Lecture Notes on Data Engineering and Communications Technologies

Volume 39

Series Editor

Fatos Xhafa, Technical University of Catalonia, Barcelona, Spain

The aim of the book series is to present cutting edge engineering approaches to data technologies and communications. It will publish latest advances on the engineering task of building and deploying distributed, scalable and reliable data infrastructures and communication systems.

The series will have a prominent applied focus on data technologies and communications with aim to promote the bridging from fundamental research on data science and networking to data engineering and communications that lead to industry products, business knowledge and standardisation.

** Indexing: The books of this series are submitted to ISI Proceedings, MetaPress, Springerlink and DBLP **

More information about this series at http://www.springer.com/series/15362

P. Karrupusamy · Joy Chen · Yong Shi
Editors

Sustainable Communication Networks and Application

ICSCN 2019

 Springer

Editors
P. Karrupusamy
Surya Engineering College
Kathirampatti, Tamil Nadu, India

Joy Chen
Department of Electrical Engineering
Da-Yeh University
Changhua, Taiwan

Yong Shi
Department of Computer Science
Kennesaw State University
Kennesaw, GA, USA

ISSN 2367-4512   ISSN 2367-4520   (electronic)
Lecture Notes on Data Engineering and Communications Technologies
ISBN 978-3-030-34514-3   ISBN 978-3-030-34515-0   (eBook)
https://doi.org/10.1007/978-3-030-34515-0

This Springer imprint is published by the registered company Springer Nature Switzerland AG
The registered company address is: Gewerbestrasse 11, 6330 Cham, Switzerland

We are honored to dedicate the proceedings of ICSCN 2019 to all the participants and editors of ICSCN 2019.

Foreword

It is with deep satisfaction that I write this Foreword to the Proceedings of the ICSCN 2019 held in Surya Engineering College (SEC), Erode, India, at July 30–31, 2019.

This conference was bringing together researchers, academics, and professionals from all over the world, experts in sustainable networking technology, sustainable applications, sustainable computing and communication technologies.

This conference particularly encouraged the interaction of research students and developing academics with the more established academic community in an informal setting to present and to discuss new and current work. The papers contributed the most recent scientific knowledge known in the field of ultra-low-power sustainable system, ultra-low-power sustainable system, sustainable vehicular ad hoc networks, Internet-enabled infrastructures for sustainability, sustainable mobility, and vehicle management. Their contributions helped to make the conference as outstanding as it has been. The Local Organizing Committee members and their helpers put much effort into ensuring the success of the day-to-day operation of the meeting.

We hope that this program will further stimulate research in sustainable big data frameworks, energy and power constrained devices, low-power communication technologies, sustainable vehicular ad hoc networks, smart transport systems, and smart data analytics techniques through this exciting program.

We thank all authors and participants for their contributions.

E. Baraneetharan
Conference Chair, ICSCN 2019

Preface

This conference proceedings volume contains the written versions of most of the contributions presented during the conference of ICSCN 2019. The conference provided a setting for discussing recent developments in a wide variety of topics including communications, networks, and sustainable applications. The conference has been a good opportunity for participants coming from various destinations to present and discuss topics in their respective research areas.

ICSCN 2019 Conference tends to collect the latest research results and applications on intelligent data communication technologies and networks. It includes a selection of 84 papers from 258 papers submitted to the conference from universities and industries all over the world. All of the accepted papers were subjected to strict peer reviewing by 2–4 expert referees. The papers have been selected for this volume because of quality and the relevance to the conference.

ICSCN 2019 would like to express our sincere appreciation to all authors for their contributions to this book. We would like to extend our thanks to all the referees for their constructive comments on all papers; especially, we would like to thank to Organizing Committee for their hardworking. Finally, we would like to thank the Springer publications for producing this volume.

E. Baraneetharan
Conference Chair, ICSCN 2019

Acknowledgements

ICSCN 2019 would like to acknowledge the excellent work of our conference Organizing Committee, keynote speakers for their presentation on July 30–31, 2019. The organizers also wish to acknowledge publicly the valuable services provided by the reviewers.

On behalf of the editors, organizers, authors, and readers of this conference, we wish to thank the keynote speakers and the reviewers for their time, hardwork, and dedication to this conference. The organizers wish to acknowledge Thiru. Andavar. A. Ramasamy, Ln. K. Kalaiyarasan, Dr. S. Vijayan, Prof. E. Baraneetharan for the discussion, suggestion, and cooperation to organize the keynote speakers of this conference. The organizers also wish to acknowledge for speakers and participants who attend this conference. Many thanks given for all persons who help and support this conference. ICSCN 2019 would like to acknowledge the contribution made to the organization by its many volunteers. Members contribute their time, energy, and knowledge at a local, regional, and international level.

We also thank all the chair persons and conference committee members for their support.

Contents

An Android Based Assessment Tool for the Visually Impaired

Hassan Abbas, Azka Rizwan, Attiya Baqai$^{(\boxtimes)}$, Daniyal Ahmed,
Taha Hussain, Munazza Naeem, and Adeel Raja

Department of Electronic Engineering, Mehran University of Engineering
and Technology, Jamshoro, Sindh, Pakistan
attiya.baqai@faculty.muet.edu.pk

Abstract. Vision processing is one of the important function human brain performs utilizing its one third of resources. Any injury or damage in the skull housing brain results in various visual impairments that can vary from severe impairment to only cognitive visual problems. Individuals with visual impairments (VI) may have intact perception of their movements. The progression of such medical conditions of patients can be identified, defined and tracked using medical assessments tools. Any development of such a functioning and cost effective assessment tool assist the medical practitioner to access the history of patient resulting in successful diagnosis and treatment, continuous observation and experimentation. In this work a new tool is developed for the medical rehabilitation and treatment of the visually impaired patients–especially young children–and provide a safety measure by recommending preventive and corrective actions from further deteriorating the impairment. It provides a measure of a stable life for the individuals by assessing them for defining characteristics of visual and/or cognitive impairments. Seven modules are designed in this research, namely color detection, object size detection, static detection, linear detection, motion detection, vision detection for static objects and vision detection for mobile objects. Each module was tested on three groups of subjects. The results from this tool are verified by ophthalmologists and neurologists. The system can identify different states of multiple subjects. This assessment tool is easily available, reliable and very cheap.

Keywords: Visually Impaired (VI) · Cortical/cerebral visual impairments (CVI) · Application Programming Interface (API) · Disability-Adjusted Life Year (DALY)

1 Introduction

The factors that affect vision or loss of acuity in people include both the human eye and the visual center (brain). We can formulate a successful diagnostic test for people with specific visual impairments. The basic thing to understand in designing an assessment tool for visual impairments is that blindness is essentially the inability to see well enough to be able to have a reasonable quality of life. People with some form of a visual impairment are NOT categorized as 'blind'. Damage to the eyes, neural network pathways or visual center of the brain Damage to the head can result in visual

© Springer Nature Switzerland AG 2020
P. Karrupusamy et al. (Eds.): ICSCN 2019, LNDECT 39, pp. 1–11, 2020.
https://doi.org/10.1007/978-3-030-34515-0_1

difficulties that can vary from severe impairment to only cognitive visual problems [1]. The analysts and medical practitioners have started to collect consonant data with respect to the symptoms and causes related to cortical visual impairments CVI. It is woth noting that researchers have different views for the complete set of causes, an opinion depending upon the medical condition or even labeling the suitable diagnosis which can describe the loss of vision by brain damage [2]. Furthermore, there is a definitive lack of medical equipment specifically catered to diagnosing patients with CVI, which further complicates the treatment process. In this research an effort is made to develop a functioning and cost effective tool in this field that assist in rehabilitation and treatment of the CVI patients.

The rest of the paper is organized in four sections. Section 2 discusses the literature review and background. Section 3 explains the formulation of the algorithms and logics for the designed modules incorporated in the form of an android app. It also explains the various groups of subjects used for testing and experimentation in this research. Section 4 gives an insight to the interpretation of results whereas Sect. 5 concludes the paper.

2   Literature Review

The clarity of vision or visual activity (VA) depends on the retinal functions including sharpness of retinal focus and the sensitivity of the brains interpretative faculty [3, 4]. A leading public health concern in the developed country is the cortical/cerebral visual impairments (CVI) among children [5–7]. In clinical terms the definition of CVI is notably due to dysfuntioning of vision which is mostly because of visual pathways perinatal injury and thier structures instead of ocular pathology alone. The most common causes of impaired maturation of key visual pathways is perinatal hypoxia and ischemia. CVI may also be caused by severe head trauma. Children with CVI often experience a variety and merger of multiple visual dysfunctions which depends upon where and upto which extent the damage is. Such dysfunctions include decreased visual acuity, visual field deficits and impairments in oculomotor, visuomotor and cognitive visual processing [8, 9]. Recovery in CVI patients is particularly challenging due to the variability in the extent and location of brain injury across individuals [10–12]. Many children with symptoms of visual perception disorders go unnoticed. These children can be studying or can be in travel without any help, but facing many challenges for which accommodations are necessary [13]. There is a difference between behaviour exhibited by CVI children and they are not same for every child so their diagnosis can be different for different children [14]. Individuals with CVI can be children who have intense visual impairment because of CVI, they can have additional disabilities, they can have useful and functional vision but having cognitive challenges, and the children working around the academic level which is expected for their peers according to their age [15, 16]. The diagnosis severity of CVI further depends on certain characteristics with help in identifying the level of cognizance in children with CVI. Some available mobile applications include Art of Glow [17], Peekaboo Barn [18], Infant Zoo Lite [19], My Talking Picture Board [20] among others. While these apps focus on improving some parameters like color, sound, light and visually guided responses in the long term, they do not have any assessment parameters to judge and determine the severity and/or type of the visual impairment the individual may have. The aim of this paper is to develop a prototype software assessment tool, which can diagnose and assess cognitive visual impairments through a set of tests to determine certain characteristics about the impairment, which a doctor can then score on a pre-defined clinical chart. It is proposed that this tool can be used to monitor and assess the progress of patients with visual impairments or learning in-capabilities.

3 Methodology

This tool consists of seven modules, each designed to test a certain characteristic of the subject's visual or cognitive state. This tool is capable of storing the results summary in the database which can be exported to the doctor or caregiver. A CVI patient cannot see clearly so a parent/caregiver has to make his user login account in which all of his testing data can be saved, also has to initiate his test by starting first module then testing of patient will continue by itself through all the modules.

3.1 Software Design

This software assessment tool is developed to run like an interactive game on touch screen devices. The primary part is the graphic subject interface, which contains the instructions in text and speech. The secondary part is the assessment running behind every module and keeps the record of the visual or cognitive characteristic being checked by that module. The working algorithm can be seen in Fig. 1. This software tool is developed using Android Studio v.3.1, Java for main programming and.xml for layout coding. The modules and assessment parameters are depicted in Table 1.

Table 1. Assessment modules

#	Module	Name	Assessment
1	One	Color detection	Which colors the subject responds to and how quickly?
2	Two	Object size detection	What object size the subject can detect?
3	Three	Static detection	Detects static objects or not?
4	Four	Linear detection	Detects linear motion or not?
5	Five	Motion detection	Detects mobile objects or not?
6	Six	Vision detection (static)	What the subject's field of vision is? (if randomly moving static objects are detected)
7.	Seven	Vision detection (mobile)	What the subject's field of vision is? (if randomly moving mobile objects detected)

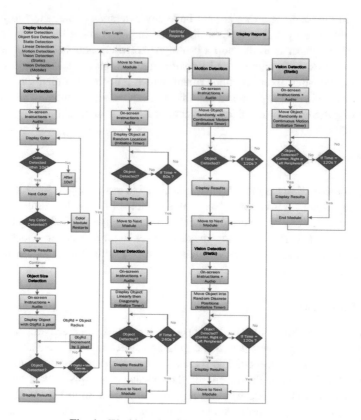

Fig. 1. Working algorithm of assessment tool

3.1.1   Module One: Response to Colors

Subjects with visual/cognitive impairments have different responses to colors than healthy people. Mostly they have a higher cognitive response to red and yellow, while subjects with defective visual acuity have a higher visual response to blue. Module one displays four primary colors on the screen as displayed in Fig. 2.

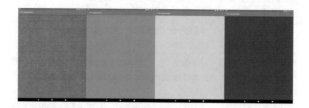

Fig. 2. Four colors used in color detection module

The subject is instructed to touch the screen as soon as they visualize the color. If the screen is not tapped within 10 s, the module moves on to the next color. This module lets the tool determine that which colors the subject responded to, and in how much time.

3.1.2 Module Two: Object Size

Subjects with CVI have different responses to different object sizes. Module two displays an object in the color to which the subject responded most quickly. The subject is instructed to touch the object when visible. The object is displayed in the center of the screen and moves to cover the whole screen shown in Fig. 3. The size of the object is calculated with ratio of pixels to screen size in inches. The smallest size from which object starts to expand is 1 pixel which is incremented by 1 pixel, the maximum possible size object reaches will depend on canvas aspect ratio of the screen.

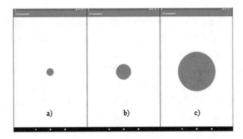

Fig. 3. Module two testing screen (a) small size (b) medium size (c) large size

3.1.3 Module Three: Static Object Detection

Some subjects with CVI are unable to detect stationary objects, and can only detect objects in motion. Module three displays an object in the color and size that the subject responded to in module 2 and 3. This is a stationary object that changes location discretely every five seconds as shown in Fig. 4. If the static object is not detected after one minute, the module notifies about no results, exits and moves on to the next module.

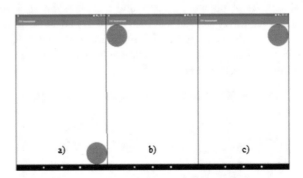

Fig. 4. Module three testing screen (a) bottom right corner (b) upper left corner (c) upper right corner

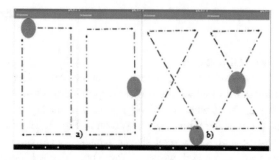

Fig. 5. Module four testing screen (a) linear motion and (b) diagonal motion

3.1.4 Module Four and Five: Motion Detection (Linear, Diagonal and Random)

Some subjects with CVI don't detect motion at all, detect complete motion or detect motion with latency. Module four displays an object in linear motion first and then in diagonal motion for a predefined time duration with a color and size to which the subject responded previously. It moves linearly from top right corner to all four corners and then will move diagonally from the same corner to all four corners of the screen for four minutes as shown in Fig. 5. If the mobile object is not detected after four minutes, the module notifies through audio and text for no results, exits and moves on to the next module. Module five displays an object in random motion with a color and size to which the subject responded previously, as shown in Fig. 6. The subject is instructed to touch the object as soon as it becomes visible to him. If the mobile object is not detected after two minutes, the module notifies through audio and text for no results, exits and moves on to the next module.

3.1.5 Module Six and Seven: Static and Continuous Field of Vision Detection

CVI patient are also unable to detect objects in all fields of vision. Figure 7 shows that humans have a 180° view normally and can see in all 180° angles without moving the head. Subjects with CVI, however, may not be able to see in all directions. For instance, some subjects with CVI have tunnel vision and can only see centered objects. Module six displays an object in the color and size that the subject responded to previously. This is a stationary object that changes location discretely (random motion) every five seconds to the defined angular vision regions as shown in Fig. 8. If the static object is not detected after two minutes, the module notifies for (audio and text) no results, exits and moves on to the next module.

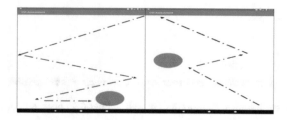

Fig. 6. Module five testing screen (random motion)

Module seven displays an object in the color and size that the subject responded to previously. This is an object in motion (random) that moves to all the defined angular vision region. If the mobile object is not detected after one minute, the module displays (audio and text) no results, exits and the program terminates. These modules let the tool determine the region of vision in which the subject responded to stationary and/or mobile objects and in how much time.

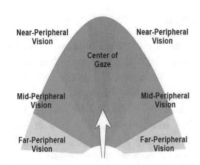

Fig. 7. Fields of vision

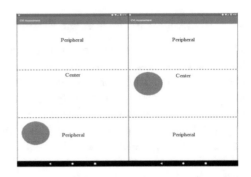

Fig. 8. Module six testing screen for field of vision

4   Results and Discussion

The beta version of the assessment tool was tested out on fourteen subjects in three groups. The 1st control group consisted of healthy subjects. They were tested upon to verify the results of the tool and form the 'normal scale' of the tool for comparison between completely healthy subjects and patients with suspected visual impairments. The subjects in the 2nd control group had a history of low visual acuity and/or diseases affecting the eye. Subjects in the 3rd control group had a history of visually related neurological disorders. All subjects were tested for the parameters i.e. response to color (s), response to object size, visual latency, field of vision, and motion detection.

4.1   Control Group One: Healthy Subjects

Table 2 displays the assessment of the 5 healthy subjects and their average responses through 5 consecutive tests to the selected parameters. The subjects in this control group had their results arranged with respect to their ages varying from 4 years of age to 19 years. The tests were performed at different times and at different places such as in a room with different wall colors and background lighting within the range of 200 Lux to 500 Lux. It was observed that almost all the subjects had similar responses and scored 10 on the CVI assessment chart (healthy), but slightly different latency results were obtained for subject 1 i.e. 0.8 s in response to blue color due to the young age factor.

4.2   Control Group Two: Low Visual Acuity/Neurological Disorder(S)

Subjects in group two and three were patients with known or diagnosed visual acuity issues, impairments or vision-neuro disorders. Five patients were assessed using the beta application. These patients ranged in ages from four to twenty years. The patients were assessed in different stimulated environments to gauge their reactions satisfactorily. Table 3 summarizes these results.

4.3   Discussion with Doctors for Results' Verification

The results of patients from group two and three were discussed with an ophthalmologist, neurologist and pediatrician, who concurred with the responses that were obtained from the patients through the developed assessment tool. It was observed that patients with issues in visual acuity were more partial to color blue and had slower response times. Patients with myopia or hyperopia have no issues in focusing or response time, but hyperopia had trouble identifying small objects. The patient with primary open angle glaucoma (POAG) had perfect response times, but had severely limited field of vision. Table 3 summarizes the patients' responses and Table 4 summarizes the discussion with doctor.

4.4   Tool Generated Results for Various Modules

The assessment tool developed generates a summary of subject as shown in Fig. 9 which compares the responses of the subject on different colors, object sizes, types of motions and field of vision. It shows that the subject responded strongly to green and yellow colors with low latency, i.e. 1 s and 4 s respectively as compared to red and blue, i.e. 6 s and 7 s respectively, object of medium size, motion of all types, and center of gaze vision. The summary is then exported to the doctor via e-mail.

Fig. 9.   Tool generated summary of a subject

Table 2. Averaged and summarized responses of the healthy subjects

Subject#	Age (years)	Response to color(s)	Average response time to colors (sec)				Vision location at	Motion detected	Object size (radius, inch)
			Red	Blue	Green	Yellow			
1	4	All 4	0.6	0.8	0.6	0.6	All points	Static, linear and random	1
2	8	All 4	0	0	0	0	All points	Static, linear and random	1
3	11	All 4	0	0	0	0	All points	Static, linear and random	1
4	14	All 4	0	0	0	0	All points	Static, linear and random	1
5	19	All 4	0	0	0	0	All points	Static, linear and random	1

Table 3. Averaged and summarized responses of subjects from group 2 and 3

Subject#	Age (years)	Disease	Average response time to colors (sec)				Vision location (most responses)	Motion detected	Object size (radius, inch)
			Red	Blue	Green	Yellow			
1	7	Myopia	0	0	0	0	All gaze points	Static and mobile	1
2	4	Partial blindness	–	3	–	3	Near-Peripheral gaze	Static	136
3	11	Hyperopia	0	0	0	0	All gaze points	Static	110
4	22	Partial blindness	6	7	1	4	Center gaze	Static and mobile	128
5	15	(POAG)	0	0	0	0	Center gaze	Static and mobile	2

Table 4. Averaged doctor's comments on patient summaries

Subject#	Age (years)	Doctor's comment (strong area)	Doctor's comment (weak area)	Doctor's comment (result)
1	7	All	None	Responded to all colors quickly, full field of vision, motion identified and smallest object detected
2	4	None	Latency, limited field vision, motion detection and size	Responded to blue color only with average latency at near peripheral vision, no motion identified and largest object size detected only
3	11	Latency and complete field of vision.	Motion detection and size	Responded to all colors quickly at all field angles, no motion identified and largest object size detected only
4	22	Complete field of vision, motion detection and size	Latency	Responded to all colors but with latency as in the case of blue with high latency, center gaze of vision, motion identified and object detected but of large size
5	15	Latency, motion detection and size	Limited field of vision	Responded to all colors quickly at center gaze only, motion identified and smallest object size detected

5   Conclusions

Testing on various subjects and the doctors feedback verified the results of this assessment tool and its vitality in the long run. These results concluded that if this tool is deployed on a wide scale and used primarily in blind schools and clinics, the patients can be tested first for any visual/cognitive diseases (primarily CVI) before beginning their long term, conditional treatment. Also, if a patient is successfully diagnosed with a specific cognitive visual impairment, measures can be taken to make stimuli in their daily life as unencumbering as possible for them.

Acknowledgement. This research is funded by Research Incubation in Public Universities (RINU) Sindh Pakistan in collaboration with Innovation and Entrepreneurship Center (IEC), Mehran UET Jamshoro Sindh Pakistan. The study presented here is performed after seeking consent of the patients and the confidentiality and anonymity of the research respondents will be respected and ensured. Moreover, all participants participated in the study voluntarily.

References

1. World Health Organization, "International statistical classification of diseases and related health problems". World Health Organization, vol. 1 (2004)
2. Jan, J.E., Heaven, R.K.B., Matsuba, C., Langley, M.B., Roman-Lantzy, C., Anthony, T.L.: Windows into the visual brain: new discoveries about the visual system, its functions, and implications for practitioners. J. Vis. Impair. Blindness 107(4), 251–261 (2013)
3. ElMaksoud, G., Gharib, N., Diab, R.: Visual-based training program for motor functions in cerebral palsied children with cortical visual impairment. Int. J. Ther. Rehabil. Res. 5(4), 265 (2016)
4. Lueck, A.H.: Comment: cortical or cerebral visual impairment in children: a brief overview. J. Vis. Impair. Blindness 104(10), 585–592 (2010)
5. Dutton, G., Bax, M.: Visual Impairment in Children due to Damage to the Brain: Clinics in Developmental Medicine. Mac Keith Press, London (2010)
6. Ospina, L.H.: Cortical visual impairment. In: Pediatrics in Review, pp. 30 (2009)
7. Shams, L., Kim, R.: Crossmodal influences on visual perception. Phys. Life Rev. 7(3), 269–284 (2010)
8. Huo, R., Burden, S.K., Hoyt, C.S., Good, W.V.: Chronic cortical visual impairment in children: aetiology, prognosis, and associated neurological deficits. Br. J, Ophthalmol (1999)
9. Kong, L., Fry, M., Al-Samarraie, M., Gilbert, C., Steinkuller, P.G.: An update on progress and the changing epidemiology of causes of childhood blindness worldwide. J. AAPOS 16, 501–507 (2012)
10. Dutton, G.N.: Types of impaired vision in children related to damage to the brain, and approaches towards their management. J. South Pac. Educ. Vis. Impair. 6(1), 14–30 (2013)
11. Good, W., Jan, J., Burden, S., Skoczenski, A., Candy, R.: Recent advances in cortical visual impairment. Dev. Med. Child Neurol. 43(1), 56–60 (2001)
12. Groenveld, M., Jan, J.E., Leader, P.: Observations on the habilitation of children with cortical visual impairment. J. Vis. Impair. Blindness 84, 11–15 (1990)
13. Hoyt, C.S.: Visual function in the brain-damaged child. Eye 17, 369–384 (2003)
14. Jackel, B., Hartmann, E., Wilson, M.: A survey of parents of children with cortical or cerebral visual impairment. J. Vis. Impair. Blindness 104(10), 613–623 (2010)
15. Merabet, L., Devaney, K., Bauer, C., Panja, A., Heidary, G., Somers, D.: Characterizing visual field deficits in cerebral/cortical visual impairment (cvi) using combined diffusion based imaging and functional retinotopic mapping: a case study. Front. Syst. Neurosci. 10, 13 (2016)
16. Rogers, M.: Vision impairment in Liverpool: prevalence and morbidity. Arch. Dis. Child. 74 (4), 299–303 (1996)
17. Ariyatrakool, N.: Art Of glow (2019). https://apps.apple.com/pk/app/art-of-glow/id38768 0399. Accessed 23 June 2019
18. Peekaboo Barn, I.: Night & Day Studios (2014). https://apps.apple.com/us/app/peekaboo-barn/id300590611. Accessed: 23 June 2019
19. treebetty LLC. Infant Zoo LITE: Visual Stimulation for Babies (2016). https://apps.apple.com/us/app/infant-zoo-lite-visual-stimulation-for-babies/id677233768. Accessed 23 June 2019
20. Little Bear Sees. My Talking Picture Board (2017). https://itunes.apple.com/us/app/my-talking-picture-board/id586535395?mt=8. Accessed 23 June 2019

RETRACTED CHAPTER: Social Media Analytics Using Data Mining Algorithms

Harnoor Anand[✉] and Sandeep Mathur

Amity Institute of Information Technology, AUUP, Noida, India
harnooranand6@gmail.com, sandeep2809@gmail.com,
smathur@amity.edu

Abstract. Social Media sites like Facebook, Twitter, LinkedIn, and Google + contain an outsized volume of unprocessed information. By analyzing this knowledge, new information will be gained. The traditional data processing techniques won't be applicable for processing the increasing number of dynamic and unstructured data. In this paper, we have a tendency to discuss data processing, social media knowledge, data processing techniques applied in social media applications. A detailed survey has been performed on the works that drained the sector of social network data processing analysis and techniques followed to perform {the knowledge the info the information} mining on the social network data. Results of this survey will function the baselines for the future data processing technologies. Data mining is evolving as a procedure for examining pre-existent databases to come up with new helpful information. The social networking sites are gaining an increased research potential for researchers as most of the people are depending upon the social network for news, and opinion of other users [1]. Internet based life mining is emerging as the strategy for speaking, dissecting and mining designs that square measure in move from information in web based life by abusing numerous strategies to overcomes the bottlenecks in web-based social networking.

Keywords: Data mining · Social media · Social network data analysis

1 Introduction

Data mining is evolving as an integral asset, which can discover examples and connections inside our information. Information mining finds concealed data from vast databases [1]. The general objective of the information mining process is to separate data from an informational index and change it into a reasonable structure so that it can be used in a wide manner. Social systems can be utilized in numerous financial exercises like expanding verbal showcasing, promoting research, General advertising, Idea age and new item advancement, Co-development, Customer administration, Public relations, Employee correspondences and for increasing the reputation.

Regulated and unsupervised calculations are utilized to distinguish the concealed examples present in the information. Directed methodologies rely upon some from their earlier information (for example class marks). Unsupervised calculations are utilized to portray information with no earlier guidance that concerns on what sorts of examples

The original version of this chapter was retracted: The retraction note to this chapter is available at https://doi.org/10.1007/978-3-030-34515-0_85

will be found by the calculation, The assortment of work can be achieved for date relating to information mining of online internet based life information, which is practiced with some rendition of either [2].

Regulated or unsupervised learning calculations. Deciding if a regulated or an unsupervised methodology would be best relies upon the informational collection and the specific inquiry being explored, Data sets can be summed up into three kinds: information with marks, information without names, and information with just a little part of names. Order is a typical regulated methodology and is fitting when the informational collection has marks or a little bit of the, information has names. Order calculations start with a lot of preparing information which tends to incorporate class names for every datum component. The calculation gains from the preparation information and fabricates a model that will consequently; arrange new information components into one of the unmistakable classes that gave the preparation information. Characterization guidelines and choice trees are instances of regulated order procedures.

Clustering is a common unattended data mining process that's helpful when explaining knowledge sets consisting of no labels. Not as classification algorithms, algorithms of clustering don't rely upon training knowledge for the development of a model. Rather, grouping calculations figure out which components in the informational collection are like each other dependent on the likeness of the information components. Similitude can be characterized as Euclidian separation for some numerical information, sets yet frequently in information related with web based life, group procedures must most likely arrangement with. For this situation, bunching strategies use watchwords that are spoken to as a vector (to speak to a report) and the cosine comparability measure is utilized to separate how, comparative one vector (information component) is to another.

Scientists have developed many data processing techniques so as to beat the issues like size, noise and dynamic nature of the social media information. Because of the massive volume of data in social media, automatic processing is required so as to research it in a given amount of time. The dynamism within the social media information ends up in the speedy evolution of sets over time; such dynamic data are often simply handled by varied data processing techniques [3].

There are various data mining techniques:

(1) **Characterization:** Characterization generalizes and summarizes the different data characteristics.

(2) **Classification:** It is a process which classifies the given data into different classes.

(3) **Regression:** Regression and classification are almost same processes. The difference is that the object to be predicted is continuous rather than discrete.

(4) **Association:** In this method, there's associate association between the objects. It discovers the association between numerous infos and also the association between the attributes of the one database.

(5) **Clustering:** In Clustering, data is grouped into new classes in such a way that it describes the data [4].

2 Literature Review

In the literature review, it's been clearly noticed that a lot of sorts of analysis works on the basis that how these interests of the users is analyzed for future conclusions and usages. there have been many ways put forwarded.

Bogdan Batrinca; consistently and strategically explains the techniques, tools, and platforms for social media analytics, a close study on information retrieval techniques is in their paper. They discuss end necessary terms like social media, sentiment analysis, scraping, opinion mining, behavior social science, NLP and numerous toolkits and code platforms. The recent analysis challenges like scraping, information cleanup, holistic information sources, information protection, information analytics, analytics dashboards, and information visual image are also mentioned. They also considered the companies' policies of limiting data access to gain financial edges [5].

Karthikeyan and Vyas; The text mining or text data processing is one of the data mining process which derives better quality info from texts. Massive media corporations like the tribune etc., by exploiting the text mining techniques so as to get the clear data and to supply higher search experiences to the readers which will consequently increase the 'stickiness' of the location and helps the location to get larger revenue [6].

Adedoyin-Olowe, Gaber, and Stahl; Sentiment lexicon is considered as a wordbook of the emotional words that are often utilized by the reviewers in their communication. It includes a listing of normal words, that helps within the improvement of the information from mining techniques once they are used for mining a sentiment in a document. Relying upon the range in subjects, numerous collections of sentiment lexicon is generated.

The sentiment words utilized in sports, as an example, are not like those which are utilized in politics. we will focus on topic-specific incidence by increasing the incidence of sentiment lexicon combined with the high manpower.

J. Bonneau, J. Anderson, and G. Danezis; Social networking sites like Twitter, LinkedIn and additional consists of users connected along with distinctive profiles. individuals victimization these social networking sites will connect with their legendary individuals and might share their content along. purchasers alter those profiles relying upon individual inclinations but some traditional information might incorporate relationship standing, birthday, associate degree email address, and main residence. it's determined by users that WHO will have authorization to their profiles. data flowing through these social networking sites has result in privacy problems and currently it's become a social group issue [7].

It is critical to ensure individual protection when working with interpersonal organization information. Ongoing productions feature the need to ensure security as it has been appeared even, anonym punch this sort of information can at present uncover individual data when best in class information investigation methods are utilized. Settings of privacy additionally can restrain the capacity of information mining applications to think about each snippet of data in an interpersonal organization. Be that as it may, some loathsome strategies can be utilized to usurp protection settings [8].

Ritu Mewari put in front her opinion that mining provides a transparent platform to get the public's mood by filtering the data's noise. It additionally provides various techniques helpful in extracting and joining a human opinion from unstructured and creaky text knowledge. Supposition mining is a consuming field of web mining. There exists a great deal of advantages of supposition mining at client and business level. A main part of information is day by day posted on sites like Facebook and twitter. User put in forward all of their sentiments within the type of reviews and feedback on daily basis. Associate in nursing opinion mining method provides the thanks to extract pearl data from it [9].

An investigation by Al-Daihani, S. M., applied; the approach in mining on an extensive tweets. The total courses of twitter of events of 10 scholarly libraries were utilized to gather the dataset for this examination. Almost 23,707 no of tweet shaped the absolute dataset having 7625 no of hashtags, 17,848 notices, and 5974 no of retweets. The essentialness of information and mining approaches re accounted for inside the examination and their motivation is to pick up knowledge with the total social information of scholastic libraries so the procedure of sic leadership and key arranging could progress toward becoming encouraged for promoting of administrations and supporter outreach [10].

Kermanidis, K.L; Said that sentiment analysis through social media usage has witnessed an enormous amount of interest from students within the previous few years. In that, the authors mentioned the influence of tweets sentiment on elections and therefore the impact of the elections' results on sentiment [11].

Ache, B., Lee, L; Millions of individuals get to web based life locales, for example, Twitter, Facebook, LinkedIn, YouTube and MySpace to scan out for data, breaking news and news refreshes. A large portion of the updates are regularly posted by new individuals they have never and may never have contact with. Thusly, data accumulated via web-based networking media is here and there used to settle on important choices. While a few gatherings are recovering data from online life destinations, others are posting data for the utilization of other web clients [12].

3   Social Network Data Mining Algorithm

3.1   Overview of the Algorithms

When mining social network knowledge it ought to be a mix of net structure mining and online page mining. Social Network Analysis is evaluating the anatomy of the social network. Social Network analysis wherever was discussed a lot in 1994 by the researchers particularly within the area of science, geography, social science, economics, biology and medical specialty, social science, economics, biology, and medical specialty, many tools are introduced within the space of informal community examination like Graph Characterization Toolkit, Tweet Hood, Net Driller, HiTS/ISAC, a cloud-based library for huge scale interpersonal organization investigation. Online page mining was a great deal of basic in selling and promoting investigation [13].

3.2 Analysis of the Algorithms

In informal organization information mining, existing information mining calculations can't be utilized legitimately in view of dynamic conduct. While breaking down the writing on informal community information mining strategies; it was discovered that every calculation has qualities and shortcomings. The accompanying area clarifies the current calculations in detail.

3.2.1 Graph Mining Algorithms

Pattern Visualization Approach. Graph mining algorithm is one of the popular techniques of data mining in Social Network Analysis. World Wide Web consists of all social networks. All the hyperlinks interconnected documents are stored in World Wide Web. Web structure investigation dependent on chart calculations has been dissected in numerous sorts of research in past years. Lahiri and Berger-Wolf (2008) have made and tried strategies joining system, quantitative, semantic, information handling, transformation, and representation based segments. They have presented another chart mining calculation "intermittent sub graph mining, or the revelation of all association designs that happen at standard time interims" keeping dynamic flow of socials in consideration. This algorithm depends on regular example mining in value-based and chart databases with intermittent example mining in one-dimensional and multidimensional arrangements [14].

Bouqu et al. introduced a system in 2009 which depends on unique diagram separation and chart bunching. This system is fit for recognizing the dynamic changes of the informal organization structure and distinguishes occasions dissecting transient measurement and uncovered order progressive systems in interpersonal organizations. The specific calculations treat the system as a chart yet it limits the bunching issues and diagram parceling issues. As an answer least spreading over trees can be utilized to distinguish clients who have comparative profiles and solid connections. Zhang et al. have directed an examination on the pertinence of usual insatiable, slope climbing and centrality-put together calculations with respect to dynamic informal community information to recognize key clients for target promoting by mapping the system to a chart in 2010. They have proposed another estimation seeking calculation dependent on the heuristics data from the above calculations.

Despite the fact that the charts map the association or the connection between the hubs it doesn't demonstrate the relationship quality. One fascinating instrument has been created called associate t, o give h recurrence data on a social relationship among various substances in the systems by utilizing a Frequent [15].

3.2.2 Classification

Classification is that the, technique of categorizing knowledge into, one among several classes. This could be applied in net data processing, to classify user profiles supported profile, characteristics. Preferred classification algorithms, in data naïve Bayesian classifier, and natural networks. Suma and Furmanek (2010) introduced, a motivating rule referred to as C &RT, combining classification and regression tree algorithms, to work out rules to spot target teams to promote this could be employed in real social network knowledge.

3.2.3 Clustering

Clustering is grouping a group of items, such how that things within the same, cluster area unit a lot of kind of like one another than to those in, different teams. These teams area unit referred to as a cluster clump is principally, employed in data retrieval in net mining. Based on, past analysis clump can increase the potency in data retrieval. Graph-based clump is comely employed in net structure mining, as explained with in the early section. Text- based clump is most typically employed in website mining weather not you produce clusters supported the content of the online document. Bartal introduced a stimulating methodology comb social network analysis and text based clump to predict the nodes of a social network would be connected next [16].

3.2.4 Associations

Affiliation rule mining is utilized to discover visit example and connection among informational index. Nancy had use affiliation standards to mine, informal organization information utilizing 100 Facebook college pages. The exploration concentrated on the detailing of affiliation rules utilizing which choices can be made and utilizes the Apriori Algorithm to determine affiliation rules [17].

3.2.5 Semantic Web and Ontology

Semantic Web, a new analysis space wherever it tends to present aspiring to internet knowledge. this permits machine and humans to connect with each other with intelligence and exchange data. Many researchers has been carried out in this file like using semantic geo catalogs and recovery of mental health information. Clarifies applying measurable learning techniques semantic web information. It has utilized an all-encompassing FOAF (companion of-a-companion) metaphysics connected as an intercession composition to incorporate Social Networks and a cross breed element compromise strategy to determine substances of various information sources. Clarifies the use of Semantic Web innovation to distinguish the relationship between different areas in a Social Network. presented a novel methodology of utilizing semantic similitude measure dependent on pre-characterized ontologies for arranging interpersonal organization information. It has built up a calculation to recover data in informal communities to recognize patterns. This Algorithm has implemented semantics to decide the pertinence of systems utilizing unstructured information. The calculation was fed on twitter messages.

3.2.6 Markov Models

Markov chains unit of measuring is additionally a mathematical rule program. The undergoes transitions from one state to a special, among a finite or denumerable form of potential states. It's a random methodology anywhere sequent state depends entirely on this state and not the sequence of events that preceded it. Markov model is furthermore utilized in internet mining to predict users next action [18] (Fig. 1).

Fig. 1. Key research issues in online social network analysis

Influence Propagation: These days, as OSNs are pulling in a huge number of individuals, the last depend on settling on choices dependent because of such locales. For instance, impact spread can help choose which motion picture to watch, which item to buy, etc. Subsequently, impact proliferation has turned into a significant instrument for successful viral showcasing, where organizations attempt to advance their items and administrations through the verbal spreads in OSNs. This further inspires the examination network to complete broad investigations on different parts of the impact proliferation issue.

Community or Group Detection: By and large, bunch discovery in OSNs depends on dissecting the structure of the system and discovering people that correspond more with one another than with different clients. Bunching a person with a certain goal in mind can additionally make an appraisal about the individual, for example, what exercises, products, and administrations, an individual may be keen on.

Expert Finding: OSNs comprise of a few specialists in a particular space and other individuals who join the system to get help from these specialists. These OSNs can be utilized to look for such specialists inside a gathering of individuals. For instance, a subject related master can be sought dependent on the investigation of the connection among creators and collectors of messages.

Link Prediction: The mass measure of information accessible in OSNs can be mined to make expectations about 'who is a companion of whom' as an individual may be just a couple of steps from an alluring social companion yet may not understand it. By social occasion helpful data about an individual, OSNs can deduce new collaborations among individuals from an OSN that are probably going to happen sooner rather than later.

Recommender Systems: Recommender frameworks (RS) give suggestions to clients about a lot of s or administrations they may be keen on. This office in OSNs has turned out to be mainstream because of the simple access of data on the Internet. Hardly any significant uses of RS are its utilization in a few sites for proposal of things, for example, motion pictures, books, contraptions, and so forth.

Predicting Trust and Distrust among Individuals: Because of the ceaseless development of networks in OSNs, the subject of trust and doubt among people in a network has turned into a matter of extraordinary concern. Past evaluations uncover that a few clients endeavor to either irritate or exploit the ordinary climate of such online networks. Thus, there emerges a need of surveying every client of an OSN people group to foresee the dimension of trust or doubt that can be processed for them.

Behavior and Mood Analysis: Because of the ceaseless development of networks in OSNs, the subject of trust and doubt among people in a network has turned into a matter of extraordinary concern. Past evaluations uncover that a few clients endeavor to either irritate or exploit the ordinary climate of such online networks. Thus, there emerges a need of surveying every client of an OSN people group to foresee the dimension of trust or doubt that can be processed for them.

Opinion Mining: OSNs have offered ascend to different survey destinations, blog storehouses, online discourses, and so forth where individuals can express their thoughts and feelings, trade learning and convictions, censure items and thoughts. Information mining of sentiments of explicit subjects permits the discovery of client prospects and needs, and furthermore emotions or responses of individuals about specific convictions, items, choices or occasions.

Data Mining Algorithms

Information mining system are unit partitioning into 2 methodologies: a right way methodology is used in forecast wherever it endeavors to foresee a condition of latest associate incentive by taking a goose at the noted qualities. The second methodology. On- direct methodology is used to acknowledge new examples by models info need to be clean and organized. The mining model is created on the related to Algorithms [19].

Association Rules – This calculation container be utilized in showcasing sordid examination similar recognizing strategically pitching chances. This takes numerous things in a solitary exchange, filters the information and checks the occasions the things show up in the exchange so it tends to be utilized to distinguish the connections in the vast informational collections.

Clustering – This calculation bunches the information as indicated by their comparative qualities. This can be utilized to distinguish the relationship of the attributes among a gathering. At the point when new information is presented, its qualities can be mapped with the connections, it very well may be utilized to foresee the conduct of the new information. Bunching can be utilized to discover abnormalities of the information too. This is usually utilized in frameworks of misrepresentation location and Customer Relationship Management.

Decision Trees – This is the straightforward and a standout amongst the most usually utilized calculation. This is utilized to anticipate discretely and proceeds with factors.

Straight relapse – This is foreseeing just proceeds with factors utilizing single various direct relapse formulae [20].

Logistic Regression – It uses a sickish network consisting of no hidden layers.

Naïve Bayes – This may be used to figure probabilities for every conceivable condition of the data property once a condition of a discerning character is-tic is given. This may be used as a result of the beginning calculation of the anticipating procedure.

Neural Networks – The calculation received from man-made consciousness. This can be utilized to scan for nonlinear useful conditions. This will achieve nonlinear changes on the information in coats from information coats to the shrouded coats lastly to the yield coat.

Sequence Clustering – This searches for cluster primarily founded models than the likeness if the data. The model uses the succession of occasions by utilizing shouted chains. The states area unit models a lattice and also the change of travelling beginning with one sate then onto successive within the cell of the grid. With these a chance for a grouping of progress will be determined by duplicating probabilities of state advances within the succession. The chains of most astounding probability will be used to show the teams.

Time Series – This can be used to conjecture payoff with factors. This can be a UNIX of 2 calculations referred to a auto egreeion trees and Auto regressive coordinated moving constellation [21].

4   Result and Discussion

Data mining techniques for social media analysis. Wherever it's shown that data processing may be a powerful tool which can facilitate to hunt out hidden patterns and also the varied relations between the information. Processing discovers hidden facts from huge databases. The general objective of information mining technique is to extract information from an enormous data set and remodel it into an evident structure for additional use.

The different data Mining techniques are

   I. Characterization.
  II. Classification.
 III. Regression.
 IV. Association.
  V. Clustering.
 VI. Change Detection.
 VII. Deviation Detection.
VIII. Link Analysis.
 IX. Sequential Pattern Mining.

Social network finds its application in several business activities like Co-innovation, Customer service, General promoting, increasing spoken promoting, marketing research, plan generation and new development, publicity, worker communication, and reputation management. Clustering [22].

The Algorithm by Marcelo Maia, for a various group of users, dissimilar interaction patterns that are not similar can be experiential.

Characterize and identify user profiles in online social networks..Display more suitable advertisements based on user behavior. Cleavage Ant.

Colony Metaphor by Al- Fayoumi

Using maximum group and subgrouping criteria social network structure are clustered. Imposing performance on a take a look at run mistreatment the quality cluster sequence knowledge significantly for the clump of the instance named as Keller six.

Graph Ct Snag Algorithm by Ediger, Jiang

To perform analysis on public data from a micro blogging network like Twitter. the focus of analysts more on a much smaller data subset.

Decision Tree Algorithm- SemanBozkur et al.

Most precise results.

Opinion Mining and Sentiment Analysis- Pak et al.

Large data set. Improved needed increasing the amount of the training data. Back Propagation.

Analysis- Priyadershini et al.

A highly effective tool for classification. Combination of the trail, learned. Web mining techniques- Raju, Sravnthi while using web mining for social Media for analysis, data sampling is a huge issue. Filtering the data is difficult. Communities Overlapping [23].

Sentiment Classification- Vinodhini, Chandrasekaran

Sentiment detection used in different applications such as identity and classified reviews, summarize the review and other real-time methods and applications.

Decision Tree Classification and K Means Algorithm- Ling , Yin, et al.

Analyze the potential mix of users. To process a huge total of data in a timely manner, decision tree-based and a k- means algorithms are used.

Social Network Models Statically and for Analysis- Some

Link analysis and deep and dark networking Is recent trends in social networks.

K-Nearest Neighbor and Native Bayes- LopamudraDey, Sanjay Chakraborty

Give better results For hotel reviews these algorithms not Suitable.

Association Rule Mining- Hemant Kumar Soni, Sanjiv Sharma, et al.

Displays and describes the association rule mining [24].

5 Conclusion

This paper gives an undeniably stream appraisal and update of relational association examination and investigation available. Composing has been assessed reliant on different pieces of casual network examination.

This study ponders the use of the techniques and thought of data burrowing for relational associations examination and overviews the related expounding on substance mining and casual associations. Casual people group examination assisted through the frameworks of Web mining is a captivating field of research.

Be that as it may, there are numerous difficulties in this exploration field to be resolved with further research progress.

References

1. Pal, S.K., et al.: Web mining in soft computing framework. IEEE Trans. Neural Netw. **13**, 1163–1177 (2016)
2. Nanno, T., et al.: Automatically collecting, monitoring, and mining Japanese weblogs. In: WWW Alt. 04: Proceedings of the 13th International World Wide Web Conference on Alternate Track Papers & Posters (2014)
3. Gross, R., et al.: Information revelation and privacy in online social networks. In: ACM Workshop on Privacy in the Electronic Society (WPES) (2015)
4. Liu,H., et al.: Toward integrating feature selection algorithms for classification and clustering. IEEE Trans. (2015)
5. Esuli, A., et al.: Sentiwordnet: a publicly available lexical resource for opinion mining. In: Proceedings of the 5th Conference (2016)
6. Maia, M., et al.: Identifying user behavior in online social networks. In: SocialNets 2008 (2015)
7. Ho, A., Maiga, A., et al.: Privacy protection issues in social networking sites (2016)
8. Dey, L., et al.: Opinion mining from noisy text data (2016)
9. Al-Fayoumi, M., Banerjee, Jr.S., et al.: Analysis of social network using clever ant colony metaphor. In: Proceedings of Wiset (2017)
10. Ediger, D., Jiang, K., et al.: Massive social network analysis: mining twitter for social good. In: ICPP (2010)
11. Priyadarshini, R.: Functional analysis of artificial neural network for dataset classification. In: IJCCT, August 2010
12. Bozkir, S., et al.: Identification of user patterns in social networks by data mining techniques. In: IMCW (2010)
13. Pak, A., et al.: Twitter as a corpus for sentiment analysis and opinion mining. In: Proceedings of the LREC (2010)
14. Baccianella, S., et al.: Enhanced lexical resource for sentiment analysis and opinion mining. In: Proceedings of LREC 2010 (2010)
15. Asur, S., et al.: Predicting the future with social media (2010)
16. Lin, K.Y., et al.: Why people use social networking sites: an empirical study integrating network externalities and motivation theory. Comput. Hum. Behav. **27**, 1152–1161 (2011)
17. Taboada, M., et al.: Lexicon-based methods for sentiment analysis. IEEE (2011)
18. Raju, E., Sravanthi, K.: Analysis of social networks using the techniques of web mining. Gearcase 2(10) (2012)
19. Vinodhini, G., et al.: Sentiment analysis and opinion mining: a survey. IJARCSSE **2**(6), 282–292 (2012)
20. Khoshnood, F., et al.: Designing a recommender system based on social networks and location-based services. IJMIT **4**(4), 41 (2012)
21. Long, X., Yin, W., et al.: Churn analysis of online social network users using data mining techniques. IJEMS (2012)

22. Vinerean, S., et al.: The effects of social media marketing on online consumer behavior. IJBM **8**(14), 66 (2013). ISSN 1833–3850
23. Some, R.: A survey on social network analysis and its future trends. IJARCCE **2**(6), 2403–2405 (2013)
24. Dalal, M.K., et al.: Automatic classification of unstructured blog text. IJILSA, 108–114 (2013)

A Blockchain Technology Solution to Enhance Operational Efficiency of Rice Supply Chain for Food Corporation of India

Alpana Kakkar[✉] and Ruchi

Noida, India
rtalwar829@gmail.com

Abstract. Rice and wheat are the staple food grains among the major commodities that are procured and securely stored by Food Corporation of India (FCI), which is the responsible body to implement the decisions made by the Ministry of Consumer Affairs, Food and Public Distribution (MCAFPD). The Government of India (GOI), alongside other state offices, attempts procurement of wheat and rice under price support scheme. Food grains are procured from farmers from mandis, and then they are moved to various FCI depots through rail or road and finally distributed to state agencies for public distribution. FCI has been able to make sure the availability of sufficient foodgrains in depots of all states by proper planning and efficient supply chain operations. For effective SCM, FCI has implemented IIFSM (Integrated information system for food grain management) and FAP (Financial accounting package) to make the increasing demand of the population and food security. We have studied the existing FCI's supply chain management for rice, and the extensive literature review was also done to understand the supply chain management in the food sector. To escalate and track the productivity of paddy and increase the security and efficiency of rice supply chain operations of FCI, we have proposed a Blockchain technology model of RSCM (Rice Supply Chain Management) in this paper.

Keywords: Blockchain · Supply chain management · Rice supply chain · FCI · Food security and procurement

1 Introduction

Food Corporation of India (FCI) is an organization working under the Ministry of Consumer Affairs, Food and Public Distribution domains. FCI was set up under the Food Corporation Act 1964 for satisfying the successful price support activities to shield the interest of farmers and, throughout the nation.

FCI has two broad objectives, ensuring Minimum Support Price (MSP) to the farmers and availability of food grains to the weaker section at a reasonable price. FCI also arbitrates in the market to ensure that the prices are conventional and legitimate and hence the overall food security of the country increases. Food Corporation of India along with state agencies look after the acquirement of wheat and paddy under the price scheme [26].

© Springer Nature Switzerland AG 2020
P. Karrupusamy et al. (Eds.): ICSCN 2019, LNDECT 39, pp. 24–31, 2020.
https://doi.org/10.1007/978-3-030-34515-0_3

Initially, before the harvest during each crop season, the Government of India (GOI) announces MSP for the procurement procedure. These prices are based on the recommendation of the Commission of Agricultural Costs and Prices (CACP) [23], These are other accommodating components in the record like the expense of different horticultural data sources and a sensible edge for the ranchers for their produce. Besides, to encourage the acquirement of nourishment grains and different state offices with the help of state and built up countless focuses at different mandis and key places. The price support operations are so vigorous, extensive and effective that they have resulted in providing a sustaining time income of farmers over a long period. Ultimately, all the stocks brought to the centers are within the Government of India's specifications and are purchased at fixed support prices. After procurement, the need is to have an assertive movement plan for the food grains. Movement is an essential process for FCI just as it fulfills the goals of nourishment arrangement and National sustenance security act. FCI initiates the development of nourishment grains to clear the stocks from surplus areas and to adhere to the deficit regions for the TPDS (Target Public Distribution center) and other schemes [25]. Also, it creates reserves of stocks in the deficient areas. For example, Haryana, Punjab, Odisha, Andhra Pradesh, and Chhattisgarh are surplus States as far as rice obtainment just as utilization. Hence, according to the TPDS requirements rice is moved to deficit states for sale as well as to create reserves.

In a year, FCI transports over 42 million tons of food grains across the country utilizing the depots, silos, railheads and FCI's. The movement of stock depends on a few factors such as availability in the surplus region and then the amount of requirement in deficit region. This process is done monthly [25]. The work and operation of the FCI have expanded and improved, even the quantity of food grains procured has increased and hence the movement and transports, but to maintain the overall efficiency on such a large scale, IT has been playing an essential role in all aspects. Food security and effective distribution of grain, pricing of grains and sustainability of farmers are supported by IIFSM (integrated info system for food grain management) and FAP (Financial Accounting Package) [24]. To reduce food wastage, increasing operational efficiency, to enhance food security and lean management in the rice supply chain. We have proposed a model for rice supply chain management using Blockchain into its functioning to increase its operational efficiency. Section 2 is about the literature review done to understand FCI's role, functioning and existing Supply Chain Management of foodgrains. We have studied research papers on Supply Chain Management of various sectors and a few specifically of food sectors. In Sect. 3, we are describing the proposed model. Conclusion is written in Sect. 4.

2   Literature Review

Commission agents receive the payment from FCI which they further pass on to the farmers after deducting their commission. Also, the paddy is distributed to rice mills and depends on the storage facility in the mills. FCI allocates the paddy according to the storage capacity of mills. The Millers start working on the stock and submit the rice back to the FCI and FCI distributes the rice to Public Distribution Center (PDS).

Figure 1 shows the traditional Rice Supply Chain which includes farmers, grain markets (anaj mandis), rice mills, distributors, wholesalers, and retailers. The operations in the Rice Supply chain management are performed in various stages. Each stage depends on the functionality of the next stage. Hence, cooperation is needed between different stages to conduct activities in customized manner-the multiple steps involved in the rice supply chain are as follows- The first stages starts with the harvesting of the paddy by farmers. Farmers are classified as "Small Land Holders" and "Large Land Holders" the classification is done based on how much land they hold large landholders grow paddy in vast amount with the goal that they can offer their paddy straightforwardly to rice mills. Acquiring paddy legitimately from small landholders isn't gainful for rice mills as their expense of coordination and getting isn't ideal to rice handling organizations. In this manner, small landholders sale their paddy to the commission agents. Now and then rice mills approach farmers and give their prerequisites ahead of time with the end goal that the buyer requests are met. After getting paddy from farmers, rice mills convert paddy to rice. Changing over paddy to rice incorporates numerous stages like husking, cleaning, production, polishing, and packing. Later on, this rice is being sold to distributors and wholesalers from where local retailers get it and further sell it to the consumer. Industrial buyers like supermarkets and restaurants directly from rice mills. In this way, rice reaches to the public after going through all the stages in the supply chain.

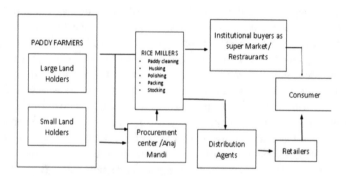

Fig. 1. Traditional rice supply chain.

Ganeshan and Harrison in 1995 defined SCM as, "A supply chain is a network of facilities and distribution options that performs the functions of procurement of materials, the transformation of these materials intermediate and finished products, and the distribution of these finished products to the customers" [16]. Lee and Corey in 1995 defined it as, "The integration activities taking place among a network of facilities that procure Raw material, transform them into intermediate goods and then final products, and deliver products to the consumer through a distribution system" [17]. Christopher in 1998 defined SCM as, "The supply chain is the network of an organization that is involved, through upstream and downstream linkages, in the different processes and activities that produce value in the form of products and services in the hands of ultimate customer" [18]. The food processing defines as the agri-industrial system aims for plantation, cutting, transport, and processing from farm to mill and mill to market [3]. Currently, technological advancements like IOT, RFID(radio-frequency identification)tags, sensors, barcodes, GPS tags, and chips, are used to track the location of products and packages [4]. This allows an enhanced, real-time tracking of goods from their origins. Food supply chain system guarantees that bringing a quality item to the customer is the best way to rebound the trust of pupils on food security in the market [6]. With the development of the internet, a lot of advanced technologies have been applied in SCM to track the chain of products between the mediators in the supply chain [5, 7]. Blockchain is denoting the start of a new period and is a revolutionary advancement in decentralized data innovation. First developed in Bitcoin's basic framework in 2008 [14, 15]. The future of blockchain technology is not going to end soon; it is being used in many types of currencies and legal contracts [10]. Blockchain is the finest technology which tracks all the transactions so that they can be easily available to each person and Blockchain keeps all the transaction secure as well [8]. Blockchain is a distributed ledger which is divided into blocks and each block in the Blockchain has its own data of transactions which is uploaded on the block by using internet, if someone tries to break into any block which is kind of impossible to break into because all the blocks are interlinked with each other and for that hacker needs hefty amount of resources, this quality of blockchain makes it unchangeable and incredibly hard to alter [9, 12]. Blockchain can be open source, can be available to many people in the globe and it can also be private with restricted users too. Blockchain has various types which are Public, Private and Consortium which depends on how the user or an organization want to use [9, 10]. Public blockchain is decentralized, open-source and anyone can write or read the data in the blockchain. Private blockchain gives authorization to one specific organization which gives access to read or write and can be accessed by highly trusted clients. Consortium blockchain is a mixture of private and public [9]. Consortium blockchain gives authorization to an organization to have full command on the blockchain except an individual to participate in the validation process on the blockchain. Hence, it's on the organization how they want and which type of information they want to transmit on their blockchain framework [9]. In 2014, second generation "Ethereum" blockchain was designed by Buterin [10]. It was created in such a way, that a person with least computer skills can also be worked and established on Decentralized-application(DApp) [20] on the blockchain. Ethereum [21] is also redistributed not governed by any central entity same as bitcoin "blockchain." It has its own currency called "Ether" and also, has a virtual machine (Ethereum Virtual Machine) EVM. "Solidity" is a programming language which is also made by Ethereum [22] and it is utilized in working if decentralized apps. "Smart Contract" is also build by Ethereum blockchain [22] and also uploaded on the blockchain so every individual can work on it. A smart contract is a self-executing way in which code is used to trigger an event that is mentioned in the contract. This contract is formed after the mutual understanding between both parties in the contract [11].They are self-executable and trigger when the condition is satisfied. It can be implemented as a way to overcome paperwork. It can cause intervention which is generally due to the involvement of the middlemen. The smart contract once deployed have their unique address [10, 11] Blockchain can also be used for making a system which is decentralized in nature, and this can also be used in different systems like SCM which can use blockchain's functionality like security, transparency and unchangeable [11]. Some SCM system uses blockchain technology and provides an efficient and effective way of surveillance in the supply chain system like production, processing, warehousing, distribution and retail [12]. Rice being mostly consumed and easy to make food, every customer deserves to have good quality rice. RSCM plays a vital work in bringing paddy from farmers, converting into the rice and bringing into the plates of every individual [13]. In our research, we are implementing Blockchain technology on "DApp (Decentralized Application) and Smart Contract as digital contract" for making the rice supply chain more efficient and immutable to frauds.

3 Proposed Model for Rice Supply Chain Using Blockchain Technology

Here, we discuss a method of how Blockchain can help in the enhancement of RSCM and help to make the process more efficient. Blockchain being immutable can be very useful in creation on the distributed ledger which will be accepted by all the nodes in the system, and a decentralized detectability framework is given to follow the supply of rice during the methods required inside the rice production network [4] (Fig. 2).

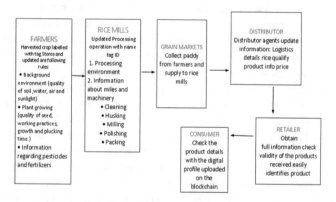

Fig. 2. Implementing blockchain technology in rice supply chain

The first step is all stakeholders involved in the RSCM needs to register themselves on the blockchain network with a new Unique-ID such that they can put the information/data related to the distribution of paddy by the farmers till the distribution of rice in the blockchain network. The proposed SCM system works in various steps like Production, Procurement, Storage, Distribution and Retailing.

3.1 Production

Production stage consists of collecting paddy and marking the paddy sacks with specific tags. Those tags will be scanned on the digital profile which is made on the blockchain network and all the information regarding paddy will be transferred on the blockchain. All the vital info like planting time, kinds of seeds utilized, sort of pesticides used, soil, season, rancher subtleties, and so on. After selling paddy to rice mills or commission agents, a digital contract is made on the blockchain which is called Smart contract, i.e., exchanged between the farmers and to those who are selling their sacks to. All the information mentioned above is sent to the network on blockchain.

3.2   Procurement

During the procurement process, the farmer brings the paddy to the procurement center and the procurement center will be responsible for the weight of paddy and then the payment process begins. With the use of a smart contract, the process will be reliable and automated for the payment carried out for the farmer. The payment to the farmers depends on the set MSP price, and quantity of paddy procured from them. A smart contract is self-executable and can be used for minimizing the human intervention in the incentive to the farmer. This can be very useful as the farmer will get the accurate price for the paddy that they have sold to the procurement center. The payment will be made online directly to the farmer's bank account without any paperwork required.

3.3   Storage

After the procurement, the stock will be stored in various locations as per need and policy. As the stock moves, the Blockchain will be updated, and stock will be having the digital signature attached where they can easily find the existence of the stock in the blockchain network. If the food stored in the warehouse for any reason is not matching the quality of the Rice which can be tested by the QA (Quality Analyst) at any point in the lifecycle of the Supply Chain to maintain the Quality Standard of the Rice, the stored Rice can be traced back to the procurement center from where the paddy harvested.

3.4   Process

After receiving rice, will read and enter information regarding rice and update it on the digital profile of paddy. The report includes all those conditions which are being done in mills like the handling and changing paddy into rice and rice goes through the cleaning process, husking, arranging, polishing and packing. While processing, the tags which were labeled on the sacks might get destroyed, but this would not affect because the digital profile of the event is being updated every time it goes to the next step in processing. Before sending it to the distributor rice sacks are labeled with the new tags, goes through scanning by tag reader and then details are uploaded on the blockchain's digital profile. All the requirements are done processed rice is sent to distributors.

3.5   Distribution

After processing, issuing of rice starts, and data regarding the creation of rice, expenses, and conveyance are being exchanged on to the Blockchain at unequivocal between times with an ultimate target that exceptional exercises can be recorded, and thing can be followed in the dispersion technique.

3.6   Retail

At the retailing, when a retailer gets rice sacks from distributors, they can almost acquire and audit every one of the information regarding the process related to RSCM. Any client who comes to the retailer with the application software which is working with blockchain technology can see the history of rice and information of rice regarding every information of RSCM which is being uploaded on the network mentioned above Also, the entire SCM system is transparent; the result of cheats and scam will be lessened to the minimum.

4   Conclusions

The idea to implement Blockchain in Rice Supply Chain will not only change the way of how the FCI manages the supply chain but also help in managing significant wastage and proper management of rice. If there is an event where the FCI finds out that a particular product is found to be contaminated which could hamper the health of an individual by consuming it.

With the proposed framework of RSCM using Blockchain technology, operational efficiency of the entire SCM is suggested to be improved manifold. With the help of Blockchain, we can view the audit trail the process from procurement till it reaches the consumer. If needed the rice stock from that specific farm or batch can be recalled rapidly. Also, the consumer in the PDS will be benefitted by actually getting the right share of a food product. Incorporating Blockchain into the RSCM may give us a recognizability system by which we can follow, review the security and nature of item all through the method. It will also fight fraud and provides transparency in the order and Creating a permanent history of RSCM on a digital platform which is immutable will make this entire SCM more efficient.

References

1. Boschi, A.A.: An exploration of blockchain technology in Supply chain management. Eur. J. Purch. Supply Manag. **7**(1), 39–48 (2018). https://doi.org/10.1109/wict.2011.6141215
2. Tribis, Y.: Supply chain management based on blockchain: a systematic mapping study. In MATEC Web of Conferences (2017). https://doi.org/10.1051/matecconf/201820000020
3. Nir, K.: Exploring blockchain technology and its potential applications for education. Int. J. Manag. (2018). https://doi.org/10.1016/j.ijnfomgt.2017.12.005
4. Tian, F.: An agri-food supply chain traceability system for China based on RFID & blockchain technology. In: 2016 13th International Conference on Service Systems and Service Management (2016). https://doi.org/10.1109/ICSSSM.2016.753842
5. Winger, Y., Meita, S., Wang, J., Rit, M.: Making sense of blockchain technology: how will it transform supply chains? Int. J. Prod. Econ. **211**, 221–236 (2018)

6. Gautam, N., Chauhan, S.S.: Supply chain management: for the India agri-food sector, vol. 3, NO. 8 (2015). ISSN 2320–2092
7. Zyskind, G., Nathan, O.: Decentralizing privacy: using blockchain to protect personal data. In: Security and Privacy Workshops (2014)
8. Acharjamayum, I., Patgiri, R., Devi, D.: Blockchain: a tale of peer to peer security. In: 2018 IEEE Symposium Series on Computational Intelligence (2018). https://doi.org/10.1109/ssci.2018.8628826
9. Nir, K.: Exploring blockchain technology and its potential applications for education. Int. J. Manag. (2018). https://doi.org/10.1016/j.ijnfomgt.2017.12.005
10. Buterin, V.: A next-generation smart contract and decentralized application platform. Ethereum White Paper (2017)
11. Sreehari, P., Nandakishore, M.: Smart will converting the legal testament into a smart contract. In: 2017 International Conference on Networks & Advances in Computational Technologies (2017). https://doi.org/10.1109/NETACT.2017.8076767
12. Saveen, A., Radmehr, P.: Blockchain, ready manufacturing supply chain, using distributed ledger. IJRET 05(09) (2016)
13. Westerkamp, M., Victor, F., Kupper, A.: Blockchain-based supply chain traceability: token recipes model manufacturing processes (2018)
14. Nakamoto, S.: Bitcoin: a peer-to-peer electronic cash system. Consulted, pp. 1–9 (2008)
15. Bogart, S., Rice, K.: The blockchain report: welcome to the internet of value (2015)
16. Ganeshan, R., Harrison Terry P.: An introduction to supply chain management. Department of Management Sciences and Information Systems (21: 1/2), pp. 71–77 (2001)
17. Lee, H.L., Billington, C.: The evolution of supply-chain-management models and practice at hewlett packard. Interfaces 25, 42–63 (1995)
18. Christopher, M.: Logistics and supply chain management: strategies for reducing costs and improving services. Int. J. Res. Eng. Technol. 05(09), 1–10 (1998)
19. Saveen, A.: blockchain ready manufacturing supply chain using distributed ledger. IJRET 05 (09) (2016)
20. Wei, C., Zehua, G.: Decentralized applications: the blockchain-empowered software system (2018)
21. Sergei, T.: Ethereum: state of knowledge and research perspectives (2017)
22. Parizi, R.M., Dehghantanha, A.: Smart contract programming languages on blockchains: an empirical evaluation of usability and security. In: Chen, S., et al. (eds.) ICBC 2018, LNCS, vol. 10974, pp. 1–17 (2018)

Web Links

23. http://fci.gov.in/
24. http://fci.gov.in/information.php
25. http://fci.gov.in/movements.php
26. http://fci.gov.in/aboutUs.php

Managing and Monitoring E-Waste Using Augmented Reality in India

Siva Rama Krishnan Somayaji[✉], Surendheran Kaliyaperumal,
and Vivek Velayutham

Vellore Institute of Technology, Vellore, India
{siva.s,surendheran.k,vivek.velayutham}@vit.ac.in

Abstract. The growing demand of digital technologies are increasing the quality of human life at an unprecedented rate, but at the same time managing e-waste from these digital technologies are emerging as one of the biggest challenges for most of the industrial and government bodies. Collection of E-waste causes various problems such as environmental effluence, resource scarcity and it also results in developing various health hazards to human population. In this paper, we propose a framework using wireless sensor network and augmented reality to detect and mark the various high pollutant zone, which would save people from being exposed to hazardous environment.

Keywords: Wireless sensor network · Augmented reality · Effluent · GPS · GSM

1 Introduction

Electronic waste, e-scrap, or Waste Electrical and Electronic Equipment (WEEE) [13] is a loose category of excess, outdated, damaged, or unwanted electrical or electronic devices which have become out-of-use due to advancement in technology, approaching the end of their useful life. This includes used electronic components and gadgets which are destined for reuse, resale, salvage, recycling or disposal. It is also alarming to note that computer wastes containing hazardous heavy metals in large quantities are amassing very fast. Health hazards for humans [14] when exposed to e-waste consist of change in thyroid function, changes in cellular expression and function, adverse neonatal outcomes, changes in temperament and behaviour, and decreased lung function. India has 178 registered e-waste recyclers, recognised by the state governments to process e-waste [15]. But many of India's e-waste recyclers aren't recycling waste at all. Whereas, some are storing it in hazardous environments, others don't even have the ability to handle such waste, as per by the report of Union Environment ministry.

Augmented reality in short 'AR' is a technology that alters the visuals of reality by adding CGI (Computer generated imagery) either in 2D or 3D and allows user interaction with them [16].

AR may be made use of in various industries, including armed forces' investigation, the video games field, medication as well as in design. It is actually ending up being progressively common as an interdisciplinary area and also an impressive

© Springer Nature Switzerland AG 2020
P. Karrupusamy et al. (Eds.): ICSCN 2019, LNDECT 39, pp. 32–37, 2020.
https://doi.org/10.1007/978-3-030-34515-0_4

modern technology [17]. Especially, the simple fact that the smart phones, tablets and PCs possess extra strong CPUs, much higher resolution capable monitors and highly-developed video cameras along with attributes including GPS, electronic compasses and accelerometers has actually brought AR into the limelight.

AR is becoming popular in recent times but it's not an entirely a new thing. AR is here among us for a while and a cinematographer – Mortan Heilig was the one who initiated the whole idea in 1950 that eventually lead to the invention of AR. He tried to immerse his viewers in the on-screen activities by providing all senses of the entire story to deliver a 'near to real-world' feeling for his viewers. In the year 1962, he came up with a motorbike simulator called *Sensorama* [16] which provided senses of visual, audio, smell and vibration.

Several researchers contributed towards the evolution of AR since that and the phenomena of AR took its proper place when Paul Milgram and Fumio Kishino tossed the continuum of AR in the year 1994. Virtual Reality completely immerses the user into the virtual world whereas the AR handles both the virtual and real-world elements in its simulation [16]. The Fig. 1. clearly shows where the AR lies in its capabilities.

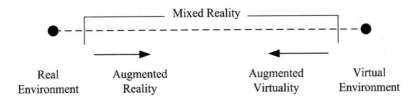

Fig. 1. Paul Milgram and Fumio Kishino AR continnum

AR has the prospective to be more highly effective and extra reliable through being incorporated along with various other modern technologies. The influence of *Pokémon Go* – An AR mobile game released in 2016, which has been creating fantastic impacts in a brief time, is actually still being reviewed. This AR game has been installed greater than one hundred million times in the first month and also made around $10 m daily [17]. At this point, the best threat associates with how users view this modern technology. When used in the right way, it will sure be a part of our daily life.

2   Literature Survey

The authors reviewed various ICT methods [1] involved in waste management. The main methods which were used to monitor and manage the solid waste was data aggregation techniques, GPS techniques, RFID and networking. The major challenge here is that the data acquired from dump yard should be accurate enough for efficient processing of solid waste.

Augmented Reality is an emerging technology [2] that deals with superimposing computer-generated imaginary over real-time visuals that are often displayed on an

HMD (Head Mounted Display). AR is a technology that combines information technology, computer generated digital images, animations and artificial intelligence.

The authors surveyed various IoT techniques [3] in real time systems. The authors compared the various network technologies such as WiFi and ZigBee based on number of nodes and packet loss. Based on the result analysis, the authors observed that the WiFi performed better than other technologies. Delay in data transmission is the huge issue in the paper discussed above.

Authors from the University of Évora, conducted a study using AR for environmental management [4]. In their study, the researchers have used augmented geo-referenced environments and found positive results in managing water purity levels, evolution of sub-soil structures, etc., For their research, they came up with a client-server based model named ANTS (Augmented Environment) and laid out a prototype for developing new AR tools that could act as an aid in areas such as environmental management.

The authors proposed a new framework [5] with the goal to improving the dump site treatment and transmission optimization in the waste management process. The wireless sensor network is used to sense the data from the dump yard and a sink would aggregate the data and send it to server for further processing. The major research challenge in this proposal is processing delay and accuracy. Real time systems such as waste dump monitoring site require minimum delay and maximum accuracy.

The authors reviewed different scenarios and case studies [6] involving IoT technologies to make the world greener. The authors also review various methods of green IoT and suggest different policies to build a green infrastructure. Furthermore, they suggest the usage of energy saving natural and organic radical batteries which do not use any heavy metals.

The authors discussed the challenges in designing a smart city [7] framework. The main components of this framework would be environment monitoring and location information services. The major challenge in this framework is to integrate the different ICT components and establishing an interface for them to communicate.

The author proposed a new method for process management in virtualization [8] which aids to reduce e-waste. The authors also surveyed the various places from where e-waste is being generated and their respective pollutant levels.

Another study that discusses MAR (Mobile Augmented Reality) analyses the pros and cons of using mobile phones [9] in combination with LBS (Location Based Services). The authors have analysed many possibilities of using AR along with the location of the users and they insist that in order to achieve convincing user/object tracking, the application interface should use more than one type of position retrieving methods. The authors also have diagnosed a few methods on how the location of a point-of-interest shall be shared with a server. They say, when a mobile device has some set of coordinates, it can be sent to the server which in return can be displayed in AR via a camera or can be mapped on a map. The authors recommend Google maps, as the service is provided for free and also comes with an API (Application Programming Interface) which is free to use as well for software developers.

Another comparative study done by the researchers of Media Studies and Research at Pillai Institute of Information Technology and Engineering [10] compared various types of AR SDK's and one section of their research compared the tracking capabilities of those SDK's. The comparison chart released by them clearly shows that both 'Metaio' and 'D'Fusion' SDK's support GPS tracking feature and they both support the Android and iOS platform for application development.

In relation with our study, a Canadian company – 'Scope', being experts in AR options for market, [10] has actually cultivated a program resource named *WorkLink* that functions through improving typical paper-based job directions into Smart guidelines. The customers are actually submerged into a 3D computer system created environments that overlay on real-life visuals. Scope's Founder, as well as Head of state David Nedohin, discusses that, in concept, this could possibly aid in the managing of risky misuses, such as atomic refuse.

Furthermore, a magazine named *"EnergyMatters: The nuclear waste opportunity"* [11] in its Autumn 2017 issue states that in spite of the intricacies that border the waste management sector, the top priority on every hazardous site is actually very clear – 'ensure our folks return home safely by the end of every day!'. Modern technology has actually lowered individual interference in the clean-up as well as rubbish administration methods however, the society can easily reduce the threats even further through welcoming immersive modern technology. The magazine further states that the immersive technologies are often helpful in training the workers and engineers in hazardous environments and 'virtually' walk them through the potential pitfalls even before they physically hurt themselves. Eye tracking technology in AR also helps tremendously in intimating the danger that lies right in front of the workers in a hazardous environment, says the author of the article.

Though the usage of AR in e-waste management is an emerging technique, AR is being extensively used for similar purposes in various fields [12]. For example, researchers from the University of the Balearic Islands successfully proposed an AR enabled module that detects the contamination level from a water source. The module makes use of an augmented smart phone camera app and scans for coloured signals that are generated by the nano particles in immunoassay which in return, yields the concentration of contamination in a simpler form which can be understood by non-technical people too.

3 Proposed Method

The proposed model consists of a high-powered drone that's fitted with various types of sensors such as gas, iconic, chemical etc., and will be operated from a remote location or somewhere safe outside the e-waste dump yard. The drone is also fitted with GSM and GPS modules for location detection and connecting with server. As the drone navigates across the dump yard, the sensors attached to it shall sense the hazardous places in the yard and flag them and send the level of hazards along with their GPS coordinates to the server. The server shall log the received information with the time stamp (Fig. 2).

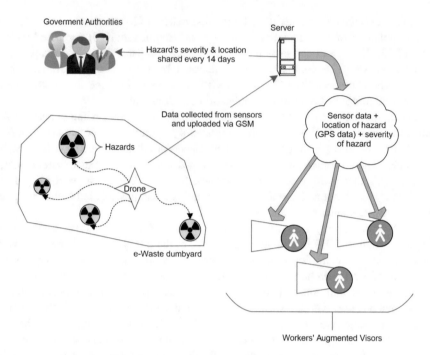

Fig. 2. e-waste remote monitoring model

The database stored in the server will be in sync (latest time stamped entry) with the augmented HMD's of the workers via internet. The HMDs will keep the worker well-informed about the anomalies even before they enter the dump yard. Once they are inside the dump yard the augmented HMDs will show them the direction and location of highly toxic area as an overlay over their real world visual so that the workers can proceed with caution or steer clear of the hazardous areas.

The database captured from the drone will be shared in the form of an electronic report with the concerned government officials every fortnight for their awareness and for necessary sanitation actions that may arise time to time. The drone scanning can be done every morning or as required to keep the safety of the workers' health.

4   Conclusions

Based on the survey carried out in this paper, it is concluded that there is not much safety measures for the workers of e-waste dump yards. They have been exposed to various health hazards as part of their job. The method proposed in this article would remotely monitor the effluent's level in an e-waste dump yard minimizing the human interference in evaluating the hazard levels and also will give the workers a chance to carry out necessary precautions well before entering the dump yard. The proposed design not only warns the workers but also would guide them via their augmented safety helmets to avoid high pollutant zones when they are inside the dump yard. This proposed design demands for further research to understand the challenges that may arise while implementing this plan especially in a developing country like India.

References

1. Hannan, M.A., Al Mamuna, M.A., Hussain, A., Basri, H., Begum, R.A.: A review on technologies and their usage in solid waste monitoring and management systems: issues and challenges. J. Waste Manag. **43**, 509–523 (2015)
2. Amin, D., Govilkar, S.: Comparative study of augmented reality Sdk's. Int. J. Comput. Sci. Appl. (IJCSA) **5**(1), 11–26 (2015)
3. Suresh, P., Daniel, J.V., Parthasarathy, V., Aswathy, R.H.: A state of the art review on the Internet of Things (IoT). In: International Conference on Science, Engineering and Management Research. IEEE (2014)
4. Romão, T., Correia, N., Dias, E., Trabuco, A., Santos, C., Santos, R., Nobre, E., Câmara, A., Danado, J., Romero, L.: Augmenting reality with geo-referenced information for environmental management. In: Proceedings of the 10th ACM international symposium on Advances in Geographic Information Systems, pp. 175–180 (2002)
5. Longhi, S., Marzioni, D., Alidori, E., Di Buo, G., Prist, M., Grisostomi, M., Pirro, M.: Solid waste management architecture using wireless sensor network technology. In: 5th International Conference on New Technologies, Mobility and Security (NTMS). IEEE (2012)
6. Arshad, R., Zahoor, S., Shah, M. A., Wahid, A., Yu, H.: Green IoT: An investigation on energy saving practices for 2020 and beyond. In: Future Networks: Architectures, Protocols, and Applications, vol. 5, pp 15667–15681. IEEE (2017)
7. Theodoridis, E., Mylonas, G., Chatzigiannakis, I.: Developing an IoT smart city framework. In: International Conference on Information, Intelligence, Systems and Applications. IEEE (2013)
8. Krishnan, S.S.R., Balasubramanian, K., Mudireddy, S.R.: Design/Implementation of a novel technique in virtualization to reduce e-waste. In: Int. J. Adv. Res. Comput. Commun. Eng. 2 (12) (2013)
9. Aurelia, S., Raj, M.D., Saleh, O.: Mobile augmented reality and location based service. Adv. Inf. Sci. Appl. **2**, 551–558 (2014)
10. https://resource.co/article/augmented-reality-comes-waste-management-11342
11. https://www.atkinsglobal.com/ ∼ /media/Files/A/Atkins-Corporate/group/sectors-documents/energy/energy%20matters/Energy-Matters-Magazine.pdf
12. Russell, S.M., Doménech-Sánchez, A., de la Rica, R.: Augmented reality for real-time detection and interpretation of colorimetric signals generated by paper-based biosensors. ACS Sens. **2**(6), 848–853 (2017)
13. http://ec.europa.eu/environment/waste/weee/index_en.htm
14. https://www.thelancet.com/journals/langlo/article/PIIS2214–109X(13)70101-3/fulltext
15. https://www.downtoearth.org.in/blog/waste/recycling-of-e-waste-in-india-and-its-potential-64034
16. Jamali, S.S., Shiratuddin, M.F., Wong, K.W.: A review of augmented reality (AR) and mobile-augmented reality (mAR) technology: learning in tertiary education. Learn. High. Educ. **20**(2), 37–54 (2013)
17. Alkhamisi, A.O., Monowar, M.M.: Rise of augmented reality: current and future application areas. Int. J. Internet Distrib. Syst. **1**(04), 25 (2013)

Optimization of Sensing Time in Cognitive Radio Networks Based on Localization Algorithm

P. Poornima[1]([✉]) and S. Chithra[2]

[1] Annai College of Engineering and Technology,
Kumbakonam, Tamilnadu, India
poornimapriya21@gmail.com
[2] SSN College of Engineering, Chennai, Tamilnadu, India

Abstract. Cognitive radio is one of the promising technologies that allows opportunistic spectrum access for unlicensed users namely the secondary users (SUs) to utilize the white spaces available in the spectrum allocated to the primary user (PU). To achieve a power efficient spectrum sensing the geometry-based localization algorithm is proposed. Here, the deployment of primary and secondary users are first considered. Then the secondary users are categorized based on their known and unknown locations. In order to locate the unknown secondary user, within the prescribed geometry, is identified by employing the Spider Monkey Optimization algorithm. To ensure optimal convergence for the given problem statement N number of iterations are done and best values are calculated using the fitness factor of the samples. Based on the best values, the optimized location of the unknown SUs is recorded and priority is given to those SUs that remain close to the PU and maintains a proper degree of separation. Since wireless devices are more vulnerable to security threats, this paper also aims in identifying malicious nodes and prohibiting them to access the white spaces by inducing denial of service through primary user to safeguard the spectrum access by outliers. Simulation results also provide a satisfactory outcome for achieving power effective spectrum sensing and performance of the proposed algorithm is studied in the presence of malicious and normal nodes.

Keywords: Spectrum sensing · Spider monkey optimization · White spaces · Malicious node detection

1 Introduction

Cognitive radio (CR) is an intelligent technology that enables efficient spectrum utilization under different circumstances. It incorporates more software-based radios that ease parameter changes such as detecting the presence of spectrum, tuning the modulators and amplifiers to the required range and so on. To accomplish these goals it classifies the users into two major groups: Primary and secondary users. The primary user (PU) is used to define those members who have licensed spectrum access. The secondary user (SU) who does not have any license but still considered as eligible users to access the unused spectrum when the PU is absent. This amendment by FCC has

© Springer Nature Switzerland AG 2020
P. Karrupusamy et al. (Eds.): ICSCN 2019, LNDECT 39, pp. 38–48, 2020.
https://doi.org/10.1007/978-3-030-34515-0_5

created a major revolution in the field of telecommunication. Cognitive radio is then proposed as the way to exploit the advantages offered by the FCC. For example in television broadcasting channels are allotted with Ultra-high and very high frequency ranges. To avoid interference, guard bands are provided between each station and the frequency in guard band remains idle. According to the FCC report, for the white space devices like the microphone/mobile phone this guard band frequency is more enough to perform its operation. To achieve better spectrum utilization, some challenges imposed by the CR configuration needs to be solved. The key challenges of CR are spectrum sensing, power utilization, and interference management. Spectrum sensing can be done by the SUs, it is coded accordingly by the professionals to configure itself to the available PU spectrum. Sometimes it may ends with false alarms, to overcome those drawbacks hybrid algorithms are proposed. Since most of the wireless networks are battery-operated, power management in CR is also challenging. Most of the power wastage occurs when the SU tries to sense the spectrum that is not in its range. Here comes the time based and geometry-based sensing phenomena. To achieve effective energy consumption, geometry-based sensing can be employed. As per the revised amendment by FCC dated 20th March 2019, more provisions are given to location based sensing schemes. Hence in this paper GDOP based localization algorithm is proposed. To avoid interference related issues SUs are provided with an optimization mechanism to access the spectrum holes effectively. To achieve this we make use of spider monkey optimization in this work.

2   Related Work

Cognitive radio is an emerging technology offers wireless personal devices for accessing the spectrum, which is one of the most valuable resources in this digital era of communication. It defines some protocols that need to be satisfied by the secondary user if it wants to access the licensed spectrum [1]. The term "white space" denotes the unused spectrum, which is intentionally left unused between different TV stations to avoid interference and related issues. It allows secondary users like unlicensed microphones and other white space devices (WSDs) to make use of this spectrum without causing any interference [2]. Some critical changes are recommended by the FCC for the operation of white space devices to ensure proper security for the primary user data and also to incorporate more WSDs to avoid the idle spectrum holes. Here we consider two of them for our objective:

(a) If the secondary user is fixed and maintains its position in the same location then it needs to register its device information with the respective primary user database.
(b) The location identification of the unknown secondary user is increased from ±50 m to ±100 m within the geographical location of the primary user.

In order to locate the actual position of the secondary user some location-based techniques like GPS can be employed [3]. As per the FCC rules for designing cognitive radio networks, we make use of the location-based algorithms for nodes deployment and opportunity is given to those nodes which satisfies the rule (b) as mentioned above. The localization of nodes enables the CRNs to effectively manage power consumption and also helps to overcome the shortcomings in hardware [6]. Location-based detection of nodes can be done in different ways namely: proximity detection, triangulation detection and scene analysis [7]. Since in CRNs most of the nodes are mobile in nature, for robust analysis proximity detection offers a high degree of accuracy. To find optimal convergence for the desired solution we find Spider Monkey Optimization (SMO) results are more efficient when compared with other optimizations like Genetic Algorithm, Ant Colony Optimization, etc., it is primarily based on the fact, the fission-fusion social behavior of animals [9]. SMO makes use of the advantage of the foraging behavior of monkeys. It converges quickly as per the objective function to validate the results with the minimum number of experiments [10]. By choosing proper fitness value, selected secondary users can avail the white spaces. After optimization, we need to calculate the selected secondary users' distance through geometric dilution of precision [8]. In the wireless scenario, privacy is always a challenging factor to ensure proper QoS for transceiver nodes. There is more possibility of collapse the operation of CRNs in terms of both sensing time and channel allocation. Adversary users make all possible ways to exploit the data corresponding to the licensed user and also try to reduce the channels allocated to the legitimate secondary users [12]. GDOP measurement is used in tracking devices for maintaining the mobile device connection with the base station to avoid interruptions in service [13]. To avoid much power consumption, it is needed to measure the receiver signal strength for locating the closer primary user [14]. In order to encourage effective sensing with limited power, secondary users divides their frame into two slots namely: sensing slot and transmission slot. The sensing slot is used to determine whether the spectrum is free and if it available for transmission, then SU starts its data transmission process [15].

Based on the above-discussed factors, the rest of the paper is organized as follows. Section 3 deals with nodes deployment and predicting the actual location of unknown secondary users. Section 4 indulges the SMO based GDOP for grouping secondary users within the transmission range of the primary user for achieving power-efficient sensing. Section 5 is dedicated to identifying malicious nodes and emulating them from the CRN. Section 6 corresponds to Simulation results that are used to demonstrate the performance of the proposed localization algorithm.

3   Localization of Nodes

The scope of this paper relies on the location of nodes present in the transmission range of the primary user. White spaces can serve more secondary users operating under different frequency ranges. Consider a Cognitive Radio Network (CRN) comprising of primary and secondary nodes, in which the secondary users are classified into two types based on their location information namely: known or fixed SU and unknown SU. For geometry-based sensing, the location of nodes is given more importance. SU which remains stationary and repeatedly accesses the spectrum allocated to the specific PU can be added as a genuine white space device to avoid frequent sensing mechanism. In terms of fixed SU, their location information is registered with the corresponding primary user database. Su which is mobile and attempts to access different spectral ranges are unknown and they need to satisfy some of the constraints to prove itself as a genuine user. To find an optimal transmission link, location information of all SUs are needed. Hence to locate the SUs, SMO based GDOP algorithm is proposed. The process of localization is done by taking the best values from multi-dimensional scaling matrix and further process is done in proximity matrix. The localization mechanism is shown in Fig. 1.

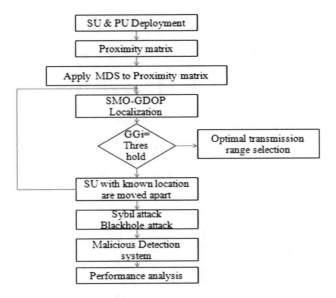

Fig. 1. Flow diagram

4 SMO Based GDOP Algorithm

By using the current position of the nodes only, we can allocate them with the necessary channels. For having update-to-date location information, geometric dilution of precision is employed. It maintains the location information of all the nodes and constantly checks with the threshold value of $GG_i \leq \Upsilon$ i.e., it should be within ± 100 m of the transmission range. If the above criteria are not satisfied, the SU is declared as a non-participating node and does not auction any channel for its use.

4.1 Parameters of the SMO Algorithm

The SMO algorithm depends on the following parameters for its proper operation: Local Leader phase (LLP), Global Leader phase (GLP), Local Leader Learning phase, Global Leader Learning phase, Local Leader Decision phase, Global Leader Decision phase. A global position update is done by choosing the best values amongst the iterations done so far.

Steps involved in the SMO algorithm:

4.1.1   Initialization of Population (N)

Here the nodes are generated randomly up to N values. Each element corresponds to the active SM member in the group. Each member is represented as follows:

$S_i = \{1, 2, \ldots, N\}$ where $i = \{1, 2, \ldots, N\}$

$$S_{ij} = S_{minj} + UD(0, 1) \times (S_{maxj} - S_{minj})$$

Where Smaxj and Sminj are bounded values of Si in the jth direction.

UD (0, 1) represents uniform distribution of variables. Random distribution of nodes is shown in Fig. 2 and are labeled as primary user, known and secondary users.

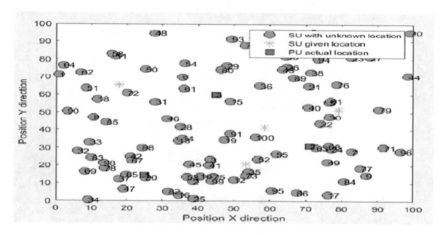

Fig. 2. Distribution of nodes in the CRN

4.1.2   Local Leader Phase (LLP)

It mainly deals with the fitness function calculation based on the past experience of each group member (GM).

If the fitness value of the new position becomes higher than that of the old position, then position update request is processed.

The position update for ith GM is given by,

$$S_{newij} = S_{ij} + UD(0, 1) \times (LLP_{kj} - S_{ij}) + UD(-1, 1) \times (S_{rj} - S_{ij})$$

Algorithm 1 Local Leader Phase Position Update:

```
for each k ∈ {1, ..., GM} do
for each member Si ∈ kᵗʰ group do
for each j ∈ {1, ..., D} do
if UD(0, 1) ≥pbr then
    S_newij = S_ij + UD(0, 1) × (LLP_kj − S_ij )+UD(−1, 1) × (S_rj − S_ij )
else
S_newij = S_ij
end if
end for
end for
end for
```

4.1.3   Global Leader Phase (GLP)

Every new position update made by the GM in the LLP needs to be reported to their Global leader for finding a better position for the desired target.

The position update made by GLP is given as follows:

$$S_{newij} = S_{ij} + U(0,\ 1) \times \left(GLP_j - S_{ij}\right) + U(-1,\ 1) \times \left(S_{rj} - S_{ij}\right)$$

Algorithm 2 Global Leader Phase Position update:

```
for k = 1 to GM do
count = 1;
GS = kth group size;
while count < Group size do
for i = 1 to Group size do
if UD(0, 1) < pi then
count = count + 1.
Randomly select j ∈ {1...D}.
Randomly select S_r from kᵗʰ group provided r ≠i .
S_newij = S_ij + U(0, 1) × (GLP_j − S_ij )+U(−1, 1) × (S_rj − S_ij )
end if
end for
if i is equal to Group size then
i = 1;
end if
end while
end for
```

4.1.4   Local Leader Learning Phase

In this phase based on the fitness value calculation, a position update is done by using the greedy algorithm.

For choosing the appropriate values, Local limit count is set to 1 for each GM and in case of position update, it is incremented by 1.

4.1.5   Global Leader Learning Phase

Now it's the time to intimate each GM about the newly found positions to reach the target and similar to the local leader learning phase it sets global leader count and initialized as 1 and depending on the position update by a local leader it is incremented by 1.

4.1.6   Local Leader Decision Phase

The local limit count values are analyzed by the Local leader and the threshold value is maintained for validating the updated positions and those values of fitness that are not in compliance with the count are labeled as not updated and retains the values of the old position.

Algorithm 3 Local Leader Decision Phase:
    for k = {1...GM} do
    if LocalLimitCountk > LocalLeader Limit then
    LocalLimitCountk = 0.
    GS = k^{th} group size;
    for i ∈ {1...Group size} do
    for each j ∈ {1...D} do
    if U(0, 1) ≥probability of random location then
    $S_{new\,ij} = S_{minj} + UD(0, 1) \times (S_{maxj} - S_{minj})$
    else
    $S_{new\,ij} = S_{ij} + UD(0, 1) \times (GLP_j - S_{ij}) + UD(0, 1) \times (S_{ij} - LL_{kj})$
    end if
    end for
    end for
    end if
    end for

4.1.7   Global Leader Decision Phase

The entire operation depends primarily on the decision made by the global leader and hence it focuses mainly on the information collected from those Local leaders with vast deviations in their members' fitness value findings. It maintains global leader limit count to ensure those values are satisfactory for maintain the position and initiates updating else old values are registered for further process is shown in Fig. 3.

Algorithm 4 Global Leader Decision Phase:

if GlobalLimitCount > GlobalLeader Limit then
GlobalLimitCount = 0
if Number of groups < GM then
Divide the population into groups.
else
Combine all the groups to make a single group.
end if
Update the Local Leaders position.
end if

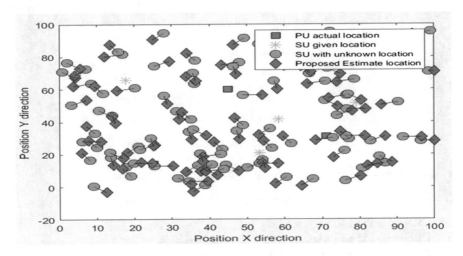

Fig. 3. Optimized location of unknown secondary nodes.

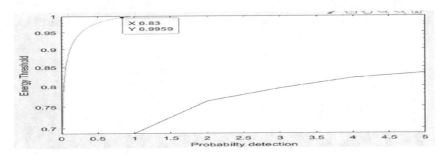

Fig. 4. Probability of unknown location of secondary nodes detection by proposed algorithm

5 Malicious Node Detection

As detailed earlier, to safeguard the valuable information of the primary use from the adversaries it becomes necessary to identify the nodes which behave suspiciously in the given CRN. Here we are analyzing the activities of every secondary user based on its RSSI and nodes that exhibit variations in its interference levels are considered malicious in nature. For the prediction of nodes, we make use of a decision system that is able to discriminate malicious nodes from legitimate nodes. Different types of malicious behavior like Sybil attack, denial of service attack and black hole attack are predicted is shown in Fig. 5.

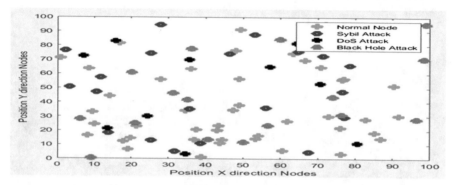

Fig. 5. Discrimination between normal and malicious nodes.

6 Performance Evaluation and Simulation Results

The proposed algorithm has experimented under various conditions like:

(a) CRN with PU and known SUs.
(b) CRN with PU and known as well as unknown SUs.
(c) CRN with normal nodes and malicious nodes.

In our proposal, we've chosen the number of nodes as 100 and the threshold value for localization as ±100 m. Experimental results have demonstrated favorable outcomes when compared with traditional localization algorithms is shown in Fig. 6. SMO based GDOP provides a greater contribution in selecting appropriate SU based on its fitness factor is shown in Fig. 4.

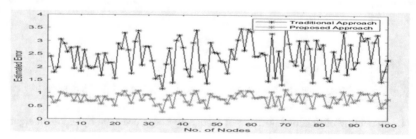

Fig. 6. Proposed algorithm vs Traditional approach.

7 Conclusion

Hence, in this paper power-efficient spectrum sensing based on SMO is proposed. The main objective of consuming less power is achieved by providing spectrum access to those SUs in the range of nearby primary user. To ensure proper security to the legitimate nodes, the detection system has been built and enables better channel utilization by the genuine nodes. The proposed algorithm abides the rules and regulations of FCC and it can be incorporated in designing CRN for real-time purposes.

References

1. Mitola, J., Maguire, G.Q.: Cognitive radio: making software radios more personal. IEEE Pers. Commun. **6**(4), 13–18 (1999). https://doi.org/10.1109/98.788210
2. https://transition.fcc.gov/oet/spectrum/table/fcctable.pdf. Federal Communications Commission Office Of Engineering and Technology Policy and Rules Division
3. https://www.fcc.gov/document/amendment-part-15-rules-unlicensed-white-spaces-devices
4. Fragkiadakis, A.G., Tragos, E.Z., Askoxylakis, I.G.: A survey on security threats and detection techniques in cognitive radio networks. In: IEEE Communications Surveys & Tutorials, vol. 15, no. 1, pp. 428–445 (2013). https://doi.org/10.1109/surv.2011.122211.00162
5. Saeed, N., Nam, H.: Energy efficient localization algorithm with improved accuracy in cognitive radio networks. IEEE Commun. Lett. **21**(9), 2017–2020 (2017). https://doi.org/10.1109/lcomm.2017.2712802
6. Guibène, W., Slock, D.: Cooperative spectrum sensing and localization in cognitive radio systems using compressed sensing. J. Sens. **2013**, 9 (2013). https://doi.org/10.1155/2013/606413. Article ID 606413
7. Farid, Z., Nordin, R., Ismail, M.: Recent advances in wireless indoor localization techniques and system. J. Comput. Netw. Commun. **2013**, 12 (2013). https://doi.org/10.1155/2013/185138. Article ID 185138
8. Yongcai, A., Bo, Z., Baozhuo, Z., Shili, W.: Change of geometric dilution of precision (GDOP) for integrated system. In: 2016 IEEE Information Technology, Networking, Electronic and Automation Control Conference, Chongqing, pp. 660–662 (2016). https://doi.org/10.1109/itnec.2016.7560443
9. Bansal, J.C., Sharma, H., Jadon, S.S.: Memetic Comp. **6**, 31 (2014). https://doi.org/10.1007/s12293-013-0128-0
10. Al-Azza, A.A., Al-Jodah, A.A., Harackiewicz, F.J.: Spider monkey optimization: a novel technique for antenna optimization. IEEE Antennas Wirel. Propag. Lett. **15**, 1016–1019 (2016). https://doi.org/10.1109/LAWP.2015.2490103
11. Clancy, T.C., Goergen, N.: Security in cognitive radio networks: threats and mitigation. In: 2008 3rd International Conference on Cognitive Radio Oriented Wireless Networks and Communications (CrownCom 2008), Singapore, pp. 1–8 (2008). https://doi.org/10.1109/crowncom.2008.4562534
12. Chen, R., Park, J., Reed, J.H.: Defense against primary user emulation attacks in cognitive radio networks. IEEE J. Sel. Areas Commun. **26**(1), 25–37 (2008). https://doi.org/10.1109/JSAC.2008.080104

13. Sharp, I., Yu, K., Guo, Y.J.: GDOP analysis for positioning system design. IEEE Trans. Veh. Technol. **58**(7), 3371–3382 (2009). https://doi.org/10.1109/tvt.2009.2017270
14. Yu, Y.H., Sun, C., Qin, N.N., Gao, K., Chen, D.Z.: CR-RSS location algorithm for primary user in cognitive radio. J. China Univ. Posts Telecommun. **21**(1), 22–25 (2014). https://doi.org/10.1016/S1005-8885(14)60264-8
15. Liu, X., Li, F., Na, Z.: Optimal resource allocation in simultaneous cooperative spectrum sensing and energy harvesting for multichannel cognitive radio. IEEE Access **5**, 3801–3812 (2017). https://doi.org/10.1109/ACCESS.2017.2677976

Analysis of POX and Ryu Controllers Using Topology Based Hybrid Software Defined Networks

K. Rohitaksha[1,2(✉)] and A. B. Rajendra[3]

[1] JSS Academy of Technical Education, VTU University, Bengaluru, India
rohithaksha.k@gmail.com
[2] Department of CSE, Vidyavardhaka College of Engineering,
VTU Research Center, Mysuru, India
[3] Vidyavardhaka College of Engineering, Mysuru, India
abrajendra@vvce.ac.in

Abstract. Software Defined Networking (SDN) enables operation and management of communication networks by decoupling the control and forwarding plane. However, replacing the existing legacy switches in the network with the SDN switches leads to the high budget and also pure SDN suffers from scalability and robustness. Hybrid SDN is a new emerging technique in which legacy switches are connected with the SDN switches in order to overcome some of the drawbacks with respect to SDN. In this paper, we compare the performance of two open source controllers namely POX and Ryu, by considering the different Hybrid SDN topologies. The results are taken by running the simulation in Mininet for different QoS requirements in all the topologies.

Keywords: SDN · Hybrid SDN · POX · Ryu

1 Introduction

The SDN model has gained significant research attention recently. In SDN architecture the control plane and forwarding plane are separated from each other. Forwarding plane consist of routers and switches which are responsible for forwarding of packets control plane is responsible for decision making which packet should be sent which direction. Deployment of Software Defined Network is financially very costly. Hybrid SDN network model integrates the existing network devices to be connected with the SDN switches. Hybrid SDN model allows for the network administrator to program only in the SDN regions. Hybrid SDN supports the legacy protocols and OpenFlow protocol. Hybrid SDN models allows the controller to forward the packets to the legacy switches and their by leaving the forwarding policies to the traditional control plane. Hybrid networks can overcome from the problem of controller failure by shifting to the time-tested convolutional mechanism. Hybrid SDN supports some of the specific features of the pure SDN combined with the benefits of the convolutional switches. With the feature of decoupling the control and the forwarding plane, SDN enables for ease of network design.

© Springer Nature Switzerland AG 2020
P. Karrupusamy et al. (Eds.): ICSCN 2019, LNDECT 39, pp. 49–56, 2020.
https://doi.org/10.1007/978-3-030-34515-0_6

Hybrid SDN models can classified based on the architecture, components and the functionality. These are the Hybrid SDN with controller only, with middleware only, with Upgrade/agent, Edge placements, SDN overlay, SDN and Non-SDN zones. In service base Hybrid SDN model nodes are placed at certain locations in the network based on the deployment policies. In case of traffic class based model SDN nodes are placed in order to monitor the heavy traffic under the SDN. Hybrid SDN overcome from the problem of congestion by enabling the legacy routing mechanism.

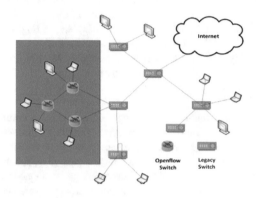

Fig. 1. Hybrid SDN Network.

Figure 1 [17] shows the example of Hybrid SDN Network. The figure is divided into two portions in which the portion in the green indicates the SDN switches enabled with the OpenFlow protocol and the rest of the network indicates the normal switches.

SDN controller plays a important role in the operation of the SDN neytwork. These controllers are: POX [1], Ryu [2], Trema [3], FloodLight [4], and OpenDaylight [5]. POX is open source controller which supports for rapid development and proto-typing. It supports for topology discovery and pythonic interface. POX provides a framework for communicating with SDN switches using either the OpenFlow or the OVSDB protocol. Ryu is open source software defined networking controller which provides software components and APIs to create network management and control applications.

The implementation is done on Mininet emulator which acts as a testing bed for the Hybrid SDN. Mininet support light weight virtualization which is reflects the real network, running real kernel, switch and application code. The analysis is done for different topologies and network parameters. The results can help to choose the best controller for Hybrid SDN that have different QoS requirements.

2   Related Work

Khondoker *et al.* [6] have analyzed the properties of open source controllers. The authors propose an Analytic Hierarchy Process to compare the properties of the controllers through the following factors like the languages it supports for writing the script, operating system, REST API, newer version of OpenFlow protocol etc. Based on the results "Ryu" is selected to be the best controller based on the requirements.

Shalimov *et al.* [7] considered different QoS requirements for comparing the performance of different open source network controllers. The test was conducted for different parameters like throughput, latency etc. according to the authors Ryu performance is better based on the test bed considered.

Kaur *et al.* [9] analyzed the performance of SDN networks with respect to language support, OpenFlow support, being open source, GUI and REST API. In this paper they have used POX controller and Mininet to emulate SDN and verified the working of network applications in POX.

Govindraj *et al.* [10] proposed in datacenters and enterprises OpenFlow enabled switches are more efficient, since this type of switches reduces the bandwidth wastage and provides load balancing feature.

Yahya *et al.* deliberates different types of OpenFlow switch including five different types of SDN emulators and simulators. Based on the survey conducted NOX, OpenDaylight supports for OpenFlow protoc version 1.3 and for GUI, EstiNet is better one.

2.1   Mininet

Mininet is a network emulator that allows us to create virtual network with switches, hosts and SDN controller. Mininet provides network testbed for developing OpenFlow applications. It supports for large number of developers to work independently on the same topology parallel. It provides command line interface CLI for executing the network tests and for debugging. Supports for Python API for creating the network and experimentation. Mininet boosts faster, scales larger and provides more bandwidth. It allows the developers to prototype the real network efficiently. Mininet does not require physical network for testing the complex topologies. Using the Mininet the user can customize the rules for forwarding the packets. Mininet provides OVS switches and OVS controllers. Mininet enables to install additional packages like D-ITF. One of the drawbacks of the Mininet is for all the virtual hosts it uses single Linux kernel.

3   Experimental Settings

We have considered three different topologies Linear and Tree topology for analyzing the performance of the controllers. The Mininet virtual machine is installed on Ubuntu 14.04 Lts operating system.

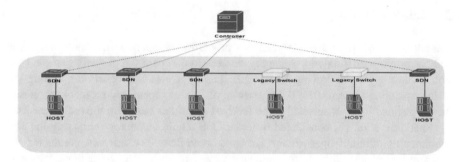

Fig. 2. Linear topology.

The Fig. 2 depicts linear topology with six hosts, four SDN enabled switches and 2 legacy switches. As indicated in the diagram the flow tables were installed on the SDN switches and were connected with the SDN controller. The test was conducted by transferring packets from legacy switch to SDN switches and results were recorded.

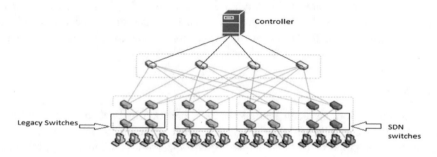

Fig. 3. Fat tree topology with K = 4.

The Fig. 3 shows the architecture of fat tree topology. The topology consists of three levels in which the last level two of the switches are legacy switches and the remaining switches are SDN switches. The results were considered by sending the packets from the Hosts connected to the normal switches to the SDN switches.

The test was conducted by running the topology using POX and Ryu controller. In each of the topology the few switch was configured as the legacy switches and other switches has OpenFlow enabled switches. Each of the controller is executed with different QoS parameters.

3.1   Performance Evaluation

The general network QoS parameters are:

Latency: The amount of time that it takes for a packet to travel along the network from source to destination.

Jitter: The variation of latency.

Loss: The percentage of packets that failed to reach their destination. Packet loss ratio is measured in percentage of total.

Throughput: The ability of network to carry a volume of data over a unit of time

Fig. 4. POX controller

Figure 4 shows the POX controller running in the background. The figure indicates the POX version is being used.

Fig. 5. Ryu controller

Figure 5 indicates the Ryu controller is up and the flow table of the intermediate switches and the port through which the packets are forwareded.

Table 1. Comparsion between Ryu and POX controller using linear and tree topolgy.

Linear topology						
	Time taken	Transfer in bytes	Jitter	Total datagrams	Lost	percentage
Ryu	427 s	10 GB	0.012 ms	7304366	61	0.00084%
POX	427 s	10 GB	0.008 ms	7304366	26	0.00036%
Tree topology						
Ryu	568 s	10 GB	0.271 ms	7304366	3621401	50%
POX	546 s	10 GB	0.050 ms	7304366	3587743	49%

Table 1 shows the results. Simulation was conducted by transmitting the 10 GB of packets from the legacy switches to the SDN switches using the UDP protocol. According to the observation the percentage of packets loss in POX controller is less than the Ryu controller for both the topologies.

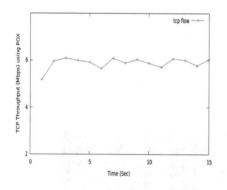

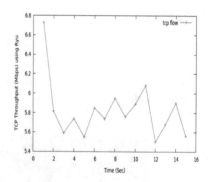

(a) TCP throughput using POX controller   (b) TCP throughput using Ryu controller

Fig. 6. Linear topology TCP throughput

Figure 6 shows the TCP throughput performance of linear topology using the POX and Ryu controller. The simulation was run for 15 s and the bandwidth was measured. By comparing the Fig. 6(a) and (b) the throughput performance of TCP using POX controller is around 5.8 Mbps and Ryu is 5.6 Mbps.

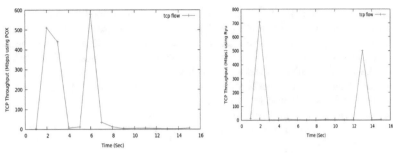

(a) TCP throughput using POX controller   **(b)** TCP throughput using Ryu controller

Fig. 7. The throughput performance of TCP in tree topology

Figure 7 depicts the throughput performance of the tree topology using Ryu controller. By comparing the the ouput of the POX and Ryu controlles we can cleary observe that the throughput is high using POX when compared with the Ryu.

4 Conclusion

İn this paper, we have proposed the methodolgy for analysing the performance of two controllers using Hybrid Software defined Network topologies. POX and Ryu are the two opensource SDN controllers. These controllers supports for modularity and also supports for the Python language. The topologies considered here are linear and tree topolgy. We have configured few nodes has the normal switches and other nodes has SDN switches. The results shown indicates that the performance of the POX controller is better when compared the Ryu controller with respect to both the topologies. The test was conducted over ten iterations and in all the iteration the results were consistent.

References

1. POX Controller. Accessed 11 June 2013. http://www.noxrepo.org/pox/about-pox/
2. Ryu. Accessed 11 June 2013. http://osrg.github.io/ryu/
3. Trema. Accessed 11 June 2013. http://trema.github.io/trema/
4. FloodLight. Accessed 11 June 2013. http://www.projectfloodlight.org/floodlight/
5. OpenDaylight. Accessed 11 June 2013. https://www.opendaylight.org/
6. Khondoker, R., Zaalouk, A., Marx, R., Bayarou, K.: Feature-based comparison and selection of software defined networking (SDN) controllers. In: 2014 World Congress on Computer Applications and Information Systems (WCCAIS), 17 January 2014, pp. 1–7. IEEE (2014)
7. Shalimov, A., Zuikov, D., Zimarina, D., Pashkov, V., Smeliansky, R.: Advanced study of SDN/OpenFlow controllers. In: Proceedings of the 9th Central & Eastern European Software Engineering Conference, pp. 1–6. ACM (2013)
8. Al-Somaidai, M.B., Yahya, E.: Survey of software components to emulate OpenFlow protocol as an SDN implementation. Am. J. Software Engin. Appl. **3**(6), 74–82 (2014)

9. Kaur, S., Singh, J., Ghumman, N.S.: Network programmability using POX controller (2014). http://sbsstc.ac.in/icccs2014/Papers/Paper28.pdf. Accessed Sept 2015

10. Govindraj, S., Jayaraman, A., Khanna, N., Prakash, K.R.: OpenFlow, load balancing in enterprise networks using Floodlight controller. University of Colorado (2012). http://morse. colorado.edu/~tlen5710/12s. Accessed Sept 2015

11. Mininet an instant virtual network on your laptop (or other PC) (2015). Mininet Team. http:// Mininet.org. Accessed Sept 2015

12. NOX. Accessed 11 June 2013. http://www.noxrepo.org/

13. Azodolmolky, S.: Software Defined Networking with OpenFlow. Packet Publishing, Birmingham (2013)

14. Mccauley, J.: Pox: A python-based Openflow controller (2014). http://www.noxrepo.org/ pox/about-pox, Accessed Sept 2015

15. Mininet an instant virtual network on your laptop (or other PC) (2015). Mininet Team. http:// Mininet.org. Accessed: September 2015

16. Dixit, A., Hao, F., Mukherjee, S., Lakshman, T.V., Kompella, R.: Towards an elastic distributed SDN controller. In: Proceedings ACM SIGCOMM HotSDN (2013)

17. Amin, R., Reisslein, M., Shah, N.: Hybrid SDN networks: a survey of existing approaches. IEEE Commun. Surv. Tutorials **20**, 3259–3306 (2018)

Automated Robot Communication System Using Swarm Intelligence

Prachi R. Rajarapollu$^{(\boxtimes)}$ and Debashis Adhikari

School of Electrical Engineering, MIT Academy of Engineering,
Alandi, Pune 412105, India
prrajarapollu@entc.maepune.ac.in,
dadhikari@entc.mitaoe.ac.in

Abstract. Swarm intelligence is best example of natural phenomenon giving of team work. Here the focus is on use of swarm intelligence to form the communication between automated robots for better and reliable results. With the main aim of exploring the concept for the progressive development of swarm robotics in an engineering field and solve the complex real time applications by setting a communication among automated robots. Currently, Swarm robotics is one of the most important application areas for swarm intelligence. Swarming behaviors of groups of organisms is called as swarm intelligence. There are many advantages of swarm intelligence like, good performance, high reliability, low cost with less complexity. Swarm robotics is the group of robots which are self-assembled and self-communicating to solve such task, which are difficult to solve individually by a single robot. Swarm robots are designed in such a way that they can self-configure and dynamically change their structure to meet environmental conditions. Section 1 will give the detail introduction of research work carried out. While coming consecutive sections will explain literature review, complete implemented system, result and references used.

Keywords: Swarm robotics · Swarm intelligence · Multi-Robot Communication · Server-Client Application · Obstacle detection

1 Introduction

There are various insects and creatures in the natural world, which follow simple law to accomplish an assignment cogently and expertly. The different cases of collective intelligence seem to arise from very simple individuals that are working together. There are many examples of collective intelligence seems to arise from very simple individuals that are working together [1]. Swarm Intelligence is the trait found throughout the animal kingdom. Swarm intelligence is an effective tool for managing complicated tasks, which comes via coordinating with each other without the leader. Application of swarm intelligence on the robots is called as swarm robotics. Swarm robotics is group of robots communicating together to perform task. This provides lot of diversity and helps in widening the capabilities of individual robot.

This swarm of simple robots coordinates their actions in a decentralized way. Robot swarming provides flexibility and robustness. Swarm Robotics is inspired from the

© Springer Nature Switzerland AG 2020
P. Karrupusamy et al. (Eds.): ICSCN 2019, LNDECT 39, pp. 57–64, 2020.
https://doi.org/10.1007/978-3-030-34515-0_7

swarming nature of organisms. Examples such as Bees, Ants [11], Schooling of fishes, Flock of birds, colony of bats, bacteria colony [9], etc. Behaviors of such beings are mirrored on the robots. Different algorithms are developed for functioning on the same concept, so that robots can replicate the functioning of swarm [2–5]. Swarm robotics provides application in the field of science such as navigation [8] in the confined area, agriculture foraging, searching for tumors in the human body; target searching, search and rescue operations, military applications. Swarm robots fix up other issues, which a single large robot faces [6]. These robots are cost efficient, power efficient [7], durable as such, if small unit of swarm fails, the swarm continue to work, multi-tasking is possible as each robot or the batch of swarm with different tasks. This paper focuses on one of the application of swarm intelligence. Swarm bots are implemented using Arduino microcontroller. Multiple robots communicate using sensors. Stepper motors are used to maneuver the bots. Path of each robot is monitored [10]. Ant colony optimization is used for identifying the shortest path for the robots [11, 12].

2   Literature Review

Where the problems that seems difficult or impossible for single individuals to resolve, can easily be solved by a group of organisms by working together. So, swarm intelligence can be seen as a mechanism in which multiple robots can be used to overcome some of their own cognitive limitations. Such system doesn't work on Master-Slave configuration, where one robot is master and others are slaves and slaves have to follow the instructions given by the master. But here each and every robot acts freely and asks for help when ever required. By this type of approach it becomes easy to solve the problems that are hard to solve individually. There are number of fields where swarm robotics is playing important role like, environment monitoring systems, oil cleaning, under water surveillance and many more.

A large number of researches are taking place continuously in the field of Swarm Intelligence. Zhu and Tang [1] gave a broad overview of swarm intelligence. They defined the term swarm intelligence in three parts namely biological basis, artificial literature and swarm engineering.

Nair, Frye, Coronado and Qin [2] has developed multi-robot system using particle swarm optimization (PSO) technique, which is used to evaluate the social characteristics of the insects. Also different approaches are discussed that are used for formation control and collaborative control of self-governing vehicles.

Patil, Upadhye, Kazi and Singh [4] have used Arduino MEGA-2560 to control robot for swarm application. In order to accomplish their objectives they have made use of target tracking algorithm.

Liyan, Sainan, Geng, Yongli and Guanyan [11] has proposed ant colony clustering algorithm based on swarm intelligence which is much efficient than ant colony algorithm. They have also proposed a new methodology of picking and dropping objects.

Lv and Zhu [10] have explained a new dynamic routing algorithm for packet switch communication networks. This algorithm is inspired by ants that explore the area as a network and find the best route to the destination.

Ravinandan, Prasad and Kumar [12] have proposed a novel way of discovering different paths/routes to the known destination and create an intellectual map using swarm intelligence. The technique is developed so that human interruption. The target of this paper is path exploration in combat and non-combative area for search and rescue operation.

3   Developed System

The system had been developed step by step by setting following goals, initially design a system of robots to follow a random path to reach a destination. In the initial phase all the robots will try to identify the path to the destination. The robot which will identify the destination first will communicate with other robots. These robots after the communication is get setup among all, will calculate shortest path to reach destination. Every robot will create its own database that will contain step values given to the stepper motors. The robots trying to find destination will share its database so that a map can be created using the step values. The database created by each robot will be stored for reference and it will be used for reaching the destination point. The system flowchart has been shown in Fig. 1.

There will be a confined area/field where two or more bots will be travelling randomly in search of an unknown destination (Black-patch). Swarm-robots travel the route randomly to reach the aim. The microcontroller used is ATmega328p. The power supply of 7 V to 12 V is provided to the micro-controller. Interfaced are the two IR sensors, Wi-Fi module (ESP8266) and two stepper motors. The system block diagram has been shown in Fig. 2.

IR (infrared sensor) used serves two purposes. One is for object detection and other is for black spot detection. IR sensor is present in circular manner above the robot for detection of objects around the robot. One IR sensor will be below the robot for black spot detection (destination identification).

Micro-controller generates random step values. These are feed to the stepper motors through the motor driver. According to the step values received by the stepper motors, the robots direct themselves in a random direction. Thingspeak.com is a very popular website used for implementation of IoT (Internet of Things) applications. The random step values generated are transmitted to thing speak server with the help of Wi-Fi module interfaced with micro-controller. Using various AT commands, a wireless connection is established between the ESP and thing speak web server. Thingspeak.com provides API keys (Write API key and Read API key) using which the data can be stored or accessed. Once the connection is established, using write API key the ESP starts sending the step values to the web server, results has been shown in Fig. 6. After successfully sending data to the server, the connection is cut-off. The send step values have accessed through a program using read API key or download the file in various formats such as.csv,.txt, etc.

The step values retrieved from the server used to generate a GUI map. The map makes understanding of robot location easier as shown in Fig. 4. If any of the robots find the destination, it will reveal its location to other robots and wait at the destination until the remaining robots arrive. The notified robots find the shortest pathway to the destination from their current location [10]. Following the shortest distance, they will reach the destination.

Once the destination is traced and notified to other robots, the shortest path to the destination of each individual robot will be calculated in GUI (software) and then the respective number of steps of the motors will be calculated and given to the robots [14].

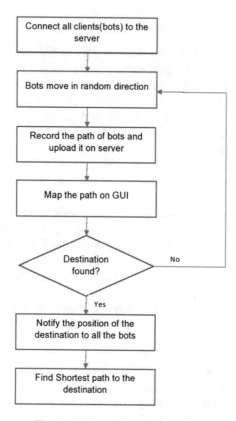

Fig. 1. Proposed system flow-chart

There will be a confined area where two robots will be travelling freely in search of a black spot. Black Spot is nothing but the destination of all the robots and all will try to find it. To detect the black spot and to communicate with other robots, IR sensors have been used. The robot that finds the black spot will notify to other robots that he have successfully identified the black spot. Initially, there is no master robot or Slave robot as shown in Fig. 3(a), (b) and (c). All the robots will travel randomly in the area and the robot that first find the black spot will act as master and order the others to find the shortest path to destination and follow it and reach the destination. As shown in the Fig. 3, there are three robots i.e. bot1, bot2 and bot3.

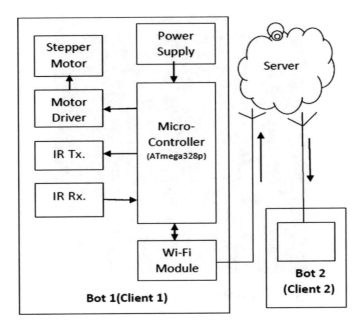

Fig. 2. System block diagram

All robots are allowed to departure from a same starting point. These three robots move in random fashion in order to find the black spot (destination). Random step values are given to stepper motor so that the robots can follow a random path. Bot2 is successful in finding the black spot and hence it will convey the message to bot1and bot3 that he has found the black spot. WiFi module is used for transmission and reception of the information between two robots. The step values of each bot will be transmitted to server using Wi-Fi module [13]. Accessing these step values a map will be created as shown in Fig. 4. Once the destination is determined the other two robots i.e. bot1 and bot3 referring the map, will find the shortest route to the destination using any shortest path algorithm which will save the time of robot to reach destination as well make the system more time efficient [15]. Following the shortest path the robots will reach the destination.

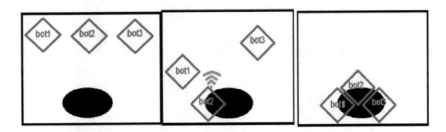

Fig. 3. Robots finding destination (black spot) with same starting point and reaching to destination by following shortest path

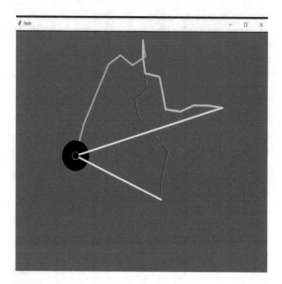

Fig. 4. Map created using fetched values from server

4   Result and Conclusion

Complete system has been implemented successfully. Results received have been shown in following Fig. 5. The designed robot is shown in Fig. 6. Interfacing of IR sensor is done to serve two purposes, destination identification and obstacle detection. Stepper motor is directing the robot in required direction. The data (i.e. step values given

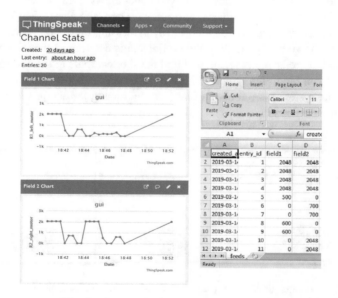

Fig. 5. Communication set up among robots

to stepper motor) is successfully sent to the server website (www.thingspeak.com) using Wi-Fi module. Client can access data from server and create a map based on step values. Robots are tracing the destination successfully. The communication has been set up successfully among robots. Experimental results received has been tested on various scale and given the satisfactory output. Bots are following the accurate path and reaching to destination with shortest path.

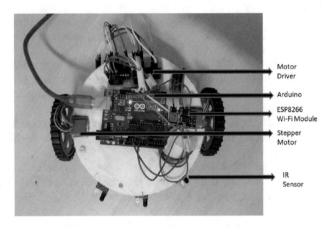

Fig. 6. Complete hardware system implemented

References

1. Zhu, Y.F., Tang, X.M.: Overview of swarm intelligence. In: 2010 International Conference on Computer Application and System Modeling (ICCASM 2010), vol. 9, pp. 400–403 (2010)
2. Corondo, E.M., Qin, Y., Nair, S.C., Frye, M.T.: Swarm intelligence for the control of group of robots. In: 10th System of Systems Engineering Conference (SoSE), pp. 205–206 (2015)
3. Robbins, N., McLurkin, J., McMullen, A., Habibi, G.: A robot system design for low-cost multi-robot manipulation. In: 2014 IEEE/RSJ International Conference on Intelligent Robots and Systems (IROS 2014), pp. 912–918, September 2014
4. Kazi, F.S., Patil, D.A., Upadhye, M.Y., Singh, N.M.: Multi robot communication and target tracking system with controller design and implementation of swarm robot using arduino. In: 2015 International Conference on Industrial Instrumentation and Control (ICIC), pp. 412–416, May 2015
5. Samsudin, K., Arvin, F., Ramli, A.R.: A short-range infrared communication for swarm mobile robots. In: 2009 International Conference on Signal Processing Systems, pp. 454–458, May 2009
6. Kottas, A., Drenner, A., Papanikolopoulos, N.: Intelligent power management: Promoting power consciousness in teams of mobile robots. In: 2009 IEEE International Conference on Robotics and Automation, pp. 1140–1145, May 2009
7. Sauze, C., Neal, M.: Long term power management in sailing robots. In: Oceans 2011 IEEE-Spain, pp. 1–8, June 2011

8. Salwani, S., Jain, S., Chandwani, V.K.: Ad-hoc swarm robotics optimization in grid based navigation. In: 2010 11th International Conference Control, Automation, Robotics and Vision, 7-10 December 2010, pp. 1553–1558 (2010)

9. Ming, L.: A novel swarm intelligence optimization inspired by evolution process of a bacterial colony. In: Proceedings of the 10th World Congress on Intelligent Control and Automation, Beijing, China, 6–8 July 2012, pp. 450–453 (2012)

10. Zhu, Y., Lv, Y.: Routing algorithm based on swarm intelligence. In: Proceedings of the 10th World Congress on Intelligent Control and Automation, Beijing, China, 6–8 July 2012, pp. 47–50 (2012)

11. Geng, T., Yongli, L., Guanyan, C., Liyan, D., Sainan, Z.: Ant colony clustering algorithm based on swarm intelligence. In: 2013 6th International Conference on Intelligent Networks and Intelligent Systems, pp. 123–126, May 2013

12. Kumar, M.V., Ravinandan, M.E., Prasad, E.V.: Adaptive path exploration and cognitive map generation using swarm intelligence. In: 2016 International Conference on Electrical, Electronics, Communication, Computer and Optimization Techniques (ICEECCOT), pp. 318–321 (2016)

13. Mai, S., Steup, C., Mostaghim, S.: Simultaneous localisation and optimisation for swarm robotics. In: 2018 IEEE Symposium Series on Computational Intelligence (SSCI), Bangalore, India, pp. 1998–2004 (2018)

14. Bozhinoski, D., Birattari, M.: Designing control software for robot swarms: software engineering for the development of automatic design methods. In: 2018 IEEE/ACM 1st International Workshop on Robotics Software Engineering (RoSE), Gothenburg, Sweden, pp. 33–35 (2018)

15. Hou, L., Fan, F., Fu, J., Wang, J.: Time-varying algorithm for swarm robotics. IEEE/CAA J. Autom. Sinica 5(1), 217–222 (2018)

Data Sharing and Privacy–Preserving of Medical Records Using Blockchain

Shraddha Suhas Kavathekar[✉] and Rahul Patil

Computer Engineering, Pimpri-Chinchwad College of Engineering, Pune, India
shraddhaskavathekar@gmail.com,
rahulpatilpink@gmail.com

Abstract. The present way to deal with patient health records disappoints clinicians and fails to fulfill the requirements of a diverse patient population. Many of these issues are raised due to absentiseem of direct control by doctors and patients over the innovation that serves them. As these innovations empower secure credentialing of authorized professionals and audit of clinical exchanges without the cooperation of a doctor's facility or state middle person, we have actualized health data innovation that is bought and controlled by the people in the clinical relationship. A Blockchain is an open Distributed Ledger that can record transactions between two parties efficiently, verifiable and permanent way. This empowers every patient and each authorized specialist to claim and totally control their open source associated health records inside a protected situation. The aim is to study concepts and techniques of blockchain and to understand how blockchain technology works for different applications in different scenarios. Our arrangement is a patient-focused way to deal with privacy assurance in cloud computing and data dependent on advanced assent models, blockchain identity, and blockchain audit. Digitally signed and encrypted transactions are verified by the peers. Public blockchains, for example, Bitcoin and Ethereum have been appeared to expel institutional control and decentralize innovation with advantages to development and restriction opposition.

Keywords: Blockchain · Smart contracts · Decentralization · Ethereum

1 Introduction

The Blockchain innovation has gone past the "development trigger" and is at "peak of inflated desires." It holds the possibility to disturb diverse ventures, for example, back, saving money, store network just as healthcare. In any case, before making a plunge directly into the subtleties of how blockchain is influencing healthcare, how about we comprehend what is it first [1].

1. Private key cryptography—In private key cryptography, a mystery key is utilized as a variable alongside a calculation to scramble and unscramble the code. The key is kept mystery not withstanding when the calculation isn't. In a blockchain, a reference of the protected advanced identity is made, in any case, the exchanges are on the open system.

© Springer Nature Switzerland AG 2020
P. Karrupusamy et al. (Eds.): ICSCN 2019, LNDECT 39, pp. 65–72, 2020.
https://doi.org/10.1007/978-3-030-34515-0_8

2. Distributed ledgers— A distributed record otherwise called a common record is alluded to as an accord of shared records. In DLT, the record is refreshed progressively and no focal expert is considered capable to keep up the record. Rather, organize members keep the record refreshed. Any progressions made in the ledgers are reflected inside seconds.
3. Authentication—Authentication is a procedure that demonstrates validity. In a Blockchain, every one of the exchanges are verified before getting added to the chain. This procedure happens through calculations that approve and confirm every one of the exchanges. When the data is encoded and carefully marked and put away, the realness is fixed.

Healthcare firms, innovation trend-setters and the individuals from generally healthcare area are paying special mind to approaches to discover what's conceivable in the present occasions and what blockchain could improve and moderate later on [2]. Blockchain has the ability to draw out a monstrous achievement in the healthcare biological system as it can without much of a stretch acquire explicit changes the healthcare the executives of the patient. With the guide of this innovation, the power will return to individuals' hands. Implying that people will be in charge of taking care of their own records accordingly, gaining the general power of their own information [5, 6].

The innovation holds the capacity to effectively enhance patient consideration quality while keeping up the assets at a sensible rate. Every one of the difficulties and impediments that happen in numerous dimension authentication can be dispensed with through blockchain. With the expanding selection rate, Blockchain has advanced toward the healthcare part. Indeed, even in its starting stage, the innovation is as a rule decidedly acknowledged by individuals in the healthcare biological system.

As indicated by an examination led by IBM, around 16% of healthcare administrators are resolved about their plans to execute blockchain arrangement in their work this year, while around 56% expected to embrace blockchain constantly 2020 [7]. The complete vision for blockchain to disturb the healthcare part in the coming occasions is resolve issues that torment the present framework. Envision a healthcare framework where all the data is effectively available by doctors, patients and drug specialists at some random time. Blockchain permits the creation and sharing of a solitary normal database of health data [4].

This framework would be available by every one of the elements engaged with the procedure regardless of which electronic therapeutic framework they use. This offers higher security and straightforwardness while permitting doctors discover more opportunity to spend on patient consideration and their treatment. Additionally, it will likewise empower better sharing of measurements of looks into which, thusly, would encourage clinical preliminaries and treatment treatments for any uncommon ailment. In a healthcare framework, smooth information sharing between healthcare arrangement suppliers can prompt exactness in determination, powerful medicines, and financially savvy biological system. The everyday development of patient information requires appropriate use of assets so as to make the best use of the experiences found through it.

2 Literature Survey

Concentrating on quality health care administrations implies guaranteeing patient health the board at a predominant dimension consistently. Be that as it may, government principles and directions are making forms much increasingly monotonous and protracted. Because of this, keeping such procedures flawless and as yet giving successful patient consideration isn't practical much of the time. The real issue in giving quality healthcare administrations is the hole among suppliers and payers. The reliance of go between in the production network aggravates it even [10].

In the healthcare part, basic patient information and data stays dispersed crosswise over various offices and frameworks. Because of this, essential information isn't open and helpfully accessible in the midst of need. The current healthcare environment can't be viewed as entire as different players in the framework don't have a framework set up for smooth process the executives. Additionally, it is likewise named as lacking for dealing with the trading of data and requires certain significant changes.

The abuse of accessible information is keeping healthcare associations from conveying fitting patient consideration and astounding administrations for better health. In spite of being genuinely proficient as far as economy, these associations are not ready to satisfy the requirements of patients. Following are a couple details from assets that underscore on this reality. Healthcare information ruptures in associations are evaluated to cost around $380 per record in the present occasions. This sum is relied upon to increment with the progression of time. Numerous healthcare offices today are as yet subject to obsolete frameworks for keeping patient records. These frameworks hold the usefulness of keeping nearby records of the patient information. This can make it troublesome for the specialist to analyze which is tedious for the specialist and dull for the patients as well. Because of this, the expense of keeping up a patient-situated business is expanded extensively [8, 9].

Issues winning in the present healthcare segment are not restricted. They continue developing with high-force with time. The requirement for an in fact propelled framework is evident. Consider the issue of medication falsifying which prompts misfortunes of around $200 million. It could be lessened drastically if a framework with exact following highlights is established in the production network. Some other tedious and dreary process that outcomes in staggering expenses in the healthcare business is Health Information Exchange. Since patients don't have any power over their information, the odds of identity burglaries, monetary information violations and spamming are expanding each day [3].

In spite of having devices like PCs and cell phones at each healthcare office nowadays, we're as yet not ready to gather, examine, secure and trade information consistently. In this manner, the healthcare framework today not just needs a development framework rather it additionally needs a framework that is smooth, straightforward, financially effective and effortlessly operable.

3 Proposed Work

3.1 Architecture

Our hypothesis is that huge numbers of these issues are because of an absence of direct control by doctors and patients over the innovation that serves them. As these advancements empower secure credentialing of authorized specialists and audit of clinical exchanges without the cooperation of a doctor's facility or state mediator, we can exhibit health data innovation that is bought and controlled by the people in the clinical relationship.

A unique and precise as well as patient-centered way to deal with privacy assurance in cloud computing and data sharing in healthcare has been created dependent on advanced assent guidelines, blockchain identity, and audit. As the cost of cloud innovation keeps on dropping and the assessment of individual information keeps on rising, preparations dependent on self-sovereign as conflicting to institutional innovation are increasingly handy By authorizing every patient and each authorized professional to privilege and totally control and give rights, their open source associated health records innovation, we can illuminate probably the most troublesome privacy problems including patient identity, admittance to social factors of health, and the emerging dangers of re-distinguishing proof of information through machine insight (Fig.1).

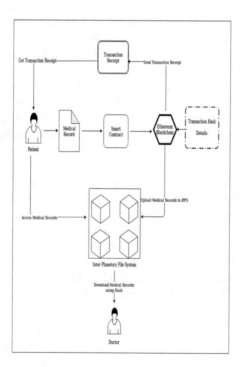

Fig. 1 System architecture

3.2 Algorithm

The below code snippet gives the main function of accessing records if and only if a patient provides their name:

```
    The below code snippet gives the main function of accessing records
if and only if a patient provides their name:
function    addName(uint256    _recordID,    string_name)    public
   patientExist(msg.sender) onlyPatient(_recordID)
   recordExists(_recordID,          msg.sender)    notEmpty(_name)
   patientNotProvidedName(_recordID, msg.sender)
{
   records[_recordID][msg.sender].providedName=t rue;
   records[_recordID][msg.sender].name=                    _name;
   addresshostpitalInRecord=
records[_recordID][msg.sender].hospital;
   mappingByName[hostpitalInRecord][_name]+=1
;
payPatient(msg.sender);
emit           NameAddedToRecords(_recordID, msg.sender);
}
```

4 Result and Discussions

Healthcare Dapp Performance and Transaction Status

By the time of compilation of this paper only frontend of project is remaining. Firstly, we will compile our solidity smart contract which is the basis of transaction between two accounts. Metamask is used to choose ropsten test network and show the interaction between accounts below Fig. 2 shows the contract deployment status (Figs. 3 and 4):

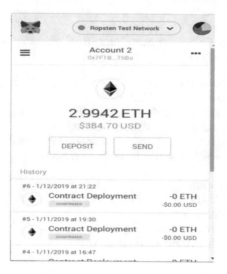

Fig. 2 Contract deployment on metamask

Fig. 3 Successful transaction view on etherscan

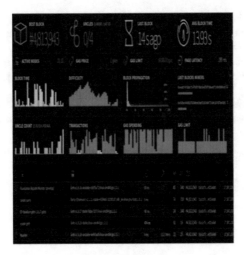

Fig. 4 Statistics

5   Conclusion

Our demonstration shows a practical implementation of a DAPP for sharing patient's data with physicians and hospitals. While this DAPP may be used for sharing patient data it assures security and integrity of patients' data. We implement a solution where the powers of institutions are restricted whereas patient rights are taken care of. In our set up two accounts, used by the patient and the physician, access the Ethereum test network. In our smart contract demonstration, we have a patient fill a form in the web app, have an additional person ask the access, and then allow or reject the access thereafter; thus, presenting the four core functions of our smart contract.

References

1. Azaria, A. Ekblaw, T. Vieira, Lippman, A..: MedRec: using blockchain or medical data access and permission management. In: IEEE 2nd International Conference Open Big Data, p. 6 (2016)
2. Zhu, H., Liu, X., Lu, R., Li, H.: Efficient and privacy preserving online medical pre-diagnosis framework using nonlinear SVM. IEEE J. Biomed. Health Inf. **12** (2016)
3. Aras, S.T., Kulkarni, V.: Blockchain and Its applications - a detailed survey. Int. J. Comput. Appl. **180**(3), 29–35 (2017)
4. Fan, K., Wang, S., Ren, Y., Li, H., Yang, Y.: MedBlock: e_cient and secure medical data sharing via blockchain. J. Med. Syst. Springer, Heidelberg (2018). https://www.readbyqxm d.com/journal/27398/7
5. Michalas, A., Weingarten, N.: HealthShare: using attribute-based encryption for secure data sharing between multiple clouds. In: 1063–7125/17 2017 IEEE 30th International Symposium on Computer-Based Medical Systems (2017)

6. Ji, Y., Zhang, J., Ma, J., Yang, C., Yao, X.: BMPLS: blockchain-based multi-level privacy-preserving location scheme for telecare medical information systems. Springer, Heidelberg (2018). https://www.readbyqxmd.com/journal/27398/

7. Kaur, H., Alam, M.A., Jameel, R., Mourya, A.K., Chang, V.: A proposed solution and future direction for blockchain-based heterogeneous medicare data in cloud environment. Springer, Heidelberg (2018). https://dblp.org/pers/a/Alam:Mohd=_Afsha

8. Li, H., Zhu, L., Shen, M., Gao, F., Tao, X., Liu, S.: Blockchain-based data preservation system for medical data. Springer, Heidelberg (2018). https://www.readbyqxmd.com/journal/27398/7

9. Lin, J., Niu, J., Li, H.: PCD: a privacy-preserving predictive clinical decision scheme with e-health big data based on RNN. In: IEEE Conference on Computer Communications (2017). 978-1-5386-2784-6

10. https://solve.care/media-digest2

11. https://ambcrypto.com/top-seven-healthcare-blockchains/

Analysis of PAPR in an OFDM Signal by Adopting Clipping and Filtering System and also by Varying Clipping Levels and Symbol Size

M. N. Geetha[1]([✉]) and U. B. Mahadevaswamy[2]

[1] Department of Electronics and Communication Engineering,
Vidyavardhaka College of Engineering, Mysuru, India
geethashekar73@gmail.com
[2] Department of Electronics and Communication Engineering,
Sri Jayachamarajendra College of Engineering, Mysuru, India
ubms_sjce@yahoo.co.in

Abstract. OFDM is the most efficient methods for the transfer of high to low data rate in the wireless communication system. It has 'N' number of advantages namely immune to multi fading, lesser inter symbol interference and high spectral efficiency. Despite the hyp, OFDM also has the disadvantage of having a very high PAPR i.e. PAPR ratio. In the view of reducing PAPR, clipping and filtering system is commonly employed and the reduction of PAPR is observed by varying clipping levels and symbol size.

Keywords: OFDM · PAPR · Clipping · Filtering · Clipping level · Symbol size

1 Introduction

In designing any contemporary communication system, one should know how the OFDM signals are realized inside the system. A well known DFT is used so as to realize this OFDM signal. One of the major application of using the so called multi-carrier system (MS) is to send as many as large and huge data streams in a constrained bandwidth. These data's are then partitioned and sent to parallel subcarriers. One should also observe that the PAPR will be very high due to non uniform distributions of peaks in the subcarriers. Since OFDM system generally uses a very high powered amplifier [1, 2], their energy efficiency is more or less exactly related to the PAPR of the signal input. Also because of non uniform distributions of these peaks in the subcarriers, the power amplifier drives to a saturation region. Due to this reason it is more important to lessen or reduce PAPR nearly possible in an OFDM signal. Different techniques or methods are available to lessen or reduce this PAPR, namely one is by making use of Signal distortion techniques (SDT) or by Signal non distortion techniques (SNDT). In SDN, there are many other techniques that can be used namely

© Springer Nature Switzerland AG 2020
P. Karrupusamy et al. (Eds.): ICSCN 2019, LNDECT 39, pp. 73–84, 2020.
https://doi.org/10.1007/978-3-030-34515-0_9

clipping and filtering [3–6], peak windowing [7, 8], companding techniques [9, 10]. Similarly in SNDT there are many other techniques namely Selective-Mapping-Technique (SMT) [11], Tone Injection (TI) [12, 13], PTS [14], Tone reservation [15–18]. There are tradeoffs in the prior mentioned methods including its data rate, BER, computational complexity and performance parameters. Clipping and filtering system is the elementary method among all other method to reduce this PAPR in an OFDM signal. In this technique, initially the Input signal i.e. OFDM signals are first clipped to the predefined level. Later, these signals are made to overleap the filter in order to mitigate the band radiations. Else ways, the deterioration of the wireless channels is minimized by conniving the wireless network. The effects in the wireless channels can be rise above or can be overcome by using OFDM. This in turn reduces the persuade of the multipath without the help of complex equalizers. The OFDM then transmits a huge data stream along N lower rate subcarriers.

In any OFDM signal model, the input signal is made to transmit and it is then modulated by making use of PSK, QPSK or QAM techniques. Later, IFFT process is performed at the end of transmitter section. As a result of this, production of orthogonal subcarriers occurs at the end of transmitter section. One of the limitation of the OFDM system is high PAPR which is due to addition of numerous subcarriers amplitude as well as phases from that system [19]. Under other conditions, the signal that is transmitted has a high peaks called as PAPR. This in turn degrades the power ampliers performance thus causing the decrease in SNR of the signal. Therefore the reduction efficiency can be calculated using CDF. This paper is formulated and explained as follows. The Sect. 1 explain the basic introduction about the OFDM signal and explain the various methods or the techniques available for the reduction of PAPR. The Section 2 explain the system model that is present in this paper and Section 3 explains the methodology used for analysis of this OFDM signal and its effect in PAPR reduction. The Section 4 explains the results and discussion that is observed in this paper using clipping and filtering method for varying clipping levels and symbol size. Finally Section 5 describes the conclusion section of the work.

2 System Model

Initially, the OFDM signal divides the wideband into numerous orthogonal subcarriers. The size of the DFT used is say 'N'. The subcarriers here convey so called data frames F and this let 'F' be denoted as follows

$$F = [f1, f2, f3, \ldots \ldots fp] \tag{1}$$

Where $f_p = [f_0, f_1, f_2 \ldots f_{N-1}]^T$.

Mathematically OFDM is indicated as

$$D(t) = \frac{1}{N} \sum_{k=0}^{N-1} C_k \, e^{j2\pi f_k t} \tag{2}$$

Also, The PAPR is expressed as

$$PAPR = \frac{max \, |x(t)|^2}{E|x(t)|^2} \tag{3}$$

Where $x(t)$ = Signal

$\qquad\qquad E[x(t)]$ = average power of the signal $x(t)$.

Similarly PAPR in decibels (dB) is expressed as

$$\text{PAPR in (dB)} = 10 \log 10 \, (PAPR) \tag{4}$$

3 Methodology

As discussed in the Sect. 1, the simplest and easiest method for mitigating the PAPR is to make an clipping and filtering system. Figure 1. Shows the method of how the signal is computed to the system and how the OFDM signals looks after performing both the clipping and filtering system. Initially the input signal is disposed to the QPSK and the outputted signal from the QPSK is fed as an input to the IFFT module. Once the inverse FFT operation is done, the signal is fed into clipping module. Clipping is an non linear process where it causes a distortion in out band as well as in band. But as the signals are analyzed in the FFT module, the BER is reduced. In this clipping method, a part of OFDM signals are clipped by considering some threshold value. Later, a high peak of this signal is generally sliced of and it is then reach through the Power Amplifier (PA). After this process, a signals are defined and it is then limited by put into action of clipper to a predefined level. This predefined level is called as clipping level. This clipping is generally located at the transmitter level.

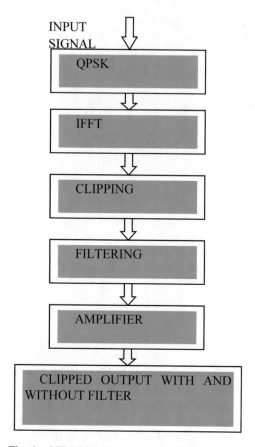

Fig. 1. OFDM Signal transmission block diagram

4 Result and Discussion

In this section, comparison of clipping performance is done for the OFDM signal. Here the different clipping levels are used and varied for a normal OFDM signals. Similarly the symbol size of 1024 is also considered. The PAPR performance is measured for both original OFDM signal as well for clipped OFDM signal. Figures 2, 3 and 4 shows the result for clipping level 1, 1.25, 1.5 with the symbol size of 1024 in each case.

In the below Fig. 2, one can observe that for a clipping level of 1 and symbol size of 1024, the PAPR ratio for original OFDM signal is 8.35 and with clipping and filtering system the PAPR is 2.98.

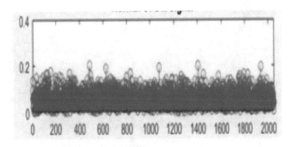

(a) Normal OFDM signal

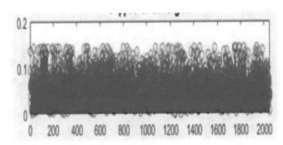

(b) Clipped OFDM Signal

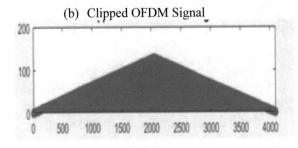

(c) Clipped and Filtered OFDM signal

Fig. 2. Clipping level = 1, symbol size = 1024

From this one can conclude that PAPR using clipping and filtering system outperforms well compare to clipped OFDM signal. The results are tabulated in Table 1.

Similarly Fig. 3 shows the output for clipping level 1.25. The PAPR for original OFDM signal stands 7.46 and with clipping and filtering system, the PAPR is 2.74. From this one can conclude that PAPR using clipping and filtering system outperforms well compare to clipped OFDM signal. The results are tabulated in Table 1.

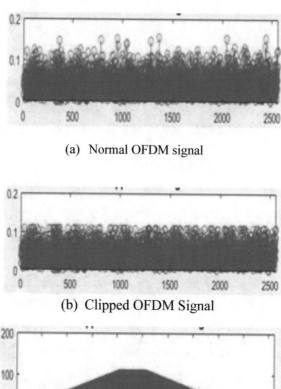

(a) Normal OFDM signal

(b) Clipped OFDM Signal

(c) Clipped and Filtered OFDM signal

Fig. 3. Clipping level = 1.25, symbol size = 1024

Figure 4 shows the output for clipping level 1.5 and symbol size of 1024. The PAPR for original OFDM signal is 7.66 and with clipping and filtering technique the PAPR is 2.63. From this one can conclude that PAPR reduction using clipping and filtering technique outperforms well compare to clipped OFDM signal. The results are tabulated in Table 1.

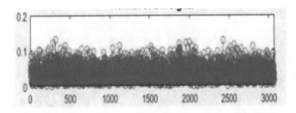

(a) Normal OFDM signal

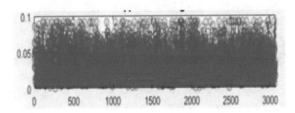

(b) Clipped OFDM Signal

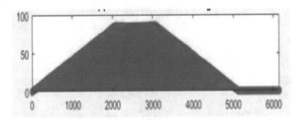

(c) Clipped and Filtered OFDM signal

Fig. 4. Clipping level = 1.5, symbol size = 1024

Table 1. Comparison of PAPR of Original and Clipped OFDM Signal with different clipping level.

Clipping levels	Symbol size	PAPR of original OFDM	PAPR of Clipped OFDM
1.00	1024	8.3547	2.9898
1.25	1024	7.4650	2.7401
1.5	1024	7.6687	2.6309

Next, Comparison of PAPR of original and Clipped OFDM Signal with different clipping levels are tabulated and are shown in the Table 1. Also, Fig. 5 shows the comparison results in terms of bar graph. From the graph one can observe that as the clipping level increases PAPR of the original OFDM and PAPR of clipped OFDM decreases.

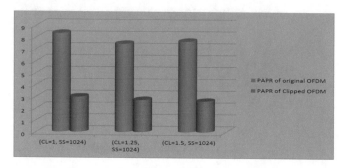

Fig. 5. Comparison of original OFDM and clipped OFDM with varying CL

Similarly, the performance of both clipping and filtering can be seen in the following below Figs. 6, 7 and 8 with varying clipping levels and also with varying symbol size.

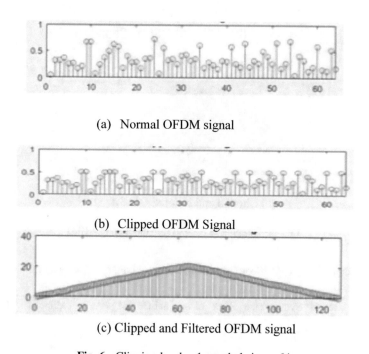

(a) Normal OFDM signal

(b) Clipped OFDM Signal

(c) Clipped and Filtered OFDM signal

Fig. 6. Clipping level = 1, symbol size = 64

The above Fig. 6 shows the output for clipping level 1 and symbol size of 64. The PAPR for original OFDM signal stands 3.5621 and with clipping and filtering system the PAPR is 2.9209. From this, one can conclude that PAPR employing clipping and filtering system outperforms well compare to clipped OFDM signal. The results are tabulated for varying symbol size in Table 2.

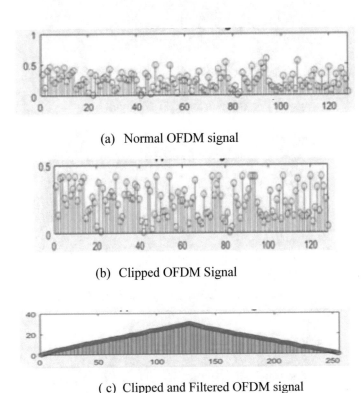

(a) Normal OFDM signal

(b) Clipped OFDM Signal

(c) Clipped and Filtered OFDM signal

Fig. 7. Clipping level = 1.5, symbol size = 128

Similarly, the above Fig. 7 shows the output for clipping level 1.5 and symbol size of 128. The PAPR ratio for original OFDM signal stands 4.7107 and with clipping and filtering system the PAPR is 3.0370.

Similarly, Fig. 8 shows the output for clipping level 1.5 and symbol size of 1024. The peak to average power ratio for original OFDM signal is 8.1457 and with clipping and filtering technique the PAPR is 3.0076.

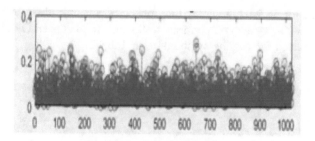

(a) Normal OFDM signal

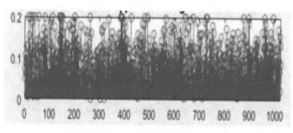

(b) Clipped OFDM Signal

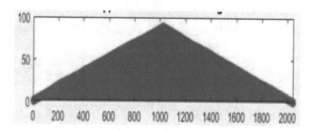

(c) Clipped and Filtered OFDM signal

Fig. 8. Clipping level = 1.5, symbol size = 1024

Next, Comparison of PAPR of original and Clipped OFDM Signal with different clipping levels and for varying symbol size are tabulated and are shown in the Table 2. Also, Fig. 9 shows the comparison results in terms of bar graph. From the graph one can observe that as the clipping level increases PAPR of the original OFDM and PAPR of clipped OFDM decreases. Figure 10 shows the overall comparison of original OFDM signal and clipped OFDM by considering both varying CL and SS.

Table 2. Comparison of PAPR of original and Clipped OFDM Signal with varying symbol size

Clipping levels	Symbol size	PAPR of original OFDM	PAPR of Clipped OFDM
1.00	64	3.5621	2.9209
1.25	128	4.7107	3.0370
1.5	1024	8.1457	3.0076

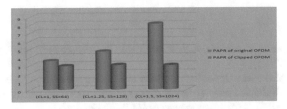

Fig. 9. Comparison of original OFDM and clipped OFDM with varying symbol size

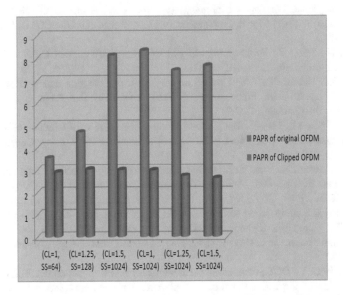

Fig. 10. Overall comparison of Original OFDM and Clipped OFDM considering both varying CL and SS

5 Conclusion

OFDM is found to be most attractive and simplest method for both wireless applications as well as wired applications owing to its very high and huge data rates. On the other hand, one of its major drawbacks is in generating the high peak to peak average ration. As discussed in this paper, there are many techniques to reduce the PAPAR. This paper highlights more on clipping and filtering technique. It has the potential to reduce the PAPR but losses its data rate substantially. From the result section it can be inferred that reduction in PAPR for the OFDM signal is done using a simple clipping and filtering method. There are many other methods and using which a PAPR can be reduced and the choice of the technique to be used depends on the system requirement.

References

1. Bhad, S., Gulhane, P., Hiwale, A.S.: PAPR reduction scheme for OFDM. Procedia Technol. **4**, 109–113 (2012)
2. Wirastuti, N.M.A.E.D.: Understanding peak average power ratio in VFFT-OFDM systems. In: Applied Mechanics and Materials, vol. 776, pp. 419–424. Trans Tech Publications, Switzerland (2015). www.scientific.net/AMM.776.419
3. Wang, L., Tellambura, C.: A simplified clipping and filtering technique for PAR reduction in OFDM systems. Signal Process. Lett. **12**(6), 453–456 (2005)
4. Ochiai, H., Imai, H.: One the clipping for peak power reduction of OFDM signal. In: Proceedings of IEEE Global Communications Conference (GLOBECOM), San Fransisco, USA (2000)
5. Ochiai, H., Imai, H.: Performance analysis of deliberately clipped OFDM signals. IEEE Trans. Commun. **50**(1), 89–101 (2002)
6. Ojima, M., Hattori, T.: PAPR reduction method using clipping and peak-windowing in CI/OFDM system. In: Proceedings of IEEE 66th Vehicular Technology Conference, September 30–October 3 (2007)
7. Cha, S., Park, M., Lee, S., Bang, K.-J., Hong, D.: A new PAPR reduction technique for OFDM systems using advanced peak windowing method. IEEE Trans. Consum. Electron. **54**(2), 405–410 (2008)
8. Chen, G., Ansari, R., Yao, Y.: Improved peak windowing for PAPR reduction in OFDM. In: Proceedings of IEEE 69th Vehicular Technology Conference, April 26–29 (2009)
9. Wang, X.B., Tjhung, T.T., Ng, C.S.: Reduction of peak-to-average power ratio of OFDM system using a companding technique. IEEE Trans. Broadcast. **45**(3), 303–307 (1999)
10. Jiang, T., Xiang, W., Richardson, P.C., Qu, D., Zhu, G.: On the nonlinear companding transform for reduction in PAPR of MCM signals. IEEE Trans. Wirel. Commun. **6**(6), 2017–2021 (2007)
11. Baig, I., Jeoti, V.: DCT precoded SLM technique for PAPR reduction in OFDM systems. In: Proceedings of the 3rd International Conference on Intelligent and Advanced Systems, Kuala Lumpur, Malaysia (2010)
12. Mourelo, J.T.: PAPR reduction for multicarrier modulation. Ph D. thesis, University of Stanford (1999)
13. Yoo, S., Yoon, S., Kim, S.Y., Song, I.: A novel PAPR reduction scheme for OFDM systems: selective mapping of partial tones (SMOPT). IEEE Trans. Consum. Electron. **52**(1), 40–43 (2006)
14. Han, S.H., Lee, J.H.: PAPR reduction of OFDM signals using a reduced complexity PTS technique. Signal Process. Lett. **11**(11), 887–890 (2004)
15. Krongold, B., Jones, D.: An active-set approach for OFDM PAR reduction via tone reservation. IEEE Trans. Signal Process. **52**, 495–509 (2004)
16. Hussain, S., Louet, Y.: Tone reservation's complexity reduction using fast calculation of maximal IDFT element. In: International Wireless Communications and Mobile Computing Conference, August 6–8 (2008)
17. Wang, L., Tellambura, C.: Analysis of clipping noise and tone-reservation algorithms for peak reduction in OFDM systems. IEEE Trans. Veh. Technol. **57**(3), 1675–1694 (2008)
18. Li, H., Jiang, T., Zhou, Y.: An improved tone reservation scheme with fast convergence for PAPR reduction in OFDM systems. IEEE Trans. Broadcast. **57**(4), 902–906 (2011)
19. Mahadevaswamy, U.B., Geetha, M.N.: A comparative survey on PAPR reduction in OFDM signal. In: 2016 International Conference on Electrical, Electronics, Communication (2016)

Stack-Based WSS Scheme for Four-Degree Network Node Module

G. R. Kavitha[1($\boxtimes$)] and T. S. Indumathi[2]

[1] Department of Electronics Engineering, Sir MVIT, Bangalore, India
kavigr@gmail.com
[2] Visvesvaraya Institute of Technology, Bangalore, India
drindumathits@gmail.com

Abstract. The existing switching modules for networking are lacked with some of the limitations associated with scalability, and configurability. Hence, this article introduces a more flexible, stack-based switching module, where some independent (1 × n) wavelength selective (WS) switches is realized on a single 4-degree network node. The stack-based WS switching module is designed in different ways such as either for transit side or add or drop operation of a Colourless, Directionless, and contentionless (CDC) ROADM (Re-configurable Optical Add or Drop Multiplexer). The ROADM design is introduced by applying stack-based WS switching scheme. The cost is analyzed through a 4-degree network node. The proposed CDC-ROADM architecture is demonstrated to recognize the cost reduction towards a minimum 35 percent in the proposed test network nodes when measured with the traditional/existing CDC-ROADM architecture based on multicasting switches and WS switches. According to experimental results, the proposed architecture can efficiently decrease the number of components in add or drop side and reduces the total cost of >70% and >80% in the proposed 4-degree network node.

Keywords: Multi-casting switches · Reconfigurable ROADM · Wavelength Selective Switch (WSS) · Wavelength contention

1 Introduction

In the past decades, optical communication networks were proposed with static paths. To each path, a big margin, usually in the manner of 3-to-5 dB for commercial systems, is allotted to accommodate the network performance and to guarantee operation beyond years/decades. Recently, by the introducing of advanced ROADMs, i.e., colorless, directionless and contentionless (CDC) ROADMs [1], the wavelength-paths can be changed dynamically based on the end-user request, which affects the network performance highly when compared with static networks. Additionally, high data-bit rates are usually lead to decrease the non-linear tolerance, and the making of large margin allocation is more challenging. Also, the current optical networks are made up of both hardware (i.e., proprietary transport equipment) and software components for network

© Springer Nature Switzerland AG 2020
P. Karrupusamy et al. (Eds.): ICSCN 2019, LNDECT 39, pp. 85–96, 2020.
https://doi.org/10.1007/978-3-030-34515-0_10

operation as well as management. To solve these issues, [2] have introduced a new ROADM scheme and designed an interoperable specification for RODMs containing optical-switch, radio signal transmitter, and pluggable element and YANG model for supervised software network. However, Wavelength Selective Switches (WSS) are the key building blocks of Reconfigurable Optical Add/Drop Multiplexers (ROADMs). This allows single/multiple wavelengths carrying channels to add/dropped through an optical fiber without the conversion of optical signals on each of the Wavelength Division Multiplexing (WDM) channels to electrical signals and revert to the optical signals. It means that ROADMs are capable of redirecting the signals from any network node to another network node based on wavelength without the conversion of optical-electrical and electrical-optical signals (O-E-O).

ROADMs with advanced features, i.e., CDC (colorless directionless and contention less) also permit the service operators to drip any wavelength-channel from any incoming direction of the network node to its transceivers or else add a wavelength channel from any transceivers to the living direction [3]. The ROADM system is composed of a pair of logical input and output ports such as In and Add (input ports) and Out and Drop (output ports). Based on their switching techniques, conventional ROADM architecture design can be categorized into three groups: such as ROADM-I, ROADM-II, and ROADM-III. The first category (ROADM-I) contains a large optical fiber switch with some Add/Drop wavelengths ($\lambda \geq 1$) example: MEMS (Micro-Electro-Mechanical System) [4] whereas second category (ROADM-II) is integrated with several small optical fiber switches arranged in parallel order example: multi-port optical fiber Bragg grating [5]. The third category (ROADM-III) consists of single optical fiber switch along with single Add/Drop wavelength, for example: vertically joined semiconductor Bragg grating ROADMs [6].

The different categories of ROADMs have different characteristics based on their device performance at physical-level [7–9]. In the proposed study, a low-cost ROADMs system design is introduced by applying stack-based Wavelength Selective Switches (WSSs) model. The WSS builds a hybrid network center for electronic (passes busty data via a packet network) as well as for optical data (passes bulk data through the optical circuit).

Since the stacked based WSSs could transmit some optical elements and contains a common alignment method, leading to smaller unit cost and smaller footprint. This process will make it commercially viable to deploy a huge number of wavelength selective switches in a single network node. This paper illustrates how to utilize a stack-based WSSs module in CDC-ROADMs. Finally, the simulation results show the comparative analysis of the proposed system with conventional techniques based on WSS modules and multicasting switches.

Section 2 overviews of related work which illustrates the existing research study being done in this area. The Sect. 3 gives description on ROADM architecture while Sect. 4 presents proposed stack-based WSSs module design followed by result discussion being carried out from the proposed study, which represents the comparative analysis of system performance in Sect. 5. Finally, Sect. 6 briefly presents the conclusion of the study.

2 Related Work

The existing researches on ROADM is discussed in this section. Homa et al. [9], have introduced a multi-level ROADM architecture which provides automated provisioning in current multichannel optical fiber networks. The motive of this architecture is to minimize the costs, improve the provisioning time, and mitigate the error from physical reconfiguration. In [10], Strasser and Wagner have measured the operational and WSSs performance requirements for ROADMs applications. Moreover, have defined the new techniques and development strategies which improve the performance and flexibility of ROADMs optical networks. Finally, the system performance was evaluated by all channels performance concerning various parameters. Bhardwaj and Soni [11], have designed an optical transmission model, was integrated with several components such as laser light, modulator, single mode optical fiber, dispersion compensation (fiber Bragg grating), amplifier, and light detector device. Also analyzed the dispersion-compensation using different lengths of optical fibers and simulation results have been measured using different parameters (i.e., input power, length of the fiber, etc.) Finally, overall simulation results were analyzed by optic system simulation at 20 GB/s transmission systems.

Ishii et al. [12] developed advanced ROADMs using WSSs and discussed some important technical issues during the implementation of the WSSs model; such as optical design, electrical monitoring and management, and so on. The proposed WSSs system model integrated with control process, voltage monitor, memory device, temperature sensor devices, and other elements which were utilized to control the proposed optical module. In [13], Simmons, have demonstrated different scenarios where wavelength-contention may occur within a ROADM and noted that different contention situations could be reduced by applying an effective algorithmic approach. Showed that contentionless feature is quite crucial by comparing with other features like as colorless, and directionless. Additionally, the author evaluated the ROADM contention property in detail and provided a clear picture regarding contentionless ROADM. Kavitha et al. [14] have proposed a new constraint-based optical fiber network. The aim was to maximize the overall throughput of the optical network by optimizing the optical signal to noise-ration exploiting ROADM. The experimental analysis was carried in a Matlab environment which evaluated the signal performance (i.e., optical signal to noise ratio). The simulation results have provided better throughput performance by comparing to existing works.

In [15], Kavitha et al. has proposed an extension model of ECON which was initially based on mathematical concept to improve the optical signal to noise ration performance while in this paper author introduced an ORPT model using ROADM,

WSSs and queuing network approach which improves the performance of the optical network in terms of behavioral in optical network. The proposed model experimented in different environments which show the efficiency of the optical routing scheme. Oda et al. [16] proposed a living network for open ROADMs. They experimentally showed that proposed living network scheme for different networks that independently maintains the records of their path level performance also improves the system estimation accuracy through multi-span BER (bit error rate) pre-test. Tang and shore [17], have presented a reconfigurable OADMs module for dynamic optical networks which evaluated the capability of wavelength routing in ROADMs. Also developed a theoretical routing energy model which is considered for various ROADM architectures. The numerical result shown that different types of ROADMs have different wavelength routing capability, which is significant, depends on dynamic traffic.

Tripathi et al. [18], discussed a comprehensive study on different kinds of ROADM architecture. Whereas, in [19], Peter and Collings have explored the colorless and directionless ROADM architectures and discussed the opportunities to enhance the ROADM architecture with colorless and directionless features which provides an add and drop switching mechanism. Current advances in wavelength selective switching method toward high density with great optical performance provides an opportunity to enhance the WSSs applications into a larger set of applications, including colorless and directionless add or drop switching scheme. Kavitha et al. [20] analyzed the ROADM model, which can increase the performance of optical channels. A new technique was adopted by applying mathematical modeling exploiting a single node to node link. The experimental analysis was carried out by adding/dropping some channels using ROADM, which enhanced the service quality and system performance in optical-network.

3   ROADM Architecture

The current and futuristic generation of networking system is much focused on optical networks where the ROADM is the main component of it. The use of ROADMs helps to add/drop operation through unique Wavelength Division Multiplexing (WDM) channel group in the network nodes. By which it can generate better optical performance, configurability, and cost-effectiveness. Previously, the ROADMs have supported line and ring architectures. However, the ROADMs developed in such a way that it supports higher degree node values and is helpful for futuristic optical networks. To do so, WSS based ROADMs are used, which offers multiple degrees and emerged as a significant technique offering flexibility and degree updation.

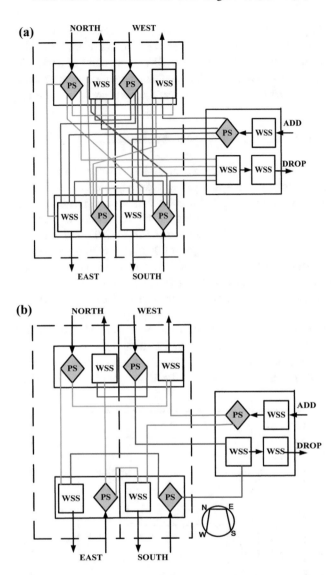

Fig. 1. (a) Symmetric 4-degree ROADM architecture (b) A 4-degree ROADM (asymmetric architecture)

Figure 1(a) indicates the 4th-degree operation of WSS based ROADM, which is held within the network. This node configuration helps to reach one node to its neighbor node positioned South (S), North (N), West (W) and East (E) directions and vice-versa. These signals received from different directions are separately split through Power Splitter (PS) and then passed to WSS positioning at the outer side of rest three directions and the local DROP module. The WSS selects and performs the integration of wavelengths obtained from rest 3-directions, from local ADD, and it is forwarded to the desired location. Using this architecture, any wavelength value entering the network node is routed at the output of one or more directions. Also, service providers need not target the investigating approach. This gives the effective asymmetric switching in nodes.

Figure 1(b) gives the asymmetric ROADM architecture of degree four and configuration is utilized 1×4 PSs, and WSSs each which precisely ensures a subset from their port travels in the other directions and rest of the ports are reserved for further scaling.

4 Proposed Stack-Based WSSs Architecture

This section illustrates a highly flexible, stack-based switching module, which significantly reduces the cost and footprint per wavelength selective switch (WSS). The proposed module is integrated with different elements for application at add or drop of a colorless, directionless and contentionless (CDC) ROADM where multiple $(1 \times n)$ WS switches is realized into a single device by exploiting 2-D beam steering. The stack-based WSSs module is composed with a degree of flexibility to either increase the rate of WS switches or to remove the few numbers of WS switches with high port-count. The cost of the switching module is estimated in the order of its component cost. Therefore, the prime objective is to significantly reduce the cost of each WSS owing to the extensive transiting of switching components and packaging. The proposed WSSs add or drop switching module for ROAD is configured by utilizing $(m \times n)$ multicasting switching approach where an array of $(m1 \times n)$ splitters is paired with another array of $(n1 \times m)$ switches. In this $(m \times n)$ multicasting switches are modeled for an m-deg network and capable of inserting (λn) n-wavelength channels into the network node in CDC format or vice versa.

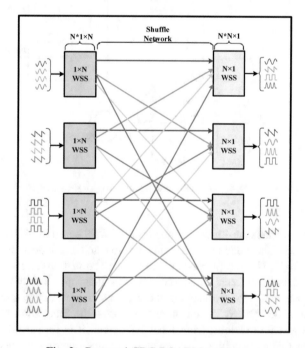

Fig. 2. Proposed CDC-ROADM architecture

In a general network node, every input optical fiber carries eighty wavelength channels, and out of them, 20% is dropped for general processing. As a result, Dropn (m × n) switches are needed for CDC drop processing in an N-degree ROADM network node, where Dropn = N × 80 × 20/N*100. Consequently, the number of port (1 × n) WSSs utilized on the transit side, and it will be (N + Dropn − 1). Similarly, the equal number of insertion channels are considered to inserted to the network node for further transmission operation, the same number of (m × n) switches are needed for CDC add/insert operation. Owing to the great loss of (m × n) multicasting switches, an array of m-amplifiers are required to be fixed at the interface between every switch and the ROADM transit side. Hence, two Drop (m × n) amplifiers are needed for CDC add or drop operation. Moreover, the CDC ROADM architecture is well-suited with transceivers based on direct detection since the wavelength switches can perform a necessary filtering operation before the fiber signals reaching the transceiver node. Though, the proposed stack based WS switching model is executed in a single architecture, with a low cost or minimum footprint.

The proposed switch based WSS module is configured as a wavelength cross-connect (WxC) model. Figure 2 illustrates the functional module of the proposed scheme where (4 × 4) WxC are used in which four arrays of (1 × 4) WS switches are connected of four arrays of (1 × 4) WS switches through a shuffling interconnection. The proposed stack-based WS switching module is integrated with six independent (4 × 4) WxC connectors, three independent (8 × 8) WxC connectors. These WxCs are applied on configurated CDC-ROADM add or drop module. This configuration is capable of interface with large-scale network node with a range of 12-degree. Nevertheless, some wavelength channels which is added/drop operation is still tough in some cases. Therefore, as a solution, a conventional static CDC architecture was introduced in [22]. Though, the wavelength contention (WC) is averted by control software, since the number of (1 × n) parallel splitters in the array surpass the degree (N) of the network node. The same wavelength of N signal channels is added or drop in N-deg network node because the optic fiber for each degree impossible to carry multiple channels at the same wavelength. Hence, the same wavelength of the signal channel could be utilized for different (1 × n) splitters with less impact on the connectivity.

5   Results and Analysis

In this section, the system analyzes the cost for the proposed Stack-based WSSs ROADM architecture. In this, 4-degree network node configuration is utilized, which is composed of 4-parallel optic fibers. The price of the traditional CDC-ROADM architecture is based on independent WSSs and multicasting switches, which are considered as a benchmark for the comparative analysis. Here, system compared the network complexity and dimensions of each component in the WSS module with their corresponding in a standalone (1 × n) WSS module. However, the price of the conventional ROAM architecture is integrated with standalone multicasting switches, and WS switches are exploited as a benchmark for comparative analysis (As shown in Table 1). When (4 × 16) multicasting switches are applied in proposed 4-deg network module, four multicasting switches are needed to add 64-channels to the proposed 4-degree network node, and an equal amount of multicasting switches are also needed for dropping operation. Equivalently, an eight (1 × 7) WS switches are required for transit side. Out of seven ports of each WS switch, three ports are attached with WS switches for other degrees (X-deg), and remaining other 4-ports are attached with equivalent four (4 × 16) multicasting switches in add/drop side. In this case, it is observed that optical signals are traveled between fibers attached to the same ROADM module.

Table 1. Cost-comparison of ROADM architecture for the four-degree network node

			Unit cost	Quantity	Component cost
Benchmark	Transit add/ drop	1 × 20 WSSs	1.00	32	32.0
		4 × 6 MSCs	0.50	32	16.00
		amplifiers	0.10	128	12.80
Conventional architecture	Transit add/ drop	81 × 32 stacked WSS module	1.50 [1.85][a]	4	6.00 [7.40]
		81 × 32 stacked WSS module	1.50 [1.85]	8	12.00 [14.80]
		1 × 4 space switch amplifiers (low amplification)	0.02	512	10.24
			0.05	64	3.24
Proposed architecture	Transit add/ drop	81 × 32 stacked WSS module	1.50 [1.85]	4	6.00 [7.40]
		481 × 12 stacked WSS module	1.50 [1.85]	4	6.00 [7.40]
		1 × 16 plitters/couplers amplifiers	0.01	32	0.32
			0.01	32	0.32

However, in the existing network node, 32 WS switches are applied in transit side and 32 (4 × 16) multicasting switches are utilized for add or drop operation. Moreover, for each (4 × 16) multicasting switches need 4-amplifiers, 128-low amplifiers are needed. Whereas in the proposed network node, (8 × 16) multicasting switches are utilized, and for each needs 8-amplifiers to recompense the addition loss. Totally, 128 −(8 × 16) multicasting switches, 210 amplifiers, and 128−(1 × 20) WS switches are required for it.

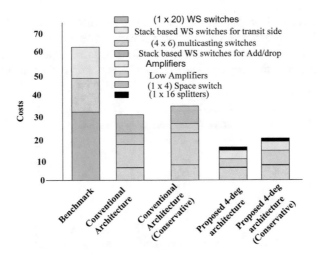

Fig. 3. Comparative analysis

However, in the following section, have briefly overviewed on ROADMs and future scope of the optical networks. This survey study provides information about how the network operators can resize their core optical networks, concentrating on multi-service reconfigurable optical networking (MRON) markets and opportunities.

The cost analysis of the proposed architecture in terms of transit cost and Add/drop cost, and the total cost is compared with benchmarked and conventional architecture (given in Table 2).

Table 2. Cost analysis

Method/cost	Conventional	Benchmarked	Proposed
Transit cost	7.4	32	7.4
Add/drop cost	28.44	28.80	10.92
Total cost	35.64	31.44	18.32

6 Future Scope of the ROADMS and Metro Optical Networks (MONs)

Currently, the ROADM scheme is not a new concept as it is part of a handful of service provider's wavelength division multiplexing (WDM) equipment in both North America and Europe since 2002 (). Long haul optic networks have been integrated with ROADMs in last three years, positioning the levels for a migration of this characteristic into local networks and metro optical networks (MONs) as cost reduction and flexibility of network requirements escalate. Therefore, today, it is very precisely to develop a ROADM model as a market space point of view. In [23], the author introduced a new concept of "Re-configurable Optical Networking Market (RONM)" which encloses with which are labeled ROADMs systems, optical add/drop switches (OADS), wavelength switching systems and which service provider refers to a Digital Optical Network (DON). In this, the similarity in network wavelengths is simple, automated fashion in local and metro region networks, along with embedded transport and network switching at both wavelength and sonnet layers in a single network component.

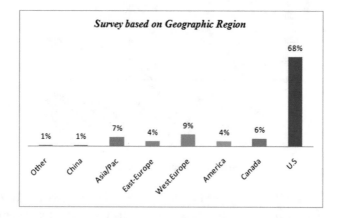

Fig. 4. Survey report based on geographic region

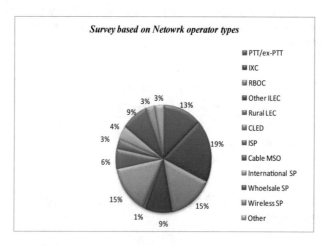

Fig. 5. Survey report based on network operator types

The Figs. 3 and 4 illustrate the survey report of service provider professionals and direct interface with a wide range of Network operators and communications with main operators in the supplier network. This report provides a complete assessment of ROADMs and opportunity for optical networking scheme also provides an exclusive result analysis of telecom service providers concerning their optical network expansion plans and expected role where ROADMs will play an important role in that expansion (Fig. 5).

7 Conclusion

The proposed study presents a stack-based WSS (wavelength selective switch) module, where some (1 × n) WSSs ca be recognized into a single device by exploiting 2-D beam steering. The proposed module is integrated with different elements for application at add or drop of a colorless, directionless and contentionless (CDC) ROADM where multiple (1 × n) WS switches is realized into a single device by exploiting 2-D beam steering.

The proposed WSSs module is configured with high degree network connectivity to either maximize or realize the amount of WS switches with the larger port count. The cost/price of the proposed WSSs module is determined based on its configured components costs. This module is significantly minimizing the cost of the WSS. However, the WSS cost analysis has been evaluated by 4-deg network node configuration. The outcome results show that from this module, approximately 40% of the cost is saved, as compared with a conventional ROADM module based on MCSs. The cost reduction in the proposed architecture is primarily owing to the use of stack-based WSS module at transit side.

References

1. Wagener, J., Strasser, T.: Characterization of the economic impact of stranded bandwidth in fixed OADM relative to ROADM networks. In: Presented at the Optical Fibre Communication Conference/National Fiber Optic Engineers Conference (OFC/NFOEC), Anaheim, CA (2006)
2. Eldada, L., Fujita, J., Radojevic, A., Izuhara, T., Gerhardt, R., Shi, J., Pant, D., Wang, F., Malek, A.: 40-channel ultra-low-power compact PLC-based ROADM subsystem. In: Presented at the Optical Fibre Communication Conference/National Fiber Optic Engineers Conference (OFC/NFOEC), Anaheim, CA, Paper NThC4, March 5–10, 2006
3. Yang, H., Robertson, B., Wilkinson, P., Chu, D.: Low-cost CDC ROADM architecture based on stacked wavelength selective switches. Commun. Netw. **9**(5), 375 (2017)
4. Yano, M., Yamagishi, F., Tsuda, T.: Optical MEMS for photonic switching-compact and stable optical cross-connect switches for simple, fast, and flexible wavelength applications in recent photonic networks. IEEE J. Sel. Top. Quantum Electron. **11**(2), 383–394 (2005)
5. Tran, A.V., Zhong, W.D., Tucker, R.S., Song, K.: Reconfigurable multichannel optical add-drop multiplexers are incorporating eight-port optical circulators and fiber Bragg gratings. IEEE Photonics Technol. Lett. **13**(10), 1100–1102 (2001)
6. Horita, M., Yazaki, T., Tanaka, S., Matsushima, Y.: Extension of operation range of semiconductor optical add and drop multiplexer. In: Proceedings of International Conference on Indium Phosphide and Related Materials, pp. 579–582 (2001)
7. Mezhoudi, M., Feldman, R., Goudreault, R., Basch, B., Poudyal, V.: The value of multiple degree ROADMs on metropolitan network economics. In: Presented at the Optical Fibre Communication Conference/National Fiber Optic Engineers Conference (OFC/NFOEC), Anaheim, CA, Paper NThA4, March 5–10, 2006
8. Hsieh, T., Barakat, N., Sargent, E.H.: Banding in optical add-drop multiplexers in WDM networks: preserving agility while minimizing cost. In: Proceedings of IEEE International Conference on Communications, vol. 2, pp. 1397–1401 (2003)

9. Homa, J., Bala, K.: ROADM architectures and their enabling WSS technology. IEEE Commun. Mag. **46**(7), 150–154 (2008)
10. Strasser, T.A., Wagener, J.L.: Wavelength-selective switches for ROADM applications. IEEE J. Sel. Top. Quantum Electron. **16**(5), 1150–1157 (2010)
11. Bhardwaj, A., Soni, G.: Performance analysis of optical communication system using fiber Bragg grating. SSRG Int. J. Electron. Commun. Eng. (SSRG-IJECE) **2**(1) (2015)
12. Ishii, Y., Ooba, N., Sahara, A., Hadama, K.: WSS module technology for advanced ROADM. NTT Tech. Rev. **12**(1) (2017)
13. Simmons, J.M.: A closer look at ROADM contention. IEEE Commun. Mag. **55**(2), 160–166 (2017). https://doi.org/10.1109/MCOM.2017.1600829CM
14. Kavitha, G.R., Indumathi, T.S.: ROADM framework for Enhanced Constraint-based Optical Network (Econ) for enhancing OSNR. In: 2015 6th International Conference on Computing, Communication and Networking Technologies (ICCCNT), Denton, TX, pp. 1–7 (2015)
15. Kavitha, G.R., Indumathi, T.S.: Novel ROADM modeling with WSS and OBS to improve routing performance in optical network. Int. J. Electr. Comput. Eng. (IJECE) **6**(2), 700–707 (2016)
16. Oda, S., et al.: A learning living network with open ROADMs. J. Lightwave Technol. **35**(8), 1350–1356 (2017)
17. Tang, J.M., Shore, K.A.: Wavelength-routing capability of reconfigurable optical add/drop multiplexers in dynamic optical networks. J. Lightwave Technol. **24**(11), 4296–4303 (2006)
18. Tripathi, D.K., Singh, P., Shukla, N.K., Dixit, H.K.: Reconfigurable optical add-drop multiplexers a review. Electr. Comput. Eng. Int. J. **3** (2014)
19. Roorda, P., Collings, B.: Evolution to colorless and directionless ROADM architectures. In: OFC/NFOEC 2008 - 2008 Conference on Optical Fiber Communication/National Fiber Optic Engineers Conference, San Diego, CA, pp. 1–3 (2008)
20. Kavitha, G.R., Indumathi, T.S.: Enhanced constraint-based optical network for improving OSNR using ROADM. Int. J. Appl. Innov. Eng. Manag. **3**(3) (2014)
21. Abedifar, V., Shahkooh, S.A., Emami, A., Poureslami, A., Ayoughi, S.A.: Design and simulation of a ROADM-based DWDM network. In: 2013 21st Iranian Conference on Electrical Engineering (ICEE), Mashhad, pp. 1–4 (2013)
22. Zong, L., Zhao, H., Feng, Z., Yan, Y.: Low-cost, degree expendable and contention-free ROADM architecture based on $M \times N$ WSS. In: Optical Fiber Communication Conference (OFC), paper M3E.3 (2016)
23. Clavenna, S.: ROADMS and the future of metro optical networks. Heavy Read **3**, 1–5 (2005)

Applying Lagrange Model to Fill Data During Big Data Streaming

Sindhu P. Menon[(⊠)]

Jain College of Engineering and Technology, Hubli, India
sindhu33in@gmail.com

Abstract. Advancements in technology have significantly reshaped the social and economic environment. Businesses are coming up with new strategies to uncover hidden information from data in order to support better prediction and analysis. Data continues to grow at a rapid rate and it has become necessary to process the quality data. In mission critical applications, streaming of data plays a very important role. Discontinuity in data stream is unaffordable as it consumes more time and money. This paper proposes a technique through Lagrange's Interpolation, which could avoid discontinuity in data streams when large data is being processed.

Keywords: Big data · Velocity · Lagrange's Interpolation · Streaming

1 Introduction

In this digital era, Data is being generated at an unprecedented rate every second. The quality of this data is questionable, resulting in poor analysis. Data containing noise costs organizations and countries a good fortune. In US it is estimated to cost about 600 billion dollars annually [14]. Data preparation phase accounts to about 70% of the time and cost. About 14% of the data contain incomplete information like missing tags, incomplete structure and so on. This makes such documents not worthy for analysis. Due to the existence of such large data, quality has become a key issue [12]. There are other features of big data like volume, veracity, value and so on [8]. To infer value from any data, the quality should be good. In [11] a methodology to handle the unstructured data is proposed. In [5], the authors have stated that the greatest issue with big data is the occurrence of errors. Errors could be committed during entry of data, or due to redundancy. This could result in inaccurate analysis and loss in revenue, thereby failing to comply with the rules and regulations put forth by the government and industry [9].

Many researchers have worked on this area. In [2], a comprehensive overview on data encryption frameworks and their impact on security measures where big data velocity became prime concern have been shown. It presents a comprehensive analysis of encryption mechanisms and also stated that encryption mechanisms are no more obstacles for big data processing due to the efficient parallel computing paradigm. Four analytical models using data encryption strategies are implemented and tested over Hadoop based big data applications. The experimental analysis was carried out considering a test bed Amazon Web Services (AWS). The test bed considered datasets of

© Springer Nature Switzerland AG 2020
P. Karrupusamy et al. (Eds.): ICSCN 2019, LNDECT 39, pp. 97–107, 2020.
https://doi.org/10.1007/978-3-030-34515-0_11

size (5–20 GB) and performance was measured in terms of processing time metric. On the same grounds [15] emphasises on an analytical methodology integrated with a high-throughput analytical platform has been designed and implemented. The tool has been explicitly incorporated into extracting, storing and analyzing high-dimensional velocity big data to meet the requirement of commercialized/ industrial usage in terms of equipment remote monitoring and diagnostic use case. A similar study has been carried out by Zhang and Shi [16] where the emphasis is on classification over big velocity data to extract significant attributes by incorporating a cross domain canonical correlation analysis (CD-CCA). The study mostly focused on the problems associated with correspondence pivot and other specific feature extraction from various areas. To handle this problem the study introduced aforementioned modelling to learn from semantic space of multi-view correspondences. The learning paradigm of proposed CD-CCA involves transferring of knowledge by reducing the dimensionality of data in a multi-view way. The extensive simulation has been carried out on 144×6 classification problems associated with 20 new groups which ensure effectiveness in terms of prediction accuracy. The study in [1] also emphasized about handling big data velocity issues by introducing a popular open-source stream processing engine. The tool targeted at obtaining real-time integration and trend in real-time social media platforms like twitter or Bitly streams [17]. The case study demonstrated that the proposed tool can effectively process high dimensional big data even for high velocity streams. Authors in [10] also briefed an insight into big data analytics and how big data can be processed into smart data for better analytics. But in spite of all these works, velocity still remains a challenge to be addressed.

Data arriving at a rapid rate has to be processed real time. This analysis is very critical for various online services. The data size could reach 40 ZB or even more by 2022. Hence the most crucial thing is to find valuable knowledge from this voluminous data to support prediction. Many applications like health care are time sensitive and for them there should be no break in the flow of data. The main objective of this work is to provide a solution to data streaming when there is a disruption in the flow of data.

2 Proposed System

2.1 Data Set

The data set used for this work is obtained from PRO-ACT database which contains 10,47,253 records of patient's details.

2.2 Methodology

The concept pertaining to velocity mostly deals with the speed of data being generated as well as pooling resources from various sources in conventional networks. Big data characteristics are not only limited to its conventional nature but has a higher impact on its speed. The evaluation of speed associated with big data determines the flow of data, produced from different computing resources including business processes, networks, machines, mobile devices, social media, etc.

Insight into these data help researchers, business leaders and IT professionals in making valuable decisions for their respective field of interest. Data packet analysis - a different form of network analysis is also an integral part of cyber security where the velocity of data stream being utilized is the prime concern. The current explosion in data has led to a situation where massive amount of data chunks are getting generated and transmitted through internet every fraction of second. Data thus generated will generate meaning only if they are continuously being analysed.

Hence there is a need for continuous flow of data or continuous data being generated. Sometimes due to hardware failures, data may stop arriving for a certain time period. In such cases, the need of the hour is to design a methodology to fill in the data for continuous analysis if there is a stoppage. The system proposed here is designed taking this requirement into consideration. When there is partial data arriving from a particular source, this system fills that data by using Lagrange's Interpolation. It will thus ensure that the analysis or prediction of information does not stop.

The foundation of the proposed system lies towards the problem of designing a cost effective integrated modelling method which can be used to control the issue of reducing the big data speed problems. The extensive study found that there is no means by which the speed of incoming data can be controlled. The only method by which this could be achieved is to store it effectively eliminating redundancy. The following are the processes involved into designing the core levels of the proposed system.

2.2.1   Pre-processing Stage

The system initially is designed in a way where it considers the input data stream generated from [13] in which the data veracity problems are efficiently handled. Algorithm 1 explains the process designed for the purpose of handling the speed problems of large volumes of data generated from healthcare.

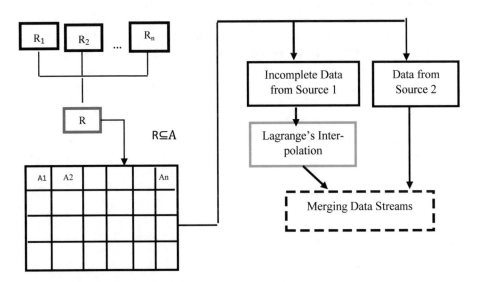

Fig. 1. Proposed system design to handle data discontinuity

The assumption here is that data comes from a variety of sources. For e.g. Blood Pressure (BP) of a patient may come from the demographic data and lab data. When the parameters are same for the same patient id, the value which arrives first will be saved and the rest discarded to avoid redundancy.

In Fig. 1, R_1, R2....R_n are the same attributes which arrive from different sources. Since the attributes correspond to the same variable, the first data represented by R is saved and the rest discarded. The entire set of attributes are represented by A where $A = \{A_1, A2........A_n\}$. This is stored in a matrix form as shown above.

2.2.2   Filling Missing Data

The next process includes filling of missing values in case there is a discontinuity in the stream of arriving data. This is achieved by using Lagrange's Interpolation as shown in Fig: 2.2. Interpolation is a technique through which we find certain values from some known data point [7]. Initilaly, compute the first x_i using mean of previous two values i.e. x_{i-1} and x_{i-2}. Then apply Lagrange's Interpolation to compute $f(x_i)$. The faster processing of data thereby increases the system throughput and plays a crucial role in the next phase of the study when the prediction needs to be carried out. The following section presents the proposed algorithm design principles and its operational aspects from a theoretical perspective.

3   Implementation

The prime concern of the proposed algorithm is to offer a cost effective computational model well capable of mitigating the issues which occur in data streaming. The existing studies are found to be very less effective to provide counter-measure to the big data velocity problems. The earlier work on velocity [14–16] have only shown how to store the data when it comes at high velocity.

The problem formulation of the system considers that the speed of data flow cannot be controlled but the improvisation to the speed factor of incoming data rate can be controlled by employing a proper data management and classification model.

The algorithm for filling data in case of loss of data is as follows:

Algorithm 1: Maintaining continuity of data stream
Input: Structured data with good quality
Output: Loss of data overcome by Lagrange's Interpolation
    1.   For continuous data aggregation
    2. $\Phi_i \leftarrow$ aggregate (),
    3.   Compute f(p) $\longleftarrow$ $[\Phi_i + \Phi_{i-1}]/2$
    4.   Check if the new f(p) is unique . If it not replace Φ_{i-1} with Φ_{i-2}
        and repeat the   process until we get a unique value
    5.   Initialise sum $\longleftarrow$ 0
    6.   For i $\leftarrow$ 1 to n-1 do
    7.   For j $\leftarrow$ 1 to n-1 do

$$\text{sum} = t(p) = \sum \left[\frac{(p-p_1)\ (p-p_2)\ (p-p_3)(p-p_4)}{(p_i-p_{i+1})(p_i-p_{i+2})(p_i-p_{i+3})(p_i-p_{i+4})} *t_j \right]$$

//p_i represents the previous value of the field (in this case the previous
value of id), t_j represents the field in the same row as p_i for which the missing
data is computed
    8.   End for
    9.End for

The study thereby incorporated two different operations namely (a) Pre-operational processing and (b) Filling of missing values of incoming healthcare big data. The algorithm also tries to reduce the cost of computation which helps in the next phase of the study where the prediction of the disease stage is performed based on its prior healthcare records.

Explanation of Algorithm 1
This work aims at substituting the missing data for a particular time period.
   Step 3: Compute $f(p) \leftarrow [\Phi_i + \Phi_{i-1}]/2$

$$\text{Pid (new1)} = (x1 + x2)/2 = (124 + 864)/2 = 494 = Y$$

Step 4: Here Pid is the patient id which must be unique
If (Y is present in x1....xn), where 'n' corresponds to the previous 1000 values, recompute Y

$$Y = \text{Pid (new2)} = (x1 + x3)/2$$

Repeat this for until a unique value is obtained. If after 1000 iterations, a unique value is not obtained, then recompute Y as ((x1 + x2)/2)++. Check for uniqueness in a similar manner.

Table 1. Prediction of values using Lagrange's interpolation

T1	PID		Age		BP		BPD		Sugar	
	212		38		310		389		114	
	182		53		144		178		132	
	200		20		120		185		280	
	412		84		80		127		260	
	864		16		240		167		450	
	124		36		220		243		140	
Predicted values	Proposed	Existing	Proposed	Existing	Proposed	Existing	Proposed	Existing	Proposed	Existing
T2	494	864	63	16	292	240	360	167	407	450
	309	124	46	36	236	220	301	243	347	140

Using Lagrange's Interpolation [6] formula we fill in the remaining missing values. The mathematical computation for the same is shown in Table 1.

Step 6, 7, 8: The Lagrange polynomial interpolation equation can be given as,

$$t(p) = \frac{(p-p1)(p-p2)(p-p3)(p-p4)}{(p0-p1)(p0-p2)(p0-P3)(p0-p4)} t_0 + \ldots\ldots + \frac{(p-p0)(p-p1)(p-p2)(p-p3)}{(p4-p0)(p4-p1)(p4-P2)(p4-p3)} t_4$$

$$
\begin{aligned}
t(494)(Age) &= \frac{(494-864)(494-412)(494-200)}{(124-864)(124-412)(124-200)} * 36 + \frac{(494-864)(494-412)(494-200)}{(864-124)(864-412)(864-200)} * 16 \\
&+ \frac{(494-864)(494-412)(494-200)}{(412-124)(412-864)(412-200)} * 84 + \frac{(494-864)(494-412)(494-200)}{(200-124)(200-864)(200-412)} * 20 \\
&= 0.55 * 36 + 0.04 * 16 + 0.32 * 84 + 0.833 * 20 \\
&= 63
\end{aligned}
$$

$$
\begin{aligned}
t(494)(BP) &= \frac{(494-864)(494-412)(494-200)}{(124-864)(124-412)(124-200)} * 220 + \frac{(494-864)(494-412)(494-200)}{(864-124)(864-412)(864-200)} * 240 \\
&+ \frac{(494-864)(494-412)(494-200)}{(412-124)(412-864)(412-200)} * 80 + \frac{(494-864)(494-412)(494-200)}{(200-124)(200-864)(200-412)} * 120 \\
&= 0.55 * 220 + 0.04 * 240 + 0.32 * 80 + 0.83 * 120 \\
&= 292
\end{aligned}
$$

$$
\begin{aligned}
t(494)(BPD) &= \frac{(494-864)(494-412)(494-200)}{(124-864)(124-412)(124-200)} * 243 + \frac{(494-864)(494-412)(494-200)}{(864-124)(864-412)(864-200)} * 167 \\
&+ \frac{(494-864)(494-412)(494-200)}{(412-124)(412-864)(412-200)} * 127 + \frac{(494-864)(494-412)(494-200)}{(200-124)(200-864)(200-412)} * 185 \\
&= 360
\end{aligned}
$$

$$
\begin{aligned}
t(494)(Sugar) &= \frac{(494-864)(494-412)(494-200)}{(124-864)(124-412)(124-200)} * 140 + \frac{(494-864)(494-412)(494-200)}{(864-124)(864-412)(864-200)} * 450 \\
&+ \frac{(494-864)(494-412)(494-200)}{(412-124)(412-864)(412-200)} * 260 + \frac{(494-864)(494-412)(494-200)}{(200-124)(200-864)(200-412)} * 280 \\
&= 0.55 * 140 + 0.04 * 450 + 0.32 * 260 + 0.83 * 280 \\
&= 406.6
\end{aligned}
$$

Next Data Set
Pid = Average (494, 124) = 309

$$t(309)(Age) = \frac{(309-494)(309-124)(309-864)(309-412)}{(494-124)(494-864)(494-412)(494-200)} * 46 + \frac{(309-494)(309-124)(309-864)(309-412)}{(124-494)(124-864)(124-412)(124-200)} * 36$$
$$+ \frac{(309-494)(309-124)(309-864)(309-412)}{(864-494)(864-124)(864-412)(864-200)} * 16 + \frac{(309-494)(309-124)(309-864)(309-412)}{(412-494)(412-124)(412-864)(412-200)} * 84$$
$$= 46$$

$t(309)(BP) = 236$, $t(BPD) = 301$, $t(309)(Sugar) = 347$.

Algorithm 1 initially reads the attributes for which the data has to be computed. In any hospital data, Patient id is a common attribute. This method starts by computing a new id based on the mean of the previous two ids. If this value is not unique, then it recomputes these values using i−2 in place of i−1. To this newly computed value, Lagrange's Interpolation is applied to fill the missing data. Literature shows that earlier such missing data is computed by just filling values from the previous record. The disadvantage with this is it leads to duplication of data with not much value added. Hence the proposed strategy is an improvement over existing systems as it eliminates redundancy.

Once the data is filled it can be used for further prediction of progression of ALS which will be done in the next chapter.

4 Experimental Results After Handling Discontinuity in Data Streams

4.1 Results

The proposed system is evaluated with ALS data set obtained from PRO-ACT [3], where the database has 10,47,253 records of patient's details. Table 2 shows the time taken to process the data, in our case processing refers to the time taken to fill the missing data.

The earlier work on velocity [14–16] have only shown how to store the data when it comes at high velocity. The work carried by us in handling velocity has not been done earlier.

Table 2. Analysis of processing time

Record count	Time taken (mins) with proposed system	Time taken (mins) with existing system	Record count	Time taken (mins) with proposed system	Time taken (mins) with existing system
5000	0.07	0.03	440000	30.47	29.38
10000	0.08	0.04	445000	31.31	30.13
15000	0.07	0.04	450000	31.34	29.12
20000	0.15	0.06	455000	32.37	27.19
25000	0.13	0.08	460000	32.21	28.04
30000	0.22	0.12	465000	33.44	28.47
35000	0.3	0.19	470000	33.17	27.5

(*continued*)

Table 2. (*continued*)

Record count	Time taken (mins) with proposed system	Time taken (mins) with existing system	Record count	Time taken (mins) with proposed system	Time taken (mins) with existing system
40000	0.4	0.25	475000	34.51	28.21
45000	0.45	0.38	480000	35.04	29.14
50000	0.53	0.44	485000	35.58	29.11
55000	0.55	0.46	490000	36.11	30.09
60000	0.6	0.32	495000	36.64	32.21
65000	0.57	0.44	500000	34.19	31.19
70000	0.42	0.51	505000	32.28	36.44
75000	0.17	0.48	510000	38.24	29.19
80000	0.32	0.33	515000	38.18	31.19
85000	0.37	0.22	520000	39.31	35.25
90000	0.42	0.56	525000	39.8	36.56
95000	0.68	0.15	530000	40.38	39.45
100000	1.03	0.57	535000	40.91	40.11
105000	1.08	1.02	540000	41.45	41.14
110000	1.13	1.08	545000	41.9	39.12
115000	1.18	1.11	550000	42.51	38.39
170000	6.9	6.25	605000	48.38	48.02
175000	8.01	7.17	610000	48.2	44.15
180000	8.53	7.37	615000	49.45	43.38
185000	8.8	7.18	620000	49.9	48.28
215000	11.34	8.12	650000	53.19	48.26
220000	12.9	9.57	655000	53.2	47.17
225000	13.23	10.24	660000	54.25	48.1
230000	13.66	11.16	665000	54.9	49.18
235000	13.9	9.19	670000	55.32	48.12
240000	14.11	9.35	675000	55.5	51.18
245000	14.45	10.12	680000	56.39	55.28
250000	14.6	10.27	685000	56.22	52.19
255000	14.8	11.17	690000	57.46	53.56
260000	15.08	12.08	695000	57.9	54.28
265000	15.72	12.35	700000	58.52	55.18
270000	15.9	12.56	705000	59.06	56.37
275000	16.13	14.41	710000	59.59	56.39
280000	16.36	14.57	715000	60.12	58.12
285000	16.59	15.12	720000	60.66	61.38
290000	16.2	15.28	725000	61.19	62.07
295000	17.02	15.49	730000	61.72	62.29
300000	17.11	16.29	735000	62.26	63.29

(*continued*)

Table 2. (*continued*)

Record count	Time taken (mins) with proposed system	Time taken (mins) with existing system	Record count	Time taken (mins) with proposed system	Time taken (mins) with existing system
305000	17.23	16.48	740000	62.9	65.19
310000	17.29	17.17	745000	63.33	60.28
370000	22.38	22.49	805000	69.13	72.12
375000	24.17	24.19	810000	70.26	77.14
380000	25.19	24.38	815000	70.8	69.37
385000	25.47	25.39	820000	71.33	72.37
390000	25.9	25.18	825000	71.83	75.27
395000	26.16	26.19	830000	72.4	77.19
400000	26.29	26.08	835000	72.53	79.16
405000	26.49	27.05	840000	73.46	71.17
410000	27.38	27.28	845000	74	70.18
415000	27.46	28.19	850000	74.53	72.32
420000	28.14	28.23	855000	75.07	70.18
425000	29.17	28.17	860000	75.6	70.05
430000	29.71	29.02	865000	76.13	71.18
435000	30.24	29.45			

The graph in Fig. 2 depicts the time taken by the proposed system. The main advantage of this work is in the methodology adapted to fill in missing values where redundancy gets eliminated and results in better prediction.

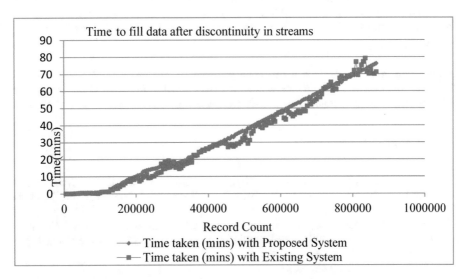

Fig. 2. Graph showing the record count v/s m processing time

The objective of this work is based on the assumption that if data is occurring at a continuous speed from time t_1, t_2....t_n and at time t_n it stops occurring due to some error like a hardware failure, then the system must replace that data until it resumes since we are dealing with critical applications in health care.

4.2 Comparison with the Existing Approach

In the existing system, when there is discontinuity in the data stream, the system automatically replicates the previous 'n' records. Even though the time consumed by the existing system is less when compared with the proposed system, it is not a good approach as the objective is not to just fill the values but to use them for further processing. When the data is just replicated, it results in redundant values which will not give efficient prediction accuracy. Another approach in existing system is to take the mean of n values to fill in the missing data [4]. The disadvantage with mean imputation is that it is problematic for multivariate analysis because it guarantees no relationship between imputed variable and measured variable. The proposed system fills the missing data with new values, hence the problem of redundant data does not occur and it results in better prediction.

A very simple model where irrespective of any speed of arrival of the incoming data, the data could be efficiently stored in our database for further processing. The algorithm takes the output of prior algorithm where incomplete data problem is addressed.

4.3 Complexity of the Algorithm

In this algorithm, Lagrange's Interpolation has been used to fill in records. This method involves a number of additions, subtractions, multiplications and divisions. The best and worst case time complexity can be computed as follows. The number of operations for determining the value of Lagrange's polynomial $L_n(a)$, is of $2(n + 1)^2$ multiplications and divisions, $n(n + 1)$ additions and subtractions. The two methods are of $O(n^2)$ time complexity.

Acknowledgement. I would like to extend my gratitude to the members of PRO-ACT who have developed the data set used in this work. PRO-ACT (Pooled Resource Open-Access ALS Clinical Trials Database) contains about 8500 records of ALS patients whose identity is hidden.

References

1. Chardonnens, T., Cudre-Mauroux, P., Grund, M., Perroud, B.: Big data analytics on high velocity streams: a case study. In: 2013 IEEE International Conference on Big Data, pp. 784–787 (2013)
2. Dupré, L., Demchenko, Y.: Impact of information security measures on the velocity of big data infrastructures. In: 2016 International Conference on High Performance Computing & Simulation (HPCS), pp. 492–500 (2016)
3. https://nctu.partners.org/ProACT
4. https://en.wikipedia.org/wiki/Imputation_(statistics)

5. Tee, J.: Handling the four 'V's of big data: volume, velocity, variety, and veracity (2013). TheServerSide.com
6. Lovett, D.L., Felder, D.L.: Application of regression techniques to studies of relative growth in crustacean. J. Crustac. Biol. **9**(4), 529–539 (1989)
7. Manembu, P., Kewo, A., Welang, B.: Missing data solution of electricity consumption based on Lagrange Interpolation case study: IntelligEnSia data monitoring. In: International Conference on Electrical Engineering and Informatics (ICEEI), pp. 511–516 (2015)
8. Menon, S.P., Hegde, N.P.: A survey of tools and applications in big data. In: 2015 IEEE 9th International Conference on Intelligent Systems and Control (ISCO), pp. 1–7. IEEE (2015)
9. Sarsfield, S.: The butterfly effect of data quality. In: The Fifth MIT Information Quality Industry SymPosium (2011)
10. Sheth, A.: Transforming big data into smart data: deriving value via harnessing volume, variety, and velocity using semantic techniques and technologies. In: 2014 IEEE 30th International Conference on Data Engineering, Chicago, IL, USA, p. 2 (2014)
11. Sindhu, C.S., Hegde, N.P.: A framework to handle data heterogeneity contextual to medical big data. In: 2015 IEEE International Conference on Computational Intelligence and Computing Research (ICCIC), pp. 1–7 (2015)
12. Sindhu, C.S., Hegde, N.P.: A novel integrated framework to ensure better data quality in big data analytics over cloud environment. Int. J. Electr. Comput. Eng. **7**(5), 27–98 (2017)
13. Sindhu, C.S., Hegde, N.P.: An approach to mitigate veracity issue in big data using regression. Int. J. Comput. Eng. Res. (IJCER) **7**(11), 51–54 (2017)
14. Eckerson, W.W.: Data quality and the bottom line: achieving business success through a commitment to high quality data. Data Warehous. Inst. **730**, 1–36 (2002)
15. Williams, J.W., Aggour, K.S., Interrante, J., McHugh, J., Pool, E.: Bridging high velocity and high volume industrial big data through distributed in-memory storage & analytics. In: 2014 IEEE International Conference on Big Data (Big Data), October 27, 2014, pp. 932–941 (2014)
16. Zhang, B., Shi, Z.Z.: Classification of big velocity data via cross-domain canonical correlation analysis. In: 2013 IEEE International Conference on Big Data, pp. 493–498 (2013)
17. Zeng, D., Gu, L., Guo, S.: Cost minimization for big data processing in geo-distributed data centers. In: Cloud Networking for Big Data, pp. 59–78. Springer (2015)

A Survey on Data Auditing Approaches to Preserve Privacy and Data Integrity in Cloud Computing

V. Kavya$^{(\boxtimes)}$, R. Sumathi, and A. N. Shwetha

Department of CSE, Siddaganga Institute of Technology, Tumakuru, India
kavyav1095@gmail.com, {rsumathi,shwethaan}@sit.ac.in

Abstract. Cloud computing is recognized as one of the prominent and rapid growing technologies in this contemporary world. Cloud allows the users to store the data without any physical infrastructure and it provides access to the users over the internet anywhere and at any time without harming its accuracy and reliability. However, on the other hand storing data in the cloud creates an inconvenience to the users by creating security issues like data integrity, privacy & availability in cloud. Security exists as one of the recent research challenges in the cloud. The data stored on the cloud can be hacked and altered by any person and sometimes even the data can be deleted by the Cloud Service Provider (CSP) for his own personal benefit. So, there exist many solutions to solve the security issues in the cloud. Data auditing is found as one of the popular approach to provide integrity and privacy using Third Party Auditor (TPA). In this paper, pros and cons of six different data auditing schemes are discussed along with the results.

Keywords: Cloud computing · Third Party Auditor · Cloud Service Provider (CSP) · Data auditing

1 Introduction

Cloud computing is an internet based technology that provides on demand services to various organizations to store the data on remote servers, manage and access the data over an internet from anywhere and anytime across the world [14]. The user has to pay only for the services what they have used. The benefits of cloud are it provides services to the large organizations to store their data, provides flexibility by the natures of backup and recovery, elasticity, independent of location, allow multiple sharing. Many of the large companies like Google, Microsoft, IBM, oracle, Amazon, Sales force make use of cloud to store and access their resources between various cloud users [11]. Some of the examples of cloud services are Gmail, Google drive, YouTube, Face book etc. The challenges of cloud computing are data security, privacy, interoperability, multi-tenancy and portability. Among these data security is considered as one of the major issue in cloud. It includes data privacy, data integrity, data confidentiality, data availability are the security challenges in cloud computing [13]. Thus, the primary objective is to maintain integrity and security of stored data in cloud due to the

© Springer Nature Switzerland AG 2020
P. Karrupusamy et al. (Eds.): ICSCN 2019, LNDECT 39, pp. 108–118, 2020.
https://doi.org/10.1007/978-3-030-34515-0_12

robustness nature of cloud and large quantities of data it takes through internet. The data integrity means protecting data on the cloud from deletion and modification by an unauthorized user. Data confidentiality is also a major aspect for users to store their private or confidential data in the cloud. Data privacy is the ability to preserve and protect the data from an unauthorized user. In order to achieve all these, data auditing approach is designed.

This paper presents a survey of different data auditing schemes to preserve privacy and integrity of data. Each scheme is discussed with pros and cons along with their results. Section 2 describes the cloud delivery and service model. Section 3 discusses data auditing, in Sect. 4 we discussed some of the existing surveys. Section 5 presents the six different auditing schemes along with results and pros and cons. Section 6 presents the summary table for auditing schemes and finally Sect. 7 concludes the paper.

2 Cloud Delivery and Deployment Models

Cloud Delivery Models: A cloud delivery model specifies the services offered to users and the applications supported. There are three basic cloud delivery models [12], listed below:

1. **Software as a Service (SaaS):** It provides service to the user's on demand. Users can access software which is readily available in cloud. The customers need to pay only for the services they have used. In this model cloud service providers can take care of the infrastructure and platforms required to run the software applications on the internet. The cloud users need not take care of its infrastructure required to run the particular software. The advantages of SaaS are it takes less time to develop a particular application and thus increases the availability of applications easily. Examples are drop box, Google apps.

2. **Platform as a Service (PaaS):** It is a type of cloud service model which offers full or partial development of tools or environment that users can access and utilize online. It provides platform to develop and test various applications without any difficulties in building and maintaining infrastructure on cloud environment. To provide scalability for the applications, PaaS provides a defined combination of OS and applications servers. Examples are force.com, Google App engine.

3. **Infrastructure as a Service (IaaS):** It provides the storage facilities, data management. The cloud users can access the resources on demand. Examples are Amazon Web Service (AWS). In this type, customers usually use the service providers (SP) and application program interface (API) to start, stop, access, modify and configure their virtual servers and storage as needed.

Cloud Deployment Models: There are three types of cloud deployments, categorized based on organization's ability to manage and secure assets as well as business needs

1. **Public cloud:** In this type of cloud, each user can access and able to alter the data of other cloud users via internet. This can be owned or operated by the third parties.

All the users use the same infrastructure to store the data with limited resources and security. It provides less security to the users. It offers services via web applications as well as web services over the internet to the public or a large group of organization. It provides an elastic and cost effective service to the users. Example for public cloud is Google, Amazon etc.

2. **Private cloud:** This type of cloud is customized to a particular users or and organizations. It can be leased or can be owned by a particular one. It is more expensive and secure and it cannot be accessed by anyone except the organization members. Private clouds are service based and very flexible compared to the public cloud. This type of cloud is maintained by the organizations like schools, hospitals, industries where the resources are managed by the organization.

3. **Hybrid cloud:** It is the combination of both public and private clouds. It allows the users to transfer data between public and private clouds. The reason for combining both public and private cloud is to control the business data on private cloud and outsource the less critical data to the public cloud. This cloud provides increase in data storage, cost-effective services.

4. **Community cloud:** It provides a shared infrastructure to the organizations to support a particular organization. Community cloud offers higher level of privacy, security and policy compliances. This type of cloud is utilized by the large group of organizations. The resources are shared and accessed among the organizations.

3 Data Auditing

Data auditing is a new approach which make sure the security of the stored data on the cloud environment [13]. Auditing is the process of collecting and evaluating the user data stored on the cloud environment which can be either carried out by a data owner or by a Third Party Auditor (TPA). The TPA acts as a verifier to examine the trustworthiness of data stored in the cloud without downloading the original data. Auditing can be done in two ways: Private auditing: Auditing is carried out by user or data owner and public auditing: Auditing is carried out by TPA.

The Fig. 1 shows the system architecture of data auditing. It consists of following entities namely: data owner, cloud service provider (CSP), Third Party Auditor (TPA). Data owner has a responsibility related to data. The data owner has to first login and register with the CSP and TPA. The new user has to fill the registration form and he should be active on the system to complete registration. Whenever data owner wants to access the data stored on the cloud he has to login to the cloud by using his own password. It includes both individual and organizations. It is dependent on CSP for maintenance of data. Cloud service Provider provides enough infrastructure and resources for storing the data to user. It is responsible for handling the users. A client logins to the cloud, CSP sends a password to the owner and he can access the data. TPA keeps on verifying the integrity of data on demand without archiving the original data from cloud service provider and notifies the owner if there is any modification in the stored data.

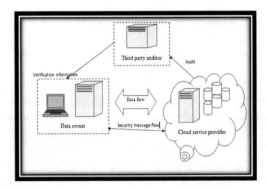

Fig. 1. System architecture of data auditing

4   Literature Survey on Data Auditing Surveys

This section presents some of the existing literature surveys on data auditing.

Rajeswari et al. [8] made a survey on data and security issues on cloud computing. They have made a survey of existing works on data security in terms of integrity, access control and attribute based encryption (ABE). An access control is used verify the policy of a user to store data and attribute based encryption is used to verify the policy's. They have presented the comparison table for different works. From this survey they have concluded that to ensure the integrity the cloud should provide less storage space and less overhead.

Sawanth et al. [9] made a survey on auditing techniques to preserve privacy of the data stored on the cloud. They made a survey of four different data auditing protocols. The first approach is Provable Data Possession (PDP) allows the users to prove the integrity of stored data on cloud without archiving it. The second approach is Cooperative provable data possession (CPDP) is used to solve the problem of multiple storage issues on cloud. The third approach is Dynamic provable data possession (DPDP) is an extension of PDP which uses a technique called rank based authenticated list which is used to evidence the integrity of the data stored on the cloud. The performance of DPDP is better than PDP. The last approach is Efficient and privacy preserving protocol (EPPAP) is proposed to reduce the burden of TPA. It uses a bilinear pairing algorithm which generates a data proof with a challenge stamp.

Bhavani et al. [10] had discussed different data auditing techniques to preserve privacy of data on the cloud and also present some of the existing privacy preserving techniques in detail. Among them are oruta which preserves privacy of data stored in a cloud using ring signatures to develop homomorphic authenticators. The second method is Authenticated File System (AFS), which is similar to the oruta system except it will check the freshness of data. In conclusion they have suggested that none of the discussed auditing support the multiple users and in future the work need to be enhanced to support multiple users.

5 Data Auditing Schemes

Some of the existing data auditing schemes to achieve integrity and privacy of data are discussed below.

More [3] proposed a data auditing scheme using AES and SHA-512 algorithm. The motivation for this scheme is that most of the existing solutions uses cloud server to check the integrity of data but it also has the responsibility of storing a large quantity of data in the server. This increases the load on cloud server. So in order to reduce the load on cloud server this system is proposed. This scheme includes three entities namely: data owner, Third Party Auditor (TPA), cloud server. The data owner is responsible for splitting the file need to be uploaded on cloud in to multiple blocks and that blocks are encrypted by using an AES algorithm, hash is generated for each blocks which are encrypted using SHA-2 algorithm and at last it combines the generated hash values and creates a digital signature for a file using RSA algorithm and sends the signature to TPA. The encryption and hashing is done in order to protect the data from the malicious attacker. Thus it achieves confidentiality of data on cloud. The cloud server is used to supervise the stored and encrypted blocks and sends the encrypted data whenever TPA requires. The function of a TPA in this scheme is to examine the trustworthiness of the data periodically or on demand without archiving the data from the server and thus maintains privacy of data on cloud. TPA checks the generated signature with the stored signature, if it matches indicates that the data is not modified and if it mismatches, it indicates that the data is modified by an attacker. After verification, TPA notifies the data owner regarding the verification process. Thus TPA is responsible for verifying the data and thus reduces the load on the cloud server. The performance of this scheme is evaluated by taking varying block size and file size. The authors noticed that the increase in block size leads to increase in hashing time and the encryption time goes on decreases.

Pros:
1. More secure.
2. Provides data integrity, confidentiality and privacy.

Cons:
1. Performance of the system mainly depends on the block size.
2. Provides good results for only fixed file size.

Jakhotia [1] proposed the data auditing scheme to achieve integrity of data by generating the proof for original data stored by client. The motivation for this scheme is that the cloud users don't know how the cloud administrative deals with stored data. Even though the data integrity is achieved, there is a possibility of data getting spoiled due to the immoral activities that had taken place and the user is unaware of it. So while the user is uploading a data, the user has to copy the file. This causes extra load on the users. Thus in order to reduce the dependency on the TPA, this system was proposed. This system is a challenge response protocol where the cloud service provider has to validate the user that the file is not harmed by any modifications or corruption of a file. The system consists of four entities namely Cloud Storage Server (CSS), Cloud Service Monitor (CSM), client and an auditor. The client stores the large amount of data on the cloud server. The CSM acts as controlling system for all the process running on the cloud. It is responsible for encryption of uploaded file using AES algorithm and hash values are generated for the encrypted data using SHA-512 algorithm and thus achieves the data integrity. It also verifies the requested tag from a CSS and generates the proof by comparing with the stored tag and thus CSM reduces the dependence on TPA. The Auditor in this scheme just compares the hash value and updates the client by sending the verified results. The CSS provides a large space to the client for data storage. The results of this scheme are performed by considering the fixed block size of 5 and 10. The proposed system gives good results compared to the existing system and it is noticed that as the block size increases the tag generation time is decreasing, and thus it provide better results as the block size increases.

Pros:
1. More secure.
2. Achieve data integrity and confidentiality.
3. Tag generation time for encryption and decryption is less.
4. Uses a unique key at only once for downloading a file. Therefore this reduces generation multiple copies for the same data.

Cons:
1. Multiple accessing of a same file is not possible
2. Backup of the data is complex means that if a client lost the downloaded file, he cannot download the file again using the same key so client should repeat the download process again this creates, an extra load on the client.

Kumar [2] proposed the data auditing scheme to achieve integrity of data using hybrid algorithm. The hybrid algorithm is a combination of Identity based Encryption (IBE) and MD5 method. The aim of this system is to improve the security trade off during data-transmission using MD5 & IBE algorithms. The system consists of TPA, user and server. The user upload the file on the server. The cloud server is used to generate a hash value for data uploaded by the user. TPA verifies the data on receiving the request from the server. The system consists of two phases: uploading and downloading phase. In uploading phase, user requests a server for storing a data on cloud by login to the cloud server using his own ID and password. After uploading a file the data is encrypted using IBE and hash is generated by using MD5 algorithm. The IBE is used by the server. In downloading phase, users send a request to the server for retrieving the data. The TPA on the cloud checks the authorization of user. If the authorization is verified then the user sends a file ID that needs to be downloaded. If the verification is done then the user is allowed to download the file. Thus it achieves integrity of data. This scheme is evaluated by considering the parameters like key size, computational time, encryption time and decryption time. The authors considered different key size for evaluating both RSA and IBE algorithm. For RSA, authors noticed that increase in key size leads to increase in key generation time. The encryption and decryption time is approximately equal for different key sizes. For IBE, key generation time is little less compared to the RSA and the computational time for decryption and encryption is same.

Pros:

1. Provides integrity.

Cons:

1. Less secure.
2. Computation time is high.
3. Does not provide confidentiality and privacy.

Rashmi [6] proposed the data auditing scheme using a remote data possession checking protocol. The motivation for this scheme is that if a user does not have a copy of data uploaded, the user is unable to access the data from the cloud. Because during integrity checking phase, verified data should compared with the original data. If the digital signature doesn't match then user is unable to download file. This problem exists in PoR scheme [1]. In order to reduce the load of user from the above mentioned problem the authors proposed this scheme. The proposed protocol is based on homomorphic hash algorithm. To know the position of the data stored on the cloud the system uses Merkle Hash Tree (MHT) algorithm. This system mainly consists of two phases: setup phase and audit phase. Setup phase is also called as preprocessing phase, where the client process the data using tag generation algorithm and stores the generated metadata at server and at TPA. The audit phase is also called as verification phase. Here TPA challenges the server to provide the verified data at the time of review. By using the ProofGen algorithm it will generate the proof and sends to the TPA. Once it receives the proof message it will verify the data using ProofVer algorithm. The authors evaluated the results for both individual users by taking different number of blocks and they have made the comparison of time with single user and a auditing with TPA (multi-user). In single auditing, user is able to check 4–5 hash value blocks with the server and takes more time and the multi user (auditing with TPA) is able to check more than 40–50 blocks and takes less computation time compared to the individual user.

Pros:

1. Achieves data integrity
2. Reduces the overhead
3. Supports dynamic operations.

Cons:

1. Doesn't provide confidentiality and privacy
2. Provides good results only with multiple users
3. Data is modified in the cloud server it has to repeat the tag generation process and thus increases the computation overhead.
4. Unable to delete the duplicate files in the cloud server and this leads to loss of data and increase the storage space.

Chakraborty [7] proposed a data auditing scheme using bilinear pairing algorithm. The motivation for this scheme is that the user has to store their data in their local server after uploading the data to the cloud server and later it is used for archiving the data. There are certain limitations for the users like limited resources, less storage space in their local machines and this causes the extra load on the user. Thus the main goal of this scheme is to ensure the integrity of the data stored on the user local machine using TPA. The TPA verifies the integrity of data by using bilinear pairing algorithm. The integrity is verified by following stages: Initialization phase where the variables are defined in this phase, Binding phase, Upload the data, Generates a hash value, Public and private key generation using bilinear algorithm, Generate a tag for encrypted data. The system mainly consists of a three entities namely client, TPA and cloud server. Client is responsible for storing data on cloud server. Cloud server provides a sufficient space for an individual users or an organization. TPA acts as a verifier between a user and CSP. This scheme uses three techniques namely fragmentation, encryption and hashing. Fragmentation technique divides a file in to a number of blocks and assigns id to the blocks and AES algorithm is used to encrypt the n blocks. The proposed scheme uses different keys for encrypting each block. Hashing is used for integrity checking using the message digest. The algorithm of this scheme consists of key generation, signature, proof and verification. The authors analyzed the results by considering different number of blocks ranges from 50–600 and calculated the mean of 20 trials. They compared the proposed scheme with the random masking technique; it is observed that the proposed scheme takes less time for proof generation.

Pros:
1. Achieves data integrity.
2. Less computational time compared to the random masking technique.
3. Less cost.

Cons:
1. Doesn't provide privacy and confidentiality
2. Requires more amount of time for proof verification.

Salunake and Patel [4] proposed an auditing scheme using replication and division algorithm. The motivation for this scheme is that when a user is archiving the data, it can be hacked or altered by an attacker this causes increase in delay while archiving the data. In order to overcome this problem the authors proposed this scheme using the concept of fragmentation and replication. This scheme uses the concept of data scattering among various nodes of cloud. The nodes are placed with certain distance using T-colouring concept where the hacker is unable to identify the next node. By using the replication concept, each data file is divided into number of fragments and the replica of those fragments is stored at different node and thus reduces the retrieval time. In order to achieve the data security, the fragments are encrypted using AES-128 algorithm and the TPA is responsible for periodically monitoring the data and report the status to user and service provider. The system mainly consists of four entities namely user, cloud server, TPA and proxy agent. Proxy agent helps in repairing the damaged fragment by always staying online while the owner gets disconnected. Once he uploads the file on to the server. Figure 2 shows the architecture of the system. The client uploads the data file and the file is splitted into number of fragments depends on file size. The fragments are aligned by using T-colouring concept. The nodes are initially assigned with open color. When the values are assigned to the nodes all the neighboring nodes with some distances are assigned with close color according to the set T. The encrypted fragments are hashed using the SHA algorithm and send a hash value to the server. While auditing, TPA checks the integrity by calculating the hash values for the fragments stored on server. If mismatch occurs, TPA sends the id of original and modified fragment to the proxy agent, who is responsible for replacing the modified fragment with the original fragment. The authors performed result analysis by comparing both AES & DES algorithm for the plain text with varying file size ranges from 512 Kb to 2.5 Mb. It is observed that encryption time for AES take less time compared to the DES algorithm.

Pros:
1. Provides integrity of data.
2. Modified data can be recovered quickly.
3. Reduces the retrieval time.

Cons:
1. In T-colouring process, if some of the nodes get lost this again leads to increase in retrieval time.
2. Provides no guarantee about placement of fragments at correct nodes.
3. Doesn't provide integrity and privacy for the stored data.

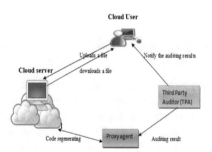

Fig. 2. Architecture of replication system

6   Summary of Data Auditing Schemes

This section presents the summary table for the above discussed data auditing schemes. From the table it is observed that auditing using AES algorithm achieves integrity, confidentiality and privacy. In Proof of Retrievability, Hybrid algorithm & division and replication scheme doesn't provide privacy of data (Table 1).

Table 1. Summary table for different data auditing schemes

Scheme	Objectives	Algorithms used	Achieves integrity	Achieves confidentiality	Achieves privacy
Auditing using AES algorithm	To develop secure and efficient auditing model, and reduce the burden on the cloud server	AES, SHA-2, RSA digital signature	Yes	Yes	Yes
Proof of Retrievability (PoR)	To reduce mistrust and dependency on the audit server	SHA-512, AES	Yes	Yes	No
Hybrid algorithm	To reduce the security tradeoff and improve the performance for transmitting a data	IBE, MD-5	Yes	Yes	No
Remote data possession checking protocol	Ensure the data integrity even the user doesn't have the original copy of uploaded file on its local machine	Homomorphic hash algorithm, Merkle Hash Tree (MHT)	Yes	No	No
Integrity checking using bilinear algorithm	To minimize overload on the user and to examine the integrity for the data stored on the user local machine	AES, bilinear algorithm	Yes	No	No
Division and replication	Reduce the retrieval time when data gets tampered	AES-128, SHA-256, T-colouring,	Yes	Yes	No

7 Conclusion

The users cache their data in the cloud and retrieve the data whenever it is required. Preserving privacy, integrity and confidentiality of data on cloud storage remains as a challenging task. This survey analyzes various research ideas on TPA schemes with various approaches to confirm the privacy and integrity of data. Hence it is necessarily required to prevent data from unauthenticated and unknown users and impartial auditing by data owners in the cloud. In this survey, six different data auditing schemes to preserve integrity and privacy are discussed in detail and finally a summary table for auditing schemes is presented. It is observed that each technique has made some improvements to provide security to the stored data on the cloud. The pros and cons of each of the proposed system is evaluated as of the security level of each scheme is compared. There is still scope for improvement to reduce the overhead and to provide the security for multiple users. All the proposed systems have their own benefits and drawbacks and in the future, researches can be conducted to overcome these shortcomings.

References

1. Jakhotia, K., Bhosale, R.: Novel architecture for enabling proof of retrievability using AES algorithm. In: ICCMC. IEEE (2017)
2. Gajendra, B.P., Singh, V.K.: Achieving cloud security using third party auditor, MD-5 and identity-base encryption. In: IEEE (2016)
3. More, S.S., Chaudhari, S.S: Secure and efficient public auditing scheme for cloud storage. In: IEEE (2016)
4. Salunkhe, S.D., Patil, D.: Division and replication for data with public auditing scheme for cloud storage. In: IEEE (2016)
5. Kaura, W.C.N., Lal, A.: Survey paper on cloud computing security. In: ICIIECS. IEEE (2017)
6. Rashmi, R.P., Sangve, S.M.: Public auditing system: improved remote data possession checking protocol for secure cloud storage. In: IEEE (2015)
7. Chakraborty, S., Singh, S.: Integrity checking using third party auditor in cloud storage. In: International Conference on Contemporary Computing. IEEE (2018)
8. Rajeswari, S., Kalaiselvi, R.: Survey of data and storage security in cloud computing. In: ICCS. IEEE (2017)
9. Sawant, N.M., Mane, N.S.: A survey on auditing techniques used for preserving privacy of data stored on cloud. In: ICEEOT. IEEE (2016)
10. Bhavani, S.D., Sudhakar, G.: A survey of auditing techniques for privacy preserving and ensure data correctness in the cloud. In: IJARCSSE (2017)
11. Kolhar, M., Abu-Alhaj, M.M., El-atty, S.M.A.: Cloud data auditing techniques with a focus on privacy and security. IEEE Secur. Priv. **15**(1), 42–51 (2017)
12. Rao, C.C., Leelarani, M., Kumar, Y.R.: Cloud computing services and deployment models. IJECS **2**(12), 3389–3390 (2013)
13. Geeta, C.M., Raghavendra, S., Rajkumar, B., Venugopal, K.R., Iyengar, S.S., Patnaik, L.M.: Data auditing and security in cloud computing: issues, challenges and future direction. Int. J. Comput. (IJC) **28**(1), 8–57 (2018)
14. Hiremath, S., Kunte, S.: A novel data auditing approach to achieve data privacy and data integrity in cloud computing. In: ICEECCOT. IEEE (2017)

Accident Avoidance System in Blind Curves and Fog Using Integration of Technologies

Adityapratap Singh$^{(\boxtimes)}$, Rajasi Ratnakar, Ankit Rajak, Neha Gurale, and Utkarsha S. Pacharaney

Department of Electronics Engineering, Datta Meghe College of Engineering, Airoli, India
singhadityajoy@gmail.com, rajasi0610@gmail.com, ankitrajak9167@gmail.com, nehagurale10@gmail.com, utk21pac76@gmail.com

Abstract. In India, the number of road accidents and deaths caused by fog and mist has steadily increased over the years. Visibility is greatly reduced due to the dense fog. Also at a blind curve, approaching vehicles tend to collide with each other due to lack of visibility over the other vehicles. Such condition increases the demand for widespread availability of different information and communication technologies over road networks to avoid vehicle accident. In this paper, we have proposed a system based on integration of various modules such as RF transmitter and receiver pair, temperature and humidity sensor, ultrasonic distance sensor of long range, ATmega microcontroller, etc., to avoid collision of vehicles due to fog and blind curve. The purpose of the system is to buzz an alarm in blind curves when vehicles approach each other and alert the driver in case of fog. A sensor set, a management unit and a monitoring platform is included in this system.

Keywords: RF communication · Vehicular network · Sensor network · Blind curves · Fog

1 Introduction

Road accident is a collision between any vehicle (e.g. car, truck, motorcycle, etc.) or it may be take place in the middle of the road with any pedestrian, animal or obstacle. Accident results in serious injury to life and property. One major road accident occurs every minute in the country and 16 die every hour on Indian roads. Measures to reduce the number of road accidents must therefore be taken. The main contributing factors to road accidents are:

1. **Road Conditions:** Damaged road, potholes, eroded road merging of rural roads with highways, illegal speed breakers, diversions, unexpected turnings, blind curves etc.
2. **Weather Conditions:** Fog, wind storms, hail storms, snow, and heavy rainfall.

P. Karrupusamy et al. (Eds.): ICSCN 2019, LNDECT 39, pp. 119–134, 2020.
https://doi.org/10.1007/978-3-030-34515-0_13

This paper focuses on two major factors leading to road accidents i.e. the occurrence of unexpected blind curves and the reduced vision problem in presence of fog. So we aim at providing a solution to the visual and the natural factors leading to road accidents.

(a) **Accidents occurring due to blind curves:**
    A blind curve is a curve or bends in the road that one cannot see around while driving a vehicle. In addition to this the narrow, steep curvy roads in hilly regions contribute to the spate of accidents. Similar situation arises at crossroads in cities where huge structures or buildings are built on the turning points. These obstacles provide either partial or no-visibility of the incoming traffic to the drivers, thus increasing the risk of collision. As study has found that the number of crashes caused by such blind spots has risen by 50% over the last two years. So a warning system that aims at minimizing chances of an accident due to blind corners is necessary.

(b) **Accidents occurring due to fog:**
    In northern India, 37% of the road accidents are caused due to bad weather. The major contributor is fog. Road fatalities on account of weather conditions like fog are increasing day by day and getting deadlier. The number of casualties due to fog related crashes increases by 20% every year. Fog hits the train and air traffic as well. On roads, car and trucks ran over one another leading to loss of lives. Normal movement of vehicular traffic gets affected. This problem is thoroughly considered in our project.

2 Literature Survey

Over the years many different solutions were proposed to address the above mentioned problems. This section gives an overview of work done to overcome this problem. Saraf *et al.* has developed a technique to measure the distance between two vehicles and determine the traffic density in that particular area. The main objective of his study is to improve road safety by providing real-time alerts to the vehicle drivers about hazards and obstacles in their path and their immediate area. This is made possible through vehicle and roadside unit intercommunication [1]. Ismail *et al.* has developed a system to alert the user of their respective vehicle while taking turns using a wireless communication system. By means of the red light indication, the system alerts users of the other side lane to either reduce the speed or stop the second vehicle to avoid collisions [2]. Leo *et al.* has proposed a system which negotiates the problems of vehicles travelling in hairpin bends in hilly areas. The system provides priorities to the vehicles near the turnings, speed control, intimation of vehicle breakdown in the hilly areas and accident alerts. It uses GPS and GSM technologies [3]. Rakul *et al.* has also tried to avert collisions between vehicles mainly occurring in short corners, blind curves and hairpin bends. When the Ultrasonic transceiver detects any vehicles at hairpin bends, it sends the signal via Wi-Fi router to the microcontroller. When the vehicle driver has installed this android application, the traffic information will be delivered as voice recognition [4]. Leo, Monisha, Sakthi and Sunder *et al.* in this paper try to lessen the burden on the drivers by bring out a technical solution by placing an interface between the man and the machine [5]. Ueki, Mori, Nakamura, Horii and Okada *et al.* in this paper describe a vehicular collision avoidance support system (VCASS), and show its performance by various experiments with two real vehicles [6]. Sengupta, Rezaei, Shladover, Cody, Dickey and Kumar *et al.* introduce the concept of cooperative collision warning (CCW) systems, The CCW idea gives alerts or circumstance mindfulness showcases to drivers dependent on data about the movements of neighboring vehicles acquired by remote correspondences from those vehicles, without utilization of any ranging sensors [7]. Kumar *et al.* has proposed a system which addresses issues due to adverse weather conditions as heavy rain and huge snow fall. Humidity sensor is used to predict the rain and fog. After detecting the adverse weather conditions, the next step is to avoid collisions [8]. MacHutchon et al. proposed a fog identification and cautioning framework in which fog finders dependent on the forward scatter principle were chosen. Following meteorological, traffic and site overviews the fog hazard index was applied to recognize reasonable destinations for the area of the detectors. To caution drivers of fog, cautioning messages will be shown at inside distinguished fog region. This would give the alerts [9]. Negru *et al.* has designed a system which uses image processing for detecting fog. The images taken from the camera of the moving vehicle are employed to calculate the visible distance. The system works in real time [10]. Singhal *et al.* has proposed an GPS based sensor system employing Information and communication technology (ICT) for collision avoidance [11]. Mueller and Trick *et al.* has proposed a method that Investigates effects of driving experience and visibility on driving behavior [12]. de Bruyne *et al.* has proposed a system with a 430 MHz transceiver in cars coupled to the 4-way flasher control switch [13]. Hancock and de Ridder et al. led inquire about on new and imaginative research technique that grants examination to be directed in a protected and compelling way, for social investigation of reaction in the final seconds and milliseconds before impact [14]. Ye, Adams and Roy et al. propose and study the effect of a 802.11 based multi-hop Mac protocol that engenders an emergency warning message (EWM) down a detachment of cars on a roadway [15].

3   Proposed System

The proposed system is a simple cost effective solution to the above mentioned problems. It includes RF transmitter and receiver pair to address the problem of blind curves and Ultrasonic sensor to address the issues in presence of fog. These two components are the most accurate, highly precise and inexpensive as well in their respective domains. So the system evolves on the best components of the two worlds. The basic block diagrams showing the working of the system are shown in Figs. 1, 2, 3 and 4. The electronic modules of the system are implemented on the road (piezoelectric sensor), in the roadside tower and in the vehicle.

3.1   Block Diagram

Figure 1 shows the sensor array which is implemented on the roads.

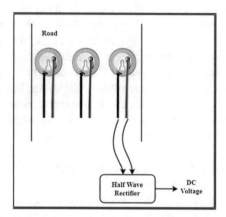

Fig. 1. The array of piezoelectric sensors

Whenever a vehicle passes on the road the Piezoelectric Sensors buried underneath will generate a voltage proportional to the pressure applied to it. This voltage will be then applied to the Half Wave Rectifier which will convert the generated AC voltage to proportional DC voltage.

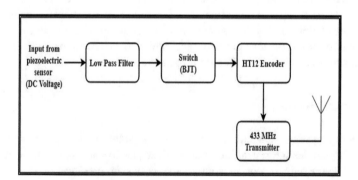

Fig. 2. Block diagram of the module placed in the tower

The module shown in Fig. 2 will be implemented in the tower. The generated DC voltage from the half wave rectifier will act as an input to the low pass filter which will filter out the unnecessary noise. Then this filtered signal will turn on the system by turning on the switch i.e. BJT. This signal will be then encoded by HT12 Encoder and then transmitted via antenna using 433 MHz Transmitter.

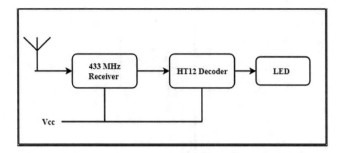

Fig. 3. Block diagram of the circuit mounted in the vehicle

Figure 3 shows the circuit mounted in the vehicle. The transmitted signal that is encoded will be received via 433 MHz receiver in the vehicle which will be then applied to the HT12 decoder which then will decode the encoded signal and will turn on the LED thus alerting the driver. Once the presence of approaching vehicle is detected the transmission occurs for around 2 s, during that period the signal stored in the tower (based on the standard conventions) is transmitted, so that driver on the other side will have sufficient time to react.

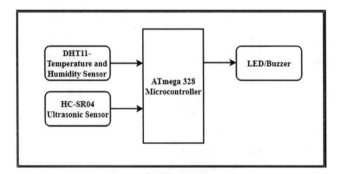

Fig. 4. Schematic used in presence of fog

The schematic shown in Fig. 4 will be used in case of fog. It consists of DHT11 sensor to sense the humidity and temperature conditions and HC-SR04 to sense and measure the distance from the vehicle. The outputs from DHT11 i.e. Temperature and Humidity Sensor and HC-SR04 i.e. Ultrasonic Sensor will act as inputs to ATmega 328 Microcontroller and which in turn will turn on LED/Buzzer (Fig. 4).

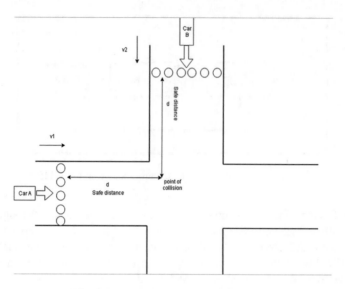

Fig. 5. Schematic diagram for blind curve

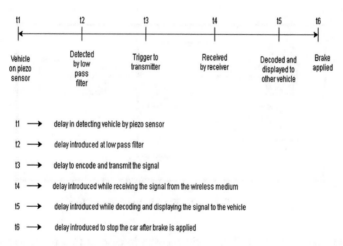

Fig. 6. Instances in blind curve

3.2   Timing Diagram and Calculations

3.2.1   In Case of Blind Curves

v1 is the speed of a vehicle (Car A)
v2 is the speed of another vehicle (Car B)
d is the Safe distance

$$\text{Total delay} = t_1 + t_2 + t_3 + t_4 + t_5 + t_6$$

$t_3 = t_5 = 9.23 \times (10^{-9})$ [since it is in the range of nano (10^{-9}) it can be neglected]

t_4 [This delay can also be neglected because radio communication is done at very high speeds]

t_6 [Delay after brake is applied along with drivers response time]

$$\text{Total delay} = t_1 + t_2 + t_6$$

$$\text{Total delay} = 0.2 + 0.22 + 2.75$$

$$\textbf{Total delay} = \textbf{3.17 s}$$

Since we are considering local road accidents we can assume an average speed of 35 km/hr

So,

$$\text{Safe distance}(d) = 3.17 * ((35 * 1000)/(60 * 60))$$
$$= 30.819\,\text{m}$$

So the safe distance where the piezoelectric sensors should be mounted is approximately 31 meters from the point of collision as seen in Fig. 5.

3.2.2   In Case of Fog

Considering safe distance between both vehicles to be d_2, so ultrasonic sound will have to travel double the distance at the speed of 346 m/s as seen in Fig. 7. At lower temperatures the speed of sound may decrease slightly, but considering it to be same (Fig. 7).

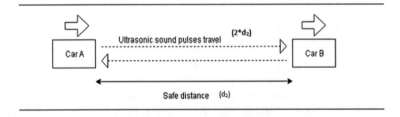

Fig. 7. Schematic diagram for Fog case

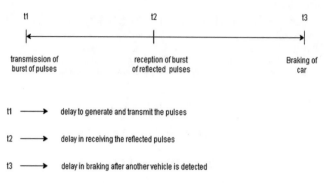

Fig. 8. Instances in Fog case

Considering the case in which vehicles are operating along the safe distance and since for close range the difference in speed of vehicle won't be of utmost priority so we consider instantaneous velocity or assume that the velocity of vehicles to be constant.

Assuming average speed of cars in fog to be 35 km/hr and Safe distance between cars when alarm is to be triggered is 28 m, so that if car in the front applies brakes suddenly, system will still be able to prevent the collision.

$$\text{Total delay} = t_1 + t_2 + t_3$$

$$\text{Total delay} = 2 \times \left(10^{-4}\right) + 0.08746 + 2.75$$

$$\textbf{Total delay} = \textbf{2.83 s}$$

This much time is required to stop the vehicle completely.

$$\text{Minimum safe distance } (d_2) = \text{Avg speed of car} * \text{Total delay}$$
$$\text{Minimum safe distance } (d_2) = ((35 * 1000)/(60 * 60)) * 2.83$$
$$d_2 = 27.51 \, \text{m}$$

So approximately a minimum distance of 28 m is necessary to avoid collision.

Here in braking time (t_3) along with delay to stop the vehicle after application of brake we have assumed one second delay for driver reaction.

3.3   Hardware Components

3.3.1   RF Transmitter and Receiver

The RF module works at radio frequency, as the name recommends. The frequency varies in the range from 30 kHz to 300 GHz. Digital data is represented in this RF system as variations in carrier wave amplitude. This type of modulation is known as the Amplitude Shift Keying (ASK) Transmission by means of RF has numerous points of interest over IR (infrared). Signals can travel over larger separations through RF, making it reasonable for long-range applications. Additionally, while IR works generally in line-of-sight mode, RF signals can travel, even when there is an obstruction between transmitter and receiver. Furthermore, RF is more reliable than IR transmission. RF communication uses a specific frequency unlike IR signals that are affected by other sources of light emitting. This RF module incorporates a RF transmitter as well as a receiver. The transmitter/receiver pair (Tx/Rx) works at 433 MHz frequency. The RF transmitter and receiver is shown in Fig. 9.

Fig. 9. RF transmitter and receiver

3.3.2   HT12 E and HT12 D

HT12E is used to encode RF Transmitter data and HT12D is used to decode RF receiver data. The HT12E & HT12D Encoder and decoder ICs for remote control system applications are a series of CMOS LSIs respectively. Encoder can encode 12 bit of information consisting of 8 bits of address and 4 bits of data. Each address and input of data can be programmed externally or fed into switches. These Encoder and decoder ICs are paired together. A pair of encoder/decoder with the same address number and data format should be selected for proper operation (HT12E is coupled with HT12D). The decoder receives the serial address and data transmitted by a carrier using an RF transmission medium and gives output to the output pins after processing the data.

3.3.3   DHT11 Temperature and Humidity Sensor

The DHT11 is a digital temperature and humidity sensor that is basic & ultra low-cost. It uses a capacitive humidity sensor and a thermistor to measure the vicinity air and spits a digital signal on the data pin (no analog input pins required). Using it is quite easy but it requires careful timing for data acquisition. This sensor's readings are 2 s old, as we can get data from it after every 2 s when using the library, it is only real downside that you can get. Its technology ensures high reliability and long-term stability. This sensor includes a resistive element and a sense of measuring devices for wet NTC temperature. It has excellent quality, quick response, ability to prevent interference, and advantages of high cost performance. The DHT11 Temperature and Humidity Sensor are shown in Fig. 10.

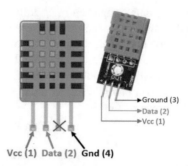

Fig. 10. DHT 11 temperature and humidity sensor

3.3.4   Piezoelectric Sensor

A piezoelectric effect is called the ability of a piezoelectric material to convert a mechanical stress into electric charge. The outer circle gives negative output voltage and positive output voltage is provided by the inner circle.

3.3.5   Ultrasonic Sensor

An ultrasonic sensor is a device that uses sound waves to measure the distance to an object. By sending out a specific frequency sound wave and listening to bounce back for that sound wave it measures distance. The distance between the sonar sensor and the object can be calculated by recording the elapsed time between the sound wave being generated and the sound wave bouncing back. The ultrasonic sensor is shown in Fig. 11.

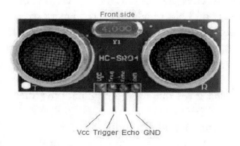

Fig. 11. Ultrasonic sensor

3.3.6   ATmega Microcontroller

AVR is a microcontroller family that Atmel developed in 1996. These are modified 8-bit RISC single-chip microcontrollers in Harvard architecture. The core of Atmel AVR combines a rich set of instructions with 32 working registers for general purposes. All 32 registers are connected directly to the Arithmetic Logic Unit (ALU), allowing access to two independent registers in one clock cycle performed instruction. The resulting architecture is more code-efficient than conventional CISC microcontrollers while achieving throughput up to ten times faster.

4 Working of the Proposed System

The prototype of the proposed system is shown in Fig. 12. The detail working of detection for both blind curve and fog is explained in this section.

Fig. 12. Final prototype implementation

4.1 In Case of Blind Curve

An array of piezoelectric sensors is embedded in the road in parallel fashion so as to detect the vehicle and give the voltage output. Piezoelectric sensors generate an electric charge proportional to the pressure applied to it. Hence when vehicles pass over it the signal is obtained. When piezoelectric sensor detects the presence of vehicle, signal is sent to this module where it is first passed through a passive low pass filter (LPF) and then given to the base of BJT which acts as the switch. Value of the capacitor plays the vital role as it determines the ON time of the switch and in turn the transmission time. Signal from previous module is obtained as a result of switching of BJT, that signal is encoded with the help of HT-12 encoder module and transmitted over RF using ASK modulation over a 18 m radius (approx). Buttons here represent different input patterns that can be transmitted and antenna is used to increase the range. On detection of vehicles presence on the crossroads, the transmitted signal is received by this module (Receiver works at a frequency of 433 MHz). The received signal is then decoded by HT-12 decoder IC and the corresponding signal is displayed on the green LED's, while the red LED will display the reception of data serially.

4.2   In Case of Presence of Fog

To detect the presence of fog, DHT11 sensor is used. Based on the pre-decided temperature and humidity threshold value (temp < 10 °C, humidity > 83 RH), the ultrasonic sensor is triggered. The ultrasonic sensor is used to detect the obstacles in front of the vehicle. The distance between the two vehicles is calculated and depending on it the alerts are provided. The alarm is triggered continuously when the distance is very less.

5   Results

This section presents the result for blind curve and fog detection.

5.1   Blind Curve

According to the analysis done on the prototype following are the observations:

1. Piezoelectric sensor is capable of detecting high speed vehicles accurately with a speed of around 30–35 km/h, For an increase in speed above this, the entire system may be recalibrated. The generated signal via piezo sensor is AC in nature so the setup is quite capable of detecting high speed signals and much more can be concluded from it.
2. If the car is within the radius (approx 18 m) of transmission, the signal is detected (in the receiver implemented on the vehicle) without much latency.

5.2   Fog

For various values of distance between the vehicles, the following results were obtained.

If the vehicle in question is closer to another vehicle below a particular defined threshold (less than 15 cm as seen in prototype) then an alarm will be triggered continuously along with the glow of red LED indicating immediate attention of the driver as shown in Fig. 13. This is the first LED in the array. (In real case 28 m is the corresponding distance as measured theoretically).

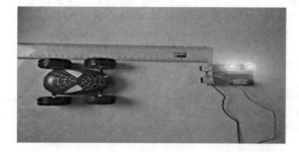

Fig. 13. Vehicle present very close to the system mounted vehicle

When the vehicle is about to enter the range of critical distance (less than 30 cm), the second LED (green) of the array glows as shown in Fig. 14. The buzzer beeps in intervals.

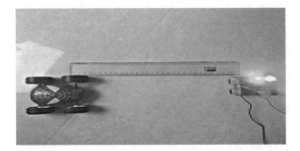

Fig. 14. Vehicle lies in the short range distance

Depending upon the distance measured by the ultrasonic sensor if the vehicle lies in the mid-range (around 40 cm), the third LED (green) of the array glows indicating less or no danger as shown in Fig. 15.

Fig. 15. Vehicle lies in the mid-range distance

When the vehicle lies away (more than 50 cm) from our system mounted vehicle, there is no danger. So the green led glows as shown in Fig. 16.

Fig. 16. The vehicle is far away from system mounted vehicle

When fog occurs, a particular threshold is tested for temp and humidity (temp < 10 °C, humidity > 83 RH) and ultrasonic sensor is activated to measure the distance of vehicles present in the front of vehicle as seen in Fig. 17.

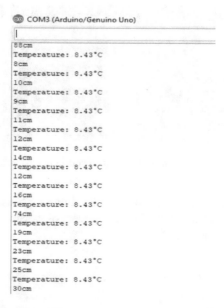

Fig. 17. The distance from the front vehicle with temperature values indication in presence of fog.

6 Conclusion and Future Scope

The main aim of the proposed accident avoidance system is to avoid accidents occurring due to blind curves and fog by proposing a low cost design with high accuracy. The key idea is establishing wireless communication between various vehicle nodes thus avoiding accidents. This system uses integrated RF and ultrasonic communication which results in providing alerts with minimal latency. The performance of the system may change depending on the operating conditions so it should be calibrated accordingly.

As we know Energy harvesting is a basic need, but the cost of generation is quite high and another issue which follows it is the storage of energy, for this purpose our project may be slightly modified in the future, the key factor for this are the piezo-electric sensors used.

Consider the following scenario:

1. Suppose piezoelectric sensor is embedded within the roads to detect the presence of vehicles, so along with that it may also helps in generation of energy because of movement of vehicles and pedestrians due to its basic property. Hence the system

may work on the energy generated by itself and the excess energy can be used or stored for other purposes.

2. In parking lot, parking is a big issue and people use mirrors to track others but it's inconvenient solution so implementing piezo sensors along with our sensors and some calibration will help solve this issue. Furthermore if the parking is in basement or indoors our project does not need any other sources to charge itself as it is required with solar chargers which are redundant in this particular scenario.

Compliance with Ethical Standards. I ADITYAPRATAP SINGH (Corresponding author) hereby declare that the below mentioned points are true & correct to the best of my knowledge & belief and the work done in this paper is our (all authors mentioned) original piece of work.

1. Disclosure of potential conflicts of interests:

Funding: This study was self funded.

Conflict of Interest: The authors declare that they have no conflict of interest.

2. Informed consent:Informed consent: "Informed consent was obtained from all individual participants included in the study."

References

1. Saraf, P.D., Chavan, N.A.: Pre-crash sensing and warning on curves: a review. Int. J. Latest Trends Eng. Technol. (IJLTET) **2**(1) (2013)
2. Ismail, M., et al.: Intersection cross traffic warning system for vehicle collision avoidance. Int. J. Adv. Res. Electr. Electron. Instrum. Eng. **3**, 13155–13160 (2014). https://doi.org/10.15662/ijareeie.2014.0311031
3. Leo, J.J., et al.: Vehicle movement control and accident avoidance in hilly track. In: 2014 International Conference on Electronics and Communication Systems (ICECS), pp. 1–5. IEEE (2014)
4. Rakul, R.S., Ravia, S., Thirukkuralkani, K.N.: Implementation of vehicle mishap averting system using arduino microcontroller. Int. J. Eng. Res. Technol. (IJERT), **5**(4) (2016)
5. Leo, J., Monisha, R., Sakthi, B.T.S., Sunder, A.C.: Vehicle movement control and accident avoidance in hilly track. In: 2014 International Conference on Electronics and Communication Systems (ICECS) (2014). https://doi.org/10.1109/ecs.2014.6892757
6. Ueki, J., Mori, J., Nakamura, Y., Horii, Y., Okada, H.: Development of vehicular-collision avoidance support system by inter-vehicle communications – VCASS. In: 2004 IEEE 59th Vehicular Technology Conference VTC 2004-Spring (IEEE Cat. No. 04CH37514). https://doi.org/10.1109/vetecs.2004.1391463
7. Sengupta, R., Rezaei, S., Shladover, S., Cody, D., Dickey, S., Krishnan, H.: Cooperative collision warning systems: concept definition and experimental implementation. J. Intell. Transp. Syst. **11**(3), 143–155 (2007). https://doi.org/10.1080/15472450701410452
8. Kumar, T.A.S., Mrudula, J.: Advanced accident avoidance system for automobiles. Int. J. Comput. Trends Technol. (IJCTT) **6**(2), 79–83 (2013)
9. MacHutchon, K.R., Ryan, A.: Fog detection and warning, a novel approach to sensor location. In: 1999 IEEE Africon. 5th Africon Conference in Africa (Cat. No. CH36342), vol. 1, pp. 43–50. IEEE (1999)
10. Negru, M., Nedevschi, S.: Image based fog detection and visibility estimation for driving assistance systems. In: 2013 IEEE 9th International Conference on Intelligent Computer Communication and Processing (ICCP), pp. 163–168. IEEE (2013)

11. Singhal, V., Jain, S.S.: A novel forward vehicle collision avoidance methodology through integration of information and communication technologies at intersections (2012)
12. Mueller, A., Trick, L.: Driving in fog: The effects of driving experience and visibility on speed compensation and hazard avoidance. Accid. Anal. Prev. **48**, 472–479 (2012)
13. de Bruyne, P.: Fog collision-avoidance warning device. In: Proceedings IEEE 31st Annual 1997 International Carnahan Conference on Security Technology. https://doi.org/10.1109/ccst.1997.626264
14. Hancock, P., De Ridder, S.: Behavioural accident avoidance science: understanding response in collision incipient conditions. Ergonomics **46**(12), 1111–1135 (2003). https://doi.org/10.1080/0014013031000136386
15. Ye, F., Adams, M., Roy, S.: V2 V wireless communication protocol for rear-end collision avoidance on highways. ICC Workshops - 2008 IEEE International Conference on Communications Workshops (2008). https://doi.org/10.1109/iccw.2008.77

Design of a Home Guide System for Dementia Using Integrated Sensor

S. Sajithra Varun[1(✉)] and R. Nagaraj[2]

[1] Department of ECE, The Oxford College of Engineering, Bangalore, India
sajithra_s@yahoo.co.in
[2] Kalasalingam Academy of Research and Education, Anand Nagar,
Krishnankoil, India
nagaraj.ramrao@gmail.com

Abstract. Kalman filtering with its amazing pattern of achieving most favourable outcome has become one of the world's widely accessible and interesting algorithms due to its optimal estimation solution. The proposed system conceptualizes the integration of two sensors Global Positioning System and inertial navigation system and aiding patients with memory loss for proper navigation. This paper aims at enhancing the technology along with the common GPS tracker by bringing back the person with Alzheimer's or dementia securely to the destination and also back home. GPS/INS integrated device interfaced with hearing aid navigates the patient by giving instructions through wireless communication framework. The initial phase of the paper deals with integration of the two sensors using MATLAB, which will be extended in developing the prototype device which secures the living of our loved ones. By combining GPS/INS along with medical aids, the proposed paper proves the significance of the integration of different fields in the healthcare domain.

Keywords: Kalman filter · Accelerometer · INS · Global Positioning System

1 Introduction

Localization has gained its own popularity in a wide variety of applications in this modern life with its huge research attention. Integration of the sensor has made a huge impact on day-to-day life on various commercial applications due to its popularity. The underlying uses of GPS tracker are many such as tracking children, as both indoor and outdoor unit. As safety is one of the major issues in society, GPS tracking has made a huge impact in the current market. Even the remote areas can be reached by humans in short span of time. Extending the research in medical field for aiding dementia both long term and short term memory loss is a challenging task. GPS tracker along with geo-fencing has gained its popularity worldwide. Interfacing the former device along with a hearing aid that communicates with the patient with respect to the embedded destination address paves a miracle by navigating the patient back to destination without the help of third person. GPS helps in tracking while INS helps in navigation and integration of both minimizes the error and gives more precise outcome by using kalman filtering [1].

© Springer Nature Switzerland AG 2020
P. Karrupusamy et al. (Eds.): ICSCN 2019, LNDECT 39, pp. 135–142, 2020.
https://doi.org/10.1007/978-3-030-34515-0_14

GPS utilizes a minimum of four satellites for triangulation to monitor the exact location of the person. The receiver with monitoring device monitors the portable device that is attached to the target such as humans, vehicles or any other belongings and it totally constitutes the tracking system. GPS sends the latitude and longitude coordinates to the device which in turn transmits the coordinates to centre so that the trajectory can be interfaced through Google maps [2]. An open source software which is capable to do the calculation in simplified method and user accessible was selected such as Android device which offers wide services is low cost [3, 4].

The heart of precise estimation is the Kalman Filter algorithm with prediction and measurement process. Extended Kalman Filter (EKF) enhances the estimation of the three dimensional rigid body by taking the non-linearity into consideration. Distance and velocity from accelerometer and gyroscope are optimized using Kalman Filter through recursive steps [5, 6]. The Position accuracy can be enhanced enormously by this innovative method of tracking technology and reaching out at any areas is made easier.

The ephemeris and almanac data from the satellite will be obtained by the corresponding GPS receiver. The GPS receiver is fed as one of the input and its data are recorded in the buffer for further analysis. The GPS receiver updates the data once in every 0.01 ms. To activate, the GSM module server has to send a message by which it gets activated [7, 8]. The kalman filter is used to interpolate the data and obtain the best and optimum estimate with the aid of GPS. The algorithm can be further extended to EKF for more optimum results. The restoration of velocity is accomplished by a proper state estimator [9]. In this paper GPS is aided with inertial navigation system and two different algorithms one the kalman filter and speed restoration algorithms. For testing purpose GPS Trimble SPS 351 with 24 active and healthy satellites were studied and taken as reference [10, 11].

2 Proposed System

The GPS receiver connected to the target captures the signal from air using its external antenna which is designed with maximum efficiency and less leakage. However the position, velocity, attitude and time obtained from the two sensors are prone to its own disadvantages and it is not sufficient for an optimum tracking. Due to change in environment as a fact of nature, GPS signal become weak as well as it is unreliable to consider accelerometer and gyroscope data in long run. The proposed system with the help of appropriate kalman filter can overcome these former disadvantages. The wireless technology used as interface for safe return is the outstanding feature of this paper and it can serve as the base for numerous medical engineering applications.

In this proposed system the Accelerometer, GPS can be used for comparing real data with ground truth data to get accurate position of the vehicle. By using this method, the system can get error when calculating position of the vehicle. So the Kalman filter is applied to reduce error while tracking position of the vehicle [12, 13]. The following steps represent the working principle of the system.

Step1: The current Analysis depends on X direction.
Step2: GPS and Accelerometer data can be used
- GPS can provide accurate ground truth for Position
- But it can provide the data at minimum sampling rate
- It takes more power to give the position
- Acceleration can provide acceleration for X, Y, Z direction
- The result of Accelerometer output may be positive and negative

Step3:
- Estimation of state using kalman filters using recursive steps.
- Calculation of kalman gain.

Step4:
- Interfacing with the hearing aid via Bluetooth for safe return to destination.

3 Dementia and Engineering

Most common among elderly adults, dementia can affect any age group due to brain damage. It's an inevitable task to safeguard our loved ones. Although huge number of trackers and geo fencing devices are widely available in the market [14, 15]; it would be a milestone if the equipment would control and guide the patient back to home by giving necessary information using some interface. This paper enhances the existing technology by embedding the destination address and giving directional commands using hearing aid. Figure 1 depicts the idea of interfacing the integrated device and hearing aid.

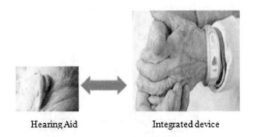

Hearing Aid   Integrated device

Fig. 1. Proposed system

4 Hardware Design

GPS module is connected to the serial port of the Arduino UNO microcontroller. Next the digital accelerometer is interfaced to i2c of the Arduino UNO microcontroller. Secondly, SD card is interfaced to SPI, which is used to record the log about filtered result and real GPS value. After that Keypad is connected to PORT pins of micro-controller, which is used to set configuration and START/STOP operation. LCD is connected to PORT pins to display the current status of the system. Blue tooth wireless technology is used for interfacing with the hearing aid. Figure 2 shows the block diagram of the hardware design.

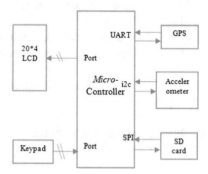

Fig. 2. Hardware design

5   Linear Systems

The implementation of kalman filter algorithm is done by considering the entire input and output characteristics as a system where the resultant parameter varies linearly with respect to the input parameter. Figure 3 depicts the design of Kalman filter. An innovative simple method of tracking any object using linear systems makes this popular and in demand to the present day scenario.

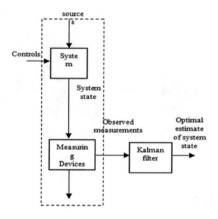

Fig. 3. Kalman filter design

6   The Discrete Kalman Filter

The prediction and measurement step in difference equation describes this algorithm. Position Estimation with sensor fusion (GPS + accelerometer) is derived in the following steps.

Step1: Read GPS value from GPS sensor, Estimates the position variation by the formula given below:

$$a = (Xgps(k) - Xgps(k-1))/(t(k) - t(k-1)) \tag{1}$$

Where

Xgps = Current Position value with minimum sample rate
Xgps (k−1) = Previous Position received from GPS by different time interval

Step2: Estimates current value based on previous GPS values, when GPS is not available.

$$E_k = a * Xgps(k-1) \tag{2}$$

Where

E_k = Position estimated by previous GPS value
A = acceleration error in present and previous value of GPS

Step3: Read the acceleration from accelerometer, Observe the current position based on previous position along with acceleration.

$$Z_k = X_k + a_k \tag{3}$$

Where

Z_k = Observed position information by accelerometer

Step4: Predict the error based on GPS deviation.

$$p_k = a * p_k - 1 * a \tag{4}$$

$$p_k = (1 - g_k)p_k - 1 \tag{5}$$

Where

p_k = Error value based on P_{k-1} and a. It is updated iteratively by previous value

Step5: Calculate gain for update the estimated result by kalman filter

$$g_k = \frac{p_k}{p_k + r} \tag{6}$$

Where

g_k = gain value which is help to update the kalman estimated result

Step6: Kalman result for filtered position (X_k) based on GPS and Accelerometer along with gain which is calculated above.

$$X_k = E_k + g_{k(Z_k - E_k)} \tag{7}$$

Where,

X_k = Kalman estimated result
r = Accelerometer Error

Step7: Update the Error iteratively for next prediction.

$$p_k = (1 - g_k)p_k \tag{8}$$

Step8: Go ahead to step1.

7   Advantages

- More precise and reliable
- Kalman filter minimizes the mean square error
- It takes less time to track position
- It doesn't require wireless network connection.
- It reduces power consumption.

8   Results

Input of GPS data value is initially 1000 samples per second which is given to the microcontroller. GPS data is down sampled by 200 when given to kalman filter. Input of Accelerometer data value is initially 1000 samples per second which is given to the microcontroller. Figures 4 and 5 shows the data from the sensors. The outcome of the kalman gain is 1 when there is zero error which is depicted in Figs. 6 and 7.

Figure 8 shows the variation in kalman gain values and Fig. 9 shows the predicted position using kalman filter.

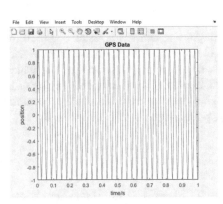

Fig. 4. GPS signal

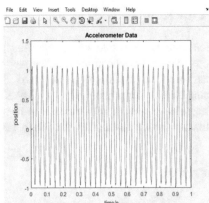

Fig. 5. Accelerometer signal

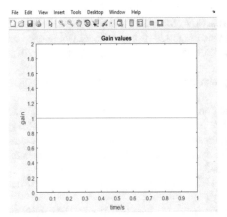

Fig. 6. Gain values of Kalman filter

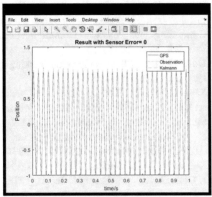

Fig. 7. Zero error output

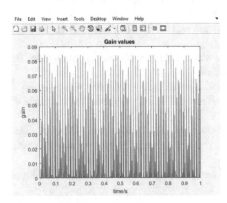

Fig. 8. Gain values of Kalman filter with error

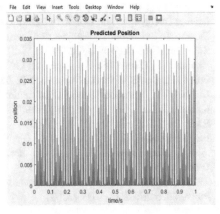

Fig. 9. Predicted position using Kalman filter

9 Conclusion

Various simulation results of kalman filter with precise accuracy is tested and used for various real time applications. Further step is to implement the hardware module with human interface which is currently used. Selection of appropriate hearing aid, which remains suitable for precise compatibility is still remaining as a major challenging task. Future enhancement on this research work will be the extension of the same concept for multitask purpose that makes the future easy, promising and giving hope to live.

References

1. Pham, V.T., Nguyen, V.T., Nguyen, D.A., Chu, D.T., Tran, D.T.: 15-state extended kalman filter design for INS/GPS navigation system. J. Autom. Control Eng. **3**(2) (2015)
2. Dafallah, H.A.A.: Design and implementation of an accurate real time GPS tracking system. In: The Third International Conference on e-Technologies and Networks for Development (ICeND2014) (2014). ISBN 978-1-4799-3166-8
3. Saha, S., Chatterjee, S., Gupta, A.K., Bhattacharya, I., Mondal, T.: TrackMe - a low power location tracking system using smart phone sensors. In: 2015 International Conference on Computing and Network Communications (CoCoNet 2015), December 16–19, 2015, Trivandrum, India (2015)
4. Uddin, M.P.: GPS-based location tracking system via Android device. Int. J. Res. Comput. Eng. Electron. **2**(5) (2013)
5. Sabatini, A.M.: Kalman-filter-based orientation determination using inertial/magnetic sensors: observability analysis and performance evaluation. Sensors **11**, 9182–9206 (2011)
6. Singhal, T., Harit, A., Vishwakarma, D.N.: Kalman filter implementation on an accelerometer sensor data for three state estimation of a dynamic system. Int. J. Res. Eng. Technol. (IJRET) **1**(6), 330–334 (2012). ISSN 2277–4378
7. Kadibagil, M., Guruprasad, H.S.: Position detection and tracking system. IRACST - Int. J. Comput. Sci. Inf. Technol. Secur. (IJCSITS) **4**(3), 19 (2014)
8. Sakre, B., Amarghade, B., Patwari, P., Suryavanshi, P., Saratkar, R.: Vehicle tracking system using GSM and GPS. Int. J. Electron. Comput. Sci. Eng. **1**(3) (2012). ISSN 2277-1956
9. Chandra, A., Jain, S., Qadeer, M.A.: GPS locator: an application for location tracking and sharing using GPS for Java enabled handhelds. In: International Conference on Computational Intelligence and Communication Systems (2011)
10. Zahaby, M., Gaonjur, P., Farajian, S.: Location tracking in GPS using Kalman filter through SMS. In: IEEE EUROCON (2000)
11. Bhutkar, G., Dhore, M.: An Improved GPS location tracking with velocity estimation. In: Proceedings of ICCNS 08, September 27–28 2008
12. Maybeck, P.: Stochastic Models, Estimation, and Control. Academic Press, Ohio (1979)
13. Kalman, R.E.: A new approach to linear filtering and prediction problems. Trans. ASME-J. Basic Eng. **82**(Series D), 35–45 (1960)
14. Ruiz-Fernandez, D., Marín-Alonso, O., Soriano-Paya, A., García-Pérez, J.D.: A tele-rehabilitation environment based on motion recognition using accelerometry. The Sci. World J. **2014**, 11 (2014)
15. Zhou, H., Hu, H.: Human motion tracking for rehabilitation—a survey. Biomed. Signal Process. Control **3**(1), 1–18 (2008)

A Comprehensive Approach to an Internet of Things (IoT) Network: Weather Monitoring System

Rahul Parthasarathy[1], Suhas R. Sharma[1],
and Prasad B. Honnavalli[2(✉)]

[1] Information Security Forensics and Cyber Resilience (ISFCR) Lab,
PES University, Bengaluru, Karnataka, India
rahul.parthal9@gmail.com, rsuhas17.sharma@gmail.com
[2] Department of Computer Science Engineering, PES University,
Bengaluru, Karnataka, India
prasadhb@pes.edu

Abstract. The overall scope of IoT sensor networks over the last decade has seen a considerable growth in technology as well as application in corporate and domestic requirement. The goal of the system proposed in this paper is a unique approach to a stable, robust, scalable, location independent, automated and efficient IoT network that is built to monitor the weather conditions at any geographic location. The data from the nodes whose positions are forethought, are transmitted one after the other in an organised, timed and precise manner to a base station where it is validated real-time by comparing values obtained from a verified online resource. Emails are sent out to the user in case of faulty behaviour of the nodes. Subsequently, this data is uploaded to a cloud platform, categorised efficiently, processed in real time and is displayed to the users on demand. Graphical visualizations of the data can be achieved using tools in the cloud platform that facilitate better understanding of the sensor data. Collision control and packet loss, an issue Wireless Sensor Networks (WSNs) face is mitigated in this model through polling techniques.

To throw light on the necessity of security in WSNs, a penetration test is performed to show one of the ways to gain unauthorized access to the network.

Keywords: IoT · Weather monitoring · Amazon web services · IoT security

1 Introduction

The Internet of Things (IoT) is a concept in computing that illustrates the thought of a plethora of electronic devices being connected to the internet, capable of identifying other devices and themselves to other devices in the network [13].

This inventive technology not only connects the devices to the internet but also offers the user plenty of features like cloud data storage, real-time data analytics,

---

R. Parthasarathy and S. R. Sharma—Equal Contribution.

P. Karrupusamy et al. (Eds.): ICSCN 2019, LNDECT 39, pp. 143–155, 2020.
https://doi.org/10.1007/978-3-030-34515-0_15

notifications based on the behaviour of the device, trigger an action from a remote location, etc. Since this technology can be leveraged in a wide range of applications, it can be implemented into almost all the industries [7].

A wide array of IoT devices connected to networks form the fundamental units of all computational devices used on a daily basis, such as baby monitors, smart TVs, smart city technologies, and others. While IoT devices have a lot of benefits to offer, they are indeed vulnerable to security threats [4].

With every passing day, the concept of IoT is becoming more ubiquitous and habitual to such a degree that it will soon revolutionize our lifestyle. Although this IoT explosion will provide a broad variety of possibilities for manufacturers and customers, it also presents glaring safety hazards. With the rise in the number of devices that are interconnected, securing each and every one of them will pose to be a huge challenge; in order to enable those IoT devices to work effectively, each of, hardware, software and the connecting network need to be secure. Without considering security, any device in the network can be compromised; once vanquished, the hackers can subjugate the functionality of that device and steal user data. Thus, securing IoT devices like all other connected devices is a force to be reckoned with.

When considering the architecture of the proposed IOT network, two topologies were taken into consideration, the wireless mesh network [1, 3] and the star wireless mesh network. Ultimately, the star wireless network being a centralised network was adopted due to the following reasons-

a. Being centralised, processing is focused at one point, thus reducing burden on the nodes where processing capabilities and available power are limited, hence reducing further reducing the cost as it reduces emphasis on more hardware.
b. Centralized processing provides better data security.

The IoT network we aim to build is a centralized network (Fig. 1), which includes multiple nodes that communicate with the base station and transmit data wirelessly in the 2.4 GHz frequency range. The data received is then processed, segregated and uploaded to a cloud platform where it is stored for further analysis and better visualisation [10].

In order to ensure accuracy of the data parameters, a comparison of the perceived parameters is made with data values extracted from legitimate source online, in case of a deviation from the expected value an intimation is sent to the user via mail to notify them of the discrepancy, thus ensuring the user is aware of the system at all times.

2   Proposed Methodology

Figures 1 and 2 represent the topology used in the system and a comprehensive representation of the model respectively. The following highlights the various segments of the system:

a. Nodes
b. Base station
c. Cloud

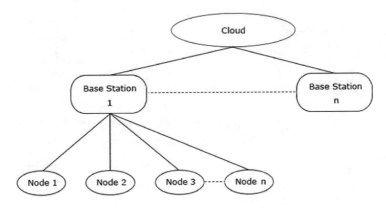

Fig. 1. Architecture

A. The nodes (refer Fig. 2. Below) act as the medium that connect the physical world and the system in context. The sensors at the each of the nodes measure environmental parameters being, temperature, humidity, CO2 levels in ppm. Each node consists of 4 main components-

a. DHT11 module
b. MQ-135 module
c. NRF24L01 + module
d. Arduino UNO

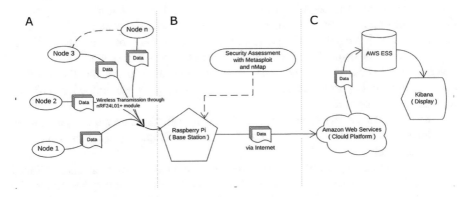

Fig. 2. The IoT system flow diagram

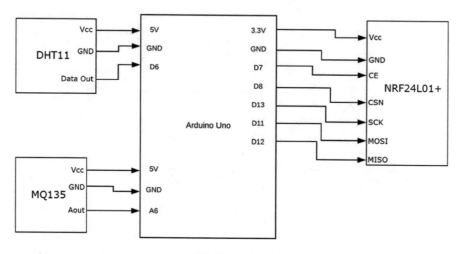

Fig. 3. Node

The DHT11 sensor extrapolates the temperature and humidity of its immediate surroundings using sensing technologies comprised of a Negative Temperature Coefficient measurement component and resistive type humidity measurement component. This extrapolated data is then converted in to digital values and transmitted to the Arduino.

The MQ-135 sensor is designed to function as an air quality monitor, for the scope of this experiment, it is used to estimate the concentration of carbon dioxide in the air. The material utilized in MQ135 is SnO2. It is a special type of material whose conductivity property deteriorates when exposed to air, whereas when exposed to an environment with a combustible gas, the conductivity of the sensor increases. Within the Arduino the change of conductivity is converted to a corresponding output signal. The gas sensor is used to measure the CO2 levels in the atmospheric air. The resulting concentration of impurities in the air is computed in parts per million (ppm) and recorded in the Arduino.

The NRF24L01+ transceiver is the module that facilitates communication between the Nodes and the Base Station in the 2.4–2.5 GHz frequency band on which the module operates.

The Arduino Uno processes the data received first-hand from the sensors mentioned above, validates it, converts the analog data [6] to the digital form and transmits the required data to the Base Station when polled for, using the nRF24L01 + module.

B. The Base station behaves as the intermediary control between the Nodes and the Cloud, by extracting sensor data from the former and pushing it to the latter. This portion consists of 2 main components:

a. Raspberry Pi 3 Model B
b. NRF24L01+ module

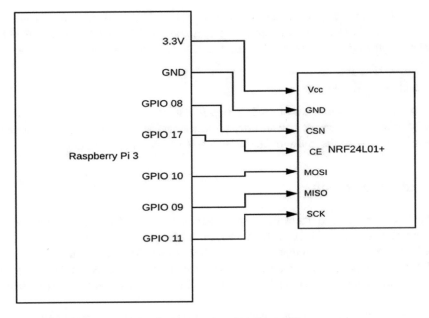

Fig. 4. Base station (Raspberry Pi)

The Raspberry Pi is deployed as the base station [9]. It polls for sensor data from each of the 'N' number of nodes utilizing the NRF24L01+ module [8], processes the data by segregating it into discrete data values and then transmits it to the cloud segment via the Internet through the 802.11n Wireless LAN adapter embedded on the board. The data is requested using distinct authentication strings that help distinguish and identify each of the nodes.

C. Amazon Web Services (AWS) was selected as the cloud platform [5]. The protocol used in AWS is MQTT (Message Queueing Telemetry Transport) [12], which is light weight and secure [2]. It makes the most of the available bandwidth and also lowers the update rates to seconds. The services utilized in AWS are as follows:

1. AWS IoT Core
2. AWS IAM
3. Amazon Elastic Search Service (ESS)
4. Kibana Plugin

The AWS IoT Core is a point in the cloud service which receives data from the base station, where data validation is performed.

The AWS IAM acts as an intermediary pipeline for data transmission from AWS IoT core to AWS ESS.

The scope of Amazon ESS for this model is for the storage of received data values and to load and query the same.

The Kibana Plugin is an extension of ESS that is used to load data from Amazon ESS and further perform graphical and statistical analysis of the same to yield a simplified visual user-end comprehension.

Scrutinizing the network on its security aspects is a vital necessity in an IOT network. In this model we attempt to obtain a reverse TCP shell targeted at the base station segment of the network. We use the following tools to conduct the test-

1. Angry IP scanner
2. Metasploit Framework

Angry IP scanner is a tool used to map IP address and host name on a network.

Metasploit framework is a security project aimed at providing information about security vulnerabilities and provide tools to assist in penetration testing.

3   Implementation

The flowchart (Fig. 5) represents the sequential flow of data through the various components of the IoT system. At the advent of the process, sensors detect the environmental parameters - temperature, humidity (using DHT11), CO_2 levels (using MQ135) at the periphery of the network. This data is then requested on-demand by the base station using the "Polling" technique. Polling is implemented in this system as follows:

The communication medium between the node and the base station makes use of a channel for the transmission path and two read and write pipes to accept and transmit data respectively. The base station requests for data using an "authentication string" which is discrete for each of the parameters in each node. Based on what sensor data is required, the base station transmits the respective authentication string along the channel. The nodes have periods of "wake" state time cycles during which they listen on the channel for any transmissions, if no data is detected within a time frame, the nodes go back to the "sleep" state. This helps regulate the power consumption, thereby increasing the power efficiency of the system. The base station broadcasts the authentication string to all nodes. On receiving a transmission request, the nodes accept the data using the reading pipe, a comparison is then made between the received authentication string and the list of authentication strings the node comprises of.

If successfully authenticated, the node transmits back the requested sensor data value using the writing pipe, else the node goes back to sleep until the next wake period. This way polling mitigates or completely removes data collision issues between the two segments. Once the data is received at the Base Station, it is sent to the cloud (here, AWS) where it is processed and displayed either graphically or in a tabular format.

The setup of AWS involves the creation of certificate initially.

Initially, a Certificate in AWS IoT Core has to be generated, which in turn provides a public and private key. This facilitates encrypted transmission of data between the base station and cloud. MQTT, a lightweight messaging protocol, is the protocol employed for transmission of data. Upon confirmation of successful transmission of sensor data to AWS IOT core, a domain is created in Amazon ESS. A rule in AWS IAM is created to push the data from AWS IoT core to Amazon ESS, and uses Kibana (a Plugin in ESS that provides customizations on the types of analysis that can be done), to perform graphical and tabular analysis as seen in Figs. 3 and 4.

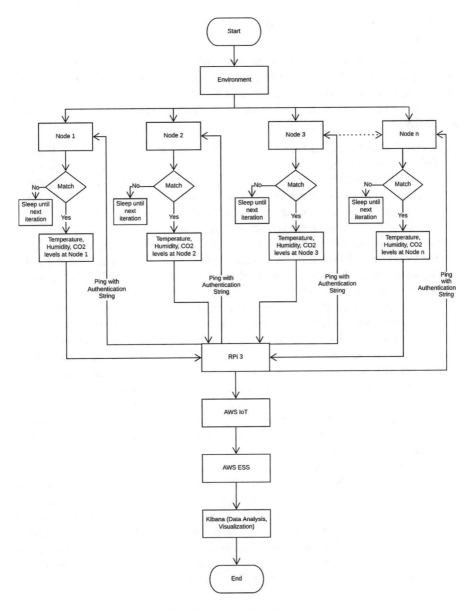

Fig. 5. Activity flow diagram

From the security standpoint of the system, the overall required outcome of this test is to obtain a reverse TCP shell from the base station segment of the model.

Angry IP scanner [14] is run to acquire the IP address of the base station. Within Metasploit Framework [11], the "exploit/multi/handler" exploit and the "python/meterpreter/reverse_tcp" payload are employed in the test. The following options of the payload must be said-

1. Lhost- The host system IP address
2. LPort- The host system port.

A payload is created using MSFVenom that has to be executed on the target system by methods such as Social Engineering techniques. Executing the respective payloads on both the host and target machines returns a reverse TCP shell of the target system.

Through the entire process, there were a few notable challenges faced that are listed as follows:

1. Energy Consumption: There would be a lot of power, energy consumed if all the nodes were active (in the "wake" state) at all times. To overcome this difficulty, an algorithm was implemented that keeps the node active for a few seconds during which it listens for requests from the base station and services them immediately, if any. In case of no requests, the node goes to the "sleep" mode after a scheduled amount of time, saving energy. The cycle repeats.
2. Identification of node - Polling: Due to the presence of multiple nodes in the network, there arises a challenge to differentiate the data from each of the nodes. Polling mechanism is implemented, where: Each parameter that is being measured in each node in the network is given a unique authentication string, the list of which is available at the base station. Cyclically, the base station polls for each parameter of each node sequentially by broadcasting the authentication string. The nodes in the network will have a short time period during which they will listen to requests from the base station. Upon receiving the authentication string, a comparison is made at each node to verify if the information requested is from that node and the particular parameter is sent to the base station if there is a match in the authentication string.
3. Time Synchronization: For the scope of this project, 3 nodes were built. Although, this number can be increased to a 120+ if needed. The time during which the base station requests data and the time during which the nodes are in active state, had to be synchronized in order to facilitate seamless transfer of data.
4. Collision Control: Since the module NRF24L01+ operates in the 2.4 GHz range and due to the presence of multiple nodes in the same network, there could be collisions that might lead to loss of data. The nodes and the base station communicate in a pre-determined channel while setup. The channel constitutes of a reading pipe and a writing pipe that is defined to avoid collision while communication. The node reads the authentication string that is sent by the base station via the reading pipe and replies via the writing pipe, therefore avoiding collision.

4 Results and Observations

The overall objective was achieved - The perceived sensor data values were successfully obtained and transmitted from the node to the cloud for analysis. The values as seen in AWS – the temperature (in degree Celsius), humidity (as a relative percentage) and CO_2 levels in the air (in parts per million (ppm)), recorded every 30 s are tabulated in Fig. 6 (below).

Time(in seconds)	Temperature(degree Celcius)	Humidity(%)	CO2 Levels(in ppm)
0	35.5	25	330.5
30	36.333	24	324.667
60	36.333	20	309.333
90	36.5	22.667	448.5
120	36.333	26.5	270.333
150	37	23.667	202
180	36.333	24	274
210	36.333	24.333	268.333
240	36.333	24.333	280.667
270	35.5	24.5	242

Fig. 6. Values as received in AWS

Figure 7 depicts the plot of Temperature with Time, where the actual temperature values received (in blue) – as seen in Fig. 6, the average temperature - calculated from the values obtained (in green) are plotted over time. The upper threshold (*) (in red) and the lower threshold (*) (in orange) denote the accuracy of the component used and is not to be mistaken for the actual limits of temperature and can be improved with the use of a component having higher precision.

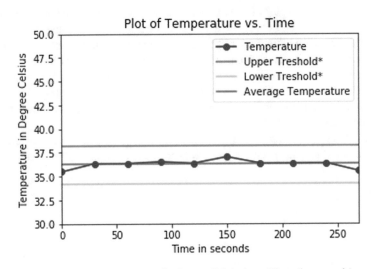

Fig. 7. Plot of temperature (in degree Celsius) vs. Time (in seconds)

Figure 8 depicts the plot of Humidity with Time, where the actual values of humidity received (in blue), the average humidity - calculated from the values obtained (in green) are plotted over time. The upper threshold (*) (in red) and the lower

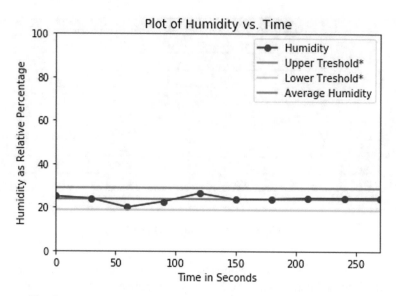

Fig. 8. Plot of humidity (as Relative Percentage) vs. Time (in seconds)

threshold (*) (in orange) denote the accuracy of the component used and is not to be mistaken for the actual limits of humidity and can be improved with the use of a component having higher precision.

Figure 9 depicts the plot of CO2 Levels with Time, where the actual values of CO2 Concentration received (in blue), the average concentration - calculated from the values obtained (in green) are plotted over time. The red line in the plot denotes the Hazardous Concentration Limit (at 1000 ppm). If the concentration values are below 1000 ppm, the air is considered to be clean and safe [15]. Any value above the limit is considered to have adverse effects on the human body. Please note that the CO2 levels measured in this trial were obtained in an indoor environment. The safe levels for outdoor ambient air is around 250-350 ppm [15].

Figure 10 shows a reverse TCP shell that is obtained and also the root directory of the base station (here, Raspberry Pi) which is accessed after successful execution of the payload.

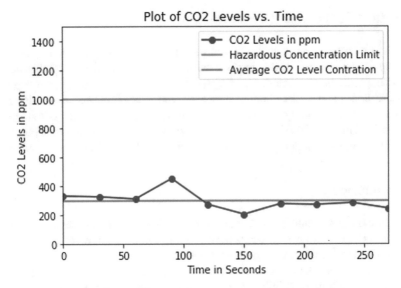

Fig. 9. Plot of CO2 levels (in ppm) vs. Time (in seconds)

```
msf exploit(multi/handler) > exploit

[*] Started reverse TCP handler on 192.168.43.149:4000
[*] Sending stage (43668 bytes) to 192.168.43.197
[*] Meterpreter session 2 opened (192.168.43.149:4000 -> 192.168.43.197:59616) at 2018-04-27 16:17:31 +0800

meterpreter > ls
Listing: /home/pi
==================

Mode                Size    Type   Last modified              Name
----                ----    ----   -------------              ----
100600/rw-------    483     fil    2018-04-25 03:17:14 +0800  .Xauthority
100600/rw-------    4810    fil    2018-04-27 14:02:14 +0800  .bash_history
100644/rw-r--r--    220     fil    1970-01-01 08:07:12 +0800  .bash_logout
100644/rw-r--r--    3523    fil    1970-01-01 08:07:12 +0800  .bashrc
40755/rwxr-xr-x     4096    dir    2018-03-22 15:57:50 +0800  .cache
40700/rwx------     4096    dir    2018-04-27 14:46:03 +0800  .config
40700/rwx------     4096    dir    2018-03-14 07:17:07 +0800  .gnupg
40755/rwxr-xr-x     4096    dir    2018-03-22 17:24:24 +0800  .idlerc
40755/rwxr-xr-x     4096    dir    1970-01-01 08:23:24 +0800  .local
40755/rwxr-xr-x     4096    dir    2018-03-22 15:57:46 +0800  .oracle_jre_usage
40700/rwx------     4096    dir    2018-03-22 01:55:45 +0800  .pki
100644/rw-r--r--    675     fil    1970-01-01 08:07:12 +0800  .profile
40755/rwxr-xr-x     4096    dir    2018-03-14 07:17:08 +0800  .themes
40700/rwx------     4096    dir    2018-04-27 14:13:03 +0800  .thonny
40700/rwx------     4096    dir    2018-03-22 15:43:50 +0800  .thumbnails
40700/rwx------     4096    dir    2018-04-25 01:34:51 +0800  .vnc
100644/rw-r--r--    209     fil    2018-03-22 01:31:03 +0800  .wget-hsts
100600/rw-------    4133    fil    2018-04-25 03:17:17 +0800  .xsession-errors
100600/rw-------    10002   fil    2018-04-25 03:17:14 +0800  .xsession-errors.old
40755/rwxr-xr-x     4096    dir    2018-04-02 18:33:23 +0800  DHT11_Python
40755/rwxr-xr-x     4096    dir    2018-04-25 02:38:00 +0800  Desktop
40755/rwxr-xr-x     4096    dir    2018-03-14 07:17:07 +0800  Documents
40755/rwxr-xr-x     4096    dir    2018-04-17 22:11:46 +0800  Downloads
```

Fig. 10. Root directory of the base station

5   Future Scope and Enhancements

Though implemented on a small scale at this stage, we believe it has the potential to either be expanded into a much larger network or be integrated into another network.

Although the network currently consists of three nodes, it is capable of supporting up to 125 nodes and a vast plurality of sensor types (moisture, gas, etc.). This project can be enhanced by adding more independent base stations and more nodes to create a bigger IoT platform of Wireless Sensors. It can also serve various other purposes in IoT, which can vary as per future demands and depends upon the users requirements.

6   Conclusion

IoT is no more just a concept, materialization of IoT is now an endeavour many R&D departments of numerous tech giants have taken up. A whole new revolutionary set of technologies are coming up in the next 2-3 years.

A significant characteristic to note is, IoT is open source and will always be open source with very little possibility of any major company monopolizing over it. Hence, IoT, just like the popular OS Linux, will witness rapid development and implementation in far parts of the globe by individuals who are enthusiastic towards tinkering and discovering the possibilities within the realm of IoT.

With the adoption of IoT comes its limitations, and the most noteworthy one to consider is its security. Cybercrime is one of the most prominent way of theft of intellectual property and monetary assets. Since IoT aims to automate day to day activities, securing these devices takes a high precedence. Thus, resources must be allocated for security fortification.

Acknowledgement. This paper and the research behind it would not have been possible without the exceptional support of our mentor, Prof. Prasad B. Honnavalli from the IFSCR Lab at PES University. We are grateful to Prof. Kedar Ramachandra for sharing his pearls of wisdom with us during the course of this research. We are also thankful to our colleagues Rishab S Kharidhi and Sharat Chandra who provided valuable inputs that assisted the research.

We are immensely grateful for their valuable comments on the earlier versions of the manuscript, although any errors are our own and should not sully the reputation of these esteemed professionals.

References

1. Math, R.K.M., Dharwadkar, N.V.: IoT Based lowcost weather station and monitoring system for precision agriculture in India. In: 2nd International Conference on I-SMAC (IoT in Social, Mobile, Analytics and Cloud) (I-SMAC)I-SMAC (IoT in Social, Mobile, Analytics and Cloud) (I-SMAC) (2018)
2. Zamfir, S., Balan, T., Iliescu, I., Sandu, F.: A security analysis on standard IoT protocols. In: Proceedings International Conference Applied and Theoretical Electricity (ICATE), Craiova, Romania, pp. 1–6 (2016)

3. Liu; Y., Tong, K.F., Qiu, X., Liu, Y., Ding, X.: Wireless mesh networks in IoT networks. In: 2017 International Workshop on Electromagnetics: Applications and Student Innovation Competition (2017)
4. Chandan, A.R., Khairnar, A.R.: Security Testing Methodology of IoT. In: International Conference on Inventive Research in Computing Applications (ICIRCA) (2018)
5. Amazon Web Services Documentation. https://aws.amazon.com/documentation/
6. Arduino Software. https://www.arduino.cc/en/Main/Software
7. Evans, G.D.: The Internet of Things—How the next evolution of the internet is changing everything. Cisco Internet Business Solutions Group (IBSG), white paper (2011)
8. Wang, J., Wang, M., Zheng, K., Huang, X.: Model checking nRF24L01-based internet of things systems. In: 2018 9th International Conference on Information Technology in Medicine and Education (ITME) (2018)
9. Raspberry Pi Documentation. https://www.raspberrypi.org/documentation/
10. Biswas, A.R., Giaffreda, R.: IoT and cloud convergence: opportunities and challenges. In: 2014 IEEE World Forum on Internet of Things (WF-IoT) (2014)
11. Kennedy, D., O'Gorman, J., Kearns, D., Aharoni, M.: Metasploit the penetration testers guide. No Starch Press, San Francisco (2011)
12. Tantitharanukul, N., Osathanunkul, K., Hantrakul, K., Pramokchon, P., Khoenkaw, P.: MQTT-topics management system for sharing of open data. In: 2017 International Conference on Digital Arts, Media and Technology (ICDAMT) (2017)
13. Waher, P.: Learning Internet of Things. Packt Publishing, Birmingham (2015)
14. Angry IP Scanner. https://angryip.org/documentation/
15. Safe levels of Carbon Dioxide. https://www.dhs.wisconsin.gov/chemical/carbondioxide.htm

Fingerprinting Based Localization with Channel State Information Features and Spatio-Temporal Deep Learning in Long Term Evolution Femtocell Network: An Experimental Approach

Manasjyoti Bhuyan$^{(\boxtimes)}$ and Kandarpa Kumar Sarma

Department of Electronics and Communication Engineering, Gauhati University,
Guwahati 781014, Assam, India
manasjyoti.b@gmail.com, kandarpaks@gauhati.ac.in

Abstract. The need for accurate indoor localization for related services and radio resource management have imparted increased research attention on fingerprinting based positioning. In this work, we present a novel spatio-temporal deep learning (STDL) based fingerprinting method for indoor localization utilizing Channel State Information (CSI). The system works in two phases, acquisition and training and tracking. Experimental evaluation of the method for LTE femtocell signal with Software Defined Radio (SDR) hardware found to outperform spatial and temporal Received Signal Strength (RSS), spatial CSI and temporal CSI based methods.

Keywords: Localization · Fingerprinting · Indoor · Femtocell · SDR · RSS · CSI · LTE

1 Introduction

Global positioning system (GPS), which is a dedicated satellite based infrastructure for positioning and navigation, requires line of sight (LOS) signal paths and fails in indoor environment because of heavy multipath, fading and shadowing [1]. Radio Frequency (RF) fingerprinting based localization has attracted the researchers from around the world as it has the potential to provide location of a mobile device accurately in indoors. Recent development in this prospect of localization, beamforming and beamspace is growing towards the implementation of machine learning algorithms [2, 3]. Machine learning based fingerprinting [4], beam training [5], beam selection [6] and beam prediction [7] have outperformed its predecessors. Machine learning has found applications in radio resource management tasks such as adaptive power allocation, adaptive resource utilization etc. [8].

Cellular networks have gained a reputation of providing poor indoor coverage resulting in inferior call quality and low data rate. Quality of Service (QoS) is even worst in urban areas because of penetration loss of radio frequencies (RF) in high raised buildings. This situation needs addressing as most of the cellular communication

© Springer Nature Switzerland AG 2020
P. Karrupusamy et al. (Eds.): ICSCN 2019, LNDECT 39, pp. 156–163, 2020.
https://doi.org/10.1007/978-3-030-34515-0_16

happens from indoors and leads to the development of Femtocells, which are small cells with typical coverage of a floor area inside buildings. Femtocells are user deployed off-the-shelf hardware which is connected to the core network through broadband ethernet. These dense deployment of 4G femtocells comes with the inevitable intra-tier interference. While intra-tier interference is a problem needing solutions, these advanced communication standards are suitable from smart antenna, massive MIMO, beamforming and space division multiplexing enabling point of view. Although beamforming and beamspace are observed from gain and spectral reuse perspective respectively, these techniques can also essentially support in reducing intra tier interference. For beamforming, beamspace or beam selection to be accurate and effective in indoors with heavy multipath and outdoors with considerable user vehicular velocity, accurate realtime (w.r.t user scheduling time or Transmit time Interval, TTI) 3D localization of target is a must. Motivated by this fact, considerable amount of research work can be seen in literature [5, 7, 9]. Mobile users with some velocity needs not only positioning but also tracking for effective implementation of such advanced techniques. Also, the knowledge of the target receiver is utilized in achieving adaptive cell size, fine grained power management, selective resource assignment to reduce intra-tier interference, smooth out handoff and other location aware services.

Indoor localization techniques can be broadly classified into two categories, triangulation based techniques and fingerprinting based techniques. In triangulation based techniques, the position of the target is estimated by collecting distance estimates from some parametric information like Angle of Arrival (AoA) [10], Time of Arrival (ToA) [11], Time difference of arrival (TdoA) [12] and mixture of these parameters [13]. This requires more than one Base Station (Access Points or eNodeBs). Fingerprint based techniques are two step methods, namely offline phase and online phase, and yields higher accuracy as compared to triangulation based techniques [14, 15]. In the offline phase data is collected through measurement or survay. In the online or tracking phase inference about the location of the user is drawn from the data available through offline phase and measurement data at tracking time. This is done by either of these three methods namely, (i) deterministic method [16], (ii) probabilistic method [17] and (iii) machine learning based methods [2, 4].

Most of the contributions in fingerprinting based indoor localization considers RSS as the measurement metric [16, 17]. While even for a stationary receiver in indoor environment the variation of RSS in one minute is upto 5 dB, RSS seems to not qualify as a good measurement metric for indoor environment [18]. In contrast, because of heavy multipath, uniqueness of CSI at each point in space in an indoor environment is high with a slowly diminishing correlation to its nearest points [19]. This renders CSI suitable metric for fingerprinting as compared to RSS.

Also, there can be a consideration regarding spatial pattern or temporal pattern of the measurement metric. Signal patterns acquired during walking in indoor are temporal patterns. Although in most of the works temporal patterns are not considered, authors in [20, 21] utilizes temporal patterns to improve and correct the fingerprinting results.

In this paper, we propose a novel spatio-temporal technique of fingerprinting based localization using learning structure for LTE signal and verify the proposed model using SDR hardware based testbed. The novelty of our work is twofold, one being the application of stacked autoencoder coupled with LSTM structure and the other being the experimental methodology utilized. In Sect. 2, we describe about the CSI acquisition procedure and SDR hardware setup. In Sect. 3 we elaborate on the STDL structure. Finally, in Sect. 4 results of localization is discussed and Sect. 5 concludes the paper.

2   CSI Acquisition and Preprocessing

CSI acquisition phase consists of generating, transmitting, receiving and decoding LTE downlink signal. LTE frequency division duplex (FDD) signal at 1.92 MHz sampling rate comprising of 6 resource block (RBs) i.e., 72 subcarrier and 41 sub-frames are generated. The samples so generated are than upsampled and zeropadded to form the final signal to be transmitted at 1870 MHz (LTE FDD band3). Before selecting the frequency band for the experimentation, the site is investigated with a spectrum analyzer for presence of any commercial transmission and interference at that band. Zeropadding to the transmit signal has been done to aid the detection phase. Figure 1. shows the steps involved in the process of data generation.

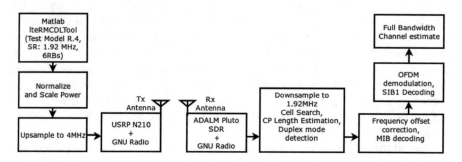

Fig. 1. CSI acquisition toolchain

The experimental setup is shown in Fig. 2. Each received packet is of duration 41 ms which is equal to the number of subcarrier in each packet. After LTE decoding for each packet we get complex valued CSI matrix of size 72 × 14 where the row and column dimensions indicate subcarriers and OFDM symbols respectively. Further, magnitude of the CSI matrices were averaged over all OFDM symbols in a packet to finally get real valued CSI vector of size 72 × 1. This averaging is done based on the assumption that the channel variation within each packet is negligible which we assume valid as indoor users move only at walking velocity. Also, this averaging reduces the effect of noise along each subcarrier. The CSI preprocessing process is shown in Fig. 3.

Fig. 2. Experimental Setup. Handheld UE on left side with ADALM Pluto, ZHL4240 RF Power Amplifier and Power Pack. Top mounted (8 feet height) USRP N210 based eNodeB on Right side. Laptops are not shown in Figure.

A total of 30 min worth of data has been captured during the measurement phase out of which 15000 CSI vectors were obtained. Signal instants with low PSS-SSS correlation is discarded. The whole measurement campaign has been done in 800 square feet of floor area comprising of part of a corridor and one room of area 300 square feet.

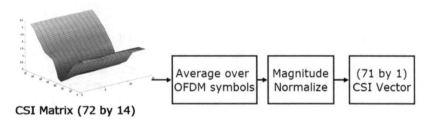

Fig. 3. CSI preprocessing steps

For labeling the data, tiles of the measurement site is numbered along the walking path and video of that path is taken. Each tile is of size 1.5 feet × 1.5 feet. Tiles numbers is than extracted from each video frame and aligned with the captured CSI. Thus, for each captured CSI one label is obtained.

3   System Model

In this work we use a stacked sparse autoencoder with two levels for dimensionality reduction coupled with an LSTM network for sequence to sequence modeling based classification. The first encoder reduces the dimension of CSI vectors from 72×1 to feature vector of size 30×1. The feature output of the first encoder becomes the input for the second encoder, which further reduces the feature dimension to 5×1. A total of 10000 samples are used to train the autoencoder network. Thus the stacked autoencoder learns to generate unique features for each unique locations. Further, we investigated the performance of the system by doing classification from the features using a softmax layer. But as a general consequence of human movement, the probability of presence of an walking user at any particular spot slowly diminishes and do not fluctuate randomly. This correlation in time is also an useful parameter and a recurrent neural network can learn that behavior easily. Hence, we use a single LSTM layer to learn from the features. LSTM layer is trained with the features for sequence classification while the target sequence being the labels extracted from the video recorded at the time of measurement. The system model is shown in Fig. 4.

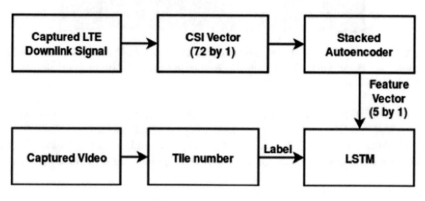

Fig. 4. System Model

The overall training of the network comprises of four steps. In first step autoencoder with 30 hidden neuron is trained. In the second step, autoencoder with 5 hidden neuron is trained with the features from the first encoder. Then in the third step, combination of the first and the second encoder with a softmax layer is trained for classification to fine tune the encoder combination. In the fourth step an LSTM layer with 150 hidden neurons is trained with the features obtained form the fine tuned second encoder to classify to one of 50 labels. The network is shown in Fig. 5.

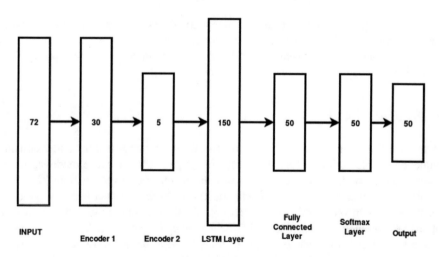

Fig. 5. Structure of the trained network

4 Results and Discussion

Mean and standard deviation of errors available from the test data for different measurement scenarios are shown in Table 1. The huge improvement around corners and inside/outside room is attributed to the LSTM which has reduced the effect fluctuating noise like behavior of CSI measurements. Although we are getting no errors in few scenarios, it should be mentioned that our quantization level for measurement error is 1.5 feet which is the dimension of numbered tiles. So we always get errors in multiple of 1.5 feet.

Table 1. Mean errors for different measurement conditions

Corridor corner		Inside/outside room		Absolute position	
Mean error (m)	Std. dev (m)	Mean error (m)	Std. dev (m)	Mean error (m)	Std. dev (m)
0	0	0	0	1.37	0.91

Feature vector size is selected empirically. Measurement accuracy for different feature vector size is shown in Table 2. Feature vector size greater than 5×1 does not provide any improvement in accuracy and hence feature size of 5×1 is selected. Unlike most of the works available in literature, proposed system is robust in terms of detecting target user in distinctive parts of the floor.

Table 2. Mean errors for different feature vector sizes

3×1		4×1		5×1		6×1	
Mean error (m)	Std. dev (m)	Mean error (m)	Std. dev (m)	Mean error (m)	Std. dev (m)	Mean error (m)	Std. dev (m)
3.2	2.7	2.29	1.83	1.37	0.91	1.37	0.91

5 Conclusion

In this paper, a fingerprinting based localization method using CSI features and spatio-temporal deep learning (STDL) is proposed. The temporal variation of CSI along with spatial variation is exploited using coupled autoencoder-LSTM configuration. Experimental results prove the effectiveness of the temporal information of CSI variation for walking indoor user. Also, as the proposed system needs only one eNodeB for positioning, is applicable in real world scenario.

Acknowledgements. Authors would like to thank Mr. Diganta Kumar Pathak for his assistance in the measurement phase of this work and TEQIP III, MHRD, Govt. of India for the support.

References

1. Djuknic, G.M., Richton, R.E.: Geolocation and assisted GPS. Computer **34**(2), 123–125 (2001)
2. Wang, X., Gao, L., Mao, S., Pandey, S.: CSI-based fingerprinting for indoor localization: a deep learning approach. IEEE Trans. Veh. Technol. **66**(1), 763–776 (2017)
3. Chen, H., Zhang, Y., Li, W., Tao, X., Zhang, P.: ConFi: convolutional neural networks based indoor Wi-Fi localization using channel state information. IEEE Access **5**, 18066–18074 (2017)
4. Vieira, J., Leitinger, E., Sarajlic, M., Li, X., Tufvesson, F.: Deep convolutional neural networks for massive MIMO fingerprint-based positioning. In: 2017 IEEE 28th Annual International Symposium on Personal, Indoor, and Mobile Radio Communications (PIMRC), pp. 1–6 (2017)
5. Va, V., Shimizu, T., Bansal, G., Heath, R.W.: Online learning for position-aided millimeter wave beam training. IEEE Access **7**, 30507–30526 (2019)
6. Anton-Haro, C., Mestre, X.: Learning and data-driven beam selection for mmWave communications: an angle of arrival-based approach. IEEE Access **7**, 20404–20415 (2019)
7. Wang, Y., Narasimha, M., Heath, R.W.: Mmwave beam prediction with situational awareness: a machine learning approach. In: 2018 IEEE 19th International Workshop on Signal Processing Advances in Wireless Communications (SPAWC), pp. 1–5 (2018)
8. Klautau, A., Batista, P., Gonz´alez-Prelcic, N., Wang, Y., Heath, R.W.: 5G MIMO data for machine learning: application to beam-selection using deep learning. In: 2018 Information Theory and Applications Workshop (ITA), pp. 1–9 (2018)
9. Salman, A., Qureshi, I.M., Saleem, S., Saeed, S., Alyaei, B.R.: Novel sensing and joint beam and null steering based resource allocation for cross-tier interference mitigation in cognitive femtocell networks. Wirel. Netw. **24**(6), 2205–2219 (2018)
10. Guo, X., Huang, Y., Li, B., Chu, L.: DOA estimation of mixed circular and non-circular signals using uniform circular array. In: 2014 7th International Congress on Image and Signal Processing, pp. 1043–1047 (2014)
11. Alsindi, N.A., Alavi, B., Pahlavan, K.: Measurement and modeling of ultrawideband TOA-based ranging in indoor multipath environments. IEEE Trans. Veh. Technol. **58**(3), 1046–1058 (2009)
12. Wang, G., Li, Y., Ansari, N.: A semidefinite relaxation method for source localization using TDOA and FDOA measurements. IEEE Trans. Veh. Technol. **62**(2), 853–862 (2013)
13. Nerguizian, C., Despins, C., Affes, S.: Indoor geolocation with received signal strength fingerprinting technique and neural networks. In: de Souza, J.N., Dini, P., Lorenz, P. (eds.) Telecommunications and Networking - ICT 2004, pp. 866–875. Springer, Berlin (2004)
14. Kaemarungsi, K.: Design of indoor positioning systems based on location fingerprinting technique. Ph. D. thesis, University of Pittsburgh (2005)
15. Lin, T.-N., Lin, P.-C.: Performance comparison of indoor positioning techniques based on location fingerprinting in wireless networks. In: 2005 International Conference on Wireless Networks, Communications and Mobile Computing, vol. 2, pp. 1569–1574 (2005)
16. Bahl, P., Padmanabhan, V.N., Bahl, V., Padmanabhan, V.: Radar: an in-building RF-based user location and tracking system. Institute of Electrical and Electronics Engineers, Inc., ACM SIGMOBILE Test-of-Time Paper Award, 2016 (2000)
17. Youssef, M., Agrawala, A.: The Horus WLAN location determination system. In: Proceedings of the 3rd International Conference on Mobile Systems, Applications, and Services, MobiSys 2005, pp. 205–218. ACM, New York (2005)

18. Wu, K., Xiao, J., Yi, Y., Chen, D., Luo, X., Ni, L.M.: CSI-based indoor localization. IEEE Trans. Parallel Distrib. Syst. **24**(7), 1300–1309 (2013)
19. Pecoraro, G., Di Domenico, S., Cianca, E., De Sanctis, M.: CSI-based fingerprinting for indoor localization using LTE signals. EURASIP J. Adv. Signal Process. **2018**(1), 49 (2018)
20. Kim, Y., Shin, H., Cha, H.: Smartphone-based Wi-Fi pedestrian-tracking system tolerating the RSS variance problem. In: 2012 IEEE International Conference on Pervasive Computing and Communications, pp. 11–19 (2012)
21. Shen, G., Chen, Z., Zhang, P., Moscibroda, T., Zhang, Y.: Walkie-Markie: indoor pathway mapping made easy. In: Presented as part of the 10th USENIX Symposium on Networked Systems Design and Implementation (NSDI 13), USENIX, Lombard, IL, pp. 85–98 (2013)

A Survey on Various Message Brokers for Real-Time Big Data

Spandana Srinivas$^{(\boxtimes)}$ and Viswavardhan Reddy Karna

Department of TCE, RVCE, Bengaluru, India
spandanackm93@gmail.com, viswavardhank@rvce.edu.in,
viswavardhan.kv.2018@ieee.org

Abstract. In the current scenario, processing the huge information is very difficult in the data pipelines. The solution to overcome the above problem is by using message brokers. This will help in collecting and delivering the huge amount of real-time data. In this work, producer client is designed using java language to fetch the information of network elements using Kafka messaging scheme. The consumer application is configured to fetch information from kafka bus. Moreover, performance metrics of kafka bus is also collected. From the results, it is observed that kafka achieved 100% throughput and latency is 1.3 s for fetching 2 lakh messages. From this, it is understood that kafka is very fast, reliable and fault tolerant messaging scheme. Moreover, from the survey it is shown that kafka is the best messaging scheme.

Keywords: Kafka · Zookeeper · Topic · Producer · Consumer

1 Introduction

In the modern web technology, message brokers plays an important role as they are middleware programs which provides asynchronous communication between different components of a software architecture using publish-subscribe paradigm. Both the receiver and sender are loosely coupled and to communicate with each other both need not be online at a time. Basically, message brokers used for

- To ensure loosely coupling between producer and consumer.
- To maintain asynchronous/non-blocking communication between producer and consumer.
- In case consumer cannot process messages as fast as producer sending

To enhance the communication layer scalability, message brokers are used. Server publishes their messages to some specific topic inside broker and if client subscribes that particular topic client receives messages that published by the server. The reverse is also possible. Messages can be of various formats (JSON, XML, text, binary etc.). There are lots of message brokers available like Apache Kafka, RabbitMQ, Apache ActiveMQ, Kestrel etc., [1].

© Springer Nature Switzerland AG 2020
P. Karrupusamy et al. (Eds.): ICSCN 2019, LNDECT 39, pp. 164–172, 2020.
https://doi.org/10.1007/978-3-030-34515-0_17

1.1 Comparision of Various Message Brokers

RabbitMQ, Kafka and ActiveMQ all serve the basic purpose but can go about their jobs differently. Some of the differences are explained in Table 1.

Table 1. Comparision of message brokers [2–7].

RabbitMQ	ActiveMQ	Kafka
Developed by Pivotal	Founders from Logic Blaze	Developed by Linkedin & now part of Apache
Uses Erlang Language	Uses Java Language	Uses Java Language
High usage of memory and resources	High usage of memory and resources	Very low usage of memory and resources
Publish rate is 20 K/s messages	Publish rate is 5 K/s messages	Publish rate is 2 millions/s
Synchronisation is very slow	Synchronisation is very slow	Ability to cluster & synchronise fast
Performance is medium	Performance is medium	Performance is high
Plugin system, Reliability, Flexible routing, Clustering, tracing are some of the features	Clustering, cross language, JMS to JMS bridge, Master-slave message groups are some of the main features	Scalable, high volume, fault tolerance, realiable, zero down time, replication and partition are some of the features

Everysystem has its specific requirements to select the message brokers. Some essential features which we consider as important are: durability, easy to configure, high availability, developer friendly, high throughput and delivery acknowledgement. According to the Table 1 kafka provides high throughput with good performance.

In this work, we propose kafka message distribution scheme which is highly reliable with high throughput and low latency. First we have created the producer client, configured the consumer and checked wether the message are parsing successfully between the producer and consumer. In order to do this, producer client is written in java language, consumer is configured to fetch the information from a proper topic. Later the performance analysis of Kafka messaging scheme is evaluated.

The organization of the paper is as follows: Sect. 1 explains about kafka messaging system and the comparision between various message brokers. Section 2 describes about architecture of Kafka and its components. Workflow of kafka and its implementation is explained in Sect. 3 followed by experimental results and analysis in Sect. 4.

2 Kafka

Kafka aims at providing unified, quality of service parameters to handle huge amount of real-time data. It's storage layer is essentially a massive scalable publish/subscriber message queue designed as a distributed transaction log, which makes it highly valuable for enterprise infrastructures to process streaming data [8].

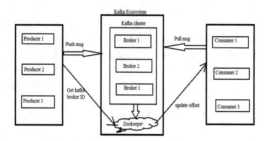

Fig. 1. Kafka-cluster architecture [13].

Architecture of Kafka cluster is shown in the Fig. 1. This architecture contains the components involved in Kafka messaging system. Some of the components of Kafka are as follows.

2.1 Broker

A Kafka brokers are the heart of the Publish/Subscribe messaging system. The Kafka cluster usually consists of several brokers to balance the load. Kafka brokers are stateless, so zookeepers are used to maintain their cluster status. The broker is responsible to receive the messages from producer, to store them in a log file and to send them to the consumer. The broker elected by zookeeper, acts as a middleman between producers and consumers. Thousands of reads and writes files can be handled by broker instance without affecting the performance. Apache Kafka broker offers a set of APIs for implementing publishers and subscribers.

1. Producer's API
2. Consumer API

Any application wishing to send a message must use the producer API to send the data to broker Kafka. The broker receives the message, sends an acknowledgment, and stores the data in a log file. When a client application wants to read the message, it uses its API to read the broker's messages [9, 13].

2.2 Zookeeper

Kafka broker is a cluster without a master. A master-slave architecture is not followed in this cluster. The services are mainly used to inform the producer and the consumer regarding the availability of a new broker in the Kafka system or a broker failure in the

Kafka system. In accordance with the notification received by the Zookeeper regarding the presence or failure of the broker, the producer and the consumer make the decision and begin to coordinate their work with another broker.

Each Kafka broker has a unique-id that has to be defined in the broker configuration file. The details of the Zookeeper connection in the broker configuration file also need to be specified. When the broker starts, it connects to the Zookeeper and creates an ephemeral node (using broker-id) to represent an active broker session. The ephemeral node remains intact as long as the broker session with the Zookeeper is active. When a broker loses connectivity with the Zookeeper for any reason, the Zookeeper automatically removes that ephemeral node. Thus, the list of active brokers in the cluster is maintained as a list of ephemeral nodes under the path/brokers/ids in the Zookeeper [10].

2.3   Producer

Transmission of messages to the brokers are done by producers. However, when a new broker instance is started, all the producers search and then transmits the data or messages to the broker. Producer does not wait for the broker's acknowledgments and sends messages as quickly as the broker can handle [9].

2.4   Consumer

An application that reads the data sent by the producer is a consumer. Kafka brokers doesn't have state, which implies that the consumer must manage the number of messages consumed using the partition offset. After receiving the message from broker, consumer processes it and then sends the acknowledgement to kafka. Next, offset will be changed to new value and updates in the zookeeper as zookeeper maintains offset values [13].

2.5   Topic

A stream of messages which belongs to particular category is named as topic. The message classification mechanism is called as Topic. The information is stored in Topics. The topics can be compared to the name of the table in the database. The producer writes the information or messages in Topic and the consumer fetches it from the Topic. The log files for each topic is created by the broker. While writing the messages, producer specifies the Topic name for which messages belongs to and the broker keeps the message in that particular log file of the topic. When consumer needs to utilize the message, it must mention the particular topic name from which it has to use the message.

2.6   Partition

A single topic can store millions of messages. It is therefore unfeasible to keep every one of these messages in a single file. Topic partitions are a mechanism for dividing the topic into smaller fragments. For Apache Kafka, a partition is nothing more than a physical directory, as shown in Fig. 2.

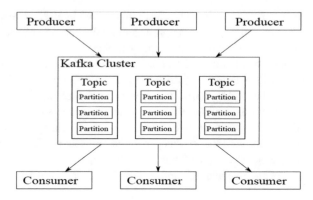

Fig. 2. Partition in topic [8].

Apache Kafka creates a separate directory for each topic partition. A partition is implemented as a set of segment files of equal sizes. While creating the Topic, if the partitioned mentioned is three, it means that kafka will create three various directories. The actual log files are stored in these directories as segmented files. Topics can have multiple partitions, which can handle an arbitrary amount of data. Each Kafka message of a partition is uniquely identified by the offset. For example, the offset for the first message of the partition would be 0000, the offset for the second message is 0001, and so on [8, 15].

2.7   Replication Factor

The replication factor specifies the number of copies to keep for each partition. The purpose of maintaining multiple copies of partitions is to provide fault tolerance. Replicas do not read or write data. They are used to prevent data loss. For Example, if the topic is created for RVCE, when creating the topic, if the number of partitions is mentioned as five and a replication factor as three. In this case, Kafka should create fifteen directories. This indicates that the whole content of the topic "RVCE" is divided into five different directories and each directory is copied three times. These directories are called as partition replica [14].

2.8   Leader and Follwer

The node which is elected as a leader will be responsible for reading and writing the files in a given partition. Other nodes in that instances will act as a follower and obeys the rules and regulations of the leader. However, in some cases, if the leader fails, one of the followers will automatically be elected as a leader. When creating the topic, if the number of partitions is set to five and then Kafka creates five directories. These five directories are called leader partitions. So leaders are created first. Then if the specified replication factor is three. This means that Kafka must guarantee three copies for each of these five partitions. One is already there (the leader), so Kafka creates two additional directories for each leader, and copies are called as followers. Followers are duplicate copies of the leader.

3 Workflow in Kafka

Both the publish-subscriber and queue based messaging schema is supported by kafka, in a fast and reliable manner. In the two cases also, the message to a kafka topic is sent by producer and it can be picked up by consumer using any type of message system based on the need. Figure 3 represents the methodology involved in kafka.

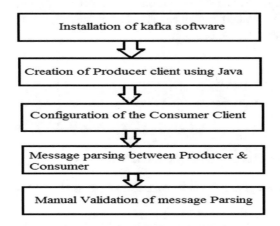

Fig. 3. Methodology

The workflow of the Publish-Subscribe Messaging is explained below:

1. First the kafka should be installed to the Virtual Machine or Linux
2. The Producer client is written in java language mentioning from which host and port the producer has to publish the information.
3. All the information are stored in the partitions configured for the particular topic by Kafka broker. Kafka broker guarantees that the messages are distributed equally among all the partitions. If six messages are sent by Kafka Producer and there are three partitions, Kafka stores two messages in each partition.
4. Consumer subscribes to the specific topic. Consumer is configured in such a way that it has to consume all the messages under the topic specified.
5. Once the consumer subscribes to a specific topic, the topics's current offset is provided by kafka to the topic application and offset will be saved in the zookeeper.
6. Kafka will be requested by the consumer at a regular instance for new messages.
7. Once Kafka receives the state change messages from producers, it forwards to the consumer application.
8. Consumer receives and process the messages received.
9. The acknowledgement is sent from consumer to the kafka broker after message processing.

10. Once the acknowledgement is received by Kafka, the offset is changed to new value and it will be updated in zookeeper. The next messages are read correctly by consumers even during server outrages, because zookeeper maintains the offsets.
11. Until the consumer stops the request, the above process will be repeated.

4 Experimental Results

The experiment is conducted to fetch the real-time information of the Network elements. The producer client is created using java language in Eclipse IDE.

The producer client fetches the information of the NE from the network and post the information of 10,000 equipment into Kafka bus. Figure 4 refers to messages posted by producer into the bus.

```
       ρυσιιπϧ mⴭϽϽaϧⴭ ϽϽϽϽ ιπ ιιⴐιaιιⴐπ ϭ
76     before postMessageToKafka
77     Sending a Kafka message to topic inventoryStateChangeIngest
78     Sending a Kafka message to topic inventoryStateChangeIngest
79     sendResult is null
30     PUBLISHED TO KAFKA TOPIC:::::::::::::{"sourceTenantId":"GlobalTenant","msgType
31     Kafka : success to send message: [B@6662974a
32     after postMessageToKafka
33     Posted count till now in the Iteration : 0is 10000
```

Fig. 4. Message of NE sent by producer to the Kafka bus (10,000 messages)

Figure 5 refers to the complete information of the ten thousand messages sent to the Kafka consumer via Kafka bus.

```
:false},{"Name":"OperationalState","Value":"test0","DisplayName":"Op
false,"ReadOnly":false,"Discovered":true},{"Name":"UpdatedByAppId","'
se},{"Name":"AdministrativeState","Value":"INSTALLED","Mandatory":fa
d","Value":"admin","Mandatory":false}]},"SureName":"GGMKS01-AVAT-03-
:"15:1","SubType":"Network Element Node STATE","DisplayName":"GGMKS(
Category":"ATTRIBUTE","CreationDate":"2018-05-25T16:52:58.745+05:30'
5T10:21:57.242+05:30","Associations":{},"Label":"State"},"Features"
By","Value":"FTPFile","Mandatory":false}]}}]},"Label":"Equipment","'
rchy":false}
^CProcessed a total of 10000 messages
```

Fig. 5. Message of NE received by Kafka consumer (10,000 messages).

From Figs. 4 and 5 it is observed that all the 10,000 messages sent from the producer is received by the consumer. Throughput can be calculated by using formula below.

$$Throughput = \frac{Number\ of\ messages\ received\ by\ consumer}{Number\ of\ messages\ sent\ from\ producer}$$

$$\text{Throughput} = \frac{10,000}{10,000} = 100\%$$

From the above equation, it is observed that throughput is 100% in Kafka.

Figure 6 refers to the latency in Kafka Messaging System when two lakhs ten thousand messages have been sent. It shows that the maximum latency it can have is 1309 ms i.e., 1.309 s which means that the over all time taken by producer to send message to Kafka and consumer is subscribe from Kafka is just 1309 ms even when they are two lakhs messages.

```
[root@mdfautomation7 kafka]# bin/kafka-run-class.sh org.apache.kafka.tools.ProducerPerformance --topic
inventoryStateChangeIngest --num-records 210000 --record-size 8000 --throughput 1500000 --producer-prop
s bootstrap.servers=localhost:9092
21267 records sent, 4253.4 records/sec (32.45 MB/sec), 789.4 ms avg latency, 1165.0 max latency.
25294 records sent, 5058.8 records/sec (38.60 MB/sec), 833.7 ms avg latency, 1112.0 max latency.
26948 records sent, 5389.6 records/sec (41.12 MB/sec), 770.2 ms avg latency, 856.0 max latency.
27000 records sent, 5400.0 records/sec (41.20 MB/sec), 751.6 ms avg latency, 839.0 max latency.
25962 records sent, 5192.4 records/sec (39.61 MB/sec), 795.4 ms avg latency, 862.0 max latency.
23810 records sent, 4762.0 records/sec (36.33 MB/sec), 829.8 ms avg latency, 1128.0 max latency.
19328 records sent, 3865.6 records/sec (29.49 MB/sec), 1092.7 ms avg latency, 1309.0 max latency.
28994 records sent, 5798.8 records/sec (44.24 MB/sec), 710.2 ms avg latency, 761.0 max latency.
210000 records sent, 5006.914310 records/sec (38.20 MB/sec), 804.01 ms avg latency, 1309.00 ms max late
ncy, 773 ms 50th, 1117 ms 95th, 1252 ms 99th, 1304 ms 99.9th.
[root@mdfautomation7 kafka]#
```

Fig. 6. Latency in Kafka messaging system.

5 Conclusion and Future Work

The Kafka is a distributed messaging system for processing of huge data with the high throughput. Kafka uses some special practical designs and offers the publish and subscribe function. Based on its architectural design, it supports both online and offline message processing at the same time. Experimental results show that Kafka is scalable and has very less latency even when large volume of data is passed. Because of its advantages, it is being used in many real-time examples like Linkedin, Twitter, Netflix, Oracle etc.

With its rapid widespread, Kafka can still be improved in some of the aspects. First, to manage the brokers in the Kafka it utilizes Zookeeper at present, if Kafka tries out some other methods so as to remove the dependency of Zookeeper then Kafka can become more integrated and reliable.

Second, there can be many various applications using the same Kafka Cluster. It is much necessary to guarantee that an application would not run out of all the sources in the system suddenly. Therefore, it is important to take an effective measure to allow each application to choose the right level of priority based on its requirement.

Thirdly, it is also suggested that performance of kafka messaging has to be compared with other messaging schemes like ActiveMQ, RabbitMQ etc. experiment is conducted.

References

1. Hiraman, B.R., Chapte Viresh, M., Abhijeet, C.K.: A study of Apache Kafka in big data stream processing. In: International Conference on Information, Communication, Engineering and Technology (ICICET), pp. 1–3, Pune (2018)
2. Ionescu, V.M.: The analysis of the performance of RabbitMQ and ActiveMQ. In: 14th RoEduNet International Conference - Networking in Education and Research (RoEduNet NER), pp. 132–137, Craiova (2015)
3. Hong, X.J., Yang, H.S., Kim, Y.H.: Performance analysis of RESTful API and RabbitMQ for microservice web application. In: International Conference on Information and Communication Technology Convergence (ICTC), pp. 257–259, Jeju (2018)
4. Klein, A.F., Ştefănescu, M., Saied, A., Swakhoven, K.: An experimental comparison of ActiveMQ and OpenMQ brokers in asynchronous cloud environment. In: Fifth International Conference on Digital Information Processing and Communications (ICDIPC), pp. 24–30, Sierre (2015)
5. He, D., Kang, Y., Su, X.: Research on data exchange platform based on JMS. In: 2nd IEEE International Conference on Computer and Communications (ICCC), pp. 110–113, Chengdu (2016)
6. Chen, G., Du, Y., Qin, P., Zhang, L.: Research of JMS based message oriented middleware for cluster. In: International Conference on Computational and Information Sciences, pp. 1628–1631, Shiyang (2013)
7. Shree, R., Choudhury, T., Gupta, S.C., Kumar, P.: KAFKA: the modern platform for data management and analysis in big data domain. In: 2nd International Conference on Telecommunication and Networks (TEL-NET), pp. 1–5, Noida (2017)
8. Wang, Z.: Kafka and its using in high-throughput and reliable message distribution. In: 8th International Conference on Intelligent Networks and Intelligent Systems (ICINIS), pp. 117–120, Tianjin (2015)
9. Bang, J., Son, S., Kim, H., Moon, Y., Choi, M.: Design and implementation of a load shedding engine for solving starvation problems in Apache Kafka. In: IEEE/IFIP Network Operations and Management Symposium, pp. 1–4, Taipei (2018)
10. Nguyen, C.N., Kim, J., Hwang, S.: KOHA: building a kafka-based distributed queue system on the fly in a hadoop cluster. In: IEEE 1st International Workshops on Foundations and Applications of Self* Systems (FAS*W), pp. 48–53, Augsburg (2016)
11. Versaci, F., Pireddu, L., Zanetti, G.: Kafka interfaces for composable streaming genomics pipelines. In: IEEE EMBS International Conference on Biomedical & Health Informatics (BHI), pp. 259–262, Las Vegas, NV (2018)
12. Kato, K., Takefusa, A., Nakada, H., Oguchi, M.: A study of a scalable distributed stream processing infrastructure using Ray and Apache Kafka. In: IEEE International Conference on Big Data (Big Data), pp. 5351–5353, Seattle, WA, USA (2018)
13. Le Noac'h, P., Costan, A., Bougé, L.: A performance evaluation of Apache Kafka in support of big data streaming applications. In: IEEE International Conference on Big Data, pp. 4803–4806, Boston, MA (2017)
14. Wang, M., Liu, J., Zhou, W.: Design and implementation of a high-performance stream-oriented big data processing system. In: 8th International Conference on Intelligent Human-Machine Systems and Cybernetics (IHMSC), pp. 363–368, Hangzhou (2016)
15. Vargas, W.V., Munoz-Arcentales, A., Rodríguez, J.S.: A distributed system model for managing data ingestion in a wireless sensor network. In: IEEE 7th Annual Computing and Communication Workshop and Conference (CCWC), pp. 1–5, Las Vegas, NV (2017)

Implementation of AOR Alarm States for Network Elements and Their Service Impact Analysis

Sujata N. Bogur$^{(\boxtimes)}$ and K. Viswavardhan Reddy

Department of TCE, RVCE, Bangaluru, Karanataka, India
sujatabogur091@gmail.com, viswavardhank@rvce.edu.in

Abstract. A Computer network consists of millions of devices and is connected together for proper communication. In networks, communication paths are very important as they transfer the data from one point to another and should be monitored continuously. An alarm should be generated when a network element and its communication link in a topology is failed. Moreover, alarms should contain the severity of the damage to the communication between the network entities. The severity will help the operator to prioritise the alarms. For efficient and continuous monitoring, the network topology should be analysed with respect to location. Hence whenever alarms are generated, the network topology can be analysed with respect to the severity of alarm and the location of the device. In this paper, we mainly focus on implementation of AOR (Automated Operation Recovery) alarm system with the severity states and location of the devices. Thus we can provide end-to-end service dashboard with tables, maps to identify, monitor and track the quality of service levels.

Keywords: AOR alarms · Severity state · Kafka · Neo 4j database · Restlet client

1 Introduction

The AOR system delivers an end to end closed loop automated fault management solution, receives events (alarms and clears) as input from different element manager systems that indicate some problem in the network. AOR solution reduces manual and repetitive supervision activities and maximizes the quality of operations and maintenance processes over the network. AOR normalizes the events received from the network elements, of any vendor, in any of the different technology areas of the network, processes and enriches these events and generates outputs as corrective actions backwards to the network elements or otherwise, trouble tickets towards the L2 maintenance teams.

AOR assures the established communication path between the network entities and monitors the states and severity of different states of network entities. Monitoring of huge number of entities in the network requires complex operations, hence AOR provides a simplified supervision system for monitoring the network topology and

© Springer Nature Switzerland AG 2020
P. Karrupusamy et al. (Eds.): ICSCN 2019, LNDECT 39, pp. 173–182, 2020.
https://doi.org/10.1007/978-3-030-34515-0_18

location of all the entities. AOR reduces the cost of service operation centre (SOC) supervisory activities.

Service Impact analysis explains the services which are effected whenever any fault is identified in the network. In a network topology all the entities are associated with a service. Hence, whenever alarms are generated, the corresponding service with respect to each entity can be analysed. There are two types of services as listed below:

- End-user services - services which are directly attached to a customer. For example, VPN (Virtual Private Network), DSL (Digital Subscriber Line), VOIP (Voice over Internet Protocol) services, ELINE (Ethernet –Line), VCPE (Virtual Customer Premises Equipment) and so on.
- Infrastructure services - services which serve multiple customers and provide a common medium for the end user services. For example, optical transport services, IP (Internet Protocol) Routing service.

End user services or infrastructure related services can either gets connected to a network element node or to a set of sub network element (that is, physical port).

The paper [1] presents a network computing model and architecture that are being used by AT&T to automate its global network management services. The model is based on client/server technology, and the architecture is designed to address the diverse requirements of an operational work centre which include a user-friendly graphical interface, connectivity to different element management systems (EMSs) for alarm tracking and performance measurement.

The paper [2] evaluates a novel cell outage detection algorithm, which is based on the neighbour cell list reporting of mobile terminals. Using statistical classification techniques as well as a manually designed heuristic, the algorithm is able to detect most of the outage situations.

Authors in paper [3] presents an automated diagnosis in troubleshooting (TS) for Universal Mobile Telecommunications System (UMTS) networks using a Bayesian network (BN) approach.

The papers [1–3] explains different methodologies for detection and generation of alarms in the network The different severity factor for the AOR alarm states is not considered in the above mentioned papers. In this paper, the generation of alarms is based on the different severities. The associations between the network entities, topology map, hierarchical view and geo map, helps to monitor the network and to locate the faults in the network.

In this paper the AOR alarm system for monitoring network health condition is explained. Network topology is created for eighty six thousand entities and different severities are defined for all the entities using cypher script in Neo 4j database. The severities are prioritized based on the degree of fault in the network. The results are analysed using hierarchical view and geo maps.

Rest of the paper is structured as follows: Sect. 2 explains about the methodology and software implementation. Section 3 explains about the results and analysis. Section 4 explains about the conclusion and acknowledgement.

2   Methodology

2.1   Overview

The Fig. 1 explains about the overview of the methodology. Initially, kafka sends 86000 entities. Kafka is a messenger which transfer the data to the database. Then the locations are created using the Restlet client tool. The created locations need to be associated with the equipments. The creation of the associations between the entity is performed using cypher query. The severities for all the network topology is defined in a logical manner. Then the network health is monitored on User Interface.

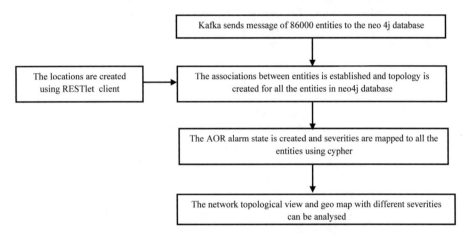

Fig. 1.   Flowchart of working principle

2.2   Kafka

Kafka is a disseminated streaming platform. In this paper, the eighty six thousand entities are sent using Kafka as a messenger bus. The data sent is stored in Neo 4j database. The creation of state and severity is done in Neo 4j database. The data is mapped to different severities and the network monitoring system is analysed on the User Interface portal.

2.3   Restlet Client

The data sent by kafka messenger bus do not have locations and endpoints. Thus the locations and endpoints are created using Rest API. For the creation of location we need to declare two important attributes i.e., latitude and longitude of the respective geographical area in which the device is situated. For creation of endpoints, the specific identification name should be declared and the association should be established between the path and two endpoints.

2.4   Topology Map

The network topology consists of path connected with two endpoints. The path is the physical connection between two entities. The path can be wired or wireless. The wired communication contains telephone networks, optical communication path etc. Each path may have number of subpaths. The subpaths are the intermediate paths between the equipments but path is defined for whole communication line between the entities. The endpoints are the virtual entities which are defined for the environment module testing. The equipments are the physical entities like, routers, switches, hubs. All the equipments are associated with location so that the network health condition can be monitored. Each path is associated with minimum one service. Service usually refers to the services used by the user for example, 3G, 4G network data, etc.

The Fig. 2 shows the network topology map. Their are different entities associated with the network topology as follows:

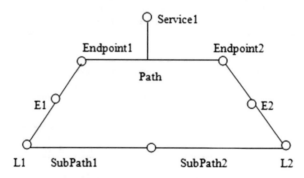

Fig. 2. Network topology view

E1, E2-Equipment 1, Equipment 2
L1, L2-Location 1, Location 2
S1-Service1
SP1, SP2-Subpath 1, Subpath 2
EP1, EP2 -Endpoint 1, Endpoint 2

All the network entities have different attributes associated for identification and for maintaining records. In this paper, the AOR alarm system for network health monitoring is explained. For eighty six thousand entities, the network topology is created and different severity conditions are assigned to all the entities with colour coding.

2.5   Severity States

Based on the degree of the damage on the network, different severities are assigned to the network entities, as follows:

- Critical ■
  This is the most important alarm. These alarms are often regarded as needing urgent attention. This event means numerous devices on the network are affected by the event. Whenever many devices fails or any communication path is blocked, then the whole network system will be effected.
- Major □
  An important alarm that needs attention, but it is not as urgent as a Critical alarm.
- Minor □
  An alarm of low importance. This alarm is generated when the device is effected with some damage but the rest of the network is working properly. This is designed to make users aware of a situation that is deemed of low significance in relation to the other Alarm types.
- Warning □
  A non-urgent alarm, that is less urgent than both the Major and Critical alarms.
- Cleared □
  For example, if the temperature of a device exceeds it's threshold value, then the device will stabilize the temperature by switching on the automatic fan, hence indicating that the device has recovered the error condition.
- Indeterminate □
  This severity explains the insufficient data regarding the degree of damage to the system device, hence the device cant be assigned with any other severity state.
- Information □
  This severity state describes about the entity details. Whenever the entity losses previous information, then an alarm will be generated with this severity. Hence the service provider will analyze the error condition and information related to the entity can be restored back.
- Condition ■
  This severity state describes about the condition of the entities. The severity explains the conditions which caused damage to the device. Their are different conditions, like environmental factors, temperature instability, pressure instability, manual negligence etc.

2.6   Neo4j Database

Neo 4j is a non relational graph database Neo 4j implements the property graph model down to the storage level. The data sent by the kafka messenger bus is stored in neo 4j database. The data do not have associations. Hence we need to create associations between the entities and create network topology for eighty six thousand entities in Neo 4j database. After the network topology is created for all the entities, the alarm severity is defined. As mentioned earlier, the severities are defined for all the entities based on different degree of damage to network. The severities need to be assigned logically, so that the most critical severity will be given highest priority in the network.

The Fig. 3 shows the flowchart of assigning severities to the entities logically. Initially the severities are defined and numbered from 1 to 8 based on the priority. Then the entity is defined with any association with other entity. The relationship is established between the entity n, m with state. Two variables are defined s1 and s2 and are aliased to state. Then the states s1 and s2 are compared and the severities are assigned based on the highest priority.

Hence based on the severities, the degree of the fault can be identified in the network and the fault can be easily resolved by the network operator.

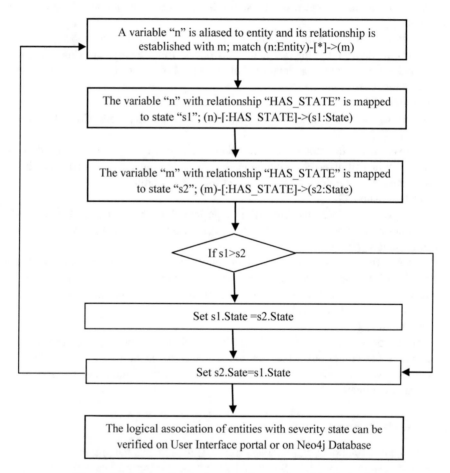

Fig. 3. Flowchart of assigning severties to the entities logically

3   Results and Analysis

The Fig. 4 shows the topology map of the network entities. From the figure it is observed that the root path has three sub paths with four endpoints as shown. The blue and yellow coloured endpoints are having severities as warning and minor respectively. The middle subpath is having two other subpaths with three endpoints. One of the subpath has two other subpaths connected to it. The lowest layer in topology has two subpaths which are having minor and critical as the severities. Since critical is the highest priority severity, thus that severity is passed on to the next higher layers of network topology. Hence the root path and its endpoints will have critical severity.

In this way it would be simple for the network operator and service provider to know the most critically effected device and fix the issue for better performance of the network.

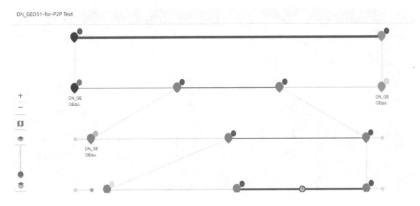

Fig. 4. Network topology view

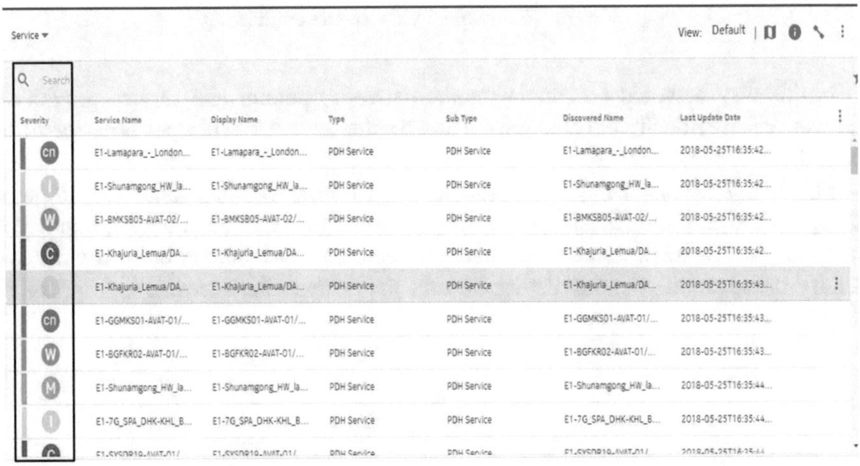

Fig. 5. Severity column for equipments on UI

The Fig. 5 shows the User Interface portal for monitoring the network entities. In the figure, it is seen that the service page is shown. The severity column is added in the display, which shows different severities associated with equipment. Different services have different severity. Similarly, the severity column is added for equipment, path, location and FCP (Functional Control Point).

The hierarchical view of the network entities is shown in the Fig. 6. Each equipment is associated with different equipments and endpoints. It is shown from the figure that the most severe alarm state is given the highest priority and it is passed on to the next level of hierarchy. As shown in the figure, the most severe state in the lowest level of hierarchy is critical, hence this severity is passed on to the next level in hierarchy. Thus the root equipment will show the most severe state, hence the root equipment is assigned with the critical condition. The severities down the hierarchy are arranged in a logical manner.

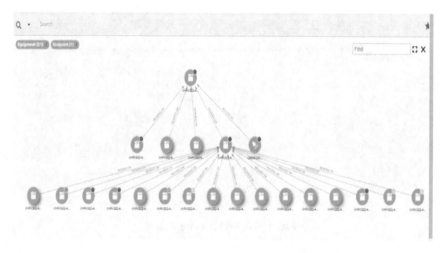

Fig. 6. Hierarchical view of the entities

In this way it would be simple for the network operator and service provider to know the most critically effected device and fix the issue for better performance of the network.

The Fig. 7 shows the geographical location of the equipments. The equipment having critical condition is situated in Bangalore and the equipment having minor state is located at Chennai. Hence, the network operator and the service provider will identify the location of faulty device and fix the error in the network to improve the performance.

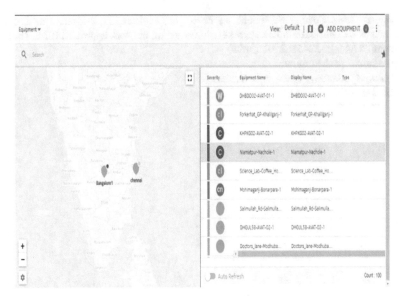

Fig. 7. Geographical map showing the severities at different locations

4 Conclusion

Monitoring huge network and detecting the faults is very difficult and not economical as it includes service operational costs. It is inefficient as, it includes error caused due to human intervention. The Automated Operational Recovery system can detect the faults in the network without human intervention efficiently and reduces service operational costs. We have defined the different severities that determines the degree of damage to the network and hence helps the network operators to find the errors easily and fix the error to improve the performance of the network.

Acknowledgement. The Authors thanks RVCE and Nokia solutions and Networks, Bangalore for the facilities and support provided for the above research work. The copyrights of the User Interface portal pictures are reserved to Nokia solutions and Networks.

References

1. Chan, P.W.: A client/server platform for automation of global network management. In: Proceedings of IEEE International Conference on Industrial Technology - ICIT 1994, Guangzhou, China, pp.355–360 (1994)
2. Barco, R., Wille, V., Diez, L.: System for automated diagnosis in cellular networks based on performance indicators. Eur. Trans. Telecommun. **16**(5), 399–409 (2005)

3. Khanafer, R.M., Solana, B., Triola, J., Barco, R., Moltsen, L., Alt-man, Z., Lazaro, P.: Automated diagnosis for umts networks using bayesian network approach. IEEE Trans. Veh. Technol. **57**(4), pp. 2451–2461 (2008)
4. https://kafka.apache.org/intro
5. https://neo4j.com/developer/graph-database
6. Nokia user guide

Performance Analysis of Digital Predistortion on Orthogonal Frequency Division Multiplexing Systems

Divya Dickshaw[✉] and P. Shanthi

Department of TCE, RVCE, Bengaluru, India
divyadickshaw285@gmail.com, shanthip@rvce.edu.in

Abstract. Wireless communication technologies are constantly evolving since the commencement of the cellular networks. Due to increase in high data rate which is observed from 1G to 4G, spectacular innovations from text to video conferencing. Orthogonal Frequency Division Multiplexing(OFDM) is one of the most promising technique that is capable to provide outrageous data rate. Drawbacks of OFDM systems is that its high Peak-to-Average Power Ratio (PAPR) and to deal with these high fluctuations of Peak-to-Average Power Ratio many amplifiers linearizing methods have been developed. One of the most commonly used linearizing technique for power amplifier to operate in linear region is digital predistortion. Digital Predistortion (DPD) technique placed preceding to power amplifier and provide the inverse transfer function of Power Amplifier (PA) and also balances the amplitude and phase harmonics. Because of its simplicity and good performance baseband band digital predistortion is preferred choice over all other linearization methods. Paper analyses the model of power amplifier which develops the impairments when compared that of input signals. Derived the DPD coefficients through general polynomial method. With PA measurements and DPD coefficients static DPD is designed. Due to the varying characteristics of power amplifier static DPD is turned to adaptive one through indirect learning architecture which uses Least Mean Square (LMS) algorithm. The advantage of using this digital predistortion technique it reduces the spectral growth, the power amplifier is operated in the linear region for longer time, efficiency is also increased. LMS algorithm provides the better reduction in spectral growth when compared with static DPD design, low complexity in designing and provide better rate of convergence. All the modelling and simulations are performed in MATLAB and SIMULINK environments.

Keywords: Power Amplifier (PA) · Orthogonal Frequency Division Multiplexing (OFDM) · Peak to Average Power ratio (PAPR) · Digital predistortion · General memory polynomial · Static DPD · Adaptive DPD · Indirect learning architecture · LMS algorithm

© Springer Nature Switzerland AG 2020
P. Karrupusamy et al. (Eds.): ICSCN 2019, LNDECT 39, pp. 183–190, 2020.
https://doi.org/10.1007/978-3-030-34515-0_19

1 Introduction

OFDM is one of the most promising technique which is capable to providing high data rate. OFDM's subcarriers are orthogonal to each other which helps in effective utilization of bandwidth [11]. Disadvantage of OFDM systems is that high PAPR and to deal with these high fluctuations of PAPR many amplifiers linearizing methods are developed. PA's are highly constrained to their efficiency and linearity. Due to limited linear characteristics of power amplifiers by increasing its efficiency PA is driven to the nonlinear region which results in distortions. To overcome these distortions many linearizing techniques have been introduced and also to deal with the high fluctuations in OFDM systems. Some of the linearization methods are feedforward linearizer, feedback linearizer, pre-distortion linearizer [8, 10]. Digital predistortion is used evolved all over other techniques. Digital predistortion suppresses the spurious emission caused by deformity of power amplifier in real wireless transmitter [7]. Digital predistortion consists of many algorithms which are to reduce the spectral growth. The main objective of digital predistortion is to add the nonlinear element before PA such that it cancels the unwanted spurious. Because of its simplicity and good performance baseband band digital predistortion is preferred choice over all other linearization of RF power amplifiers. As digital signal processing technique DPD is popular because to its good accuracy and flexible in implementation. PA develops the impairments when compared that of input signals. After analysing PA input/output are logged out on to the workspace. Later static Digital predistortion model is designed with the PA measurements and DPD coefficients. Here DPD coefficients are nothing but the inverse value of power amplifier. Static DPD is designed by general memory polynomial model [5] which helpful for derivation of the DPD coefficients. Finally, due to the variations in the power amplifiers characteristics over the time one can modify the static DPD design to adaptive one. By the use of Least Mean square algorithm, the coefficients which are derived through this method adaptive Digital predistortion can be designed by adapting indirect learning architecture.

The advantage of using this digital predistortion technique it reduces the spectral growth, the power amplifier is operated in the linear region for longer time, efficiency is also increased. LMS algorithm provides the better convergence rate and low complexity in designing it is preferred for designing adaptive DPD design. In the following paper, Sect. 2 describes about digital predistortion architecture, Sect. 3 describes about proposed methodology. Section 4 describes about analysis performed and Sect. 5 describes about the results and finally concluded with conclusion and future scope in Sect. 6.

2 Architecture Digital Predistotion

The basic principle behind indirect learning architecture [3] is same as observed in predistortion techniques there also predistorter is placed prior to the power amplifier as seen in the open loop configuration. Apart from predistorter this architecture introduces the post predistorter [6]. Its schematic is shown in Fig. 1 which involves X(n) input to

PD, Y(n) is I/P to PA and also output to PD, Z(n) is the PA's normalized O/P, Y^(n) is the output of the post predistorter, E(n) is the error signal is given by E(n) = Y(n) − Y^(n).

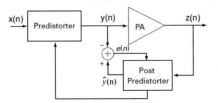

Fig. 1. Architecture of digital predistotion

The predistorter and post predistorter estimation block have the same nonlinear transfer function. Post predistorter block generates the parameters of PD, error signal E (n) is minimized parallelly. The estimated parameters are copied to predistorter when power amplifier is linear X(n) = Z(n) and thus Y(n) = Y^(n).

3   Proposed Methodology

Figure 2 shows the flow chart of methodology sequence the objective is to design a PA model which develops harmonics as increase in input power and adaptive DPD is designed to nullify those impairments.

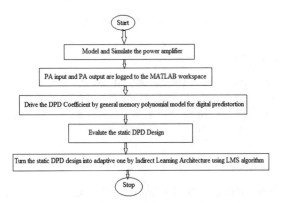

Fig. 2. Proposed methodology

Firstly, power amplifier model consists of Saleh model which is in series with digital filter. Along with the Saleh model and digital filter prior to that white gaussian noise is placed with low pass filter output observed are power amplifier spectrum along with the PA measurements. By general polynomial model [5] and PA measurements

DPD coefficients are derived. Through the DPD coefficients the static DPD is designed and simulations and spectrum of power amplifier along with the DPD are obtained. As characteristics of PA vary continuously the designed DPD is enhanced to Adaptive. In adaptive DPD coefficients are derived through LMS algorithm.

4   Analysis Performed

The designing effort is divided into following parts

4.1   Modelling and Simulating the Power Amplifier

Figure 3 shows the model of power amplifier in which baseband signals passed to the PA drive level followed by PA reference which involves Saleh model [1] in series with the digital filter.one can observe the PA I/O measurements and distortions.

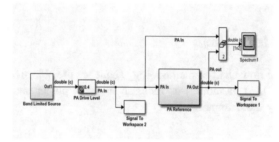

Fig. 3. Model of power amplifier

4.2   Fixed DPD Coefficients

DPD coefficients are nothing but the inverse of the PA measurements which are calculated by general memory polynomial [5]. Such that intermodulation products which are 180 degree out of phase are cancelled and power amplifier can be operated in the nonlinear region to some extent.

4.3   Static DPD Design

In the Fig. 4 DPD is added prior to the PA, by DPD coefficients and PA measurements static DPD is evaluated. Simulated design describes about reduction in the spectral growth when DPD is added prior to PA.

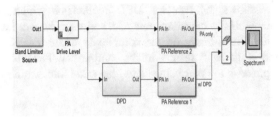

Fig. 4. static DPD design

4.4 Adaptive DPD Design

Figure 5 describes the adaptive DPD design, even though static DPD design shown significant promise but it not well suited for the adaptive implementation due to the matrix inverse and to many delays required equations which are not versatile to implement in hardware. This method, consists of DPD and adaptive DPD coefficient calculation subsystem. This adaptive DPD coefficient subsystem samples the PA I/O performs matrix coefficients to derive the DPD coefficients set and this subsystem coefficients applied to memory polynomial and outputs a predistorter waveform. For adaptive DPD implementation no inverse matrix is required.

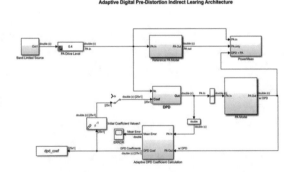

Fig. 5. Adaptive DPD design

5 Results and Discussion

Figure 6 shown below, describes the Power amplifier spectrum with pa_in and pa_out with PA output shows the 20-dB gain when compared to the input but with the expense of significant in-band distortion and spectral regrowth.

Figure 7 below shows the complex DPD coefficients which consists real and imaginary components. If the measurements of x and y are known i.e. nothing but I/O values of power amplifiers, based on the complexity of the PA K and M is chosen considering M and K to be 5 results in 25 complex coefficients. From Fig. 7 red coloured points represents the real values and blue one indicates the imaginary values (Fig. 8).

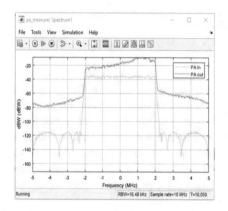

Fig. 6. Spectrum with pa_in and pa_out

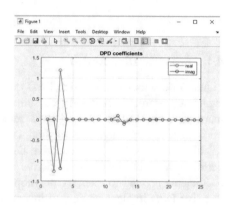

Fig. 7. DPD coefficinets

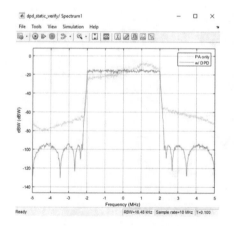

Fig. 8. Static DPD spectrum

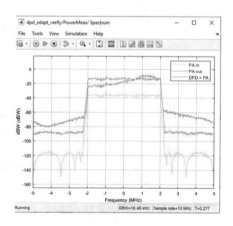

Fig. 9. Adaptive DPD spectrum

This plot compares the effectiveness of LMS algorithm which is been used for finding DPD coefficients with the I/O of power amplifier. Through this algorithm reduction in the spectral growth is observed compared with I/O power amplifier. It is better approach for hardware implementation. Shown in Fig. 9. As seen spectral regrowth of PA O/P is is −40 dBm, −120 dBm of static DPD and −120 dBm of adaptive DPD.

6   Conclusion and Future Scope

Digital predistortion mainly focused on higher efficiency, spectral growth reduction, reduction in high PAPR, low rate of convergence. In this paper, analysis is performed when power amplifier is operated in nonlinear region, distortions and spectral growth observed in output of the PA. To reduce the harmonics DPD coefficients are derived through these coefficients static DPD designed. Static and Adaptive DPD designs are performed and experimented in Simulink environment and DPD coefficients are obtained through MATLAB code. Simulation results are observed involves the spectrum of power amplifier when base band signal is passed through it. With the DPD coefficients and PA measurements obtained from the design of power amplifier static DPD spectrum is observed which indicates slighter reduction of spectral regrowth. finally designed the adaptive DPD spectrum is observed between PA I/O and DPD reduction in the spectral growth is better when compared with static DPD. The future scope of this project is totally based the algorithm which will be implemented on the adaptive digital predistortion, Recursive least mean square (RLS) and Affine Projection (AP) Algorithms are enhanced version of LMS algorithm provides the better spectral regrowth reduction. The proposed technique is limited to simulation so for practicality it can be extended to hardware implementation. Digital predistortion technique is to be implemented in 5G in the RF layer.

References

1. Shammasi, M.M., Safavi, S.M.: Performance of a predistorter based on Saleh model for OFDM systems in HPA nonlinearity. In: 2012 14th International Conference on Advanced Communication Technology (ICACT), PyeongChang (2012)
2. Duc, H.L., Feuvrie, B., Pastore, M., Wang, Y.: An analysis of adaptive digital predistortion algorithms in rf power amplifier. In: 2018 25th International Conference on Telecommunications (ICT), St. Malo (2018)
3. Paaso, H., Mammela, A.: Comparison of direct learning and indirect learning predistortion architectures. In: 2008 IEEE International Symposium on Wireless Communication Systems, Reykjavik (2008)
4. Guo, X.B., Jin, L., Tang, W., bo Chen, H.: A novel predistorter model for digital predistortion of RF power amplifiers. In: The 2012 International Workshop on Microwave and Millimeter Wave Circuits and System Technology, Chengdu (2012)
5. Morgan, D.R., Ma, Z., Kim, J., Zierdt, M.G., Pastalan, J.: A generalized memory polynomial model for digital predistortion of RF Power Amplifiers. IEEE Trans. Signal Process. **54**(10), 3852–3860 (2006)
6. Hu, Y., Boumaiza, S.: Joint RF pre-distortion and post-distortion linearization of small cell power amplifiers. In: 2017 IEEE Topical Conference on RF/Microwave Power Amplifiers for Radio and Wireless Applications (PAWR), Phoenix, AZ (2017)
7. Hussein, M.A., Venard, O., Feuvrie, B., Wang, Y.: Digital predistortion for RF power amplifiers: state of the art and advanced approaches. In: 2013 IEEE 11th International New Circuits and Systems Conference (NEWCAS), Paris (2013)

8. Bertran, E., Gilabert, P.L., Montoro, G., Berenguer, J.: Overview of power amplifier linearization based on predistortion techniques. In: 8th WSEAS International Conference on Simulation, Modelling and Optimization (SMO 2008), Santander, Cantabria, Spain, September 2008
9. Woo, Y., et al.: Adaptive digital feedback predistortion technique for linearizing power amplifiers. IEEE Trans. Microwave Theory Tech. 55(5), 932–940 (2007)
10. Singh, H., Sappal, A.S.: Comparative study of power amplifier linearization techniques. Int. J. Eng. Res. Devel. 12(3) (2016)
11. Jian, W., Yu, C., Wang, J., Yu, J., Wang, L.: A digital adaptive predistortion method of OFDM power amplifier. In: 2009 International Conference on Networks Security, Wireless Communications and Trusted Computing, Wuhan, Hubei (2009)
12. Tarasov, K.N., McDonald, E.J., Grayver, E.: Power amplifier digital predistortion - fixed or adaptive? In: MILCOM 2008 - 2008 IEEE Military Communications Conference, San Diego, CA (2008)
13. Bondar, D., Budimir, D., Shelkovnikov, B.: Linearization of power amplifiers by baseband digital predistortion for OFDM transmitters. In: 2008 18th International Crimean Conference - Microwave & Telecommunication Technology, Sevastopol, Crimea (2008)

A Taxonomy and Survey of Manifold Resource Allocation Techniques of IaaS in Cloud Computing

Saurabh Bhosale[✉], Manish Parmar, and Dayanand Ambawade

Department of Electronics and Telecommunication Engineering,
Bharatiya Vidya Bhavans' Sardar Patel Institute of Technology, Mumbai, India
saurabh.bhosale@spit.ac.in

Abstract. In cloud computing environment there are Cloud Service Providers (CSP)/Vendor, and Cloud User/Client. CSP provides application, infrastructure, and/or software. Cloud user demand for a service to the CSP via the internet which is accounted on a pay-per-usage basis. Resource allocation related parameters are optimization, cost efficiency, security, quality of service (QoS), reliability, compatibility, efficiency, and delay. In this survey, we have reviewed resource allocation algorithms and mechanisms used by researchers in the recent past and classified these techniques according to the parameters considered in the approach. According to the survey, we noticed that few parameters are well addressed by many of the researches while some are yet not much investigated. The survey will guide the researchers to achieve more vision in the field of resource allocation for IaaS in Cloud Computing.

Keywords: Cloud computing · Resource allocation · Infrastructure as a Service (IaaS) · Cloud Service Provider (CSP) · Public/private cloud

1 Introduction

Cloud computing basically is a technology which assigns computing resources to the users remotely in any network using the internet to store, manage and process their data online; in a convenient manner based on the demand of users. The Cloud Service Providers (CSP) provides services to the users for which users pay based on as-needed or pay-per-use. Cloud computing possesses few characteristics as mentioned below:

- On-demand Service Availability: A user gets the service such as server time, storage space it needs any time whenever he/she needs without contacting CSPs and any human interaction.
- Remotely Accessible Network: Cloud services are available over a network that is internet (broadly) or through LAN (privately). Additional security can be also offered through firewalls and Virtual Private Network (VPN). It can be accessed on various devices such as PCs, mobiles, laptop, etc.
- Resource Pooling: Some cases rise wherein there is uncertainty in demand of the users which means some users do not use resources to its peak capability or sometimes different applications are used by users needing full access at the same

© Springer Nature Switzerland AG 2020
P. Karrupusamy et al. (Eds.): ICSCN 2019, LNDECT 39, pp. 191–202, 2020.
https://doi.org/10.1007/978-3-030-34515-0_20

time. In such cases, the resources are allocated to users based on pooling so that it varying demand of users do not affect the performance of different applications.

- Elasticity and Scalability: It may also sometimes happen that suddenly the number of users for the same application increases and hence the number of servers to provide that service has to be increased. So the application scales itself automatically to provide the service to all the users.
- Managed Accounting: The CSPs use metering for managing and optimizing services to charge the users as per policies pre-defined.

Cloud computing thus provides various benefits as can be summarized from above that are cost saving, scalability/ flexibility, reliability, easy maintenance, and mobile access.

Along with the advantages, cloud computing also possesses certain challenges that may hinder the resource allocation and performance of the applications. Some of the challenges are as follows:

- Security: Multi-factor authentication has to be ensured when a number of people use the same cloud.
- Compliance: The Sarbanes-Oxley Act (SOX) in the US and Data Protection directives in the EU are just two among many compliance issues affecting cloud computing, based on the type of data and application for which the cloud is being used.
- Lack of standards: Most of the clouds are not interoperable.
- Vendor lock-in: It happens when other options are either expensive or just not possible.

2 Taxonomy

2.1 Cloud Services

After the cloud service is established, they are deployed in terms of business models based on the requirement. The service models are stated below:

1. Infrastructure as a Service (IaaS): This service provides infrastructure for computing resources such as servers, storage, and networks on which the users can run their software's and other applications. Examples: Amazon Web Services, Microsoft Azure, etc.
2. Platform as a service (PaaS): This cloud service model is used to deliver a platform to users with the help of which they can develop, initialize and manage applications. PaaS is similar to IaaS but it also includes DBMS, BI services, etc. Examples: Apprenda, Red Hat Openshift, etc.
3. Software as a service (SaaS): The CSPs provide software and applications running on the cloud. The user can access this service through a web browser or any program interface. This service is charged either according to consumption or simply on monthly charges. Examples: Google Applications, Salesforce, etc. (Fig. 1).

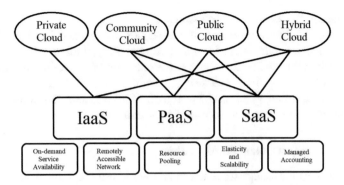

Fig. 1. Taxonomy of Cloud Computing

2.2 Deployment Models

The four major deployment models for cloud computing is identified according to the user's demand:

1. Private cloud: It is also known as the internal cloud. This model allows accessing the systems and services for a specific institution or organization. A private cloud allows only authorized users and ensures higher security, privacy, and reliability.
2. Public cloud: It is open for use. This cloud is available for the public mostly for commercial use. It is most suitable for business where a large amount of data has to be maintained.
3. Community cloud: This type of model is deployed when cloud infrastructure is shared among different organizations belonging to the same community or area.
4. Hybrid cloud: It is a combination of two or more clouds server's i.e. public, private and/or community clouds.

3 Algorithm/Mechanism

The algorithms covered in this section are only those which have used a maximum number of parameters as described by various authors from 2015 till 2018.

3.1 Markov Decision Process (MDP) and Vickrey-Clarke-Groves

3.1.1 Stochastic Matching Service

This algorithm is based on a stochastic scheduling rule based on MDP to achieve long-term efficiency. As per the literature review in [6] an MDP-based online mechanism was proposed that targeted matching of individual consumer and provider. The major drawback of this method was that it could not scale itself when the number of users increased. In order to reduce this complexity, Stochastic Matching Service mechanism can be used as it makes matching more practical by clustering the users according to their performance and then matching them within the clusters formed.

3.1.2   Dynamic VCG Auction Mechanism

The evaluation of the performance of the auction mechanism involves four major properties which are efficiency, incentive compatibility, individual rationality, and budget balance. Vickrey-Clarke-Groves is the only family of auction mechanisms that can simultaneously achieve efficiency, incentive compatibility, and individual rationality [6]. Dynamic VCG mechanism uses MDP for allocating multiple items to multiple traders in one shot and considers long term expected value.

3.2   Hybrid Adaptive Genetic Algorithm

HAGA overcomes the scalability issues in real-world problems that exist in non-linear economic models. The non-linear economic models require combinatorial optimization process in order to select an optimized composition of requests. Brute force gives best results for compositional optimization but the scalability problem is not resolved. HAGA is a hybrid of Genetic algorithm and Ant system. Mathematically it can be stated as; for N requests having arrival times as MAP $(S_N) = \{<U_1, \text{Arrival}_1>, <U_N, \text{Arrival}_N>\}$, an optimal subset MAP (S_K) has to be found where $k < N$; maximizes the fitness of the composition [9].

3.3   Nondominated Sorting Genetic Algorithm (NSGA)

This is the only algorithm that achieves scalability. The authors in [13] have proposed a modified NSGA which takes into consideration subscriber's requests and their constraints. Figure 2 shows various processes in the modified NSGA algorithm. They have addressed the weakness such as:

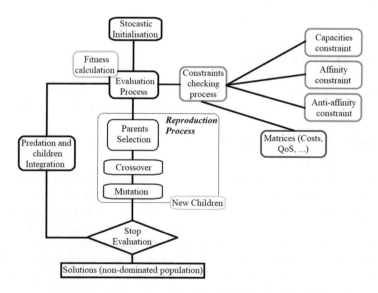

Fig. 2. Various process of modified NSGA algorithm [13].

- Exclude the subscribers which do not have constraints.
- Fix faulty subscribers using repair process.
- Preserve authorized individuals using special operators.
- Modify search space and guide the algorithm throughout.

3.4 Novel Online Auction Mechanism

This mechanism of online auctioning consists of two parts as mentioned below:

3.4.1 The Price-Based Allocation Rule and the Payment Rule

In the price allocation rule, each of the subscribers satisfying the demands is allocated with time-varying VM fleet to the cloud. The price of the VMs allocated in the entire VM fleet is lower than the unit valuation multiplied by the capacity of that VM for each time-slot. Hence, these become the cheapest VMs. The payment rule decides how much a cloud provider should charge each of its subscribers.

3.4.2 Defining the Marginal Price Function and Deriving Constant Competitive Ratio

This component focuses to define marginal price function which would benefit cloud provider. This mechanism also intends to achieve a nontrivial competitive ratio with respect to the profit of the cloud provider.

3.5 Nash Equilibrium

Nash Equilibrium can be explained with the help of game problem. All the available IaaS services provided by the cloud provider are limited and price dependent. The price allocation depends upon the subscriber's bids. Each of the subscriber bids sensibly so as to get correct VM instances for completion and implementation of their tasks ensuring maximal individual profit. Thus we can say that the bidders have a non-cooperative and competitive relationship and so this auction mechanism is modeled as if it is a non-cooperative game problem. The state of each bidder in order to fulfill his requirement and win the bidding using its optimal strategy corresponds to Nash Equilibrium [20].

3.6 Cost-Aware Adaptive Software and Hardware (CASH)

With the increase in the number of subscribers, the configurability becomes more fine-grained and the optimization space becomes more non-convex further increasing difficulty in optimization [25]. This optimization problem is overcome using CASH mechanism. It provides a simple interface to meet the performance needs of the subscriber.

This mechanism guarantees QoS and minimizes the cost for IaaS customers. It is a combination of hardware architecture and runtime management system.

To create an optimized core for any specific application, CASH uses ALU, FPU, fetch units and caches. These cores are not fixed and therefore it is not easier to access the remote cores at the runtime.

The QoS guarantee and non-convex space optimization are done using reinforcement learning and control theory.

3.7 IaaS Resource Allocation Using Column Generation Formulation (RA-IaaS-CG)

The IaaS resource allocation in the cost-efficient model is done using a pool of virtual resources which is formed by actual partitioning of physical resources. This mechanism uses Column Generation Technique wherein resource allocation is divided into a master problem and pricing problem [38].

The restricted number of columns is formed called as Restricted Master Problem and the pricing problem is evaluated based on the number of columns that iterate to give significantly reduced cost. The column represents the feasible allocation of virtual resources. The linking of both of the above-mentioned problems forms constraints for this technique due to the use of dual variables.

3.8 Federated Cloud

This algorithm offers a technique that reduces power consumption. It has been divided into two stages. In the first stage, all the subscribers are arranged in descending order based on the CPU usage. This ensures that the host having a higher workload gets maximum utility. In the second stage, appropriate virtual machines are allocated to the hosts based on the power consumption and size of CPU and RAM. This would help in saving and utilization of larger virtual machines. The policies for assigning the virtual machines are pre-defined. In case they do not run for particular hosts then they are rejected and sent to the adjacent providers. Thus helps in reducing the power consumption and increasing the profit [37].

4   Comparative Study

Figure 3 shows a graph which gives the basic scenario about how much work has been done for each parameter for resource allocation in the recent past. From the graph, it can be clearly interpreted that in 2015 most of the papers have emphasized on Compatibility, in 2016 on Cost Effectiveness and Quality of Service (QoS), in 2017 on Reliability, Efficiency, and Delay and in 2018 emphasis is on Optimization and Security.

The graph also confirms that in 2015 most of the papers have not considered Delay, in 2016 Optimization and Reliability, in 2017 Security and in 2018 most of the papers have not considered Cost Effectiveness, Efficiency, Compatibility, and Quality of Service (QoS).

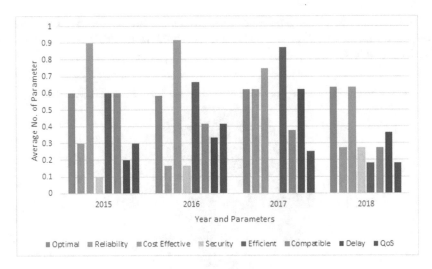

Fig. 3. Summary of parameters considered for resource allocation techniques in recent past (2015–18).

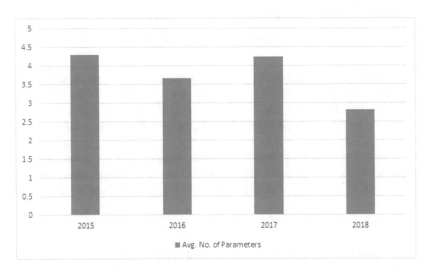

Fig. 4. Average number of parameters vs. years.

Figure 4 illustrates a graph which explains the average number of parameters referred to in papers in recent past. According to the graph, the average number of parameters considered in a paper for the year 2015 is the highest while in 2018 is lowest. From this one can infer that in 2015 the authors had considered a number of parameters for evaluation/verification. Also, the authors who published the papers for the year 2018 have concentrated on a specific number of parameters.

The average number of parameters considered in recent past is 3.54. Therefore, the years 2015, 2016 and 2017 are above the threshold while 2018 is way below the threshold (Table 1).

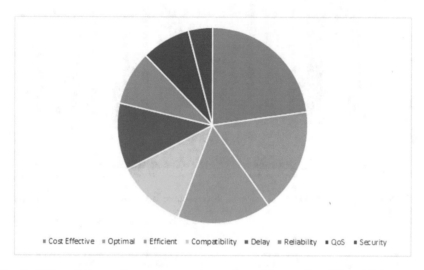

Fig. 5. Differentiating each parameter on the number of times referred from 2015 to 2018.

Figure 5 provides the number of time each parameter has been studied in all papers. The pie-chart summarizes that most authors have considered Cost Effectiveness, Optimization, and Efficiency, few have considered Compatibility and Delay while a very less number of papers have considered Reliability, Quality of Service (QoS), and Security.

If we examine the average number of parameters studied per paper, we find that Cost Effectiveness, Optimization, and Efficiency are over the threshold while Compatibility, Delay, Reliability, Quality of Service (QoS), and Security are below the threshold.

Table 1. Comparative study of resource allocation algorithm/mechanism

Algorithm/ mechanism	Parameters							
	Optimal	Reliable	Cost effective	Secure	Efficient	Compatible	Delay	QoS
Gale-Shapley Algorithm [1]	✗	✓						
Relaxing Binary Variables and Heuristic Algorithm [2]		✓	✓	✓				
Hybrid Firewall [3]		✓		✓				
Auto-Scaling Framework for Micro services [4]	✓		✓					
Dominant Strategy Incentive Compatible [5]	✓		✓	✓				

(continued)

Table 1. (*continued*)

Algorithm/ mechanism	Parameters							
	Optimal	Reliable	Cost effective	Secure	Efficient	Compatible	Delay	QoS
Markov Decision Process and Vickrey-Clarke-Groves [6]	✓		✓		✓	✓		
Dynamic Load Balancing [7]	✓						✗	
Genetic Algorithm Based on Novel Elasticity Testing [8]	✓		✓			✓		
Novel Dynamic Metaheuristic Optimization [9]	✓		✓				✗	✓
K-means and Neural Network [10]					✓		✓	✓
Skewness-Avoidance Multi-Resource Allocation [11]			✓			✓	✗	
Over-Commit Approach by Profiling Mechanism [12]	✓	✓			✓		✗	
NSGA-III [13]	✓	✓	✓		✓		✓	
Dyna-f and Q-Learning [14]	✓		✓		✓			✓
EXFed [15]		✓			✓	✓	✗	
MigrateFS [16]			✓		✓	✓	✗	
Monkey Search Algorithm [17]		✓	✓				✗	✓
Novel Online Auction Mechanism [18]	✓		✓		✓	✓	✗	
Primal-Dual Framework of Fenchel Duality [19]	✓	✓	✓		✓			
Nash Equilibrium [20]	✓		✗		✓	✓	✓	✗
Dynamic Hadoop Cluster on IaaS [21]	✓	✓			✓			
VMP based on dynamic programming strategy [22]	✓		✓		✓			
Genetic Algorithm Based on Heuristic Approach [23]	✓		✓					✓
Inter-Cloud Broker and Federation [24]	✓		✓	✗				✓
Cost-aware Adaptive Software and Hardware [25]	✓		✓			✓	✗	✓
Reciprocal Resource Fairness [26]			✓		✓		✗	
Fogbow [27]			✓		✓			
QuARAM [28]		✓	✓		✓	✓		
Healthy Threshold [29]			✓			✓		
StackAct [30]			✓		✓	✓		✓
PaaSage Prototype [31]	✓		✓	✓	✓		✗	
RA-IaaS-CG [32]	✓		✓		✓	✓	✗	✓
Bazaar-Extension [33]						✓		
MCARD [34]	✓		✓		✓			
Cooperative Resource Management [35]	✓	✓	✓			✓		
G-Cloud [36]	✓	✓	✓			✓		
Federated Cloud [37]		✓	✓	✓	✓	✓		
MILP-2P-IaaS [38]			✓		✓	✓	✗	✓
HMM and ARIMA [39]	✓		✓					✓
Fine-Grained Pricing [40]	✓		✓		✓			
2P-IaaS [41]			✓		✓			

5   Conclusion and Future Direction

The paper grants a systematic survey of resource allocation algorithms and mechanisms that are referred by researchers and the parameters they have considered. According to the survey, we conclude that cost efficiency, optimization, and efficiency are well addressed by many of the researches while compatibility, reliability, and delay are moderately considered. The research needs to be focused on parameters like QoS, and security. This paper would benefit CSP/vendors, user/client, and researchers, who are interested in doing further research in resource allocation for the cloud computing environment.

It is obligatory for a cloud computing service model to take into consideration the various request generated by the users, to meet user's demands. Hence, to satisfy the users need resource allocation techniques through IaaS Cloud Computing should be improvised. The survey presented in this paper aims to deliver insight for resource allocation in IaaS cloud computing. Additionally, we would like to suggest that an extensive investigation is required for QoS and security parameters of resource allocation schemes.

References

1. Jain, S., Purini, S., Reddy, P.V.: A multi-cloud marketplace model with multiple brokers for IaaS layer and generalized stable. In: IEEE/ACM 11th International Conference on Utility and Cloud Computing (UCC) (2018)
2. Raugust, A.S., de Souza, F.R., Pillon, M.A., Miers, C.C., Koslovski, G.P.: Allocation of virtual infrastructures on multiple IaaS providers with survivability and reliability requirements. In: IEEE 32nd International Conference on Advanced Information Networking and Applications (2018)
3. Singh, G.B., Jaafar, F., Butakov, S.: Analysis of overhead caused by security mechanisms in IaaS cloud. In: 5th International Conference on Control, Decision and Information Technologies (CoDIT18) (2018)
4. Prachitmutita, I., Aittinonmongkol, W., Pojjanasuksakul, N., Supattatham, M., Padung-weang, P.: Auto-scaling microservices on IaaS under SLA with cost-effective framework. In: Tenth International Conference on Advanced Computational Intelligence (ICACI), 29–31 March 2018, Xiamen, China (2018)
5. Halabi, T., Bellaiche, M., Abusitta, A.: Cloud security up for auction- a DSIC online mechanism for secure IaaS resource allocation. In: 2nd Cyber Security in Networking Conference (CSNet) (2018)
6. Jiang, C., Chen, Y., Wang, Q., Liu, K.J.R.: Data-driven auction mechanism design in IaaS cloud computing. In: IEEE Transactions on Services Computing, vol. 11, no. 5, September–October 2018
7. Paul, S., Adhikari, M.: Dynamic load balancing strategy based on resource classification technique in IaaS cloud. In: IEEE 7th International Conference on Advances in Computing, Communication and Informatics (ICACCI) 19–22 September 2018, Banglore, India (2018)
8. Liu, J., Qiao, J.: How to buy cloud resource better for IaaS user- from the perspective of cloud elasticity testing. In: IEEE 24th International Conference on Parallel and Distributed Systems (ICPADS) (2018)

9. Mistry, S., Bouguettaya, A., Dong, H., Qin, A.K.: Metaheuristic optimization for long-term IaaS service composition. IEEE Trans. Serv. Comput. **11**(1), 131–143 (2018)

10. Patel, E., Mohan, A., Kushwaha, D.S.: Neural network based classification of virtual machines in IaaS. In: 5th IEEE Uttar Pradesh Section International Conference on Electrical, Electronics and Computer Engineering (UPCON) (2018)

11. Wei, L., Foh, C.H., He, B., Cai, J.: Towards efficient resource allocation for heterogeneous workloads in IaaS clouds. IEEE Trans. Cloud Comput. **6**(1), 264–275 (2018)

12. Wu, C.-H., Lee, Y.-H., Huang, K.-C., Lai, K.-C.: A framework for proactive resource allocation in IaaS clouds. In: Meen, P.L. (ed.) IEEE International Conference on Applied System Innovation IEEE-ICASI 2017 (2017)

13. Ecarot, T., Zeghlache, D., Brandily, C.: Consumer and-provider-oriented efficient IaaS resource allocation. In: IEEE International Parallel and Distributed Processing Symposium Workshops (2017)

14. Ren, J., Pang, L., Cheng, Y.: Dynamic pricing scheme for IaaS cloud platform based on load balancing- a Q-learning approach. In: International Conference on Engineering, Technology and Innovation (ICE-ITMC) (2017)

15. Pucher, A., Wolski, R., Krintz, C.: EXFed- efficient cross-federation with availability SLAs on preemptible IaaS instances. In: IEEE International Conference on Cloud Engineering (2017)

16. Tsakalozos, K., Verroios, V., Roussopoulos, Delis, A.: Live VM migration under time-constraints in share-nothing IaaS-clouds. IEEE Trans. Parallel Distrib. Syst. **28**(8), 2285–2298 (2017)

17. Gupta, P., Tewari, P.: Monkey search algorithm for task scheduling in cloud IaaS. In: 4th International Conference on Image Information Processing (ICIIP) (2017)

18. Li, J., Zhu, Y., Yu, J., Long, C., Xue, G., Qian, S.: Online auction for IaaS clouds- towards elastic user demands and weighted heterogeneous VMs. In: IEEE INFOCOM – IEEE Conference on Computer Communications (2017)

19. Zhang, X., Huang, Z., Wu, C., Li, Z., Lau, F.C.M.: Online auctions in IaaS clouds: welfare and profit maximization with server costs. IEEE/ACM Trans. Netw. **25**(2), 1034–1047 (2017)

20. Wei, Y., Pan, L., Yuan, D., Liu, S., Wu, L., Meng, X.: A distributed game-theoretic approach for IaaS service trading in an auction-based cloud market. In: IEEE TrustCom-BigDataSE-ISPA (2016)

21. Wang, B., Tao, D., Lin, Z.: A load feedback based resource scheduling algorithm for IaaS cloud platform. In: International Conference on Consumer Electronics, Taiwan (2016)

22. Chang, Y., Gui, C., Luo, F.: A novel energy-aware and resource efficient virtual resource allocation strategy in IaaS cloud. In: 2nd IEEE International Conference on Computer and Communications (2016)

23. Govindaraju, Y., Hector D.-L.: A QoS and energy aware load balancing and resource allocation framework for IaaS cloud providers. In: IEEE/ACM 9th International Conference on Utility and Cloud Computing (2016)

24. Hamze, M., Mbarek, N., Togni, O.: Broker and federation based cloud networking architecture for IaaS and NaaS QoS guarantee. In: 13th IEEE Annual Consumer Communications Networking Conference (CCNC) (2016)

25. Zhou, Y., Hoffmann, H., Wentzlaff, D.; CASH: supporting IaaS customers with a subcore configurable architecture. In: ACM/IEEE 43rd Annual International Symposium on Computer Architecture (2016)

26. Liu, H., He, B.: F2C: enabling fair and fine-grained resource sharing in multi-tenant IaaS clouds. IEEE Trans. Parallel Distrib. Syst. **27**(9), 2589–2602 (2016)

27. Brasileiro, F., Falco, E.: Federation of private IaaS cloud providers through the barter of resources. In: IEEE 36th International Conference on Distributed Computing Systems (2016)

28. Soltani, S., Elgazzar, K., Martin, P.: QuARAM service recommender: a platform for IaaS service selection. In: IEEE/ACM International Conference on Utility and Cloud Computing (2016)

29. Cheng, S., Cao, C., Yu, P., Ma, X.: SLA-aware and green resource management of IaaS clouds. In: IEEE 18th International Conference on High Performance Computing and Communications, IEEE 14th International Conference on Smart City, IEEE 2nd International Conference on Data Science and Systems (2016)

30. Bruschi, G.C., Spolon, R., Pauro, L.L., Lobato, R.S., Manacero, A., Cavenaghi, M.A.: StackAct- performance evaluation in an IaaS cloud multilayer. In: 15th International Symposium on Parallel and Distributed Computing (2016)

31. Kritikos, K., Magoutis, K., Plexousakis, D.: Towards knowledge-based assisted IaaS selection. In: IEEE 8th International Conference on Cloud Computing Technology and Science (2016)

32. Metwally, K., Jarray, A., Karmouch, A.: A cost-efficient QoS-aware model for cloud IaaS resource allocation in large datacenters. In: IEEE 4th International Conference on Cloud Networking (CloudNet) (2015)

33. Pittl, B., Mach, W., Schikuta, E.: A negotiation-based resource allocation model in IaaS-markets. In: IEEE/ACM 8th International Conference on Utility and Cloud Computing (2015)

34. Dou, H., Qi, Y., Chen, P.: A novel approach to improving resource utilization for IaaS. In: 12th Web Information System and Application Conference (2015)

35. Tran, G.S., Nghiem, T.P.: Cooperative IaaS resource management- policy and simulation framework. In: 7th International Conference on Knowledge and Systems Engineering (2015)

36. Liu, T., Ji, T., Yue, Q., Tang, Z.: G-cloud: a highly reliable and secure IaaS platform. In: International Conference on Network and Information Systems for Computers (2015)

37. Bagheri, B., Abadi, C., Arani, M.G.: Improving resource management of IaaS providers in cloud federation. In: 2nd International Conference on Knowledge-Based Engineering and Innovation (KBEI), 5–6 November 2015, Tehran, Iran (2015)

38. Metwally, K., Jarray, A., Karmouch, A.: MILP based Approach for Efficient Cloud IaaS resource allocation. In: IEEE 8th International Conference on Cloud Computing (2015)

39. Mistry, S., Bouguettaya, A., Dong, H., Qin, A.K.: Predicting dynamic requests behavior in long-term IaaS service composition. In: IEEE International Conference on Web Services (2015)

40. Jin, H., Wang, X., Wu, S., Di, S., Shi, X.: Towards optimized fine-grained pricing of IaaS cloud Platform. IEEE Trans. Cloud Comput. 3(4), (2015)

41. Metwally, K.M., Jarray, A., Karmouch, A.; Two-phase ontology-based resource allocation approach for IaaS cloud service. In: 12th Annual IEEE Consumer Communications and Networking Conference (CCNC) (2015)

Automatic Traffic E-challan Generation Using Computer Vision

Manisha R. Dhage, Gaurav V. Patil$^{(\boxtimes)}$, Shrayank J. Mistry,
Prathamesh N. Tambe, and Parag H. Nankar

Department of Computer Engineering,
Sinhgad College of Engineering, Pune, India
mrdhage.scoe@sinhgad.edu, ptgauravl0@gmail.com,
shraymist@gmail.com, tprathaml1198@gmail.com,
paragnankar@gmail.com

Abstract. The Automatic recognition of license plate is the basis of effective management in traffic, the automatic detection and localization of license plate is an important part. License plate detection and contain how to extract or segment the license plate region from the license plate image a new deep learning network structure was designed, and designed network structure was used to detect and locate the license plate automatically. The system proposed by us involves automatic detection of vehicles that break the traffic rules at respective signals and registration number for every vehicle is recognized. The vehicle number detected is searched in the database for type of vehicle and owner's information. This information is used to generate e-challan in the name of the person who owes the vehicle directly and instantly and send appropriate fine message to the owner. So it will be more efficient and will require less human intervention.

Keywords: E-challan · Otsu · Contour · Tesseract

1 Introduction

Automatic Traffic E-challan (Electronic-challan) Generation using Computer Vision this system is to use to automate e-challan generation when vehicles cross zebra crossing during traffic signal. The system is based on detection of the vehicles that have broken the rule i.e. the vehicles which stop in front of zebra line, license plate detection of the vehicle breaking rule and effective challan generation. We will have a database server it has information of all the vehicle registered. The rule violation on the streets of the city is on the rise and zebra crossing violation is one of the rule that is broken very often. People crossing the roads majorly rely on zebra crossing for to safely cross the roads. If vehicles are on the zebra crossing it is inconvenient and hazardous. Are system which can detect this violation instantly will help the people as well as vehicle owners to be safer. Above all, we hope to provide a smooth, easy and hassle-free system for the traffic authority.

P. Karrupusamy et al. (Eds.): ICSCN 2019, LNDECT 39, pp. 203–213, 2020.
https://doi.org/10.1007/978-3-030-34515-0_21

2   Related-Works

In India, recently the working system is where two people (Traffic Police Officers) are involved in generation of E-challan.

One person looks at the video and takes images when signal is red. After that he checks for the vehicle which has broken the rule. The vehicles which are on Zebra crossing, number of those vehicles are taken by that person. He gives the number to the second person.

Second person takes the number and fills the form of violating traffic rule by that vehicle, by inserting that number in the form. And then the challan is generated by that system.

3   System-Implementation

The system being developed has two main features:

(1) Detecting the vehicles that have violated the rule.
(2) Generating e-challan for the vehicles that have broken the rule.

3.1   Image Preprocessing

Image is taken by the camera situated above the signal when it goes red. Image captured is full of noise and unwanted data. The image above the zebra crossing is cropped and noise is removed to make it ready for license plate detection.

3.2   License Plate Detection

Image obtained after preprocessing has not one but Multiple License plates in them. Each license plate of the vehicles that have broken the rules will be detected by the system.

3.3   Feature Extraction

After all registration plates are detected the exact number of each vehicle needs to be recognized and extracted. The alphabets and numbers on the number plate are extracted and recognized.

3.4   Database Verification

The obtained registration number is searched in the database entire database until a match is found. Once a match has been found entire details of the user is obtained.

3.5   E-challan Generation

Using the user information obtained an e-challan will be generated in the name of the owner and notification will be sent to owner of vehicle that has violated the rule via SMS or mail.

4   Methodology and Flowchart

4.1   Pre-processing

Gaussian blurring is highly effective in removing Gaussian noise from the image (Figs. 1, 2 and 3).

Fig. 1. Input image

Fig. 2. Crop image

Fig. 3. Blurring of image

ImgBlurred = cv2.GaussianBlur (img, (5, 5), 0)
Convert RGB to Gray Image -
RGB - Gray: $Y \leftarrow 0.299 * R + 0.587 * G + 0.114 * B$
Gray - RGB: $R \leftarrow Y, G \leftarrow Y, B \leftarrow Y, A \leftarrow$ Max (Channel range)
This is done with the help of cvtColor () from CV.
Threshold - Sobel operator is discrete differentiation operator. It compute an approximation of the gradient of an image intensity function.

$$G = sqrt (Gx^2 + Gy^2)$$

4.2 Recognize License Plate

- *Get Structuring Element (Rectangle)*
  Element = cv2.getStructuringElement (shape = cv2.MORPH_RECT, ksize (17, 3))
- *Closing*
  Used for closing small holes in object
  cv2.morphologyEx (sic = thresholding, op = cv2.MORPH_CLOSE, kernel = element, dust = morph_img_threshold) (Figs. 4 and 5)

Fig. 4. Closing of image

- *Find Contours*
  Contours indicate the boundary pixels of objects
  Contours = cv2.findContours (morph_img_threshold, mode = cv2.RETR_EX-TERNAL, method = cv2.CHAIN_APPROX_NONE)

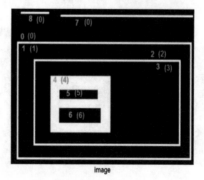

Fig. 5. Contours

1. First parameter is source image (morph_img_threshold)
2. Second parameter is contour retrieval mode mode = cv2.RETR_EXTERNAL, using this mode only outer edges are returned.
3. Third is contour approximation method.

- *Bounding Rectangle*

    x, y, w, h = cv2.boundingRect(cnt)

- (x, y) top left corner co-ordinates of plate
- w = width
- h = height
- Plate image can be obtained by adding co-ordinates and width and height. Hence number plate is detected

4.3 Number Plate Detection

Img = img.transform ((new_width, height), Image. AFFINE, (1, m, -xshift if m > 0 else 0, 0, 1, 0), Image.BICUBIC) (Figs. 6 and 7)

Fig. 6. Detected number plate

Fig. 7. Number plate

- *Creating Threshold Image*

  Extract the Value component from the HSV (Hue, Saturation, Value) color space and apply otsu thresholding to reveal the characters on the license plate.

  gaus = cv2.adaptiveThreshold (image,255, cv2.ADAPTIVE_THRESH_GAUSSIAN_C, cv2.THRESH_BINARY,115,1)

- *Creating Black and White Image*

  ```
  ret, labels1 = cv2.connectedComponents(gaus)
  # Map component labels to hue val
  label_hue = np.uint8(179 * labels1/np.max(labels1))
  blank_ch = 255*np.ones_like(label_hue)
  labeled_img = cv2.merge([label_hue, blank_ch, blank_ch])
  # cvt to BGR for display
  labeled_img = cv2.cvtColor(labeled_img, cv2.COLOR_HSV2BGR)
  # set bg label to black
  labeled_img[label_hue ==0] = 255
  labeled_img[label_hue! = 0] = 0
  ```

- *Extracting numbers and alphabets*

  ```
  ,cnts, hierarchy = cv2.findContours (lthresh, cv2.RETR_EXTERNAL, cv2.CHAIN_APPROX_SIMPLE)

  for c in cnts:
  area = cv2.contourArea(c)
  x, y, w, h = cv2.boundingRect(c)
  rect_area = w*h
  extent = float(area)/rect_area
  if (rect_area > 1000 and rect_area < 9999):
  print(rect_area)
  cv2.rectangle (labeled_img, (x, y), (x + w, y + h), 255, 0)
  pixelpoints = np.transpose(np.nonzero(mask))
  cv2.drawContours(labeled_img, cnts, −1, (255, 255, 0), 1)
  ```

Fig. 8. Processing number plate

- *Pixel Matching-*
  We check pixels if they are in Contours or not (Fig. 8)

- Pixels which are present in contours are kept as it is.
- Pixels which are not present in contours are turned to black.

Finally we get a clean image which is passed to OCR (Optical Character Recognition) engine.

- *Optical Character Recognition*

  Using Tesseract python library

```
import Image
from tesseract import image_to_string
print image_to_string(Image.open('test.png'))
print image_to_string(Image.open('test-english.jpg'), lang = 'eng') (Fig. 9)
```

Fig. 9. Processed number plate

Output: -
AXX 6850

4.4   Database Verification

Information regarding user details can be obtained from the database of vehicle owners.

4.5   E-challan Generation

Challan is generated by system using details obtained about user and SMS is send using api (Application program interface) given below. Api for SMS sending: way2sms\

5 Activity Diagram

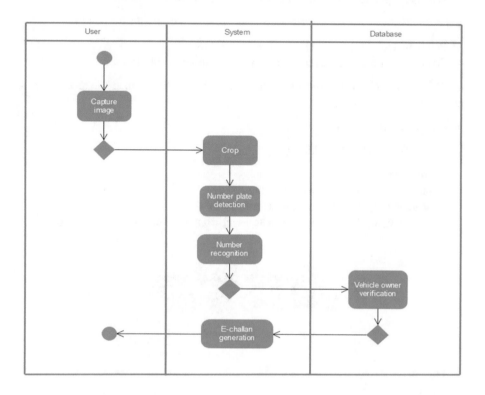

6 Results and Discussions

6.1 Main GUI Snapshots

See Figs. 10, 11, 12 and 13.

6.2 Discussions

Traffic E - Challan generation is totally python based application. It collects data from camera and detect the violated vehicle's license number. On this application, first user will get list of violated incidents. Details of incidents is displayed on this application with all necessary information and image. After user approves, this application sends message to vehicle owner when vehicle breaks the rule.

Fig. 10. Initial window

Fig. 11. After clicking on User 1

Fig. 12. After clicking on User 2

Fig. 13. E-challan generated to User 1

7 Conclusion

The objective of the Government to automate rule violation detection, recognise owner of vehicle, and generate e-challan can be obtained using the system proposed in this paper. The Automatic e-challan generation is a tool for the Government which can help her perform all duties that are currently done manually in an efficient and a smarter way and completely automatically. The system will make vehicle owner more aware of the rules as well as make it much safer for pedestrians.

Compliance of Ethical Standards. All author state that there is no conflict of interest. We used our own data. No animals and humans are not involved in this work.

References

1. Tang, Y., Luo, B., Tang, J., Wu, F.: New heuristic algorithm of license plate location based on statistical feature. Comput. Eng. Appl. **51**(6), 154–158 (2015)
2. Abolghasemi, V., Ahmadyfard, A.: An edge-based color-aided method for license plate detection. Image Vis. Comput. **27**(8), 1134–1142 (2009)
3. Comelli, P., Ferragina, P., Granieri, M.N.: Optical recognition of motor vehicle license plates. Trans Veh. Technol. IEEE **44**, 790–799 (1995)

4. Si, Q.: The design and implementation of license plate recognition system based on robust illumination. Nanjing University of Posts and Telecommunications, Nanjing (2013)
5. Deneen, K.M.V.: An algorithm for license plate recognition applied to intelligent transportation system. IEEE Trans. Intell. Transp. Syst. **12**(3), 830–845 (2011)
6. Ahmad, C.J., Shridhar, M.: Recognition of handwritten numerals with multiple feature and multistage classifier. Pattern Recognit. **2**(28), 153–160 (1995)
7. Kim, K.K., Kim, K.I., Kim, J.B., Kim, H.J.: Learning-based approach for license plate recognition. In: Proceedings of the IEEE Signal Processing Society Workshop, pp. 614–623 (2000)
8. Rahman, A., Badawy, W., Radmanesh, A.: A real time vehicle's license plate recognition system. In: Proceedings of the IEEE Conference Advanced Video Signal Based Surveillance, pp. 163–166 (2003)
9. Rice, S.V., Jenkins, F.R., Nartker, T.A.: The Fourth Annual Test of OCR Accuracy, Technical Report 95-03, Information Science Research Institute, University of Nevada, Las Vegas, July 1995
10. Smith, R.W.: The Extraction and Recognition of Text from Multimedia Document Images, Ph.D. thesis, University of Bristol, November 1987
11. Smith, R.: A simple and efficient skew detection algorithm via text row accumulation. In: Proceedings of the 3rd International Conference on Document Analysis and Recognition, vol. 2, pp. 1145–1148. IEEE (1995)
12. Rousseeuw, P.J., Leroy, A.M.: Robust Regression and Outlier Detection, Wiley-IEEE (2003)
13. Rice, S.V., Nagy, G., Nartker, T.A.: Optical Character Recognition: An Illustrated Guide to the Frontier, pp. 57–60. Kluwer Academic Publishers, USA (1999)
14. Schneider, P.J.: An algorithm for automatically fitting digitized curves. In: Glassner, A.S., (ed.) Graphics Gems I, Morgan Kaufmann, pp. 612–626 (1990)
15. Shillman, R.J.: Character Recognition Based on Phenomenological Attributes: Theory and Methods, Ph.D. thesis, Massachusetts Institute of Technology (1974)

Websites

1. https://www.python.org
2. https://www.github.com
3. https://www.tutorialspoint.com
4. https://www.opencv.org
5. https://www.sourceforge.org

Smart Electric Car Management System

Anvitha Shesh, R. Kavya, Y. B. Sanju$^{(\boxtimes)}$, K. S. Vaishnavi, and M. N. Thippeswamy

Department of Computer Science and Engineering,
Nitte Meenakshi Institute of Technology, Yelahanka, Bangalore 6429, India
anvithapuchu@gmail.com, kavyagreen72@gmail.com,
sanjuybl011@gmail.com, vaishnaviks1997@gmail.com,
mntswamynmit@gmail.com

Abstract. The electric vehicle is gaining significant research importance because of its eco-friendly frameworks. This paper proposes an application based charging facilities for the users in the cities. The main aim is to provide an online scheduling scheme to make it more convenient for every Electric Vehicle owner as well as the user to monitor the battery status of their electric vehicles at any point of time and from anywhere. It also indicates the user with the nearest charging stations and helps the user to book the slots. The administrator can keep track of user's details in a well-defined database.

Keywords: Charging stations infrastructure · Electric vehicles (EV) · Online scheduler

1 Introduction

An electric car is a plug in based automobile which runs with the help of electricity that is produced by one or more electric motors, using energy typically stored in rechargeable Li-ion batteries. Electric cars are considered to be the next disruptive market force for transportation and technology due user friendly nature, low overall cost of ownership and manufacturing along with better benefits for the environment. Considering this fact with the proposed system an attempt has been made to make use of EV for the benefit of the user. Making use of renewable energy like solar or wind or geothermal for electricity generation and charging an EV can be nearly be emission free. The World which is facing a lot of issues like pollution, global warming and ozone depletion can be reduced to some extent by making use of electric vehicles. The effective cost of refueling an electric car is a fraction of the cost of making use of sources like petrol and diesel. Thus, the electric cars can have a lower cost of ownership compared to the existing system. The initial investments on electric vehicles have to be made indeed for the betterment of the nature. This paper also aims to develop System architecture with an online scheduling scheme [9], to make it easy for the user of the car to access the battery percentage via an android application. It also indicates the user the nearest charging stations and books the slot in the electric charging station. The user also has an additional option of contacting the mobile charging station providers, who will in-turn; arrive at their location to charge their vehicle.

P. Karrupusamy et al. (Eds.): ICSCN 2019, LNDECT 39, pp. 214–223, 2020.
https://doi.org/10.1007/978-3-030-34515-0_22

In the proposed system the aim to provide the smart management of electric car and display the battery level of the car on the LCD which is affixed inside the car and the controller communicates the same to the smart phone through wireless unit such as GSM. If the battery energy of EV drops below a particular threshold or limitation the user gets an indication of the low battery status and will show the electric power stations nearby using Google map and hence obtaining the service from the power station through look ahead functionality. The driver can book a slot for his electric car in the charging station. The electric car can also be provided with backup energy. The backup energy can either be either the solar energy or petrol. There is also central controller that manages the entire system in an effective way. The proposed system is designed by developing a mobile application in android platform for user end system. The sensor data of charging level is intimated to the users through the mobile application and also the nearest charging station is detected by making use of current location and the user is intimated with the parameters such a distance from current location, approximate time required to reach the place. The user data is maintained in the local storage or cloud storage system depending on the usage of the system. The system (as referred to in Fig. 1) enables the opportunity to the controller of the system to manage and get the information about the user/driver from their place through the application. Implementation of our paper can be used in real time and make it convenient for the daily use of the user. Admin keeps track of all the data of the drivers/user and can access the data.

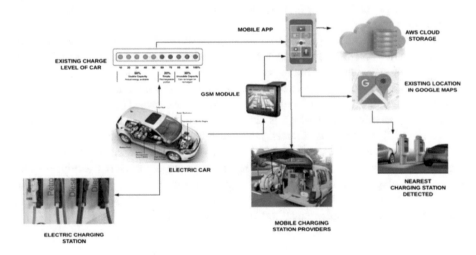

Fig. 1. System overview

2 Related Background

In the field of Electric Vehicles and Electric Vehicle Charging stations a lot of research is being carried out to provide effective solutions for the clean environment. To prevent excess of charging of Electric Vehicle the concept of smart grid is devised [1, 5], where the electricity gets limited for charging and can also be transferred to different entities to safeguard the electricity usage [11]. Maintaining the battery values by using the concept of cell balancing is found to be very effective. The networks are also designed as to where the users can charge their electric vehicles and at what time they can charge them effectively [3, 4].

Research in the field of battery is also being carried out extensively as to what kind of batteries have to be used in the upcoming days such that it is cost efficient as well as reliable for usage in long term [2]. An automated system is developed to approximate the cost of battery charging based the electric vehicles they own. Smart charging stations are devised in public that has adapters to charge various kinds of electric vehicles available in market [10].

3 Proposed Method

In the proposed system the aim is to provide the smart management of electric car. The battery level of the car is identified by the sensors and communicated to smart phone through wireless such as Wi-Fi/GSM/4G. If the battery level of EV drops below a certain threshold the gateway of the device interacts with the users application through an interface indicating the low battery status and showing the electric power stations nearby using Google map and hence obtaining the service from the power station. The driver can book a slot for his electric car in the charging station of his choice according to criteria such as distance, availability etc. It also has a central controller that manages the system.

The proposed system (as referred in Fig. 2) shows the interconnection of the Battery, Arduino and GSM module which send a voice alert to the driver's smart phone when the battery is low. The micro-controller used for this prototype is Arduino Uno. This is chosen due to its multi-functionality nature where all the modules are built-in and can be accessed via the I/O ports that are readily available. The proposed protocol has a voltage divider circuit which is used to change the battery level and the output device the LCD screen is used to see the battery value. If the battery percentage falls below the threshold level a voice alert is sent to the driver's smart phone.A mobile application is developed to monitor the battery percentage and alert the driver if the battery percentage is below a threshold level and book slots for charging in the nearest charging station.The system also have an administrator who keeps tracks of all the registrations and the details related to charging information.

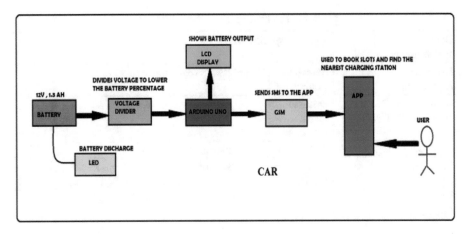

Fig. 2. Block diagram of the system

4   Novel Contribution

After studying the systems that have been deployed, the system can further be developed by adding new few features to the existing system.

1. The Smart Electric Car management system is designed using Battery, Arduino, GSM module, LCD, Voltage divider circuit and battery. The initial values from the Arduino, regarding the battery level is fetched and reported through a mobile app to the user.
2. An Android application is developed. When the voltage drops below the threshold that is 30%, a voice note in sent to the mobile application where the user has registered their mobile number via the GSM module. The application shows the current location of the driver and the nearest charging station using the latitude and longitude and also displays the direction to it in Google maps.
3. The application automatically books the slots for charging in the respective charging stations. In case the user cannot reach the charging station they can contact the mobile charging station providers, who will arrive at their location to charge the vehicles.
4. The system as a whole is integrated with the car and is able to identify the battery parameters automatically. Thus, the system is user friendly as the battery value is intimated to the phone irrespective of where the driver is present. Navigation to the charging station is obtained without doing multiple searches on Google maps. The mobile charging station providers can be contacted easily with just a button click.

5   Hardware and Software Components

Table 1 consists of the hardware and software tool that are used to implement the proposed work.

Table 1. The hardware and the software too

Hardware specifications	Software specifications	
Hardware requirements	Hardware requirements	Software requirements
Arduino Uno Board 16 × 2 LCD Display	Processor speed: 2.71 GHz	Operating System: Windows 10
12 V DC Battery	RAM: 8 GB	Platform: IOT, Cloud
GSM Module	Disk Space: 585 MB	IDE: Arduino Studio, Eclipse-Kepler, AWS Management Console, MySQL Workbench
Voltage Divider Circuit		Technologies used, JAVA, SQL

6 Implementation and Result

In this section, the implementation of the proposed work is discussed. The components are listed in Table 1 and are used to design the prototype as shown in Figs. 3 and 4. The Electric Car Management System can be classified into different phases of working such as Battery Measurement Phase, Communication phase and the Output phase.

Fig. 3. Model review

Fig. 4. System review

1. Battery Measurement Phase

   This phase is responsible for accessing the data related to the Electric Car Battery level. The data here is the voltage level of battery, which is in analog form, is converted into the digital data. The Voltage Divider circuit helps to measure the effective battery voltage which is then converted into digital data (as in Eq. 1). The input analog value is converted into the digital voltage value from 0–5 V.

$$Float\ voltage\ =\ sensor\ value\ *\ (5.0/1023.0) \tag{1}$$

Equation 1. Converting voltage value from analog to digital.

2. Communication Phase

The communication phase is the most important phase as it helps to coordinate the data from one end of the system to the other. The message from GSM (Global System for Mobile) is used to communicate between the hardware and the application. The data flow in different phase of execution is as depicted in Fig. 5.

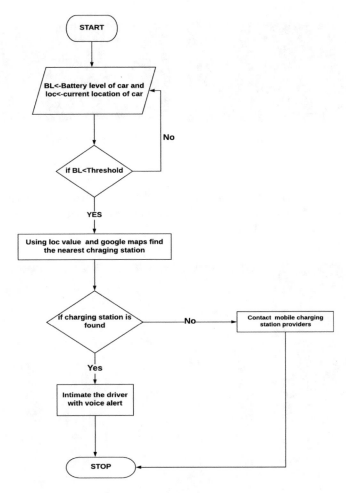

Fig. 5. Flowchart of the system

3. Output Phase

The System uses the data that is received from the communication unit to inform the user about the battery status of the car. Once the battery level reaches the threshold a voice output is enabled in the user's android application. Next, the nearest Charging station available to charge the car is detected and the slots are booked for doing the same.

7   Result

Once the battery level from the car is been detected following results will be obtained

i. LCD Display

The prior value is set in Arduino based on the assumption of battery usage, to display the battery level changing consequence as follows;

Fig. 6. Battery level full indication

Fig. 7. Battery level low indication

In Fig. 6, When the battery value is greater than 13 V the alert is initialized on the LCD saying that the battery is full When the battery level reaches the threshold that is 30% it is indicated on the LCD (in Fig. 7).

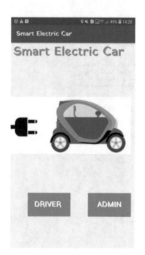

Fig. 8. Home Page   **Fig. 9.** Registration Page   **Fig. 10.** Login page

ii. Android App

The home page (In Fig. 8) of the android application is as shown. There are two options available one for the driver and the other for the administrator which will redirect to the registration page (In Fig. 9). The registration page is made use to collect the users credentials. The common login page (In Fig. 10) is used to securely login by authenticating the user name and password.

iii. Results Page:

Fig. 11. Nearest charging station is found out.

Fig. 12. Navigation in Google maps to reach the charging station.

In Fig. 11 When the location of the user is changed, it detects the nearest bunk according to which direction the user/driver is moving. On clicking of the 'Charge Bunk' button by the longitude and the latitude difference the bunk which is near to the user location will be selected and the slots will be booked automatically. There is another button for navigation that appears when the driver needs help to locate the charging station. The direction and the kilometer to the destination is also displayed (in Fig. 12) using Google maps.

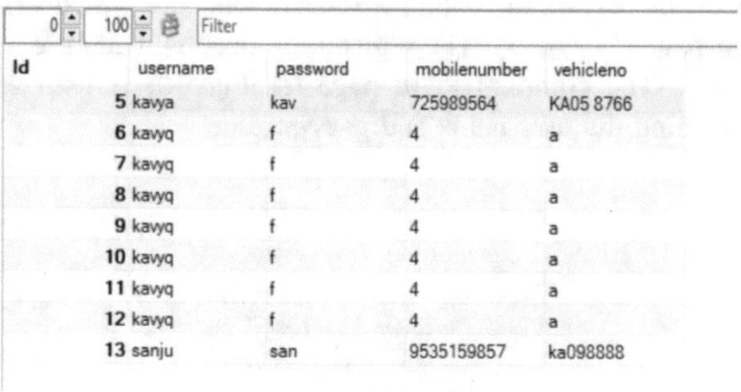

Fig. 13. List of signed in user

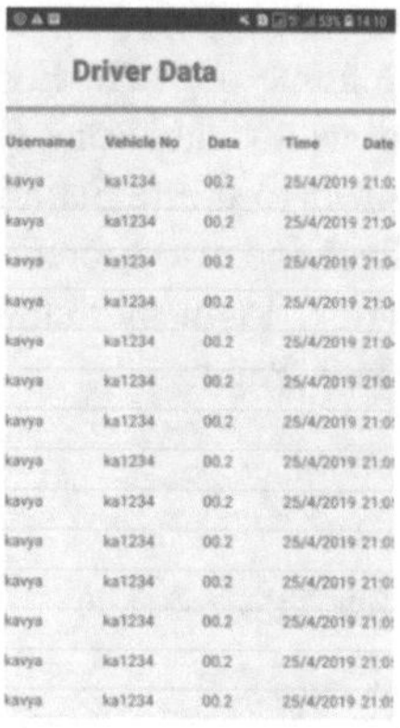

Fig. 14. User data displayed in admin page

The database of all the registered users (as in Fig. 13) can be checked, so that the administrator can get in track of the signed in users. Additionally the user data can been viewed in the application (as in Fig. 14).

8 Conclusion

A smart electric car based android application to monitor the battery level and notify the user for the same is being developed. The developed prototype has a hardware controller system that is affixed into the electric car along with android application to monitor various battery parameters. The application has two divisions that is the administrator login and user login. Initially the notification about the battery percentage is specified as a voice alert. The administrator can view the entities of the logged in user via the tables stored in the database. In a nutshell, application helps the user to find the nearest charging station and pre book the definite slots. Alternative approach is to contact the mobile charging providers to arrive at the current destination.

9 Future Enhancement

The Smart Electric Car Management application is proposed to provide navigation to the users. The speed and distance with which the user has to travel will be based on the acceleration of electric car. The system can be integrated with electric cars for working in real time. In forth coming days, iOS-based application for apple phones will also be developed.

References

1. Ferreira, J.C., Monteiro, V., Afonso, J.L., Silva, A.: Smart electric vehicle charging system. In: IEEE Intelligent Vehicles Symposium (IV), Baden-Baden, pp. 758–763 (2011). https://doi.org/10.1109/ivs.2011.5940579
2. Rahimi-Eichel, H., Ojha, U., Baronet, F., Chow, M.: Battery management system: an overview of its application in the smart grid and electric vehicles. IEEE Ind. Electron. Mag. 7 (2), 4–16 (2013)
3. Ul-Haqq, A., Brucella, C., Cecity, C., Khalid, H.A.: Smart charging infrastructure for electric vehicles. In: International Conference on Clean Electrical Power (ICCEP), Alghero, pp. 163–169 (2013). https://doi.org/10.1109/iccep.2013.6586984
4. Cui, Q., Weng, Y., Tan, C.: Electric vehicle charging station placement method for urban areas. IEEE Trans. Smart Grid. https://doi.org/10.1109/tsg.2019.2907262
5. Mukherjee, J.C., Gupta, A.: a review of charge scheduling of electric vehicles in smart grid. IEEE Syst. J. 9(4), 1541–1553 (2015). https://doi.org/10.1109/jsyst.2014.2356559
6. Huang, X., Xu, C., Wang, P., Liu, H.: LNSC: a security model for electric vehicle and charging pile management based on blockchain ecosystem. IEEE Access 6, 13565–13574 (2018). https://doi.org/10.1109/access.2018.2812176
7. Djalili, M., Nisar, H., Salih, Y., Malik, A.S.: An algorithm for vehicle detection and tracking. In: 2010 International Conference on Intelligent and Advanced Systems, Manila, pp. 1–5 (2010). https://doi.org/10.1109/ICIAS.2010.5716189
8. Manhua, W., Bing, X.: A real-time android-based monitoring system for the power Lithium-Ion battery used on EVs. In: 10th International Conference on Intelligent Computation Technology and Automation (ICICTA), Changsha, pp. 245–249 (2017). https://doi.org/10.1109/ICICTA.2017.62
9. Sallabi, F., Shuaib, K., Alahmad, M.: Online scheduling scheme for smart electric vehicle charging infrastructure. In: 2017 13th International Wireless Communications and Mobile Computing Conference (IWCMC) (2017)
10. Mustafa, M.A., Zhang, N., Kalogridis, G., Fan, Z.: Smart electric vehicle charging: security analysis. In: 2013 IEEE PES Innovative Smart Grid Technologies Conference (2013)
11. Mathew, S.A., Prakash, R., John, P.C.: A smart wireless battery monitoring system for electric vehicles. Department of Electronics, Cochin University of Science and Technology

An Effective Congestion Control System for Vehicular Adhoc Networks Using Multidimensional Data Clustering

Mohammad Pasha[1]($\boxtimes$), Huda Banu Baig[1], Mohd Umarfarooq[1],
and Khaleel Ur Rahman Khan[2]

[1] Muffakham Jah College of Engineering and Technology, Hyderabad, India
muhammed.pasha@gmail.com, huda.mirza96@gmail.com,
umarfarooq.mohd@gmail.com
[2] ACE Engineering College, Hyderabad, India
khaleelrkhan@gmail.com

Abstract. VANETs are proved to be the adequate solution to High traffic problems. Controlling congestion will effectively reduce the delay, data loss and enhance the stability of ITS. This proposal introduces an intelligent congestion control strategy which deals with the three phases to detects and control congestion. First phase detects congestion by measuring the Channel Busy Time. In the second phase the information is collected, filtered and clustered using K-means machine learning algorithm which classifies information based on type and deadline. The third phase is to set right priorities for each cluster and send notifications to the High speed vehicle group with greater priority. The Strategy focuses on controlling congestion by sending right cluster of messages to appropriate vehicles. Results from runtime highlights that the projected strategy master existing in terms of delay, throughput and also improves the efficiency of the system.

Keywords: VANETS · Intelligent transportation system · Quality of service · Traffic congestion · Machine learning · Multidimensional K means algorithm

1 Introduction

Vehicular adhoc networks (VANETs) offer automobile interactions with affordable and reliable sharing of data. The vehicle interactions are utilized to reduce congestion and accident situations, decrease traveling time, consumption of fuel and provide information about the state of roads. VANET [11] consists of an intelligent interaction mechanism comprising of Vehicle to Vehicle and Vehicle to Infrastructure Communication. They comprise of Roadside Units (RSU) and On Boarded Units, located in on the roads and on vehicles respectively.

VANET's research [15] has attracted many researchers due to heavy vehicular traffic detection. Vehicular research system should not only develop traffic monitoring application but also other applications to controls congestion and accidents by spreading information, suggesting alternate driving route etc. In VANets, Technologies of Information and Communication are integrated with transport network, vehicles and

© Springer Nature Switzerland AG 2020
P. Karrupusamy et al. (Eds.): ICSCN 2019, LNDECT 39, pp. 224–230, 2020.
https://doi.org/10.1007/978-3-030-34515-0_23

users, to improve safety and reliability of the system. Distributed traffic information scheme is the concept where vehicles exchange data about congestion or hazardous situations. In VANets Congestion happens when channel overloading occurs in dense network conditions which results in performance reduction of VANets and also results in the increase in packet loss delay [7]. In our paper, Congestion Control strategy based on Machine Learning [1] is presented which uses RSUs to control congestion. In this system, the data is filtered and clustered using algorithm based on machine learning at every RSU freely. Then right priorities for all clusters are set on the base deferral and criticalness for exchanging the messages between various classes. Next, the notifications are forward to various vehicles to prevent accident or congestion situations. Controlling network congestion accordingly has a positive impact in the performance of VANets.

2 Related Work

The measurement-based strategies are used to detect congestion by calculating parameters i.e., count of messages holding up in lines [5], Busy Time and medium usage level. Predefined threshold is used to detect congestion by comparing it with the measured parameters. In this quality of the detection depends on initial values of threshold.

   These strategies are divided into various types including Rate, Carrier-sense multiple access with collision avoidance, Priority-based, Power and Hybrid [4, 14]. Rate-based strategies control congestion by decreasing transmission rate. It's not an effective system in VANets due to high rate and topology changes. It may burdens the channel and increase crashes in network. In power associated approach, broadcast range is tuned to control the loads in medium. Santi [4] proposed Distributed Fair Transmit Power Algorithm for VANets (D-FPAV), the method tunes the transmission go for beacon messages based on vehicular traffic. Using the strategy each vehicle requires status of the entire neighboring vehicle based on this it adjusts conduction range for the Beacon messages so that the event driven messages have higher priority than the beacon. In default Congestion scheme [12], jamming is handled by using exponential back off Algorithm. The strategy adjusts the access ability of channel by tuning the parameters such as AIFS and Contention window size. The mechanism is not capable for high carrying data rate in large volume of vehicles and lead to dropping the packets before transmission. The Scheduling-based procedures [3], portray priorities for all received messages and calendars those in Control and Service channels to such an extent that urgent messages is given top priorities and are delivered on time [2]. Jung [2] presents Context Awareness Beacon Scheduling approach which schedules beacon dynamically. This approach packs the information in beacon and assigns a unique time to each vehicle. CABS improve the delay and throughput for beacon messages. The threshold-based solutions [8] also face some challenges because its performance is inclined by the initial value of the thresholds. The problem in the existing system is that it is a localized strategy which only controls congestion when the vehicles stops on the red light signal. The proposed methodology solves such problems using RSUs which detects and control congestion. It sets the right priorities [10] for every cluster on basis of message urgency with respect to its deadline. It independently detects, prevents and controls the jamming by prioritizing the messages and also prevents traffic congestion or accident situations by grouping the vehicles based on their speeds and sending the messages to the vehicles with the higher speed. This methodology comprises of three phases: (1) Congestion Detection Phase, (2) Data Control Phase, and (3) Congestion Control Phase.

3 Proposed Methodology

A **Congestion Detection Phase**

Measurement based strategy will be utilized to identify the occurred collision in the network. In this phase Channel is sensed and Channel Busy Time value is read and compared with the predefined thresholds for detecting the collision occurrence [5]. In our proposal, we consider the messages waiting in the queue to be delivered. If the Busy Time value surpasses the threshold, phase assumes that the data overcrowding occurred.

B **Data Control Phase**

In data control phase, for controlling the data flow a three-crease method is followed and they are gathering, filtering of data and clustering. In data Gathering messages passed between vehicles are collected. In filtering segment, the repeated messages gotten by RSU from various vehicles are erased to diminish the additional processing. Each message received consists of the message content, Message ID and Senders Address. In the case of rebroadcasting of the messages by different vehicles if the RSU receives the same message then it will mark that as duplicate. Data in clustering component, data is grouped on basis of multiple dimensions utilizing K means algorithm which is a solid method for arranging and grouping the substantial information sets because of various favorable circumstances such as short operating time, automatically detecting pattern, predicting future records and effectively handles expansive information.

Data that is collected from the vehicles are classified based on message type and deadline and the vehicles are clustered based on their speeds. The messages consist of emergency and non emergency messages along with the deadline. For every data set, K-means algorithm [9] follows various steps: (1) Initializing initial centroids (2) Calculating squared Euclidean separation of each record to the centers of clusters; (3) Computing latest centers by calculating mean value to find closest centroids. Repeat 2nd and 3rd steps until the group individuals never again change. K-means calculation has three sources of inputs including features, number of groups, and introductory centroids. The features consist of message, validity of messages, type of message and speed of vehicles. The intent of clustering is to reduce the difference between the cluster centroids, c and the message X_i. This can be calculated as $d(f_j, c_k)$ i.e., distance of the message x_j to c_k.

$$d(f_1, c_k) = \sqrt{(f_{j,1} - C_{k,1})^2 + (f_{j,2} - C_{k,2})^2.. + + (f_{j,z} - C_{k,z})^2} \qquad (1)$$

Where f_j is the feature that represents the message x_i and **centroid** of each cluster.

C *Congestion Control Phase*

This phase controls network congestion by setting the right priorities for the clusters. Post the cluster formation they are analyzed and the fitting parameters are set for every bunch so that crises messages which are closer to its deadline are transmitted to the prioritized high speed vehicle cluster with high transmission range and rate. This ensures that the emergency messages are communicated with less delay and the right set of messages is communicated to the right set of vehicles with lower delay and packet loss ratio. This controls network congestion and prevents any hazardous situations like accidents in future just by notifying the right set of vehicles.

4 Flow of the Proposed System

Figure 1 depicts the pictorial representation of Proposed Strategy that represents the overall flow of the system which starts with the gathering of data by RSU from vehicles, performing of filtering and Clustering and rebroadcasting of the data by assignment of right priorities to the clusters in order to control network congestion and accidents.

5 Parameters and Metrics

For the performance estimation of the system, combination of network simulators has been employed. To imitate vehicle mobility we have used Simulation of Urban Mobility. For the implementation of VANets OMNET++ has been used. An urban scenario is projected for evaluating the proposed Strategy and the pattern of highway road was taken into consideration. The communication protocol be used is IEEE 802.11p and MAC layer transmission strategy is CSMA/CA [13]. The parameters used in proposed methodology have been tabulated in Table 1.

6 Results and Discussions

In this section, various simulations are carried to analyze the system performance on basis of various metrics like Average of Delay, Proportion of total packet delivered and Average Throughput. A lot of recreations was conveyed to check the impact of number of vehicles and different run times on the presented execution metrics and then compared with the existing systems [6] for performance analysis.

Figure 2 compares the average delay obtained from the existing and proposed congestion control systems. Graph depicts that with the increase in the vehicle number average delay of the proposed strategy is far less than the existing.

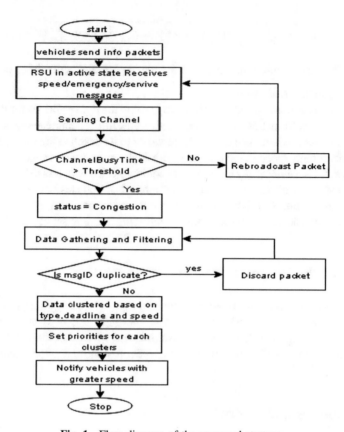

Fig. 1. Flow diagram of the proposed strategy

Table 1. Simulation parameters opted in highway road system

Specifications	Range
Size of the road	3 km.
Number of lanes	4 lanes
Speed of the vehicles	0–25 m/sec
Transmission power	3–27 Mbps
Transmission range	300 mt.
Size of each message	1024 Bytes
MAC type	IEEE 802.11p

Figure 3 presents overall throughput obtained from the proposal is more than the existing strategy [6]. The results highlights that the average throughput increases from 1.98 to 11 Mbps. Figure 3 demonstrates that, by expanding the vehicle density, the number of total packet losses is significantly less than the existing system as the data is transmitted in controlled condition [6]. The results clearly depicts that the proposed strategy is relatively better than the existing [6].

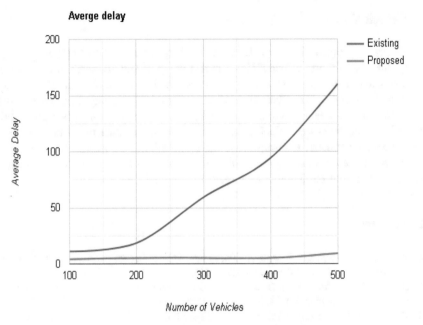

Fig. 2. Change in average delay v/s vehicle density

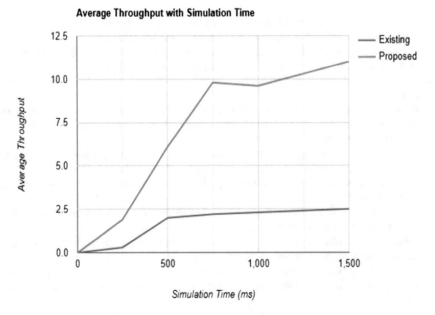

Fig. 3. Average throughput v/s simulation time (ms)

7 Conclusion

In this paper, proposal is to use an effective congestion control strategy to improve safety and reliability in VANets. It is a smart strategy. In our approach, measurement of channel Busy Time helps in detecting the congestion when the values surpass the predefined threshold. The data flow is controlled by filtering of messages and is clustered using K-means machine learning algorithm. Finally, it sets the right priorities for each cluster. RSU transmits the messages closer to deadline to the prioritized high speed vehicular cluster. Consequently, the overloading in channels medium is controlled and the priority message sets are conveyed in less time and loss in packets. Our proposed algorithm comparatively outperforms the existing.

References

1. Vijayakumar. V., Suresh Joseph, K.: Fuzzy based clustering for effective congestion control in vanet, vol. 119, no. 14, pp. 117–128 (2018)
2. Bai, S., Oh, J., Jung, J.-I.: Context awareness beacon scheduling scheme for congestion control in vehicle to vehicle safety communication, vol. 3 (2016)
3. Taherkhani, N., Pierre, S.: Prioritizing and scheduling messages for congestion control in vehicular ad hoc networks (2016)
4. Santi, P., Torrent-Moreno, M., Hartenstein, H.: Distributed fair transmit power adjustment for vehicular ad hoc networks. In: Ad Hoc Networks, vol. 515, pp. 32–125 (2006)
5. Taherkhani, N.: Improving dynamic and distributed congestion control in vehicular ad hoc networks (2016)
6. Taherkhani, N., Pierre, S.: Centralized and localized data congestion control strategy for vehicular adhoc network using machine learning algorithm (2017)
7. Gupta, N., Prakash, A., Tripathi, R.: Adaptive beaconing in mobility aware clustering based MAC protocol for safety message dissemination in VANET (2017)
8. Taherkhani, N.: Thesis on Congestion Control in Vehicular Adhoc Network, November 2015
9. Hmaerly, G., Elkan, C.: Alternatives to K means algorithm that find better clusterings (2002)
10. Suthaputchakun, C.: Priority-based inter-vehicle communication for highway safety messaging using IEEE 802.11 e (2009)
11. Karagiannis, G., Altintas, O., Ekici, E., Heijenk, G.: Vehicular networking: a survey and tutorial on requirements, architectures, challenges and solutions (2011)
12. Bouassida, M.S., Shawky, M.: On the congestion control within VANET. In: WD 2008 1st IFIP Wireless Days 2008, pp. 1–5 (2008)
13. Hsu, C.-W., Hsu, C.-H., Tseng, H.-R.: MAC channel congestion control mechanism in IEEE 802.11 p/WAVE vehicle networks (2011)
14. Sattari, M.R.J., Noor, R.M., Ghahremani, S.: Dynamic congestion control algorithm for vehicular ad hoc networks, vol. 7, pp. 95–108 (2013)
15. Golestan, K., Jundi, A., Nassar, L., Sattar, F., Karray, F., Kamel, M., et al.: Vehicular ad-hoc networks (VANETs): capabilities, challenges in information gathering and data fusion (2012)

Android Based e-Voting Mobile App Using Google Firebase as BaaS

Urmil Bharti, Deepali Bajaj$^{(\boxtimes)}$, Tulika, Payal Budhiraja,
Meghna Juyal, and Sushmita Baral

Shaheed Rajguru College of Applied Sciences for Women,
University of Delhi, New Delhi, India
ubharti@hotmail.com, deepali.bajaj@rajguru.du.ac.in,
payalbudhiraja09@gmail.com, ktulika8@gmail.com

Abstract. Making choices and choosing from options are always a part of life
and everyone wants options to choose from. Similarly when it comes to voting
and elections, it gives power of making choices to the people. Voting is a
democratic way of making decisions. Counting Ballots takes a long time that
causes delayed results. Furthermore calculating results could be biased and time
consuming which causes voters to wait for the results. In today's scenario as
everything is pacing up and new ideas and inventions are always appreciated,
our mobile app "Adhikaar" is also one of them. This Android voting app is more
efficient and convenient to conduct and manage elections as compared to traditional manual methods. The app has a simple and interactive GUI for voting
system and maintains its database using Google Firebase platform. Firebase is a
Backend-as-a-Service—BaaS that lets users build more powerful, secure and
scalable apps. Adhikaar enables user to cast their vote from anywhere, anytime
without waiting in long queues. This app is purely eco-friendly since no paper is
required. Election results are calculated automatically and declared instantly
thus reducing human effort and chances of human errors.

Keywords: Google Firebase · BaaS · Android studio · Mobile app · Online
e-Voting · JSON

1 Introduction

Voting is one of the most crucial ways that allow individuals to participate in gov-
ernmental decision-making. It is a method that allows electorates to actively participate
in decision making or express their opinion. In a democracy, a government is chosen
by voting in elections. By voting in elections people have the right to choose their
representatives, be it a local official or the prime minister of a country. Paper-based and
machine voting techniques are the traditional voting methods. Electronic Voting
Machines (EVMs) are being used in Indian General and State Elections from 1999 to
till date. EVMs have replaced paper ballots in few election categories, but there are still
a number of scenarios in our everyday life where traditional voting practices like paper
ballots are still in use. Voting in these typical scenarios can be conducted by online
mobile app based voting system. For instance, in many colleges, student council

P. Karrupusamy et al. (Eds.): ICSCN 2019, LNDECT 39, pp. 231–241, 2020.
https://doi.org/10.1007/978-3-030-34515-0_24

elections, departmental council elections, staff association elections, hostel elections are done with the help of paper ballots. The process of electing office bearers in RWA elections, club elections or professional body elections are still being carried out with the help of traditional paper ballot procedure.

The method of using paper ballots is often more cumbersome, time-consuming and prone to human biases. There are a number of factors like going to polling stations, long waiting queues, bad weather conditions, traffic etc. make traditional elections difficult for voters as well as for the election officials. These factors often results in decrease of voters' participation in the election process.

Online e-Voting is a critical step for the evolution of democracy [10]. It is an ideal means for elections of associations, councils, clubs, trade unions, educational institutions and other organisations etc. [11].

Our goal is to develop a mobile-based app that allows users to vote online via mobile phones. In this paper, a mobile app titled Adhikaar is developed with an aim to improve the voting procedure and make it more efficient. This app provides a novel method of casting votes and managing elections thus helps in reducing time, cost and human effort.

Adhikaar is developed for college students to cast their votes anytime and from anywhere using android devices. The aim of this app is to provide convenience to voters as well as election officers who monitor the voting process. The appgives election results accurately and instantaneously.

The app has an easy to use interface "Login Screen" for voters where they can login with their Google account ids. Each voter's data is stored in database containing their essential information such as Google account id, name, college roll number and a voting status which stores whether voter has voted or not.

After login, voters have access to voting interface "Voting Screen" that allows them to select candidates for the given positions and submit their voting data securely. Once a voter submits his/her response, then re-submission will not be permissible. Thus, Adhikaar ensures that there is no bogus or fake voting in the elections.

The complete voting procedure is under System Admin' s control. Admin has to sign-in with the predefined Gmail id. After successful sign in, admin gets access to "Admin Screen" where options of enabling/disabling of voting lines and viewing results are available.

Adikaar uses Google Firebase as Backend as a Service real time database that allow us to build mobile apps without server side programming [5]. It is No SQL document store database. It is cloud-based and schema free. Firebase supports OAuth2 as a built in email/password authentication system [6]. Firebase itself takes care of storing data, users verification and access rules implementation. Data is stored as JSON and synchronized in real-time to every connected client.

2   Application Features

Adhikaar is a real time e-voting mobile app. It has been developed using Android Studios version 3.1.1 (minimum android version supported is Jellybean) [1] and Firebase platform. Android Studio is Android's official IDE and its purpose is to help developers to build high quality apps for every Android device [2]. Firebase is a next-generation app-development platform on Google Cloud Platform [7]. This platform allows programmers to develop web and mobile applications without using any server side programming language [3]. Firebase allows storing user's data on its real-time database and also syncing data among other users in no time [13–15].

Some distinguished features of the app are discussed below:

1. Remote Voting: One of the biggest drawback of secret ballot-based voting system is that a voter has to be physically present to cast vote. This might lead to low voter turnout for any election. With Adhikaar app a voter can directly cast his/her vote from any location by using an Android mobile. So there is no need for any voter to visit the polling booth to cast their vote.
2. Real Time Results: With Adhikaar app, final results of any election are available just after the expiry of allotted voting time slot configured by the system administrator of the app. This app provides the real-time results of an election in tabular form as well as in easy to understand graphical display. As there is no need of physical counting of votes, so the complete voting process executes without any manual intervention, hence making it a very efficient procedure.
3. Private Poll: System administrator of the app can upload required data of all valid voters in the system. Only valid users will be allowed to vote online using app. If a user tries to sign in then credentials will be validated from Firebase and in case of failure that person won't get access to app. So this app is highly secure and validation checks are in place to restrict invalid users to vote. Only valid users will be permitted to vote online.
4. Security: For any online app, security of users' data is always an important design consideration for developers. To prevent security vulnerabilities in the app, Google Sign-in feature has been used for user verification. All valid users need to sign-in using their Gmail id to get voting access in app.
5. One User one Vote: Adhikaar has been designed to allow a user to cast vote only once. Once an election date and time has been scheduled by system administrator in app then all valid voters are allowed to vote online as per the election schedule. If a voter has already exercised his voting right using Adhikaar app then he/she will not be allowed to vote again as the voting interface will be disabled. This feature will help in mitigating risk of fake voting in an easy and efficient manner.

3 Methodology

1. Conducted an exhaustive survey to understand the current scheme of election in various colleges, schools and other institutions.
2. Identified various limitations and short comings in conventional voting systems used in elections and prepared a requirement specification document for a more advanced, efficient and convenient alternate solution.

3. Analyzed various open source databases available at present. Performed a strategic comparison with the traditional databases and finally Google Firebase database was selected to store app's data. Firebase is an efficient platform to build mobile backend services [4].

4. Collected required students data for app in a spreadsheet. Only college roll number and Gmail id of a student required to be stored for voting procedure. Also populated a spreadsheet having data for all candidates who have filled their nomination against various posts to be elected.

5. Our next task is to import voters and nominees' data from spreadsheet into Google Firebase In Firebase, data can be imported from external files or inserted manually in a structured format. Firebase supports only JSON format files for import. As the students' data was voluminous and was available in spreadsheet, we converted it to a JSON file through appropriate tool and then imported. Spreadsheet with data entries are first converted to CSV format and then this CSV data is converted to JSON using [12], appropriate column is selected as parent key node and rest columns as child. Now the converted JSON file is imported as shown in Fig. 1.

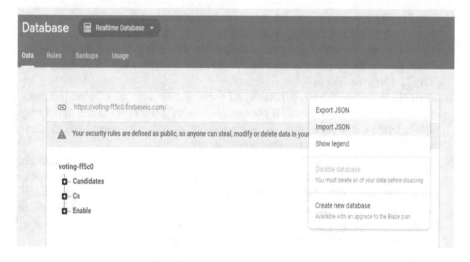

Fig. 1. Importing data in Firebase

Data in Firebase is stored as nested tree structure. The "Data" tab in Firebase dashboard interface gives a visual representation of JSON database tree as shown in Fig. 2.

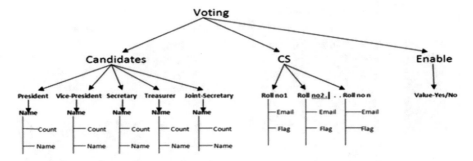

Fig. 2. Nested structure of data in firebase

6. Android Studio provides the fastest tools for building high quality and performance apps that can run on any type of Android device like phones and tablets etc. So it has been chosen as a platform for development of Adhikaar mobile app [7].
7. Prepared high level design document and was finalized after two rounds of technical review.
8. All requirements specified in SRS document and its associated algorithms were implemented using JAVA and XML.
9. Conducted unit tests for all independent modules of the app which performs different functions like displaying dynamic radio buttons, checking for individual vote count, correct updates in real-time database etc.
10. Multiple rounds of integration and system testing were conducted.
11. The app can be installed through shareable APK's by the target users who were supposed to participate in voting procedure.
12. Bugs were detected and fixed during the alpha and beta testing in real time environment.
13. Results were also recorded manually and verified with the app's result.
14. User's reviews were taken and statistical analysis was performed on the effectiveness of app.

Figures 3, 4 and 5 shows the screen flow diagram of Adhikaar app. User verification flow is shown in Fig. 6.

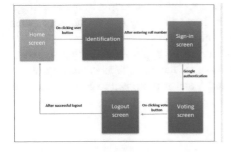

Fig. 3. Screen flow diagram for sign-in

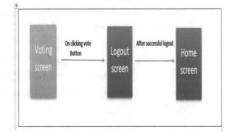

Fig. 4. Screen flow diagram for voting

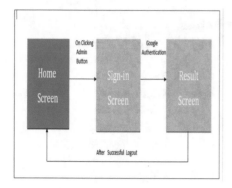
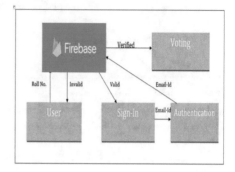

Fig. 5. Screen flow diagram for admin   **Fig. 6.** Screen flow diagram user verification

4   Implementation

The app can be installed in the Android enabled smart phones of voters by sharing the Android Application Package (APK) file [8]. The application provides two separate interfaces for (1) system admin and (2) voters.

1. **Interface for system admin–**

   1.1 **Authentication**-For authentication of system admin, Gmail id is used. After successful sign-in, admin can choose to enable or disable the election voting or can view the result on the result screen.

   1.2 **Result**-Results are displayed in the sorted order which is calculated automatically by the app. Thus, it is less time-consuming, more cost effective and less prone to human errors.

   1.3 **Control duration of poll**-Admin can restrict the duration of voting by enabling or disabling the voting lines. Voters can vote only when the voting lines are enabled on the polling day.

   1.4 **Sign out**-A sign out button is provided for signing out from the application.

2. **Interface for Voters/Electorate**

   2.1 **Authentication–**Voter authentication is done with the help of unique college roll number. Voters have to use their Gmail id to sign in to the app. After successful sign in, voter is directed to voting screen.

   2.2 **Cast Vote–**This is the main functionality of the application. Voting screen is presented to voters consisting of names of all the candidates contesting in the elections along with the name of the post. Voter is prompted to cast his/her vote and the final response is submitted.

2.3 **Sign out**–After successful voting, voter can sign out from the app.

Figure 7 shows few selected interface snapshots of app.

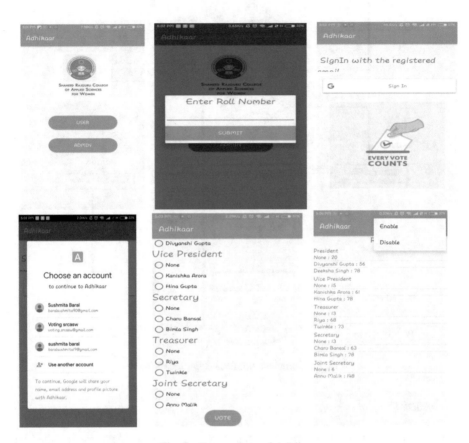

Fig. 7. Screen shots of Adhikaar

Figure 8 Shows the complete process flow of the application. It depicts different paths that can be taken by user either in system admin role or voters role.

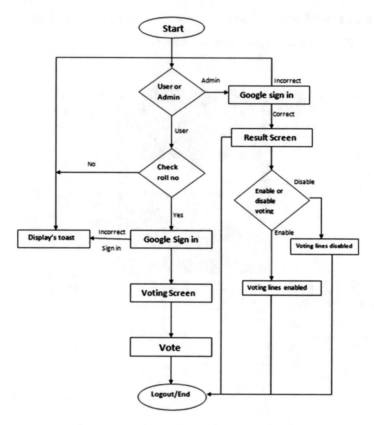

Fig. 8. Complete Adhikaar app process flow

5 Results

Adhikaar was used in the departmental elections of our college. App was installed on Android mobile phones of target students [9]. In order to evaluate the effectiveness of the app, some students cast their votes from the college campus and some from outside or their home. Figure 9 shows the results in the form of pie chart-

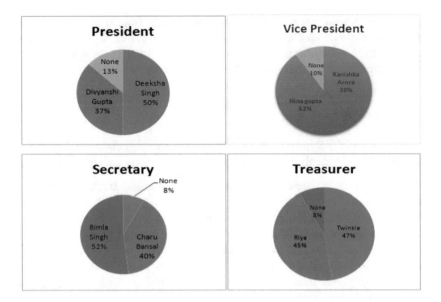

Fig. 9. Graphical representation of the final results

6 Innovations Shown by Project

App is designed to reduce overhead of its user and admin and ensures that each and every student of the college gets a chance to vote. This initiative is in the direction of smart and eco-friendly elections. The innovative features of the application are as follows:

1. The application has a unique feature of restricting the time duration of voting. Admin can enable/disable voting lines in accordance to the schedule of voting. User can vote only when voting lines are enabled.
2. The results are calculated and announced instantaneously. It has overcome the overhead of calculating result manually which saves a lot of time and human effort.
3. User identification is done on the basis student's roll number and Gmail id registered in Firebase so that only a valid student can vote. Each student can vote only once.
4. Students can cast their vote from anywhere and physical presence at college is not a constraint.
5. Interface of app is highly customized keeping in mind the need and ease of its user.

7 Conclusion

Physical presence is the biggest challenge of traditional booth based voting system as that leads to stumpy voter turnout. Everyone should be allowed to vote despite of geographical distances, work commitments, critical health concerns or adverse weather conditions etc. Adhikaar allows voters to vote directly from their Android enabled smart phones without sacrificing their home comforts within few seconds. Thus it ensures increased voting turnout %, improved over all security, reduces cost of conducting elections, and reduces time and manpower needed to conduct elections.

8   Future Scope

This project is an initiative towards a new way of conducting elections. The following features may be integrated in future for extending the functionalities of the app-

1. The app can be personalized as per user requirements (Post, candidates, user identification).
2. For securing the database application, we can use algorithms for data encryption/decryption while storing or retrieving data from database.
3. We may restrict voters from participating in the voting process with the help of geofencing. Geo-Fencing is a functionality that allows admin to restrict voters from certain geographical areas only to be able to vote. People outside that geographical area would not be able to expertise their franchise.
4. To incorporate better result visibility, admin may authorise others also to view the results. He can make it visible to all including the voters or he can decide that results would be visible to him only.

References

1. Rogers, R., Lombardo, J., Mednieks, Z., Meike, B.: Android application development: programming with the Google SDK. O'Reilly Media, Inc. (2009)
2. Bajaj, D., Yadav, A., Jain, B., Sharma, D., Tewari, D., Saxena, D., Ray, P.: Android based nutritional intake tracking application for handheld systems. In: 2017 8th International Conference on Computing, Communication and Networking Technologies (ICCCNT), pp. 1–7. IEEE, July 2017
3. Singh, N.: Study of Google firebase API for android. Int. J. Innov. Res. Comput. Commun. Eng. 4(9), 16738–16743 (2016)
4. https://firebase.google.com/
5. Plangi, S.: Overview of backend as a service platforms
6. Lane, K.: Overview of the backend as a service (BaaS) space. API Evangelist (2015)
7. Gropengießer, F., Sattler, K.U.: Database backend as a service: automatic generation, deployment, and management of database backends for mobile applications. Datenbank-Spektrum 14(2), 85–95 (2014)
8. Zapata, B.C.: Android studio application development. Packt Publishing Ltd. (2013)
9. Studio, A.: The official IDE for android. Android Studio (2016). https://developer.android.com/studio/index.html
10. Felsenthal, D.S., Machover, M.: The measurement of voting power. Books (1998)
11. Stradiotto, C.R., Zotti, Â.I., Bueno, C.O., Bedin, S.P., Hoeschl, H.C., Bueno, T.C., Mirapalheta, V.O.: Web 2.0 e-voting system using android platform. In: 2010 IEEE International Conference on Progress in Informatics and Computing (PIC), vol. 2, pp. 1138–1142. IEEE, December 2010

12. www.convertcsv.com
13. Stonehem, B.: Google Android Firebase: Learning the Basics vol. 1. First Rank Publishing (2016)
14. Rahmi, A., Piarsa, I.N., Buana, P.W.: FinDoctor-interactive android clinic geographical information system using firebase and Google maps API. Int. J. New Technol. Res. **3**(7) (2017)
15. Costa, I., Araujo, J., Dantas, J., Campos, E., Silva, F.A., Maciel, P.: Availability evaluation and sensitivity analysis of a mobile backend-as-a-service platform. Qual. Reliab. Eng. Int. **32** (7), 2191–2205 (2016)

Directed Acyclic Graph Inherited Attacks and Mitigation Methods in RPL: A Review

P. S. Nandhini[1,2(✉)] and B. M. Mehtre[2]

[1] Department of CSE, Kongu Engineering College, Perundurai, India
nandhinisvl@gmail.com
[2] Centre of Excellence in Cyber Security (COECS), IDRBT,
Established by RBI, Hyderabad, India
bmmehtre@idrbt.ac.in

Abstract. RPL (Routing Protocol for Low Power and Lossy Network) is designed for Low Power and Lossy Network (LLN). In RPL, both the nodes and links have resource constraints. There will be many RPL instances in the network. The operation of RPL requires bidirectional links that exhibits asymmetric properties. LLNs do not have predefined network topology. Directed Acyclic Graph (DAG) is organised as a topology by RPL. It uses control packets for associating the data packets with a RPL instance and for validating the routing states. The manipulation done in the control packets and constrained resource of RPL leads to various attacks such as rank attack, DIS attack, Version attack, etc., In this paper, we have discussed the operation of RPL, classification of attacks and various mitigation methods. We have also concentrated on DAG inherited attacks because it degrades the overall performance of the network.

Keywords: RPL · Bidirectional · Asymmetric · DAG · LLN · Instance · Rank attack · Version attack · Control packets

1 Introduction

Internet of Things (IoT) is a collection of heterogeneous devices or things that are uniquely identified and connected to the Internet. The smart devices can be a wide variety of devices such as smart phones, RFID tags, actuators and sensors. The origin of IoT has led to the connection of devices, people, services and many objects [1]. It plays a major role in security sensitive areas also. The IoT has transformed traditional things or objects to smart device with the help of enabling technologies such as embedded devices and softwares, communication technologies, wireless sensor network, Internet protocols and applications [2].

The devices are heterogeneous in IoT network and so they use various standards. There are various protocols for IoT. IPv6 over Low Power Wireless Personal Area Network (6LoWPAN) is one of the protocols. It is a simple low-cost communication network. It allows connectivity in applications where throughput is relaxed and the resource is constrained [3]. LoWPAN include devices that co-ordinate together for connecting the physical devices to real time applications. The architecture of 6LoWPAN is shown in Fig. 1.

© Springer Nature Switzerland AG 2020
P. Karrupusamy et al. (Eds.): ICSCN 2019, LNDECT 39, pp. 242–252, 2020.
https://doi.org/10.1007/978-3-030-34515-0_25

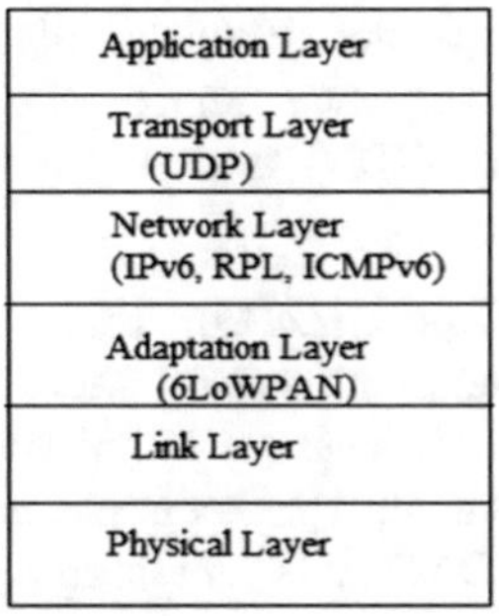

Application Layer
Transport Layer (UDP)
Network Layer (IPv6, RPL, ICMPv6)
Adaptation Layer (6LoWPAN)
Link Layer
Physical Layer

Fig. 1. Architecture of 6LoWPAN

The inclusion of a new layer called Adaptation Layer in between the Link layer and the Network Layer is the major difference regarding the architecture [4]. The functions of the adaptation layer [5] are compression and decompression of UDP and IPv6 header, fragmentation and reassembly of packets and routing of packets.

The paper is organized as follows: Sect. 2 overviews the operation of RPL protocol and the control packets used for the construction of DAG. In Sect. 3, the DAG inherited attacks and the mitigation methods are discussed. In Sect. 4, we have provided the concerns, challenges and opportunities related to RPL. The conclusion is in Sect. 5

2 RPL Overview

RPL is designed for resource constrained networks. It is a Distance Vector Routing protocol. The LLNs like Radio Networks don't have predefined topologies i.e., wires between nodes. So, RPL protocol has to find links and select nodes efficiently which is done with the help of control packets [6]. It is a source routing [7] protocol.

2.1 RPL - Control Messages

If the value of the type field in ICMPv6 message is 155 then it signifies the RPL control message. There are five RPL control messages. The control message of RPL can be identified from the code field. The control message has the base field that depends on the code.

DODAG Information Solicitation (DIS): This is used for soliciting DIO control message from nodes in RPL. It is similar to the router solicitation which is used in the neighbor discovery in IPv6. A node explore its neighbor with the help of DIS message to identify the nearby DODAGs.

DODAG Information Object (DIO): It holds the information that are needed by the node to identify an instance of RPL, learn the configuration parameters, selection of the DODAG parents and the maintenance of the DODAG. This is used by the root of the DODAG for the construction of a new DAG. It is multi-casted in the DODAG. The base object fields of this message are RPLInstanceID, Version number, Rank, Destination Advertisement Trigger sequence number (DTSN), grounded flag, mode of operation, DODAG preference. DTSN is used for the maintenance of downward routes.

Destination Advertisement Object (DAO): This is used for the propagating the information about the destination in the upward direction. In the storing mode of RPL, for the selection of the parent DAO message will be unicasted by the child. In the non-storing mode of RPL, the DAO control packet is sent to the root of the DODAG. DAO message is acknowledged by (DAO-ACK) message to the sender.

Destination Advertisement Object Acknowledgement (DAO-ACK): This control packet is delivered as a response to an unicast DAO message. The response is also unicasted. The message of recipient could be a parent of DAO or root of the DODAG.

Consistency Check (CC): It is delivered as a secured message. It is used for checking and synchronising the message counters or timestamps between each pair of nodes. It is used for issuing challenge-response.

2.2   DODAG Construction and Maintenance

DODAG construction consists of two steps: (i) DIO control message is broadcasted from the root down to the client to construct routes in the downward direction. (ii) DAO message is unicasted for constructing routes in the upward direction. It is issued by the client to the DODAG root. The construction of DODAG is depicted in Fig. 2.

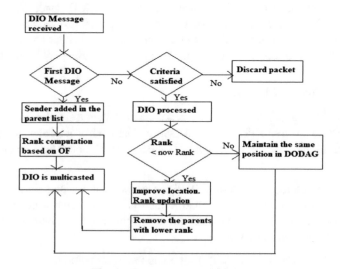

Fig. 2. Construction of DODAG

For the construction of the DODAG: (i) DIO message is broadcasted. It contains DODAGID, rank information [8], Object Function [9, 10]. (ii) DIO message can be received by a node which is willing or not willing to join the DODAG (It may also be a member of DODAG already). If a node wants to get added in the DODAG: (i) Adds the address of the sender of the DIO message to the parent list (ii) Calculate rank based on OF (iii) Forwards the updated rank in DIO.

The node then chooses a node as a parent from the parent list, through which the traffic passes. If a node is already a DODAG member and if it obtains another DIO message, it can be processed in different ways: (i) Based on the criteria of RPL, it can discard the DIO message (ii) Process the DIO message to retain its location in the DODAG where it is a member already (iii) If the computed rank is lower then it churns its location. If a rank of the node is changed, it must remove all the nodes in the parent list to avoid loops.

Grounded DODAG will offer connectivity to the nodes that are needed for obtaining the goal of the application. Floated DODAG will not satisfy the goal but it provides the routes to nodes in the DODAG (Example: During repair, it is used to maintain interconnectivity). The DODAG is maintained by the trickle timer. It is used for optimizing the frequency of message transmission depending on the condition of the network. The duration of the timer increases exponentially whenever the timer is fired.

2.3   Repairing DODAG

The repairing of the DODAG can be done in two ways: (i) Global Repair: The root of the DODAG performs a global repair by increasing DODAG version which leads to a new version of DODAG. The nodes can take a new position in the newly formed DODAG. (ii) Local Repair: This is performed within the DODAG version. The parameters necessary for configuration are specified in the DIO message. DODAG loops occur when a node gets detached from the DODAG and tries to reattach to a device in its sub-DODAG where its attached. If DIO message is missing, then loop is encountered. Mostly, this looping problem occurs during local repair.

3   Classifiaction of RPL Attacks

Due to resource constraints and manipulation of RPL control packets, RPL is prone to various attacks such as wormhole attack, blackhole attack, rank attack, version attack, etc. The attacks can be classified as address based attacks and DAG based attacks. The classification of attacks is depicted in Fig. 3. Since there are huge number of attacks, DAG inherited attacks and its countermeasures have been reviewed.

3.1   Rank Attack

In RPL, the rank of a node gets incremented from DODAG root to the child node. If the value is altered, the attacker (i) Attracts the child node for selecting the parent (ii) Improve the metrics (iii) Attract large amount of traffic to flow through the attacker.

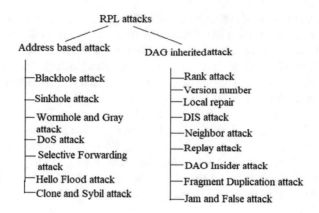

Fig. 3. Classification of RPL attacks

In Fig. 4, the black color node is an attacker. Its rank is 3 but it manipulates its rank to be 1 and tries to attract the neighbor nodes which are blue in color. According to [11], the child node will receive the information about parent through the control messages. If there is an attacker, it takes a bad quality route. There are different types of rank attack. It will degrade QoS parameters. If rank attackers are deployed in crowded area, the performance of the network is degraded.

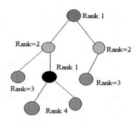

Fig. 4. Rank attack

According to authors [12], the attack aimed at rank property will have various effect on the RPL performance. The authors have depicted that the rank of node has increased to the maximum rank of its neighbors after running the simulation for particular time. The consequences of this is that (i) There are loops between root nodes and its child (ii) The network is not stable and many control packets are generated for optimizing the topology. According to [13], the rank attack is done to attract and manipulate the traffic in the network. The rank attack is done by compromising rank information. The node will not change itself.

Kamble et al. [14] proposed three types of rank attacks. They are (i) Increased Rank attack (ii) Decreased Rank attack (iii) Worst parent attack.

Mitigation Methods:

Airehrour et al. [15] proposed SecTrust-RPL. It is a secure trust-aware RPL for IoT. SecTrust -RPL is used for the detection of Sybil and Rank attack. It isolates the attacker from the network. The trust system is incorporated in the RPL protocol. It evaluates trust by examining the successful delivery of packets between the nodes. The computed trust value is used for making optimal routing decision, while isolating the attacker from the network. No re-integration of battery drained pre-trusted nodes after the recovery of battery power.

Shafique et al. [16] proposed SBIDS. It is Sink based Intrusion Detection System. It provides high detection of rank attacker. All the detection processes are carried out in the sink node and so less computational overhead. More routing metrics including energy, throughput, ETX, etc., can be included along with rank in the proposed algorithm for enhancing the detection rate of the rank attacker.

Semedo et al. [17] proposed Vulnerability Assessment of Objective Function of RPL Protocol for IoT. In this paper, the authors have induced the rank attack by altering the objective function (OF0 and MRHOF). The effect of energy depletion and delivery ratio of packet in the presence and absence of attacker is analysed. The rank attack removes a node completely from the network. The recent advancement techniques such as Machine Learning, Artificial intelligence can be deployed to detect the rank attack.

Stephen et al. [18] proposed Energy based Validation and Verification (E2V). It is a technique for detecting and identifying Rank Inconsistency Attack (RInA) in IoT. E2V detects the inconsistency in the rank based on the node's energy level. The calculation of rank, identification and elimination of malicious node are the mechanism integrated for providing a secured communication in IoT networks. The E2 V mechanism is included in the root node for the identifying the attacker. The energy of the root node gets depleted quickly.

Sahay et al. [19] proposed Attack Graph - based Vulnerability Assessment of Rank Property in RPL-6LoWPAN in IoT. The consequences of violation of the properties of rank also results in various attacks. More energy consumption and high traffic overhead are the consequences of Increased Rank attack. Large interruption in the network and DODAG formation is the consequence of decreased attack. Increased end-to-end delay and the network sub-optimization are the consequences of worst parent selection.

3.2 Version Number Attack

Version number is one of the fields in each DIO message. When there is a global repair, it is increased by the DODAG root. If the value is not updated, then the older value signifies that it can't be used as a parent. The attacker node can modify the version number, when DIO message is forwarded to its neighbors. This will lead to the rebuilding of DODAG unnecessarily. The consequence of this attack is that (i) Many loops are created (ii) Loss of data packets (iii) Overhead of control message (iv) Congestion in network.

Mitigation Methods

Mayzaud et al. [20] performed the version attack to identify the effects of it on the performance of the networks. It is identified that there is 18 times increase in overhead. The attack has doubled the delay and the delivery ratio is reduced by 30%. The attacker position is the main concern that determines the network performance.

Perrey et al. [21] proposed TRAIL (Topology Authentication for RPL). The proposed method relies on (i) DAG root is the trust node (ii) The nodes interconnect to the root in a hierarchy. The strong cryptographic mechanism is used for the detection of topology attacker. TRAIL is scalable and reliable.

Luchi et al. [22] proposed a scheme for selection of parent node in construction of route. The scheme will exclude the attacker node from the network. If there are multiple parents for a node then the node will choose a parent after denying the best candidate. The attacker will publish the lower rank than the good ones. From the simulation, it is found that the attackers had no effect in parent selection of node in the proposed scheme. It is achieved by reducing the total number of child nodes attached to the attacker.

3.3   Local Repair Attack

The attacker node will periodically send local repair message. When the node near to it, it hears this message, it starts its local repair message. On comparing with other attacks, local repair attack has more effect on the packet delivery ratio. It generates more control packet. The delay is also increased. The node energy will get exhausted due to this attack.

Mitigation Methods

TRAIL [21] discussed in the mitigation method of version number attack can be extended to local repair attack. The IDS is the best solution for mitigating local repair attack.

3.4   DIS Attack

DIS control packet is used by a node to obtain the topology information. During this attack, the attacker sends DIS message periodically. On receiving this message, the neighbor nodes will reset the DIO timer. This happens because the node assumes that there is some disruption in the topology. If the attacker node unicasts the DIS control packet, the destined neighbor node provides response to it by sending DIO. This implies that the receiver is ready to join the network. There is no impact on delivery ratio. It is a multicast attack and the delay is increased. This attack creates overhead due to control packets. The energy of the node gets depleted.

Mitigation Methods:

There is no specific mitigation method for this attack. Specification based IDS is a good solution for mitigating DIS attack.

3.5   Neighbor Attack

In this attack, the DIO control packet which is received by the attackers is broadcasted to its neighbors without any manipulation. If the sender is not in the transmission range, the attackers try to create an image that the sender is also in the transmission range. The victim node could select the parent that is not in its range. It is identical to the wormhole attack. This attack is created by forwarding the DIO control packets selectively. The consequences of this attack are: (i) QoS parameter is affected (ii) Packet delivery ratio has no effect (iii) End-to-end delay is slightly increased (iv) Slight topology disruption. If this is combined with other attacks, it is difficult to mitigate.

Mitigation Methods:
According to [3, 23], there is no specific method to mitigate this attack. It is hard to identify this kind of attack. Location based mitigation method can be used to detect this kind of attacks. Some IDS can also be extended to mitigate this kind of attack.

3.6   Replay Attack

This attack is created by replaying the routing information. The malicious node saves the routing information from other nodes and forwards it later. The receiver node will update their routing table. Thus, the table is updated using outdated routing information. The topology gets altered. This will lead to the disruption of topology and routing paths. The DIO VersionNumber or DAO message path sequence is used to confirm the freshness of the routing information.

Mitigation Methods:
According to [6], Winter suggested that integrating MIC (Message Integrity Code) is not sufficient to prevent this kind of attack.

Perazzo et al. [24] proposed that MAC layer encryption can be used to distinguish DIO message from other routing messages and data messages. This alone is not sufficient to prevent this kind of attack. Along with this, the replay protection mechanism can be used. Replay protection mechanism allows the good nodes to detect the false DIO message. So, replay protection technique can also be used for mitigation of this kind of attack. For dynamic RPL network, there is no specific mechanism for mitigation so far.

3.7   DAO Insider Attack

To build the routes in downward direction, DAO message is used. The RPL specification does not reveal when and how the DAO messages are transmitted. The attacker transmits the DAO message repeatedly to create overhead in the network. According to [25], DAO message is transmitted based on DIO trickle timer. The child unicasts the DAO message to its parent on three situations (i) On receiving a DIO message from its preferred parents (ii) On changing the parents (iii) On detecting specific errors.

Mitigation Method:

Ghaleb et al. [26] proposed SecRPL for addressing the DAO Insider attack. SecRPL limits the DAOs forwarded by a parent. The restrictions are applied in 2 ways: (1) Restricting the number of forwarded DAO messages regardless of the node that sent the DAO message intially. (2) Restricting the number of forwarded DAO control messages for each destination. The deployed malicious node triggers the DAO attack by sending DAO control message to its parent. The DAO attack varies from DIS attack because DAO packets are broadcasted from end to end. This type of attack is implemented without the need for compromising the security keys from legitimate nodes (Table 1).

Table 1. Review of the mitigation methods of DAG Inherited Attacks

Attacks	Effects on network performance	Mitigation methods	Review on the mitigation methods
Rank attack	Formation of loops, less packet delivery ratio, generation of unstable paths, packet delay. It affects the performance of the network	SecTrust-RPL, SBIDS, Vulnerability Assessment of Objective Function, E2 V, Graph – based Vulnerability Assessment	The attack aimed at rank property will create multiple impact on performance of RPL. In all the proposed mechanism the resource constraints of the nodes are considered
Version number attack	Control packet overhead is increased, less delivery of packets, more delay	Topology Authentication for RPL, Secure Scheme for parent selection	The control packet overhead is reduced and the delivery ratio is increased
Local Repair attack	Topology formation disruption due to overhead in control packets	TRAIL and IDS	It is a reliable and scalable mechanism
DIS attack	More resource consumption	Specification based IDS	Not implemented so far
Neighbor attacks	False route, route disruption, more resource consumption	Location based mitigation method	Not implemented so far
Replay attacks	Routing and topology disruption	Replay protection technique	No specific mechanism for dynamic RPL
DAO insider attack	Overhead in the network due to large number of DAO messages	SecRPL	It limits the number of DAO message sent by the parents

4 Issues, Concerns and Challenges

From the review, it is clear that vast number of researches have been conducted for preventing and detecting attacks in RPL. There are certain challenges that need to be addressed. They are

- The security features of RPL remains unexplored
- Many options are available in RPL that remains unused. This can be explored to provide security
- IDS is one of the mitigation methods for encountering the attacks. The resource constraint nature of the node should be considered while implementing IDS
- Cooja is the most widely used simulation for evaluation. Many tools and testbeds can be explored for evaluation

5 Conclusion

In this paper, DAG inherited attacks and their mitigation methods are reviewed. The performance of the network is degraded significantly, due to the manipulation of RPL control packets. The attacks, effects of the attacks on the RPL, mitigation methods and the significance of mitigation methods are tabulated. Thus, this review will provide a platform for the future researches who conduct research on RPL based attacks and mitigation methods.

References

1. Conti, M., et al.: Internet of Things security and forensics: challenges and opportunities, pp. 544–546 (2018)
2. Al-Fuqaha, A., et al.: Internet of Things: a survey on enabling technologies, protocols, and applications. IEEE Commun. Surv. Tutorials 17(4), 2347–2376 (2015)
3. Pongle, P., Chavan, A.: A survey: attacks on RPL and 6LoWPAN in IoT. In: 2015 International Conference on Pervasive Computing (ICPC). IEEE (2015)
4. Pai, V., Shenoy, U.K.K.: 6LowPan—performance analysis on low power networks. In: International Conference on Computer Networks and Communication Technologies. Springer, Singapore (2019)
5. Garg, R., Sharma, S.: A study on need of adaptation layer in 6LoWPAN protocol stack. Int. J. Wirel. Microwave Technol. (IJWMT) 7(3), 49–57 (2017)
6. Winter, T., et al.: RPL: IPv6 routing protocol for low-power and lossy networks. No. RFC 6550 (2012)
7. Clausen, T., Herberg, U., Philipp, M.: A critical evaluation of the IPv6 routing protocol for low power and lossy networks (RPL). In: 2011 IEEE 7th International Conference on Wireless and Mobile Computing, Networking and Communications (WiMob). IEEE (2011)
8. Vasseur, J.P.: Terms used in routing for low-power and lossy networks. No. RFC 7102 (2014)
9. Deshmukh-Bhosale, S., Sonavane, S.S.: A real-time intrusion detection system for wormhole attack in the RPL based Internet of Things. Proc. Manuf. 32, 840–847 (2019)

10. Verma, A., Ranga, V.: Evaluation of network intrusion detection systems for RPL based 6LoWPAN networks in IoT. Wirel. Pers. Commun. 1–24 (2019)

11. Le, A., et al.: The impact of rank attack on network topology of routing protocol for low-power and lossy networks. IEEE Sens. J. **13**(10) 3685–3692 (2013)

12. Xie, W., et al.: Routing loops in dag-based low power and lossy networks. In: 2010 24th IEEE International Conference on Advanced Information Networking and Applications. IEEE (2010)

13. Dvir, A., Buttyan, L.: VeRA-version number and rank authentication in rpl. In: 2011 IEEE Eighth International Conference on Mobile Ad-Hoc and Sensor Systems. IEEE (2011)

14. Kamble, A., Malemath, V.S., Patil, D.: Security attacks and secure routing protocols in RPL-based Internet of Things: Survey. In: 2017 International Conference on Emerging Trends & Innovation in ICT (ICEI). IEEE (2017)

15. Airehrour, D., Gutierrez, J.A., Ray, S.K.: SecTrust-RPL: a secure trust-aware RPL routing protocol for Internet of Things. Future Gener. Comput. Syst. **93** 860–876 (2019)

16. Shafique, U., et al.: Detection of rank attack in routing protocol for Low Power and Lossy Networks. Ann. Telecommun. **73**(7–8) 429–438 (2018)

17. Semedo, F., Moradpoor, N., Rafiq, M.: Vulnerability assessment of objective function of RPL protocol for Internet of Things. In: Proceedings of the 11th International Conference on Security of Information and Networks. ACM (2018)

18. Stephen, R., Arockiam, L.: E2 V: techniques for detecting and mitigating rank inconsistency attack (RInA) in RPL based Internet of Things. J. Phys.: Conf. Ser. **1142**(1) (2018). IOP Publishing

19. Sahay, R., Geethakumari, G., Modugu, K.: Attack graph—based vulnerability assessment of rank property in RPL-6LOWPAN in IoT. In: 2018 IEEE 4th World Forum on Internet of Things (WF-IoT). IEEE (2018)

20. Mayzaud, A., et al.: A study of RPL DODAG version attacks. In: IFIP International Conference on Autonomous Infrastructure, Management and Security. Springer, Heidelberg (2014)

21. Perrey, H., et al.: TRAIL: topology authentication in RPL. arXiv preprint arXiv:1312.0984 (2013)

22. Iuchi, K., et al.: Secure parent node selection scheme in route construction to exclude attacking nodes from RPL network. In: 2015 21st Asia-Pacific Conference on Communications (APCC). IEEE (2015)

23. Le, A., et al.: Specification-based IDS for securing RPL from topology attacks. In: 2011 IFIP Wireless Days (WD). IEEE (2011)

24. Perazzo, P., et al.: DIO suppression attack against routing in the Internet of Things. IEEE Commun. Lett. **21**(11), 2524–2527 (2017)

25. Dunkels, A., et al.: Contiki: the open source OS for the Internet of Things. 13 October 2012 (2015)

26. Ghaleb, B., et al.: Addressing the DAO Insider Attack in RPL's Internet of Things networks. IEEE Commun. Lett. **23**(1) 68–71 (2019)

Design and Implementation of Smart Book Reader for the Blind

Gayathri Rajendrababu$^{(\boxtimes)}$ and Rajesh Kannan Megalingam

Department of Electronics and Communications Engineering,
Amrita Vishwa Vidyapeetham, Amritapuri, India
gayathrirajendrababu95@gmail.com,
rajeshkannan@ieee.org

Abstract. This research work suggests a technique to execute an OCR based reader for the substantial number of visually debilitated. Considering how there is a large number of such people out there, we need to come up with a novel method that can be affordable as well as provides a good output. These book readers can be implemented in various ways. Here we talk around an OCR based system that is executed utilizing the product Tesseract. Later the recognized text becomes audio using TTs engines. The module utilizes a Raspberry Pi model 3B and a camera. The camera is for getting an image of the content. The image is then preprocessed before getting loaded into OCRs. The preprocessing section fuses binary images, noise removal, skew correction, division and feature extraction.

1 Introduction

The conversion of pictures into textual content is the simple explanation to Optical character recognition (OCR). It gives output as text of printed or physically composed words. In recent times, a lot of research is being carried out in OCR as it has a lot of applications and room for improvement in terms of performance.

As shown by the World Health Organization (WHO), around 28,50,00,000 humans far and wide are assessed to be ostensibly debilitated, in which 90% are from under developed countries. Keeping this in view, there's a need to help these people by getting them the necessary aid to overcome their impairment. If we look into the statistics, we can see that there is a large population, around the world, that is battling some form of visual problems on a daily basis. The actual numbers are quite alarming because out the world's population nearly 188.5 billion have some form of visual issue such as melow vision, uncorrected refractive errors or cataracts. Out of this, nearly 70–80% issues can be corrected if these people have access to proper treatment. The sad reality is that a large portion of this population cannot afford the necessary treatment and majority of them fall in the 50 above age group [4]. Visually challenged does not only refer to the people with absolute 0 vision. It also refers to the people who suffer from low vision, eye diseases etc. The evaluated number of individuals outwardly disabled on the planet is 286 million. It has been established that majority of these people fall in the 50 years and above category. As observed it demonstrates the quantity of individuals outwardly disabled, with poor vision and visually impaired per million populaces. The reasons for visual im-pairment are waterfall, 51percent,

© Springer Nature Switzerland AG 2020
P. Karrupusamy et al. (Eds.): ICSCN 2019, LNDECT 39, pp. 253–261, 2020.
https://doi.org/10.1007/978-3-030-34515-0_26

glaucoma, 8percent, AMD,5percent, youth visual impairment and corneal related issues, 4percent, refractive mistakes, tracoma etc.

Our endeavour is intended to help these poor people by being reliable as well as affordable. We have tried to reduce the hardware requirements as much as possible so as to make it portable. It contains a camera to capture the image and the Raspberry Pi 3B where all the processsing takes place. A suitable OCR engine is chosen, here we have taken Tesseract. This will convert the image of the book into text and later this is converted to sound which can be heard by the user through a headphone or speaker as per his convenience. In the software part, we have also made use OpenCV written using Python language. This was chosen due to its simple structure and convenience. Compared to the existing systems available in the market, we have tried to improve the quality of output along with speed and high accuracy.

This work has been divided into the following sections: Sect. 2 is Literature survey where similar works have been analysed. Section 3 gives the motivation behind doing this research work. Section 4 discusses the working of this research work along with the hardware and software architecture. Since image pre-processing is a major part of this research work, Sect. 5 discusses the image processing techniques implemented here. The results of the experiments conducted are given in Sect. 6 and future scope is discussed in Sect. 7.

2 Literature Survey

In [1], a smart robot is presented by the authors that can be used to guide the handicapped people - mainly visually handicapped people. It is presented as an alternative to animals. The bot can traverse and retrace multiple paths, even those which can not be accessed via GPS. It makes use of an Arduino Mega [ATMEGA1280] along with various other sensors and motors like ultrasonic sensor, IR sensor, vibration motor, micro servo motor etc. The robot is designed to work in 3 modes: 'learn, retrace and free moving'. The mode is chosen by the user using a joystick and keypad. The authors have focused on creating a cost effective bot that can guide as well as ensure the user's safety effectively. In [2] the authors provide an alternative to the touchscreen display of smartphones, which are an essential part of today's life. Touchscreens are a hindrance for the blind and available alternatives are costly. The prototype is an inexpensive wireless keyboard that makes use of Braille. It is developed using Arduino nano microcontroller and bluetooth low energy modules. In this paper, they have tested this against a standard touch keyboard for performance and have got excellent results.

[5] Discusses about an ANN (Artificial Neural Network) based optical character and handwriting recognition system. This paper also talks about some other different techniques for the same like matrix matching and fuzzy logic. Here the authors have used multi-level-perception model as the required neural network model.

[7] Talks about another device for the visually impaired that can be worn by user for comfort. This technology uses MSERs (Maximally Stable Extremal Regions) as the base for text recognition. The proposed system provides results in real time and utilises the hierarchical structure from MSERs. The outputs obtained are more steady compared to previous outputs. Results obtained show that this device out-shines other existing techs and gives excellent text detection rates.

A smart reader developed on the Raspberry Pi is discussed in [8]. The simulations are carried out using Matlab. The obtained images are binarized, skew corrected and linearised. Image segmentation is also carried out. Audio is obtained by optimal TTS software. Reader proposed in [9] is also developed on Raspberry Pi, and comes with features like text extraction, spell correction etc. It's developed on the Raspberry pi and uses OpenCV libraries for functioning which are based on Linux and Python coding. The device is gives output but with high error rates. Here portability and accuracy are an issue. Blind people come across a lot of different objects with labels in day-to-day lives. This is of extreme inconvenience. So [10] presents a framework that can help them read the labels on cylindrical objects. This is done by combining perspective projection algorithms along with feature extraction techniques. Adaboost based text classifier is used for text recognition. The device is gave satisfactory results. Contributions from [11] include text recognition frame by frame using multiple frames based on STC. These are required in route when perusing the street signs. The proposed framework is perfect with portable applications. Multiple STC forecasts in numerous edges improve the exhibition of acknowledgment. [12] talks about a system that can be used by the visually challenged to read text. The required object is isolated from background noise and region of interest is selected. A motion based algorithm is used for this. In the extracted ROI, text extraction and processing is done. Experimental results provided in the paper show state of the art results. [13] implements OCR for Devnagiri script using ANN. OCR is implemented using matrix matching techniques and fuzzy logics. [14] discusses about optical character recognition in general and talks about various methods to implement the same. [15] talks about the Tesseract software used in this project to implement OCR. It talks about how the software works and about its many applications. Even after having a slow but steady progress in the development of visual aids, [19] talks about a study that was conducted to understand how much these people are accepting the visual tools provided to them.

3   Motivation

The common language for blind people is Braille and the number of documents available today in braille is very limited. So as we can see, they have limited access to books. Also, converting each text or document into this language can be time consuming and expensive and is practically not feasible. The existing devices come with a lot of shortcomings such as cost and too many hardware components. These cannot be afforded by majority of people and so here, we come up with a device that is cheap and provides excellent results.

4   Working Principle

The working is based on a set of simple steps. Once all the required GPIO pins are accessed by the system, a message is sent to the user informing him that the system is online. Next he is required to capture an image using the camera module involved. This image undergoes skew correction noise removal etc. which are discussed later. Then OCR is implemented in this image which is then later converted to audio. The synthesized audio is heard by the user through a speaker/headphone. The hardware is kept to a minimum so that user can carry it around easily. It was also tested using a Pi cam which gave excellent results. This can be easily understood from the pictorial representation given in Fig. 1.

4.1 System Architecture

The system level architecture of our system can be explained as follows: When the process is initiated, all the sub-processes and GPIO connections are imported by the program. A notification is heard which explains that the reader is online and that the camera is on. The user is then asked to press the switch which will capture an image from the video feed. This is taken as the input image. This image is processed which includes binarization, de-blurring, skew correction etc. the output of this stage is converted to text (OCR) and then to audio. This can be heard by the user through the earphones or speaker.

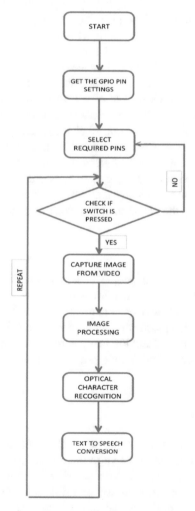

Fig. 1. Process flow

4.2   Requirements

The basic structure of the proposed reader is the raspberry pi 3B board [16]. The raspberry pi B+ is a solitary board PC which has 4 USB ports, 64-bit ARM processor, Bluetooth and Wifi, 40 GPIO pins for yield, CSI camera interface, HDMI port, DSI show interface, SOC (framework on a chip), LAN controller, SD card space, sound jack, and RCA video attachment and 5 V miniaturized scale USB connector.

The raspberry pi uses an operating system called RASPBIAN which processes the changes. The audio yield is taken from the sound jack. The net is accessed over the built in WiFi module. The page to be read is put on a base and the camera is adjusted to capture the picture. The picture is converted by the OCR engine inside the pi. After OCR conversion, the text is converted to audio. Here we used head-phones to listen to the audio. This can be replaced with a speaker for convenience.

5   Image Processing

Output obtained from the camera does not always have to be a clear image. This decreases the performance of the reader and output given to the user can be incorrect. So overcome this image processing/pre-processing can be done to increase the clarity of obtained image [18].

5.1   Binary Conversion

Obtaining a BW image where each pixel can only have two values - either 1 or 0 from RGB image is talked about here. A simple algorithm is used which can be explained as follows: First we have to transform RGB image to greyscale. Next apply suitable thresholding. And finally we get every pixel in the image is either 255 or 0.

5.2   Skew Correction

Since the target audience of our device are blind people, we need to consider the factor of angle correction. For this the following method is applie: First, detect the required text and calculate the approximate angle from mean position. Then rotate the image using a suitable algorithm (Fig. 2).

Fig. 2. Image of our proposed reader

5.3   De-Blur

In case of fuzzy or noisy images, de blurring techniques can be used to get a clean image. We apply Otsu threshold followed by Guassian blur.

6   Experiment and Result

The reader was tested for different samples with different features. Experiments were conducted taking different sample sets, each set having 20 samples. Each set was tested for different features. It was found that the reader showed an overall accuracy of 83.3% and error rate of 16.6%.

As in Table 1, the first 20 samples taken consisted of proper English words. In this case the reader gave 100% accurate output with absolutely no error. It gave clear audio output for every sample. Next 20 samples taken were combinations of random numbers where the accuracy rate was 80% since 0's were not recognised. On testing the reader with random alphanumeric combinations, it was found to have an accuracy of 85%. Error rate was highest when we tried with different fonts. The reader almost always gave correct outputs while detecting small letters. The reader had no difficulty in recognizing random combinations of letters of different sizes. Next the reader was tested to determine accuracy in a more practical situation. Pages from different books were chosen at random and number of words in the chosen paragraph were determined. It was seen that the reader worked very well and there was an average of 99% accuracy. This can be seen in Table 2.

Table 1. Experimental results - 1

Classifier	Accuracy (%)	Error rate(%)	Precision(%)
Scenario I	100	0	100
Scenario II	80	20	80
Scenario III	85	15	85
Scenario IV	60	40	60
Scenario V	80	20	80
Scenario VI	100	0	100
Overall	83.3	16.6	83.3

Table 2. Experimental results – 2

Words in paragraph	Words identified successfully	Accuracy(%)
78	78	100
56	55	98.21
43	43	100
108	105	97.22
94	94	100
15	15	100
74	72	97.29
54	53	98.14
32	32	100
19	19	100

7 Scope

This research work is in a promising field and hence there is scope for modifications. Our reader currently gives very poor outputs for handwritten texts. Thus handwriting recognition can also be further added to enhance the reader along with recognition of other languages other than English (Fig. 3).

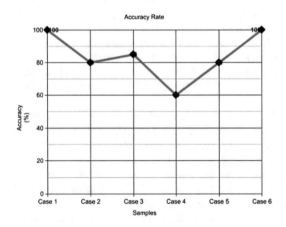

Fig. 3. This graph gives the accuracy rates for the different scenarios as seen in Table 1

8 Conclusion

So as we have seen so far, our prototype has given a very good yield in terms of accuracy and overcomes certain drawbacks of existing devices.

Compliance with Ethical Standards. All procedures performed in this study involving human participants were in accordance with the ethical standards of the institutional and/or national research committee and with the 1964 Helsinki declaration and its later amendments or comparable ethical standards. For this type of study formal consent is not required.

References

1. Megalingam, R.K., Vishnu, S., Sasikumar, V., Sreekumar, S.: Autonomous path guiding robot for visually impaired people. Adv. Intell. Syst. Comput. **768**, 257–266 (2019)
2. Rao, S.N., Dsilva, C., Parthasarathy, V.: Evaluation of a smartphone keyboard for the visually challenged. In: 2nd IEEE International Conference on Electrical, Computer and Communication Technologies 2017 (ICECCT 2017) (2017)
3. Sonth, S., Kallimani, J.S.: OCR based facilitator for the visually challenged. In: 2017 International Conference on Electrical, Electronics, Communication, Computer and Optimization Techniques (ICEECCOT) (2017)
4. Pascolini, D., Mariotti, S.P.: Global estimates of visual impairment. Br. J. Ophthalmol. 96, 614–618 (2012). https://doi.org/10.1136/bjophthalmol-2011-300539, https://bjo.bmj.com/content/96/5/614
5. Rao, N.V., et al.: Optical character technique recognition algorithms. J. Theo. Appl. Inf. Technol. 83(2) (2016)
6. Baboo, D.S.S., et al.: Embedded optical braille recognition on tamil braille system using raspberry pi. Int. J. Comput. Technol. Appl. **5**(4), 1566–1574 (2007)
7. Merino-Gracia, C., et al.: A head-mounted device for recognizing text in natural scenes. In: International Workshop on Camera-Based Document Analysis and Recognition, pp. 29–41. Springer, Heidelberg, (2011)
8. Velumurugan, D., et al.: Hardware implementation of smart reader for visually impaired using raspberry pi. IJAREEIE **5**(3), 2055–2063 (2016)
9. Nagaraja, L., et al.: Vision based text recognition using raspberry pi. Int. J. Comput. Appl. **0975**, 8887 (2015). National Conference on Power Systems Industrial Automation(NCPSIA 2015)
10. Ye, Z., Yi, C., Tian, Y.: Reading labels of cylinder objects for blind persons. In: 2017 IEEE International Conference on Environment and Electrical Engineering, pp. 1–6 (2017)
11. Rong, X., Yi, C., Yang, X., Tian, Y.: Scene text recognition in multiple frames based on text tracking, pp. 1417–1422
12. Yi, C.,, Tian, Y.: Portable camera-based assistive text and product label reading from hand-held objects for blind persons. In: IEEE/ASME Transactions on Mechatronics Student Member, vol. 19, pp. 1083–4435. IEEE (2014)
13. Singh, R., Yadav, C.S., Verma, P., Yadav, V.: Optical character recognition (ocr) for printed devnagari script using artificial neural network. Int. J. Comput. Sci. Commun. **1**(1), 91–95 (2010)
14. Sabu, A.M., Das, A.S.: A survey on various optical character recognition techniques. In: Proceedings IEEE Conference on Emerging Devices and Smart Systems (ICEDSS 2018), March 2018
15. Smith, R.: An overview of the tesseract ocr engine. In: Ninth International Conference on Document Analysis and Recognition (2007)
16. The MagPi magazine (1999). Accessed 30 Sept 2010. https://www.raspberrypi.org/magpi/raspberry-pi-3-specs-benchmarks/

17. The World Health Organisation statistics (2018). Accessed 10 Oct 2018. https://www.who. int/news-room/fact-sheets/detail/blindness-and-visual-impairment
18. Arrahmaha, A.I., Rahmatika, A., Harisa, S., Zakaria, H., Mengko, R.: Text-to-speech device for patients with low vision. In: 4th International Conference on Instrumentation, Communications, Information Technology, and Biomedical Engineering (ICICI-BME), November 2015
19. Paajala, I.J., Kernen, N.: Study for acceptance on new navigation assistance by visually impaired people. In: 2015 9th International Symposium on Medical Information and Communication Technology (ISMICT), September 2015

Using Machine Learning to Help Students with Learning Disabilities Learn

Francis Dcruz[⊠], Vijitashw Tiwari, and Mayur Soni

Department of Computer Engineering, St. Francis Institute of Technology,
Mumbai, India
dcruzfrancis04@gmail.com, vijitashw10@gmail.com,
mayursoni299@gmail.com

Abstract. The concept behind this learning modal is to connect education with technology to meet the different needs of each student. The main aim of personalized learning is to help students with disabilities.

Students with a disability often need subject matter presented through different methods, therefore it is imperative that these technological advances benefit all students with different learning styles. Machine Learning opens up new ways to help students with disabilities. Children with autism which is a neurological disorder need a personalized development system for their daily activities. Technology can play a substantial part.

The system includes 4 parts: (i) To predict the learning level of the user. (ii) Generating multimodal learning materials using web mining. (iii) User preferences are associated with the result. (iv) Personalized contents for users delineated with an intelligent interface.

Keywords: Multimodal learning material · Special Needs Children · Web mining · Machine learning

1 Introduction

Neurological disorders are diseases of the brain, spine and the nerves that connect them. Children facing neurological disorders have different growth, learning, understanding rate. To make these children learn a model is developed.

The principle inspiration driving our model is how keen and usable innovations can be explained to the kids with Neurodevelopment issue and how noteworthy materials from the Internet can help Children for better understanding. The primary point of this work is to furnish a framework that helps kids with special requirements to comprehend materials educated in the classroom with the assistance of media components (pictures and video cuts) gathered from the web. The entire framework can be isolated into three stages which are: (a) Fetch keywords that speak to the story, (b) Fetch media components from the web as indicated by child's learning level and connect them with the keywords fetched, (c) lastly apply potential filters on the recovered components to

P. Karrupusamy et al. (Eds.): ICSCN 2019, LNDECT 39, pp. 262–269, 2020.
https://doi.org/10.1007/978-3-030-34515-0_27

evade immaterial and improper components. The proposed plan is relied upon to helps kids with an extraordinary need to take in by gathering multimodal components from web identified with their content stories. AI calculation is utilized to recognize the learning dimension of a child. Distinguished dimension is utilized to create customized learning materials that upgrade the learnability of the kids. Different refining choices have been utilized to guarantee the fittingness of the material for Children.

2 Literature Survey

The clever framework accepts the child's story as information and discovers multi-modal substance from the web with story key extraction, web mining, web refining, etc. After the hunting task is performed, URL (Uniform Resource Locator) refining, space separating and web content investigation based refining strategies are utilized to evacuate undesirable materials [1–3].

The research aims to design, develop and evaluate a mobile application providing a better learning environment with the help of usable user interface design and effortless input (children's stories) system for the generation of multimodal learning materials. Children can easily acclimate with modern computing devices and learn lessons from multimodal items of Internet resources (texts, audios, videos, and images) [4, 12, 13].

A kid once determined to have learning incapacity, the kid is additionally ordered into various sorts of learning handicaps like dyslexia, dysgraphia, dyscalculia, and dysphasia. The model will likewise recognize the dimension of Learning Disability and furnish the understudy with concentrate material for improvement [5, 8].

Foreseeing and estimate mentally unbalanced Children execution by applying data mining calculations in Weka tools. In reference to that paper recognized the learning ability relies upon a few general regular qualities like language aptitude, talking expertise, composing aptitudes and so on. Besides, gathered information over those properties, connected. [6, 7] Kids' interface for seeking and pursuing has concentrated basically on two subjects: (a) Children's interface for looking and perusing (b) Multi-modal Learning Materials for Special Needs Children. The framework proposes a great interface that gives favorable circumstances to both Boolean seeking and easygoing perusing [7, 10].

Data is filled with a lot of important stories and hidden patterns. These stories and pattern can be understood by data visualization and later can be used to present it to others [11].

3   Challenges Identified

A. *How web mining techniques can be used for personalized search*

The arrangement includes first concentrates the catchphrases from the plain content gave utilizing word co-events strategy. The co-event circulation between regular terms and different terms in a sentence can assist us in finding the overall significance of a term. Removed watchwords are utilized to look for the multimodal components (writings, pictures, and video cuts) and other comparable stories over the web.

B. *Can Machine learning calculation distinguish the learning extent of a youngster*

The mental imbalance kids learning aptitudes are affected by different components like consideration, script, spelling, language, perusing, composing, understanding abilities and result from the diverse kids. Tests were taken for usage. It examines the information and recognizes examples and in this manner distinguishes the dimension of understudy.

C. *Can distinguished dimension be utilized to create customized learning materials which improve the learnability of the Children*

Initially, a sample data stored in the database is used for training and testing the system. The machine-learning algorithm uses this data to cluster students based on previous related data. The algorithm which gives the best results is selected and the process is carried on. Once students are classified into various clusters, we provide them with study materials based on the level of Learning Disability which is the same for all the students in the same cluster.

D. *How acute interfaces can improve the connection of kids with extraordinary necessities*

The principal object is to configuration, create and test model for Children with unique needs offering a UI with adaptability and ease of use. For designing and developing the prototype UCD (User-Centered Design) approach was adopted. This prototype contains almost all sorts of such required elements that conform to a high standard of usability.

4   Problem Definition

They find it hard to learn and adapt things by the traditional method and see things from other people's perspectives. Personalized learning will be modified learning programs per their learning and progress curve and this can help the teacher track the improvement as well.

5   Proposed System Methodology

5.1   Flowchart

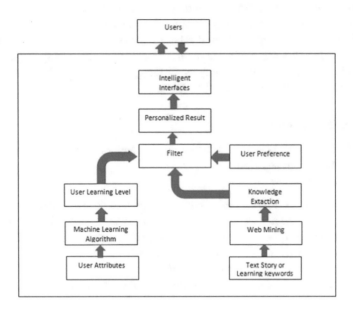

Fig. 1. Flowchart

There are three primary regions of the model as pursues:

- Web information mining, searching and separating calculation, to produce customized learning and improvement assets.
- An AI calculation to recognize the learning level and prescribe learning material for the kids.
- A multimodal, wise and usable interface for the kids thinking about their necessities (Fig. 1).

The steps in the above flowchart are explained below:

At first, the child's interfaces with the framework to scan for the required questions over the web. Here he will give a content story or learning catchphrases for tracking. Catchphrases would be separated from the narratives. The calculation utilized for watchword extraction depends on Keyword Extraction from a solitary archive utilizing Word Co-event Statistical Information calculation. Separated catchphrases that are being mined by the web then remove learning from the World Wide Web (WWW).

A separate calculation is utilized to channel the withdrawn outcomes. In our proposed framework we utilized content arrangement to distinguish whether the website page is sheltered or not for kids from where pictures and recordings will be separated. The outcome got is connected with the child's desires.

AI calculation is utilized to effectively perceive and foresee the child's level so it would help to give the child's learning materials as their dimension. First, the information is gathered from their parents and educators about learning crippled understudies. The information gathered is put away in a database in an organized organization. Information accumulation may incorporate a sort of learning inability, the dimension of handicap in different parts of talking, perusing, tuning in, language and so on. The information put away in the database is isolated for preparing and testing the framework. When the preparation is done, we start with the testing stage. When the learning dimension of an understudy is distinguished we furnish them with study materials suitable to its dimension of learning.

5.2   Proposed Algorithm

The algorithm used for keyword extraction in web data mining is Word Co-occurrence Statistical Information algorithm
Steps are as follows:

1. *Select the successive terms and dispose of high-recurrence words which are not a piece of things and action words, for example, "the", "an", "an", "in".*
2. *Clustering frequent terms with the Jensen-Shanon divergence method.*
3. *Calculate the expected probability for the term.*
4. *Calculate co-occurrence frequency for the term*
5. *Output watchwords: Choose a given number of terms having the biggest co-event recurrence esteem.*

In our proposed system the sites which the system uses to fetch images and videos are authenticated for safe use by the children using text classification.
Steps are as follows:

1. *Check URL against a rundown of genuine URLs.*
(Sites for Children' sleep time stories, learning history, book recordings for children, infant television and so on can be spared in a database table which is later used to check against the URL of the pictures and recordings recovered from the Google)

2. *Analyze the web content*
(Filtering through pictures/recordings from risky sites, and safe outcomes are brought to the child.)

In our proposed system for determining the learning level of a child, the Multilayer perceptron algorithm gives better results (Fig. 2).

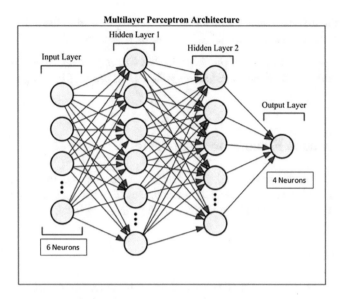

Fig. 2. Multilayer perceptron architecture

Steps followed in MLP are as follow:

1. *Determine the number of nodes in the input layer and the output layer.*
2. *Find the no. of neurons in the hidden layer by assessing the execution of the calculation.*

For most issues, we can get fair execution by setting the hidden layer setup utilizing only two guidelines:

   a. Number of hidden layers equivalent to one
   b. The quantity of neurons in the hidden layer is the mean of the neurons in the output and input layers

3. *Forward transmission:*

Initially, all weights in the network are randomly assigned. We now calculate the output V of the hidden layer node by applying some activation function.

$$V = f(X1 * W1 + X2 * W2 + X3 * W3 + \ldots \ldots)$$

where f is some activation function,

X1, X2, X3….. are input values from the dataset,
W1, W2, W3……are assigned weights

Correspondingly, the yield of different neurons in the hidden layer can likewise be determined.

A hidden layer yield is taken as a contribution for the output layer. This enables us to compute the likelihood of qualities on various outputs.

Out of the first determined probabilities are a long way from the normal likelihood, so the system is considered as "false yield or false output."

4. *Now, if we want to predict the learning level of the student depending upon the input values can be done by calculating the output probability of different learning level.*

6 Performance Evaluation Parameters

The execution of the model and proficiency can be checked by an educator who assesses how well the system calculation recognizes the learning dimension of students and how well the multimodal are mapped to their learning level.

7 Conclusion

In this paper, we tend to provide every child with personalized results, as well as a training system with information of multiple users that helps to foresee the child's learning level. The proposed approach used the data sets and collected personal information of children. Personalized general materials will generate web mining techniques for efficient transmission.

References

1. Wagley, A., Akhter, P., Bhuiyan, M., Dahal, K., Hossain, A.: Web mining to generate multimodal learning materials for children with special needs. In: The 8th International Conference on Software, Knowledge, Intelligent Management and Applications, SKIMA 20I4, Dhaka (2014)
2. Pretschner, A., Gauch, S.: Ontology based personalized search. In: 1999 Proceedings of the 11th IEEE International Conference at Tools with Artificial Intelligence, Chicago, IL (1999)
3. Pretschner, A., Gauch, S.: Personalized search based on user search histories. In: Proceedings of the 2005 IEEE/WIC/ACM International Conference on Web Intelligence (2005)
4. Bhuiyan, M., Miraz, M.H., Banik, L.: Automated generation of learning materials for children with special needs in converged platforms using android. In: The 2nd International Symposium on Advanced and Applied Convergence (ISAAC 2014) (2014)
5. Shah, M., Shah, M., Shirke, A., Deulkar, K.: Providing personalized study material for learning disability using machine learning. Int. J. Res. Sci. Eng. (2017). e-ISSN: 2394–8299 Special Issue 7-ICEMTE March 2017
6. Mythili, M.S., Shanavas, A.R.M.: A novel approach to predict the learning skills of autistic children using SVM and decision tree. (UCSIT) Int. J. Comput. Sci. Inf. Technol. (2014)
7. Hutchinson, H., Bederson, B.: Interface design for children's searching and browsing. U. of MDHCIL Technical report, HCIL-2005–25 (2005)
8. Hutchinson, H.: Children's interface design for hierarchical search and browse. ACM SIGCAPH Comput. Phys. Handicap. **75**, 11–12 (2003)

9. Marsh, J.: Young children's play in online virtual worlds. J. Early Child. Res. **7**(3), 1–17 (2010)
10. Marsh, J.: Young children's literacy practices in a virtual world: establishing an online interaction order. Read. Res. Q. **46**(2), 101–118 (2011)
11. Few, S.: Data visualization for human perception (2013). http://www.interactiondesign.org/encyclopedia/data_visualization_for_human_perception.html
12. Marsh, J.: The techno-litaracy practices of young children. J. Early Child. Res. **2**(1), 52–66 (2004)
13. Standen, P.J., Brown, D.J., Cromby, J.J.: The effective use of virtual environments in the education and rehabilitation of students with intellectual disabilities. British Journal of Educational Technology **32**(3), 289–299 (2001)
14. Attardi, G., Gulli, A., Sebastiani, F.: Automatic Web page categorization by link and context analysis. In: Proceedings of THAI, vol. 99, no. 99, pp. 105–119 (1999)
15. Kim, Y., Nam, T.: An efficient text filter for adult web documents. In: The 8th International Conference on Advanced Communication Technology. ICACT 2006, vol. 1, pp. 3-pp. IEEE, February 2006

Lightweight Elliptical Curve Cryptography (ECC) for Data Integrity and User Authentication in Smart Transportation IoT System

Jyoti Joglekar[1]([✉]), Simran Bhutani[1], Nisha Patel[1], and Pradnya Soman[2]

[1] KJSCE, Vidyavihar, Mumbai, India
{jyoti.joglekar,simran.bhutani,nisha.cp}@somaiya.edu
[2] KJSIEIT, Sion, Mumbai, India
pradnya.soman@somaiya.edu

Abstract. Data integrity and data security are major issues while transferring data in any Internet of Things (IoT) application. The sensor node as an edge node or end node collects information from the physical environment which is to be preprocessed and sent to the nearby server or cloud server for storage or for further processing. Because of the constrained resources at the edge node, it is essential to perform edge analytics with methods which utilize less power or can be performed with less computational complexity with limited memory. In this scenario, the data security related methodology should use lightweight encryption method for providing data security to avoid the attacks on data integrity while transmitting. In our Smart Goods Transport IoT system, we are using lightweight encryption method Elliptical Curve Cryptography (ECC) using ElGamal encryption scheme to provide strong authentication and data security while in transit.

Keywords: Constrained resources · IoT · Elliptical Curve Cryptography (ECC) · Data security · Edge analytics

1 Introduction

In some IoT systems strong authentication mechanism is needed for avoiding attacks such as impersonation, eavesdropping and Man-in the middle attack. These kinds of attacks are threats to data integrity as well as availability of data to the legitimate user. Hence strong encryption and secure protocols are needed in IoT systems. Though many end to end encryption schemes such as Transportation Layer Security (TLS) assure data integrity while data in transit, the major challenge in IoT system is to use light weight encryption due to limited resources of IoT end devices. Lightweight cryptography provides an efficient method which is comparable to other proven encryption standards such as Advanced Encryption Standard (AES) for data security at the end node before sending the data from end node to nearby server for further analytics or storage.

© Springer Nature Switzerland AG 2020
P. Karrupusamy et al. (Eds.): ICSCN 2019, LNDECT 39, pp. 270–278, 2020.
https://doi.org/10.1007/978-3-030-34515-0_28

Lightweight encryption is a necessity for IoT edge devices due to constrained resources available at the edge node.

The IoT devices have a limited RAM size, limited available energy due to chip size, battery power and cost constraints resulting in limited computational power. There is always a trade-off between security level and computational complexity of the algorithm. There is a need for end-to-end security while transmitting data collected by the end nodes for securing data integrity. For such applications lightweight algorithm is useful that works in a constrained environment for optimum use of resources. Elliptical curve cryptography is one of the public key cryptosystems useful for providing security of the required level.

Some important IoT security problems are insufficient authentication, lack of encryption while data in transit and privacy concerns. In our Smart Goods Transportation application, while allocating a vehicle to the customer as per his/her requirements, the server software sends a 4-digit OTP to the customer for user authentication. The vehicle which is assigned to the customer reaches the source location. Then the customer enters this OTP in the application of the driver system. Thus by entering the OTP the customer is verified and authenticated by the software. In this application we are applying Edge Analytics at customer's app by ECC decryption to extract the ECC encrypted OTP received from the server software. When customer enters the OTP in the driver vehicle app, the app at driver vehicle encrypts the entered 4-digit OTP by ECC and sends to the backend server. The server decrypts it using ECC decryption method and compares the received OTP for authentication.

2 Recent Work

A lot of work is presented on secure authentication using ECC. An ECC point multiplication for 1.3 MCyclesto employ the secp160r1 curve is described in [1], the authors implemented cycle-accurate clones using embedded processors 8-bit Atmel ATmega, 16-bit Texas Instruments MSP430, and 32-bit ARM Cortex-M0+ for software ECC implementations. In [2], the authors showcased a hardware support with feature of five prime field-based NIST ECC curves. Optimized hardware/software design approaches are presented by the authors in [3], for fast rate ECC calculations with resource constrained environment. Many other ECC algorithms were implemented by the work of authors in [4–6]. The ECC based ElGamel encryption scheme is explained in [7]. The efficient Software Implementation of Finite Fields and its applications in cryptography are presented in [8].

3 Methodology

In our proposed Smart Goods Transport IoT system customer authentication is needed before confirming the selected vehicle for goods transport by customer. After customer authentication the vehicle is assigned to the customer by the backend software.

This authentication is performed in our system using One Time password (OTP) mechanism. A 4-digit OTP is generated by the backend software and encrypted using Elliptical curve cryptographic algorithm based on Elgamel encryption scheme [9–12]. Figure 1 below shows the block diagrams for the architecture of our work.

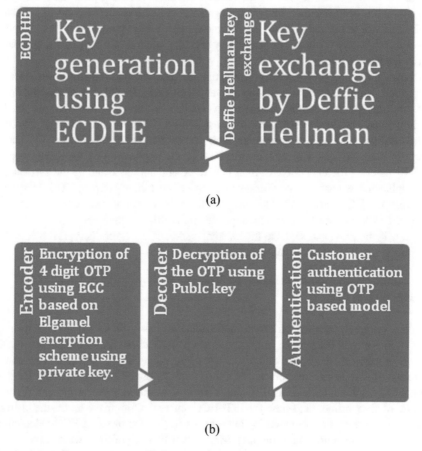

(a)

(b)

Fig. 1. Block diagram of the ECC based authentication mechanism (a) Key exchange using ECDHE algorithm (b) Encryption and decryption using Elgamel ECC algorithm

As shown in the block diagram of Fig. 1, private and public key pair generation is done. The keys are securely exchanged using Elliptic Curve Diffie-Hellman (ECDHE) where Ephemeral keys are used for the key exchange mechanism. The 4-digit OTP is encrypted using ECC and sent to the customer app. Customer app decrypts the OTP to receive the plaintext using decoder on the customer app. This 4-digit OTP is entered in the driver application to authenticate the customer. When the OTP is entered in the driver's app this OTP is encrypted by the ECC encoder at driver's app and the encrypted OTP is sent to the server software for customer's authentication. The server software decrypts the OTP and compares it with the sent OTP to authenticate the customer. After successful authentication the start of the journey message is sent to the driver's app.

3.1   Elliptic Curve Diffie-Hellman Ephemeral (ECDHE)

Elliptic-curve Diffie–Hellman Ephemeral (ECDHE) is a key agreement protocol that generates public–private key pair for communication of two sides. The secret information is to provide data security to avoid the man in the middle attack. It is a modified Diffie–Hellman protocol using elliptic-curve cryptography [6]. The derived key by using this protocol is used further to encrypt the information before sending it for communications. The public keys are either static or ephemeral. Ephemeral keys are temporary and not authenticated always. If one of the party's public keys is static, then man-in-the-middle attacks can be avoided. Holders of static private keys should validate the other public key. An example is explained below.

A cryptographic key when ephemeral, it is generated as a new one for each execution of a key generation process. Sometimes ephemeral keys are used for more than one times in a single session where one ephemeral key pair is generated for every message and the private key is combined separately with each recipient's public key.

In our modified encryption architecture, the key generation is done by using public key cryptosystem ECDHE which is the elliptical curve variant of Diffie- Hellman algorithm, as shown in the first block. The randomness is introduced making it strong for cryptanalysis attack, as the key is temporary (ephemeral) for every key generation [13–15].

The ECC field points are generated based on the randomly chosen parameters such as a chosen prime number p, the elliptical curve equation E, order of the field n and primary key d and k [9]. A new mapping table for the digits 0 to 9 as well as a new key pair is generated for every new OTP generation. The Elliptical field points are chosen randomly from generated many points for the mapping table. The OTP and the mapping table is encrypted by ElGamel ECC cryptosystem by introducing randomness by ephemeral parameters and sent to the user [6].

When A wants to provide a shared key with B over an insecure channel, initially, the domain parameters of ECC encryption (p, k, G, n) are agreed upon. Also, each party must have a key pair that suits the elliptic curve cryptography implementation that consists of a private key d and a public key as elliptical curve point Q.

Generation of Private Key d: The private key d is a randomly selected integer in the interval [1, n−1].

Generation of Public Key Q: The public key is generated by the following equation [3].

$$Q = dG \tag{1}$$

Both the communicating parties must know each other's public key prior to execution of the protocol [3].

A computes the shared secret as

$$S = d_A Q_B \tag{2}$$

B computes the shared secret as

$$S = d_B Q_A \tag{3}$$

The shared secret calculated by both parties is equal, because of the following condition of ECDH algorithm.

$$d_A Q_B = d_A d_B G = d_B d_A G = d_B Q_A \tag{4}$$

No party other than A and B can compute the shared secret. Hence the private key is kept secure and is only computed at the receiver end. Also, without the knowledge of both the private keys the message cannot be decrypted.

Pseudo code implemented: ECDHE

Input:
Elliptic curve domain parameters (p, k, G, n), Cipher Text (C_1, C_2) and Public Key Q_A

Output:
Computed private key d_A and encrypted Message M
-Compute Shared Secret $(S = d_B Q_A)$
-Compute Private Key $(d_A = S / Q_B)$
-Decrypt Cipher text using Private Key d_A $(C_1, C_2) \rightarrow M$

In the Smart Transportation IoT system, ECDHE is used as the key agreement protocol. The source will encrypt the OTP using ElGamel Encryption. The encrypted cipher text along with the public key is published as the shared secret. The receiver thus has access to the cipher text as well as the public key from which the private key is derived and subsequently used for decryption.

Thus the OTP is securely shared and obtained at the receiver end for further authentication.

4   Results

The results for ECDHE public private key generation and exchange are shown below for two sets of parameters as P: Published Point on the elliptical curve, d: Generated Private Key and Q: Public Key on both sides.

Figure 2 below shows the output of the ECDHE key generation and exchange.

Example 1:

```
ECDH Ephimeral Parameters fo key generation:
P = (1,12)          #Public Point on elliptical-curve
d1 = 190            #Sender's Private key
d2 = 10             #Receiever's Private key
Q1 = (190,2280)     #Sender's Public key which is published
Q2 = (10,120)       #Receiver's Public key


Receiver computes sender's private key d1 by using the sender's public key Q1
Computing d1 at receiver : d1 = d2*Q1/Q2
d1 = 190            #sender's private key is derived and used for decryption at receiver's end
```

Example 2

```
ECDH Ephimeral Parameters fo key generation:
P = (3,17)          #Public Point on elliptical-curve
d1 = 308            #Sender's Private key
d2 = 20             #Receiever's Private key
Q1 = (924,5236)     #Sender's Public key which is published
Q2 = (60,340)       #Receiver's Public key


Receiver computes sender's private key d1 by using the sender's public key Q1
Computing d1 at receiver : d1 = d2*Q1/Q2
d1 = 308            #sender's private key is derived and used for decryption at receiver's end
```

Fig. 2.   ECDHE output for two sets of parameters

The ElGamel ECC encryption scheme is used for encryption. Decryption of cipher text is done using published public key and generated private key at the decoder. For two set of parameters the output is presented below in Fig. 3

The Elliptical curve cryptography is used for the Smart Goods Transport IoT system using ElGamel encryption scheme for customer authentication. An implementation results are presented for ECC based OTP encryption and decryption is given in Fig. 3. As shown in the Fig. 3, the field points are generated using the ElGamel encryption scheme for the chosen parameters such as elliptical curve equation, prime number for which the field points are to be generated etc. For enhancing the security, a stochastic nature is introduced by choosing the field points randomly for generating the mapping table.

Encryption: Parameter Test1 1

```
The selected parameters for ECC-Elagamal are Elliptical Curve: y^2+y=x^3
 Public point on the curve: 1,12 Prime No:5, N: 358, d: 190 K: 28
The Map Table is:
0-0,10
1-1,2
2-1,7
3-0,4
4-0,5
5-0,15
6-0,0
7-0,19
8-0,14
9-0,9
The OTP generated by the Server side is 4699
0,5 is mapped to number4 in mapping table
0,0 is mapped to number6 in mapping table
0,9 is mapped to number9 in mapping table
0,9 is mapped to number9 in mapping table
Cipher Text (C1, C2)
 For digit 4 Cipher text 1: (28,336) Cipher text 2: (5320,63845)
 For digit 6 Cipher text 1: (28,336) Cipher text 2: (5320,63840)
 For digit 9 Cipher text 1: (28,336) Cipher text 2: (5320,63849)
 For digit 9 Cipher text 1: (28,336) Cipher text 2: (5320,63849)
```

Decryption:

```
The Map table is:
0-0,10
1-1,2
2-1,7
3-0,4
4-0,5
5-0,15
6-0,0
7-0,19
8-0,14
9-0,9
Cipher text 1: (28, 336) Cipher text 2: (5320, 63845)
Cipher text 1: (28, 336) Cipher text 2: (5320, 63840)
Cipher text 1: (28, 336) Cipher text 2: (5320, 63849)
Cipher text 1: (28, 336) Cipher text 2: (5320, 63849)
The points are
Message point[0] -> (0,5)
The points are
Message point[1] -> (0,0)
The points are
Message point[2] -> (0,9)
The points are
Message point[3] -> (0,9)
The decrypted OTP is 4699
```

Fig. 3. ECC ElGamel based encryption and decryption of the OTP for two different set of parameters E, n, P, d

Encryption: Parameter Test 2

```
The selected parameters for ECC-Elagamal are Elliptical Curve:
y^2=x^3+2x+4
 Public point on the curve: 3,17 Prime No:7, N: 341, d: 308 K: 12
The Map Table is:
0-1,7
1-0,16
2-0,9
3-2,3
4-0,5
5-0,2
6-0,19
7-1,14
8-0,12
9-1,0
The OTP generated by the Server side is 6003
0,19 is mapped to number6 in mapping table
1,7 is mapped to number0 in mapping table
1,7 is mapped to number0 in mapping table
2,3 is mapped to number3 in mapping table
Cipher Text (C1, C2)
 For digit 6 Cipher text 1: (36,204) Cipher text 2: (11088,62851)
 For digit 0 Cipher text 1: (36,204) Cipher text 2: (11089,62839)
 For digit 0 Cipher text 1: (36,204) Cipher text 2: (11089,62839)
 For digit 3 Cipher text 1: (36,204) Cipher text 2: (11090,62835)
```

Decryption:

```
The Map table is:
0-1,7
1-0,16
2-0,9
3-2,3
4-0,5
5-0,2
6-0,19
7-1,14
8-0,12
9-1,0
Cipher text 1: (36, 204) Cipher text 2: (11088, 62851)
Cipher text 1: (36, 204) Cipher text 2: (11089, 62839)
Cipher text 1: (36, 204) Cipher text 2: (11089, 62839)
Cipher text 1: (36, 204) Cipher text 2: (11090, 62835)
The points are
Message point[0] -> (0,19)
The points are
Message point[1] -> (1,7)
The points are
Message point[2] -> (1,7)
The points are
Message point[3] -> (2,3)
The decrypted OTP is 6003
```

Fig. 3. (*continued*)

5 Conclusion

In this paper a lightweight ECC-based authentication solution with a novel architecture of combination of ECDHE and ElGamel ECC is presented for smart goods transportation system for resource constrained systems. The employed authentication technique uses light weight ElGamel encryption scheme is also made ephemeral by randomly choosing new parameters and generating the mapping table from the elliptical curve point for the digits 0 to 9. Encryption of a 4-digit OTP is generated and encrypted OTP is sent to the customer app for authentication. The key exchange is done by using Elliptic Curve Diffie-Hellman with Ephemeral keys (ECDHE). Alight weight implementation of this algorithm is performed for the key exchange in our system as the proposed technique is working for the resource constrained environment of android-based application on a smart phone. Thus, a secured and lightweight authentication system is presented in this work that takes into account the limitations of the resource constrained devices. In our future work will perform studies on the attack-based evaluation and analysis of the implemented authentication solution and its performance comparison with other state of art methods.

References

1. Wenger, E., Unterluggauer, T., Werner, M.: 8/16/32 shades of elliptic curve cryptography on embedded processors. In: Progress in Cryptology INDOCRYPT 2013. Lecture Notes in Computer Science. Springer International Publishing, vol. 8250, pp. 244–261 (2013)
2. Alrimeih, H., Rakhmatov, D.: Fast and flexible hardware support for ECC over multiple standard prime fields. IEEE Trans. Very Large Scale Integr. (VLSI) Syst. **99**, 1–14 (2014)
3. Höller, A., Druml, N., Kreiner, C., Steger, C., Felicijan, T.: Hardware/software co-design of elliptic-curve cryptography for resource-constrained applications. In: 51th ACM/EDAC/IEEE Design Automation Conference (DAC), June 2014
4. Aigner, H., Bock, H., Hütter, M., Wolkerstorfer, J.: A low-cost ECC coprocessor for smartcards. In: Cryptographic Hardware and Embedded Systems (CHES). Lecture Notes in Computer Science, vol. 3156, pp. 107–118. Springer, Heidelberg (2004)
5. Hein, D., Wolkerstorfer, J., Felber, N.: ECC is ready for RFID a proof in silicon. In: Selected Areas in Cryptography. Lecture Notes in Computer Science, vol. 5381, pp. 401–413. Springer, Heidelberg (2009)
6. Plos, T., Hutter, M., Feldhofer, M., Stiglic, M., Cavaliere, F.: Security enabled near-field communication tag with flexible architecture supporting asymmetric cryptography. IEEE Trans. Very Large Scale Integr. (VLSI) Syst. **21**(11), 1965–1974 (2013)
7. Koblitz, N.: Introduction to Elliptic Curves and Modular Forms. Springer, New York (1984)
8. Lang, S., Trotter, H.: Primitive points on elliptic curves. Bull. Amer. Math. Soc **83**, 289–292 (1977)
9. Hankerson, D., Menezes, A.J., Vanstone, S.: Guide to Elliptic Curve Cryptography. Springer Professional Computing. Springer, Heidelberg (2004)
10. Guajardo, J., Kumar, S., Paar, C., Pelzl, J.: Efficient software implementation of finite fields with applications to cryptography. Acta Appl. Math. **93**(1–3), 3–32 (2006)
11. Akishita, T., Hiwatari, H.: Compact hardware implementations of the 128-bit Blockcipher CLEFIA. In: Proceedings of Symposium on Cryptography and Information Security – SCIS 2011 (2011). (in Japanese)
12. Diffie, W., Hellman, M.E.: New directions in cryptography. IEEE Trans. Inf. Theory **22**(6), 644–654 (1976)
13. Poschmann, A.: Lightweight Cryptography – Cryptographic Engineering for a Pervasive World. In IACR ePrint archive 2009/516 (2009)
14. Satoh, A., Morioka, S.: Hardware-focused performance comparison for the standard block ciphers AES, Camellia, and Triple-DES. In: Boyd, C., Mao, W. (eds.) Proceedings of ISC 2003, vol. 2851, pp. 252–266, Springer, Heidelberg (2003)
15. Eisenbarth, T., Kumar, S., Paar, C., Poschmann, A., Uhsadel, L.: A survey of light weight cryptography implementations. IEEE Des. Test Comput. **24**(6), 522–533 (2007)

To Scrap the LinkedIn Data to Create the Organization's Team Chart

Sandeep Mathur[(⊠)] and Shally Sharma

Amity Institute of Information Technology, Amity University, Noida, UP, India
Sandeep2809@gmail.com, smathur@amity.edu,
shallysharma2409@gmail.com

Abstract. From past decades, LinkedIn appears as the professional connection site for both the freelancers and the recruiters. LinkedIn users are ought to post jobs, connect different industries and updates the people with current events. The goal of the paper is to create a report on the structure of the Organization to provide a smooth and efficient reporting hierarchy which involves data Analytics on the LinkedIn Data. So it is required to create an organizational hierarchy. Web Scraping is performed on the LinkedIn site for this intended purpose.

Keywords: Data analysis · Component · Formatting · Style · Styling

1 Introduction

Web Scraping is termed as Screen Scraping, Web Data Extraction and Web Harvesting. It is a procedure utilized to extricate a lot of information from sites whereby the information is removed and spared to a nearby document in your PC or to a database in a table design. Web scraping is a popular technique which converts the entire internet into your database. Web scraping is the popular part of data analytics. The focus is set on showing how straightforward it is today to set up a data scraping pipeline, with minimal programming effort, and answer several practical needs. Web scraping is the collection of methods would get some relevant data from a website instantaneously rather than copying it automatically. A Web scraper's goal is to scan, produce and index certain kinds of information into other new Web pages. Scrapers focus on the regeneration and inventory to unstructured data in defined data sets [1, 2]. Data analytics is gaining perspective into a large volume of the data available in structured and unstructured form. Data analytics in any industry and business is the need of the hour. A large amount of data available in the form of text, audio, pictures, videos, notes and the internet data. From a point of view, hardly anything stops the viewing of qualitative advertising as a kind of Web recommendation, with both the aim of embedding the much more applicable linguistic ads on a Website page [3–5]. To derive meaning and gain knowledge out of data-data analytics is used. Various cleaning, transformations are performed over the data. The Information showed by most sites must be seen utilizing a web program. They don't offer the usefulness to spare a duplicate of this information for individual utilize. The main choice at that point is to physically duplicate and glue the information - an exceptionally repetitive occupation which can take numerous hours or now and then days to finish. Web Scraping is the procedure of computerizing this procedure so that rather than physically duplicating

© Springer Nature Switzerland AG 2020
P. Karrupusamy et al. (Eds.): ICSCN 2019, LNDECT 39, pp. 279–285, 2020.
https://doi.org/10.1007/978-3-030-34515-0_29

the information from sites, the Web Scraping programming will play out a similar undertaking inside a small amount of the time. We can explicitly recognize that references from lexically related pages are more effective than connections between irrelevant pages for iterative processing. Such conclusions prove and conceptualize the underlying assumption of the existing link-based methods which further elucidate our understanding of link actual proof. A thorough understanding is influential in introducing new methods of connecting [6–8]. The Programming language used in this project- Python, which is a popular language for web scraping. Python has various modules which make it a powerful yet easy language for web scraping [9, 10]. Figure 1 depicted the general web scraping working methodology.

in
Web Data
Web Scraping Software
Excel
Application
Database
web page

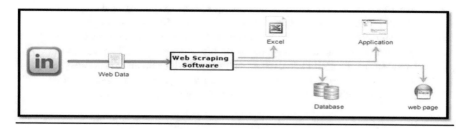

Fig. 1. Web scraping architecture

Terminologies Used:

1. Scrap/Targeted Data Source: This could be a web page, web service, etc.,
2. Data Extractor: This component extracts the required data from the response retrieved from the scrap data source Response.
3. Response: The data gathered from the source of the scrap may be an HTML, XML, simple text depending on the source of the targeted information. For instance, if you scrap a web page, you may be reluctant to get the HTML as your response.
4. Extracted Data: This is determined by you. This could just be a text, link, any crucial information, etc.
5. Web Crawling: Crawling relates to large amounts of data where you produce get your crawlers (or bots) which crawl to the depths of the web sites.

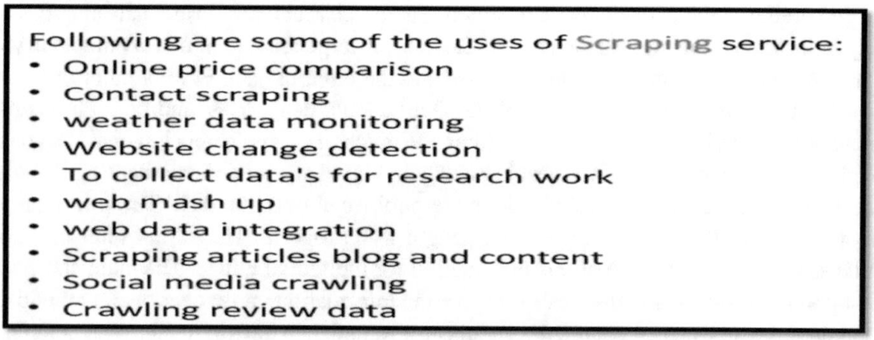

Following are some of the uses of Scraping service:
• Online price comparison
• Contact scraping
• weather data monitoring
• Website change detection
• To collect data's for research work
• web mash up
• web data integration
• Scraping articles blog and content
• Social media crawling
• Crawling review data

Fig. 2. Application of the web scraping

Figure 2 lists the web scraping major application. Understanding the objective to scrap the LinkedIn data and how it can be useful to prepare the team charts of the organizations. This can be helpful to grow the network between organizations and creating corporate Hierarchy. This implementation has various applications – Human Resource department for recruiting purposes. For an organization to grow the network among themselves and collaborate to work with different teams. To do this task two companies were chosen having an employee size of 100–200: Zepo and WebEngage.

Zepo Services-
Gets the Business Online Services-

- Creating websites
- Online Marketing
- Courier Products Anywhere
- Sell on Market Place

WebEngage Services–
Customer Engagement tool and Marketing services:-

- Personalization
- Segmentation
- Cross-channel Engagement

Team charts were prepared for both the companies by searching through the LinkedIn.

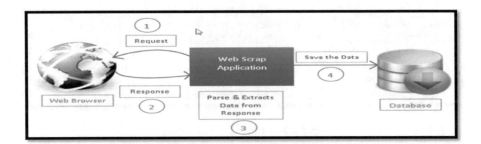

Fig. 3. Web scraping procedure

Figure 3 depicts the working principle of web scraping. Thought this procedure based on the linked in available profiles and data team charts will be searched for both sample companies.

2   Data Extraction and Analysis

In Fig. 6 the working principle of this data extraction through the LinkedIn database and analysis process has been explained by a flow chart.

ZEPO.IN

NAME	DESIGNATION	LOCATION
Nitin Purswani	Founder	Mumbai
Vikas Potta	Chief Operating Officer	Mumbai
Rajaverma K E	Finance manager	Chennai
OPERATIONAL HEADS		
Sumesh Kumar	operation officer	Pune
Raunak Kejriwal	operation officer	kota
Shubham Chandrayan	Logistics Operations Officer	Jharkhand
Prahaladh Vasudevan	Operations Officer	Mumbai
Sumeet Gadodia	product manager	Mumbai
Nitin Chaurasiya	On-Boarding Manager | Operations Officer	Mumbai
Pratik Shah	on-boarding manager	Mumbai
Shravan Kamat	ecommerce consultant	Mumbai
Nagarjun K	operation payment gateway	Bengaluru
Deepesh Kothari	Product manager and process movement coordinator	Mumbai
Abhishek Suman	operations manager	Mumbai
kunal poyekar	logistics and operations management	Mumbai
Kashif Khan	escalation manager	Mumbai
Amey Jaju	Operations Coordinator	Pune
BUSINESS TEAM		

Fig. 4. Team chart for Zepo

WEBENGAGE

Dhruvil Vaghela	Software Engineer	Mumbai
Mehul Jain	UI Designer	Mumbai
Arpit Goyal	Front-end Engineer	Mumbai
Prashant Acharekar	Programmer	Mumbai
Sneha Waykar	Application Support Specialist	Mumbai
Jharna Moorpana	Application Support Executive	Mumbai
Happy Kumar	software engineer intern	Mumbai
MARKETING TEAM		
Prasenjit Karmakar	marketing head	Mumbai
Divya Sharma Dixit	digital marketer specialist	Mumbai
Ajit Singh	Digital Marketing Manager	Mumbai
Avantika Pandey	Community manager and digital Marketer	Mumbai
SALES TEAM		
Apoorv Sood	Manager Sales (global)	Mumbai
Nilay Kohaley	Enterprise Sales Manager	Mumbai
Samir Trivedi	Enterprise Sales Manager	Mumbai
Poornima Sethumadhavan	Sales Engineer	Los Angeles,calfornia
Praneet Bhave	enterprise sales	Mumbai
HUMAN RESOURCE TEAM		
Ankita Jha	Retention Solution Manager	Mumbai

Fig. 5. Team chart for WebEngage

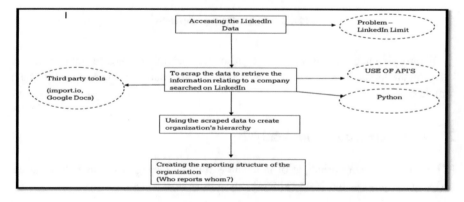

Fig. 6. Flowchart for the paper

Here accessibility from the LinkedIn accounts can search and explore the possible of interlinked accounts from same company and may find out the possible employees of one organization if they have LinkedIn accounts. It is a technology implementation that perpetually develops and alters, develops and strengthens its role as a vital part of the existences. This doesn't play a slight role in society at all now. It is a major challenge to deal with the implications of this and one that can influence everybody personal and professional. Therefore, it is vital that all in the computing industry make wise decisions regarding their conduct and make best possible use of this data analysis technique for the decision making and other processes [11, 12]. The workflow chart describes the course of action and activity to be performed at each stage. The problem encountered at the initial stage of collecting data is the LinkedIn limit to access the profiles of the individuals as it has put the limits to access the number of profiles. Even with premium accounts there are limits.

3 Python

Python is a valid and influential software application for objects analogous to Perl, Ruby, Scheme or Java. Python has powerful packages which makes it most suitable for web scraping. Anaconda package which is Python's prominent data sharing in data science platform. Anaconda's open source variant is a high - performance Python and R dispersion that includes further than 100 of the most successful data science applications. By using these tools code snippet has been designed to solve the intended original problem.

```
----------------------------------------------------------------------------------------------------------
"from bs4 import BeautifulSoup as soup
import time
from selenium import webdriver
from selenium.webdriver.common.by import By"

"driver=webdriver.Chrome("C:\\Users\\poorti
sahni\\Downloads\\chromedriver_win32\\chromedriver.exe") "
driver.get('http://linkedin.com');
time.sleep(1) # Let the user actually see something!
login_box = driver.find_element_by_name('session_key')
login_box.send_keys('poortisahni@gmail.com')
password_box = "driver.find_element_by_name('session_password')
password_box.send_keys('p886093!')"
login_box.submit()
time.sleep(1) # Let the user actually see something!
driver.get("https://www.linkedin.com/search/results/people/?keywords=snapdeal&origin=SWITCH_SE
ARCH_VERTICAL")
time.sleep(1) # Let the user actually see something!
----------------------------------------------------------------------------------------------------------
```

```
while True:
    result_ul = driver.find_element_by_class_name('results-list')
    people_list = result_ul.find_elements_by_xpath("//li[starts-with(@id,'ember')]")
    for p in people_list:
        p_id = p.get_attribute("id")
        if p_id is not None:
    p_description = p.find_element_by_xpath("//p[contains(@class,'level-1')]")
            if p_description is not None:
                print("Description: %s" % p_description.text)
            else:
                print("Description: NULL")
            p_area = p.find_element_by_xpath("//p[contains(@class,'level-2')]")
            print("place: %s" % p_area.text)
            p_link= p.find_element_by_xpath("//a[contains(@data-control-name,'srp')]")
            print("proflie-link %s" % p_link.get_attribute("href"))
            nextPageLink = driver.find_element_by_xpath("//div[contains(@class,'next')]")
    if nextPageLink is None:
        break
else:
    nextPageLink.click()
    time.sleep(3)
time.sleep(10000) # Let the user actually see something!
#driver.quit()
        print("================================")
        #print("Profile ID: %s" % p_id)
    p_name = p.find_element_by_xpath("//span[contains(@class,'actor')]")
    print("Name: %s" % p_name.text)
```

Using the above-mentioned code to apply web scraping we get the desired organizational charts of the candidate organization. These are shown in Figs. 4 and 5 respectively.

4   Conclusion

In today's world data scientists are using multiple strategies to analysis large volume of data coming from various sources. Through web scraping and coding in python for a data analytics domain will generate a desirable outcome to prove the significance of the mentioned process. Hence, this concept can be applied to various real time data analysis problem to obtain the fruitful outcomes.

5   Future Scope

Data analytics remains as the backbone of any organization and industry. Internet provides access to large data sets and information. Extracting meaningful information and relevant data for any intended purpose can be solved by data analytics itself. Social media, various websites are providing data in various formats. Businesses, government, non-private organizations are using web data analytics to become more connected to the customers, provide better services, better job opportunities and for further growth and expansion.

References

1. Glez, D., et al.: Web scraping technologies in API world. Brief. Bioinform. **15**(5), 788–797 (2014)
2. Vargiu, E., Urru, M.: Exploiting web scraping in a collaborative filtering based approach to web advertising. Artif. Intell. Res. **2**(1), 44–54 (2013)
3. Cowan, G.: Statistical Data Analysis. Oxford university press, Oxford (1998)
4. Mitchell, R.: Web Scraping with Python
5. Armano, G., Vargiu, E.: A unifying view of contextual advertising and recommender systems. In: Proceedings of International Conference on Knowledge Discovery and Information Retrieval, pp. 463–466 (2010)
6. Koolen, M., Kamps, J.: Are semantically related links more effective for retrieval? In: Proceedings of the 33rd European Conference on Advances in Information Retrieval, pp. 92–103. Springer, Heidelberg (2011)
7. Anagnostopoulos, A., Broder, A.Z., Gabrilovich, E., Josifovski, V., Riedel, L.: Just-in-time contextual advertising. In: Proceedings of the Sixteenth ACM Conference on Conference on Information and Knowledge Management, pp. 331–340. ACM, New York (2007)
8. Lacerda, A., Cristo, M., Gonçalves, M.A., Fan, W., Ziviani, N., Ribeiro-Neto, B.: Learning to advertise. In: Proceedings of the 29th Annual International ACM SIGIR Conference on Research and Development in Information Retrieval, pp. 549–556. ACM, New York (2006)
9. www.pythonprogramming.net
10. www.realpython.com
11. Berry, M.W.: Survey of Text Mining. Springer, New York (2003)
12. Adams, A., McCrindle, R.: Pandora's Box: Social and Professional Issues of the Information Age. Wiley, Hoboken (2008)

Physical Data Acquisition from Virtual Android Phone Using Genymotion

Sumit Sah, Animesh Kumar Agrawal, and Pallavi Khatri[✉]

Department of CSE, ITM University, Gwalior, India
sumitsah18@gmail.com, akag9906@gmail.com,
pallavi.khatri.cse@itmuniversity.ac.in

Abstract. With the large scale proliferation of mobile technology and smart phones, there has been an exponential rise in the cases of digital frauds and cybercrimes. With smart phones becoming powerful day by day with enhanced security features, smart phone forensics has become very important for every organisation. But the expensive forensic tools are a big impediment which leads to a no of crimes going undetected. Towards this there is a need to undertake mobile forensics without the use of commercial software's. This paper proposes manual data extraction technique using virtual phones created in an emulator (Genymotion) for android OS. Deleted data recovery using physical acquisition is done with the help of open source tool Autopsy and then a comparative analysis is done using another forensic tool Foremost.

Keywords: Information security · Android forensic · Genymotion · Virtual android device · Autopsy · Foremost · Mobile forensic

1 Introduction

Increase in digitalization all over the world has led to a large increase in digital crimes. Digital crimes are done by using smartphones to send offensive messages, sharing inappropriate contents or done by computers to gain access into another computers and to store inappropriate and illegal content. By using forensic approach investigators are able to gather the crucial evidence. Mobile forensics is an approach to retrieve the data from smartphones. It involves acquisition then examination and analysis of retrieved data from a smart phone. Same approach is used in this work.

The most crucial consideration in mobile forensics is that the data already present in the device should not be tampered be with under any situation during forensic investigation, otherwise the investigator cannot present the evidence in the court of law.

Android operating system was developed by Google in 2007 and now it has become the most popular operating system for mobile devices. Millions of smartphones are coming with android operating system since its open source architecture helps in ease of creating applications.

From last few years every social networking companies like WhatsApp, Hike, Facebook, etc. have started migrating to mobile applications because smartphones are being used extensively as compared to any other computing device. In addition, many other apps like Paytm, Tez, etc. have host of features which makes uploading photos,

© Springer Nature Switzerland AG 2020
P. Karrupusamy et al. (Eds.): ICSCN 2019, LNDECT 39, pp. 286–296, 2020.
https://doi.org/10.1007/978-3-030-34515-0_30

exchanging messages and finding lost friends very easy. The user database of these applications is very huge and that's why it makes them susceptible to digital crime and frauds. In the case of forensic investigation, a huge amount of sensitive personal information can be obtained by analysing the content of the phone. That's why the forensic approach is used for analysing the ROM of android device in this work.

In this paper a full forensic analysis of an already rooted android phone ROM is performed. The focus of this analysis is to do physical extraction and extract digital artifacts like mp3 files, pdf files, email id, png file and users Facebook id and contact number from an android ROM using a virtual environment.

This work presents forensic analysis of a rooted android ROM using physical extraction approach to obtain existing and deleted data from the mobile in virtual environment using Genymotion [1]. The experiment has been done with two tools Autopsy and Foremost to retrieve the existing and deleted digital artefacts from the device.

2   Related Work

Android OS was developed in 2007 by google and now it becomes most popular operating system for android devices. Android shares the market of 74.85% in the world [2]. Due to widely use of android device it is possible that digital crimes are done by using android devices. Hence, it is crucial to perform the forensic analysis of an android device and extract digital artifacts from an android-based mobile device. While doing forensic investigation in a digital device, care has to be taken that integrity of data is not lost. If integrity of data is lost then it can be presented before the court of law [3]. During digital analysis a forensic investigator has to strictly follow the guidelines [4] as stated by Associate of Chief Police Officers (ACPO). Some early works are already done in this field by Mellars in [5] worked on mobile device forensic in which the main focus was on retrieval of text messages and call logs from a mobile. Then Lessard and Kessler in [6] proposed a simpler examination of android mobile in which logical extraction was performed and obtained physical image using dd command. For doing this, they performed rooting of a device as a privileged super user and then obtain the physical image of the partition of the device. For analysis the physical image author used Cellebrite Universal Forensic Extraction Device (UFED). Then Vidas et al. in [7] gives new approach to avoid alteration of data for which author booted the device in recovery mode so that only recovery partition is affected. Then work proposed by Sylve et al. in [8] in which the author worked on volatile memory acquisition by using the most popular tool called Volatility. TCP connection and Memory dump of the phone SD card was used by the author to acquire the memory. The work done by Srivastava and Tapaswi in [9] author uses custom recovery image instead of stock Android recovery. Android Debug Bridge ADB is used in this work to connect to mobile device. Volatile memory of the device is being acquired during the work and AF logical tool is used for analysing the data dump. Hoog in [10] worked on android forensic where various methods for logical and physical data extraction from a mobile are being discussed. Author work on both paid and open source tools like Cellebrite UFED, MOBILedit and AFLogical etc., The author also focuses on analysing a SD card and embedded multimedia controller (EMMC) storage which are found on mobile devices. Smudge attack for bypassing the pattern-locked devices is also mentioned in this work. Guido et al. in [11] discussed about the methodology for data acquisition using physical acquisition with least data retrieval time. In this approach the author decreased the acquisition time by 20–40% by doing this they used Hawkeye algorithm. Also the work done by Li et al. in [12] was very compulsive. The author recovered deleted data from a rooted device. They recovered the deleted data from internal NAND storage and EMMC card of the device. For doing all of this process they developed a recovery tool using C++ language. The work done by Sathe and Dongre in [13], logical and physical acquisition techniques were used to exploit the digital evidences and also a comparative study between various techniques is being done to find a better approach for acquiring the digital evidence on mobile phone. They also performed an additional experiment on Samsung Galaxy Grand duos GT-19082 android smartphones and tried to acquiring the digital evidence. Raji, Wimmer and Haddad in [14] was worked on analysing data from smartphones and they also compare two forensic tools that are autopsy and Paraben E3: DS and by this work they mentioned that Paraben E3:DS performed better than open source tool Autopsy. Very interesting work is being done by Agarwal and Tapaswi in [15]. They performed forensic analysis of Google Allo messenger on Android platform. The main focus in this work was on the analysis of data stored by Allo application in internal file storage of an android device with the help of some third party application like ADB tools, db browser etc.

3   Background

The architecture of a commercial mobile analysis tool is not open source, primarily to protect the commercial interests of the manufactures. Hence, an investigator or a researcher is unable to capture the data flow between the tool and the mobile device, the memory map of the device and other finer details which can help him in gathering the data from the point of carrying out forensics. However, all tools use simple android based commands in the backend, which are nothing but Linux commands to access the mobile. In simple terms, an android device (older versions) can be treated like a memory card connected to a computer from which photos need to be accessed. However, the difference is that in case of an android mobile connected to a forensic workstation, it does not open an auto play window to give access to the treasure stored inside it. This information has to be manually extracted through android commands from it. Towards this, the android architecture which is Linux based as depicted in Fig. 1.

Fig. 1. Android architecture [*Source*: https://en.wikipedia.org/wiki/android-operating-system]

Mobile forensic is basically the branch of digital forensics. Mobile forensics means recovering the digital evidence or data under forensically sound conditions. This data acquisition process contains two methods, logical and physical. In logical analysis obtain the user files like (sms, contact, calllogs, media and app data) by connecting phone with PC and obtaining information using forensic tools like (XRY, Latern, etc.) but deleted data cannot be recovered using logical analysis because running system has no information of the deleted files, but logical analysis is fast, easy and reliable.

Physical analysis (Hex-dump analysis) is the static acquisition of mobile device forensic it is done by either connecting mobile device to PC or unmounting the flash memory and connecting with forensic hardware devices like (UFED). Data obtained by this method is basically present in raw format and it is converted into binary format by the tools used in physical analysis. By this method deleted data, files and hidden files are obtained and can be recovered along with user files like (sms, contact, call logs, media and app data).

4 Methodology

The main purpose of this research is to extract digital evidence from an android ROM using physical acquisition technique under forensic approach and also trace the deleted data. The experiment is done under forensically sound environment and during this the integrity of data is maintained. The integrity of data is a major requirement while doing digital data extraction [16]. If the integrity of data is tampered, then it is hard to find the digital evidence. The System specification used in this study is listed below in Tables 1 and 2.

Table 1. Specification of Workstation

CPU	Sixth-Gen i3 core processor
RAM	4 GB DDR4, 2133 MHz
GPU	Intel Integrated HD 520
OS	Windows 10 Professional
HDD	1 Tb SATA hard drive

Table 2. Specification of Android custom ROM

OS	Android custom 8.0
RAM	2 Gb
ROM	16 GB

The Software and tools used during this study are Genymotion, Virtual Box, Autopsy tool, and Foremost tool for analysis and extraction of deleted data. Disk dump (dd) image has been created for analysis purpose. The work in this experiment is divided into four phases as shown in the Fig. 2. The first phase is device setup phase, where we have prepared the virtual mobile device by loading and downloading some data in the device. Facebook application is installed on the device and is used for social connectivity. Some data from the device is deleted and now mobile device is put to forensic study where this experiment has tried to retrieve existing data/application, deleted data and data from Facebook page. The second phase is acquisition phase where a disk dump (dd) image of virtual mobile device is taken and is used for analysis. The dd image created is copied on the host machine for analysis. In third phase we have done the analysis of acquired dd image. During analysis, whole disk partition is analysed to get current data, deleted data from the virtual mobile device. Results retrieved out of analysis farewell discussed in the fourth and final discussion phase of the experiment.

Fig. 2. Flowchart of the methodology

5 Experimental Setup

In forensic investigation the first step is to seize the device and that device is taken into inspection. Theseizing process is done by forensic expert for preventing the evidence from tampering. This work has been performed on a virtual Android Rom using genymotion emulator so we did not need to seize the device. Android ROM is installed into genymotion and then some data set is inserted into it for investigation purpose.

Figure 3 shows the virtual android device in Genymotion. Window 10 operating system has been used for this experiment.

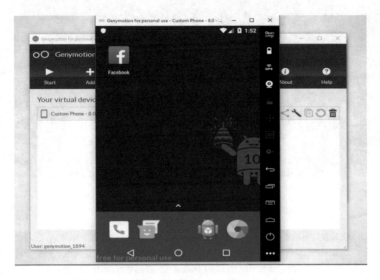

Fig. 3. Virtual android device using Genymotion

A data set consisting of pdf, jpeg files is created on the host machine and inserted into the virtual machine made on the same host for testing purpose as shown in Fig. 4. ADB (Android Debug Bridge) is used to connect with the android shell.

```
1|vbox86p:/mnt/shared/Downloads # cp data\ set -r /mnt/sdcard/
vbox86p:/mnt/shared/Downloads # cd ..
vbox86p:/mnt/shared # cd ..
vbox86p:/mnt # ls
appfuse  asec  expand  media_rw  obb  runtime  sdcard  secure  shared  user
vbox86p:/mnt # cd sdcard
vbox86p:/mnt/sdcard # ls
Alarms  Android  DCIM  Download  Movies  Music  Notifications  Pictures  Podcasts  Ringtones  data set
vbox86p:/mnt/sdcard #
```

Fig. 4. Copy data from shared memory to sd card.

After successfully completion of device setup phase now device is ready for acquisitionphase. After connecting with adb server next step is to find the path of data partition, the method of finding the path of data partition is mentioned below in Fig. 5.

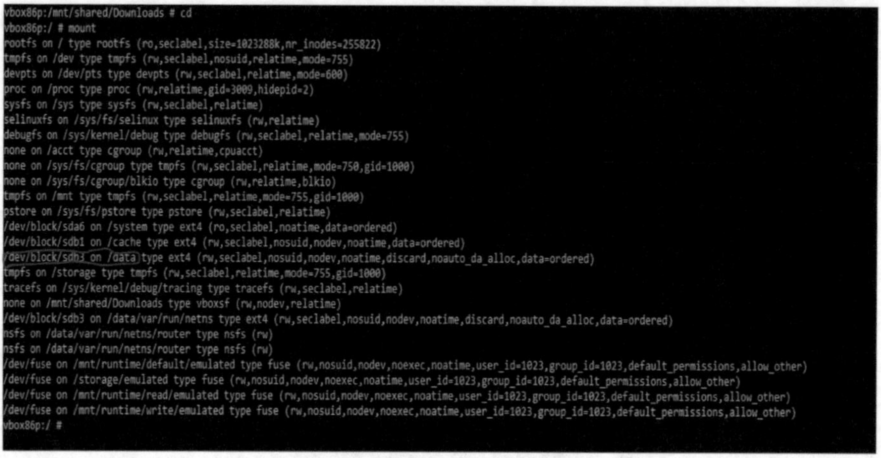

Fig. 5. Path of data partition

Path of data partition located was /dev/block/sdb3 as shown in Fig. 5. Next step is to create a disk dump (dd) image of this partition and save it into host machine for further analysis. Disk image is created in forensic investigation so that the device is not tampered during investigation. In this experiment the disk image was created to data partition after finding the path of data partition using disk dump (dd) command. Use dd command to create a disk dump of data partition and save it into host machine as shown in Fig. 6.

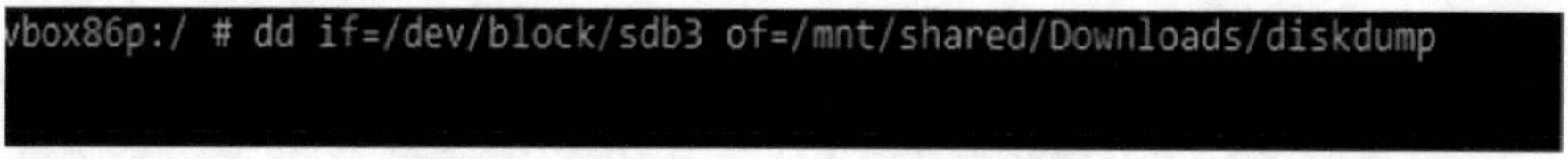

```
vbox86p:/ # dd if=/dev/block/sdb3 of=/mnt/shared/Downloads/diskdump
```

Fig. 6. Creating the dd file of an android ROM

After completion of acquisition phase, the disk dump file was automatically saved into host machine folder, the image file was then analysed using forensic approach with the help of autopsy - the sleuth kit tool and Foremost tool. Autopsy is GUI (Graphical User Interface) tool which is use for forensic analysis. It analyses whole disk to find the deleted files then categorise them into a proper manner. Figure 7 shows the analysis of image file using forensic tool called autopsy. Foremost is data carving tool which is used to recover forensic data using the header, footer, data structure of a file by using a process called carving. This tool is freely available and comes inbuilt kali Linux operating system. This is a CLI based tool and for using it some arguments are required, In Fig. 8 the analysis and recovery process of a disk dump file is shown using foremost tool.

Fig. 7. Analysis of image file using autopsy

```
root@kali:~# foremost -i /root/diskdump -o /root/foremost -t all -T
Processing: /root/diskdump
|foundat=
foundat=Core/configupdater-all.tar.lz00
foundat=Core/defaultetc-common.tar.lzLZIP
foundat=Core/defaultframework-common.tar.lzLZIP
```

Fig. 8. Analysing and recovering using foremost tool

6   Results and Discussion

This section discusses the various results obtained during this experiment. After analysing dd image file of an android ROM we could retrieve deleted as well as stored pdf, mp3, email id, png, jpeg files using autopsy tool and Foremost tool. Facebook profile was exploited to Facebook data like timelines photos, user email id, phone number could also be retrieved using autopsy tool (Fig. 9).

Source File	Keyword	Keyword Regular Expression	Keyword Preview
f0503384.elf	arjun.com11@gmail.com	(\.(?)[a-zA-Z0-9%+_\-]+(\.[a-zA-Z0-9%+_\-]+)*(\)?)]@X[...	60960232819575222J@<arjun.com11@gmail.com
Unalloc_11136_6258688_1476395008	arjun.com11@gmail.com	(\.(?)[a-zA-Z0-9%+_\-]+(\.[a-zA-Z0-9%+_\-]+)*(\)?)]@X[...	60960232819575222J@<arjun.com11@gmail.com
pluscontacts.db	arjun.com11@gmail.com	(\.(?)[a-zA-Z0-9%+_\-]+(\.[a-zA-Z0-9%+_\-]+)*(\)?)]@X[... 1	6bf2117d0f00504a <arjun.com11@gma

Fig. 9. Email id

After doing comparative analysis the result of the comparative analysis showed that Autopsy tool retrieves data far better than foremost tool. In most the cases autopsy finds large number of files. Foremost is able to acquire few pdf and some files A comparative analysis of the two approaches is shown in Table 3 (Figs. 10, 11 and 12).

Fig. 10. Shows the facebook profile information like email id, phone number

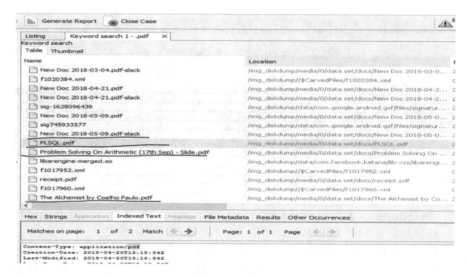

Fig. 11. Deleted pdf

audit.txt	860.0 kB
bmp	2 items
gif	10 items
htm	25 items
jar	3 items
jpg	266 items
mov	23 items
mp4	10 items
pdf	10 items
png	14,340 items
zip	29 items

Fig. 12. Extracted data using Foremost tool

Table 3. Comparison of proposed methodology with existing approach

Parameter considered	Data extraction using foremost tool	Data Extraction using autopsy
Analyse dd file	Yes	Yes
Compatible for analysing android OS file	No	Yes
Recovery of deleted files	Yes	Yes
Find Email id	No	Yes
Find social media information	No	Yes
Find Email id	No	Yes
Find social media information	No	Yes
Find Mp3 file	No	Yes

7 Conclusion

This work presents a work that attempts to obtain the physical image of virtual android device and then perform physical data extraction for forensic purpose. It also does a comparative study between autopsy and foremost tool for analysing the dump of dd image file of a virtual android device ROM. This approach can be used to carry out forensic analysis in a small organisation which does not have enough budget to buy expensive forensic tools.

References

1. https://www.genymotion.com/
2. http://gs.statcounter.com/os-market-share/mobile/worldwide
3. Induruwa, A.: Mobile phone forensics: an overview of technical and legal aspects. Int. J. Electron. Secur. Digit. Forensics 2(2), 169 (2009)
4. New ACPO guide for forensics: Computer fraud and security, vol. 2007, no. 7, p. 20 (2007)
5. Mellars, B.: Forensic examination of mobile phones. Digit. Investig. 1(4), 266–272 (2004)
6. Lessard, J., Kessler, G.: Android forensics: simplifying cell phone examinations (2010)
7. Vidas, T., Zhang, C., Christin, N.: Toward a general collection methodology for android devices. Digit. Investig. 8, S14–S24 (2011)
8. Sylve, J., Case, A., Marziale, L., Richard, G.G.: Acquisition and analysis of volatile memory from android devices. Digit. Investig. 8(4), 175–184 (2012)
9. Srivastava, H., Tapaswi, S.: Logical acquisition and analysis of data from Android mobile devices. Inform. Comput. Secur. 23(5), 450–475 (2015)
10. Hogg, A.: Android mobile forensic
11. Guido, M., Buttner, J., Grover, J.: Rapid differential forensic imaging of mobile devices. Digit. Investig. 18, S46–S54 (2016)
12. Li, Z., Xi, B., Wu, S.: Digital forensics and analysis for android devices, computer science and education (ICCSE). In: 2016 11th International Conference on IEEE, pp. 496–500 (2016)
13. Sathe, S.C., Dongre, N.M.: Data acquisition techniques in mobile forensics. In: 2018 2nd International Conference on Inventive Systems and Control (ICISC) (2018)

14. Raji, M., Wimmer, H., Haddad, R.J.: Analysing data from an android smartphone while comparing between two forensic tools. In: SoutheastCon 2018 (2018)
15. Agrawal, V., Tapaswi, S.: Forensic analysis of Google Allo messenger on Android platform. Inform. Comput. Secur. **27**(1), 62–80 (2019)
16. Ayers, R.P., Brothers, S., Jansen, W., Materese, R.: Guidelines on mobile device forensics (2016). www.nist.gov/node/5634

A Systematic Review on Load Balancing Issues in Cloud Computing

Mohammad Haris$^{(\boxtimes)}$ and Rafiqul Zaman Khan

Department of Computer Science, AMU, Aligarh, India
mohammadharis.amu@gmail.com

Abstract. Cloud computing is gaining popularity day by day due to its various qualities. It offers web services and computing resources on demand basis at a very low cost. Demand for cloud services is increasing everyday which brings a lot of load traffic with it, which need to be tackled carefully to maintain the performance of the cloud. To avoid overloaded situations, Load balancing is used to distribute load to all the available nodes efficiently. However, there are several Load balancing issues which requires special attention to give best possible solution to cloud technology. This survey highlights load balancing issues which need to be handled carefully for better performance of cloud computing. Also, we discussed the goals, need and techniques of cloud load balancing.

Keywords: Cloud computing · Load balancing · Goals of load balancing · Load balancing techniques · Issues of load balancing

1   Introduction

Cloud computing is an internet-based architecture for providing unlimited resources and services. It uses the concept of "pay-as-per-use" where user needs to pay only for what they actually use. Cloud computing creates an environment for accessing computing resources in a demand way. With the help of cloud computing, resources can be granted and removed very quickly and easily without a need of service provider interaction. It also increases the availability of resources and reduces the cost of hardware and software. Most importantly, the key focus of cloud computing is to create data center more powerful so that they provide dynamic and flexible services to the cloud users [1].

As cloud computing is growing fast, more services and better results are requested by cloud users. So, load balancing becomes a very important part of the cloud. If any component of cloud failed, load balancing helps in continuation of the services efficiently by enabling provisioning and de-provisioning of instances of applications with proper utilization of resources. Thus, Load Balancing is a process for dispersing the workload dynamically across all the nodes presented in the workspace. It avoids the scenario of overloading and underloading of nodes at any instant of time. Load balancing improves overall performance of the cloud by maximizing resource utilization [1, 2]. Figure 1 shows the distribution of workload without load balancing in which job request are assign randomly to any node without checking that node is heavily loaded

© Springer Nature Switzerland AG 2020
P. Karrupusamy et al. (Eds.): ICSCN 2019, LNDECT 39, pp. 297–303, 2020.
https://doi.org/10.1007/978-3-030-34515-0_31

or underload. Hence, it degrades the performance of cloud. Therefore, proper load balancing is required.

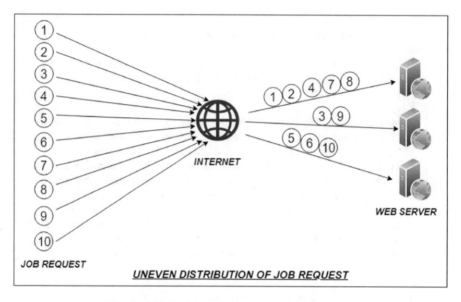

Fig. 1. Distribution of load without load balancing.

The aim of this paper is to provide overview on cloud load balancing, discusses various load balancing techniques and addresses current load balancing issues in a cloud environment.

2 Cloud Load Balancing

Load balancing is playing a vital role in cloud computing. It is a mechanism of dispersing the workload equally among all the available nodes to achieve high resource utilization ratio and user satisfaction [3]. Load balancing uses a technique to divide workload on the multiple virtual machines so that every virtual machine does equal amount of work to achieve efficient resource utilization which increases throughput and reduces response time. It minimizes the waiting time of the resources and also avoids resource overloading [1]. However, efficient allocation of resources by using proper load balancing helps in reducing costs. Hence, improving resource utilization and the performance of a distributed system ultimately will reduce the energy consumption and carbon emission which also contribute in achieving Green computing [4]. Figure 2 describes the working of proper load balancing which first calculate the load on every node before assigning the task.

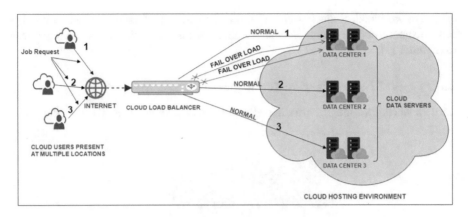

Fig. 2. Load balancing in cloud computing.

2.1 Goals of Load Balancing

Efficient Load balancing must consider following objectives for better performance of cloud systems [5, 6].

1. Efficient resource utilization.
2. Maximum throughput.
3. Minimum response time.
4. Avoiding overload.
5. Maintain system stability.
6. Build fault tolerance system.
7. Accumulate future modification.
8. Increases user's satisfaction.
9. Improve overall performance.

2.2 Need of Load Balancing

As the demand for cloud resources and services increases, load balancer is required for managing the increasing excessive workload. The strategy of balancing the load between the cloud resources will directly effect on the performance of cloud. The load imbalance across the cloud resources results in system overloading and causes to the crashing of cloud systems. This degrades the performance of cloud. The distribution of load on cloud computing is basically a process of allocation of task and job on dedicated resource such as virtual machine. In cloud computing, the distribution of load is based on distributed computing. Load balancing is the mechanism of assigning and re-assigning the load evenly across all the servers which avoid the server's overloading and server crashes. Thus, the effective load balancing is needed for cloud computing environments [7, 8].

Energy consumption and carbon discharge by the cloud resources are increasing day by day as the demand of computing resources getting high. Load balancing avoids over heating of virtual machines by minimizing the amount of energy consumed due to excessive workload. Reducing energy utilization using load balancing will automatically decrease carbon emission and for that reason also helps in achieving Green Computing [9].

2.3 Types of Load Balancing Techniques

Load Balancing techniques are divided into three categories [1, 3, 10, 11] (Fig. 3).

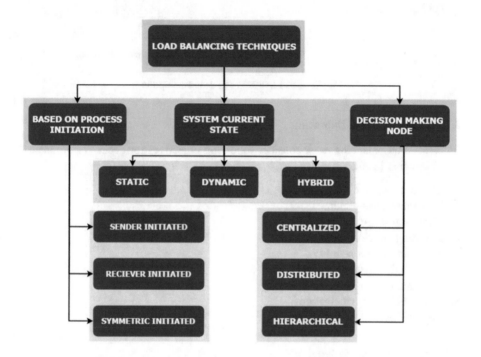

Fig. 3. Load balancing techniques used in cloud computing

- Load balancing based on process initiation:
  1. Sender Initiated: The sender initializes the request to find the underloaded nodes.
  2. Receiver Initiated: In this technique, receiver proceed the load balancing.
  3. Symmetric: It combined both types of load balancing i.e. sender and receiver initiated.

- Using state of the system, load balancing technique can be subcategories as follows:
  1. Static Load Balancing: This technique doesn't consider the system's current state. It requires prior knowledge of system like memory, communication time etc. It is simple and easy to simulate but not suitable for heterogenous environment.

2. Dynamic Load Balancing: This type of load balancing take decision based on current state of a system. It allows transferring the load from overloaded node to underloaded node in run-time. It is difficult to implement but highly adaptable.
3. Hybrid Load Balancing: Hybrid method used both static and dynamic load balancing technique. It is best for balanced distribution of computing task and helps in minimizing the communication cost of distributed computing nodes (Table 1).

Table 1. Comparison between static, dynamic and hybrid load balancing techniques [9].

Parameters	Static load balancing	Dynamic load balancing	Hybrid load balancing
Nature	Compile time	Run time	Depends on system requirement
Performance	Less	More	High
Reliability	Less	More	High
Cost	Low	Moderate	High
System info	Past knowledge	Current knowledge	Required both past and current
Response time	Maximum	Moderate	Minimum
Implementation	Easy	Difficult	Difficult
Nature of model	Homogenous	Heterogenous	Heterogenous
Complexity	Less	High	High
Fault tolerance	No	Yes	Yes

- Cloud computing nodes are highly distributed. Therefore, nodes take a decision which techniques has to be used. Depending upon which node is responsible for balancing of load in a cloud environment, load balancing techniques categories as:

1. Centralized Load Balancing: In this category, single nodes works as a central node which keeps all the information of the entire cloud network and takes decision on behave of all the nodes. It takes minimum time to analyses the resources condition but in case of failure, it takes time to recover.
2. Distributed Load Balancing: Multiple nodes takes part in Distributed load balancing. Every node takes local information of the cloud network and guarantee efficient distribution of task. If one node fails accidently, all other nodes works smoothly that makes it fault tolerance.
3. Hierarchical Load Balancing: This approach works on the bases of master-slave mode. A tree structure is created to represent levels of clouds in which nodes works under the master/parent nodes. Parent node takes decision based on information collected by slave node.

3   Load Balancing Issues in Cloud Computing

Cloud Load balancing is one the serious topic which required special attention. The current load balancing issues are as follows [11–17].

1. Energy Management: One of the critical issues in the cloud environment is reducing the energy consumption of virtual machines in data centers. Data centers directly effect on both the energy resources and the environment. Demand of cloud services increasing so fast that it required more computing resources for processing. Cloud resources must be tackled in an energy efficient manner which required a proper load balancing technique for cloud virtual machines so that it optimize resources utilization not only to satisfy Quality of Service (QoS) but also helps in achieving green computing.

2. Development of small data centers in cloud computing: Building up small data-centers around the world is beneficial for cloud users. It is cheaper than large data centers and also consumed less energy. Load balancing become a global issue to guarantee an adequate response time with an optimal distribution of resources.

3. Single point of failure: Many centralized algorithms offer effective and efficient mechanisms for processing load balancing in a dynamic way. But the problem is entire system is controlled by central node. In such condition, if the center controller crashes, then the entire system fails. For this reason, the algorithm must be designed in such a manner that it avoids any single point of failure.

4. Distribution of the cloud nodes globally: The cloud nodes are distributed globally. Some load balancing techniques are designed for smaller distance which don't consider factors such as communication delay, network delay, distance between user and resource etc. But the nodes which are present at distant locations degrades the performance of cloud as there is no load balancing techniques which works efficiently for faraway nodes.

5. Virtual Machines Migration: Virtualization is the most important feature of cloud which creates many virtual machines on a single physical machine. The aim of VM is to distribute load dynamically to other VM during execution when physical machine overloaded. So, the load balancing technique must work in an effective way to avoid bottlenecks in cloud computing.

6. Load Balancer Scalability: Cloud computing provide its services on a demand basis where user can access services anytime without a need of service provider inter-action. So, the good load balancer should accept the quick changes given by cloud users to provide quality of services to users.

7. Algorithm Complexity: Load balancing algorithms should be less complex in operations and easy to implement. For monitoring and controlling the algorithms during implementation, a complex algorithm can delay in higher communication which degrades the performance of cloud.

4 Conclusion

Cloud computing is one of the traditional computing technology for proving quality of service (QoS) to users at anytime and anywhere on pay-per-use basis. The users of cloud computing are increasing everyday as there is good demand for cloud services. In such situation, load balancing will play a crucial role in providing efficient and reliable services to users. In this work, we have identified load balancing issues which must be addressed to provide efficient load balancing in cloud technology. Also, we overview several load balancing techniques used for distributing the workload among nodes. Considering these issues will be helpful in the near future for implementation of effective load balancing techniques so that it improves the performance of cloud computing.

References

1. Mishra, N.K., Mishra, N.: Load balancing techniques: need, objectives and major challenges in cloud computing-a systematic review. Int. J. Comput. Appl. **131**(18), 11–19 (2015)
2. Haris, M., Khan, R.Z.: A systematic review on cloud computing. Int. J. Comput. Sci. Eng. **6** (11), 632–639 (2018)
3. Balagoni, Y., Rao, R.R.: Importance of load balancing in cloud computing environment: a review. Int. J. Adv. Trends Comput. Sci. Eng. **3**(5), 77–82 (2014)
4. Kansal, N.J., Chana, I.: Cloud load balancing techniques: a step towards green computing. IJCSI Int. J. Comput. Sci. Issues **9**(1), 238–246 (2012)
5. Pathak, R.: A survey of load balancing in cloud computing: challenges and algorithms. Int. J. Innov. Res. Comput. Commun. Eng. **5**(1), 740–745 (2017)
6. Joshi, S., Kumari, U.: Load balancing in cloud computing: challenges & issues. In: 2nd International Conference on Contemporary Computing and Informatics (IC3I), pp. 120–125. IEEE (2016)
7. Shaikh, S.J., Rathod, S.B.: A QoS load balancing scheduling algorithm in cloud environment. Int. J. Comput. Trends Technol. (IJCTT) **30**(1), 1–5 (2015)
8. Dangi, K., Gaud, N.: A brief review of load balancing issue in cloud computing environment. Int. J. Comput. Appl. **154**(3), 16–20 (2016)
9. Khan, R.Z., Ahmad, M.O.: Load balancing challenges in cloud computing: a survey. In: Proceedings of the International Conference on Signal, Networks, Computing, and Systems, pp. 25–32. Springer (2016)
10. Pandey, S.: Cloud load balancing: a perspective study. Int. J. Eng. Comput. Sci. **6**(6), 21602–21611 (2017)
11. Kumar, P., Kumar, R.: Issues and challenges of load balancing techniques in cloud computing: a survey. ACM Comput. Surv. (CSUR) **51**(6), 120 (2019)
12. Tripathi, A., Singh, S.: A literature review on algorithms for the load balancing in cloud computing environments and their future trends. Math. Comput. Model. **21**(1), 64–73 (2017)
13. Ghomi, E.J., Rahmani, A.M., Qader, N.N.: Load-balancing algorithms in cloud computing: a survey. J. Netw. Comput. Appl. **88**, 50–71 (2017)
14. Al Nuaimi, K., Mohamed, N., Al Nuaimi, M., Al-Jaroodi, J.: A survey of load balancing in cloud computing: challenges and algorithms. In: 2012 Second Symposium on Network Cloud Computing and Applications, pp. 137–142. IEEE (2012)
15. Mesbahi, M., Rahmani, A.M.: Load balancing in cloud computing: a state of the art survey. Int. J. Mod. Educ. Comput. Sci. **8**(3), 64–78 (2016)
16. Sreenivas, V., Prathap, M., Kemal, M.: Load balancing techniques: major challenge in cloud computing-a systematic review. In: 2014 International Conference on Electronics and Communication Systems (ICECS), pp. 1–6. IEEE (2014)
17. Khan, R.Z., Ahmad, M.O.: A survey on load balancing algorithms in cloud computing. Int. J. Auton. Comput. **2**(4), 366–383 (2017)

Semantic Analysis of Big Data in Hierarchical Interpretation of Recommendation Systems

R. Lavanya[(⊠)] and B. Bharathi

Department of Computer Science and Engineering,
Sathyabama Institute of Science and Technology, Chennai, India
lavanyaconf@gmail.com, bharathi.cse@sathyabama.ac.in

Abstract. In today's scenario where Big Data is being used at its maximum potential, showing most of its influence on searching from the internet. It gives out large amount of data to the user which becomes too overwhelming for the user to analyze and understand. This led to the introduction of Recommendation Systems whose main purpose is to give relevant datasets according to the user's preference which makes it easy for the user to understand and analyze the best option among the limited options he/she has received from the system. Recommendation Systems exhibit some kind of implicit hierarchy based on either users or items to give the best recommendation to users. But it has been noticed that these systems produce a lot of ambiguities. Hence, leading to a lot of repeated results. This paper investigate various ways to understand the implicit working of hierarchical structures and make some improvisations on the same with the help of semantic analysis under collaborative filtering approach.

Keywords: Big data · User preferences · Recommendation system · Collaborative filtering · Semantic analysis

1 Introduction

In the world where technology is getting much faster and smarter than human brains, our generation is coming across completely new aspects of data storage and analysis. To make the technology efficient, it is very important to store and analyze each and every little detail. All these little attributes when combined together introduces us with the term Big Data. Big Data is a term which introduces us with large volume of data that is produced and helps run a commerce scheduled a day-to day base. Big Data is just not about the fact that the data is large but it also about how efficient results are obtained. Thus, if analyzed properly it would give us intuitions that top to improved conclusions and calculated trade changes. In Big Data we come across four V's: Volume, Variety, Velocity and Variability. Big Data has applications all around the world. It is used for profit strategies, aircrafts, business, etc.

One of the most prominent applications of Big Data is Recommendation System. Recommendation Systems can be used in many scenarios where the user is exposed to a huge volume of items. These systems filter out the results which will not be preferred by the user and recommends the items which will gain interest of the user. This system makes it easier for the user to receive and analyze limited dataset which he/she can

© Springer Nature Switzerland AG 2020
P. Karrupusamy et al. (Eds.): ICSCN 2019, LNDECT 39, pp. 304–310, 2020.
https://doi.org/10.1007/978-3-030-34515-0_32

handle individually and help them decide on appropriate items. Every individual has a set of different opinions on the same item, recommendation system provides user with the facility to understand user's preference, tastes and with the help of that it finds desirable contents for the user. While every individual can have different preferences on different items, the basic job of a recommendation system is to find a pattern which finds similar patterns in the different opinions in order to create suggestions to both new as well as old users proposed by Qian et al. [1]. Recommendation System follows three algorithms to determine user's preference:

(1). Content Based Filtering
(2). Collaborative Based Filtering

- Model-Based CF
- Memory-Based CF

(3). Hybrid Based Filtering.

1.1 Content Based filtering

Content based filtering purely gives recommendations for worker centered happening the preferences made by the user in the past. In further disputes, these procedures attempt to recommend substances that are alike to those that a customer adored in the earlier Liu et al. [2]. In precise, numerous contender substances remain associated by substances beforehand valued by the customer and the best-matching items are recommended. It is usually considered to be a good algorithm but there are various drawbacks to this algorithm that is, it cannot be used on a type of platform where the rate of arrival of new users is very high because then there will be no history of past experiences.

1.2 Collaborative Filtering

The problem faced by content based filtering is solved by Collaborative Filtering algorithm. As we know that the term Collaborative means collective or together, so it tends to follow the approach where when a user asks for recommendation on a particular item this algorithm looks for the users rating of preference on that same item and then recommends the new user accordingly. This algorithm is not only user based, but its item based too, as it also considers the ratings of individual items as compared to other items and recommends user with items which has received the best ratings. This approach is mainly classified into two types:

- Model-Based approach: This procedure makes a paradigm for customer product rating openly together suggests the same through information;
- Memory-Based approach: This procedure makes a paradigm to find the marks not provided by the customer for alike products or friends.

1.3 Hybrid Based Filtering

It is the combination of both content based and collaborative based filtering which helps to overcome the shortcomings of procedures and gives us a better outcome or recommendation in this case. A more effective method to obtain better recommendations from any system is devised i.e. Matrix Factorization method. Matrix factorization algorithms decompose the user-item interaction matrix into product ofatwo lower dimension rectangular matrices. Simon Funk reported about this method during Netflix prize challenge in 2006. Many methods of matrix factorization exist such as the following:

- Funk SVD: It is the original algorithm which was proposed by Simon Funk. It factorizes the customer- product rating dimension as the multiple of all.
- There are some disadvantages, such as this method is not *model-based* which means that if a new user is added, the algorithm is incapable of adjusting it unless the whole model is retrained. This is an example ofaa cold- start problem.
- Asymmetric SVD: It combines the advantages of SVD++ while being a model based algorithm; therefore it considers new users with a few ratings without needing to retrain the whole model.

Hybrid Matrix Factorization: To deed the ever swelling quantity and assortment of offered communication data and use cases, these algorithms are capable of combining overt and implied communications or both contented and collaborative data.

2 Literature Work

2.1 Recommendation Systems

Recommender schemes show an vital part in aiding wired customers catch applicable info with demographic contours with chronological statistics proposed by Tang et al. [3]. Collective Filtering (CF), necessitates previous assessments to foresee unidentified assessments, covered more. Normally, collaborative filtering might categorized like— (1) memorial-centered procedures, largely focus on customer or objects in the customer-product ranking environment for reference; and (2) ideal- centered approaches, which typically accept that an fundamental ideal oversees the mode customer's rank. The informations unavailability delinquent damages recital of recommender schemes. First tactic to improve informations unavailability is graph-centered recommender schemes. Alternative tactic with assuaging informations unavailability is best N suggestions. LIke, Wang et al. [4] offered a Bayesian adapted grade on best N suggestions. Nasraoui et al. [5] assimilated public setups to medium novelization like associates ensure alike comforts thereby efficiency increases. Cai et al. [9] probed the disbelief/rival associations in community grids for approval and recognized as the same as advantageous with belief/associate suggestions. Assessment matters utilized in enlightening system performance. In recent times, categorized arrangements, becoming supporting statistics with recommender schemes.

2.2   Hierarchical Interpretation for Movie Recommendation

Bamshad et al. [6] have suggested web usage mining methodology entitled Profile Accumulations. In that procedure Grouping is through on databank by equivalent type of trades plus page outlook grouping messy to forecast the alike sheets in individual matter. Cho et al. [7] suggested prime tree initiation technique, to unravel tricky of scarcity and adaptability in collective filtering technique. Mohamed et al. [8], recommends a Tag recognized collective filtering Reference method for individual learning Environs. Here 16 dissimilar tags grounded collective Filtering procedures are employed and related in relations of accurateness and consumer indulgence. <Customer-product> Comparative transformed as <customer, piece, label> relative. The consequence displays as article centered K-means grouping Procedure stretches greatest recital in disconnected assessment. The consequence of assessment displays diversity amid excellence of customer familiarity and great approval correctness restrained by numerical quantity.

Hu et al. [9], proposes a personalized Recommendation approach grounded on Three Social Influences, Personal interest means customer–product association and relational impact and interactive concern likeness as customer-product connection of communal systems. Probablistic medium Factorization cast-off Tests remain complete on datasets Movielens and yelpp. The tactic eradicates the delicate of cold start and data scarcity. Yuan et al. [10], suggests a recommendation scheme for land selling sites concerns consumers in acquiring fresh assets. Recommendation scheme is established on inclusion instance centered perceptive (ICP) and Ontology. Former schemes cares solitary distinctive probe schemes with scheme provision multivalued quest scheme.

Customer quest behaviors are planned and a information vile is primed. Formerly the semantic gist of qualities and association amid is distinct by method. Outcome displays that this method is well-organized and rational for covering exploration in land selling sites. Cai et al. [11], suggests a cross collective filtering reference method centered on customer inclinations and article structures. Customary collective filtering endorsement training has trials like (1). Data scarcity (2).adaptability (3). Duplicity-: alike substances can have dissimilar terms and sense so recommendation scheme give diverse stuffs.

Naseem et al. [12], suggests a novel tactic of content centered Recommendation scheme grounded on Assignment knowledge. Mojtaba et al. [13], offered a cross recommendation methodology for e- learning situation. Two kinds of characteristics are measured for knowledge possessions 1. Obvious qualities alike topic and term of the issuer 2. Implied qualities are obtained through historic scores of beginners. Sumathi et al. [14], proposed an EAB-RS,a Learner preference tree (LPT) is prepared for each learner from historical logs and ratings then similarity between learners is find out and rating prediction for resources is given by NNCF Algorithm. In IAB-RS, implicit attributes are extracted using Genetic algorithms(GA) then similar users are find out and lastly score forecast is completed by NNCF. This recommendation method determinations the tricky of Data scarcity, cold start and deliver added varied approval incline.

Mobasher et al. [15] need cast-off collective technique and content withdrawal filtering methods for online suggestion. Also Fuzzy C-Mean and Ant colony grouping practices utilized for disconnected practice. Liu et al. [16] obligate cast-off the data organizations like Net- Concern Medium, Customer-Concern Medium, and Frequent-concern Medium for suggestion and personalization of websites centered on the customer concern. Hence article reference is proven on the vile of customer curiosity. Garcin et al. [17] obligate perspective hierarchy procedure for improved recommendation of update and tiers. The scheme affords enhanced recommendation about source of customer's concern. So amended forecast exactness and recommendation excellence. It suggested a grouping technique to yield improved and rapid approvals for final customer. For grouping semantically intelligible groups utilized. Too for reference Purview ontology grounded on the key concepts mined from the net insides were sited. Another one ought to grouping and relationship rule mining by web usage mining for improved approval. Hemalatha et al. [18] cast-off consecutive outline mining procedure for forecasting the following web sides. At that stage archetypal base filtering procedure casted-off, to handle successive outlines, thereby recommendation directions cohort.

2.3   Concerns in Recommendation System

Data Scarcity. It occurs while availing fewer scores for stuffs leads to problematic in endorse substances to the customer. CF schemes grieve from this tricky.

Cold Start. Provided two ways of cold start delinquent, fresh customer and first-hand article tricky. There is no evidence roughly on fresh customer and novel article makes suggestions unpredictable.

Scalability. It is the capability of approval scheme to knob increasing quantity of evidence. Evidence around the customer and article raises quickly on online. CF structures turn out to be exclusive to grip increasing quantity of evidence and stretches unclear references.

Over Specialization. Approval schemes endorse stuffs grounded on preceding account. Customer's does not acquire expanded recommendations as suggested by Praveena et al. [19]. The overall issues were narrated in the Table 1 given below.

Table 1. Comparison of techniques of recommendation system.

Sr. No.	Title of the paper	Methods involved	Algorithms and approaches	Pros	Cons	Future scope
1.	Bamshad et al.	Content based filtering	RAPPOR-perturbation algorithm	Satisfies differential privacy and utility guarantees,	(i) limited to simple data analysis (ii)can't compute aggregates on numeric attributes,	To increase accuracy with increased number of attributes
2.	Shin et al.	Content based filtering	Matrix factorization algorithms under local differential privacy (LDP) dimensionality reduction technique	Protects both user's items and ratings and Ensures privacy of both from recommender	Utility issues	Communication slide from the server to users to be reduced further
3.	Mohamed et al.	CF based	K-means clustering, Apriori, NBC, KNN, etc.	Reduces data sparsity to some extent	User experience not relate high approval accuracy	To increase the accuracy on user experience
4.	Liang Hu et al.	CF based	Probabilistic matrix factorization	Eliminate cold start and data sparsity	Gives diversified results	To increase recommendation efficiency
5.	Yuan et al.	Knowledge based	Case based reasoning and ontology	Gives a multiattribute search system	Gray Sheep problem, cold start problem	To Improve prediction quality
6.	Qian et al.	Hybrid CF method	TF-IDF model	More accurate than other CF methods	Scalability problem	To Improve efficiency, Scalability
7.	Tang et al.	Content based	BGM method	Solve the problem of data sparsity	Only for the trustworthy customers	To Improve prediction quality
8.	Cai et al.	Hybrid Approach	Nearest neighbor CF	Remove cold start and data sparsity issue	Scalability issue	Improve recommendation

3 Conclusion

Currently numerous recommendation schemes grounded on collaborative filtering, content based filtering and cross recommendation procedure ought to been projected. To solve the scalability issue connected to recommendation scheme and also giving improved recommendation, Collaborative filtering technique (CF) concerned as the optimized scheme. This paper concerns all collaborative based recommendation methods for giving improved suggestions. Also studied collaborative filtering procedure along with Cross reference for greater forecast excellence of recommendation scheme. Thus the assessment aided in diagnosing the challenges in recommendation scheme.

References

1. Qian, X., Feng, H., Zhao, G., Mei, T.: Personalized recommendation combining user interest and social circle. IEEE Trans. Knowl. Data Eng. **26**(7), 1763–1777 (2014)
2. Liu, Q., Chen, E., Xiong, H., Ge, Y., Li, Z., Wu, X.: A cocktail approach for travel package recommendation. IEEE Trans. Knowl. Data Eng. **26**, 278–293 (2014)
3. Tang, X., Zhou, J.: Dynamic personalized recommendation on sparse data. IEEE Trans. Knowl. Data Eng. **25**, 2895–2899 (2013)
4. Wang, Z., Liao, J., Cao, Q., Qi, H., Wang, Z.: Friend book: a semantic-based friend recommendation system for social networks **16** (2014)
5. Nasraoui, O., Petenes, C.: An intelligent web recommendation engine based on fuzzy approximate reasoning. In: Proceedings of the IEEE International Conference on Fuzzy Systems - Special Track on Fuzzy Logic and the Internet (2003)
6. Mobasher, B., Dai, H., Luo, T.: A personalized recommender system based on web usage mining and decision tree induction, vol. 21, pp. 426–435 (2001)
7. Cho, Y.H., Kim, J.K., Kim, S.H., Nakagawa, M.: Improving the effectiveness of collaborative filtering on anonymous web usage data. Expert Syst. Appl. **23**, 329–342 (2002)
8. Chatti, M.A., Dakova, S., Thu, H., Schroeder, U.: Tag-based collaborative filtering recommendation in personal learning environments **6**(4), 337–349 (2013)
9. Hu, L., Song, G., Xie, Z., Zhao, K.: Personalized recommendation algorithm based on preference features. Tsinghua Sci. Technol. **19**, 293–299 (2014). ISSNll1007-0214ll08/11ll
10. Yuan, X., Lee, J.-H., Kim, S.-J., Kim, Y.-H.: Toward a user-oriented system for real estate websites. Inf. Syst. **38**, 231–243 (2013)
11. Cai, Y., Leung, H.-F., Li, Q., Min, H., Tang, J., Li, J.: Typicality-based collaborative filtering recommendation. IEEE Trans. Knowl. Data Eng. **26**(3), 766–779 (2014)
12. Biadsy, N., Rokach, L., Shmilovici, A.: Transfer learning for content-based recommender systems using tree matching. IEEE Trans. Learn. Technol. **36**, 223–236 (2013)
13. Salehi, M., Kamalabadi, N.I., Ghoushchi, M.B.G.: Effective recommendation framework for personal learning environments using a learner preference tree and a GA. IEEE Trans. Learn. Technol. **6**(5) (2013)
14. Sumathi, C.P., Padmaja Valli, R., Santhanam, T.: Automatic recommendation of web pages in web usage mining. Int. J. Comput. Sci. Eng. **2**, 3046–3052 (2004)
15. Mobasher, B., Dai, H., Luo, T., Sun, Y., Zhu, J.: Integrating web usage and content mining for more effective personalization. Proc. EC-Web **20**, 165–176 (2000)
16. Liu, H., Xing, H., Zhang, F.: Web personalized recommendation algorithm incorporated with user interest change. J. Comput. Inf. Syst. **8**, 1383–1390 (2012)
17. Garcin, F., Dimitrakakis, C., Luo, T., Nakagawa, M.: Improving the effectiveness of collaborative filtering on anonymous web usage data, vol. 21, pp. 426–435 (2001)
18. Hemalatha, D., Bharathi, B.: Literature survey on existing analytical schemes to optimize the mining results using incremental map reduce. In: Proceedings: 3rd IEEE International Conference on Science Technology, Engineering and Management (2018)
19. Praveena, A., Bharathi, B.: A survey paper on big data analytics. In: Proceedings of International Conference on Information Communication and Embedded Systems (2017)

Dynamic Controller Deployment in SDN Networks Using ML Approach

Hemamalini Thiruvengadam, Ramya Gopalakrishnan$^{(\boxtimes)}$,
and Manoharan Rajendiran

Department of Computer Science and Engineering, Pondicherry Engineering
College, Pondicherry, India
{hemamail96, ramya028, rmanoharan}@pec.edu

Abstract. The Software Defined Networks (SDN) architecture deploys the programmable network by decoupling the data plane and control plane from the existing network architectures. Control activities are put into a software called controller. This new architecture, utilizes programmable controllers, enhances the intelligence of the networks' operations and enables network engineers to serve their business requirements more efficiently. One of issues in SDN is, estimating the required number of controllers needed and placing it in optimal locations. Many works have been proposed to place controllers in its optimal locations. In most of the works, the controller placement was based on some mathematical formulations, or by heuristic approach and number of controller required was given as an input parameter. In this work, a Traffic Engineering (TE) based controller deployment is proposed. For placing controllers K-Medoid algorithm was used and ANN model was created for analysing and predicting the traffic.

Keywords: SDN · Controller · Network traffic · Prediction · ANN · Controller placement

1 Introduction

In recent times, due to the development of intelligent mobile devices and network technologies, data is growing rapidly. Network devices now-a-days are heterogenous and complex to manage. In order to manage network in an efficient manner, Software Defined Networking is introduced [1]. The major goal of SDN is to isolate data and control plane from existing network architecture and put the controller logic in the centralized device called 'controller'. The controller monitors the current network state and manage the network activities. If the Traffic Engineering is implemented properly, based on which controllers can be dynamically placed.

The primary goal of TE is to route traffic from source to destination, by computing and selecting the optimal path. TE also focuses on analysing network traffic, optimizing the network performance and provides Quality of Service (Qos). It is a repetitive process in which traffic is monitored continuously to identify any congestion or packet loss in the network. TE performs both capacity management and traffic management [2]. Capacity management deals with allocating network resources with proper

© Springer Nature Switzerland AG 2020
P. Karrupusamy et al. (Eds.): ICSCN 2019, LNDECT 39, pp. 311–318, 2020.
https://doi.org/10.1007/978-3-030-34515-0_33

planning and routing. Traffic management deals with analysing traffic. The existing traffic analysing methodologies are based on IP based traffic classification and analysis, MPLS based mechanisms, port-based classification. Each of the methods has its own advantages and disadvantages. Recently, with the emergence of Machine learning (ML) in various fields have solved complex problem easier. The basic nature of ML is, it constructs models which are capable enough to make decisions directly from the given data without any human interaction. Models are defined with set of rules to be followed for processing the data. In general model are created to perform classification or prediction.

The use of ML techniques in SDN opens many doors for research community. Networking community can be benefited from ML techniques. ML can be used in network for traffic management, allocating resources, predicting the network performances, security related issues.

Traffic classification and traffic prediction are the two major areas in network traffic management. Traffic classification tries to classify the recorded traffic based protocol based traffic classification, IP based classification, port based traffic classification or payload based classification. Whereas, traffic prediction deals with analysing the network traffic and estimates the future traffic. It also helps in dynamic allocation of resources, dynamic bandwidth allocation, network planning, and security.

In SDN controller is responsible for all network activities like topology discovery, routing, scheduling, traffic management and other network management. Since, controller holds the global view of the network and aware of all traffic in the network, incorporating ML for TE in SDN may provide better traffic management than the traditional TE mechanisms [3].

The objective of this work is to predict the network traffic and allocate controllers dynamically in network. An Artificial Neural Network (ANN) model was created to predict traffic and controllers were placed accordingly.

The rest of the paper is organised as follows. Section 2, deals with the traditional traffic prediction mechanisms. Section 3, explains proposed approach for traffic prediction, Sect. 4 discusses results and finally ends with conclusion.

2   Related Work

For traffic classification and analysing network traffic, the network traffic must be captured. There are some tools for capturing the traffic. Internal behaviour of network can be captured by Network Tomography [4]. Network traffic is predicted based on the statistical values of the parameters selected for prediction. There are various traffic prediction mechanisms available. ARMA/ARIMA and Holt-winters are the commonly used prediction models.

There are various models available for network traffic prediction. Based on the historical data, traffic prediction is performed. In [5], linear models, for example ARMA, ARAR and Holt-Winters and non-linear approach like neural network are studied. Through experimental results, it is evident that non-linear models performs better than the linear approaches. In [6] FARIMA predictors using RNN was proposed.

In [7], a comparative study of different prediction model was studied. In [8], the authors proposed a ML approach for adaptive network management.

In [9], DevFlow improves scalability of network and performance by keeping PACKET_IN in the data plane of SDN without any loss in centralized view of network. In [10], a hash based ECMP was proposed for traffic engineering mechanism in SDN. In [11], a tag based traffic classification was proposed to classify network traffic. Based on the applications the packets were given tags. Atlas [12], C5.0 machine learning tool was implemented to classify traffic. Most of the approaches for controller placement follows either a mathematical approach or a heuristic approach. There are few works which addresses traffic engineering mechanism in SDN. Few papers addressed classification of traffic and few analysed the traffic. But no work has been proposed for controller placement based on network traffic in SDN.

3 Proposed Approach

The controller is responsible for deploying the flow table entries containing routing information or packet decision (forward/drop) to each and every switches connected to it. So, it is important to analyse and predict controller traffic for better controller placement. Considering that, this work proposes a TE based controller placement in SDN. The proposed approach consists of two phases. In the initial phase, controllers are placed using K-Medoids algorithm and in the second phase, controller traffic is analysed and predicted using ANN model. The Fig. 1 depicts the proposed architecture. Initially, the graph topology is given as input to the K-Medoids algorithm to identify the controller placements. The K-Medoids algorithm initally selects k points to start with. From that point, it associates, each point to its nearest medoids. Usually, the closest medoid is calculated by the distance between the two points.

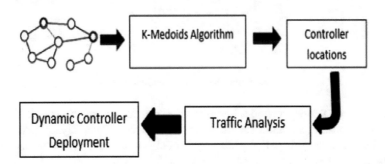

Fig. 1. Proposed architecture

In the context of controller placements, the distance between the controller and switches are calculated. The controllers are connected to the nearest switches. Since the controller installs the flow table entry of each and every switches, the shortest distance between controller and switch may reduce flow installation time which ultimately reduces delay. The number of medoid is taken as number of controllers required for that network. The controller placement algorithm is shown in Fig. 2. The main aim of this paper is to analyse traffic and allocate controllers accordingly.

Controller Placement Algorithm

Input: Graph (Topology)

Output: Controller locations

1. Initialize: randomly select k of the n data points as the medoids
2. Associate each data point to the closest medoid.
3. For each medoid m:
    a. For each non-medoid data point o: Swap m and o
    b. compute the total cost of the configuration
4. Select the configuration with the lowest cost.
5. Repeat steps 2 to 4 until there is no change in the medoid

---

Fig. 2. K-Medoids algorithm

The traffic is collected from the flow table of each switches and are taken for analysis. The collected traffic are given as input for Artificial Neural Network for traffic analysis. The following Fig. 3 shows the traffic prediction phase.

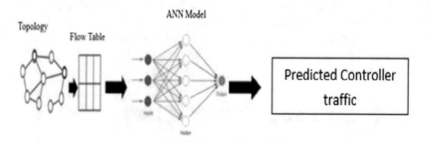

Fig. 3. Controller traffic prediction phase

ANN are or neural networks are computational algorithm. It simulates the biological nature of human brains (neurons). ANN consists of Input layer, Hidden layers, and an output layer [14]. ANN sequential model was created which consists of one input layer, multiple hidden layers and an output layer. The generated traffic was given as input and the predicted controller traffic was returned.

4 Experimental Setup and Results

In this section, the experimental scenario and results are discussed. To analyse the performance of the proposed approach, real time topologies are taken from the Internet Topology Zoo [15]. The topologies taken for experiment are LambdaNet, IRIS and

Sprint. The controllers are placed using K-Medoid algorithm. The controller placement algorithm was written in python and controller placement was accomplished using Mininet Emulator [13]. Mininet is an exclusive emulator for Software Defined Networks. The default controller in Miniet is Pox. D-ITG, TCP dump, Iperf, Curl are used to generate network traffic.

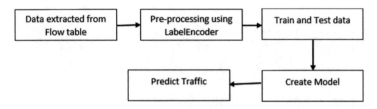

Fig. 4. ANN process

The generated traffic is collected from the switches and are sent to ANN model for further processing and it is shown in Fig. 4. ANN model was created using Jupyter Notebook in Ubuntu and written in Python3.

The attributes of flow table are n_packets, n_bytes, nw_src, nw_dst, switch, controller, duration, time. Based on the values generated the traffic is labeled as low and high. In the dataset, a new column was introduced as Traffic for training the model. In this model 0 refers to low traffic and 1 refers to high traffic.

Since the data file consists of string and other data types, all the data are converted data preprocessing was performed using LabelEncoder to convert the string and other categorical values into numerics. After the pre-processing, the data was splitted into train (60%) and test sets (40%). Then the model which is used to classify the data is imported. Now the model has to be built, compiled and trained. The Sequential model is created and the layers are added to the model.

The input taken to train the model was source id, destination id, controller port, time, switch value, number of packets, number of bytes, and traffic.

```
model = models.Sequential()
model.add(layers.Dense(10, activation = "relu", input_shape=(9, )))
model.add(layers.Dropout(0.3, noise_shape=None, seed=None))
model.add(layers.Dense(10, activation = "relu"))
model.add(layers.Dropout(0.2, noise_shape=None, seed=None))
model.add(layers.Dense(10, activation = "relu"))
model.add(layers.Dense(1, activation = "sigmoid"))
model.summary()
model.compile(
  optimizer = "adam",
  loss = "mse",
  metrics = ["accuracy"]
)
results = model.fit(
  X_train, y_train,
  epochs= 1500,
  batch_size = 40,
  validation_data = (X_test, y_test)
)
```

Fig. 5. Sequential ANN model

First the input layer gets the shape of the input, next two hidden layer are fixed with activation function rectified linear unit and finally the output layer is fixed with sigmoid activation function and model was trained with 1500 epochs and it is shown in Fig. 5. The accuracy obtained after training and testing the model is 91% and is shown in Fig. 6. The model is compared with compared with Naïve Bayes and Support Vector Machine.

```
In [8]: print("Test-Accuracy:", np.mean(results.history["val_acc"]))
        print("Test-Loss:", np.mean(results.history["val_loss"]))

Test-Accuracy: 0.9105416636807577
Test-Loss: 0.08942153296959546
```

Fig. 6. Test accuracy and test loss of ANN model

Thus, the model created is now saved as a separate model. In a new file, another new unseen traffic data is imported to predict the traffic level in the network. The created ANN model is loaded into the new file and the traffic among the controllers are predicted.

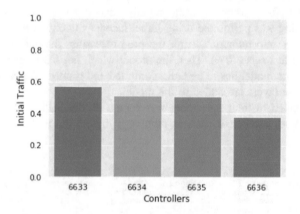

Fig. 7. Initial traffic of controllers

The initial traffic of the controllers placed in the network is shown in Fig. 7. The values 6633, 6634, 6635, and 6636 represents the controller ports. These are the default controller ports assigned to each controller placed. The Predicted traffic is shown in Fig. 8. From the prediction, the controller with port number 6636 may expect high traffic in near future.

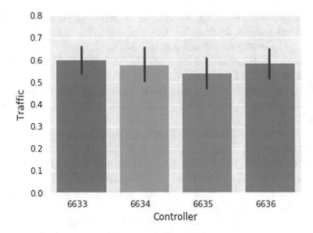

Fig. 8. Predicted traffic

The proposed model returns the future traffic of all the controllers placed in the network. From that predicted value, either number of controllers can be increased or decreased. The decision of number of controllers will be sent to the controller placement algorithm for placing controllers in the optimal location.

5 Conclusion

In SDN type of networks, all the control activities are shifted to the centralized unit called controller. The placement of controller is an important issue in SDN networks. In this work, a network traffic-based controller placement was proposed. The network traffic is analysed and predicted using ANN model. The controllers were placed using K-Medoid algorithm. The generated traffic was sent to ANN model to predict traffic. From the experimental results, it is evident that the proposed model outperforms the other ML approaches. Based on the results, the controller can be placed dynamically. In future, this work can be extended to include load balancing mechanism.

References

1. Xie, J., Yu, F.R., Huang, T., Xie, R., Liu, J., Wang, C., Liu, Y.: A survey of machine learning techniques applied to software defined networking (SDN): research issues and challenges. IEEE Commun. Surv. Tutor. **21**(1), 393–430 (2018)
2. Lu, J., Zhang, Z., Tao, H., Yi, P., Lan, J.: A survey of controller placement problem in software-defined networking. IEEE Access **7**, 24290–24307 (2019)
3. Wang, M., Cui, Y., Wang, X., Xiao, S., Jiang, J.: Machine learning for networking: workflow, advances and opportunities. IEEE Netw. **32**(2), 92–99 (2017)
4. Zhang, Y., Roughan, M., Duffield, N., Greenberg, A.: Fast accurate computation of large-scale ip traffic matrices from link loads. In: ACM SIGMETRICS Performance Evaluation Review, vol. 31, no. 1, pp. 206–217. ACM (2003)

5. Zhani, M.F., Elbiaze, H., Kamoun, F.: Analysis of prediction performance of training based models using real network traffic. Int. J. Comput. Appl. Technol. **37**(1), 472–479 (2010)
6. Wen, Y., Zhu, G.: Prediction for non-gaussian self-similar traffic with neural network. In: Intelligent Control and Automation In: 2006 The Sixth World Congress on WCICA 2006, vol. 1, pp. 4224–4228. IEEE (2006)
7. Wen, Y., Zhu, G.: Prediction for non-gaussian self-similar traffic with neural network. In: 2006 6th World Congress on Intelligent Control and Automation, vol. 1, pp. 4224–4228. IEEE (2006)
8. Gojmerac, I., Ziegler, T., Ricciato, F., Reichl, P.: Adaptive multipath routing for dynamic traffic engineering. In: Proceedings of the Global Telecommunications Conference GLOBECOM 2003, San Francisco, CA, USA, vol. 6, pp. 3058–3062 (2003)
9. Curtis, A.R., Mogul, J.C., Tourrilhes, J., Yalagandula, P., Sharma, P., Banerjee, S.: DevoFlow: scaling flow management for high performance networks. ACM SIGCOMM Comp. Commun. Rev. **41**(4), 254–265 (2011)
10. Jain, S., Kumar, A., Mandal, S., Ong, J., Poutievski, L., Singh, A., Venkata, S., et al.: B4: experience with a globally-deployed software defined WAN. In: ACM SIGCOMM Computer Communication Review, vol. 43, no. 4, pp. 3–14. ACM (2013)
11. Farhadi, H., Nakao, A.: Rethinking flow classification in SDN. In: 2014 IEEE International Conference on Cloud Engineering, pp. 598–603. IEEE (2014)
12. Qazi, Z.A., Lee, J., Jin, T., Bellala, G., Arndt, M., Noubir, G.: Application-awareness in SDN. In: ACM SIGCOMM Computer Communication Review, vol. 43, no. 4, pp. 487–488. ACM (2013)
13. Kaur, K., Singh, J., Ghumman, N.S.: Mininet as software defined networking testing platform. In: International Conference on Communication, Computing & Systems (ICCCS), pp. 139–142 (2014)
14. Basheer, I.A., Hajmeer, M.: Artificial neural networks: fundamentals, computing, design, and application. J. Microbiol. Methods **43**(1), 3–31 (2000)
15. Knight, S., Nguyen, H.X., Falkner, N., Bowden, R., Roughan, M.: The internet topology zoo. IEEE J. Sel. Areas Commun. **29**(9), 1765–1775 (2011)

Speech Separation Using Deep Learning

P. Nandal[(✉)]

MSIT, GGSIP University, Delhi, India
priyanka.lathwal@gmail.com

Abstract. In general, humans communicate through speech. The target speech which is known as the speech of interest is degraded by reverberation from surface reflections and extra noises from additional sound sources. Speech separation means separating the voices of various speakers or separating noises (background interference) from the original audio signal. Speech separation is helpful for a bountiful of applications. It is an extremely challenging task to build an automatic system for this purpose. The information about the speaker or the source of the sound and the background noises are learned by training the machine with different data using supervised machine learning. The research work presented here is primarily partitioned into 3 parts i.e. mixing the audio files, applying the algorithm to isolate the different audio files and clean the noise from them, and at last, representing the isolated output in the form of graphs and hereafter it has been endeavored to convert the graphs into the audio signal.

Keywords: Speech separation · ICA · JADE

1 Introduction

Human voice has various distinct features like pitch, frequency, quality, etc. which makes every voice different from others. The human ear has the amazing capability of splitting one sound from the other or to isolate one sound source from a blend of different sources. Using automated techniques is a very challenging task to be employed for the speech separation. The target speech is separated from the mixture of sounds which includes other speeches or the back ground interference such as noise. For example, noise is split from speech signals to enhance automatic speech recognition (ASR) accuracy [1, 2]. Singing voices are separated from music to enhance chord recognition [3]. Other applications include mobile telecommunication, hearing prosthesis. Monaural speech separation uses single microphone recordings and is very challenging task. Generally, speech separation was considered as a signal processing problem. Recently, there is a great advancement in the speech separation using supervised learning. Training data is used to learn the patterns of speech for different people, background noise and the speakers. In the recent years, several supervised algorithms have been developed. Practically, speech separation using deep learning has promoted the research recently dramatically. Deep learning techniques have been

© Springer Nature Switzerland AG 2020
P. Karrupusamy et al. (Eds.): ICSCN 2019, LNDECT 39, pp. 319–326, 2020.
https://doi.org/10.1007/978-3-030-34515-0_34

utilized to address the problems like ideal binary mask estimation and speech enhancement [2, 4, 5]. Existing speech separation algorithms still lag the human capability to separate speech.

Cherry in his well known paper [6] addressed the speech separation problem as "cocktail party problem" because of the complex listening scenarios due to multiple concurrent speakers and the reflections of their voices from the surfaces in physical space. Deep clustering techniques also boosted speech separation research. Supervised separation of speech has three predominant components: machine learning, training targets, and acoustic features. Speech separation has different categories like *speech-nonspeech separation* (*denoising*) used for *speech enhancement* or *speaker separation* where multiple speakers are concurrently speaking. Spectral subtraction approach is most widely employed for enhancing speech. In this method the power spectrum of the required speech is elicited from the noisy speech [7]. Miller and Heise conducted the first study on speech separation [8]. They concluded that a signal is split into two streams with two sine wave tones. Supervised speech separation was originally stimulated by the notion of time-frequency (T-F) masking in computational auditory scene analysis (CASA). A two dimensional mask (weight) is applied to the time frequency illustration of a source mixture in order to split the intended source [9–11]. The idea behind CASA is an ideal binary mask (IBM) [12]. The speech intelligibility is improved to a great extent by IBM in noisy conditions [13–16]. Speech separation is a form of binary classification in IBM which is known an elementary form of supervised learning. There are many other supervised learning methods present now-a-days.

SHOGUN is a machine learning toolbox which can be used for both supervised and unsupervised learning [17, 18]. Shogun toolbox supports various programming languages and technologies and also includes their scientific computation. Shogun is a standout amongst the most settled and greatest open-source machine learning community with hundreds of developers contributing every day. Shogun machine learning toolbox can be employed in speech separation. Shogun is an open-source machine learning library which provides an extensive range of proficient and unified machine learning methods. Shogun incorporates a broad collection of standard and cutting-edge algorithms.

This paper is structured as follows. Section 2 represents the proposed work and the algorithm used. Section 3 is devoted to the results and discussion. Section 4 is conclusion and future work.

2 Proposed Work and Algorithm Used

Estimates of clean speech signals can be obtained by frequency domain blind source separation (BSS). Mixed signals can be separated using BSS without having much knowledge about the original signal or how they were mixed with each other. Independent component analysis approach is utilized to have an estimation of the mixing system. The isolated signs are just approximations of the source signals. When this unmixed system is applied to the signals, estimations of the small time spectrum of the speech signals are yielded. Independent component analysis is executed using JADE (Joint Approximate Diagonalization of Eigen Matrices) algorithm. In the present work, three individual audio files are taken and mixed together using their matrices. This mixed audio file is then used as an input to be applied to JADE algorithm. Shogun toolbox is used for speech separation for the speakers using deep learning models.

2.1   Independent Component Analysis (ICA)

In ICA, a mixed signal is isolated into the constituted subparts computationally, which in turn helps in signal processing [19]. This task is finished by accepting with the aim of the subcomponents are non-Gaussian signs and that they are isolated of one another statistically. After that, ICA tries to divide a varying signal into autonomous non-Gaussian signals. For instance, the sound is a signal that is made up of the numerical addition, at each time t, of signals from distinct sources. When the analytical independence assumption is precise, blind ICA isolation of a combined signal displays excellent output. Moreover, it can also be utilized for analysing the signals which aren't presumed to be accomplished by blending.

To describe ICA model an assumption is made so as n linear mixtures $a_1, a_2, \ldots a_n$ of n independent components for all j are observed.

$$a_j = x_{j1}s_1 + x_{j2}s_2 + \ldots + x_{jn}s_n \tag{1}$$

In ICA model, the time index t is dropped, and it is presumed that each mixture a_j, including each component s_k, is a arbitrary variable, rather than a proper time signal. The inspected values $a_j(t)$ are sample of this indicated random variable which is used to represent the microphone signals in the cocktail party problem. It is also expected that the independent components and the mixed variables have zero mean. To make zero mean, the sample mean can be subtracted from the observed variables a_i to make a zero mean model. Vector-matrix notation can be used conveniently instead of sums to denote the independent components. The random vector a is used to denote the mixture of elements $a_1, a_2, \ldots a_n$ and the random vector s can be used to denote $s_1, s_2, \ldots s_n$. X denotes the matrix with elements x_{ij}. In general, bold lower case letters denote vectors and bold upper case letters are used to express matrices. Vectors used are column vectors. The mixing model presented above is expressed as

$$a = Xs \tag{2}$$

At times the columns of matrix X are needed; denoted by x_j then the model can be represented like

$$a = \sum_{i=1}^{n} x_i s_i \tag{3}$$

This analytical model given in the above statements is called ICA model. It is a creative model. It portrays the procedure of the generation of data from the mixed components. The constituent pieces are independent latent variables, which imply that the components cannot be precisely observed. Additionally, an assumption is made that the mixing matrix is unknown. An arbitrary vector a is observed, which is used to evaluate both X and s. This process must be done under as far as possible general assumptions. In ICA, It is presumed that the components s_i do not depend upon each other statistically, at the very initial stage. According to the observations, it is to expect that these independent components must have a non-Gaussian distribution.

In any case, it is not expected that these distributions are known in the fundamental model, and in the event that they are, it will streamline the issues exceedingly. X is the mixing matrix which is unknown and the size of the matrix is m by n. Some preprocessing of input data is done to save some computational costs. After making the assumption, we can calculate the inverse, W, of the matrix X and achieve those component which are independent in nature simply by:

$$s = Wa = WXs \qquad (4)$$

Here s is the estimated independent source signals. The sources are recovered when W becomes the inverse of X. ICA can also be related with the method called blind signal separation or Blind Source Separation (BSS). In BSS, the significance of the word "Blind" is like there is limited information of the mixing matrix. Minute suppositions are made about the source signals. An original signal is signified by the "source". A source signal represents an independent component, similar to the speaker in a cocktail party problem. ICA is one of the extensively employed solutions for accomplishing BSS. There are many algorithms which can be used to employ the ICA method. Joint Approximate Diagonalization of Eigen Matrices (JADE) is one of them.

2.2 JADE Algorithm for Speech Separation

JADE is an established tool for Blind Source Separation (BSS) [20]. It uses approximate joint diagonalization (AJD) of fourth-order cumulant tensor to isolate observed mixed signals into source signals. It is also called cumulant tensor for simplicity. Cumulant tensor can be simply termed as a 4D array the entries of which are defined by fourth-order cross cumulants of data. Cross cumulant is one which has two different random variables:

$$\text{Cum}(a_1, a_2, a_3, a_4) = E[\bar{a}_1 \bar{a}_2 \bar{a}_3 \bar{a}_4] \qquad (5)$$

$$\text{Here,} \, \bar{a}_i = (a_i - E[x_i]) \qquad (6)$$

Where a_i represents measured mixture of source signals, i = 1...n is number of measured mixtures and E[ ] is expectation of data which is orthogonal matrix of eigen vectors. Next preprocessing of the data is performed which is known as the whitening of the data. It is described as follows

$$J = Ha \qquad (7)$$

$$C_x = E\{xx^T\} = EDE^T \qquad (8)$$

$$H = D^{-1/2}E^T \tag{9}$$

J represents the new white matrix and H is called the whitening matrix. This step is done so that the mixed signals become uncorrelated. The main aim is to find the linear transformation W represented in Eq. (4) so that the output signals are independent of each other. Here $D = diag(\lambda_1 \ldots \lambda_n)$ is diagonal matrix, the diagonal elements are eigen values of C_x, $E = (e_1, e_2, \ldots, e_n)$ represents the orthogonal matrix, elements of which are eigen vectors of C_x and λ is a scalar eigenvalue.

Diagonalization can be viewed as shown in Eq. (8). Here diagonal matrix is obtained when covariance matrix is diagonalized by the eigen vectors. The matrix W is used to diagonalize F(M) i.e., $WF(M)W^T$ is diagonal. Here F matrix represents linear composition of terms of the type $w_i w_i^T$. The w represents the column of W. Different combination of M_i are used to make $WF(M_i)W^T$ as diagonal, where i = 1....k. The sum of the squares of off-diagonal elements, $\sum_{k \neq l} q_{kl}^2$ is the measure of the diagonality of matrix $Q = WF(M_i)W^T$. To maximize the sum of squares of diagonal elements or to minimize the sum of squares of off- diagonal elements are equivalent process.

$$J_{JADE}(W) = \sum_i \left\| diag\left(WF(M_i)W^T\right) \right\|^2 \tag{10}$$

Here $\|diag()\|^2$ denotes sum of squares of the diagonal. J_{JADE} can be maximized to obtain Joint approximate diagonalization of $F(M_i)$. Eigen matrices of the cumulant are taken in the matrix M_i. Altering the above equation algebraically,

$$J_{JADE}(W) = \sum_{ijkl \neq iikl} cum\left(s_i, s_j, s_k, s_l\right)^2 \tag{11}$$

The minimization of above equation results in the minimization of the sum of squares of cross-cumulants of s_i. The important part is the selection of set of matrices M_i which is to take the Eigen matrices of the cumulant tensor.

3   Results and Discussion

The mixed signal wave consisting of three audio signals which represent three speakers is shown in three different colors in Fig. 1. Since it is a complicated process to work on plenty of features therefore the number of features is reduced and the result is presented in the next part of the same figure. Next the true sources are presented and the ICA recovered signals are also shown in the last part of the figure. Figure 2 represents the original signal waves recovered after the application of the algorithm. The recovered graphs are then converted into audio signals and the audio signals are represented in Fig. 3. By listening to the waves represented in Fig. 3 the voice of the original input audio signals can be clearly figured out.

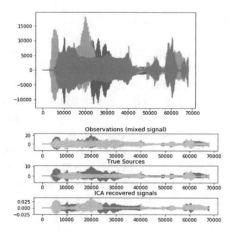

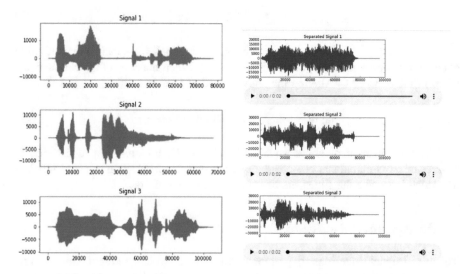

Fig. 1. Graphical representation of .wav file of input mixed audio followed by reduced features mixed signal, true sources and recovered signal from mixed signal using ICA algorithm

Fig. 2. Original signals wave.   **Fig. 3.** Output signal with wav player

4   Conclusion and Future Work

SHOGUN machine learning toolbox is used in this work for speech separation. JADE algorithm is used to perform the ICA. The mixed audio file consisting of mixed speakers' voice is input to the JADE algorithm. Due to very high number of features, audio files used are of maximum 3 s. The training time increases with increase in number of features and graphs of each file start to emerge using the matrix values as the files are processed. These matrix values are used to recreate audio files. It is also observed that there can be a trade-off among the number of features and the size of audio file if there are space and time constraints. In future, more research on JADE algorithm and tinkling with features can prove handy in speech separation. New techniques can also be implemented such as adaptive filter algorithms.

References

1. Vinyals, O., Ravuri, S.V., Povey, D.: Revisiting recurrent neural networks for robust ASR. In: Proceedings of the IEEE International Conference on Acoustics, Speech and Signal Processing, pp. 4085–4088. IEEE, Kyoto (2012)
2. Maas, A., Le, Q.V., O'neil, T.M., Vinyals, O., Nguyen, P., Ng, A.Y.: Recurrent neural networks for noise reduction in robust ASR. In: INTERSPEECH (2012)
3. Huang, P.S., Chen, S.D., Smaragdis, P., Hasegawa-Johnson, M.: Singing-voice separation from monaural recordings using robust principal component analysis. In: Proceedings of the IEEE International Conference on Acoustics, Speech and Signal Processing, pp. 57–60. IEEE, Kyoto (2012)
4. Narayanan, A., Wang, D.: Ideal ratio mask estimation using deep neural networks for robust speech recognition. In: Proceedings of the International Conference on Acoustics, Speech and Signal Processing, pp. 7092–7096. IEEE, Vancouver (2013)
5. Wang, Y., Wang, D.: Towards scaling up classification-based speech separation. IEEE Trans. Audio Speech Lang. Process. 21(7), 1381–1390 (2013)
6. Cherry, E.C.: Some experiments on the recognition of speech, with one and with two ears. J. Acoust. Soc. Am. 25(5), 975–979 (1953)
7. Boll, S.: Suppression of acoustic noise in speech using spectral subtraction. J. Acoust. Soc. Am. 27(2), 113–120 (1979)
8. Miller, G.A., Heise, G.A.: The trill threshold. J. Acoust. Soc. Am. 22(5), 637–638 (1950)
9. Lyon, R.: A computational model of binaural localization and separation. In: Proceedings of the IEEE International Conference on Acoustics, Speech, and Signal Processing, pp. 1148–1151. IEEE, Boston (1983)
10. Wang, D.: Time-frequency masking for speech separation and its potential for hearing aid design. Trends Amplif. 12(4), 332–353 (2008)
11. Wang, D., Brown, G.J.: Computational Auditory Scene Analysis: Principles, Algorithms, and Applications. Wiley-IEEE Press, Hoboken (2006)
12. Hu, G., Wang, D.: Monaural speech segregation based on pitch tracking and amplitude modulation. IEEE Trans. Neural Netw. 15(5), 1135–1150 (2004)
13. Anzalone, M.C., Calandruccio, L., Doherty, K.A., Carney, L.H.: Determination of the potential benefit of time-frequency gain manipulation. Ear Hear. 27(5), 480–492 (2006)
14. Brungart, D.S., Chang, P.S., Simpson, B.D., Wang, D.: Isolating the energetic component of speech-on-speech masking with ideal time-frequency segregation. J. Acoust. Soc. Am. 120 (6), 4007–4018 (2006)
15. Li, N., Loizou, P.C.: Factors influencing intelligibility of ideal binary-masked speech: Implications for noise reduction. J. Acoust. Soc. Am. 123(3), 1673–1682 (2008)
16. Wang, D.L., Kjems, U., Pedersen, M.S., Boldt, J.B., Lunner, T.: Speech intelligibility in background noise with ideal binary time-frequency masking. J. Acoust. Soc. Am. 125(4), 2336–2347 (2009)
17. SHOGUN-TOOLBOX Homepage. http://www.shogun-toolbox.org. Accessed 21 April 2019

18. Sonnenburg, S.Ć., Henschel, S., Widmer, C., Behr, J., Zien, A., Bona, F.D., Binder, A., Gehl, C., Franc, V.: The SHOGUN machine learning toolbox. J. Mach. Learn. Res. **11**(Jun), 1799–1802 (2010)
19. Delfarah, M., Wang, D.: Features for masking-based monaural speech separation in reverberant conditions. IEEE/ACM Trans. Audio Speech Lang. Process. **25**(5), 1085–1094 (2017)
20. Cardoso, J.F., Souloumiac, A.: An efficient technique for the blind separation of complex sources. In: Proceedings of the IEEE Signal Processing Workshop on Higher-Order Statistics, pp. 275–279. IEEE (1993)

Concoction of Steganography and Cryptography

Karthikeyan Balasubramanian, Pavan Sai Komara$^{(\boxtimes)}$,
Nischel Karanam, and Sree Nanda Sai Dasari

School of Computing, SASTRA Deemded to be University, Thanjavur, India
pavansai.komara@gmail.com

Abstract. In recent year, the seuciryt of digital applications require significant research attention, When the usage and number of users increases. This in turn makes it easy to steal the sensitive information. This paper describes how the data is encrypted and hidden in an image. Initially the data is taken and encrypted it with Burrows Wheeler Transform Algorithm and then modified Double Advanced Encryption Standard encryption is applied and the encrypted data is hidden in image. Each character of the data is represented as 8-bit binary format obtained from its ASCII value out of which first three bits is replaced with last three bits of pixel value in R Matrix and the next three bits are replaced with last three bits of pixel value (in binary format) in the G matrix and the last two bits are replaced with last two bits of pixel value in the B matrix of the image. Here hiding is achieved by process of replacing bits in image matrix. After hiding the data in the image, Mean Square Error (MSE), Peak signal to noise ratio (PSNR) values of original image and the stego image (data hidden image) are tabulated and the graph is plotted for them.

Keywords: Burrows Wheeler Transform (BWT) · Steganography · Advanced Encryption Standard (AES) · Mean Square Error (MSE) · Peak Signal to Noise Ratio (PSNR) · Stego image

1 Introduction

The security has been enhanced with the combination of cryptography and steganography. For encryption burrows wheeler transform is used and modified double Advanced encryption standard i.e. AES encryption is applied once with a key and the keys first half characters are interchanged with the last half and AES encryption is applied with the new key. Then the message is hid in the image, first three bits is replaced with last three bits of pixel value in R Matrix and the next three bits are replaced with last three bits of pixel value (in binary format) in the G matrix and the last two bits are replaced with last two bits of pixel value in the B matrix of the image. Data is hidden in image and known as stego image. MSE and PSNR values are calculated and tabulated. The behavior of MSE and PSNR values is inferred with the increase in the length of the message. From stego image, encrypted data is extracted. Inverse burrow wheeler transform is applied and Modified double AES decryption is applied to get plain text. Even if someone tries to sneak into the data, even if he extracted the message from the image he can't be able to understand the data since it is encrypted.

© Springer Nature Switzerland AG 2020
P. Karrupusamy et al. (Eds.): ICSCN 2019, LNDECT 39, pp. 327–334, 2020.
https://doi.org/10.1007/978-3-030-34515-0_35

2 Literature Survey

Information security has became the most important concern in Networks [1]. Combination of classical and quantum cryptography can produce powerful techniques which can provide security and reliability for the data. But, each technique has its own advantages and disadvantages. So, select a technique which suits our necessities in a better way.

Visual cryptography is a cryptographic technique which encrypts the data in the form of images [2]. In this data is encrypted into n transparencies and shared to participants. Decryption follows superimposition technique and it is simple because it follows the same pattern followed by human visual system.

The important purpose of a cryptographic technique is to maintain the secrecy of data and to share it only to authenticated individuals [3]. Inherent cardinal is used in the encryption so that it will be wearisome for the unauthorized person to decrypt the message which makes this technique powerful.

Information security has become the most important concern in networks. Advanced [4] Encryption Standard, a block cipher method is used to transform the text into another form. Decryption can also be done in this system. This method is used to prevent the hackers from accessing the data.

Advanced Encryption Standard is a block cipher method used for encryption and decryption of data. To maintain [5] Interoperability structure of AES should remain intact. Advanced Encryption Standard is key dependent which makes it dynamic in nature.

[6] Advanced Encryption Standard has become a vital cryptographic technique which can be used for many applications like encrypting data, electronic banking, i.e. exchange along various fields because of its less memory requirements and less power consumption.

Improvement of any algorithm is necessary [7] to improve Burrow Wheeler Transform can sometimes generate similarity measure, if more the intermixed symbols are present the more similar strings are formed. In this article two new algorithms are generated to calculate similarity measure.

Optimization improves the processing time [8] traditional AES takes more time to compute. So instead of mix columns, shift rows and changed S Box is used which is reconstructed that is used to optimize the algorithm.

To improve the statistical properties pseudo random numbers can be used as cryptographic keys [9]. Chen Chaotic system is used to design an algorithm to create pseudo random number generator which can be used to generate keys.

Medical images have images with the pixels of less potency so hiding a data in this might cause underflow problems [10]. The risk of underflow is minimized as it has huge embedding rate in the medical images.

Cryptography is a study which is used to encrypt a data such as message or an image, [11] Coupled map Lattice (CML), sub key based substitution, private key and confusion algorithm is used to secure image. Control parameters of initial chaotic system and CML are derived using 280 bit key.

The art of hiding a data in an image is known as steganography. [12] provides the information to design best system such that it is more secure than the existing models.

The data which is being transmitted by any electronic way is at high risk of attack. In such cases security plays a major role to safeguard the information [13]. AES is modified, these changes enhance the security of the information when it is compared to classic AES algorithm. In this sequence repeater is used.

Algorithms in cryptography are present to safeguard the private data but these algorithms are affected by noise. In the case of AES even the small change can affect the data and the original message is destroyed and cannot be retained. The mean variance method and global variance method is used to rectify it [14].

S-box gyration for key independent is introduced. S-box is key-dependent and it makes AES strong. Classic AES features and Cipher architecture are same, only S-box is made dependent on the key [15].

3   Discussion

3.1   Encryption

In Fig. 1, a detailed flow of data or message is processed from original message to stego image. Initially the plain text is taken, if the total number of characters is not multiple of 16, add some padding characters to it so that it becomes multiple of 16 since AES works with 16 bytes (128 bits). Then burrows wheeler transform is applied to it. Result from burrow wheeler transform is divided into blocks such that each block consists of 16 bytes (each character is considered as byte). Then modified double Advanced Encryption Standard encryption is applied to each block. 16 bytes key (k1) is taken for AES encryption and decryption. In modified double AES, encryption and decryption process is done twice. For first encryption process, key generation is done with actual 16 bytes key but for second encryption process key generation is done with 16 bytes key (k2) that is obtained by interchanging first 8 bytes and last 8 bytes of actual 16 bytes key. Now, the resultant cipher blocks, that is obtained after applying modified AES encryption to each block, are combined to give cipher text. An image is used to hide the resultant cipher text. During hiding each character of cipher text is first converted to 8-bit binary format from its ASCII value. The first three bits and the next three bits and the next two bits are replaced in last three bits of each pixel value (in binary format) in R matrix and last three bits of each pixel value in G matrix and last two bits of pixel value in B matrix respectively. The resultant image is the stego image.

3.2   Decryption

According to Fig. 2, from stego image, cipher text should be extracted. Each character of cipher text can be extracted by taking last three bits from each pixel in R matrix and last three bits from each pixel in G matrix and two bits from each pixel in B matrix. All

8 bits extracted is in binary format, decimal value of 8-bit binary format is ASCII value of character, thereby extracting a single character. Likewise all characters are extracted and put together as cipher text. After extracting the complete cipher text, modified double AES decryption is applied to cipher text. For first decryption process k2 key is used for key generation and for second decryption k1 key is used for key generation. After AES decryption is applied inverse Burrow Wheeler Transform is applied to get the plain text. Hence plain text is obtained from stego image. The plain text is obtained with padding characters. Figures 3 and 5 are the sample images, Figs. 4 and 6 are the stego images that are obtained after the encryption process. From these stego images, the decryption process is done to get the data that is hidden.

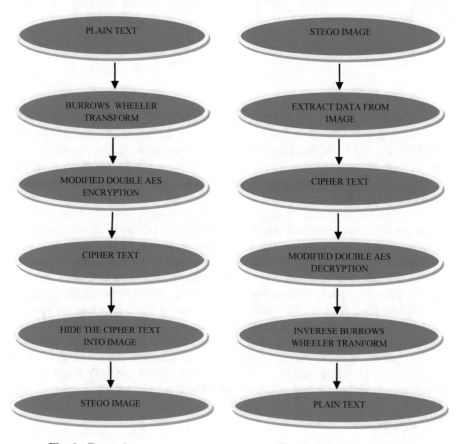

Fig. 1. Encryption process   **Fig. 2.** Decryption process

4   Figures and Tables

Fig. 3. Image 1

Fig. 4. Image 1 after 300 characters are hidden

Fig. 5. Image 2

Fig. 6. Image 2 after 1000 characters are hidden

Fig. 7. Image 3

Table 1. The table represents the MSE and PSNR values of each image with different length of strings

Filename	Length	MSE	PSNR
100.txt (image1)	100	0.005	70.478
100.txt (image2)	100	0.006	70.001
100.txt (image3)	100	0.003	72.933
300.txt (image1)	298	0.030	63.344
300.txt (image2)	298	0.031	63.129
300.txt (image3)	298	0.012	67.242
1000.txt (image1)	1000	0.112	57.600
1000.txt (image2)	1000	0.115	57.499
1000.txt (image3)	1000	0.041	61.979

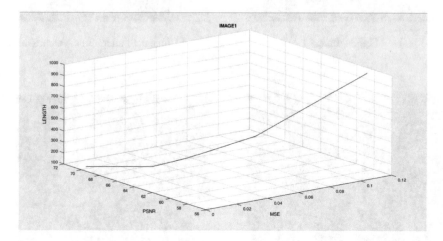

Fig. 8. Graph for Image1 (length vs PSNR vs MSE)

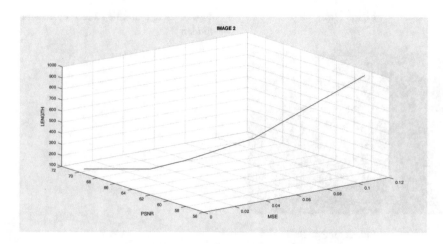

Fig. 9. Graph for Image2 (length vs PSNR vs MSE)

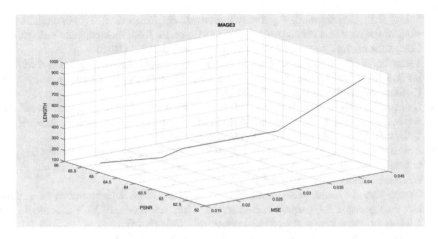

Fig. 10. Graph for Image3 (length vs PSNR vs MSE)

5 Conclusion

Hence with the integration of cryptography and steganography the security for a data can be increased. The more data is hidden into an image, the more will be the mean square error (MSE) value. As MSE value increases, the noise increases thereby decreasing peak signal to noise ratio (PSNR) value. Therefore, size of the message or data to be hidden is directly proportional to mean square error (MSE) and inversely proportional to peak signal to noise ratio (PSNR) i.e. if the size of the data is increased the MSE value increases and PSNR value decreases which can be illustrated with the scatter 3D plots drawn for length vs PSNR vs MSE for each image. Figures 8, 9 and 10 shows the 3D graphs plotted for Image 1 (Fig. 3), Image 2 (Fig. 5) and Image 3 (Fig. 7) respectively for different length of data. All the 3D plots clearly illustrates that MSE value increases and PSNR value decreases as size of data increases. All the MSE and PSNR values for three sample images with different lengths of data are tabulated in Table 1. During the process of encryption, stego image is obtained which is almost as same as original image and cannot be easily identified with naked eye. The data (plain text) that is initially hidden in original image is same as data obtained from stego image (but with some padding characters). Since the last bits of each pixel are replaced, there is less probability to find the difference between the original images and stego images with the naked eye. Security is enhanced as multiple encryption algorithms are applied (i.e. burrow wheeler transform and modified double AES) in encryption and decryption process along with steganography.

References

1. Pawar, H.R., Harkut, D.G.: Classical and quantum cryptography for image encryption and decryption. In: International Conference on Research in Intelligent and Computing in Engineering (RICE) (2018)

2. Shiny, R.M., Jayalakshmi, P., Rajakrishnammal, A., Sivaprabha, T., Abirami, R.: An efficient tagged visual cryptography for color images. In: IEEE International Conference on Computational Intelligence and Computing Research (ICCIC) (2016)
3. Vimarsha, H.M., Swetha, R.L.: Contrivance of image cryptography using inherent cardinal. In: 2nd IEEE International Conference on Recent Trends in Electronics,Information and Communication Technology (RTECIT) (2017)
4. Dumane, R., Narole, N.G., Wanjari, P.: Design of advanced encryption standard on soft-core processor Amrutha. In: published in World Conference on Futuristic Trends in Research and Innovation for Social Welfare (Startup Conclave) (2016)
5. Altigani, A., Hasan, A., Barry, B., Shamsuddin, S.M.: Key-dependant advanced encryption standard. In: IEEE International Conference on Computer, Control, Electrical and Electronics Engineering (ICCCEEE) (2018)
6. Nasser, Y.A., Bazzoun, M.A., Abdul-Nabi, S.: AES algorithm implementation for a simple low cost portable 8-bit micro controller. In: Sixth International Conference on Digital Information Processing and Communications (ICDIPC) (2016)
7. Louza, F.A., Telles, G.P., Gog, S., Zhao, L.: Algorithms to compute the Burrows-Wheeler similarity distribution. Theor. Comput. Sci. **782**, 145–156 (2019)
8. Riyaldhi, R., Rojali, Kurniawan, A.: Improvement of advanced encryption standard algorithm with shift row and S. Box modification mapping in mix column. In: 2nd International Conference on Computer Science and Computational Intelligence (ICCSCI) (2017)
9. Hamza, R.: A novel pseudo random sequence generator for image-cryptographic application. J. Inf. Secur. Appl. **35**, 119–127 (2017)
10. Bharadwaj, R., Aggarwal, A.: Hiding clinical information in medical images: an encrypted dual-image reversible data hiding algorithm with base-3 numerical framework. Optik **181**, 1099–1112 (2019)
11. Kumar, S., Kumar, M., Budhiraja, R., Das, M.K., Singh, S.: A cryptographic model for better information security. J. Inform. Secur. Appl. **43**, 123–138 (2018)
12. Kadhim, I.J., Premaratne, P., Vial, P.J., Halloran, B.: Comprehensive survey of image steganography: techniques, evaluations, and trends in future. Neurocomputing **335**, 299–326 (2018)
13. Zodpe, H., Sapkal, A.: An efficient AES implementation using FPGA with enhanced security features. J. King Saud Univ.-Eng. Sci. (2018)
14. Islam, N., Shahid, Z., Puech, W.: Denoising and error correction in noisy AES-encrypted images using statistical measures. Signal Process. Image Commun. **1**, 15–27 (2015)
15. Juremi, J., Mahmod, R., Sulaiman, S.: A proposal for improving AES S-box with rotation and key-dependent. In: Proceedings Title: 2012 International Conference on Cyber security, Cyber Warfare and Digital Forensic (Cybersecurity) (2012)

Enhanced Multipath Communication Procedure for Speed Transmission of Video Streams

V. Asanambigai[1(✉)] and S. Suganthi Devi[2]

[1] Department of Computer Engineering, Government Polytechnic College,
Perambalur, Tamilnadu, India
tradingbaskeran@gmail.com
[2] Department of Computer Engineering, Srinivasa Suubbaraya Polytechnic
College, Puthur, Nagapattinam, Tamilnadu, India
suganthidevi@yahoo.com

Abstract. Finding Alternate path communication is the solution for arbitrary video communication to build the data transfer capacity of the organization and increment the video stream. On the other hand, arbitrary multipath transmission strategies are not proper for conversational video administration conveyance because of the high measure of copied packets that origins the interruption in communication. The proposed methodology, communication procedure is utilized to guarantee the diminished measure of interruption with the little packet crash. Packet loss is the principle issue of communication methodology. An Enhanced Multipath Communication Procedure (EMC) is proposed to improve the execution of communication. EMC investigates and assesses the delay and data transfer capacity of each packet. EMC will contrast for different models of packet communication failure. The Experimental outcomes demonstrate that EMC is an effective method to build the Quality of Service for the communication of traditional video.

Keywords: Communication · Packet interruption · Packet failure · Multipath

1 Introduction

In the past years, the chatting strategy has improved quickly. On the other hand, the communication for traditional video communication is a requesting task assignment to raised data transmission [1]. Multipath communication is measured as a fruitful way to deal with advancement with the component of video communication [2]. The transfer capacity procedure is critical to guarantee communication parameter. Arbitrary procedures didn't maintain alternate paths consecutively [3]. To limit the issue, a few strategies, for example, SCTP [4], MPTCP [5] has been utilized to manage the cost of stable correspondence. A fundamental repetitive communication procedure is to simultaneously communicate video stream and its copies in abundance of various ways, and after that the receiver disperses the video communication appropriately [6].

Continued communication may root a massive measure of surplus information packets and the irregularity of calculation will be incremented. Communication

© Springer Nature Switzerland AG 2020
P. Karrupusamy et al. (Eds.): ICSCN 2019, LNDECT 39, pp. 335–342, 2020.
https://doi.org/10.1007/978-3-030-34515-0_36

procedure is a substitute strategy to build the reliability. For the circumstance of various ways, the lost video stream should be communicated with a different conceivable way to stay away from more measures of packets misplaced on the initial stage. The real trouble of communication strategy is the packet interruption [7]. Every continued communication and normal communication system may possibly origin the information crash. In traditional Video communication, the recognizable video casings must be encoded utilizing 3 sorts of frames, for example, X_Frame, Y_Frame and Z_Frame for continuous multimedia communication [8]. The organizer advises the client and important server of doled out way information. Each component of navigation table consists of two diverse communication video outlines for the speed transmission [9, 10].

2   Related Work

The multipath transport decreases the method of interchange ways that may manual for quality issue [11, 12]. RG-CMT [13] increases the communication in a quick manner to lessen the failure of packet in VANET [14] by increasing the packet transportation period and expanding data transfer capacity. FMTCP [15] has an answer for mismatched blocking issue by utilizing the created code to change over the communication images with high clash dimension of broadcasting. CMT-DA [16] proposed to diminish the video distortion as indicated by the greatest use of basic hypothesis. Out of request gathering system has proposed to lessen the packet delay conveyance [17].

CMT-QA [18] Proposed a restructuring decrease for delay conveyance for successive multipath exchange. EDPF [19] calculation is fit for progress the outline of a few similar boundaries with procured data transfer capacity. Multipath RTP (MPRTP) [20] is developed for reducing of the misplaced packet and increases the dependability within the end systems and the goal to launch video information files. Then again, in HMTP [21], the dispatcher keeps up transmitting and sending packets anticipating the beneficiary completes the process of translating and advances a stop warning. MPLOT [22] takes advantage of later grouped ways to advance the bandwidth of portable specially appointed systems and lessening parcel mending dormancy.

The main objectives of the proposed work are

1. The bandwidth is computed by utilizing the status of every path.
2. To reduce the video data.
3. Alternate path is used to send the data packets.
4. Secure communication has been done using the encoding and decoding procedure.

3   Proposed Work

3.1   Computation of the Video Frames

By alternate procedure video communication method is decreased by video spilling calculation. Dispatcher favors an information assignment calculation and update requirements. The encoded information packets must be adjusted dependent on information bundles entry in the recipient side. Video communication is utilized for communicating discussion video administration within the different ways.

Each video outline is transmitted with an interpreting positions, the beneficiary gets the successful transfer of frames consistently. While information is gotten by the beneficiary, the underlying generation is to approve the information packet. In the event that the video outline isn't having the interpreting design, the data packet will be ended. The development of organizing information application can be utilized to disentangle and produce the relevance.

3.2   Algorithm- Construction of Frames

```
Begin Procedure frame_construction
while true do
if a data packet pi is received then
if a data packet pi is arrived before its decoding then
        save packet pi in flow_sequence_number;
        else
                remove the packet pi;
        end if
        for every data packet qi in the buffer do
if the data packet qi's frame value equals the requirement
then
        if data packet qi has the same frame value with the
fragment fi then
        remodify fragment fi with the data packet
qi using transmission sequence number;
                else
        Forward the data packet qi to the buffer and increase the
frame value;
                end if
                end if
                end for
modify the requirement with frame value;
end while

End Procedure
```

3.3   Algorithm- Output Generation

In this algorithm, the initial node partitions the frames from the first edge and dispenses video edges to various ways. All through the telecom, the server distinguishes the communicated video outlines from the beneficiary. For the video outline is registered one period, the calculation multifaceted nature is recognized. At the beneficiary locale, video outlines are in the rising request by Flow_Seq_Num. The average deferral for initial class and adjusted class must be determined by a beneath mean telecom delay. Output Formation Algorithm uncovered the procedure a model of input is made. At the beneficiary side, reaction procedure associates a criticism control, a common reaction counter for all data and a substance outside contradict for each way. At the point when the underlying information stream is communicated, the beneficiary creates input manage. On the off chance that the criticism control closes, the beneficiary tosses input to the source node.

Begin Procedure Output_Generation
    Reception_counter = 0;
    Reception_End = 0;
    initialize threshold_value0, threshold_value;
    for every active path pi;
        set pi = 0;
        out_of_order = 0;
    while <> Reception_End do
    Generating an active path from path i to path j;
    Reception_Counter := Reception_Counter + j.value;
    if Reception_Counter > threshold_valuer then
    end = 1;
    end if
    if packet j is an out_of_order_packet then
    packeti.out_of_order_packet :=
packet$_i$.out_of_order_packet + 1;
        if packeti.out_of_order ≥ threshold_value0
then
            Reception_End := 1;
        end if
        end if
        if packet j is not an out_of_order_packet
then
    packet$_i$.out_of_order_packet := packet$_i$.out_of_order - 1;
        end if;
        if the timer of feedback ends then
            Reception_End := 1;
        end if
        end while
            transmit a feedback to the sender
End Procedure

3.4   Bandwidth Computation

Bandwidth demonstrates communication capacity of several related ways and computes a basic communication capacity of the multimedia. Consistency of path i for communication data is one of the elementary way for computing bandwidth, bandwth$_i$ illustrates the regular path i.

$$\text{consicy}_i = \frac{\text{rec\_pkt\_size}_i}{\text{sd\_pkt\_size}_i} \tag{1}$$

$$\text{avg\_consicy} = \frac{\sum_{j=1}^{n} \text{rec\_pkt\_size}_j}{\sum_{j=1}^{n} \text{sd\_pkt\_size}_j} \tag{2}$$

$$\text{bandwth}_i = \text{bandwth}_i \text{x} \frac{\text{consicy}_i}{\text{avg\_consicy}} \tag{3}$$

$$\text{bandwth}'_i = \text{bandwth}'_i + \lambda \text{ x bandwth}_i \tag{4}$$

4 Performance Evaluation

The performance of the proposed work is analyzed using several parameters. Initially, the Source and Destination is connected within the network using Multipath model. The video file can be separated into the video frames and the encoding and the decoding process is completed within the video frames. The size of every frame is fixed as 1280.0 pixels per frame. The resolution of the frame is 640×380 pixels per frames. The delivery rate is 50.0 frames per second. The HEVC encoding and the decoding format is used for the video frame formation. Every single frame is again separated into fragments.

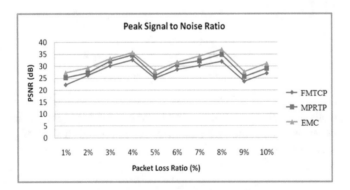

Fig. 1. PSNR

The similarity within the images are computed using the parameter SSIM. The proposed method EMC is compared with MPRTP and FMTCP. Figure 2 demonstrates the comparison for the packet loss rate with SSIM.

The proposed method EMC is compared with MPRTP and FMTCP. Figure 1 demonstrates the comparison for the packet loss rate with PSNR. The EMC is compared with MPRTP and FMTCP in Fig. 3 that the total amount of dropped frame of the proposed method is decreased compared to the related methods of FMTCP and MPRTP.

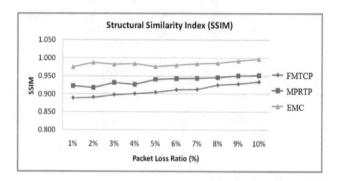

Fig. 2. SSIM

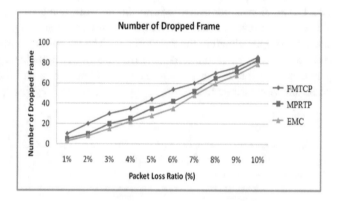

Fig. 3. Total frames dropped

For the analysis of this proposed method, the overhead of data packet ($Pt_{overhead}$) for minimized error is an important parameter for improved quality of transmission. The $Pt_{overhead}$ is computed as

$$Pt_{overhead} = \frac{\delta 1_{comm} + \delta 1_{info}}{\delta 1_{comm} + \delta_{info} + \delta_{stream}} \tag{5}$$

Where $\delta 1_{info}$ is the separated data packets must be communicated in bytes. $\delta 1_{comm}$ is the data packets for communication and δ_{stream} is the total amount of streams.

Figure 4 illustrates that the EMC is compared with correlated algorithms of MPRTP and FMTCP. The results proved that the proposed algorithm accomplished the reduced amount of the packet overload.

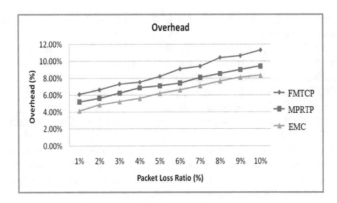

Fig. 4. Overhead

5 Conclusion

EMC actualizes a strategy to discover packet misfortune appropriate. Various ways with EMC to rebroadcast the lost packets by means of an alternate way with diminished traffic delay, avoiding ensuing loss of packet on typical way. Interpreting utmost of each casing and communication delay is measured as the metrics to pick whether to communicate the misplaced packet. The exploratory outcomes demonstrate that EMC improves similar arrangements and is proper for dispersing conversational video communication. The proposed EMC might be exhaustive for consolidated multimedia communication. In larger part of multimedia communication procedures a communication part is locked in to acquire accepting boundary and forward it to various beneficiaries.

References

1. Cisco Visual Networking Index: Global mobile data traffic forecast update. 2015–2020 white paper, p. 1 (2016)
2. Robinson, Y.H., Julie, E.G.: SMR: a synchronized multipath rebroadcasting mechanism for improving the quality of conversational video service. Wirel. Pers. Commun. 1–25 (2018)
3. Thiyagarajan, V.S., Ayyasamy, A.: Privacy preserving over big data through VSSFA and MapReduce framework in cloud environment. Wirel. Pers. Commun. **97**(4), 6239–6263 (2017)
4. Iyengar, J., Raiciu, C., Barre, S., Handley, M.J., Ford, A.: Architectural guidelines for multipath TCP development. RFC6182 (2011). https://doi.org/10.17487/rfc6182.2
5. Stewart, R.: Stream control transmission protocol. RFC4960 (2007). https://doi.org/10.17487/rfc4960.3
6. Robinson, Y.H., Julie, E.G., Saravanan, K., Kumar, R., Son, L.H.: FD-AOMDV: fault-tolerant disjoint ad-hoc on-demand multipath distance vector routing algorithm in mobile ad-hoc networks. J. Ambient Intell. Hum. Comput. 1–18 (2018)
7. Ayyasamy, A., Venkatachalapathy, K.: Context aware adaptive fuzzy based QoS routing scheme for streaming services over MANETs. Wirel. Netw. **21**(2), 421–430 (2015)

8. Thornberry, K.: Separate video file for I-frame and non-I-frame data to improve disk performance in trick play. US Patent 8,855,466.5 (2014)
9. Robinson, Y.H., Balaji, S., Julie, E.G.: Design of a buffer enabled ad hoc on-demand multipath distance vector routing protocol for improving throughput in mobile ad hoc networks. Wirel. Pers. Commun. 1–26 (2018)
10. Zhang, W., Lei, W., Liu, S., Li, G.: A general framework of multipath transport system based on application-level relay. Comput. Commun. **51**, 70–80 (2014)
11. Maxemchuk, N.F.: Dispersity routing in high-speed networks. Comput. Netw. ISDN Syst. **25**(6), 645–661 (1993)
12. Robinson, Y., Rajaram, M.: Energy-aware multipath routing scheme based on particle swarm optimization in mobile ad hoc networks. Sci. World J. **2015**, 1–9 (2015)
13. Huang, C.M., Lin, M.S.: Fast retransmission for concurrent multipath transfer (CMT) over vehicular networks. IEEE Commun. Lett. **15**(4), 386–388 (2011)
14. Gurumoorthi, E., Ayyasamy, A.: Wirel. Pers. Commun. (2019). https://doi.org/10.1007/s11277-019-06610-9
15. Kwon, O.C., Go, Y., Park, Y., Song, H.: MPMTP: multipath multimedia transport protocol using systematic raptor codes over wireless networks. IEEE Trans. Mob. Comput. **14**(9), 1903–1916 (2015). https://doi.org/10.1109/TMC.2014.2364042.10
16. Wu, J., Cheng, B., Yuen, C., Shang, Y., Chen, J.: Distortion-aware concurrent mul- tipath transfer for mobile video streaming in heterogeneous wireless net- works. IEEE Trans. Mob. Comput. **14**(4), 688–701 (2015). https://doi.org/10.1109/TMC.2014.2334592.11
17. Robinson, Y.H., Julie, E.G., Balaji, S., Ayyasamy, A.: Energy aware clustering scheme in wireless sensor network using neuro-fuzzy approach. Wirel. Pers. Commun. **95**, 703–721 (2016). https://doi.org/10.1007/s11277-016-3793-8
18. Xu, C., Liu, T., Guan, J., Zhang, H., Muntean, G.-M.: CMT-QA: quality-aware adaptive concurrent multipath data transfer in heterogeneous wireless networks. IEEE Trans. Mob. Comput. **12**(11), 2193–2205 (2013). https://doi.org/10.1109/TMC.2012.189.13
19. Chebrolu, K., Rao, R.R.: Bandwidth aggregation for real-time applications in het- erogeneous wireless networks. IEEE Trans. Mob. Comput. **5**(4), 388 (2006)
20. Ahsan, S., Eggert, L., Singh, V., Karkkainen, T., Ott, J.: Multipath RTP (MPRTP), Internet-Draft draft-ietf-avtcore-mprtp-03 (2016). https://tools.ietf.org/html/draft-ietf-avtcore-mprtp-03.17
21. Hwang, Y., Obele, B., Lim, H.: Multipath transport protocol for heterogeneous multi-homing networks. In: Proceedings of ACM CoNEXT Student Workshop, p. 5 (2010)
22. Sharma, V., Kalyanaraman, S., Kar, K., Ramakrishnan, K., Subramanian, V.: MPLOT: a transport protocol exploiting multipath diversity using erasure codes. In: Proceedings of 27th IEEE INFOCOM, pp. 121–125 (2008)

Divisible Load Scheduling from Single Source in Distributed Environments

Murugesan Ganapathy[(⊠)]

Department of Computer Science and Engineering, St. Joseph's College
of Engineering, Chennai 600 119, Tamil Nadu, India
murugesh02@gmail.com

Abstract. Divisible loads are computing loads that can be partitioned arbitrarily in to number of fractional loads and each fractional load can be independently processed in a parallel manner. Scheduling such type of loads on distributed heterogeneous environment like grid and cloud environment is a challenging task. The problem addressed here is to find the size of the fractional load to be assigned by the root processor to the child processor in a tree shaped network to minimize the computation time of a divisible load. This paper aims to develop a mathematical model to find the size of a load fraction for a divisible load with an objective of minimizing the finish time with budget and deadline as the constraints. The model was developed and solved with sample values specified in the literatures with mild assumptions. Experimental result shows that the proposed approach has obtained better solution than existing model with respect to time and cost.

Keywords: Divisible Load Scheduling · Linear programming · Resource allocation · Task scheduling · Single source scheduling

1 Introduction

Scientific problems such as image processing, signal processing applications, matrix computations and database searching etc. require large amount of computing power and also such type of problems belongs to divisible load model. Due to the nature of divisibility of these types of loads, which can be broken into an arbitrary size of load fractions and each load fraction (also called as task) can be processed independently. The term load and workload are alternatively used in this paper to specify the divisible load. While processing such type of loads, it takes a long time by a single processor to process the load. So the processing loads can be partitioned into a number of tasks which can be distributed into a network based computing system so that the total computation time can be considerably reduced. The complexity of scheduling single divisible load, to a set of processors connected through a communication link to be analyzed here.

The key players in the scheduling process are the users (customers) who submit the workload into the system, the scheduler (originator) who collects the workload from the user and a set of nodes or processing elements. The main problem to be addressed is, how to attain load balancing while distributing the task between the processors

© Springer Nature Switzerland AG 2020
P. Karrupusamy et al. (Eds.): ICSCN 2019, LNDECT 39, pp. 343–350, 2020.
https://doi.org/10.1007/978-3-030-34515-0_37

(or nodes in the network) in which the computation process is accomplished in short span of time. This balancing can take place either at starting (static) or on the fly (dynamic) of the scheduling process.

Resource specified in this paper is referred as a root processor and a set of child processors which are directly connected to the root processor. The originator (root processor or master processor) collects the workload from the user in the form of application with the scheduling decision parameters such as deadline; the time to complete the entire workload and the budget; the amount of money to be utilized for the completion of the workload processing. We used the term root processor and originator alternately. Followed by the originator there is a collection of processing elements (child processors) called as processors which are connected to the originator. There is no interconnection between the processors; the entire communications takes place through the originator. The root processor can be equipped with front-end processor to perform both computation and communication processes simultaneously.

This paper analyzes workload distribution in single level tree or star topology. The root processor is equipped with a front end processor so that it will frequently update the status information about the child processors. The role of root processor is to collect the workload from the user and identify the suitable child processor from the resource pool (cluster) to complete the computation process of the workload with respect to the decision parameters given by the user.

2   Related Works

The divisible load sharing model has been extensively used in quite few years and is very popular after the publication of the book Divisible Load Theory (DLT) authored by Bharadwaj et al. [6] in the year 1996. DLT has been applied in many scientific applications such as pattern matching [14], processing of images [1, 13, 15], broadcasting of video and audio [2, 3], large matrix processing [7], searching operation in large databases [4] and large distributed file processing [8, 18] and so on. An extensive survey on DLT was published in [5] by Beaumont et al. A comprehensive list of literatures concerning DLT and DLS is published on [9, 12, 17, 18, 21]. In [10, 19] divisible load theory framework have been carried out aims to determine the optimal fractional load which are distributed to the child processors in a network so that the processing time to be minimized.

Divisible Load Scheduling (DLS) is the process of partitioning the divisible load into fractional load, allocating the fractional load into appropriate processor to execute the fractional load, if necessary sequencing or ordering the fractional load between processors [11, 22]. An economy model for resource allocation for grid application as a mathematical model was proposed in [16]. A second order divisible computational load with non-blocking mode of communication in a single level tree network was proposed in [18, 20] and the computation of discrete wavelet transform on bus network been discussed in [8]. Most of the works discussed in the literature are based on divisible load theory principle, it is applicable mostly in offline; the workload should be available before the scheduling process is started, and also it is not suitable for dynamic nature of resources: anytime resource can join or leave from the scheduling process.

A number of unrealistic assumptions to achieve optimal solution such as all nodes finish computation process simultaneously, sequence of load allocation and also result return time is not considered.

3 Load Distribution Model

There is a need to devise an efficient load distribution strategy to schedule the divisible load in a distributed network to minimize the total computation time to complete the process of entire load. The network resources of computing power and channel bandwidth for communication have to be utilized effectively by the load distribution strategy. The divisible load distribution process undergoes as follows: the processing load arrives at one of the node in the network called originator in the tree shaped network depending on the architecture under considered. It is assumed that the root processor has the capability of multiport communication model to achieve concurrent communication and computation process and the child processors are equipped with single port communication model. So all the child processors can communicate to the root processor simultaneously and vice-versa.

Figure 1 shows the target computing platform as a single level tree network in heterogeneous platform, with N workers/child nodes labeled as P_1, P_2,..., P_N. The root processor R_0 sends out a chunk or task to its child processor over the network. It is assumed that the root processor is equipped with multiport communication model and which has a front-end processor, so that the root processor can perform both communication and computation task simultaneously.

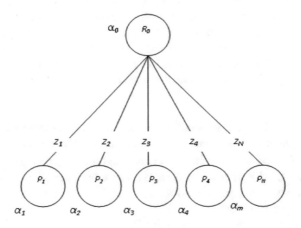

Fig. 1. Workload distribution model

This section presence the problem of scheduling divisible loads from single source in a tree shaped heterogeneous network. The objective is to develop a load distribution model for a scheduler that can adopt to minimize the total execution time (both computation and communication time) of the workload submitted. The workload to be processed could be stored in the buffer of the root processor and now the root processor act as a scheduler. Whenever the workload is received by the root processor, partition the entire workload into chunks/tasks then it distributes the task to its child processor as in the form of a single message.

All communications will takes place through the root processor and the child processors has no direct link between them. All the child processor will start computation whenever they receive their workload portion from the root processor R0. All the child processors have their own communication and computation hardware to perform both communication and computation process which allow simultaneous communication and computation process independently.

4   Design of Single Source Model

Let us assume that the amount of workload submitted by the user to the root processor is W_t. The entire workload of W_t is divided into at most $N + 1$ distinct portions (α_0, α_1, α_2,. . ., α_N).

$$\alpha_0 + \alpha_1 + \alpha_2 + \ldots + \alpha_N = W_t \tag{1}$$

$$\sum_{i=0}^{N} \alpha_i = W_t \tag{2}$$

The Eqs. (1) and (2) shows the sum of individual workload portion is equal to the total workload. In tree network there is a possibility of changing the sequence or order of load distribution among the child processors. There are $N!$ sequences are possible with N processors connected to the root processor. Our aim is to find the optimal sequence of load allocation so that the completion time will be minimized. Most of the work carried out in this area is that the order of load distribution is based on decreasing link speed. The following list shows the notations used to define the mathematical model:

c_k^p - Cost incurred to process a unit of workload by the child processor P_k
c_k^z - Cost incurred to send a unit of workload to the child processor P_k
B   - Budget (Amount to be utilized to complete the entire process)
D   - Deadline (total time taken to complete the entire process)
e_k - End time for the child processor P_k
N   - No. of child processor
s_k - Start-up time for the child processor P_k
t_k^z - Time taken to submit a unit of workload to the child processor P_k
t_k^p - Time taken to process a unit of workload by the child processor P_k
t_0^p - Time taken to process a unit of workload by the root processor
W_t - Total workload size

x_{ik} - Binary variable

     0; i^{th} workload portion is not allotted to the kth child processor

     1; i^{th} workload portion is allotted to the kth child processor

α_0 - Portion of workload assigned to the root processor

α_i - i^{th} portion of workload

Here the objective is to minimize the total execution time of the workload (application) assigned to the root processor. So the next step is to define the objective function. Initially, let us assume that the root processor is equipped with the front-end processor which means that both communication and computation process takes place concurrently by the root processor. In this paper assumed that the result compilation time is very small and it is negligible. The total execution time of the workload is same as the time required to process the task assigned to the root processor, because the root processor is the first processor to get the task first and also to finish the process last.

Now we can formally define the mathematical model with an objective of minimizing the finish time for the entire workload. The complete model can be expressed as follows:

Minimize

$$Z = \alpha_0 t_0^p$$

Subject to

$$\alpha_0 t_0^p \leq D$$

$$s_k + \sum_{i=1}^{m} (t_k^p + t_k^z)\alpha_i x_{ik} \leq D; \quad \forall k$$

$$s_k + \sum_{i=1}^{m} (t_k^p + t_k^z)\alpha_i x_{ik} \leq Z; \quad \forall k$$

$$s_k + \sum_{i=1}^{m} (t_k^p + t_k^z)\alpha_i x_{ik} \leq e_k; \quad \forall k$$

$$\sum_{i=1}^{m} (c_k^p + c_k^z)\alpha_i x_{ik} \leq B; \quad \forall k$$

$$\sum_{i=1}^{m} x_{ik} = 1; \quad \forall k$$

$$\sum_{k=1}^{N} x_{ik} = 1; \quad \forall i$$

$$\sum_{i=0}^{m} \alpha_i = W_t$$

$$\alpha_i \geq 0$$

$$x_{ik} = \{0, 1\}$$

5   Computational Experiments and Result Analysis

To test the mathematical model, we used the same sample data of [10], and which is given in Table 1. The first column specifies the various parameters used in the model and the other four column values are corresponds to the parameter values for each child processors take parts in the scheduling process. It is assumed that there are four number of child processors (N = 4) are connected with the root processors and the total workload considered for the scheduling process is $W_t = 20$.

Table 1. Input values on the scheduling process (time in sec. and cost in $)

Parameter/processor	P_1	P_2	P_3	P_4
Computation time	2	0.5	1	2
Communication time	1	0.1	2	2
Communication start-up time	1	1	1	2
Computation start-up time	0	1	1	0
Available time	10	20	30	200
Release time	0	10	20	20
Computation cost	1	5	3	2
Communication cost	0.5	1	0.3	1

The Table 2 shows the allocation of processors to the given workload for the both existing [10] and proposed model. From the table infer that when the processing cost is less than $24.13, the solution becomes infeasible and the processing cost is greater than $25.77, there is no change in the finish time, but the workload of the individual processors may change. We used the same input parameters to test the proposed model and infer that the model solution is feasible from the processing cost of $23.51. But in the existing model it is feasible only from the processing cost of $24.13. Also at a processing cost of $24.13 the finish time corresponding to the existing model becomes 60.66 s, but in the proposed model has become only 30 s. The finish time when compared to the existing model is nearly 50% less. So the proposed model gives the better solution than the existing model.

Table 2. Allocation of task to the processors

Processing cost ($)	Existing model					Proposed model				
	Size of the task allocated to the child processor				Finish time (Sec.)	Size of the task allocated to the child processor				Finish time (Sec.)
	P_1	P_2	P_3	P_4		P_1	P_2	P_3	P_4	
≥ 25.77	3	10	5.33	2	26.33	3	11.66	5.33	0	26.33
24.25	3	5	7.5	4.5	41.5	3	9.5	6.5	0	27.55
24.13	3	0.003	7.67	9.33	60.66	3	9.5	7.35	0.15	30
23.51	Infeasible (from <24.13)					3	8.44	8.55	0	33.52
<23.51						Infeasible				

It is observe that increase in schedule length result with decreases in cost and the budget is greater than 25.77 then budget (processing cost) constraint is inactive. When the processing cost B $\leq$ 23.51 the problem is infeasible. In the existing work, to reach the optimality, the author has enforced the condition that all the processors have to participate in the scheduling process even if the usage cost of the processor is high.

According to the existing model solution all the processors should participate in the scheduling process, but in the proposed model only one time the fourth processor is used due to higher communication and computation time. Also the rate of data transfer (communication time t_k^z) and the computation time of the second processor is very less compared to the other processors. So it can be used for processing large amount of data. But in the proposed model it is used to the maximum due to its availability, computation and communication time. The processor P_1 is utilized in the full schedule period, because it is the only processor available during the entire scheduling period (from 0 to 10). The next maximum time utilized processor is the second processor P_2 due to its availability (during the period 10 to 20), no other processors are available during that period. The other processors P_3 and P_4 are available from the schedule period 20 onwards so that processors are not fully utilized.

6 Conclusion

This paper proposed a solution to process a divisible workload received from a single source by optimally partition the divisible load into chunks and which are allocated to appropriate processor with respect to the given budget and deadline to minimize the computation time to process the entire workload. To partition and process the workload, a mathematical model was developed with the objective is to minimize the completion time to process the entire workload through linear programming technique and the model was solved with some sample values used in the literatures with some minor assumptions by using LINDO software package. Results shows that the proposed model outperforms well compared with the other popular approaches discussed in the literatures. This work can be extended with multiple sources and/or multi-instalment to speed up the finish time to process the entire workload.

References

1. Aali, S.N., Shahhosseini, H.S., Bagherzadeh, N.: Divisible load scheduling of image processing applications on the heterogeneous star network using a new genetic algorithm. In: Proceedings of the 26th Euromicro International Conference on Parallel, Distributed and Network-based Processing, UK (2018)
2. Altilar, D., Paker, Y.: An optimal scheduling algorithm for parallel video processing In: Proceeding of the IEEE International Conference on Multimedia Computing and Systems, pp. 245–248 (1998)
3. Altilar, D., Paker, Y.: Optimal scheduling algorithms for communication constrained parallel processing. In: Proceeding of the Euro-Par 2002. LNCS, vol. 2400, pp. 197–206. Springer (2002)

4. Blazewicz, J., Drogdowski, M., Markiwicz, M.: Divisible task scheduling – concept and verification. Parallel Comput. **25**(1), 87–98 (1999)
5. Beaumont, O., Casanova, H., Legrand, A., Robert, Y., Yang, Y.: Scheduling divisible loads on star and tree networks: results and open problem. IEEE Trans. Parallel Distrib. Syst. **16** (3), 207–218 (2005)
6. Bharadwaj, V., Ghose, D., Mani, V., Robertazzi, T.: Scheduling Divisible Loads in Parallel and Distributed Systems. IEEE Computer Society Press, Washington (1996)
7. Chan, S., Bharadwaj, V., Ghose, D.: Large matrix-vector products on distributed bus networks with communication delays using the divisible load paradigm: performance and simulation. Math. Comput. Simul. **58**(1), 71–92 (2001)
8. Chin, T.T., Bharadwaj, V., Jia, J.: Handling large-size discrete wavelet transform on network-based computing systems: parallelization via divisible load paradigm. J. Parallel Distrib. Comput. **69**(2), 143–152 (2009)
9. Chen, C.-Y.: Scheduling divisible loads on heterogeneous linear networks using pipelined communications. In: Proceedings of the Fuzzy Systems Association and 9th International Conference on Soft Computing and Intelligent Systems, Japan (2017)
10. Drogdowski, M., Wolniewicz, P.: Optimum divisible load scheduling on heterogeneous stars with limited memory. Eur. J. Oper. Res. **172**(2), 545–559 (2006)
11. Ghatpande, A., Nakazato, H., Beaumont, O., Watanabe, H.: SPORT: an algorithm for divisible load scheduling with result collection on heterogeneous systems. IEICE Trans. Commun. **91**(8), 2571–2588 (2012)
12. Lawenda, M.: Multi-instalment divisible loads scheduling. A thesis, Poznan University, Poland (2006)
13. Lee, C., Hamdi, M.: Parallel image processing applications on a network of workstations. Parallel Comput. **21**(1), 137–160 (1995)
14. Legrand, A., Su, A., Vivien, F.: Minimizing the stretch when scheduling flows of biological requests. In: Proceedings of the SPAA 2006, pp. 103–112. ACM Press (2006)
15. Li, X., Bharadwaj, V., Ko, C.: Distributed image processing on a network of workstations. Int. J. Comput. Appl. **25**(2), 1–10 (2003)
16. Murugesan, G., Chellappan, C.: Resource allocation for grid applications: an economy model. In: Lecture notes on Machine Learning and Systems Engineering, pp. 439–449 (2010)
17. Robertazzi, T.G.: A product form solution for tree networks with divisible loads. Parallel Process. Lett. **21**(1), 13–20 (2011)
18. Suresh, S., Mani, V., Omkar, S.N., Kim, H.J., Sundararajan, N.: A new load distribution strategy for linear network with communication delays. Math. Comput. Simul. **79**(5), 1488–1501 (2009)
19. Suresh, S., Cui, R., Robertazzi, T.: Scheduling nonlinear divisible loads for a single level tree network. J. Supercomput. **61**(3), 1068–1088 (2012)
20. Suresh, S., Cui, R., Kim, H.J., Robertazzi, T., Kim, Y.: Scheduling second order computational loads in master-slave paradigm. IEEE Trans. Aerosp. Electron. Syst. **48**(1), 780–793 (2012)
21. Tong, W., Xiao, S., Li, H.: Fault-tolerant scheduling algorithm with re-allocation for divisible loads on homogeneous distributed system. IAENG Int. J. Comput. Sci. **45**(3) (2018)
22. Wu, F., Cao, Y., Robertazzi, T.: Optimal divisible load scheduling for resource-sharing network. J. Distrib. Parallel Clust. Comput. (2019, submitted)

Deployment of Fog and Edge Computing in IoT for Cyber-Physical Infrastructures in the 5G Era

Sultan Ahmad$^{(\boxtimes)}$ and Mohammad Mazhar Afzal

Department of Computer Science, Glocal University, Saharanpur, UP, India
sultan.14nov@gmail.com, mazhar@theglocaluniversity.in

Abstract. To bridge the gap between existing cloud-based ICT architectures and large-scale Internet of Things (IoT) deployment, fog and edge computing (FEC) is becoming promising and anticipating paradigm to facilitate computing and communication for the next generation cyber-physical infrastructures. The key features of fog and edge computing include low latency, locality, scalability, security and privacy. Those features form a solid foundation to meet the service requirements of smart cyber-physical systems. However, although there are stand-alone, solutions exist already in this domain; many aspects are still unexplored yet in practical context also. In addition, we lack a coherent edge computing architecture that embeds these principles and can effectively enhance our cyber-physical infrastructures. In this research paper, we investigate how to enhance many expected features such as self-adaptiveness and resilience in cyber-physical systems in 5G era and IoT deployment. The focus is on fog and edge computing, which encompasses computing, communication, data analytics, security and privacy.

Keywords: Fog Computing · Cloud Computing · Edge Computing · Security and privacy · IoT · Cyber-physical infrastructure

1 Introduction

The cloud computing has become innovative and emerging paradigm in information technologies. Our society and industry are migrating from traditional system and adopting this most innovative strategies [1]. In cloud computing all data are being saved to the data center from network edge for processing. This leads latency. This also needs more use and requirement of bandwidth capacity. The industries are concern about some specific data privacy and security issues as data are stored onsite. Also, cloud servers can work and communicate only on IP. However, IoT is using countless Protocols. Therefore, these all requirement of the current system leads us to develop Fog and Edge Computing (FEC) architecture that overcome the current problems for big data services in 5G era. The strong reason behind the motivation to adopt Fog and Edge Computing (FEC) architecture is the latency issue related to the cloud-centric management for IoT ecosystem. Fog Computing is also famous as Real-Time Edge Computing. It is a logical extension from Cloud Computing to the edge of the network. These new computing paradigms distribute the tasks from the cloud to the edge

© Springer Nature Switzerland AG 2020
P. Karrupusamy et al. (Eds.): ICSCN 2019, LNDECT 39, pp. 351–359, 2020.
https://doi.org/10.1007/978-3-030-34515-0_38

resources such as sensors, embedded devices, routers, switches or the other IoT devices. Fog and Edge computing (FEC) become as a distributed computing perspective paradigm in which computation of data is being performed near by the data generating devices or close to the IoT devices. It focuses on devices and technologies that are attached to the entity or physical object (things) in the Internet of Things (IoT).

The increasing number of connected devices and the various networking issues (e.g. network latency, unstable bandwidth, unstable connectivity, etc.) will influence the whole performance of the system of IOT in cyber physical infrastructure. Especially when the big data service needs to provide the timely responses. Therefore, the system needs FEC architecture that distributes certain computational tasks to the edge devices in the edge network of IoT system, where the data source devices located at, towards enhancing the overall performance.

The IoT and FEC technologies promise to make our live easier and commodious by changing each physical object and entity in our living environment and surrounding into a smart object. These physical objects are able to sense other things in the environment and able to perform expected tasks, compute and communicate with rest of things. They have a unique identifier, an embedded system and the ability to transfer data over a network. These computing technologies help to go one-step forward in developing Smart Cities, Smart Healthcare, Smart Transportation, Smart Energy, automotive, robotics and industrial automation. That makes our live more convenient.

2   Related Works

These new Computing paradigms might possibly help to meet QoS requirements and recent applications' security. As regards of edge computing, resource provisioning and the ongoing Horizon 2020 RECAP project [2] intends an integrated cloud–Fog-Edge architecture directed towards solving application placement, capacity provisioning and infrastructure management. Cloud/Edge infrastructure monitoring is endowed with application, workload models and infrastructure that are in response give an optimized system, which arranges applications and progressively configures the infrastructure.

The research undertaken by Xu et al. [3] suggested a utility-aware resource allocation method for edge computing, known as Zenith. Service providers can establish resource sharing contracts with edge infrastructure providers with Zenith acknowledging a latency-aware resource provisioning algorithm to plan edge tasks in a way that complies with their latency requirements. ENORM) [4], Edge Node Resource Management is a structure that handles edge/fog node resources. Further auto-scaling edge nodes is enabled by monitoring application. Static priority-assignment prioritizes specific applications. Based on linear search, Provisioning and auto-scaling mechanisms are proportionately simple implementations. When the sources themselves are mobile, the edge cloud paradigm is also feasible [5]. Garcia-Valls has investigated in the context of systems with varying non-functional requirements likes cyber-physical systems, cloud and mobility [6]. In their work, they have revealed the context, describing the main challenges and choose the contemporary contributions on the areas of middleware design for cyber-physical systems, cloud computing as well as middleware for mobility support within cyber-physical systems. Elshenawy et al. [7], in other research work, have suggested a coordination and integration framework to enable Intelligent Transportation Systems of Systems (ITSoS) designing and implementing operations of intelligent Transportation Systems (ITS) in smart cities in the context of the Internet of Things. Chiefly their proposed framework is based on three-pillars: ontological semantic knowledge representation, integrated service planning and integrated service execution. The writers in [8] have furnished the notation of Smart e-Health Gateway at the edge of the network in the context of Internet-of-Things based healthcare systems, especially at smart homes or hospitals. There have been beneficial knowledge of these gateways and constructive control over both the sensor infrastructure network and the Internet for transmitting data and services such as local storage, embedded data mining, real-time local data processing etc. The writers have also propounded the idea of Fog Computing in healthcare IoT systems and proposed a Geo-distributed intermediary layer of intelligence between sensor nodes and Cloud to cope with many challenges in healthcare systems likes energy efficiency, mobility, reliability and scalability issues. Firouzi et al. [9] have evaluated the fast-pace development in the field of the Internet of Things (IoT) and Big Data Analytics for the next generation of eHealth and mHealth services.

This evolution includes big data analytics, wearable sensors, edge cloud, fog computing and body sensor networks. They examined the main concerns in IoT technologies that pertain to smart sensors for healthcare applications including smart healthcare wearable sensors, body area sensors, advanced pervasive healthcare systems, and Big Data analytics to provide eHealth services to individuals for healthier lifestyles.

Franco Cicirelli provided a set of methodological guide lines exploitable for the plan and execution of smart city applications by properly using iSapiens [10]. Ghosh has focused on a combination of dimensions essential for IoT likes reducing latency for streaming data flows across edge and Cloud, while conserving energy and bound by the compute capabilities of the devices [11]. He explored application quality parameters, such as latency and throughput; limited system characteristics like CPU, network, and power; or specific architectures such as Cloud-only, Mobile-Cloud, and Fog Computing. Psaras [12] has given his view that edge computing needs infrastructure decentralization. Infrastructure decentralization requires a decentralized trust mechanism and a distributed model of Internet governance scheme. He discusses the features of edge-computing ecosystem and decentralized IoT and catalogue of the elements that require to be planned, as well the challenges that need to be addressed.

Gupta et al. [13] in their study present Fog computing paradigm in which cloud services are implemented in the network edge to successfully decrease the delay and network congestion. Several open issues and problems exist concerning the real implementation of fog computing, which are summarized by Vaquero and Rodero-Merino [14]. Application synchronization and discovery, Storage limitation, and Management mechanisms related problems are discussed in their work.

3   Fog and Edge Computing (FEC) Architecture

The integration of fog/edge computing paradigm into the IoT architecture is very important aspect of new cyber physical infrastructure of next generation 5G era. FEC devices can be linked and connect to create a mesh network to share data on network and that will be able to do load balancing and fault tolerance. They may reduce the communication between the cloud and things in IoT eco-system. They can support resilience and self-adaptiveness feature of system. Architecturally, the devices of fog/edge may able to communicate vertically as well as horizontally in the ecosystem of IoT. The distributed system of FEC paradigm can be very much helpful and elaborate the IoT requirements. As till now, there is no fix general accepted architecture for IoT. The IoT architectures and the requirements can only be seen in presentation and illustration given in different authors view and framework. The related works enlighten authors view and ideas also. A general integration of Fog and Edge Computing (FEC) and Cloud Computing in IoT eco-system can be shown in given figure below (see Fig. 1).

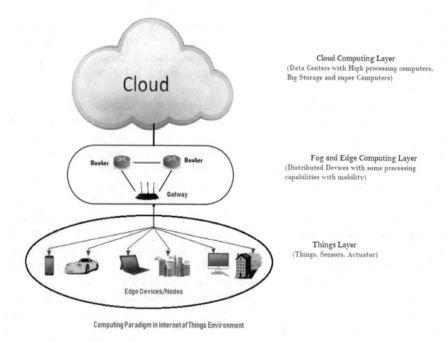

Fig. 1. Fog and edge computing with cloud architecture in IoT environment

Fog computing devices always reside near by the edge of network but it may not necessary to be at the network edge. On the other hand, edge computing devices generally place at network edge and so it becomes first approach point to connect for devices of IoT. In general, Fog devices are generally near by the IoT end-devices, but devices related to Edge Computing are more close to the IoT devices. In many situations and based on requirements of works, fog computing and edge computing are adopted interchangeably. Some people considers EC as a constituent of the FC and microdata center paradigm for IoT [15]. Fog Computing devices and Edge Computing devices are located close to the IoT end users. Edge Computing services are in edge-devices close to end users, on the other hand Fog Computing services are in network-edge devices that are few hops away from end-user. The limitation of Edge Computing is its confined storage and energy. This is the reason to list it in constrained devices category. Greater requirement of resources and more latency are occurred due to the increasing IoT applications and devices [16]. EC demand is higher than FC, due to its proximity to the end-user devices. FC requirements inclined towards infrastructure domain and EC towards things domain.

Thus in order to integrate Fog Computing and Edge Computing, certain aspects of both must be considered. Autonomy, hierarchical structure, Security, openness, scalability and reliability, agility feature and programmable feature are keys. These are inherent features of FC and EC. The peculiarities between both is the main reason of motivation of integrating FC and EC.

4 Characteristics of Computing Paradigm

Fog Computing and Edge Computing (FEC) and Cloud Computing, all are comprising three building aspects. The three are compute, storage and networking. But there are many features and characteristics that make them distinguish from each other [see Table 1]. Network edge location, low latency, location awareness are unique attributes of Fog and edge computing. Geographical and architectural distribution in FEC contrast to Cloud Computing that provide centralized services in data centers.

Table 1. Comparison of cloud, fog and edge computing.

Characteristics	Edge Computing	Fog Computing	Cloud Computing
Latency and jitter	Low	Low	Highest
Service location	Network edge	Local network	Within internet
Bandwidth utilization	Very less	Less	High
Response time	Very less	Less	High
Storage capacity	Less	Less	High
Server overhead	Very less	Low	High
Enery consumption	Less	Less	High
Congestion in networks	Low	Less	Highest
Scalability	High	High	Medium
Quality of service and experience	High	High	Medium

Cloud Computing drives toward on demand big scalable data center [17], whereas Fog and Edge Computing push further on scalability. Cloud does not support for mobility. In Fog Computing and Edge Computing, the fundamental tasks such as compute, store and communicate can be performed in mobility. All sources of data, consumers, processing devices and storage resources are being working in mobile state. With respect to end-users, attainment of Quality of Experience (QoE) is foremost requirement, which can be fulfilled and achieved by minimizing response time and throughput in Fog and Edge Computing. Real-time computation and analysis of stream data is main objective of FEC. In contrast to the batch processing, FEC should always work in real time without delay or latency problem. Data analysis is being done always in motion not in rest. Fog server is generally installed with Local Area Network. Data is intermediary processed and filtered before sending and storing to the cloud server.

Cloud Connectivity bases on the internet with wireline technologies (Gigabit Ethernet), but in FEC, the things are connected over wireless in order to make mobility possible. Further, in Cloud Computing paradigm, the data center is deployed on homogeneous resources to make less operational charges and complexity as a particular cloud provider is nothing but a particular business entity. On the other hand, FEC architecture is managed by different business entities. That's why resources are heterogeneous and vary in capacity, form factors and operating environment.

Therefore, FEC is most sought architecture to run real time applications and internet of things smart services. It is highly demandable for smart IoT Solutions in many sectors (see Fig. 2).

Fig. 2. IoT verticals

5   Deployment of Fog and Edge Computing (FEC)

Until now, approx. 20 billion of IoT devices have been deployed and used on the world internet. Everyday this number is being increased exponentially. It is expected that some 30 billion objects may be connected to the IoT by 2020. The IoT will evolve in technologies and grow in size over time. Gartner published its Hype Cycle for Emerging Technologies report in year 2015 [18]. This is the longest-running annual Hype Cycle, providing a cross-industry perspective on the technologies (see Fig. 3). These all connected devices or things are able to sense, communicate, compute and acute. They are having great feature likes intelligence, multimodal interfaces and different attributes.

Fig. 3. Gartner 2015 hype cycle for IoT emerging technologies

FEC are generally deployed to provide computing facilities for these IoT infrastructures. The huge data generated by IoT devices represent a vast volume of data generated from the IoT networks. However, increasing number of connected devices and the various networking issues (e.g. network latency, unstable bandwidth, unstable connectivity, etc.) influence the whole performance of the system. Especially when the big data service needs to provide the timely responses [19]. So this need to deploy FEC paradigm to process these in real-time. Edge computing becomes a helping trigger in this big demand of data analysis in real-time. Edge devices are being deployed in the base station of the network. Data streams are being transmitted through edge devices. They perform lightweight computing locally in real-time and transfer the data streams to the cloud for batch processing. This new combination of edge and cloud with IoT is also known as Fog Computing.

The mechanism related to deployment control of Fog and Edge Computing allows client to deploy customizable software program in dynamic mode. Also clients can do configuration and setting for execution control of Fog and Edge nodes. They can control, which and when should a node execute. FEC network topology as a service can be provided by a FEC, so that client can transfer his program from one node to another node in FEC. To meet the optimal performance of applications, client can also control multi nodes using deployment control process.

6   Conclusion

Nowadays, IoT is everyplace and it is being implemented in various ways in different service sectors. This leads need of better technologies and policies for IoT applications. Cloud Computing is one of the leading technologies that providing cloud resources and cloud services to IoT. Fog Computing and Edge computing (FEC) are also emerging as an attractive solution for the problems and situations that requires data processing in real-time manner for IoT. However, need to provide more effective framework and effective services for upgrading services to the growing smart IoT applications. These computing technologies should be fully tested for high demand and feasibility in the exponential growing IoT and its applications. This makes the whole system more smart, less costly and optimized. A lot of problems and issues need to be explored and addressed in near future consequently.

References

1. Ahn, G., Park, Y.-J., Hur, S.: Probabilistic graphical framework for estimating collaboration levels in cloud manufacturing. Sustainability **9**, 277 (2017)
2. Ostberg, P.O., Byrne, J., Casari, P., Eardley, P., Fernandez Anta, A., Forsman, J., Kennedy, J., Duc, T.L., Mariño, M.N., Loomba, R., et al.: Reliable capacity provisioning for distributed cloud/edge/fog computing applications. In: Proceedings of the 2017 European Conference on Networks and Communications (EuCNC), Oulu, Finland, 12–15 June 2017, pp. 1–6 (2017)
3. Xu, J., Palanisamy, B., Ludwig, H., Wang, Q.: Zenith: utility-aware resource allocation for edge computing. In: Proceedings of the IEEE International Conference on Edge Computing (EDGE), Honolulu, HI, USA, 25–30 June 2017, pp. 47–54 (2017)
4. Wang, N., Varghese, B., Matthaiou M., Nikolopoulos, D.S.: ENORM: a framework for edge node resource management. IEEE Trans. Serv. Comput. (2017)
5. Liu, H., Eldarrat, F., Alqahtani, H., Reznik, A., de Foy, X., Zhang, Y.: Mobile edge cloud system: architectures, challenges, and approaches. IEEE Syst. J. **12**, 1–14 (2017)
6. Garcia-Valls, M., Bellavista, P., Gokhale, A.: Reliable software technologies and communication middleware: A perspective and evolution directions for cyber-physical systems, mobility, and cloud computing. Future Gener. Comput. Syst. **71** (2017)
7. Elshenawy, M., Abdulhai, B., El-Darieby, M.: Towards a service-oriented cyber–physical systems of systems for smart city mobility applications. Future Gener. Comput. Syst. **79**, 575–587 (2018). https://doi.org/10.1016/j.future.2017.09.047. Part 2. ISSN 0167-739X
8. Rahmani, A.M., Gia, T.N., Negash, B., Anzanpour, A., Azimi, I., Jiang, M., Liljeberg, P.: Exploiting smart e-Health gateways at the edge of healthcare internet-of-things: a fog computing approach. Future Gener. Comput. Syst. **78**, 641–658 (2018). https://doi.org/10.1016/j.future.2017.02.014. Part 2. ISSN 0167-739X
9. Firouzi, F., Rahmani, A.M., Mankodiya, K., Badaroglu, M., Merrett, G.V., Wong, P., Farahani, B.: Internet-of-things and big data for smarter healthcare: from device to architecture, applications and analytics. Future Gener. Comput. Syst. **78**, 583–586 (2018). https://doi.org/10.1016/j.future.2017.09.016. Part 2. ISSN 0167-739X
10. Cicirelli, F., Guerrieri, A., Spezzano, G., Vinci, A.: An edge-based platform for dynamic Smart City applications. Future Gener. Comput. Syst. **76**, 106–118 (2017). https://doi.org/10.1016/j.future.2017.05.034. ISSN 0167-739X

11. Ghosh, R., Simmhan, Y.: Distributed scheduling of event analytics across edge and cloud. ACM Trans. Cyber-Phys. Syst. **2**(4), 28 (2018). https://doi.org/10.1145/3140256. Article 24
12. Psaras, I.: Decentralised edge-omputing and IoT through distributed trust. In: MobiSys 2018: The 16th Annual International Conference on Mobile Systems, Applications, and Services, 10–15 June 2018, Munich, Germany, p. 3. ACM, NewYork (2018). https://doi.org/10.1145/3210240.3226062
13. Gupta, H., Vahid Dastjerdi, A., Ghosh, S.K., Buyya, R.: iFogSim: a to olkit for modeling andsimulation of resource management techniques in the internet of things, edge and fog computing environments. Softw. Pract. Exp. **47**, 1275–1296 (2017)
14. Vaquero, M., Rodero-Merino, L.: Finding your way in the fog: towards a comprehensive definition of fog computing ACM SIGCOMM. Comput. Commun. Rev. **44**, 27–32 (2014)
15. Aazam, M., Huh, E.N.: Fog computing: the cloud-IOT/IoE middleware paradigm. IEEE Potentials **35**(3), 40–44 (2016)
16. Hu, P., Dhelim, S., Ning, H., Qiu, T.: Survey on fog computing: architecture, key technologies, applications and open issues. J. Netw. Comput. Appl. **98**, 27–42 (2017)
17. Zamani, A.S., Akhtar, M.M., Ahmad, S.: Emerging cloud computing paradigm. Int. J. Comput. Sci. **8**(4), 304 (2011). paper id 'IJCSI-2011-8-4-164
18. Gartner: Gartner says 6.4 billions connected things will be in USA in 2016. http://www.gartner.com/newsroom/id/3114217
19. Klonoff, D.C.: Fog computing and edge computing architectures for processing data from diabetes devices connected to the medical Internet of Things. J. Diab. Sci. Technol. **11**(4), 647–652 (2017). https://journals.sagepub.com/doi/full/10.1177/1932296817717007

RETRACTED CHAPTER: An Innovative Selection of Road Side Unit Emplacement in VANETs

K. Suresh Kumar Reddy[1]([⊠]) and V. Tamizhazhagan[2]

[1] Department of CSE, Annamalai University, Chidambaram, TN, India
kambamsureshkumarreddy@gmail.com
[2] Department of Information Technology, Annamalai University, Chidambaram, TN, India

Abstract. Travelling plays an essential role in our daily lives. Most of the accidents will happen in unknown places where we use to visit. VANET improves the safety of our intelligence transportation system by providing sensible and proficient data dissemination about actions like accidents, traffic jams and road environment along driver's information. To reduce the accidents in VANET, they use On board Units and RSU for communication of vehicular networks. While V2I selection of RSU is plays an important role. Existing methods will place the RSU in Dense areas of vehicles or at the Traffic signal junctions. We proposed a new way of placing and selecting RSUs in District and urban connecting highway roads.

Keywords: VANET · V2I · V2X · RS

1 Introduction

A standout amongst the best business utilizations of portable specially appointed systems is Vehicular impromptu systems (VANETs). The primary objective of VANET is improved safety security and productivity of transportation. FCC (government correspondences commission) has dispensed 76 MHz of for committed short–extend interchange. In current pattern, the creative advancement in both transportation and correspondence innovations made data trade conceivable on streets utilizing savvy vehicles. The correspondence between vehicles in meager vehicular adhoc systems are described with extremely towering communication delays, it unfavorably influences the nature of interchanges.

VANET is made out of a lot of highly versatile hubs, which results in as often as possible changing system topology with irregular availability. There are two potential approaches to convey in VANET: vehicle to vehicle, vehicle to foundation or both. At the point when a resource vehicle identifies a danger or a mishap, it can create a security note to the succeeding vehicles. V2I plays a high effect on unwavering quality and data trade. In this paper we center on street side unit (RSUs) position. It will go about as passages. We examined and broke down the current chips away at setting a

---

The original version of this chapter was retracted: The retraction note to this chapter is available at https://doi.org/10.1007/978-3-030-34515-0_85

© Springer Nature Switzerland AG 2020, corrected publication 2024
P. Karrupusamy et al. (Eds.): ICSCN 2019, LNDECT 39, pp. 360–367, 2020.
https://doi.org/10.1007/978-3-030-34515-0_39

RSU. To improve the general inclusion of vehicular system they are framework hubs of fixed base stations conveyed along street side.

Moreover, traffic variances are constrained by the basic street system and street systems don't change as frequently as traffic does. The majority of roadside units are in stopping position. At the same time as other roadside units will travel so as to meet the changes. Since real streets depend on an elevated moving limit, they will in general be exceptionally well known courses and provides regular possibility for getting the stationary street side units.

This paper is organized in five segments: In the Sect. 2, we presented most significant positions of RSU, in Sect. 3, we examine our proposed investigation of setting RSUs, in Sect.4, we talk about recreation and results, and finally in Sect. 5 we conclude the paper.

2 Related Work

RSU situating is a covering area improvement issue. In wrapping obstacles, aim is to find the ideal positions to cover each of the Vehicles, whereas considering a few imperatives, for example, sending cost and applications nature of administrations prerequisites. The most of effort is done in the direction of cover the field of RSUs enhanced sitting. In [1–3, 6–9], the creators intend to put RSUs ideally so as to upgrade network. They consider crossing points as best areas in urban situations since the thickness in convergences is normally higher and data can spread every which way. In [4], the creators expect to improve information spread in a urban territory. Present model in area issue as a Maximum Inclusion with Time entry Problem. To ideally send RSUs will identifies the greatest node in the street, they utilize a hereditary heuristic to unravel the MCTTP and get the best RSUs situation. In [5], the objective is to locate the ideal positions for RSUs to cover the vehicles in metropolitan streets and decrease the deferment of security info proliferation. The creators consider convergence positions as potential areas to convey RSUs.

In [8], the creator recommends a cost-proficient RSUs organization conspire. They intend to restrest safety endorsements in urban region inside short defer in usage of RSUs. In this plan, they estimate that every means of transportation will converse with RSUs in limited duration whatever its genuine position, and the additional overhead time utilize for modifying courses to refresh testaments. They model the issue as a set covering issue and utilize the P-time estimation calculation called "Covetousness Set Cover" [12] to illuminate it. In [9], the creators propose a RSUs sitting system for record downloading on account of urban situation. The principle objective is to ensure document downloading achievement proportion and postpone necessity with the most reduced arrangement cost. They likewise use crossing points as RSUs starting arrangement and consider just V2I correspondences. To improve record downloading achievement proportion, they first model the between gathering time among vehicles and RSUs as a period nonstop standardized Markov with two states (detached and associated with RSU). At that point, they find the target work that relates RSUs organization and record download proportion achievement and postponement. From that point forward, they model a weighted undirected diagram of the street arrange as where each border speaks to the ordinary sitting back on the relating street.

Compare to most computing systems VANETs operate in different location. Vehicles are moved in ordered path, it will make the path as predictable. Though, the fast moving vehicles reduces the available time for swapping the messages. Most of the challenges in VANET are bandwidth estimation, unseen and uncovered problems. Traffic is highly depending on the no of vehicles and capacity of roads.

The current situation, they use best effort scheme [18] to believe evaluated framework in realistic settings to locate the best hole between each two progressive RSUs. In [11], the's creators will probably limit the regular broadcast time of data to a given RSU. To successfully gather information in roadway situation, they propose to utilize a uniform beginning RSUs conveyance (every twice radio ranges remove). To lessen the sending cost and get the best RSUs situation, they build up a heuristic dependent on inflatable extension technique. Be that as it may, the creators in [17] and [11] propose their answer immediately to Private Street. Similarly, the standardized appropriation isn't the best beginning circulation, as it won't reflect on the size of the vehicles in the street and raise the passage deferral of data. In [2], the creators model the scheme accessibility uses a gooey model and a stochastic model.

The proposed RSUs aim is a utilization of their form. The gooey model uses to register the system size. The stochastic model identifies arbitrary conduction of vehicles. This model is not quite the same as the Poisson-landing area model (PALM) portrayed in [16]. Precision be told, this model is planned for metropolitan situations; it considers traffic light and associations between vehicles. It allows deciding the stage of network in a given street. In [2] the creators contemplate the network in VANETs dependent on consequences of permeation hypothesis. They think about autos thickness, the rate of prepared vehicles, and correspondence inclusion to examine the circulation of separated hubs and the effect of putting the RSU in crossing points. The creators demonstrate that RSUs arrangement in intersection does not affect fundamentally the extent of disconnected vehicles. With the interrelated work, we can see that the issue of RSUs sending relies upon the application. Numerous works consider just V2I correspondence for example the vehicle are in the RSUs' radio coverage to impart communication. What's more, the creator uses crossing points on account of urban situation and uniform circulation on account of parkway situation.

3 Proposed Implemetation

Our work is enlivened from all situations of RSUs and our proposed examination is cement the upgrade of propelled existing work. Keeping RSUs on thick regions of natural roadways likewise plays wellbeing yet our tale investigation is actualized dependent on street foundation of concerned regions and spots. We made an overview on unplanned inclined zones of different places in Andhra Pradesh. In those studies we found most hazardous infrastructed streets in Madanapalle region in chittoor region. We make sense of the a portion of the coincidental areas of around there (Fig. 1 and Table 1).

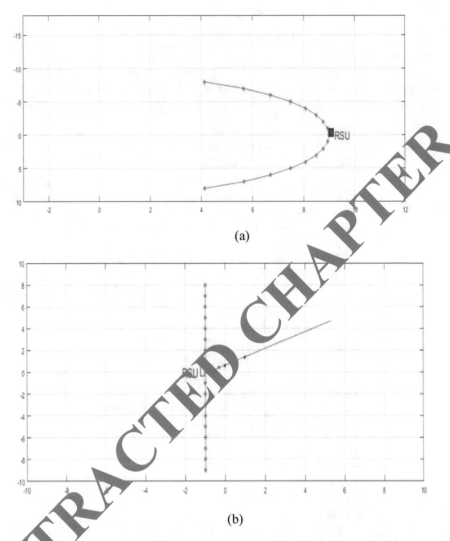

Fig. 1. (a) Sharp edge curve (b) "Y" junction roads

The following are the list of cases where we can find the accidents zones, (Tables 2 and 3).

Table 1. Notations

Variable	Notation	Variable	Notation
Location of vehicle on path A	X_A	Inclination of vehicle on path B	M_B
Location of vehicle on path B	X_B	Inclination of vehicle on path C	M_C
Location of vehicle on path C	X_C	Location of Road side Unit (sharp edge curve)	$LoRSU_{sec}$
Inclination of vehicle on path A	M_A	"Y" Junction Road side Unit	$YJunRSU$

Table 2. Algorithm (where to placeRSUs)

Path 1	Path 2
Step 1: Start	Step 1: Start
Step 2: find X_A	Step 2: find X_B
Step 3: find M_A	Step 3: find M_B
Step 4: find $RSU_{sec}(A)$	Step 4: find $RSU_{sec}(B)$ intercep
Step 5: Close	Step 5: Close

If the values of $RSU_{sec}(A) == RSU_{sec}(B)$ then pla
"RSU Equipment"

Table 3. Algorithm (where to place RSUs)

Path 1	Path 2
Step 1: Start	Step 1: Start
Step 2: find X_A	Step 2: find X_B
Step 3: find M_A	Step 3: find M_B
Step4: find JunRSU(A)	Step 4: find YJunRSU(B) intercept
Step 5: Close	Step 5: Close

Path 3
Step 1: Start
Step 2: find X_C
Step 3: find M_C
Step 4: find YJunRSU(C)
Step 5: Close

If the value of $YJunRSU(A) == YJunRSU(B) ==$
$YJunRSU(C)$ then place "RSU Equipment"

CASE 1: SHARP EDGE CURVE ROADS

$$LoRSU_{sec}(i) = M_{i\times} X_i + K \text{ where } i = A, B, C\ldots K = \text{Constant} \quad (1)$$

CASE 2: "Y" JUNCTIONS:

$$YJunRSU(i) = M_{i\times} X_i + K \text{ where } i = A, B, C\ldots K = \text{Constant} \quad (2)$$

4   Simulation and Results

We calculated and implement a proposed work of placing RSU using MATLAB Software using (R2012a).

At Sharp Edge Curves
Figure 2 gives the results of placing RSUs at the sharp edge curve with the Novality of RSU_{sec} method (Table 4).

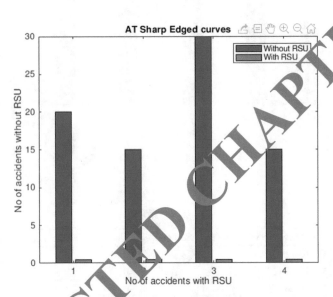

Fig. 2.   The results of RSU_{sec}

Table 4.   Results

S. no	Duration	No of accidents without RSU	No of accidents with RSU_{sec}
1	WEEK 1	20	0
2	WEEK 2	15	0
3	WEEK 3	30	0
4	WEEK 4	25	0

AT "Y" Junctions
Figure 3 gives the results of placing RSUs at the "Y" junctions with the Novality of YJunRSU Process (Table 5).

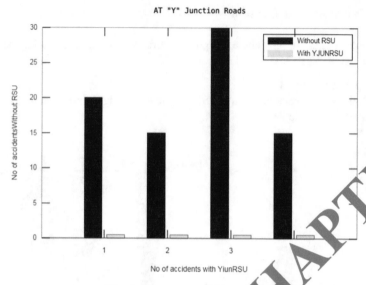

Fig. 3. The results of YJunRSU

Table 5. Results

S. no	Duration	No of accidents without RSU	No of Accidents with YjunRSU
1	WEEK 1	20	0
2	WEEK 2	15	0
3	WEEK 3	30	0
4	WEEK 4	25	0

5   Conclusions

We calculated and implement a well-organized, strong, safe and economical method of placing RSU in Critical infrastructure roads. The results show that our novel method reduces the accidents at sharp edge curves and at Y junctions. In future this work will be very useful for traffic analysis and to analyze various security issues.

References

1. IEEE 802.11P. http://www.ieee802.org/11/Reports/tgp_update.htm
2. Liu, J., Li, J., Zhang, L., Dai, F., Zhang, Y., Meng, X., Shen, J.: Secure intelligent traffic light control using fog computing. Future Gener. Comput. Syst. **78**, 817–824 (2018)
3. Giordano, L.G., Reggiani, L.: Vehicular Technologies - Deployment and Applications. InTech (2013)
4. Almeida, M.: Advances in vehicular networking

5. Maric, M.: An efficient genetic algorithm for solving the multi-level uncapacitated facility location problem. Comput. Inform. **29**(2), 183–201 (2010)
6. Cormen, T.H., Stein, C., Rivest, R.L., Leiserson, C.E.: Introduction to Algorithms, 2nd edn. McGraw-Hill Higher Education, New York (2001)
7. Cavalcante, E.S., Aquino, A.L.L., Pappa, G.L., Loureiro, A.A.F.: Roadside unit deployment for information dissemination in a VANET. In: Proceedings of the Fourteenth International Conference on Genetic and Evolutionary Computation Conference Companion - GECCO Companion 2012, p. 27 (2012)
8. Aslam, B., Amjad, F., Zou, C.C.: Optimal roadside units placement in urban areas for vehicular networks. In: 2012 IEEE Symposium on Computers and Communications (ISCC), pp. 000423–000429 (2012)
9. Sun, Y., Lin, X., Lu, R., Shen, X., Su, J.: Roadside units deployment for efficient short-time certificate updating in VANETs. In: 2010 IEEE International Conference on Communications (ICC), no. January 2014, pp. 1–5 (2010)
10. Liu, Y., Niu, J., Ma, J., Wang, W.: File downloading oriented roadside units deployment for vehicular networks. J. Syst. Archit. **59**(10), 938–946 (2013)
11. Trullols, O., Barcelo-Ordinas, J.M., Fiore, M., Casetti, C., Chiasserini, C.-F.: A max coverage formulation for information dissemination in vehicular networks. In: 2009 IEEE International Conference on Wireless and Mobile Computing, Networking and Communications, pp. 154–160 (2009)
12. Altınel, İ.K., Aras, N., Güney, E., Ersoy, C.: Binary integer programming formulation and heuristics for differentiated coverage in heterogeneous sensor networks. Comput. Netw. **52**(12), 2419–2431 (2008)
13. Aloisio, A., Izzo, V., Rampone, S.: FPGA implementation of a greedy algorithm for set covering. In: 14th IEEE-NPSS Real Time Conference, p. 5 (2005)
14. Xiong, Y., Ma, J., Wang, W., Niu, J.: Optimal roadside gateway deployment for VANETs. Prz. ELEKTROTECHNICZNY 7, 273–276 (2012)
15. Apollonio, N., Simeone, B.: The maximum vertex coverage problem on bipartite graphs. Discret. Appl. Math. **165**, 37–48 (2014)
16. Lochert, C., Scheuermann, B., Wewetzer, C., Luebke, A., Mauve, M.: Data aggregation and roadside unit placement for a vanet traffic information system. In: Proceedings of the Fifth ACM International Workshop on VehiculAr Inter-NETworking, pp. 58–65 (2008)
17. The Network Simulator - ns-2. http://www.isi.edu/nsnam/ns/

Identical Destination Based Community in Internet of Vehicles (IDCIoV) for Optimal Path Identification

S. Suguna Devi$^{(\boxtimes)}$ and A. Bhuvaneswari

Department of Computer Science, Cauvery College for Women,
Trichy 620018, India
{s.sugunadevi.it,
bhuvaneswari.it}@cauverycollege.ac.in

Abstract. IoV is an emerging field which involves Internet of Things where the things are vehicles. The problems such as efficient neighbor identification with optimized data routing transmission which could not be solved using VANET can be solved by IoV. This paper focuses on navigation based community for neighbor identification using IoV. The current work is classified into two phases. They are Location Identification using navigation based community with Indian Regional Navigation Satellite (IRNS) System and Optimal path Identification using Acyclic tree transformation. The proposed research work provides a solution for the data to be transmitted through multiple communities without link breakage during data transmission. The path reliability in neighbor selection affects the Packet Delivery Ratio and Packet Delay Ratio which has been simulated using VANET MOBISIM.

Keywords: IoV · Routing · QOS · Neighbor identification · PDR

1 Introduction

VANET is a type of network which uses vehicles as nodes for its data transmission. The vehicles in IOV are interconnected with more complex systems and are known as Smart vehicles. The data from the vehicle are collected from via the sensors. Inorder to communicate with other vehicles or internet various network devices are used. An embedded software platform is mandatory to process status information and to control all devices. Eventhough variety of sensors are employed, the functionality related to an IoV remains a simple one. In IoV, vehicles play a dual role: The vehicles acts as clents and peers. As a peer it performs distributed computing which communicates with the internet. IoV is a combination of peer-2-peer and client-server computing technology. The file sharing and integrated driving technology comes under peer to peer paradigm. Vehicles uses the data from internet servers which can be a ordinary computing node or a cloud data center. They are classified into three types based on networks which are V2V, V2Infrastructure, Vehicle to Internet. In V2V the data is transmitted between vehicles through On Board Unit. In V2Infrastructure the data from vehicle to another vehicle is transmitted via infrastructure which is the roadside unit. The third category is

© Springer Nature Switzerland AG 2020
P. Karrupusamy et al. (Eds.): ICSCN 2019, LNDECT 39, pp. 368–378, 2020.
https://doi.org/10.1007/978-3-030-34515-0_40

the Internet of Vehicles (IoV) where the data is transferred from the infrastructure to the internet in which the vehicles act as things. In the proposed algorithm the location of the emergency vehicle is identified with the help of IRNS by constructing a community of vehicles which travels along the same direction. After identification of position of vehicles the optimal path to the emergency vehicle is determined by the acyclic tree transformation technique.

The Navigation community based algorithm is used to identify optimum path between the transmitting and receiving vehicles. Generally, these routes share some common intermediate nodes either Road side unit or vehicles. Communities are a flexible tool to group the identical directional vehicles. An individual community value can be assigned to a single route or multiple routes. A route can be assigned a single community value or multiple values. Networks use the community attribute to assist in implementing identical community administrative routing policies. A route's assigned value can allow it to be accepted into the network, or rejected from the network, or allow it to modify attributes.

2   Existing System

The small world feature is incorporated into IOV. The community aware mechanism proposed by Qiu et al. [1] is used to build a list of contact for long distance communication. This is realized by the community detection algorithm which divides the smart phones into different communities. After the vehicles are connected to Smartphone, the greedy algorithm is used to construct the small world model so as to find the importance of vehicle which are calculated using position information as shown in Fig. 1. The node with shortest distance is selected. If more than one node has the same distance to the destination then the path which connects in straight lines to the destination is selected. The local importance of node is calculated in order to evaluate the position of the vehicle. The local importance is evaluated by first and second importance degree. The first importance degree of vehicle v is equal to its degree after staining. The number of neighboring vehicle of v is denoted as kv. The second importance of vehicle v is the sum of kv neighboring vehicles first importance degrees.

By adjusting the local importance, the algorithm tends to the first or second importance degrees. A Coefficient of proportionality has been used to find a trade off between the two important degrees. Compared to greedy perimeter stateless routing (GPSR) protocol and connectivity- aware routing (CAR) protocol, the results show that the proposed scheme performs better in terms of data packet delays and packet delivery ratios. Hence the existing system has some issues which are addressed in the proposed system.

Issue1: Which node reveal the location information to the source.

Issue2: On what basis the vehicles are classified into community?

Issue3: Who identifies the existence of nearby smart phones.

Issue4: When a packet is transferred from one community to other and if no intermediate intersection node occurs between the community then it does not work.

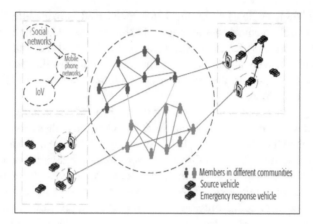

Fig. 1. Community aware algorithm (Source: (community-aware data propagation with small world feature for internet of vehicles. pg 85)

Hence the Identical navigation community based neighbor aware routing has been proposed in order to address the above issues.

3   Literature Review

The main goal of IoV is to realize the communication and relationship between human vehicles and reduces cost, increases the efficiency. IoV also satisfies the human expectation to enjoy with their vehicles. From this, VANET is only the solution which is a sub network for IoV. In vehicle Telematics, The vehicle transfers the electronic data and provide information services and this works with more complex technologies.

Cheng et al. [2] explored a survey on routing protocols in IOV and algorithms of routing to evaluate the approaches. In this paper, the authors described five various routing protocols and categorized them into three based on the transmission. After the transmission is completed, the protocols were classified into four types depending upon the information required for routing process and analyzed the various dimensions applicable to their process. Finally the target network was selected.

Marzak et al. [3] have proposed his own classification and divided into six classes like topology, position, cluster, geocast, broadcast and mobicast. Comparison was made in between the protocols that reviewed on a few metrics such as count of hops to achieve the goal, overhead frequency, status of the link in the network and communication environment.

Harous et al. [4], provided a review on routing protocols in IpV network. In this paper they discussed about various research data used for routing protocols in IpV and evaluated a good literature survey which explained routing protocols. According to this criteria, their classification was made based on transmission strategy. The proposed work described the pros and cons of these five types and detected many challenges while designing the routing protocols for IoV.

Dhandala et al. [5] provided various solution to IoV based traffic management problem which is faced in our daily life. In this work, variety of communication techniques were discussed to manage the vehicles and traffic with the help of IoV in traffic management.

Chen et al. [6], proposed a novel architecture of a cognitive IoV which is designed to overcome the existing problem in IOV networks. The overview of cognitive IoV with its evolution associated technologies and architecture are also discussed in this paper. The proposed work mainly focused on communication technologies and reveals how to mine the useful information from both physical and network layer to augment the safety of transportation and network security.

Xu et al. [7] described a detailed study on IoV and Big data, and the relationship between these two in vehicular environment. Additionally, they provided how IOV manages the storage and transmission computing on big data. It also describes the IOV benefits using big data with performance evaluation and discussed the growing issues of big data which enabled in IoV.

Bujari et al. [8] discussed the basic behaviors of topologies based and level based routing protocols with various graphs on IoV. In this process, a well- known simulator was used and displayed the process outcome in terms of delivery rate and the path dilation was achieved by the implementation of various routing protocols.

Zhang et al. [9] discussed the problems which raised when modeling the scalable IoV system through defining carstream. In this paper, the carstream was collected and processed various types of driving data and also provided various services based on the gathered data. This study revealed an idea of how to design the carstream depending upon massive data and mentioned the challenges faced due to designing and mobility of carstream.

Arora and Yadav et al. [10] proposed an algorithm for transferring secured data in IoV framework. This data transfer algorithm authenticated and secured the framework which made by using block chain technology to ensure positive communication between nodes and thus the block chain improved the efficiency.

Wang et al. [11] developed a traffic management framework based Crowd sensing and is a distributed framework. In this work, the uploading of messages for vehicles is based on dynamic reward model and effectiveness of the framework is increased due to conduct analysis based simulator. This paper explored the overview of IOV and essential role of traffic management in SIOV which is a significant area for research and challenges.

Khoueiry et al. [12] proposed a novel graph based encoding and decoding techniques to communicate two vehicles simultaneously together. The proposed method achieved and rate pair print without utilizing the time sharing mechanism.

Haouari et al. [13] proposed segment based approach. This approach solved the problem occurring while loss of internet connection or lowest–infiltration rate of vehicles on the internet. Therefore the proposed approach improved the accuracy with the low overhead during the maximum transmission range of vehicles.

4 Proposed System

The proposed system are classified into two phases

- **Location identification in navigation based community using IRNS.**
- **Optimal path identification using acyclic tree transformation.**

This algorithm consists of a location or navigation based community. Each navigation services which travel in the same direction are divided into community.

(a) **Location identification in navigation based community using IRNS.**

The Access Point (AP), Transmission vehicles, receiving vehicles, Road side terminal Unit are the main constituents of IOV Network. At the intersection Road Transmission Unit is connected with Access point. The vehicle moving along the same destination forms a identical community. Multiple community subnets are those RTU which are connected to multiple Access Points. Since the transmission range of a vehicle is considered as a rectangular area where the central node is the vehicle whose location is to be identified. The beacon messages which are received by the IRNS satellite forms a rectangular area which is responsible for identifying the location of the emergency vehicle with the help of Receiving Signal Strength Indicator(RSSI) as shown in Fig. 2. Interact with Beacon nodes to get atleast four RSSI values and calculate x and y using Eqs. 1 and 2. If it has particularly large RSSI then use history position to calculate the distance.

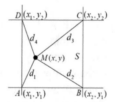

Fig. 2. Location Identification using IRNS

The vehicles position (x.y) is calculated using the following mathematical equation

$$\begin{cases} d_1^2 - (x - x_1)^2 = d_2^2 - (x_2 - x)^2 \\ d_4^2 - (x - x_1)^2 = d_3^2 - (x_2 - x)^2 \end{cases} \tag{1}$$

$$\begin{aligned} x &= \frac{1}{2} \cdot \left[x_1 + x_2 - \frac{(d_1^2 + d_4^2) - (d_2^2 + d_3^2)}{2(x_1 - x_2)} \right] \\ y &= \frac{1}{2} \cdot \left[y_1 + y_2 - \frac{(d_1^2 + d_2^2) - (d_3^2 + d_4^2)}{2(y_1 - y_2)} \right] \end{aligned} \tag{2}$$

Although a community of blind node can be obtained by local node positioning but the decision may not satisfy the requirement for the applications like IoT. However, by increasing the node community range to gain a high precision is not a good choice which needs a number of reference nodes, which has become an obstacle for Multi-directional vehicles applications. To overcome this problem, IRNS based community allocation is proposed. It designs a fine-grain community separation method for blind node in a IoV network, as depicted in Fig. 2, by using a compact algorithm which can easily implement two-directional vehicles localization with a regular deployment of reference signal from the satellite. In Fig. 3, by exploring Destination requested by source (ΔAMB) and Destination requested by Destination (ΔCMD), set of equations given in Eq. 1 is obtained. The community coordinates is calculated using Eq. 1 and common community is also identified which is represented as in Eq. 2.

$RSSI_{(i)}$ [4]; where i is neighbour nodes // identification process

Max $_{[RSSI]}(x_j\ y_j)$ [4] ; where(j = 1,2,......n)

  If [(L(D,A) == L(C, B)) / (L(D,C) == L(A, B))]

    Calculate X (x_1, x_2)

    Calculate Y (y_1, y_2)

  ElseIf (RSSI$_{(i)}$ > RSSI$_{(j)}$)

    Distance = H (x_j, y_j) // Position estimation using history

    Max $_{[RSSI]}$ (x_j, y_j) [2]; where (j = 1, 2 ,......n)

    Calculate X (x_1, x_2)

    Calculate Y (y_1, y_2)

  Else

Reinitiate identification process

(b) **Optimal path identification using acyclic tree transformation.**

The networks between vehicles are represented as Lossy Links (LL) because of high mobility. The process of identifying optimum route between the Tx-vehicle and Rx is called routing in IoV. Routing protocols need to satisfy the requirement of LL. Routing protocol for Low Power Lossy networks (RPL), is one of the widely used routing protocol. In RPL the network between the vehicle node and RTU nodes are established using Acyclic Tree formation. The proposed algorithm improves the routing path selection process in IoV inorder to achieve the reliability. The proposed algorithm classifies vehicle nodes into three categories namely, Identical Community node, Non-Identical community node and Static nodes. The network is Loopless uni-Directional towards the AR node (LDA). The LDA construction is carried out using three types of Beacon messages such as Re-LDA, RP-LDA and RA-LDA respectively. In LDA construction friend hop selection is one of the important processes. It is carried out into two phases universal construction, Local construction. The process of LDA construction is depicted in Fig. 3 shows R1, R2 and R3 are identical since they are travelling in same direction.

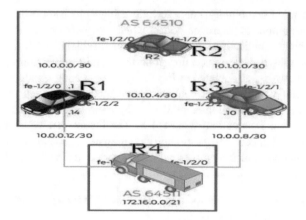

Fig. 3. Acyclic tree transformation for optimum path identification

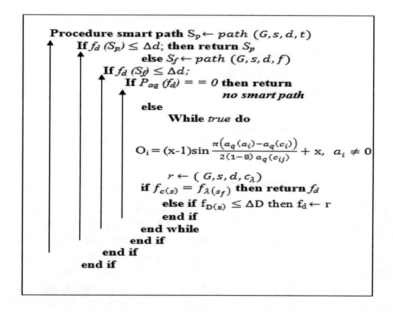

The optimum path function is identified using the following algorithm smart path.

Where G is the graph in which s is the source, d is the destination δ,d is the delay variation which is the minimum delay and f is the forwarding attack path. The range 1 to 8 which specifies the minimum and maximum hop count for a forwarding path. When the delay variation is less than the selected node then that node is selected as the next forwarding node for transmission.

5 Performance Evaluation

In proposed algorithm, the real time vehicle communication and traffic is simulated using VanetSim. It is used to analyze the vehicle communication in realistic urban environment. The realistic road map of kallanai road starts with (x = 7259124, y = 150832) and ends at (x = 7248028, y = 1507727). The roadside unit with communication radius of 500 m is used to communicate between vehicles. The parameter constraints that are used to examine the p proposed algorithm is tabulated in Table 1

Table 1. Performance Evaluation Parameters.

Simulation map area	W = 9,900,780 H = 3896,172
Enable recycling vehicle	Yes
Communication interval	160 ms
Beacon interval	240 ms
Fall back only flooding messages	Yes
Speed max	80 kmph
Speed min	20 kmph
Direction	Two way

Packet Loss Ratio
Packet Loss ratio is the number of packet lost while receiving at the destination to the total number of packets sent by the Source.

$$\text{Packet Loss Ratio} = \frac{\text{Number of Packets lost}}{\text{Total Number of Packets sent}}$$

PDR is defined as the ratio between the number of packets received by the destination (DP) and the packets generated by the source. The packet Loss is compared between CIoVS and IDCIoV and it is decreased when the community density is increased as shown in Fig. 4.
Packet Delivery Ratio

$$\text{Packet Delivery ratio} = DP/SP$$

It is defined as the number of packets recieved at the destination to the number of packets sent at the source. After Simulation Packet delivery ratio is found to be high in IDCIoV due to high community density as shown in Fig. 5.

Path Reliability
Path Reliability is defined as the probability that an operational path existed for a specific source and destination pair during the entire time interval (0, t) and is also called as terminal reliability.

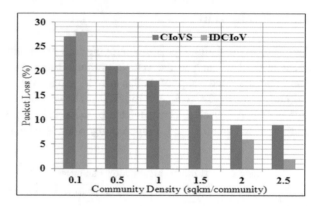

Fig. 4. Comparison of CIoVS with IDCIoV in terms of Packet loss

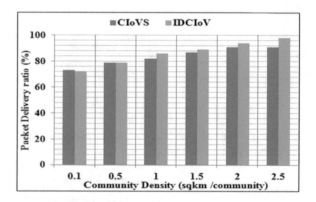

Fig. 5. Comparison of CIoVS with IDCIoV in terms of packet delivery ratio

If a specific source destination pair is of special interest, we define the *path reliability*—sometimes called *terminal reliability* as the probability that an operational path has existed for this source destination pair during the entire interval [0, t].

The packet delivery ratio is increased with respect to increase in the community density and hence the packet loss is considerably reduced. Further the path reliability is high when compared to the existing CIoVS as shown in Fig. 6. Due to reliable path the packet loss is reduced and the PDR is increased. From simulation it has been identified that with 100 nodes and 10 similar community the bandwidth is 98 for CIoVS and 95 for the proposed IDCIoV. The Overhead occurred in the same set up for IoVS is 75 and IDCIoV is 83.

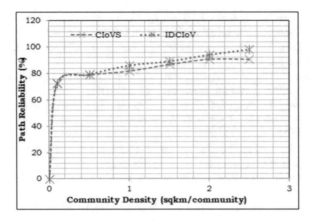

Fig. 6. Comparison of CIoVS with IDCIoV in terms of Path Reliability

6 Conclusion

Intelligent transportation system gains a vital attention by using the concept of Internet of Vehicles. The proposed Identical Destination based community algorithm creates a community based on the same forwarding direction and the location of the node is identified using the received signal strength by forwarding beacon nodes. Then the optimum path is identified with respect to hop count for forwarding data to the Emergency vehicles. The simulation results has been shown by comparing the IDCIoV's with CIoVS and the packet loss is reduced when the node density inside the community is increased. The simulation is carried out under heterogeneous environment. The advantage of using identical direction community is viewed in the form of delay of packets during delivery. The navigation community routing can be further improved by the communication between vehicle and road side unit, thus will subsequently reduce the end to end delay and the overhead. This paper has given solution to problems such as identification of location information and classification of community.

References

1. Qiu, T., Liu, X., Li, K., Hu, Q., Sangaiah, A.K., Chen, N.: Community-aware data propogation with small world feature for internet of vehicles. IEEE Commun. Mag. **56**(1), 86–91 (2018)
2. Cheng, J.J., Cheng, J.L., Zhou, M., Liu, F., Gao, S., Liu, C.: Routing in internet of vehicles: a review. IEEE Trans. Intell. Transp. Syst. **16**(5), 2339–2352 (2015)
3. Marzak, B., Toumi, H., El Guemmat, K., Benlahmar, A., Talea, M.: A survey on routing protocols for vehicular ad-hoc networks. Indian J. Sci. Technol. **9**(1), 1–20 (2016)
4. Senouci, O., Aliouat, Z., Harous, S.: A review of routing protocols in internet of vehicles and their challenges. Sens. Rev. **39**(1), 58–70 (2019)

5. Rawat, D.B., Garuba, M., Chen, L., Yang, Q.: On the security of information dissemination in the internet-of-vehicles. Tsinghua Sci. Technol. **22**(4), 437–445 (2017)
6. Chen, M., Tian, Y., Fortino, G., Zhang, J., Humar, I.: Cognitive internet of vehicles. Comput. Commun. **120**, 58–70 (2018)
7. Xu, W., Zhou, H., Cheng, N., Lyu, F., Shi, W., Chen, J., Shen, X.: Internet of vehicles in big data era. IEEE/CAA J. Autom. Sinica **5**(1), 19–35 (2018)
8. Bujari, A., Gaggi, O., Palazzi, C.E., Ronzani, D.: Would currentad-hocrouting protocols be adequate for the internet of vehicles? A comparative study. IEEE Internet Things J. **5**(5), 3683–3691 (2018)
9. Zhang, M., Wo, T., Xie, T., Lin, X., Liu, Y.: CStream an industrial system of big data processing for internet-of-vehicles. Proc. VLDB Endow. **10**(12), 1766–1777 (2017)
10. Arora, A., Yadav, S.K.: Block chain based security mechanism for internet of vehicles (IoV) (2018)
11. Wang, X., Ning, Z., Hu, X., Ngai, E.C.-H., Wang, L., Hu, B., Kwok, R.Y.: A city-wide real-time traffic management system: enabling crowdsensing in social Internet of vehicles. IEEE Commun. Mag. **56**(9), 19–25 (2018)
12. Khoueiry, B.W., Soleymani, M.R.: An efficient NOMA V2X communication scheme in the Internet of vehicles. In: 2017 IEEE 85th Vehicular Technology Conference (VTC Spring), pp. 1–7. IEEE (2017)
13. Haouari, N., Moussaoui, S., Senouci, S.M., Boualouache, A., Guerroumi, M.: Enhanced local density estimation in internet of vehicles. IET Commun. **11**(15), 2393–2401 (2017)

A Survey on Omnipresent HTTPS and Its Impact on Network Security Monitoring

Madhu Khurana$^{(\boxtimes)}$ and Priyanka Malik

Noida, India
iamkhurana669@gmail.com, pmalik2@amity.edu

Abstract. HTTPS is gaining widespread popularity for performing secure transactions. Most popular sites have made default choice as HTTPS. Therefore, this paper makes a survey through various study done in the area and it has comprehensively explored the various tools, technologies, and mechanisms to deal with secured network in a robust way. We make a complete analysis and evaluation of HTTPS protocol–is it ensuring security or are we entering into a vicious cycle of finding weaknesses and trying to fill the gaps in Network security Monitoring. The gaps like Man In The Middle, active and passive attacks on device, compromising response time or accessibility at the cost of security are among the most important ones that has been explored and proposed solutions for various research studies. Data collected from couple of up to date research works and their conclusion has been discussed to provide a brief overview so as to provide the reader with global understanding of the research progress in this area.

Keywords: HTTPS · MITM · Network security monitoring

1 Introduction

Emerging unprecedented increase in online activity is posing new challenges through Innovation and Diversity in terms of secured network communication. HTTPS is the safe version of HTTP. HTTPS is the de facto standard for safe sessions. A recent report shows an average 40% growth of HTTPS usage every six months. In the present day, we are extensively relying on HTTPS for all the transactions namely e-banking, e-commerce, e-procurement, home banking among others. It ensures end to end safe exchange of monetary and other sensitive data. HTTPS is being used about 4 times today than what it was being used in 2015, according to French ISPs. HTTPS traffic is increasing ever since at an unfathomable pace, owing to increasing concern about privacy and security. It works in the Transport Layer Protocol of OSI model. It is not a protocol as such, yet a product of layering the HTTP on top of various versions of SSL or TLS protocols. Digital Certificates on the basis of ITU X.509 standard guarantee the authenticity of the communicating ends. In 1995 when the first version of HTTPS was released on Secure Socket Layer (SSL) version2, major security issues were detected and hence SSL version 3 released in 1996 and even after three major protocol revisions, SSL Version 2 was recommended to discontinue by IETF in 2008 which was banned in 2011. Transport Layer Security (TLS) was declared as standard in 1999 by The Internet

© Springer Nature Switzerland AG 2020
P. Karrupusamy et al. (Eds.): ICSCN 2019, LNDECT 39, pp. 379–388, 2020.
https://doi.org/10.1007/978-3-030-34515-0_41

Engineering Task Force (IETF). This layer is the one that makes HTTP secure and hence named HTTPS. The basis of HTTPS is asymmetric Cryptography to cipher a secret key which is then used for encrypting the message. Therefore, the triad of Confidentiality, integrity and authentication can be achieved by using HTTPS [1].

An HTTPS Uniform Resource Locator (URL) depicts that the webpage will be downloaded by browser through a separate default port (443) and there lies an extra TLS encryption/authentication layer after HTTP. Therefore, HTTPS-based data exchange is considered safe, and symbol of "lock" indicates that their sensitive and confidential data in transit is safe. This paper discusses how attackers through innovative tools technologies and procedures available and widespread in cyberworld can successfully intercept the data transfer and invade and exploit the safety and security of communication. We will discuss how using innovative tools technologies newer and better ways to secure our system is being worked at.

2   Contributions

In this paper, we present a complete mechanism that features the current research work followed in the HTTPS implementation of network security Monitoring. A survey on the related literature is done, placing each work in the specific category and briefly describing them so as to provide the main idea of the suggested work. In this way, the scholar can understand the research development in the said area and highlight the burning issues. The aim is to help the scholar to understand and derive the security management issues with particular emphasis on HTTPS. Finally, a comparative discussion of different techniques prevalent today is done, to highlight their strengths and weaknesses. The detailed categories in the paper is organized as follows: Category A present emerging cyber threats in HTTPS, the gaps and the proposed solutions through various research in the respective area, category B describes Current Research trends to highlight the various up to date surveys and their conclusions, followed by Section C that covers the various mitigation techniques the tools techniques and suggested solution already implemented or still being worked and explored. Category D addresses the overall opinion and future Directions and stresses on current scenario and the issues and the important aspects that need immediate attention and points to work on in future. Subsection E addresses evaluation of current state of the art of topic discusses the way things are being handled at the moment.

2.1   Emerging Cyber Threats

On one side, there is danger of the attackers taking advantage of the assumption that if data exchange is HTTPS based, it is exploitable through Man in the middle (MITM) attacks where HTTPS server sends a certificate with its public key to the web browser and the entire path becomes vulnerable if this certificate isn't trustworthy. A MITM attack replaces original certificate authenticating HTTPS server with a modified certificate. A negligent user, if ignores the warning notification of the browser, attack is guaranteed. Simply by compromising few features of Local Area Networks (LANs) and inexperienced users' attitudes, certificates can be spoofed and powerful encryption through HTTPS in the Network security becomes potentially hazardous [3].

On the other hand, middle boxes or Proxies that are important devices at network or application layer. These devices play crucial role by contributing to the tasks like caching, Intrusion Detection System (IDS), URL filtering, and compression. It becomes difficult to operate Middle boxes with HTTPS being all over, since they inspect monitor and modify the traffic. It is specifically a problem in the organization' Bring Your Own Device (BYOD) context. Various organizations in the modern world permit employees to bring their own personal devices for official use, namely laptop, smartphones. Middleboxes in such scenarios perform the tasks like virus detection, compression, URL filtering and user traffic being monitored by such middleboxes, With HTTPS in place, Middleboxes is an impossible task and hence a major issue [4].

In particular, HTTPS brings the issues of security management of ciphered traffic. There are couple of vacuums that exist along with encrypted data, namely firewalls, IDS becoming blind to data being exchanged, enhancing vulnerabilities by making it tough to implement security policy and probability of malicious activities hiding in the ciphered traffic. Most of the current techniques namely Deep Packet Inspection (DPI) to port based to Internet Protocol (IP) address to Domain Name System (DNS) to Server Name Indication (SNI) filtering is prone to be ineffective in front of HTTPS traffic. The emphasis has been upon the need for new ways to explore the expanding HTTPS volume with security breaches to cover new challenges related to Network Security Monitoring [5].

Though passive attacks become difficult once the network traffic is encrypted. Despite HTTPS in place, information regarding operating system, and browser can be probed at. Desktops and laptops can be easily compromised. The attacker can get information about user and can collect data on groups of users for potential attack. In the passive attack, where the victim doesn't interact actively, he can compromise on the network volume of device from the network [7].

Notably, HTTPS and its growing popularity is a double-edged sword, it ensures secured transmission, protects against malicious attacks, yet enhances the burden of analysis of ciphered traffic for authenticity. Analyzing ciphered traffic is tough. The secure application is difficult to comprehend in terms of traffic and hence malicious activity. The Attackers can invade taking advantage of the fact and can lead to security issues [6].

With HTTPS being synonymous for end to end security, it brings overheads of extra cost both to host and servers, extra burden on server, deteriorating response time, performance and hence negatively impacting user experience. Cryptography is the main point behind all the cost and performance issues. This implies performance along with response time cannot coincide with security. Simultaneous co-existence of the two very heart of communication seems impossible [8].

The long distance for signals in satellite connection leads to large latency despite its high capacity. This high latency makes it difficult to use HTTPS as it is encrypted. With the ever-increasing websites switching to Https as default, this performance degradation becomes a serious issue posing several threats in Satellite networks. In simple words, satellite networks with high bandwidth and high latency has difficult co-existence with secured access and performance. Satellite networks in tandem with low latency network can enjoy secured access but it defeats the basic purpose. The issue is important as satellite networks are very popular and catering to their needs is crucial.

Notably, during the traversal of a webpage from publisher to user's browser, any illegal content can be inserted into this page by one of the various ends namely Internet service Protocol (ISP) to network operator to Wireless Access Point to some middle box like proxy or firewall. These contents are usually misused for advertising. This in turn deteriorates user overall experience and it is highlighted that unsolicited content injection gaining popularity is affecting user experience and overall impact along with being a security threat [9].

Interestingly with HTTPS in place, keyboard or signature matching through machine learning and data mining techniques is not possible by eavesdroppers and attackers, although personal data of user is maliciously being used by ISPs. The providers are using middle boxes to capture internet traffic aiming at dealing security attacks through techniques like DPI, in practice collecting user data which is further used for analysis and marketing uses.

With the increasing and growing traffic over https, organizations finding it hard to analyze the encrypted traffic against malware and ensuring implantation of security policies in an encrypted environment is very challenging. Confidential information can be deliberately being sent out from organization in an encrypted environment, as it is a challenge to filter and analyze this data There is a need for new tools and methods to monitor and filter https traffic to ensure security and privacy in a Network [10].

Notably, a taxonomical evaluation of the HTTP security headers' exposure to attackers was tested in 240 Mozambique websites. http security vulnerabilities were examined and the ways to decrease the security risks on websites were worked upon. It was concluded that there is a lack of Network security Monitoring in national strategy particularly in https. The Mozambique government is integrating Information and Communications Technology (ICT) to enhance communication and information in public. Since World wide web is prone to attacks and malicious activities, most people actively providing data is not in the interest of citizens. It further highlights the threat of our sensitive data at the mercy of malicious activities and prone to potential attacks and security issues. There are gaps at various levels and many issues need robust planning before such actions can be rolled out.

2.2   Current Research Trends

It presents an up-to-date research on security of most popular websites, ranked as top one million as per Alexa. It is a survey of current web security policies prevalent through HTTP and concludes that HTTP response headers are growing consistently but still its adoption isn't wide spread. HTTPS sites seem to implement the policies more than HTTP only websites. Also, the popular sites implement it more than not so popular sites. The popular ones are prone to HTTPS over HTTP. Overall, security policies on HTTP header is low, well below satisfactory rates. The paper analyses the current adoption rate of security policies on the popular sites. A study on the correlation between rates of adoption and variables like HTTPS usage and the positioning of the ranking is another notable aspect [14].

It is proposed that with mobile office becoming popular and fixed physical location on legacy network fading away. SSL Virtual Private Network technology is emerging with its powerful capabilities like auditing, logging and reporting. The technology ensures security to users. They can access sensitive data remotely with speed and security intact. The data can be accessed on mobiles also. The overhead of processor and memory consumption in turn deteriorates performance [11].

Cloud computing is new area for research because of the popularity of technology, anywhere anytime platform. Its un parallel capability to access services as and when required is contributing to its unprecedented growth. U.S Department of defense has mandated High Level Architecture (HLA) as facto for DoD modeling and simulation technology. Run Time architecture (RTI) is a middleware software. Cloud Simulation Platform is a modeling & Simulation platform which combines virtualization technology, pervasive computing, cloud computing and high-performance computing. It is gaining wide attention from all communities, industry and academia HLA/RTI has been suggested for this on the basis of HTTPS and public Key Infrastructure (PKI) techniques [12].

Notably, wearable computing devices trust the smartphone to ensure security in communication. As most of them are supported by wireless connection, security cannot be effectively supported. It is proved through experiments that wearables can have secured communication and the additional HTTPS cost has same effect on smartphones and wearables. Wearables namely smart watches, smart glasses and fitness bands are becoming highly in demand similar to smartphones. The sensors in these commodities have not only personal data of users but it also captures physical surrounding data and pattern and behavior of online activities. This data is highly sensitive and cannot be controlled through any other method. Considering the confidential aspect of data, data transferring must take place through secure communication protocol such as https. Whereas wireless communication (Wi-Fi) capabilities and cellular networking becoming widespread security in such wearables is a big concern and needs immediate attention. It is emphasized at this stage that secure communication protocols are there and able to provide efficient services to such wearables through cellular networking as most of them are connected through Wi-Fi. The technical support in terms of providing and ensuring secured data exchange and transmission is the need of the day [13].

A survey on Alexa top 20000 Sites, .jpg sites and 120 bank association sites and online banking focused on probing areas namely, HTTP/HTTPS usage rate analysis, Server- side certificates, banking login services. It concluded that about half of them had FQDN mismatch. It also revealed that design guidelines of HTTPS sites are declared as countermeasure against the problem. A big percentage is still using HTTP as against HTTPS increasing the possibility of malicious activities and other security issues [14].

2.3 Attack Mitigation Techniques

The paper highlights the all-time concern for HTTPS namely Man-In-The-Middle attacks. Since HTTPs has weakness of not being able to tolerate to unauthorized access using SSL Man in the Middle technique. This is a big security hole in the performance of HTTPS. HTTPS sensitive data can be easily captured and maliciously used.

It is prone to attacks and hijacks. Suggested solution highlights to fill the gap through one of three techniques, namely Static ARP at the switch or using ARP.

Watch to alert the administration and uses the technique of correlating the IP and MAC address together and Anti-sniffing technique that provides a function to scan a machine capturing information by analyzing Network Interface Card running in promiscuous mode, thereby enhancing the security of Network System. All these three techniques are simple that can be enforced in practice [15].

It is proposed that the popular security vacuum Man in the Middle (MITM) attacking techniques to evade HTTPS can be easily achieved through SSL Stripping. This captures HTTPS traffic and directs it to HTTP only and maliciously hijack the crucial data. It is difficult to comprehend such an activity for a novice user. This is carried out with such ease as the mechanism and guidelines for SSL Stripping is available easily. Moreover, SSL Stripping can be achieved by any MITM techniques. Suggested solution ISAN-HTTPs provides mechanism against SSL Stripping attacks by implementing a connection to HTTPS, that determines URL and implements HTTPS data exchange for ensuring against SSL Stripping attacks [16].

Noticeably, Electronic Mail (email) is the de facto communication and data exchange mode of the modern world. As per an estimate, it will be 2.9 billion in 2019 against 2.6 billion in 2015. Also, as per research, 90% of the emails exchanged every second is either a spam or a virus. Spoofing is yet another security area which is menace to handle. It is a serious problem in which a malicious person, called spoofer sends an email to a user to extract sensitive data as a forged identity, disguising the identity as an authentic source. The proposed solution Web-based Anti- spoofing email application is suggested. In this technique not only you can find out, rather analyze and further control such attacks with help of triggering and filtering the message through spoof filter. Also, an option of notifying the spoofer is there to take the complete command of the situation in active mode [17].

As HTTPS is the preferred and default option for most popular sites, the volume of encrypted data and consequently the traffic is increasing at an unprecedented pace. There is a pressing need to analyze ciphered application traffic thereby serving as the base for several mitigation techniques and mechanisms, namely Quality of Service, implementation of security policy, firewall, middle box and network security. A machine learning based approach is suggested in a holistic view to reveal the effectiveness of approach as a robust mitigation tool to investigate further [18].

It is proposed that Android Platforms are the most wide-spread globally, yet highly insecure. There are security issues in most applications, making it vulnerable to malicious activities, thefts and attacks. These security vacuums that are prevalent in applications and libraries, makes HTTPS benefits null and void. The suggested technique emphaSSL focuses on ensuring network security in Android applications to mitigate various attacks in this front [19].

HTTPS is the default option for most popular websites namely Google, Amazon, Yahoo and the likes. These sites use encrypted protocol only for a single aspect user authentication and the rest of the session is on HTTP, increasing the probability of sensitive data to stealing and malicious activities and attacks. The crucial session data copying such as cookies, identifiers or hijacking session. Recent research shows that Session hijacking is a big challenge of internet age. The tools techniques and guidelines for this malicious activity are freely available. Therefore, to use new ways without falling back on User agent data is the need and the same is pointed through technique called browser fingerprinting to tackle the issue to a great extent [20].

Interestingly, Short Messaging System (SMS) has zero-day security as it can be compromised by Man In The Middle attacks. A mitigation technique to handle this big issue is required and hence the suggested framework uses HTTPS for enhancing security providing confidence and Authentic SMS through encryption [21].

The popularity of mobile computing, multimedia sources and social networks has led to the development of Content Distribution Network (CDN). The need for streaming and sharing content has increased with technologies and hence various security issues and concerns. With data available, illicit unauthentic evaders become active. It focuses on two crucial and parallel web technologies, namely HTTPS and CDN (Content Delivery Network) and the way the two work in tandem. Wide data studied concludes issues namely invalid certificates, private key sharing and not secure back end communication. Few of these issues being operational and others due to multiple parties in service. The study will be a platform to probe mitigation solutions on Network security management with respect to HTTPS and CDN in future [22].

It is presented that Online social networking (OSN) giving organizations to show their presence online and users to connect with others when physically and logistically connect is difficult. OSN poses serious threats to crucial data as it is prone to malicious and potential hijacks and leakage. Existing methods are not equipped to handle this vulnerability. Hence a technique called shutter which is an information protection system. The system is based on domain gateway for OSN. It filters traffic in a fine grain approach to assess http requests and hence enhances the parsing phase of OSN. It also takes advantage of rule matcher to achieve high matching throughput and speed up the rules insertion and deletion operations [23].

2.4   Evaluation of Current State of Art in Topic

We have come a long way from HTTP to HTTPS, from SSL Version2, Version 3 to TLS and now we are moving ahead with lot of tools and technologies and methods in place to work in tandem with these, taking care of all the vacuums lying in the protocol with known and unknown vulnerabilities. We need to learn few lessons from the past so that we move ahead in a robust and firm way to security. Documents leaked by ex-NSA Contractor Edward Snowden has confirmed that HTTPS is a preferred target for malicious cyberattacks. Compromising gaps in web security owing to HTTPS leads to risks at global cyber security level and is ethically a question mark. On the one hand, there is a challenge to be compatible with legacy protocols and software with known vulnerabilities, pace and decay of traditional protocols. On the other hand, responding slow in the web security ecosystem with emerging technologies namely cloud computing, VPN, Wearable Android application, mobile computing will have negative impacts too.

Encryption is the language of secured network but even after encryption are we secured. The added overhead in terms of cost, performance, efficiency and most importantly time we are bearing but the question is are we safe. At what level and how much security we are asking for. With sensitive data everywhere in the network, just by adopting HTTPS can I sit back in terms of security. Who has the onus, in case of leakage and attacks. Even after all these security in place, are we not having and had incidents in the past of the likes of Heartbleed and Poodle bleed.

Middle boxes are there to analyze encrypted data and traffic and protect the end user's sensitive information but what if the websites and ISPs are themselves stealing this data and selling to vendors. The important information about user identity and online preferences and patterns, all these are compromised. Who needs to address these critical aspects.

HTTPS is truly a double-edged sword, since it is secured everyone is using it, increasing traffic at an unbelievable pace. Due to heavy traffic, it needs to be analyzed and comprehended. Voluminous traffic makes this task highly difficult and seemingly impossible.

The way ahead is in terms of users not required to be scared and alert and with the expectation of being tech savvy to safeguard themselves from malicious attacks and hijacks and sensitive data copying, where people are compulsorily using ICT for all the transactions, authorities to take care of a secured and highly robust network.

Last but not the least, robust techniques for detection, analysis and control all three together are required to ensure minimizing the attacks in HTTPS. Cybercrimes and Cyberlaws need to work in tandem strictly to achieve the secured computer network.

2.5   Overall Opinion in Future Directions

Public key Cryptography implementation in end to end security architecture that allows to create a full secure web server attack namely SSL, HTTP and efficiency of connections. Solutions like Elliptic Curve Cryptography (ECC) and sizzle are in the exploratory stage.

Since HTTPS has a weakness of insecure requests and responses leading to Man In The Middle attacks and security vulnerability, mechanisms to detect unauthorized modifications is required. Projects like Steganography are in its initial stages and a lot needs to be done.

Another vulnerable area to deal with is the existence of fresh pages at URLs at places where nothing should exist. Analyzing and finding such URLs can save various attacks which go undetected even by experienced professionals. Some solution based on lexical features is in its early stages and yet to be worked upon.

Notable, session hijacking attacks are gaining popularity and robust solutions must be worked at in future. The prevalent approach of forcing crucial data and functionality over ciphered data, and allowing others to be accessed through HTTP is a burning issue that can lead to various issues further ahead.

3   Conclusion

The aim of this survey paper is to highlight the major contributions in the paper previously done on HTTPS, why it is gaining popularity and reliance on it for secured network. Through various papers, it is highlighted that although HTTPS is increasingly becoming popular and de facto standard for data transmission and communication on

Internet. There are various gaps like MITM, usage of Middleboxes and BYOD implementation, ciphered traffic monitoring, cost, effort and skill needed that makes the security of network through HTTPS a difficult task. Yet, sophisticated new techniques ECC, Steganography are being devised and worked upon, through cyberlaws and policies, robust methods for detection, reporting, logging and controlling all working in tandem will pave the path ahead towards a secure cyber world.

Acknowledgment. The work described in this paper was supported by my mentor Mr. Kieran McLaughlin. The author pays gratitude for openhandedly giving his time and expertise, in helping on the challenges within cyber security to be viewed from a new standpoint, and in giving the insight into the investigation into cyber security monitoring as a psychological as well as a security question.

References

1. Larsson, E., Sigholm, J.: Papering over the cracks: the effects of introducing best practices on the web security ecosystem (2016). 978-1-5090-1724-9/16 ©2016 IEEE
2. Ouvrier, G., Laterman, M., Arlitt, M., Carlsson, N.: Characterizing the HTTPS trust landscape: a passive view from the edge. IEEE Commun. Mag. **55**, 36–42 (2017)
3. Ford, R., Howard, M.: Man-in-the-middle attack to the HTTPS protocol. IEEE Computer Society, IEEE, January 2009
4. Liu, X., Qian, F., Qian, Z.: Selective HTTPS traffic manipulation at middleboxes for byod devices (2017). 978-1-5090-6501-1/17/ © 2017 IEEE
5. Shbair, W.M., Cholez, T., Francois, J., Chrisment, I.: A Multi-level framework to identify https services. In: 2016 IEEE/IFIP Network Operations and Management Symposium (NOMS) (2016)
6. Husák, M., Cermák, M., Jirsík, T., Celeda, P.: Network-based HTTPS client identification using SSL/TLS fingerprinting. In: 2015 10th International Conference on Availability, Reliability and Security (2015)
7. Muehlstein, J., Zion, Y., Bahumi, M., Kirshenboim, I., Dubin, R., Dvir, A., Pele, O.: Analyzing HTTPS encrypted traffic to identify user's operating system, browser and application. In: 2017 14th IEEE Annual Consumer Communications and Networking Conference (CCNC) (2017)
8. Yan, L., Deng, H., Chen, X., Ye, X.: Service differentiation strategy based on user demands on https web servers. In: SERA 2018, 13–15 June 2018, Kunming, China (2018). 978-1-5386-5886-4/18/ ©2018 IEEE
9. Fu, J., Xie, M., Wang, Y., Mei, X.: An empirical study of unsolicited content injection into a website. In: 2017 International Conference on Networking and Network Applications (2017)
10. Shbair, W.M., Cholez, T., Goichot, A., Chrisment, I.: Efficiently bypassing SNI-based HTTPS filtering. In: 2015 IFIP/IEEE International Symposium on Integrated Network Management (IM2015): Experience Session Paper (2015). 978-3-901882-76-0 @2015 IFIP
11. Song, Y., Li, H., Cheng, L., Xiang, M., Cai, J.: SSL VPN resources log optimization techniques based on Bloom Filter algorithm (2016). 978-1-4673-9194-8/16 © 2016 IEEE
12. Zhang, Z.H., Chai, X.D., Hou, B.C.: System security approach for web-enabled HLA/RTI in the cloud simulation environment. In: 2011 6th IEEE Conference on Industrial Electronics and Applications (2011). 978-1-4244-8756-1/11/ © 2011 IEEE
13. Kolamunna, H., Chauhan, J., Hu, Y.: Kanchana Copyright Information. https://doi.org/10.1109/cicsyn.2012.50. 978-0-7695-4821-0/12 © 2012 IEEE

14. Fowdur, T.P., Veerasoo, L.: An email application with active spoof monitoring and control. In: 2016 International Conference on Computer Communication and Informatics (ICCCI - 2016), 07–09 January 2016, Coimbatore, India (2016). 978-1-4673-6680-9/16/ ©2016 IEEE
15. Foroushani, V.A., Zincir-Heywood, A.N.: Investigating application behavior in network traffic traces. In: 2013 IEEE Symposium on Computational Intelligence for Security and Defense Applications (CISDA) (2013). 978-1-4673-5911-5/13/ c 2013 IEEE
16. Wei, X., Wolf, M., Guo, L., Lee, K.H., Huang, M.-C., Niu, N.: emphaSSL: towards emphasis as a mechanism to harden networking security in android apps (2016). 978-1-5090-1328-9/16/ ©2016 IEEE
17. Unger, T., Mulazzani, M., Fruhwirt, D.: SHPF: enhancing HTTP(S) session security with browser fingerprinting. In: 2013 International Conference on Availability, Reliability and Security (2013). https://doi.org/10.1109/ares.2013.33. 978-0-7695-5008-4/13 © 2013 IEEE
18. Khan, M.M., Bakhtiari, M., Bakhtiari, S.: An HTTPS approach to resist man in the middle attack in secure SMS using ECC and RSA. In: 2013 13th International Conference on Intelligent Systems Design and Applications (ISDA) (2013). 978-1-4799-3516-1113/ ©2013 IEEE
19. Komatineni, S., MacLean, D., Hashimi, S.Y.: Pro Android 3, ser. Apress. Paul Manning, April 2011. http://www.apress.com/9781430232223
20. Elgin, B.: Google buys android for its mobile arsenal. Bloomberg Businessweek, August 2005. http://www.businessweek.com/technology/content/aug2005/tc20050817\0949\tc024.htm
21. Liang, J., Jiang, J., Duan, H., Li, K., Wan, T., Wu, J.: When HTTPS meets CDN: a case of authentication in delegated service. In: 2014 IEEE Symposium on Security and Privacy. IEEE (2014). https://doi.org/10.1109/sp.2014.12
22. Wu, T., Li, J., Wu, N., Ou, T., Yang, B., Li, B.: Shutter: preventing information leakage based on domain gateway for social networks. In: 2014 IEEE International Conference on Ubiquitous Intelligence and Computing (2014). https://doi.org/10.1109/uic-atc-scalcom.2014.121. 978-1-4799-7646-1/14 © 2014 IEEE

A Survey on Different Approaches for Malware Detection Using Machine Learning Techniques

S. Soja Rani[✉] and S. R. Reeja

Dayananda Sagar University, Bangalore, India
soja.naveen@gmail.com, reeja-cse@dsu.edu.in

Abstract. Malwares are increasing in volume and variety, by posing a big threat to digital world and is one of the major alarms over the past few years for the security in industries. They can penetrate networks, steal confidential information from computers, bring down servers and can cripple infrastructures. Traditional Anti-Intrusion Detection/Intrusion prevention system and anti-virus softwares follow signature based methods which makes the detection of unknown or zero day malwares almost impossible. This issue can be solved by more sophisticated mechanisms in which, static and dynamic malware analysis can be used together with machine learning algorithms for classifying and detecting malware. Through this paper we present a survey on the different techniques for concealment and obfuscation used to make sophisticated malware as well as the different approaches used in malware detection and analysis.

Keywords: Malware analysis · CyberSecurity · Machine learning

1 Introduction

Data breaches, ransomware attacks, targeted botnet as well as malware attacks are in the daily news these days. Brain, the first PC virus in 1986 to the highly obfuscated Wannacry in 2017, the malware epidemic is continuing its alarming expedition. Due to the advancements and increased use in technology, we can observe a continuous evolution of malwares in volume, variety and velocity. Malicious software or in short, Malware, can be code, scripts, or any other content that are designed to interrupt the normal operation or gather information illegally that leads to loss of privacy, gain unauthorized access to system resources, and other abusive behaviour. Computer viruses, spyware, Trojan horses, worms, adware, botnets, rootkits etc. come under the huge umbrella of malware which forms the integral component of almost all the data breaches. Attackers are also using more tools, like polymorphic malware and zero-day malwares, to evade the current malware detection tools. The wide spread use of World Wide Web is also an inevitable reason behind the increase in threat from malware. There are multiple doors for the adversary to enter the enterprise network which are guarded by perimeter security tools such as firefalls, antivirus, network based and host based intrusion detection/prevention tools which at times may not be able to distinguish between genuine user and an adversary. Once the adversary enters the enterprise

© Springer Nature Switzerland AG 2020
P. Karrupusamy et al. (Eds.): ICSCN 2019, LNDECT 39, pp. 389–398, 2020.
https://doi.org/10.1007/978-3-030-34515-0_42

network and gains knowledge about the network, they unwind their plan of action and attack the target. During their course of action in the network, the leave behind some changes in the data or signal which can be detected using data science based tools that can raise alerts.

2 Malware Camouflage Evolutions

The advancements in the malware code by concealing the appearance have become a serious challenge for the antivirus companies. On the basis of the concealment technology used malwares can be classified as Encrypted, Oligomorphic, Polymorphic and Metamorphic Malwares.

2.1 Encrypted Malware

Encrypted malware has two basic sections in their structure: a decryption loop and main body. Decryption loop or the Decryptor code is responsible to decrypt and encrypt the program of the main body, the actual malware, which is meaningless otherwise. After getting into the host, virus begins its action where the decryptor loop executes first to decode the main body into machine executable code. Since the decryption module remains the same, these viruses are detectable by analyzing the decryption module. Different keys are used for each infection to hide the signature and to make the detection process harder.

2.2 Oligomorphic Malwares

Next step in the concealment tactics for defending the short comings of encrypted malware led to the development of Oligomorphic malwares where the decryptors are mutated from one variant to other. This type of malware can generate no more than few hundreds of decryptors which are randomly chosen for a new victim but still remains detectable with signatures. Once signatures of all the decryptors are made, these malwares cannot evade signature based detection techniques.

2.3 Polymorphic Malwares

In order to overcome the limitations of Oligomorphic malwares the next concealment technique came into prevalence is the polymorphic malwares which can create countless number of distinct decryptors. Here the presentation of the code is made unique constantly from one variant to other. When the malware executes, a new decrytpor is generated which joins with the encrypted malware body to construct a new malware variant [1]. Polymorphic malwares utilize different code obfuscation approaches such as substitution of instructions to mutate its decryptor or insertion of junk codes etc. to build a new variant for a new victim which is done using a mutation engine or obfuscation engine. Even though these malwares are able to hide from signature matching techniques, their body after decryption appears the same as well as behaviour which can be used as the source for detection. The detection tools for the same adopts the emulation technique where the malware is executed in an emulator and signatures can be constructed efficiently and can be detected using the conventional detection mechanisms.

2.4   Metamorphic Malwares

The metamorphic malware is the most novel approach in the 2nd generation of malwares. Different from the previous camouflage generations, metamorphic virus has no encrypted part because of which it does not need decryptor part, but employs a mutation engine which mutates the whole malware body rather than the decryptor. Each new copy of this type may have different size, code sequence, structure and syntactic properties, but the behaviour remains the same. Since a professional metamorphic malware does not leave a single pattern vulnerability to make it detectable, the defence software should be highly sophisticated and built on heuristics and behaviour based analysis and detection techniques.

2.5   ZeroDay Malware

ZeroDay Malware is the one which invades a software vulnerability for which there is currently no available defence or fix. Malware makes use of the vulnerability present in the system to perform adversarial actions on the system which can compromise the confidentiality, integrity, or availability of the system. Zeroday malwares cause significant damage by exploiting the delay that can happen between the delivery of the malware and the development of the counter-measures.

3   Approaches for Malware Analysis

Malwares can be analyzed through various methods and can be broadly categorized into two – static and dynamic analysis. Malware analysis gives a detailed and well understanding about the functioning of the malware as well as what can be done in order to eliminate the threats of the malware.

3.1   Static Analysis

It is the priliminary malware analysis technique where malware code is decompiled and examined. Static analysis examines malware binary without actually running it or without viewing the actual code or instructions. Different techniques and tools can be used to provide information about its functionality and collect information to produce simple signatures which is the unique identification for the binary file. Commonly used signatures include file name, DLLs called by the malware, URLs accesses, MD5 checksums or hashes, file type, file size etc.

3.2   Dynamic Analysis

As the complexity and sophistication of malware increases, it becomes hard to analyse the malware using the static signatures. Dynamic analysis (also known as behavior analysis) executes malware in a controlled and monitored environment to observe its behavior. During dynamic analysis the Malware is executed in a controlled, isolated virtual environment like Cuckoo Sandbox for studying the malware thoroughly and to understand its functionality with out damaging your system. After the execution some features or indicators which can be used in detection are extracted. The features revealed with basic dynamic analysis can include IP addresses, domain names, registry keys, file path locations or can be any additional files located on the system or network. Many tools are available in order to make the dynamic analysis efficient and safe like Netcat, Wireshark, Regshot InetSim, ApateDNS, Procmon, etc.

4 Approaches for Malware Detection

The available collection of the techniques for malware analysis and detection inclusive of those adopted by the industries and those that are not can be categorized into four approaches - Static Signature based approach, Static Behavior based approach, Dynamic Signature based approach, Dynamic Behavior based approach.

4.1 Signature Based Detection

Unique sequences of bytes called signatures are extracted, once the malware is identified and then added to a database which may contain hundreds of millions of signatures that are already identified as malicious objects. The techniques scans the file in the system to find the defined malware signature, if found, an alert of the presence of malware is sent. Signatures can be efficiently and quickly scanned and identified by algorithms. Most of the commercially available anti-malware products use this method as the primary technique for identifying malicious objects. The signature based malware detection deploying machine learning can be learned by analyzing the assembly features or the binary features.

Signature based detection can generally identify only previously known and identified malware. Even though it is easy to use, scanning and pattern matching becomes costly as the malware signature database is increasing exponentially [2]. Another major reason being today's advanced malware can alter its signature to avoid detection. The technique is reactive in nature and thus is unable to identify threats or attacks from the new malwares. The Cisco 2017 Annual CyberSecurity Report states that 95% of malware files they analyzed weren't even 24 h old, indicating the prevalence of zero day attacks.

4.2 Behaviour-Based Malware Detection

Behaviour-based malware detection analyses the given sample based on its behaviour or actions. Abnormal and unauthorised actions and run time activities of the given sample in the sandboxed environment make it malicious or at least suspicious. There Installing rootkits, registering for auto start, disabling security controls, attempt to discover a sandbox environment are all behaviours that can point towards a malicious behaviour. Behaviour based detection can either use static analysis or dynamic analysis. If the malicious behaviour is evaluated as it executes, then it is called *dynamic analysis*. *M*alicious intent can also be evaluated by *static analysis* where the system is worried only about the structure and program of the malware. Behaviour based detection is the leading technology today as it can detect metamorphic and zero day malwares in near real time. Machine learning based behaviour-based detection systems either use API calls or assembly features deployed by the malware for learning.

There are still a few important limitations of behaviour based detection. If malware understands that it is running in a sandbox or virtual environment, it will try to hide its malicious intensions by curtaining its malicious activity. Second drawback is that the malware may take its own time to unleash the malicious activity which is limited in a virtual environment resulting in latency in dynamic analysis. Behaviour based systems are prone to false alarm [3], which may make the system more vulnerable by taking the real malware as another false alarm. It has been demonstrated that anomaly detection-based techniques are susceptible to mimicry attacks.

4.3   Hybrid Malware Detection Approach

Both signature and behaviour-based malware detection has its own pros and cons. Even though static and dynamic analysis are strong mechanisms that can stand alone to for malware analysis, another genre which can reap the benefits is the hybrid approach which uses both static features and dynamic features for analysis. The best security will come from utilizing both technologies. This technique is more beneficial to reverse engineer complex malwares. True intentions and capabilities of a malware can be analyzed better using a hybrid approach.

Information security industry is often misled by vendors promising next-generation firewalls and other sophisticated security software. Most of these sophisticated anti-malware systems rely on the decades old signature-based approach for malware detection that is not up to the level of detecting today's metamorphic malware and zero-day attacks. Behaviour based malware detection strategies should be used to augment the signature based systems so that the sensitive data or critical operations of the organizations remain protected. Researchers propose the use of hybrid approach to augment the results of standalone detection techniques so as to reduce high probability of false alarm [4].

5   Review of the Malware Detection Approaches

Here few current researches in malware detection are analyzed together with the machine learning algorithms used. Automating the CyberSercurity industry leveraging data science using machine learning algorithm is gaining popularity in recent years. By making use of learning algorithms in anti malware industry we can not only detect the known malwares but also act as knowledge for the detection of new variants of mal-ware including the polymorphic and zeroday malwares. This technique does not have to replace the existing standard detection methods, but surely can act as an add-on improvising the detection probability. Since machine learning techniques are more computationally demanding when compared with the standard existing systems, it may not be suitable for end users but can be promisingly implemented at the enterprise gateway level to act as a central anti-malware engine. Even though the infrastructure can turn out to be costly, but it can help in an efficient and promising way by protecting valuable enterprises data from the security threat and can prevent immense financial damages.

5.1   Review of the Selected Signature-Based Detection Approaches

Signature based detection approaches reduces the overhead of the system as well as the execution time taken for malware detection or prediction. This approach can be used in all three target environments including Windows based systems, embedded systems and smart phones.

Wang and Wang [5] presented a machine learning based automatic malware recognition framework using support vector models (SVMs). Behavioral signatures were used to train the SVM classifier. The system performed good and the classification error decreased with increasing size of the test data. For different sizing (N) of malware samples, the prediction accuracy of malware detection goes up to 98.7% with N = 100.

Another strategy was proposed by Santos et al. [6] to identify malware files. Frequency of occurances of opcode groupings was the feature used in this study. The importance and the recurrence of each opcode grouping were evaluated. The paper also presented the result and validation which proved that the new system was able to detect unknown malware files as well.

Martín et al. [7] in their paper illustrated the use of third part calls to ignore the effects of the different camoufalging strategies used by the different malware families for obfuscation. They combined clustering and multi target advancement to create a classifier in view of specific behaviours carried out by outsider or third party call groups. This analyzer, named MOCDroid, achieves a precision of 95.15% in test with 1.69% of false positives which is much better than the commercial antivirus engines from VirusTotal.

Hellal and Ben Romdhane [8] in their research utilized static analysis to form a technique for mining data from graph to recognize variations of malware. MCFSM, Minimal contrast frequent sub graph miner, a novel algorithm was proposed by them for identifying the seen malicious behavioural patterns which can distinguish between malicious and benign programs. The proposed method displayed high malware detection rates combined with low false positive rates. The behavioural signatures requirement was also limited for the proposed method.

Fan et al. [9] proposed a system to detect malicious files based on instruction sequences extracted from the data sample to build a sequence mining algorithm. The identified patterns were used to train an All-Nearest-Neighbor (ANN) classifier for malicious and benign classification. The method utilizing the pattern mining appended by ANN classifier proved to be successful even for polymorphic and zeroday malware samples.

Boujnouni et al. [10] proposed a new approach which was very effective against attacks like polymorphism and metamorphism which are the techniques used by malware creators in order to obfuscate the code. The approach used N-grams to train an improved version on Support Vector Domain Description. Experimental results were evaluated after the successful classification of several hundreds of malware and benign files to confirm the accuracy and feasibility of the proposed method.

5.2   Review of the Selected Behavior-Based Detection Approaches

Dynamic analysis can performed by monitoring various runtime execution features including API call hooking, binary hooking, data flow analysis, instruction trace, multiple path execution, running in sandbox or virtual machine, using machine learning, etc. Even though the approach can be used in all the three target environments, most of the research studies have been conducted in the smart phone environment.

Ye et al. [11] put forward a heterogeneous deep learning framework for autonomous malware detection. They monitored the runtime portable executable files and extracted the Windows API calls which served as the feature for the learning system. Deep learning framework comprised of an AutoEncoder which was driven by multi-layer Boltzmann machines together with a layer of associative memory. The method took advantage of both supervised and unsupervised learning so as make use of the unlabeled data sets in the training phase which trained the deep neural network. Data was collected from Comodo Cloud Security Center on which experiments was done and the results proved better precision and low false positives compared with shallow learning systems.

Bayer et al. put forward TTAnalyze, a tool for analyzing Windows executables. The approach monitors the behaviour by running the binary in an open source PC emulator, Qemu and used Windows native system calls and Windows API function calls as features for the learning system [12].

In his work [13] Willems et al. made use of API hooking and dynamic linked library injection techniques in their anti malware tool, CWSandbox. The software was built for malware detection for Win32 family of OS. Behavioural analysis was performed on features like registry manipulation, network communications, file system changes and operating system interactions. The first five bytes of the application programming interface was replaced with unconditional jump for implementing the hook.

Mohaisen et al. [14] addressed the shortcomings in the then existed systems through the introduction of AMAL, an automated and behavior-based malware analysis and labeling system. The system has to parts. The first subsystem, AutoMal runs the files in virtualized environments and has a collection of tools to collect the behavioural features that characterize the malware like file usage, memory, network, and registry. MaLabel makes use of the collected features to build classifiers trained by data samples which can be used to classify the files into groups of similar behaviour. Experiments was conducted on medium sized samples of size 4000 as well as on large scale samples of size 115,000 over 13 months and the results was promising a precision of 99.5% and recall of 99.6% for certain families' classification, and more than 98% of precision and recall for unsupervised clustering.

Norouzi [15], in their proposal for malware detection presents a data mining classification approach which relies on behavioural features of the malicious software using dynamic analysis. The executive XML file is used to extract features and provides input to the WEKA tool. The performance is evaluated using a real case study data set using WEKA tool to illustrate the performance efficiency as well as training data and test.

5.3   Review of the Selected Hybrid Detection Approaches

Eskandari et al. [16] introduced an hybrid approach called HDM-Analyzer, which uses both the major approaches to analyze an executable file. While static analysis analyze the file in the program source code level, dynamic analysis extracts the features of the file by observing program's activities during its execution time. The proposed HDM-Analyzer takes into account the advantages associated with both types of analysis thus reducing the analysis and detection time as well as maintaining the required level of precision and accuracy of classification using Bayesian network, Naive Bayes, Lazy K-Stare.

Yuan et al. [17], in their paper proposed to combine both the static and dynamic analysis features of Android apps and used deep learning to automatically detect the malicious ones called DroidDetector which was successful in classifying apps as malicious and benign. The experimental results proved that deep learning turned out to be efficient for detecting and classifying malware and that the efficiency of the software increased with the variety and volume of data. 96.76% detection accuracy was achieved which concluded that DroidDetector was much efficient than the traditional machine learning techniques.

Dali et al. [18], in their paper is making use of machine learning algorithms to learn the difference in features between malicious and non malicious apps automatically. They proposed a deep learning based application, DeepFlow, for detecting malware in which data flows in the Android applications is analysed. The experimental results showed high detection F1 score of 95.05%, which outperformed the classic machine learning based approaches.

Ding et al. [19], in their paper proposes an Application Programming Interface based association mining rules for identifying malicious files. One of the approaches is to use efficient rule quality by removing the APIs that cannot become frequent items. Finding association rules with strong discrimination ability is another approach for the same. The results proved that the use of these new approaches resulted in faster running of OOA and that the time complexity for data mining is reduced by 32% and the time complexity for classification is reduced by 50%.

A hybrid approach with both signature and behaviour based approaches driven by machine learning algorithms for detecting malicious apps was introduced by Rehman et al. [20] in their research. The Android apps are reverse engineered to extract manifest files, and binaries. Classification algorithms including Decision trees, SVM, KNN and W-J48 were used to perform classification experiments. Results showed that SVM was the better classifier in case of binaries and KNN in case of.xml files. The proposed system was tested on benchmark datasets and results proved better accuracy in malware detection.

6 Conclusion

This paper presents a thorough study on the evolution of the concealment techniques as well as obfuscation methods employed in the generations of malware. The paper also focuses on the two analysis methods of malware – static analysis and dynamic analysis. The paper also concentrated on Signature based malware detection and Heuristics or Behaviour based malware detection, the two detection strategies prevailing in the industry. Owing to the advantages and disadvantages in both the detection strategies, a combination approach has also been evolved, the hybrid detection strategy. The paper also tries to review a few literatures of the malware detection approaches using machine learning. The reviewed papers are classified into the above mentioned 3 categories - (1) signature-based, (2) behaviour-based approaches and (3) Hybrid approach.

References

1. Digital Object Identifier: The effects of traditional anti-virus labels on malware detection using dynamic runtime opcode. https://doi.org/10.1109/ACCESS.2017.2749538
2. Beaucamps, P.: Advanced polymorphic techniques. Int. J. Comput. Sci. **2**(3), 194–205 (2007)
3. Wong, W., Stamp, M.: Hunting for metamorphic engines. J. Comput. Virol. **2**, 211229 (2006)
4. Govindaraju, A.: Exhaustive statistical analysis for detection of metamorphic malware. [MS Project], San Jose State University, US (2010)
5. Wang, P., Wang, Y.-S.: Malware behavioural detection and vaccine development by using a support vector model classifier. J. Comput. Syst. Sci. **81**, 1012–1026 (2015)
6. Santos, I., Brezo, F., Ugarte-Pedrero, X., Bringas, P.G.: Opcode sequences as representation of executables for datamining-based unknown malware detection. Inf. Sci. **231**, 64–82 (2013)
7. Martín, A., Menéndez, H.D., Camacho, D.: MOCDroid: multi-objective evolutionary classifier for Android malware detection. Soft. Comput. **21**, 7405–7415 (2017)
8. Hellal, A., Romdhane, L.B.: Minimal contrast frequent pattern mining for malware detection. Comput. Secur. **62**, 19–32 (2016)
9. Fan, Y., Ye, Y., Chen, L.: Malicious sequential pattern mining for automatic malware detection. Expert Syst. Appl. **52**, 16–25 (2016)
10. Boujnouni, M.E., Jedra, M., Zahid, N.: New malware detection framework based on N-grams and support vector domain description. In: 2015 11th International Conference on Information Assurance and Security (IAS), pp. 123–128 (2015)
11. Ye, Y., Chen, L., Hou, S., Hardy, W., Li, X.: DeepAM: a heterogeneous deep learning framework for intelligent malware detection. Knowl. Inf. Syst. **54**, 265–285 (2017)
12. Bayer, U., Moser, A., Krugel, C., Kirda, E.: Dynamic analysis of malicious code. J. Comput. Virol. **2**(1), 67–77 (2006)
13. Willems, C., Holz, T., Freiling, F.: Toward automated dynamic malware analysis using CWSandbox. IEEESecur. Priv. **5**(2), 32–39 (2007)
14. Mohaisen, A., Alrawi, O., Mohaisen, M.: AMAL: high-fidelity, behavior-based automated malware analysis and classification. Comput. Secur. **52**, 251–266 (2015)
15. Norouzi, M., Souri, A., Samad Zamini, M.: A data mining classification approach for behavioral malware detection. J. Comput. Netw. Commun. **2016**, 9 (2016)

16. Eskandari, M., Khorshidpour, Z., Hashemi, S.: HDM-analyser: a hybrid analysis approach based on data mining techniques for malware detection. J. Comput. Virol. Hacking Tech. **9**, 77–93 (2013)
17. Yuan, Z., Lu, Y., Xue, Y.: DroidDetector: android malware characterization and detection using deep learning. Tsinghua Sci. Technol. **21**, 114–123 (2016)
18. Dali, Z., Hao, J., Ying, Y., Wu, D., Weiyi, C.: DeepFlow: deep learning-based malware detection by mining Android application for abnormal usage of sensitive data. In: 2017 IEEE Symposium on Computers and Communications (ISCC), pp 438–443 (2017)
19. Ding, Y., Yuan, X., Tang, K., Xiao, X., Zhang, Y.: A fast malware detection algorithm based on objective-oriented association mining. Comput. Secur. **39**(Part B), 315–324 (2013)
20. Rehman, Z.-U., Khan, S.N., Muhammad, K., Lee, J.W., Lv, Z., Baik, S.W., Shah, P.A., Awan, K., Mehmood, I.: Machine learning assisted signature and heuristic-based detection of malwares in Android devices. Comput. Electr. Eng. **69**, 828–841 (2017)

Classification of Walkers Based on Back Angle Measurements Using Wireless Sensor Node

Ramandeep Singh Chowdhary[(✉)] and Mainak Basu

GD Goenka University, Gurgaon, Haryana, India
raman85friends@gmail.com, mainak.basu87@gmail.com

Abstract. High end technology enabled devices are being used these days to perform classification and analysis of walking styles of athletes and patients for therapeutic applications. Hence it has become an encouraging step to carry out research in related domain. Various sports have significant health benefits which contribute to muscular, heart, and mental health. However, there is high risk of having injuries while playing outdoor sports and are very common in athletes. Relative excessive loading and impulsive impact on the muscle tissues causes almost all type of basic and severe injuries. To study the phenomenon of injury occurrence, avoidance, improvement in training techniques, and therapeutic applications, an open source electronic device has been fabricated using microcontroller and gy-521 sensor module. The designed system was used to study the effect of lower back movement of persons while walking and was able to classify subjects based on the lower back deviation angle. This result shall form the basis of designing customized training sessions suited for athletes to minimize injuries and suggesting of physiotherapy for patients with lower back pain. The designed system can be used as a reliable evaluation device for lower back analysis in various field environments without any constraints. The device could support injury management, performance enhancement, and rehabilitation of lower back pain patients.

Keywords: Gyroscope · Lower back analysis · MPU6050 · Sports analytics · Wearable sensors

1 Introduction

Various physical activities along with sports have important mental, health, and psychological benefits which also include cardiovascular and musculo-skeletal benefit that are used by many people. It had also been seen that lower, upper, and middle back musculo-skeletal injuries happen in large number of sport activities because of muscular stress. These muscular stress conditions occurs because of relative excessive loading on bones and muscle tissues, sudden impact or excessive stretching of the muscular tissue i.e., high muscle strain values relative to actual strength of tissues. This is the main cause of having injuries while playing any sport. One significant reason this muscular stress and strain is walking technique. Bertoli et al. [1] developed a method which uses an inertial sensor on the lower back, to perform a quantitative gait analysis in presence of $180°$ and $360°$ turns in freezers and non freezers in Parkinson's Disease

© Springer Nature Switzerland AG 2020
P. Karrupusamy et al. (Eds.): ICSCN 2019, LNDECT 39, pp. 399–411, 2020.
https://doi.org/10.1007/978-3-030-34515-0_43

patients. Chong et al. [2] developed a method to select sensors based on mutual information, as well as learning a prediction model for lower limb joint kinematics. Kam et al. [3] developed plastic optical fiber based sensor and used intensity interrogation technique to analyze back bone bending in sagittal plane. Esfahani et al. [4] designed and developed a new device to record planar low back (thorax vs. pelvis) and shoulder motions using printed textile sensors on an undershirt. Xu et al. [5] showed in their research work, the feasibility of using a sensor system to measure human joint motions. The sensor in [5] can provide accelerometer and gyroscope measurements at 100 samples per seconds. In most of the cases all the tasks are completed in a controlled laboratory conditions under proper supervision since motion analysis based systems (known as bio-mechanics systems) are camera-based systems which captures the body movement and should remain spatially fixed during the testing session.

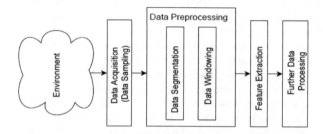

Fig. 1. Steps required for sports activities classification

One of the factor which negatively affect this technique is the change in ambient light. Figure 1 shows the steps required for activity classification in case of sports and therapeutic applications. Classification is performed on the extracted features from different data segments obtained from sensors. These windows or segment imparts a significant input to calculate accuracy in classification of events. Importance of sampling frequency in activity classification is a well known fact among researchers. On the other hand issues such as segmentation methodology and size of sampling window with respect to extracted features are not covered in the referred research work. Similarly, computational load (CL) is another important factor which should be considered when data classification is performed mostly in case of embedded systems for real time recognition and classification.

2   Human Walking Mechanism

Human walking style is also known as ambulation. Walking is analyzed as a pendulum in inverted orientation. In this pendulum model the body moves over the stiff limb with each forward step. Human walking has a different model from running model. An efficient way to differentiate walking and running techniques is to find height of subject's center of mass (CoM) by using cameras for recording motion. While walking, at middle of the stance, CoM attains maximum height. On the other hand, during running it attains minimum height. This change in height of CoM is valid for experiments performed over flat or nearly flat ground. Human walking resembles double pendulum model in inverted orientation. During forward stance, it is observed that the leg that looses the grip from the ground swings in forward direction from the hip (It represents the first pendulum). In second stance the leg makes contact with the flat surface touching heel first and then rolls through to the toe (It represents the second inverted pendulum). At any instance of time during walking, one foot always touches the ground. Running generally has a phase similar to that of a projectile. In this the subject performing running experiment has both the feet in air. It's a known fact that differences in human walking patterns based on gender of the person. Walking techniques of females has smaller step width as compared to males. Females also have more pelvic movement while walking, hence, human walking analysis generally takes gender into consideration. Walking analysis is also used to design customized training sessions for sport persons and for rehabilitation of patients with lower back pain. It also plays an important role in sports bio-mechanics. Using this technique athletes can improve their style of walking which further identifies movement-related problems caused due to injuries. The bio mechanics of walking is affected by many parameters, and deviation from normal walking style can be temporary or permanent. These parameters are given as, External parameters (such as terrain, footwear, and clothing), internal and physical parameters (such as sex, age, height, and weight), psychological parameters (such as emotions and personality), physiological parameters (such as anthropometric characteristics of a person i.e. measurements and proportions of body), pathological parameters (musculo-skeletal anomalies, neurological diseases, trauma, and psychiatric disorders).

3 Hardware Design

The implemented hardware was based on Hub-Node architecture and uses NRF communication chip (nRF24L01). It works on a frequency band of 2.4 GHz - 2.4835 GHz. The base band communication engine on which nRF24L01 works is known as Enhanced ShockBurst^{TM}. This wireless communication chip operates on ultra low power hence can support long battery life. It consumes only 12 mA of current while transmitting wireless data. It operates on a minimum voltage of 1.9 V to maximum voltage of 3.6 V. All input/output pins available on nRF24L01 communication chip are 5 V tolerable and it can be connected to Arduino directly without the need of level shifters. The implemented design based on nRF24L01 communication chip has 126 channels in total and reasonably higher data transfer rate starting from 250 Kbps and going to the maximum value of 2 Mbps. When working as a receiver it uses an on chip filters for noise reduction. The Enhanced ShockBurst^{TM} communication engine has dynamic payload sizes starting with minimum of 1 byte going up to maximum of

32 bytes. It also has six pipes which were used to increase throughput of the system by implementing data pipelining. This pipelining feature is known as MultiCeive™ and enables multiple reception of data from 6 different transmitters and can form a 1:6 star network.

(1) NRF Node Design: The circuit of wireless node is explained in this section. It consists of MPU6050 sensor which was used for data collections. After data acquisition from the sensor it transmits the data to the hub for creating log files. Figure 2 shows the circuit of wireless NRF node. The design was based on Arduino Nano CPU which performs calibration of MPU6050 sensor module before logging data for further processing. One important consideration to be noted is that the sensor should remain stationary for 5 s as precise calibration is required. After calibration was done, the sensor node starts collecting lower back deviation angle data and after primary processing, it was transmitted to the Hub for data logging. In the implemented design the baud rate for wireless communication was configurable and was set at 115200 to achieve high speed communication. An on-board red LED was available on the node which indicates calibration completion status. A blinking functionality was given on the node which will make the LED blink 5 times when the calibration of MPU6050 sensor was done. A voltage divider circuit was designed to keep track of battery discharging. Status LED would be in OFF state if the battery voltage was more than the 75% of maximum value and it would be in ON state otherwise. The Node was designed in a modular pattern which was interfaced with ready to use modules. A double sided PCB was fabricated which was having all the modules on a single plane along with battery for power supply. To make the node wearable using a wearable belt, a cut out was also made in the PCB.

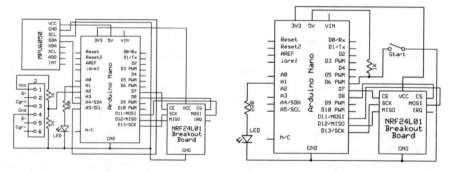

Fig. 2. Schematic diagram for wireless NRF node.

Fig. 3. Schematic diagram for wireless NRF hub.

(2) NRF Hub Design: Hub was designed and programmed as a receiver which collects all the data coming from the node side. It is having a NRF based transceiver module which can send and receive wireless data. Sending of data from hub to the node was required for sending acknowledgments which helps in avoiding overlapping of data and data loss at high speed wireless communication. An Arduino Nano was used in Hub design which receives data from the node and creates log files in the laptop. All the received data was used for lower back angle analysis and the subjects were classified after performing this analysis. The logging of data was performed with the help of Java based software. The Java software creates CSV file for multi axis sensor data coming from the node. Figure 3 shows the schematic diagram for Hub which has a synchronous switch (denoted as "Start") for introducing synchronization in Hub- Node architecture.

4   NRF Sensor Node Orientation

Precise sensor orientation is a major factor in classification of subjects in sports domain and for therapeutic application. This is important as it alters sampled data which significantly changes the accuracy of classification. Proper sensor orientation also allows coaches and doctors to perform accurate monitoring and analysis of athlete's lower back body movement techniques in both training and evaluation sessions. Accurate sensor orientation can be maintained while walking using wearable inertial sensors which have the upper hand with respect to non wearable sensors of being self sufficient as an independent module. In this way the data measurement was not affected by movement, surrounding conditions, and location while walking in training session or in the field. Three dimensional spacial analysis can also be performed of different body movements by using tri-axial accelerometers and gyroscopes which can measure movement in x, y, and z axis.

5   Classification Parameters

Diversity should be followed at the time of parameter selection which will result in better analytical and classification results. Various parameters plays an important role in achieving high accuracy and low computation time in activity classification out of which three parameters viz. sampling rate, segmentation method and window size are the most important ones.

(1) Sampling Rate: One of the most important steps in activity classification is data acquisition from the sensors as performing experiments again with the same test subjects is not always possible. Under-sampling and oversampling should be avoided at the time of data acquisition as under-sampling leads to information loss and over-sampling can result in noise introduced into collected data. Oversampling involves more computational load (CL) for analysis of classification problem as data along with noise is required to be processed. As stated in sampling theorem $f_{sampling}$ is dependent on f_{max} which is part of the data signal and is given as $f_{sampling} \geq 2f_{max}$.

Table 1. Sampling frequency for daily activity classification

Freq. (Hz)	Authors	Activity	Remark
512	Huynh et al. [6]	Daily activity	Highest $f_{sampling}$
256	Sekine et al. [7]	Accel. data	Data acquisition
76.25	Bao et al. [8]	2- axis accel.	1/3 of [7]
64	Preece et al. [9]	Daily activity	1/4 of [7]
50	Wang et al. [10]	Daily activity	Re-sampling
52	Casale et al. [11]	3-axis accel.	–
50	Ravi et al. [12]	3- axis accel.	–
50	Parkka et al. [13]	2- axis accel.	–
50	Maurer et al. [14]	Light sensor	Tested 1 to 30 Hz
10/15	Antonsson et al. [15]	–	99% FFT spectrum
20	Bouten et al. [16]	Diff. activities	Sufficient $f_{sampling}$
5	Gjoreski et al. [17]	–	Lowest $f_{sampling}$

The main point extracted from literature review is that $f_{sampling} \approx 50$ Hz is the most suitable value for $f_{sampling}$. Antonsson et al. [15] stated that 98% of the amplitude of the fast fourier transform spectrum is obtained with $f_{sampling} = 10$ Hz, and 99% below the value of 15 Hz. This is in accordance with the findings of Bouten et al. [16] stated that a $f_{sampling} = 20$ Hz is enough for successful activity classification. Table 1 shows comparison of sampling frequency in literature review for daily activity classification.

(2) Segmentation Method: As in most of the cases data acquisition is performed using live streaming of data from sensors. It becomes tough to decide which data points should be considered from the incoming live data for pre- processing which is performed after data acquisition. There are various segmentation methods which are used to break the incoming data segments into smaller data chunks for pre-processing of data. Various features which were extracted from the dataset and the accuracy of classification are the two main issues which depend on the selected segmentation technique as it impacts them effectively. Hence it can be said that performance of any classification algorithm will become weak if there are non-differentiable extracted features as given by Pietka et al. [18]. In real time various segmentation techniques puts a strong impression on different capabilities as complicated segmentation techniques increases computational load and might improve accuracy of classification results. In the literature review several different segmentation techniques has been identified which were used in various research projects, such as: Fixed-size Overlapping Sliding Window (FOSW) [18, 19], Fixed-size Non-overlapping Sliding Window (FNSW) [18, 19], Top-Down (ToD) [19], Bottom-Up (BUp) [19], Symbolic Aggregate approXimation (SAX) [18], Sliding Window And Bottom-up (SWAB) [19], String Matching (SM) [18], Reference-based Windowing (RbW) [20], Dynamic Windowing (DWin) [21] and Variable-size Sliding Window (VSW) [22]. The main difference between various segmentation techniques is the online (segmentation can be done before the complete data is collected) and offline (complete dataset should be available before data segmentation is performed) capabilities of these segmentation techniques.

Generally for real time applications online techniques are used. [19] states that online algorithms can have poor approximation results under constraint environment but have a relatively good performance on noisy data.

(3) Window Size: Many researchers who have used fixed size window segmentation methods mostly use changing window sizes. Table 2 gives the details of different window sizes used for activity classification by various researchers. Furthermore, Ortiz et al. [22] showcased some alterations in selecting fixed window sizes and tried to dynamically change the window size based on different events in received data from various sensors as different activities have different time frames. It was stated that longer window sizes can cover more than one activities whereas small windows could split an activity into multiple parts. Both these cases leads to non optimal information for activity classification algorithm.

Table 2. Different window sizes in various activity classification research work

Window (s)	Authors	Remark
1	Casale et al. [11]	Short window size
2	Preece et al. [9]	Small daily activities
3	Preece et al. [9]	Gain in accuracy
2	Nyan et al. [23]	Short window size
2.56	Wang et al. [10]	Short window size
4	Maurer et al. [14]	Short window size
5	Parkka et al. [13]	Short window size
6.7	Bao et al. [8]	Short window size
60	Kasteren et al. [24]	Long window size
74	Patterson et al. [25]	Long window size

6   Technique Analysis

There were various steps required for classification of walkers based on back angle measurement. These steps were data sampling, data segmentation, data processing, classification and result analysis. Back angle measurement while walking could be in three planes viz. coronal/frontal, sagittal, and transverse planes. Lower back deviation angles were measured using a wireless sensor node attached at the lower back of the subjects. Angular movement data of lower back was sensed and transmitted to the wireless hub which captures and stores data for creating logs. Hence, maximum sector span angles were measured for lower back movement of subjects while walking. The analysis technique is given in Fig. 4. Data sampling was performed using Arduino and MPU6050 sensor. The sampling rate of 300 Hz (i.e. 300 samples per seconds) was used as it was beyond the sufficient criterion for activity classification and captures minute details of deviation angles of subjects while walking. This sampled data were stored in the excel files in CSV formats to make them compatible with data handling using pandas in python. First all the necessary packages which were required for the creation of data set and data analytics were imported into the python environment. These packages makes it possible to create data set from the CSV files of all twenty subjects. The created data sets were processed to get the deviation angles for all subjects in degrees around the vertical Z-axis coming out of the back of the subjects. These processed data files were used as input to the random forest classifier, decision tree classifier, and support vector machine classifier to perform classification. The deviation angles were used as input for walker classification which produced different target classes as outputs. Predicted classes were analyzed using confusion matrix which was generated after the completion of random forest classification process.

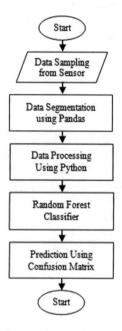

Fig. 4. Flow diagram for running style classification

7 Designed Experiments

Evaluation of the designed hardware was performed by fabricating an experimental setup i.e. a test bed goniometer. It was used to check and validate the working of the designed wireless node. Sensor data acquisition was performed using wearable gyroscope based inertial sensor. Subjects having different walking proficiency levels were selected for this research work which encourages to analyze variations in subjects (i.e., speed and walking techniques).

A. Experimental Protocol and Setup
Subjects were addressed to perform experiment which had a walking task at self selected and normal walking pace. All subjects performed experiments for a duration of 15 s and walked on a pre defined experiment path. This setup was followed as

mentioned in previous research work which authenticates and supports data acquisition while walking. It also recommended to have experiment duration of greater that 30 steps to get reliable experimental data. This setup also eliminates the spatio-temporal rhythm while performing continuous walk which were evident with repeated single trials. The NRF wireless sensor node was placed on lower back of the subjects and it continuously gather data for the full test duration.

B. Experiment with Subjects

The accelerometer/gyroscope sensor was placed on the lower back of the runners as given in Fig. 5. Sensor position was selected in such a manner to avoid thick muscles; as deformations of soft tissues occur due to contractions in muscles and impact of feet on ground while walking can affect sudden accelerations which further can affect accuracy of classification. The sensor was mounted on the subject's body with fabric belts having velcro stitched on it for firm mounting. The fabric had elasticity which will not restrict the subject's walking and method of technique analysis. Further, the subjects were asked to walk for 15 s with normal self selected walking pace.

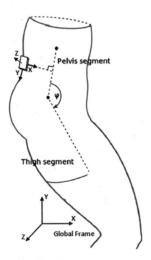

Fig. 5. Placement of inertial sensor on the lower back.

Data acquisition were performed from MPU6050 gyroscope sensor which were transmitted wireless to remote NRF hub. All the sampled data which was received by the NRF hub was converted into CSV files which was then handled in python using pandas to generate training and testing datasets. Data sensing from the sensor is independent of any event, so a dedicated signal was used for starting point synchronization in the data stream of collected data. This synchronization signal was generated by a push type switch available at the hub. The synchronization need to be send every time before starting walking experiment with different subjects.

8   Experimental Results

Table 3 gives the measured values for 20 subjects after performing walking experiment. It was found that the average sector span angle for male subjects (AVG_{SSAm}) was 9.3° and that for female subjects was 13.2°. Acceleration values were also analyzed and it was found that average acceleration in X-axis for male subjects was 9.83 and for female subjects was 9.46. Acceleration values in Z-axis were found to be more for female subjects as they were having greater sector span angle than male subjects while performing walking experiments. It was observed that average acceleration in Z-axis for male subjects was 2.56 and for female subjects was 3.42.

Table 3. Data collected for 20 subjects after performing walking experiment.

Feature	Value	Remark
AVG_{SSAm}	9.3	Average sector span angle for male subjects
AVG_{SSAf}	13.2	Average sector span angle for female subjects
AVG_{AXm}	9.83	Average acceleration in X-axis for male subjects
AVG_{AXf}	9.46	Average acceleration in X-axis for female subjects
AVG_{AZm}	2.56	Average acceleration in Z-axis for male Subjects
AVG_{AZf}	3.42	Average acceleration in Z-axis for female subjects

Figure 6 shows the periodic lower back movement. It shows the deviation angles captured from the gyroscope around Z-axis (i.e. axis perpendicular to the lateral/coronal plane). The performed experiments shows that there was movement of lumbar spine in both directions around the spine. The measured values also indicate that the angular values were not same in both directions around the axis but varies on the body posture while walking. This type of study can be used as meaningful data for predicting posture correction techniques while playing and recovering from back pain problems.

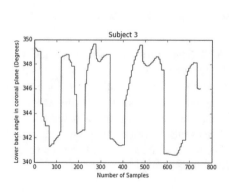

Fig. 6. Angle measurement in coronal plane.

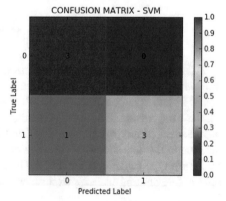

Fig. 7. Confusion matrix

Figure 7 shows the confusion matrix for predicted classes using support vector machine (SVM) classification algorithm. In this the target class was set to "1" for female subjects and it was set to "0" for male subjects. There were true labels (0 and 1) for actual class and predicted labels (0 and 1) for predicted classes. It was seen from Table 4 that SVM classifier had an accuracy of 85.71% which was highest when compared with random forest and decision tree classifiers which were having an accuracy of 71.42% and 57.14% respectively. Figure 7 also shows that there were one false predictions when the target class was falsely predicted as 0. The complete set of data was divided in two sets namely training set and test set. In total seven records were to be tested using three classification algorithms hence sum of predicted labels in the confusion matrix is seven. Three more performance metrics were calculated for all classifiers i.e. F1-score, sensitivity, and computation time and is shown in Table 4. The classification of walkers was based on back angle measurements and the results are given in Table 4. It was seen that support vector machine got the highest prediction accuracy of 85.71% and decision tree classifier had the lowest prediction accuracy. It was also analyzed that SVM was the fastest classifier with least computation time of 0.001 s.

Table 4. Performance metrics of different classifiers

Type	Acc.	F1	Sen.	Time
SVM	85.71%	0.857	0.75	0.001
DT	57.14%	0.400	0.25	0.002
RF	71.42%	0.666	0.50	0.101

9   Conclusion

The research work resulted in designing of a new device and presented a new way of classification of different types of walkers based on back angle measurements. It was found that the device was capable of performing classification of subjects just by analyzing data received from sensor mounted at the lower back of the subjects. In the implemented system sampling frequency of 300 Hz was used which was much beyond the minimum criterion for getting reasonable classification accuracy. It was concluded that SVM classification algorithm got the highest accuracy of 85.71% with F1 score of 0.857 and sensitivity of 0.75. It was also found that SVM was having the least computation time of 0.001 s as compared to random forest and decision tree classifiers.

Acknowledgement. I would like to mention that a non-invasive wearable belt was designed and fabricated for the purpose of performing experiments. Because of the non-invasive nature of the belt no ethical committee was formed by the University but informed consent was taken from all participants before starting the experiment. However information with regards to participant's confidentiality has been maintained and has not been promoted online or in any other form for any other purpose.

References

1. Bertoli, M., Cereatti, A., Croce, U.D., Mancini, M.: An objective assessment to investigate the impact of turning angle on freezing of gait in Parkinsons disease. In: Proceedings of IEEE Biomedical Circuits and Systems Conference (BioCAS), pp. 1–4, 19–21 October 2017
2. Chong, E., Choi, T., Kim, H., Kim, S.J., Hwang, Y., Min Lee, J.: Informative sensor selection and learning for prediction of lower limb kinematics using generative stochastic neural networks. In: Proceedings of 39th Annual International Conference of the IEEE Engineering in Medicine and Biology Society (EMBC), pp. 1–4, 11–15 July 2017
3. Kam, W., O'Sullivan, K., Mohammed, W.S., Lewis, E.: Low cost portable sensor for real-time monitoring of lower back bending. In: Proceedings of SPIE, 25th International Conference on Optical Fiber Sensors (OFS), pp. 1–4, 24–28 April 2017
4. Esfahani, M.I.M., Nussbaum, M.A.: A smart undershirt for tracking upper body motions: task classification and angle estimation. IEEE Sens. J. **18**(18), 7650–7658 (2018)
5. Xu, W., Sanchez, C.O., Murray, I.: Measuring human joint movement with IMUs. In: Proceedings of 15th IEEE Student Conference on Research and Development (SCOReD), pp. 172–177, 13–14 December 2017
6. Huynh, T., Schiele, B.: Analyzing features for activity recognition. In: Proceedings of Joint Conference on Smart Objects and Ambient Intelligence: Innovative Context Aware Services: Usage and Technologies, Grenoble, France, pp. 159–163, 12–14 October 2005
7. Sekine, M., Tamura, T., Fujimoto, T., Fukui, Y.: Classification of walking pattern using acceleration waveform in elderly people. In: Proceedings of the 22nd Annual IEEE International Conference of the Engineering in Medicine and Biology Society, Chicago, MI, USA, pp. 1356–1359, 23–28 July 2000
8. Bao, L., Intille, S.: Activity recognition from user-annotated acceleration data. In: Pervasive Computing, pp. 1–17. Springer, Berlin (2004)
9. Preece, S.J., Goulermas, J.Y., Kenney, L.P., Howard, D.: A comparision of feature extraction methods for the classification of dynamic activities from accelerometer data. IEEE Trans. Biomed. Eng. **56**, 871–879 (2009)
10. Wang, N., Ambikairajah, E., Lovell, N.H., Cellar, B.G.: Accelerometry based classification of walking pattern using time frequency analysis. In: Proceedings of 29th Annual International Conference of The IEEE Engineering in Medicine and Biology Society, Lyon, France, pp. 4899–4902, 22–26 August 2007
11. Casale, P., Pujol, O., Radeva, P.: Human activity recognition from accelerometer data using a wearable device. In: Pattern Recognition, Image Analysis, pp. 289–296 (2011)
12. Ravi, N., Dandekar, N., Mysore, P., Littman, M.: Activity recognition from accelerometer data. In: Proceedings of 7th Conference on Innovative Applications of Artificial Intelligence, Menlo Park, CA, USA, pp. 1541–1546, 9–13 July 2005
13. Parkka, J., Cluitmans, L., Ermes, M.: Personalization algorithm for real-time activity recognition using PDA, wireless motion bands, and binary decision trees. IEEE Trans. Inf. Technol. Biomed. **14**, 1211–1215 (2010)
14. Maurer, U., Smailagic, A., Siewiorek, D.P., Deisher, M.: Activity recognition and monitoring using multiple sensors on different body positions. In: Proceedings of International Workshop on Wearable and Implantable Body Sensor Networks, Boston, MA, USA, pp. 113–116, 3–5 April 2006
15. Antonsson, E.K., Mann, R.W.: The frequency content of gait. J. Biomech. **18**, 39–47 (1985)
16. Bouten, C.V., Koekkoek, K.T., Verduin, M., Kodde, R., Janssen, J.D.: A triaxial accelerometer and portable data processing unit for the assessment of daily physical activity. IEEE Trans. Biomed. Eng. **44**, 136–147 (1997)

17. Gjoreski, H., Gams, M., Chorbev, I.: 3-Axis accelerometer activity recognition. ICT Innov. 51–58 (2010)
18. Pietka, E.: Expert systems in parameter extraction of ECG signal. In: Proceedings of the Annual International Conference of the IEEE Engineering in Medicine and Biology Society, New Orleans, LA, USA, pp. 165–166, 4–7 November 1988
19. Keogh, E., Chu, S., Hart, D., Pazzani, M.: An online algorithm for segmenting time series. In: Proceedings of the International Conference on Data Mining, San Jose, CA, USA, pp. 289–296, 29 November–2 December 2001
20. Chu, C.: Time series segmentation: a sliding window approach. Inf. Sci. **85**, 147–173 (1995)
21. Kozina, S., Lustrek, M., Gams, M.: Dynamic signal segmentation for activity recognition. In: Proceedings of the International Joint Conference on Artificial Intelligence, Barcelona, Spain, pp. 15–22, 16–22 July 2011
22. Ortiz, J., Olaya, A.G., Borrajo, D.: A dynamic sliding window approach for activity recognition. In: Proceedings of the International Conference on User Modeling, Adaptation and Personalization, Girona, Spain, pp. 219–230, 11–15 July 2011
23. Nyan, M.N., Tay, F., Seah, K., Sitoh, Y.Y.: Classification of gait patterns in the time-frequency domain. J. Biomech. **39**, 2647–2656 (2006)
24. Van Kasteren, T.L.M., Noulas, A., Englebienne, G., Krose, B.J.: Accurate activity recognition in a home setting. In: Proceedings of the Conference on Autonomous Agents and Multi Agent Systems (AAMAS 2007), Seoul, Korea, pp. 1–9, 21–24 September 2007
25. Patterson, D.J., Fox, D., Kautz, H., Philipose, M.: Fine-grained activity recognition by aggregating abstract object usage. In: Proceedings of International Semantic Web Conference (ISWC), Galway, Ireland, pp. 44–51, 18–21 October 2005

Sensitivity Factor Based Congestion Management of Modified 33-Bus Distribution System

Santanu Chakraborty[1](✉), Sougata Koley[1], Subham Mandal[1],
Rajat Kumar Mandal[1], Birendra Krishna Ghosh[1], Mainak Biswas[1],
Rupamit Dutta[1], Piyali Ganguly[2](✉), and Soumyadip Roy[3]

[1] Department of Electrical Engineering, Techno International New Town,
Rajarhat, Kolkata 700156, India
[2] Department of Electrical Engineering, Seacom Engineering College,
Jala-Dhulagori, Howrah 711302, India
piyali.march@gmail.com
[3] Department of Electrical Engineering, Camellia School of Engineering
and Technology, Baichi 712134, West Bengal, India

Abstract. In modern power system we are very much concern about the power quality of the existing network, i.e. generating ends, distribution ends and consumer end. Now a day's power congestion in generation end i.e. transmission line can be control by using FACTs devices, DGs placements. To manage congestion in distribution system, implementing these conventional methods, become challenging task. In simple word, Congestion happens due to shortage of production which is overcome by DG placements and in some case due to transmission network failure, can be overcome by using FACTs devices. But that needs proper identification of congested line or nodes. It is also shown that problem solving at the point of occurrence is effective as well as economic. So it becomes necessary to control the congestion in distribution line of the power system. This paper is based on DG placement in congested node of distribution system with the help of forward backward sweep method and Sensitivity factor (LS_g). In modified IEEE 33 bus distribution sys power loss at each bus is mathematically calculated by FB sweep method and also compared with simulated data depending on these LS_g is calculated to identify the congested line then placing DGs to overcome the critical situation.

Keywords: Congestion · Distributed Generator · Congestion management · Deregulation

Note: This work is based on our earlier work- Sinha, Aman Annu, Mukesh Kumar Arya, Ankit Kumar, Rjiat Kumar Mandal, Mainak Biswas, Birendra Krishna Ghosh, and Soumyadip Roy. "Linear Sensitivity Based Congestion Management of IEEE 30 Bus System Using Distributed Generator." In 2018 2nd International Conference on Trends in Electronics and Informatics (ICOEI), pp. 691–696. IEEE, 2018.

© Springer Nature Switzerland AG 2020
P. Karrupusamy et al. (Eds.): ICSCN 2019, LNDECT 39, pp. 412–423, 2020.
https://doi.org/10.1007/978-3-030-34515-0_44

1 Introduction

The ultimate motive for every power industry is to transmit power at minimal cost and how the power loss in mainly transmission and distribution lines can be minimized to transfer bulk amount of power from the generation to the load side. Mainly the congestion occurs in a power system network, when the load connected to the distribution line requires more power than the power transmitted through the transmission line. Simply, congestion refers to an overloading the transmission and distribution line when the thermal bounds or other material constants are violated. The physical and system limitation of the distribution line can lead to the congestion of distribution network. Physical limitations such as thermal limitation of distribution line or distribution transformer lead to the congestion which may affect the line compensating devices (capacitor banks). These effects will change the voltage profile of a node, power angle calculation and transient stability as well as steady state stability will also be affected and these will cause a system disturbance which may permanently blackout the total network, as the whole system is interconnected [1].

Now a days, in distribution system when demands become very high at load end i.e. more than one substation or load end needs more power at a time but that amount of power generation is not possible at that instant to match the demand, which creates congestion. DG placement is the most accepted way to solve this race around condition but proper placement is required to maintain the quality power flow and economic balance. This work is very much fruitful for modern Distributed Energy Management System.

As the increase in population along with developments and infrastructures require huge amount of power, congestion occurs. So, the congestion management is the most important in terms of safety, security and reliability of the distribution network. Many researchers have been done to find the solutions for the congested transmission and distribution lines, so that the electricity demand by the loads can be fulfilled. Different methods are being used for controlling the congestion in distribution line, such as Optimal Power Flow (OPF), Generator Re-scheduling (GR), Load shedding, Distributed Generation, Flexible Alternating Current Transmission System (FACTS) devices, Parallel distribution lines [2, 3].

2 Proposed Methodology of Congestion Management

There are several techniques acquired to manage the congestion in transmission line. But directly those methods can't be implemented on distribution system due physical and system limitation. Out of those different methods for handling the congestion, Linear Sensitivity based congestion management can be implemented in distribution system merging with 'Forward-Backward sweep'. With one simple example the method discussed below.

2.1   Single Ended 33 Bus Distribution System

Figure 1 describes the single line diagram of 3–Bus distribution system with single end feeding [4].

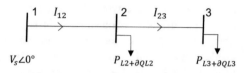

Fig. 1. Single line diagram of 3–bus distribution system

$$V_1 = V_s \angle 0° , \ V_2 = V_s \angle 0° , \ V_3 = V_s \angle 0°$$

V_s is sending end voltage V_1, V_2, V_3 are different node voltages. Initially all are assume as V_s with zero angle.

In First Iteration:- Using the equations shown below line current are calculated.

$$I_2' = \left(\frac{P_{L2} + \partial QL2}{V_s \angle 0°} \right) ; \ I_2' = \left(\frac{P_{L3} + \partial QL3}{V_s \angle 0°} \right)$$

In Backward Sweep I_{23}' and I_{12}' is replaced as –

$$I_{23}' = I_3'$$
$$I_{12}' = I_2' + I_{23}'$$

In forward Sweep the node voltages are modified as-

$$V_2' = V_s \angle 0° - Z_{12} I_{12}'$$
$$V_3' = V_{23}' - Z_{23} I_{23}'$$

Now in Second Iteration line currents and node voltages are further modified as:-

$$I_2^2 = \left(\frac{P_{L2} + \partial QL2}{V_2'} \right) ; I_3^2 = \left(\frac{P_{L3} + \partial QL3}{V_3'} \right)$$
$$I_{23}^2 = I_3^2$$
$$I_{12}^2 = I_2^2 + I_{23}^2$$
$$V_2^2 = V_s \angle 0° - Z_{12} I_{12}^2$$
$$V_3^2 = V_{23}^2 - Z_{23} I_{23}^2$$

At K_{th} Iteration node voltages can be expressed in general form as, V_2^k V_3^k and Voltage at $(K-1)_{th}$ Iteration is,

$$V_2^{k-1} = V_3^{k-1}$$

Let the Error of Voltage at node (2) at K_{th} Iteration is,

$$e_2^k = \left| v_2^k - v_2^{k-1} \right| \quad \& \quad e_3^k = \left| v_3^k - v_3^{k-1} \right|$$

So max Error will be

$$e_{max}^k = \max\left(e_2^k, e_3^k \right)$$
$$e_{max}^k \leq \varepsilon\,(\text{tolerance, threshold value})$$

In Fig. 2 the SLD of Nth Bus Distribution system is shown. By using this example the mathematical calculation is performed [5, 6].

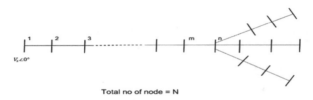

Total no of node = N

Fig. 2. SLD of Nth bus distribution system

Step 1: Iteration of voltages

$$V_j^{(0)} = V_s \angle 0° \quad \text{for, } j = 2, 3 \ldots N$$

Step 2: Iteration count initialization k = 1

Step 3: $I_J^{(k)} = \left(\frac{P_{Lj} + J_{Lj}}{V_j^{(k-1)}} \right) \quad \text{for, } j = 2, 3 \ldots N$

Load Current Calculation

Step 4: Backward Sweep:-

$$I_{mn}^k = I_n + \Sigma(\text{of all the currents of brunches emerged from bus} - N)$$

Step 5: Forward Sweep:-

$$V_n^{(k)} = V_m^{(k)} - 2 \quad I_{mn}^{(k)} \quad \text{for, } n = 2, 3 \ldots N$$

Step 6: Error in each branch

Error bus in J_{th} bus in K_{th} Iteration

$$e_j^{(k)} = \left| v_j^k - v_j^{k-1} \right| \quad \text{for, } n = 2, 3, \ldots N$$

Step 7:

$$e_{max}^k = \max of$$
$$\left(e_2^{(k)}, e_3^{(k)}, e_4^{(k)}, \ldots \ldots \ldots \ldots \ldots e_n^{(k)} \right)$$

Step 8:

If,

$$e_{max}^k \leq \varepsilon \, (\text{tolerance})$$

Step 9:

Calculate LS_g using

$$LS_g = \frac{(|V_m - V_n|)(\operatorname{Im} n)^*}{(p_{L_k} \pm \varDelta p_{L_k})}$$

Mathematically, LS for brunch m-n for power change in K_{th} line can be written as [7] above expression, where, $(|V_m - V_n|)(\operatorname{Im} n)^*$ define as Power loss in "m, n" bus and $(p_{L_k} \pm \varDelta p_{L_k})$ is power change in K^{th} Line.

Step 10:

Check $LS_g > 1$
If true, Print bus no as Congested Bus.
Else, update iteration count k = k + 1 and go to Step 3.

3 System Specification

Modified IEEE 33-Bus System is describes in Fig. 3 using MATLAB simulation tool. Though it is not possible in existing stable distribution system some abnormality is forcefully created by changing the line data like resistance susceptance and load demands. For this real power and reactive power is different at different Buses. In Table 1 Sl. No is define as Bus no and P, Q are corresponding power demand.

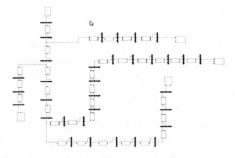

Fig. 3. Simulated diagram of modified 33-bus distribution system

Table 1. Bus data of modified IEEE 33-bus distribution system

Sl. No.	Real power (P) P.U.	Reactive power (Q) P.U.
1	0.00132	0.02525
2	0.70711	0.42427
3	0.6364	0.28285
4	0.84854	0.56569
5	0.42427	0.21213
6	0.42427	0.14142
7	1.41423	0.70711
8	1.41423	0.70711
9	0.42427	0.14142
10	0.42427	0.14142
11	0.3182	0.21213
12	0.42427	0.24749
13	0.42427	0.24749
14	0.84854	0.56569
15	0.42427	0.07071
16	0.42427	0.14142
17	0.42427	0.14142
18	0.6364	0.28285
19	0.6364	0.28285
20	0.6364	0.28285
21	0.6364	0.28285
22	0.6364	0.28285
23	0.6364	0.35356
24	2.96988	1.41423
25	2.96988	1.41423
26	0.42427	0.17678
27	0.42427	0.17678
28	0.42427	0.14142
29	0.84854	0.49498
30	1.41423	4.24268
31	1.06067	0.49498
32	1.48494	0.70711
33	0.42427	0.28285

4   Result and Discuss

Congestion in distribution system is based on some assumptions like extreme overload condition, out of transformer capacity and some economic factors. In this paper some fictitious loading and parametric changes has been done to make the some node or Bus congested (forcefully) to show that placement of the distribution generator can improve this chaotic situation [8, 9]. Values of Line sensitivity index or factor (L_{gs}) shows more than 1 for two nodes (15 and 23) before DG placed at node 15 and 23. Table 3 shows the simulation result where LS_g is 1.088 for Bus 15 and 1.1609 for 23 Bus. Figures 4 and 5 shows that the voltage profile of congested Buses. It is clear from this plots that when the line is congested the voltage level decreases which denotes the quality of the system detrains and after placing DGs the voltage level reaches it steady (Table 2).

Table 2. Line data of modified IEEE 33-Bus Distribution system

Sl. No	From Bus	To Bus	R (p.u.)	X (p.u.)
1	1	2	0.0922	0.0470
2	2	3	0.4930	0.2511
3	3	4	0.3660	0.1864
4	4	5	0.3811	0.1947
5	5	6	0.8190	0.7070
6	6	7	0.1872	0.6188
7	7	8	0.7114	0.2351
8	8	9	1.0300	0.7400
9	9	10	1.0440	0.7400
10	10	11	0.1966	0.0650
11	11	12	0.3744	0.1238
12	12	13	1.4680	1.1550
13	13	14	0.5416	0.7129
14	14	15	0.5910	0.5260
15	15	16	0.7463	0.5450
16	16	17	1.2890	1.7210
17	17	18	0.7320	0.5740
18	2	19	0.1640	0.1565
19	19	20	1.5042	1.3554
20	20	21	0.4095	0.4784
21	21	22	0.7089	0.9373
22	3	23	0.4512	0.3083
23	23	24	0.8980	0.7091
24	24	25	0.8960	0.7011
25	6	26	0.2030	0.1034
26	26	27	0.2842	0.1447
27	27	28	1.0590	0.9337
28	28	29	0.8042	0.7006
29	29	30	0.5075	0.2585
30	30	31	0.9744	0.9630
31	31	32	0.3105	0.3619
32	32	33	0.3410	0.5302

It is also found that depending on power loss the congestion occurs in the bus as Fig. 8 shows. At Bus No 15 and Bus No 23 both sensitivity factor (blue one) and Power loss (red one) has a pick and both of them crosses the average level of their own [10, 11]. Graphical representation of Sensitivity factor for different Buses is shown in Fig. 6 and Comparison of Power loss obtained from simulation and mathematical analysis for different Buses is shown in Fig. 7.

Table 3. Comparison of simulated & mathematical result

Sl. No.	Simulink calculation				Mathematical calculation			
	V_{Bus}	I_{Branch}	Power loss	LS_g	V_{Bus}	I_{Branch}	power loss	LS_g
1	$1\angle0°$	$0.01495\angle151.76°$	0.02164	0.785	$1\angle0°$	$0.014785\angle155.76°$	0.01964	0.4899
2	$1.1510\angle-0.15929°$	$0.58032\angle149.03°$	0.0761	0.509	$1.2310\angle-0.1529°$	$0.5952\angle154.03°$	0.0866	0.4908
3	$1.4715\angle-0.99293°$	$0.02931\angle143.61°$	0.0201	0.126	$1.44815\angle-0.9993°$	$0.0313\angle135.61°$	0.0139	0.368
4	$1.5899\angle-1.6219°$	$0.02515\angle144.27°$	0.0291	0.210	$1.5599\angle-1.6219°$	$0.0455\angle156.27°$	0.0231	0.245
5	$1.6965\angle-2.0707°$	$0.024\angle1455.25°$	0.0221	0.351	$1.7665\angle-2.0707°$	$0.0254\angle155.80°$	0.0358	0.348
6	$1.9660\angle-1.0869°$	$0.11764\angle157.14°$	0.01765	0.189	$1.7660\angle-1.0869°$	$0.1356\angle177.14°$	0.0208	0.231
7	$2.0147\angle0.60576°$	$0.089694\angle157°$	0.00246	0.028	$2.847\angle0.60576°$	$0.09878\angle167°$	0.0031	0.540
8	$2.0817\angle0.45690°$	$0.0063941\angle159°$	0.0034	0.033	$2.0347\angle0.45690°$	$0.007891\angle131°$	0.0037	0.5421
9	$2.1603\angle0.98666°$	$0.057418\angle158°$	0.0013	0.011	$2.3203\angle0.98666°$	$0.06354\angle156°$	0.0013	0.2137
10	$2.2321\angle1.4068°$	$0.051402\angle157.8°$	0.0023	0.0251	$2.4221\angle1.4068°$	$0.03676\angle144.8°$	0.0022	0.3212
11	$2.2427\angle1.3797°$	$0.04632\angle158.75°$	0.0034	0.1704	$2.2927\angle1.3797°$	$0.049876\angle158.75°$	0.0019	0.4313
12	$2.2609\angle1.3458°$	$0.04005\angle169.89°$	0.0053	0.0012	$2.54609\angle1.3458°$	$0.0588\angle169.89°$	0.0011	0.2341
13	$2.3325\angle1.8745°$	$0.03415\angle161.5°$	0.0034	0.00186	$2.6525\angle1.8745°$	$0.03978\angle161.5°$	0.0248	0.2714
14	$2.3585\angle2.2671°$	$0.02209\angle166.64°$	0.034	0.927	$2.3485\angle2.2671°$	$0.0334\angle166.64°$	0.0475	0.7102
15	**$2.3746\angle2.4547°$**	**$0.17665\angle16.29°$**	**0.0943**	**1.088**	**$2.4346\angle2.4547°$**	**$0.21445\angle16.29°$**	**0.08613**	**1.182**
16	$2.3901\angle2.5755°$	$0.01264\angle163.43°$	0.0121	0.589	$2.7601\angle2.5755°$	$0.0144\angle163.43°$	0.012	0.658
17	$2.4125\angle2.9344°$	$0.00758\angle161.5°$	0.0054	0.4109	$2.4525\angle2.9344°$	$0.01248\angle161.5°$	0.0001	0.209
18	$2.4193\angle2.9830°$	$0.51862\angle161.77°$	0.02155	0.2589	$2.5693\angle2.9830°$	$0.5352\angle161.77°$	0.02848	0.2589
19	$1.2581\angle2.1501°$	$0.27158\angle168.06°$	0.0534	0.715	$1.3581\angle2.1501°$	$0.3148\angle168.06°$	0.077	0.687
20	$1.7629\angle10.551°$	$0.17107\angle168.95°$	0.0345	0.213	$1.7929\angle10.551°$	$0.2345\angle168.95°$	0.0235	0.435
21	$1.8588\angle12.102°$	$0.08138\angle170°$	0.00776	0.1124	$1.4588\angle12.102°$	$0.0964\angle170°$	0.0013	0.232
22	$1.9419\angle13.539°$	$0.25153\angle155°$	0.0456	0.8211	$1.7619\angle13.539°$	$0.2789\angle155°$	0.0321	0.7241
23	**$1.6069\angle-0.10697°$**	**$0.2174\angle156°$**	**0.0356**	**1.1609**	**$1.6969\angle-0.10697°$**	**$0.1986\angle156°$**	**0.0777**	**1.0335**
24	$1.8488\angle1.842°$	$0.1026\angle156.95°$	0.0308	0.921	$1.9688\angle1.842°$	$0.1234\angle156.95°$	0.0102	0.7903
25	$1.9621\angle2.6114°$	$0.1174\angle130°$	0.0357	0.383	$1.9621\angle2.6114°$	$0.1009\angle136°$	0.0202	0.523
26	$1.9912\angle-1.3727°$	$0.10981\angle128°$	0.00389	0.234	$2.0912\angle-1.3727°$	$0.1875\angle128°$	0.0037	0.352
27	$2.0234\angle-1.7673°$	$0.10264\angle126.32°$	0.00576	0.158	$2.0765\angle-1.7673°$	$0.1446\angle126.32°$	0.0057	0.345
28	$2.1660\angle-2.4688°$	$0.09713\angle125°$	0.00534	0.290	$2.2345\angle-2.4688°$	$0.0976\angle125°$	0.0032	0.243
29	$2.2674\angle-3.0160°$	$0.08554\angle121.42°$	0.00456	0.391	$2.4455\angle-3.0160°$	$0.07876\angle128.42°$	0.0043	0.455
30	$2.3102\angle-3.5960°$	$0.03826\angle150°$	0.0023	0.1124	$2.5678\angle-3.5960°$	$0.04678\angle155°$	0.0011	0.243
31	$2.3604\angle-3.2179°$	$0.02698\angle148°$	0.0067	0.7210	$2.1234\angle-3.2179°$	$0.037688\angle148°$	0.008	0.561
32	$2.3714\angle-3.1143°$	$0.0058412\angle140°$	0.011	0.324	$2.4567\angle-3.1143°$	$0.0064512\angle143°$	0.021	0.254
33	$2.3747\angle-3.0849°$	$0.10264\angle126.32°$	0.0057	0.0518	$2.6986\angle-3.0849°$	$0.1264\angle126.32°$	0.057	0.0518

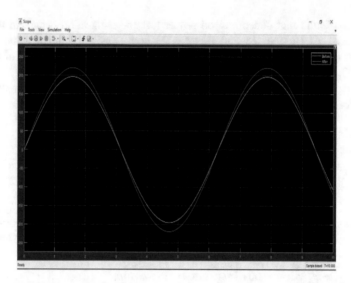

Fig. 4. Voltage profile diagram of bus no. 15 before and after congestion

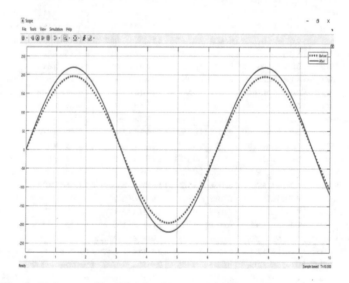

Fig. 5. Voltage profile diagram of bus no. 23 before and after congestion

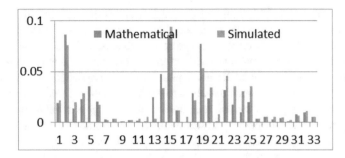

Fig. 6. Graphical representation of sensitivity factor

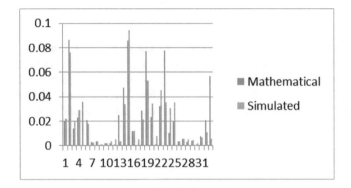

Fig. 7. Comparison of power loss at different Bus

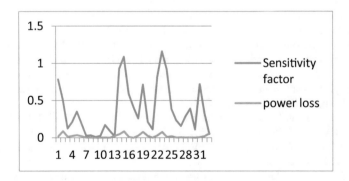

Fig. 8. Comparison between Sensitivity factor and power loss

5 Conclusion

In this paper the power congestion management method has been tested on 33 bus system in Simulink and also the data from the Simulink is compared to the data comes from the mathematical formula, which are almost same.

The method used to find out the most sensitive and congested lines in this modified IEEE 33 bus distribution system is Linear Sensitivity Analysis. Power in each line is examined and analyzed thereby recognizing the congested lines in the system. The congestion management approaches are enforced to sensitive lines to keep the power demand not beyond its critical limits. Thus, it can be concluded that in distribution system congestion can be successfully managed. Basically a very small system in contrast reflects nearby values of sensitivities shows that a reliable power 33 Bus network system. The DG placement has compact and dominating impact on the congested line to control steady power flow and system stability. In RMU System this method can also be applicable on the basis of some simplifications and real time assumptions.

6   Future Scope

By controlling or overcoming the problem of congestion from the line [12], we can subdue a number of problems that are regularly faced by the system operators. We will be able to provide a continuous supply of power to different consumers. The commonly faced issue of overloading can easily be dealt with. Power losses will be minimized and the idea of stable and uninterrupted supply of power can be achieved.

References

1. Nayak, A.S., Pai, M.A.: Congestion management in Restructured Power System using an optimal power flow framework. PSERC Publication, 02–23 May 2002
2. Dutta, S., Singh, S.P.: Optimal rescheduling of generators for congestion management based on Particle Swarm Optimization. IEEE Trans. Power Syst. 23(4), 1560–1569 (2008)
3. Patel, A.K., Parekh, B.R.: Transmission congestion management using linear sensitivity factors for electrical power system. In: 2nd International Conference on Latest Trends in Engineering, Science, Humanities and Management (ICLTESHM-2017). Indian Federation of United Nations Associations, New Delhi, India, February 2017. ISBN 978-93-86171-31-3
4. Singh, K., Yadav, V.K., Padhy, N.P., Sharma, J.: Congestion management considering optimal placement of distributed generator in Deregulated Power System Networks. Electr. Power Compon. Syst. 42, 13–22 (2014)
5. Sharma, A., Jaswal, R.A.: Congestion management by TCPST in IEEE 9 Bus System using Matlab Simulink. IJESRT (2013). ISSN: 2277-9655
6. Electric Grid Test Case Respitory: IEEE 30-Bus System. https://electricgrids.engr.tamu.edu/electric-grid-test-cases/ieee-30-bus-system/
7. Tauseef Khan, Md., Siddiqui, A.S.: Congestion management in deregulated power system using FACTS device. Int. J. Syst. Assur. Eng. Manag. 8(1), 1–7 (2017)
8. Charles Raja, S., Venkatesh, P., Manikandan, B.V.: Transmission congestion management in restructured power system. In: 2011 International Conference on Emerging Trends in Electrical and Computer Technology (ICETECT), pp. 23–28. IEEE (2011)
9. Singh, S.N., David, A.K.: Optimal location of FACTS device for congestion management. Electr. Power Syst. Res. 58(2), 71–79 (2001)

10. Shivkumar, S., Devaraj, D.: Congestion management in deregulated power system by rescheduling of generators using genetic algorithm. In: 2014 International Conference on Power Signals Control and Computations (EPSCICON), pp. 1–5. IEEE (2014)
11. Sinha, A.A., Aayushi, M.K.A., Kumar, A., Mandal, R.K., Biswas, M., Ghosh, B.K., Roy, S.: Linear sensitivity based congestion management of IEEE 30 bus system using distributed generator. In: 2nd International Conference on Trends in Electronics and Informatics (ICOEI), pp. 691–696 (2018)
12. Tabandeh, A., Abdollahi, A., Rashidinejad, M.: Security-based congestion management considering power system components' uncertainty. In: 2016 21st Conference on Electrical Power Distribution Networks Conference (EPDC), pp. 172–179. IEEE, April 2016

Efficient and Accurate Property Title Retrieval Using Ethereum Blockchain

Saee M. Joshi[✉] and K. Rajeswari

Department of Computer Engineering,
Pimpri Chinchwad College of Engineering, Pune, India
saee.joshi46@gmail.com,
kannan.rajeswari@pccoepune.org

Abstract. A simple Decentralized Application (Dapp) of Land Registry based on the Ethereum Blockchain can be used as an alternative to the current Land Registry System. Here the owner registers the land by providing necessary land details like the survey number of land, market value of the land/property etc. The Registration process is carried by a Superadmin who indeed acts as registrar or government authority. The Superadmin is the only responsible person who can register the land coming under particular area into the system. The owner has to transfer his full piece of land or any particular property to the buyer and no partial transaction will be allowed is specified by the Smart Contract. Even though the entire registration is done by the Superadmin acting as Government authority/Registrar still the transaction happens in a very transparent manner between the seller and the buyer without any Mediator.

Keywords: Blockchain technology · Solidity · Smart contracts · Land Registry · Metamask · Ganache · Remix IDE

1 Introduction

1.1 Blockchain Technology

Blockchain is a hotly debated issue right now and it's essential to really understand it. In conventional money related transactions, there is a "trusted" (brought together) outsider who is in charge of confirming the transaction. For instance, on the off chance that we send cash to another person it will be confirmed by a bank and the bank will charge an expense for doing this [1]. In any case, blockchain is open-source innovation that evacuates the requirement for a go-between or "middleman" when making transactions. Rather, information on each transaction is put away in "blocks" which are time- stepped and connected together to frame a "chain [7]." The majority of the information put away in the chain can't be changed or expelled, which implies that it is secure and permanent until the end of time [4]. It is a record of transactions that, rather than being put away in one explicit (incorporated) area, is put away and encoded on thousands of PCs around the globe. This trusted advanced record is unmistakable to all members and it demonstrates each component of the transaction. This evacuates the need to depend on a middleman to make secure transactions [12]. There's no

© Springer Nature Switzerland AG 2020
P. Karrupusamy et al. (Eds.): ICSCN 2019, LNDECT 39, pp. 424–438, 2020.
https://doi.org/10.1007/978-3-030-34515-0_45

compelling reason to confide in a bank. For whatever length of time that the transaction is recorded, everybody can see it and go to an agreement to check it.

Ethereum, an open programming stage that enables designers to fabricate and convey blockchain applications (called DApps, in type of smart co tracts). It makes building blockchain applications increasingly effective. In October 2017, Michael Arrington TechCrunch) made blockchain history when he utilized an Ethereum smart contract to buy a loft in the Ukraine. It was viewed as one of the principal times ever a real resource other than a cryptographic money was exchanged on the blockchain [5]. This was a critical initial phase in mechanizing the procedure of real estate buys and disposing of the requirement for one single amazing element that controls the whole framework. In the event that a smart contract can dispose of the requirement for an outsider when sending cash, it can likewise disturb a scope of different ventures where mediators are as of now required to give trust [15]. This has a tremendous measure of potential for the real estate industry. The primary advantages of blockchain speed, trust and expelling a focal authority can all apply to the real estate showcase.

Right now real estate is the biggest resource class on the planet and it is worth around dollar 217 Trillion consistently. Notwithstanding, the market is divided, which makes it troublesome for financial specialists to buy a bit of a real estate property [11]. Real estate possession has recently been very elitist [9]. Most common individuals can scarcely stand to buy one home, not to mention put resources into numerous properties. So as to meet all requirements to purchase a home, buyers must have fantastic credit and have the capacity to exhibit their strong budgetary standing as a great job and a hold of advantages. Tokenization of real estate property dispenses with the issue of illiquidity. It is conceivable to change over real estate esteem into"tokens" that speak to a benefit, (for example, an individual real estate offer) and exchange and move those tokens rather, at any minute in time [13]. Tokenization is when delicate information is supplanted with a special image to distinguish it, which has the majority of the fundamental data without trading off security. In this way, basically resource sponsored tokenization implies changing over the rights to an advantage into a computerized token [10]. When a property is tokenized, a token is produced on a smart contract and then an incentive is connected with that Token which compares to the real resource [8].

1.2 Ethereum

Internet has a foundation of new phase called Ethereum:

- Money and payments are built in Ethereum.
- Users can own their data and your apps cannot be tampered with using Ethereum.
- Every user has access to an open source financial system using Ethereum.
- Ethereum is an platform that is open source to every user which is completely neutral and is not controlled by any third party like a company or person (Fig. 1).

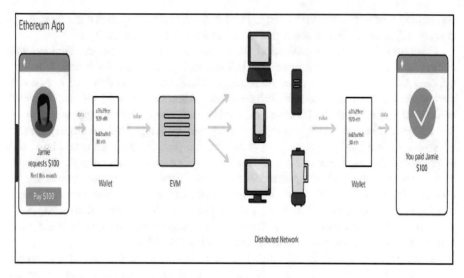

Fig. 1. Working of ethereum [17]

1.3   Smart Contracts

Smart contracts are contracts that help any form of transaction without the interference of third party users. The owner has to transfer his full piece of land or any particular property to the buyer and no partial transaction will be allowed is specified by the Smart Contract (Fig. 2).

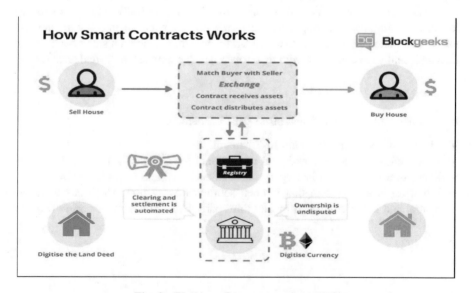

Fig. 2. Working of an smart contract [18]

1.4 Truffle Framework

The life of Ethereum developer is becoming easier by using Truffle Framework which is development environment and is also used for testing and compiling the smart contracts used for building DApps. In the Ethereum Community Truffle is most widely used IDE. For running the Decentralized Application (DApp) truffle helps in compile and migrating the smart contracts so that we come to know that smart contract functions do not consist of any errors. Once Truffle compile and migrate is done a contract address is generated for that particular smart contract which is used to run and test the DApp.

1.5 Metamask

Metamask is a bridge that allows you to access a distributed web in your own web browser. Without running a full Ethereum node Metamask allows to run the Ethereum DApps in your own web browser. Metamask has a secured identity vault and manages the identities provided by Metamask user interface on different sites and also helps in signing different blockchain transactions. It also allows to store Ether's that is a type of cryptocurrency in the Metamask wallet so that the Ether's can be used by both the sellers and buyers in order to make the transaction successfull.

2 Literature Survey

2.1 The Land Registry in the Blockchain, Mats Snall, Jorgen Sannagard, Susanne Forsberg Tobias Lundberg, Testbed 2017

This article [1], was to attempt to separate the hype from the hyperbole of blockchains and how they might relate to land administration. There is no doubt that in countries embarking on a mission to embed the notion of title and have it attached to identified individuals may do worse than consider the advantages that the blockchain concept may potentially offer. With a largely blank canvas from which to work, the blockchain might well, to borrow a visual image, stack up against other central forms of land registries once some of the uncertainties in terms of its application are ironed out. By operating without any need for long, ongoing central involvement (if this is possible), the effect on the public purse is greatly diminished, and with the increased flexibility offered by a distributed ledger and the ability to have a true peer-to-peer verification and validation service, greater financial flexibility is achieved. Such advantages may be significant in countries where there is no central enforceable evidence that one has a secured and legally binding title to land. With this possibly affecting some two-thirds of the worlds population, it has been estimated that unleashing the value of certain property rights could stimulate the worlds economy by some US$20 trillion.

2.2   A Blockchain Based Property Ownership Recording System, Alex Mizrahi, ChromaWay (2017)

This paper [2], describes major three categories of Blockchain like Public, Private and Permissioned Blockchain. Different consensus algorithms can be used for these three major categories of Blockchain. A public blockchain means that every user is valid to use this type of open source blockchain. Every user can participate in network such that they become nodes in network and can obtain rewards by performing their responsibilities in the network. PoW (Proof-of-Work), PoS (Proof-of-Stake) are the consensus algorithms used for Public Blockchain. As compared to public blockchain, the private blockchain has characteristics of modification and low cost for transactions. Private blockchain owner has a very high authority to change the information and the other nodes have limited access to read the information and access it for their own needs.

2.3   Blockchain for Property a Roll Out Road Map for India, Vishnu Chandra, Baladevan Rangaraju, Copyright 2017 India Institute

This paper [3],describes that two-third of civil cases pending in Indian courts deal with land related disputes, most of which revolve around establishing ownership. The existing legal framework, based on the system of presumptive ownership, lends itself to litigation as land and property undergo several mutations over generations that are not always captured on public records. Moreover, data related to a specific parcel of land is stored in siloed government departments, and in formats that vary substantially from one state to another. Access to these records is time- and cost-intensive, as they involve frequent visits and bribes to government agents. For some time now, there have been suggestions to improve this abysmal state of land records by introducing blockchain technology. The popularity of cryptocurrencies has brought forth this new concept of storing and sharing of information between participants.

2.4   Comparative Analysis of Blockchain Consensus Algorithms, L.M. Bach, B. Mihaljevic and M. Zagar, MIPRO 2018

This paper [4], describes that blockchain uses cryptocurrencies which is a new virtual currency and has become popular having the strong technology such as Blockchain behind it. Blockchain which is known as distributed ledger helps in verifying each and every transaction using Peer-to-Peer network(P2P). This paper gives a comparative analysis of various consesnsus algorithms that are currently used in modern blockchains. The earlier consensus such as PoW and PoS can be utilized to make more hybrid and new consesnsus algorithms for more secured implementations.

2.5   A Review on Consensus Algorithm of Blockchain, Du Mingxiao, Ma Xiaofeng, Zhang Zhe, Wang Xiangwei, Chen Qijun, IEEE 2017

This paper [5], describes that Bitcoin has a basis of Blockchain Technology which is more valuable and appreciable and so is attracting more attention in many areas.

Decentralization, Stability, Security, non-modifiability are all characteristics of Blockchain. Maintaining safety and efficiency are major roles of Blockkchain using various Consensus algorithms. This paper provides basic principles and characteristics of various consensus algorithms that helps in analyzing the performance and applications of different consensus.

3   System Architecture

The system includes two major tasks (Fig. 3):

A. Registration
B. Transaction

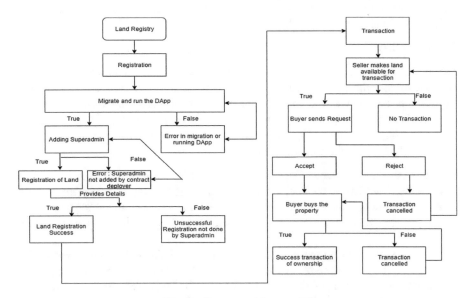

Fig. 3. System architecture [6]

A. Registration

Here the user provides the land details to the government authority who is registered as the superadmin. The land which is going to be registered should be in the same area as the superadmin who is going to register the land. The superadmin varifies the details with the existing records and enters into the Dapp.

The details that are enrolled into the Dapp are:

- state
- district
- village
- survey number
- owner address
- market value.

Along with these, an ID generated from the first four details of the land is also passed in. This ID is generated in the function computeId() using SHA256. The values entered in the registration form of the UI is passed in to the function Registration() and the details are mapped using the ID generated from above. Later on this mapping allows searching for a land easier. There is another field in the registration page, which is, adding the superadmin. There the address of the superuser and the village in which he/she is working is provided. The village is mapped in to the address so that it becomes easier to check that only the superadmin assigned to a village is able to register the details of a land in that village.

B. Transaction

The transaction of a property has several stages involved. The algorithm is designed in such a way that there is no need for any central authority to varify the transaction process. It is important to note that the owner of a property can sell the land as a whole, i.e., there is no partial transaction of the property. This is just to simplify the problem in hand. Later on, while improving the dApp, more of these functionalities can be added.

The following are the steps involved:

- Making the land available: Once the buyer and seller agrees to make the transaction, the seller should make the land available to buy. The land owner passes the property ID to the function makeAvailbale() and the function varifies the account of owner and changes the value of is Available to true which implies that the land is open to buy.
- Sending request to land owner: When the land is available to buy, the buyer sends a request to the land owner to buy the property. The ID of the land is feeded into the function requestToLandOwner(). The function 'isAvailable' verifies whether the land is available to buy by checking its value. The buyers address is stored if the value is true, inside requester which was initially 0 address. The value of isAvailable is then set to false so that no more request can be sent and request status is changed from default to pending.

The above two functions are important in the transaction process because if there is no makeAvailbale function then any one can send request to the land owner which will overwrite the request of the original buyer. If there is no request function then anyone who sents the exact amount to buy the property can actually get the property.

- Viewing the request: The function viewRequest takes the property ID as the input and returns the address of the requester. The function is for checking the address of the buyer.
- Processing the request: Once the seller views the requester address and if it is the right one, then the seller can process the request by puting property ID, request status to the function processRequest. The function, as usual, varifies whether the input is done by the owner of the land and process the request. If the requester address is not of the original buyer then the seller can reject the request and the function changes the value inside requester to 0 address and request status as default. This is for reverting the states to the original and starting the transaction process from zero.
- Buying the property: Once the request is approved the buyer can buy the property. The buyer enters the land ID to the function buyProperty. The function check whether the request status is approved or not and if it is approved, then it checks if the amount given is greater than the sum of market value and 10% of market value which goes as the land tax. If the conditions are satisfied then the amount is transferred to the land owners account. The functions then changes the ownership of the land to the buyer. Removing the ownership of the previous owner is done by calling another function removeOwnership. This function is called after the transfer of amount is complete. removeOwnership removes the property from the last owners asset list.

4   Experimental Results

The dApp is created to make the land registration and land transaction transparent and decentralised. Angular is used to implement the front-end and ethereum solidity contract is used in the back-end. Angular is a JavaScript framework which is used to build single page applications. The main feature is that the application can be build in in such a way that will help in reducing code repetition and will become easier to debug. It also gives a way to modify the HTML elements dynamically which makes it easier to create real-time interactive pages. The ethereum solidity contract in the back-end makes the dApp decentralised and transparent.

A. Ganache-cli

The latest version of TestRPC IS Ganache-cli. It is a fast and customizable blockchain emulator. Making calls to the blockchain is made easier ith Ganache-cli without running an actual Ethereum node. Transactions can be mined instantly. Accounts are recycled, reset and instantiated with a fixed amount of Ether. Modified Gas price and mining speed can be obtained.

Below Fig. (4), gives an idea about how the Ganache accounts are been used for transactions on Metamask by installing ganache with the command: npm install-g ganache-cli and then for accessing those accounts you can simply type ganache-cli on your terminal to get a list of available accounts.

Fig. 4. Ganache-cli accounts

B. Metamask

By using the Mnemonic in Ganache-cli a Metamask ac- count is to be imported as shown in Fig. (5) which consists of 10 accounts to be used as Owner account or buyer account which consists of 100 Ether's each as shown in below Fig. 5.

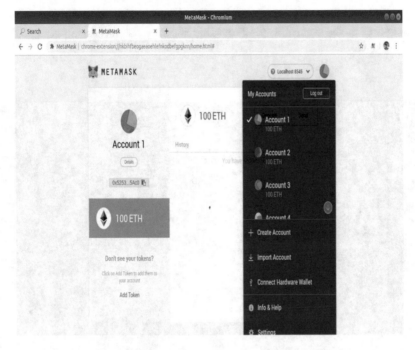

Fig. 5. Metamask accounts created successfully

C. Smart Contract Compilation and Migration

Figure (6), shows that the Setup includes installation of some dependency packages on visual studio terminal as npm, node js, truffle and after installing dependencies truffle migrate command will deploy the smart contracts as well as compile and Migrate it which will be displayed on terminal.

Fig. 6. Smart contracts migration

Figure (7), shows that Ng serve command compiles all modules in the project folder and takes the control to the Front-End showing the successfull compilation of SmartContract.

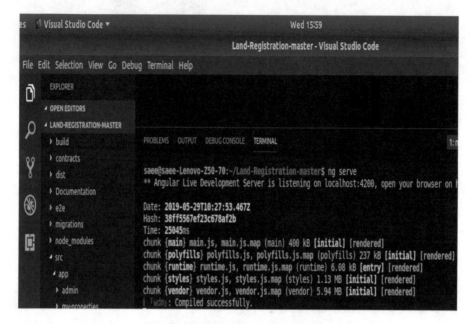

Fig. 7. Compilation successfully done

5 Testing With Remix IDE (integrated Development Environment)

- Integration Testing: Remix-tests is also a CLI, it can be used in a continuous integration environment which support node.js. The elements of the architectural design corresponding to larger groups of tested software are integrated and tested until the software works as a system.

Below Figs. (8, 9, and 10) are some Screen-shots taken while testing Land Registration Smart Contracts on Remix IDE (Integrated Development Environment).

Figure (8), is showing how the Smart contract is deployed for testing on the Remix IDE Environment. Once Smart contracts are been deployed then they can be compiled by using different compiler versions available and after successfull compilation the contract can be deployed in order to get Etherscan test report.

Figure (9), shows the number of compiler version available to compile contracts and needs to be selected same as the pragma solidity version used to write the smart contract.

Whereas, Fig. (10), shows the detailed Etherscan Report after testing the smart contracts and provides details such as the gas fee, block number, transaction hash, gas limit etc.

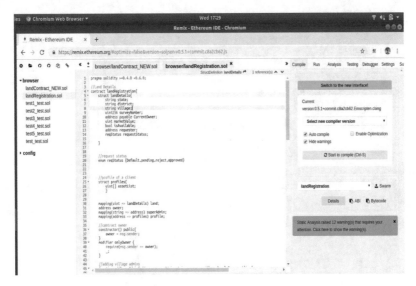

Fig. 8. Smart contracts deployed on remix IDE

Fig. 9. Selecting compiler version according to smart contract

After completing the testing phase successfully the below Etherscan Report is generated giving all the transaction details indicating that the deal between two parties is been successfully completed.

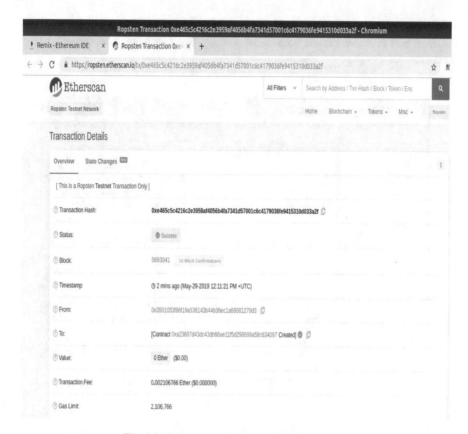

Fig. 10. Transaction details report after testing

6   Conclusion

The aim of this study was to see how Blockchain could be used as a transfer proof technology to solve Security issues in Property Registration, Ownership and Transfer from one party to another. Blockchain as a distributed ledger can help conduct Property Title Retrieval in a transparent manner, which can also be validated. Secure Storage on Blockchain can form the basis for strengthening the good governance principles of ethical conduct. Thus Blockchain Technology can be used to reduce reliance on trusted third parties, reduce costs and reduce number fraud and errors.

This technology brings Security to Real Estate transactions because there is no central point of failure. From Security point of view the ledger is distributed among many computers so attacker needs to attack 51% of network which will be next to impossible. Real Estate development should assess whether and when Blockchain can be used as a technology in their organization. Hence this project work eliminates the third party like Real Estate Broker as in Existing System and totally relies on Block-chain Platform to maintain transparency and make transfer-proof transactions.

7 Future Work

In our research attempt, we have focused on how Blockchain technology brings security to Real Estate transactions because of its transparency and no central involvement. Thus, our future work would be to extend the use of Blockchain Technology to facilitate an accessible, stream-lined Indian Real Estate Market making it more efficient and effective by employing faster or simpler working methods. Current system of this project is able to do Registration and Transfer using Public Ethereum Blockchain of the Property but is not able to view the previous transactions of that Property/Land. Future enhancement is to show all the previous owners of that current land that means a history of records/transaction made on the Land/Property till date should be available to the person interested in buying the property. In Future Proof-of-Authority (PoA) can also be used so that the property details can only be viewed by the authorized person and not by anyone else.

References

1. Snall, M., Sannagard, J., Forsberg, S., Lundberg, T.: The Land Registry in the blockchain. Testbed (2017)
2. Mizrahi, A.: A blockchain based property ownership recording system. ChromaWay (2017)
3. Chandra,V., Rangaraju, B.: Blockchain for Property a Roll Out Road Map for India. Copyright 2017 India Institute (2017)
4. Bach, L.M., Mihaljevic, B., Zagar, M.: Comparative analysis of blockchain consensus algorithms. MIPRO 2018
5. Du, M., Ma, X., Zhang, Z., Wang, X., Chen, Q.: A review on consensus algorithm of blockchain. IEEE (2017)
6. Krishnan, A.: Blockchain Projects that Will Change the Real Estate Industry, Invest in Blockchain (2018)
7. Crosby, M., Nachiappan, Pattanayak, P., Verma, S., Kalyanaraman, V.: BlockChain Technology Beyond Bitcoin. Fairchild, University of California (2015)
8. Updating the Land Registration Act 2002, @ Crown Copyright (2016)
9. Registering property the paths of digitization, Doing Business (2016)
10. Lian, H., Yu, Y.: Reference of land registration system from countries in Asia Pacific region: comparison of the United States, Canada, Russia, South Korea, Japan and Australia. IOSR J. Bus. Manag. (2016)
11. Akingbade, A.O.: Improvement of Availability of Land Registration and Cadasrel Information in Ondo State Nigeria (2005)
12. Porat, A., Pratap, A., Shah, P., Adkar, V.: Blockchain consensus: an analysis of proof-of-work and its applications
13. Nicholsland, S.: Registration: Managing Information for Land Administration, University of New Brunswick

14. Mulholland, G., Pontesthe, L.: Technology behind the integrated cadastre/land registry solutions in the Canadian Provinces of New Brunswick and Nova Scotia. In: Symposium on Innovative Technologies for Land Administration Madison WI, June 2005
15. Land Titling and Credit Access Understanding the Reality, USAID (United States Agency International Development)
16. Ethereum: ethereum.org. https://ethereum.org. Accessed 28 June 2019
17. Working of Ethereum - Google Search. Accessed 28 June 2019
18. Smart Contracts - Google Search. Accessed 28 June 2019

RETRACTED CHAPTER: Internet of Things Based Air Pollution Monitoring and Forecasting System

P. Gokul[1(✉)], J. Srikanth[1], G. Inbarasu[1], Kamalraj Subramaniyam[1], and G. K. D. Prasanna Venkatesan[2]

[1] Department of ECE, Karpagam Academy of Higher Education, Coimbatore, India
gokulkarthick39@gmail.com, srikanthsrij86@gmail.com, inbarasu37@gmail.com, kamalrajece@gmail.com
[2] Faculty of Engineering, Karpagam Academy of Higher Education, Coimbatore, India
prasphd@gmail.com

Abstract. The quality of air plays a predominant role in human health and has greater influence on life expectancy. In our project, five different gas sensors are used to evaluate the quality of air. The quality of air is classified into good, moderate, unhealthy, and hazardous. For instance, the work has been conducted in Coimbatore (smart city). The presence of harmful gases determines the nature of the air and this classification can be monitored periodically by measuring the amount of SO2 and NO2, which acts as a vital parameter. These measured and classified data can be transferred to public access and decision can be taken, so that we have developed an open access website to store and display the values of classified air quality. This information to the people can be used to improve quality of life by improving the quality of air.

Keywords: Gas sensors · PIC micro controller · IOT · Air quality

1 Introduction

Pollution (contamination) is defined as unwanted particle mixture in the globe and it affects the environment. The major kinds of contamination are air contamination, water contamination, light contamination etc. This paper deals with air pollution. Air pollution is term of toxic gases being mixed to the air. It is the contamination of the atmosphere with harmful gaseous emission such as smoke, smog and other toxic gases rendered in the air. The harmful effects in human health due to the toxic gases are heart diseases, cancer, pneumonia, bronchitis, and asthma.

World health organization predicted that around 90% of people inhale polluted air. In the year 2016, survey reports that 4.2 million individuals demise caused by ambient air contamination. Around 3.8 million individual demise caused by household air contamination. Air contamination level remains hazardously high in many places around the world. New database from WHO shows that 9 out of 10 people inhale air

The original version of this chapter was retracted: The retraction note to this chapter is available at https://doi.org/10.1007/978-3-030-34515-0_85

which contains high level of pollutants [2]. In Times of India a survey in 2017 states that air pollution causes about 12.4Lac people death in India. Around 77% of India's contamination was presented to outside air contamination levels over the national surrounding air bore benchmarks safe breaking point. Indian therapeutic board has given a review of normal future in India have being 1.7 years higher if the air contamination level [2]. In Tamil Nadu, top three pollution city are Chennai, Trichy, Coimbatore [1].

PPM (Part Per Million) is a mathematical term used to measure the gas level using sensors In residential areas Co, Co2 emits more than 350PPM and is considered as deadly range. In industrial areas So2 and No2 gas range is maximum and also in garbage station methane gas range is too high, PPM level will be varied from the gas sensor.

In Fig. (1) among them, CO2 is (65%) of the major pollution in environment, next to it CH4 (16%) is moderately polluting the environment and So2 of (2%) creates low level of pollution.

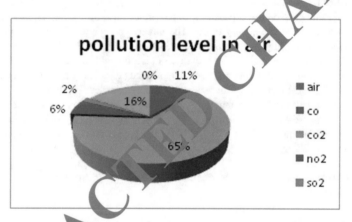

1. Pollution level in air worldwide

In Fig. (2) represents that, The main pollutants are emitted by industry, transport, agricultural fields, external garbage, commercial usages, and electric power department. In which majorly transportation make 58% pollution, 16% pollution in electric power, 13% of pollution in industrial areas, 9% of pollution in residential area and 4% pollution in commercial areas.

In Fig. (3) represent the pollution effects between 1970 and 2000 people increasing every year. So that it can be predicted that in the year 2000 and 2030 is expected as drastic changes and increment in the air pollution causes death in upcoming years.

2   Literature Review

Gurjar et al. [13] discussed the SO2, NO2, CO2, CO, CH4, MH3, NMBOC and TSP gases and its measures in Delhi. It is center point hard to gadget fitting outflow methodologies for NO2 fixations for a traffic intersection like ITO. DCE being a generally spotless site far from major roadways does not feel the vehicular effect and

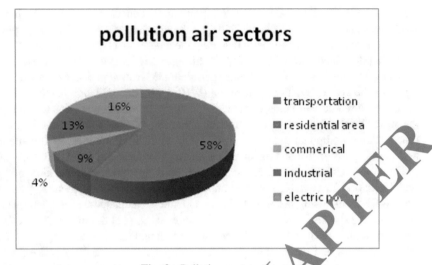

Fig. 2. Pollution sector

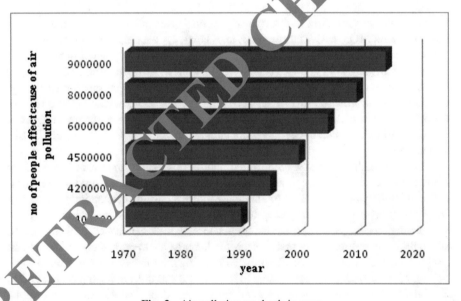

Fig. 3. Air pollution vs death in year

demonstrates a relentless diminishing pattern. Configuration estimations of SO2 uncover that focuses at all locales are generally well underneath the standard. Industrial emanations are the essential supporters of SO2 in Delhi. Goyal et al. [12] discussed SO2, NO2 and small particle matter that are measured and its awareness among people. Ngoc et al. [14] describes about micro scale air quality monitoring in urban areas based on two major technologies: wireless sensor networks and service-oriented architecture.

Biswas et al. [9] described about the pollutant's concentration changes higher in week days and lower in weekends and also the analysis seasonal changes in pollution higher concentration in winter and lower concentration in monsoon (rainy) Tudose et al. [24] described about CO, ammonia, nitrogen oxides, gasoline and diesel exhaust, natural gases, methane, CO2 and other petroleum derivative gases. As indicated by the report of the World Health Organization (WHO), air contamination is a critical hazard factor for different well-being conditions including skin and eye diseases, bothering of the nose, throat and eyes. It additionally causes genuine conditions like coronary illness, lung malignant growth, pneumonia, bronchitis, trouble in breathing and hacking because of irritated asthma. Bales et al. [8] discussed NO2, O3, CO polluted gas in atmosphere monitoring and sharing within their local communication and online social network. Xi et al. [25] worked on about pollution, weather, chemical component from WRF model using the So2, No2, Co, O3 and is calculated by the Pm10. The AQI and IAQI are the kits used in this paper. Fuertes et al. [11] describes about CO, CO2 gases measure using API developed in C++ language. Dexterous strategies, for example, scrum and extraordinary writing computer programs are utilized so as to guarantee programming quality. Chen et al. [26] discussed the air pollution monitoring and forecasting in real time value in IOT.

Raipure and Mehetre [18] proposed the utilization of AVR ATmega-32 Micro-controller and sensor matrix to distinguish the sensor esteems from an alternate sensor like parameters MQ5, MQ7, temperature, and dampness dataset. The recreation results demonstrate that the execution of the nature of administration expanded in the system. Sumithra et al. [23] work connected IOT checking air or water quality, climatic or soil conditions to screen air contamination utilizing sensors alongside distributed storage and enormous information investigation. Rybarczyk and Zalakeviciute [20] this work attempted to address the topic of how to anticipate fine particulate issue given a mix of climate conditions by Predicting the air contamination with dataset utilizing choice tree calculation.

Shah and Mishra [21] IoT empowered natural observing framework for checking temperature, relative dampness, and CO2. Taneja et al. [22] endeavored to use information mining apparatus WEKA to anticipate future alongside calculation for example straight relapse and multilayer preceptor on. Basher et al. [4] examined about O3, SO2, NO2 contaminate gases and it has been checked in urban zones utilizing an AI calculation with bolster vector machines, M5P demonstrate trees, and counterfeit neural system. The execution assessment estimated utilizing the forecast pattern. Gómez et al. [5] this work utilized, audience sensors to gather data and the administration utilized as an arbiter to the capacity that recover and conveyance data. The AI calculation can be utilized to anticipate for the future cool to convey exact data to take speedy estima-tions. Ayele and Mehta [7] portrayed checking air poisons of a specific region and characterizing its air gauge. Paulraj, Subramaniam et al. [15] designed a data recording system to record the event related evoked potentials of the somatosensory stimuli. Paulraj, Subramaniam et al. [16] designed a system to determine the hearing perception level using the evoked potentials using machine learning. Chittilappilly, Subramaniam [10], designed an industrial fault detection system by utilizing the SVM for classifying the extracted features. Augustine and Subramaniam [6], designed a LPG monitoring and booking system to determine the daily consumption of the gas and intimate by

SMS. Ramakrishna, Kasthala and Prasanna Venkatesan [19], proposed an Arduino based heavy vehicle monitoring and control system to determine the fuel leakage, engine control, speed control. Swaminathan and Prasanna Venkatesan [3], designed an embedded traffic control system using wireless adhoc sensor system. Pachiyaannan and Prasanna Venkatesan [17], presented a MSP based wireless sensor nodes. The literature papers [3, 6, 10, 15–17, 19] are used as a reference for future enhancement of the system with the help of new sensors and classifiers.

In our project, we mainly concentrate on five human life affect harmful gases (CO_2, CO, SO_2, NO_2, CH_4). The sensors are used to detect the gases in the envi- ronment. The data sent through the PIC micro-controller and it converts the value in analog to digital and data will be stored into the IOT. The quality of air is classified in to four class good (CO-50PPM, CO2-350PPM, CH4 < 1000PPM, SO2-0.1PPM, NO2-< 0.1PPM), moderate(CO-200PPM, CO2-1000PPM, CH4-1000PPM, SO2-1PPM, NO2->5PPM), unhealthy(CO-800PPM, CO2-2000PPM, CH4-5000PPM, SO2-3PPM, NO2-3PPM), and hazardous (CO-6400PPM above, CO-5000PPM above, CH4-150000 above, SO2-above 5PPM, NO2 10PPM above). result will be updated every three minutes. The project main used of awareness of the people.

3 Methodology

4 Working

The gas sensors are used to discover the current status in the air. The sensors are connected to the PIC micro-controller. The controllers acquire the value and transform into digital form and those values are modernize in the web server through RS 232 modem. The non-remittance values are already stored in the web server. The current data will be proposed with the non-remittance values to analyze the result. The pollution status will be uploaded in the web server. So, the user can view anywhere through internet. When there is pollution that occurs this shows the total amount of gases which is present on the particular locality example CFC (chlorofluorocarbon) carbon-monoxide and many toxic gases can identified by the amount of percentage which is contemporary on it. The main working principle behind this is IOT which collects documentation from the cloud which consists of documentation about the pollution status which is contemporary in our environment. The microcontroller which is used in this device is that PIC (16F877A) microcontroller which consists of 5 outputs and 5 inputs so that many sensors can be clubbed in collaboration which totally sums up together as a pollution detector and monitoring using an IOT device (Fig. 4).

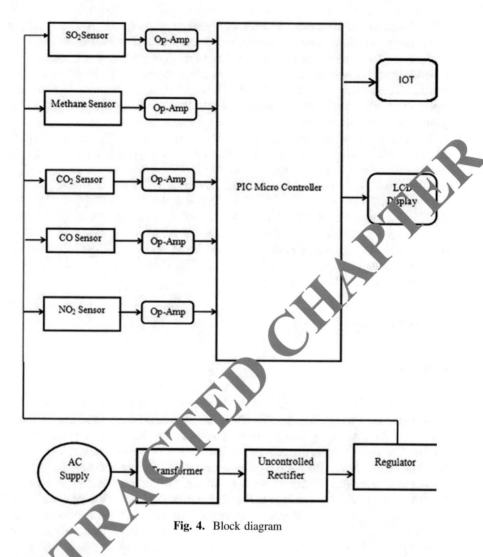

Fig. 4. Block diagram

4.1 Rs232

RS-232 (Recommended Standard - 232) is a broadcast communications standard for paired sequential interchanges between gadgets.

MAX232

The MAX232 from Maxim was the principal IC which in one bundle contains the fundamental drivers (two) and beneficiaries (additionally two), to adjust the RS-232 flag voltage levels to TTL rationale.

ORC Protocol

This specification documents the ORC network protocol in its entirety for the purpose of enabling its implementation in other languages.

5   Working Module

See Fig. 5.

Fig. 5. Working module pollution monitoring

6   Tabulation

This result was taken in gandhipuram bus stand and echanari junction (Fig. 6 and Tables 1, 2, 3, 4, 5 and 6).

Fig. 6. LCD display output of pollution

Table 1. PPM level in pollution gases

S.No	Gases	Range good	Moderate	Unhealthy	Hazardous
1	CO gas	0 PPM	0–100 PPM	200–1000 PPM	>1000 PPM
2	CO2 gas	250–35 PPM	350–1000 PPM	1000–5000 PPM	40000 PPM
3	CH4 gas	0 PPM	1000 PPM	50,000 PPM	1,50,000 PPM
4	SO2 gas	1–5 PPM	6–10 PPM	10–20 PPM	>20 PPM
5	No2 gas	0 PPM	0.1–5 PPM	5–10 PPM	>10 PPM

Table 2. CO gas PPM level

CO gas PPM level characteristic	
Range	Health effects
0 PPM	Good and fresh air
0 < 100 PPM	Headache
200–1000 PPM	Unconsciousness
>1000 PPM	Death

Table 3. CO2gas PPM level

CO2 gas PPM level characteristic	
Range	Health effects
250–350 PPM	Normal air
350–1000 PPM	Air exchange level is low
1000–5000 PPM	Heart rate and s nausea
40000 PPM	Permanent brain damage, coma, even death

Table 4. SO2 gas PPM level

SO2 gas PPM level characteristic	
Range	Health effects
1–5 PPM	Normal air
6–10 PPM	Irritation of eye, noise and throat
10–20 PPM	Threshold of toxicity
>20 PPM	Death

Table 5. CH4 gas PM level

CH4 gas PPM level characteristic	
Range	Health effects
0 PPM	Normal level of air
0–1000 PPM	NIOSH 8–hours TLV
5000–15000	Potentially explosive
500000	Asphyxiation

Table 6. NO2 gas PPM level

NO2 gas PPM level characteristic	
Range	Health effects
0 PPM	Good and fresh air
0.1–5 PPM	Irritation for noise, eye and throat
5–10 PPM	Noise, eye and throat irritation
>10 PPM	High noise, eye and throat irritation

240	Methane:0010	3/19/2019 10:24:39 AM
239	CO:0021	3/19/2019 10:24:33 AM
238	CO2:0210	3/19/2019 10:23:53 AM
237	HO2:0026	3/19/2019 10:23:47 AM
236	NO2:0065	3/19/2019 10:23:41 AM
235	Methane:0010	3/19/2019 10:23:33 AM
234	CO:0021	3/19/2019 10:23:29 AM
233	CO2:0210	3/19/2019 10:22:47 AM
232	HO2:0026	3/19/2019 10:22:41 AM
231	NO2:0069	3/19/2019 10:22:34 AM
230	Methane:0010	3/19/2019 10:22:29 AM
229	CO:0021	3/19/2019 10:22:21 AM
228	CO2:0220	3/19/2019 10:21:42 AM
227	HO2:0026	3/19/2019 10:21:38 AM
226	NO2:0071	3/19/2019 10:21:29 AM
225	Methane:0010	3/19/2019 10:21:23 AM
224	CO:0021	3/19/2019 10:21:16 AM
223	CO2:0240	3/19/2019 10:20:36 AM
222	HO2:0026	3/19/2019 10:20:30 AM
221	NO2:0077	3/19/2019 10:20:23 AM
220	Methane:0010	3/19/2019 10:20:17 AM
219	CO:0021	3/19/2019 10:20:10 AM
218	Sulphur_dioxide_concentartion_is_high	3/19/2019 10:19:55 AM
217	CO2_concentartion_is_high	3/19/2019 10:19:35 AM
216	Sulphur_dioxide_concentartion_is_high	3/19/2019 10:19:34 AM
215	CO2:0770	3/19/2019 10:19:31 AM
214	HO2:0026	3/19/2019 10:19:24 AM
213	NO2:0092	3/19/2019 10:19:18 AM
212	Methane:0010	3/19/2019 10:19:11 AM
211	CO2_concentartion_is_high	3/19/2019 10:19:10 AM

Fig. 7. Result from air pollution monitoring

7 Result and Discussion

The result of our study indicates that CO2, CO, NO2, SO2 and CH4 are predominant air pollutants that changes every second. The overall results show that SO2 & NO2 had smaller impact in atmosphere rather than CO, CO2 and CH4 like gases. The pollution range measured also depicted in online website. The studies have been taken in Gandhipuram and echanari and the results have been tabulated (Figs. 7, 8 and 9).

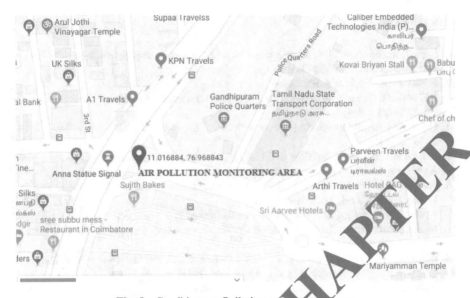

Fig. 8. Gandhipuram Pollution monitoring system

Fig. 9. Echanari pollution monitoring system

8   Conclusion

In this project we have developed an air quality monitoring system using IOT. In this, we have used five sensors for acquiring data from any location or place .data based on the threshold value of the gas the quality of air is classified into 4 classes. Classes are good, moderate, unhealthy and hazardously thus the values are uploaded in IOT so that it is on able to monitor the level of pollution in the location.

References

1. https://timesofindia.indiatimes.com/city/trichy/Trichy-2nd-most-polluted-city-in-TN-WHO-survey/articleshow/52275878.cms
2. https://www.who.int/airpollution/en/
3. Embedded traffic control system using wireless ad hoc sensors. Middle-East J. Sci. Res. **20**(2), 225–227 (2014)
4. Urban air pollution monitoring system with forecasting models. IEEE Sens. J. **16**(8), 2598–2606 (2016)
5. IoT for environmental variables in urban areas. In: ELSEVIER on International Conference Ambient Systems, Networks and Technologies, ANT-2017, vol. 109, pp. 67–74 (2017)
6. Augustine, I., Subramaniam, K.: LPG consumption, monitoring and Booking system. Int. J. Appl. Eng. Res. **4**, 3993–3998 (2016)
7. Ayele, T.W., Mehta, R.: Air pollution monitoring and prediction using IOT. In: 2nd International Conference on Inventive Communication and Computational Technologies (2018)
8. Bales, E., Nikzad, N., Quick, N., Ziftci, C., Patrick, K., Griswold, Citisense: mobile air quality sensing for individuals and communities (2012)
9. Biswas, J., Upadhyay, E., Nayak, M., Yadav, A.K.: An analysis of ambient air quality conditions over Delhi, India from 2004 to 2009. Atmos. Cl. Sci. **1**, 214–224 (2011)
10. Chittilappilly, A.J., Subramaniam, K.: SVM based defect detection for industrial applications. In: 2017 4th International Conference on Advanced Computing and Communication Systems (ICACCS), pp. 1–5 (2017)
11. Fuertes, W., Carrrera, D., Villacis, C., Toulkeridis, T., Galarraga, F., Torres, E., Aules, H.: Distributed system as Internet of Things for a new-low-cost, air pollution wireless monitoring on real time. In: 2015 IEEE/ACM 19th International Symposium on Distributed Simulation and Real Time Applications (2015)
12. Goyal, S.K., Ghatge, S.V., Nema, P., Tamhane, S.M.: Understanding urban vehicular pollution problems vis-a-vis ambient air quality-case study of mega city. Environ. Monit. Assess. **199**, 557–569 (2006)
13. Gurjar, B.R., Aardenne, Lelievelda, J., Mohan, M.: Emission estimates and trends for mega city Delhi and implication (2004)
14. Ngoc, T.Q., Lee, J., Gu, K.J.: An ESB based micro scale urban air quality monitoring system. In: Fifth IEEE International Conference on Networking, Architecture, and Storage (2010)
15. Paulraj, M.P., Subramaniam, K., Yaccob, S.B., Adom, A.H.B.: Application specific multichannel EEG data acquisition system for portable biomedical application. In: Proceedings of the Conference on MUCET, pp. 15–28 (2012)
16. Paulraj, M.P., Subramaniam, K., Yaccob, S.B., Adom, A.H.B., Hema, C.R.: A Machine learning approach for distinguishing hearing perception level using auditory evoked potentials. In: IEEE EMBS Conference on Bio-Medical Engineering Sciences, pp. 991–996 (2014)
17. Pachiyaannan, M., Venkatesan, G.P.: Designing of Wireless Sensor Node using MSP (2013)
18. Raipure, S., Mehetre, D.: Wireless Sensor Network Based Pollution Monitoring System in Metropolitan Cities. In: IEEE on International Conference (2015)
19. Ramakrishna, G., Kasthala, S., Venkatesan, G.P.: Arduino based automatic vehicle control system. Int. J. Sci. Res. Comput. Sci. Eng. Inf. Technol. **2**(6) (2017)
20. Rybarczyk, Y., Zalakeviciute, R.: Machine learning approach to forecasting urban pollution. IEEE J. (2016)

21. Shah, J., Mishra, B.: IOT enabled environmental monitoring system for smart cities. In: International Conference on Internet of Things and Applications (IOTA) (2016)
22. Sharma, N., Taneja, S., Sagar, V., Bhatt, A.: Forecasting air pollution load in Delhi using data analysis tools. In: International Conference on Computational Intelligence and Data Science (2016)
23. Sumithra, D.A., Ida, J.J., Karthika, K., Gavaskar, D.S.: A smart environmental monitoring system using Internet of Things. IJSEAS J. **2**(3) (2016)
24. Tudose, D.S., Patrascu, T.A., Voinescu, A., Tataroiu, R., Tapus, N.: Mobile sensors in air pollution measurement. In: 8th Workshop on Positing Navigation and Communication, pp. 166–170 (2011)
25. Xi, X., Wei, Z., Rui, X., Wang, Y., Bai, X., Yin, W., Don, J.: A comprehensive evolution of air pollution prediction improvement by a machine learning method. In: IEEE International Conference on Service Operations and Logistics, and Informatics (SOLI) (2015)
26. Xiaojun, C., Liu, X., Xu, P.: IOT - based air pollution monitoring and forecasting system. In: International Conference on Computer and Computational Sciences (2015)

Survey of Detection of Sound Level Using Internet of Things

Aditi Dande[✉] and D. M. Bhalerao

Savitribai Phule Pune University, Pune, India
aditidande009@gmail.com, dsmbhalerao1@gmail.com

Abstract. Internet of Things (IoT) is known as a platform where objects deal with the actuators, processors and sensors wherein they communicate with one another to serve for a meaningful purpose. This paper surveys about the various methods, different protocols, and applications in the new emerging area. This survey paper highlights some IoT technologies, which deals with some of the most important technologies, and makes use of some applications that leads to huge difference in human life especially for the senior citizens and special people. If compared with same survey papers in this area, this paper is much more comprehensive in its coverage and it covers most major technologies from sensors to applications.

1 Introduction

Noise pollution causes very serious problems to the humans. Noise can be defined as over-loud or disturbing sound. Sound levels are measured in decibels (dB). It is a unit used for expressing the relative intensity of sound which measures on a scale from 0 to 130. Any sound which is at or above 85 decibels can damage hearing. Noise pollution can cause serious problems like hypertension, high stress levels, hearing loss, sleep disturbances, and other harmful effects. Sound becomes unwanted when it disturbs normal activities such as sleeping, conversation, or disrupts one's quality of life.

In this paper, noise level detection using audio system has been implemented. This system includes Mic sensor, node MCU at transmitter side and buzzer, node MCU at receiver side. Mic sensor senses the surrounding noise as analog values. Node MCU converts that analog values to digital values and those digital values can be displayed to firebase. At the receiver side, digital values can be seen. In this survey paper, we have implemented that if noise level is below 10 than buzzer will be OFF. Or else if noise level is above 10, buzzer will be ON. This will give indication that the particular area has more noise so accordingly further action can be taken. Taking into consideration this application, comparison of protocols like HTTP, IPv6, 6LowPAN has been studied. In this survey paper, we studied about 6LowPan in detailed.

P. Karrupusamy et al. (Eds.): ICSCN 2019, LNDECT 39, pp. 452–460, 2020.
https://doi.org/10.1007/978-3-030-34515-0_47

2 Literature Survey

This paper highlights the implementation of an Internet of Things (IoT) application which focuses on performance to monitor the sound level. The phonometer which is used to measure sound level is connected on Arduino MEGA combined with ESP8266 Wi-Fi module [1].

This paper suggests the importance of IP for Smart things which tries to expand the use of IP into asset obliged gadgets over a wide scope of low-control interface advancements – IEEE 802.15.4 speaks to one such connection. Stretching out IP to low-control, remote individual zone systems (LoWPANs) was once viewed as unrealistic in light of the fact that these systems are exceptionally obliged and should work unattended for multiyear lifetimes on unobtrusive batteries [2].

This report portrays the casing group for transmission of IPv6 bundles and the strategy for framing IPv6 connect residential areas stateless autoconfigured addresses on IEEE 802.15.4 systems [3].

This paper suggests that IPv6 (Internet Protocol) (6LowPAN), has turned into the concept of low-rate remote individual territory systems. IPv6 is considered as the connecting platform for IEEE802.15.4, 6LowPAN innovation has been pulled in broad consideration. This paper presented IEEE802.15.4 standard and 6LowPAN innovation, talked about 6LowPAN's focal points, broke down its key advancements, and the utilizations of 6LowPAN were investigated [4].

Transmitting of IPv6 data packets over Low-control Wireless Personal Area Networks (6LoWPAN) has been taken into consideration for this paper. Also, this paper suggests that the measure of IPv6 bundles is much greater than the size parcel of the IEEE 802.15.4 information connection layer [5].

This paper highlights, various sorts of 6LoWPAN headers are studied. By understanding the 6LoWPAN header exemplification, it is important to examine the new header in 6LoWPAN parcel to accomplish complete steering usefulness [6].

In this paper, based on IoT Air Pollution Monitoring System is developed where Air Quality on webserver can be examined with the help of internet and will trigger an alarm indicating that air quality has gone down after a certain level. This paper proposes an air quality and also sound pollution to monitor system that allows us to monitor and check live quality of air and also the pollution in a particular area through IoT [7].

This paper studies that the IEEE 802.15.4 standard is a minimal effort, low-rate, low-control wireless personal area network innovation. IPv6 over low-control wireless personal area network (6LoWPAN) are determined by the IETF committee [8].

6LowPAN is a low-control remote individual zone network (LoWPAN) which includes the IEEE 802.15.4-2003 standard. In [I-D.ietf-6lowpan-format], the interface identifier [RFC4291] for a 6LoWPAN gadget is dependent on the EUI-64 [EUI64] location, where various RFC files has been described [9].

This paper suggests that 6LoWPANs are framed by gadgets that are good with the IEEE 802.15.4 standard. For any reason neither the IEEE 802.15.4 standard nor the 6LoWPAN organization can characterize particularly how work topologies can be kept up. This implies that how 6LoWPAN arrangement and multi-jump directing can be kept upheld [10].

Some of the common issues which deals with wireless sensor networks that analyzes the environmental noise which includes usage of more power, more cost, and less scalability. In this paper, a novel approach to develop an autonomous system for collecting environmental noise information was developed. So the system which is highly scalable, easy to use, low-cost, and low-powered was designed to encourage its widespread adoption [11].

3 Components Used

ESP8266 Wifi Module:
Figure 1 shows the ESP8266 WIFI Module. This device is one of the leading IoT devices in the world. It is very cheap and effective to use. The hardware connections required to connect to the ESP8266 module are straight-forward but there few important items to consider which are related to power:

Fig. 1. ESP8266 Wifi module [5]

- The ESP8266 requires 3.3 V power. It should not power with 5 volts
- The communication for ESP8266 takes place via serial at 3.3 V and does not have 5 V tolerant inputs. So, a level conversion to communicate with a 5 V microcontroller like Arduinos need to be used.

Mic Sensor:
Figure 2 shows the Mic sensor. It is a microphone which is acoustic to electric transducer or sensor that detects sound signals and converts them into an electrical signal. When the moving plate (diaphragm) vibrates in time with the sound wave, the distance between the plates and hence the capacitance is changed. The changes in capacitance can then be converted to an electrical signal.

Fig. 2. MIC sensor [7]

Buzzer:
Figure 3 shows the diagram of buzzer. It is an audio signaling device, which can be mechanical, electromechanical, or piezoelectric. Uses of buzzers include alarm devices, timers. The vibrating disk which is present in magnetic buzzer gets attracted to the pole with the help of magnetic field. When an oscillating signal is moved through the coil, a fluctuating magnetic field is produced which leads to the vibration of the disk which is at a frequency same as that of the drive signal.

Fig. 3. Buzzer [7]

4   Proposed System

To discuss 6LowPAN protocol of Internet layer we consider the example of transferring of audio signal over internet. The system consists of transmitter and receiver side. At transmitter and receiver side Node MCU has been used, with Wi-Fi module ESP8266 package. User can record voice using mic connected to system. The output of mic is given as input to an audio amplifier. Audio amplifier amplifies audio signal to sufficient high level so that it can be transferred without interference.

Figure 4 shows Node MCU is connected to both sides i.e. at transmitter and receiver side. Mic sensor senses the surrounding noise and sends those values to firebase server through Node MCU. Node MCU uses 6LoWPAN internet protocol. With the help of this protocol information is being send to the server. Node MCU converts Analog noise signals to digital signals. Digital signals can be seen on Firebase system. For this survey paper, we have taken into consideration digital Nosie level as 10. If noise value is below 10, then buzzer will be OFF and if noise level goes above 10, buzzer will be ON. This will give clear indication that noise level area has reached the danger level and accordingly further action can be taken. This system is standalone system and can be implemented near temples, hospitals where noise pollution can cause harmful effects to the common man.

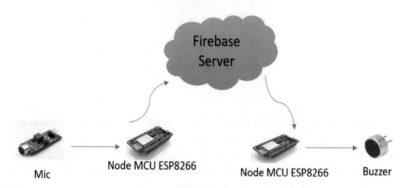

Fig. 4. Proposed system

Figure 5 shows the experimental setup of the proposed system. Where, Mic is connected to input side of node MCU and buzzer works as an output. Firebase server will display the noise level from the surrounding network. Configuration of this system is based on the hotspot of the mobile network. Where authentication details are provided in 6LowPan code.

Fig. 5. Experimental setup

Figure 6 shows the buzzer is ON when the value of Mic sensor reaches above 10. This value is depending upon the surrounding noise levels. As the buzzer is ON, this will give indication that surrounding area has more sound pollution and accordingly further action can be taken.

Fig. 6. Buzzer is ON when Mic value exceeds 10

5   Results

Figure 7 shows the Firebase result. The values sensed from the surrounding network will be sensed from the Mic sensor. Noise will be analog values. Those values will be converted into digital values through Node MCU and digital values will be seen on Firebase system. And at the receiver side if the values of Mic sensor is increased then accordingly buzzer will be ON. These values will be sent through mobile to the noise detecting committee of that area indicating that noise has reached its maximum value. Those people will receive notification on their mobile for such information so that they can take further steps.

Fig. 7. Firebase result

Table 1 shows the various comparison parameters among three protocols namely 6LowPAN, IPv6 and HTTP. Based on the above parameters which protocol will be best suited for this application can be decided.

Table 1. Comparison parameters.

Parameter	6LoWPAN	IPv6	HTTP
Time to respond (sec)	1.2	1.2	5.2
Throughput per 1000 trials	1000	998	993
Data Loss			
Overall system power consumption (mW)	859	897	984
Parameter	6LoWPAN	IPv6	HTTP
Time to respond (sec)	1.2	1.2	5.2
Throughput per 1000 trials	1000	998	993
Data Loss			
Overall system power consumption (mW)	859	897	984

Figure 8 shows the output of proposed system. Comparison of three protocols is studied and based on the parameters like Time to respond, throughput, data loss it can be seen that 6LowPan protocol consumes least power and this protocol has maximum throughput.

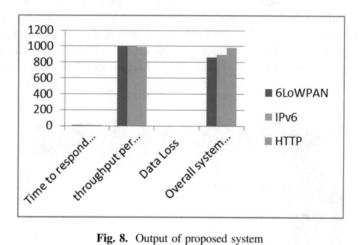

Fig. 8. Output of proposed system

6   Conclusion

Keeping the embedded devices in the environment monitors the protection (i.e., smart environment) to the environment. For this implementation, it is necessary that the sensor devices are to be deployed in the environment so that data can be collected, and further analysis can be done. After deploying sensor devices in the environment, environment can be brought into real life which means that it can be useful to interact with different objects which are available through the network. After this, data which is collected, and further results will be provided to the destination user through the Wi-Fi.

Noise level monitoring system with Internet of Things (IoT) concept can be used to monitor surrounding pollution. It can also send the sensor parameters to the cloud (Google Spread Sheets). This data can prove helpful for future analysis and also this data can be easily shared to other destination users.

This paper surveys the design and implementation of audio signal system on Node MCU. The system has a MIC as an input device and a buzzer as an output device. Based on this application comparison of protocols namely 6LowPan, IPv6 and HTTP has been done and 6LowPan is studied deeply. Depending on the results obtained it is estimated that 6LoWPan protocol consumes least power and has least chances of data loss. In this paper 6LoWPAN protocol of internet layer of IoT is discussed using example of audio transfer through internet. This system can be used where noise level is above the danger level and accordingly further action can be taken.

References

1. Likic, M.: An IOT solution for secured and remote sound level monitoring. In: 2019 18th International Symposium INFOTEH -JAHORINA (INFOTEH) (2019)
2. Jamil, M.S., Jamil, M.A., Mazhar, A., Ikram, A., Ahmed, A., Munawar, U.: Smart environment monitoring system by employing wireless sensor networks on vehicles for pollution free smart cities. In: Humanitarian Technology: Science, Systems and Global Impact 2015, HumTech (2015)
3. Guthi, A.: Implementation of an efficient noise and air pollution monitoring system using Internet of Things (IoT). Int. J. Adv. Res. Comput. Commun. Eng. 5(7), 237–242 (2016)
4. Khan, W.Z., Xiang, Y., Aalsalem, M.Y., Arshad, Q.: Mobile phone sensing systems: a survey. IEEE Commun. Surv. Tutorials 15(1), 402–427 (2013)
5. El-Bendary, N., Fouad, M.M.M., Ramadan, R.A., Banerjee, S., Hassanien, A.E.: Smart environmental monitoring using wireless sensor networks. In: K15146_C025.indd (2013)
6. Chowdhury, A.H., Ikram, M., Cha, H.-S.: Route-over vs Mesh-under Routing in 6LoWPAN. ACM (2009)
7. Ee, G.K., Ng, C.K., Noordin, N.K., Ali, B.M.: A review of 6LoWPAN routing protocols. In: Proceeding of Asia Pacific Advanced Network (2010)
8. Lehmann, G., Rieger, A., Blumendorf, M., Dai, S.A.: A 3-Layer architecture for smart environment models. In: A Model-Based Approach/Labor Technische University Berlin, Germany. IEEE (2010). ISBN 978-1-4244-5328-3/10 ©
9. Chang, J.M., Chi, T.-Y., Yang, H.-Y., Chao, H.C.: The 6LoWPAN adhoc on-demand distance vector routing with multi-path scheme (MLOAD). IET (2010)
10. Kim, E., Kaspar, D., Gomez, C., Bormann, C.: 6LoWPAN routing requirements. IETF, draft-ietf6lowpan-routing-requirements-04
11. Hui, J.W., Culler, D.E.: Extending ip to low-power, wireless personal area networks. IEEE Internet Comput. 10, 37–45 (2008)
12. Liu, H., Bolic, M., Nayak, A., Stojmenović, I.: Integration of RFID and wireless sensor networks, School of Information Technology & Engineering, University of Ottawa, Ottawa, Canada, K1 N 6N5
13. Weber, S., Andrews, J.G., Jindal, N.: An overview of the transmission capacity of wireless networks. IEEE Trans. Commun. 58, 3593–3604 (2010)

14. Valero, X., Alías, F., Oldoni, D., Botteldooren, D.: Support vector machines and self-organizing maps for the recognition of sound events in urban soundscapes. In: Proceedings of the 41st International Congress and Exposition on Noise Control Engineering, New York, NY, USA, 19–22 August 2012
15. Porambage, P., Braeken, A., Gurtov, A., Ylianttila, M., Spinsante, S.: Secure end-to-end communication for constrained devices in IoT-enabled ambient assisted living systems. In: Proceedings of the IEEE 2nd World Forum on Internet of Things (WF-IoT), Singapore, 5–8 February 2015

Big Data Performance Analysis on a Hadoop Distributed File System Based on Modified Partitional Clustering Algorithm

V. Santhana Marichamy[1(✉)] and V. Natarajan[2]

[1] Department of Computer Applications, SRM Valliammai Engineering College,
SRM Nagar, Kattankulathur, Chennai 603203, Tamil Nadu, India
sandalswamy72@gmail.com
[2] Department of Instrumentation Engineering, Anna University, MIT Campus,
Chennai 600044, Tamil Nadu, India
natraj@mitindia.edu

Abstract. This paper proposes a Big Data Performance Analysis based on a modified Partitional Clustering Algorithm (PCA) on a Hadoop Distributed File System (HDFS) which is commonly used in various business applications. This paper has utilized an improved K-means clustering algorithm, which selects the initial clustering centers based on the density parameters. After calculating the density parameter, the data with largest density parameter is selected as the first initial clustering center point, all the left data in the field is deleted from the dataset. By repeating the above phases, K initial clustering centers are found. A new method to improve the precision and packing effect of the K-means computation is needed as there is a poor assurance of finding an initial centers. The proposed approach does not select the initial clustering algorithm randomly, so the stable K value can be obtained by calculating Variance based Cluster Validity Index (VCVI). The performance of the proposed method is evaluated with the parameters Precision, Clustering time and Recall. The experimental result shows that the proposed approach reduces the complexity along with various parameters are compared with existing methods.

Keywords: HDFS · PCA · VCVI · K-means cluster · Density parameter · Clustering time · Recall

1 Introduction

The Corporate world generates tons of data every single day and makes use of sensors for gathering information about the climatic conditions, audios, videos and images in the public network sites and play a major role [1]. In particular, medical field has a tremendous amount of data which has to be stored in a highly secured warehouses for privacy preservation, such kind of data is known as "BIG DATA". Many successful techniques on data mining, extraction of patterns and knowledge from big data set were used by the cooperators for an efficient decision making to earn customer satisfaction [2].

© Springer Nature Switzerland AG 2020
P. Karrupusamy et al. (Eds.): ICSCN 2019, LNDECT 39, pp. 461–468, 2020.
https://doi.org/10.1007/978-3-030-34515-0_48

Hence, it is necessary to have access to the data of interest effectively. This is achieved by clustering the big data set in a compact format to have a good quality of clustered data. It is an unsupervised learning technique, in which each cluster will have the same kind of data and they are dissimilar from supplementary clusters [3].

In this paper, it is proposed to have health care insurance information in the Hadoop Distributed File System [4] without presenting a calculation and correspondence overhead between the private and the public distributed file system. Here, at first an improved K-means partition clustering algorithm [5] is utilized for distributing the high-dimensional information where each partition is considered as a cluster.

2 Related Works

Numerous research works have previously existed in literature which was based on the big data clustering techniques and schemes. Some of the works are reviewed here.

Eluri et al. [6] has their primary goal was to characterize information into clusters with the end goal that objects are gathered in a similar group when they are comparable rendering to specific measurements and not at all like objects of different gatherings. From the AI viewpoint clustering can be seen as unsupervised learning of ideas. Hadoop is a dispersed document framework and an open-source execution of MapReduce managing huge information. Apache Mahout bunching calculations are actualized over Hadoop utilizing MapReduce worldview. In this paper three clustering calculations were portrayed: K-implies, Fuzzy K-Means (FKM) and Canopy clustering executed by utilizing Apache Mahout just as giving an examination. What's more, they underlined the bunching clustering calculations that are the prevalent performing for big data.

3 Proposed Method

3.1 Problem Statement

This proposed approach for this big data performance analysis based on a Hadoop Distributed File System (HDFS) on a modified Partitional clustering algorithm (PCA) technique which are reduced the complexities with the help of various parameter analysis.

3.2 Our Contribution

In this paper, the proposed methodology would contribute in decrease of an excessive amount of capacity in a database sector, if we ultimately store the whole sensitive data on the storage sector. Different commitments incorporate the decrease of message overheads and the delay presented by interchanges among the application and database.

3.3   The Proposed Framework for HDFS Based Modified PCA

The proposed work is appeared in Fig. 1. The Hadoop Distributed File System (HDFS) is the essential information storing framework utilized by many applications [9].

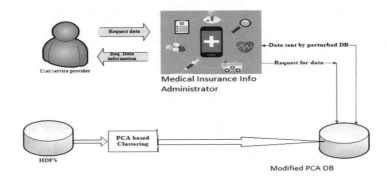

Fig. 1. The data process on HDFS based modified PCA

At the point, Big Data in the HDFS system, separates the data into isolated groups and disperses them into various blocks in a cluster, therefore empowering exceedingly proficient by parallel organizing. Then the clustering procedure is carried out by an improved K-means clustering algorithm i.e., PCA is applied to the sub partitioned blocks/groups.

3.3.1   Partitional Clustering Algorithm (PCA)

The main objective of this algorithm is to group the similar data based on their features and they were clustered. In this paper we have utilized an improved K-means clustering algorithm, which selects the initial clustering centers based on the density parameters.

After calculating the density parameter the data with largest density parameter is selected as the first initial clustering center point, all the left data in the field $\in$ is deleted from the dataset. The proposed approach does not select the initial clustering algorithm randomly, so that the stable K_{opt} value can be obtained by calculating variance based cluster validity index (VCVI) [7].

In Euclidean space R^m, a dataset comprising of $D = \{x_1, x_2, x_3........x_n\}$ is given, among that based on the clustering algorithm the dataset is subdivided into K clusters $C = \{c_1, c_2, c_3........c_k\}$. The variance based clustering algorithm is evaluated by equation

$$VCVI(K) = \frac{1}{n}(K^2 G + nT)$$

(1)

Where, G is the inter cluster variance, which is the mean square sum of the distance between the centers of each cluster to global center. Where, T_i is the sum of squared distance of intra cluster of the ith cluster? The optimal clustering number K_{opt} can be calculated by using the formula (2)

$$K_{opt} = \min 2 \leq K \leq n - 1\{VCVI(K)\} \tag{2}$$

From the above equations, the value of K is optimal when the VCVI index value reaches the minimum value or else it will at least reach near the optimal number of clusters is described in Fig. 2.

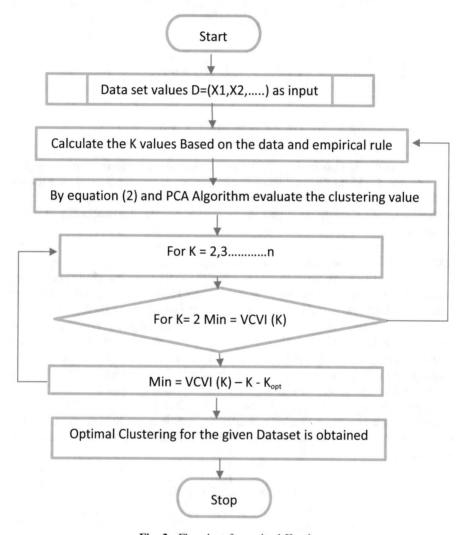

Fig. 2. Flowchart for optimal K value

4 Implementation and Analysis

In this section, the performance metrics used for evaluation of the proposed privacy preserving algorithm are presented. These metrics are Precision, Clustering time and Recall. The proposed technique is implemented in Java Net beans. For that we are taking health care Dataset in https://www.academia.edu/ [8].

5 Result and Discussion

In Table 1 and Fig. 3, for all the document sizes, the proposed calculation is commonly speedier than the existing clustering algorithm technique, in light of the way that, the proposed procedure has no emphasis to annoy the information anyway the current strategy takes numerous rounds. For a document size of 10 MB the exactness rate of existing methodology is of about 88.888%, comparably for same record measure the proposed methodology is of about 90.45%. Consequently, the proposed modified Partitional based K means clustering methodology gives approximately 100% information accuracy and it is more powerful than the existing K – Means clustering strategy.

Table 1. Precision

File size (MB)	10	20	30	40	50
Existing K-means clustering algorithm (%)	88.8888	90.7103	91.9117	93.3333	95.7894
Proposed partitional clustering algorithm (%)	90.4534	92.6644	93.43545	95.34223	96.4555

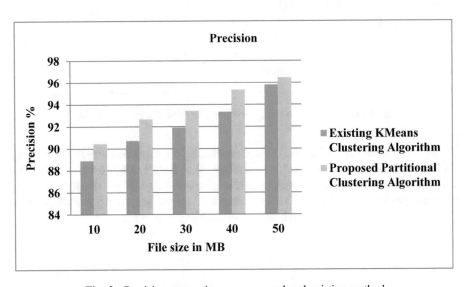

Fig. 3. Precision comparison on proposed and existing method

Clustering is the task of grouping a set of objects in such a way that objects in the same group called a cluster or to those in other groups and time taken for this process is said to be clustering time.

In Table 2 and Fig. 4 the clustering time taken by proposed and existing clustering algorithm is shown with respect to file size. Here at file size of about 30 MB the clustering time taken by the existing approach is about 23.3457 s and for the proposed approach is about 20.5464 s, which is better than that of the existing approach.

Table 2. Clustering time

File size (MB)	10	20	30	40	50
Existing K-means clustering algorithm (Sec)	11.346	18.435	23.3457	29.645	37.454
Proposed partitional clustering algorithm (Sec)	10.5567	16.4354	20.5464	26.443	35.234

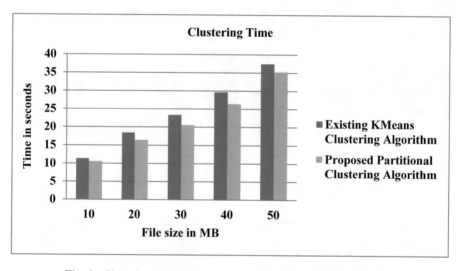

Fig. 4. Clustering time comparison on proposed and existing method

In Table 3 and Fig. 5 the recall rate percentage by proposed and existing method by clustering algorithm is shown with respect to file size. Recall is the part of genuine sets that were recognized, here at document size of around 30 MB the recall rate by the existing approach is about 92.2333% and the recall for the proposed approach is about 93.232432%, which is better than that of the existing approach.

Table 3. Recall

File size (MB)	10	20	30	40	50
Existing K means clustering algorithm (%)	92.1223	91.8855	92.2333	93.88775	95.8876
Proposed partitional clustering algorithm (%)	93.112	93.3534	93.232432	94.83232	96.83223

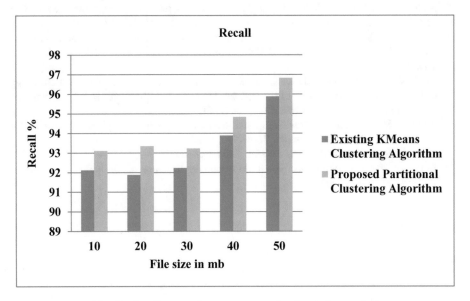

Fig. 5. Recall comparison on proposed and existing method

6   Conclusion

The conventional K-means calculation irregularly chooses the number (K) of initial clustering centers. In this paper, the proposed approach does not select the initial clustering algorithm randomly, so that the stable K optimal value can be obtained by calculating variance based cluster validity index (VCVI). This approach has executed and shown in the comparative results of parameters like precision, cluster execution time and recall. These results shown better performance compared with the existing method. In future, the Hadoop platform should be further directed to improve its performance and efficiency. This work can be extended further to execute Geometric Data perturbation (GDP) to improve the execution of information recovery processing.

References

1. Craven, M., Page, C.: Big data in healthcare: opportunities and challenges. Big Data **3**(4), 209–210 (2015)
2. Wolfert, S., Ge, L., Verdouw, C., Bogaardt, M.: Big data in smart farming – a review. Agric. Syst. **153**, 69–80 (2017)
3. Liang, F., Yu, W., An, D., Yang, Q., Fu, X., Zhao, W.: A survey on big data market: pricing, trading and protection. IEEE Access **6**, 15132–15154 (2018)
4. Mellor, J., Stone, M., Keane, J.: Application of data mining to "big data" acquired in audiology: principles and potential. Trends Hear. **22**, 233121651877681 (2018)
5. Anwar, M., Oakes, M.: Data mining of audiology patient records: factors influencing the choice of hearing aid type. In: BMC Medical Informatics and Decision Making, vol. 12, no. 1 (2012)
6. Eluri, V., Ramesh, M., Al-Jabri, A., Jane, M.: A comparative study of various clustering techniques on big data sets using Apache Mahout. In: 2016 3rd MEC International Conference on Big Data and Smart City (ICBDSC) (2016)
7. Zhu, E., Ma, R.: An effective Partitional clustering algorithm based on new clustering validity index. Appl. Soft Comput. **71**, 608–621 (2018)
8. https://www.academia.edu/17510514/Big_Data_Hadoop_In_Health_Care
9. Borthakur, D.: The hadoop distributed file system: architecture and design. Hadoop Project Website, pp. 1–14 (2007)

Efficient Network Security Architecture for Heterogeneous Smart Meter Environment

Senthil Perumal$^{(\boxtimes)}$, Sandanalakshmi Rajendiran,
and Sivadarishini Souprayen

ECE Department, Pondicherry Engineering College, Puducherry, India
{senthilmjk, sandanalakshmi, sivadarishini176}@pec.edu

Abstract. In smart grid communication network, Smart meter (SM) to DCU communication is currently Homogenous in standard. As the trend of SM to DCU communication is approaching towards heterogeneous nature, Intrusion detection will be a challenging task, leading to security issues. A novel Network security scheme is proposed for Heterogeneous SM to DCU environment. Security is established in the network by using and Particle Swarm Optimization (PSO) and Bilinear Diffie-Hellman (BDH) The PSO algorithm helps in finding an access points (AP) best position selection method to forward SM data to DCU through local centralisation method. The BDH gives the best identification scheme is secure against the attacks. The simulation results show an increase of 15% Packet delivery ratio, 7% increase in the end to end delay and 20% increase in throughput is achieved with PSO algorithm for Heterogeneous SM environment with respect to AODV in Homogenous SM enviornment.

Keywords: Heterogenous · Smart grid · Network security · Key exchange

1 Introduction

Smart grid alludes to future generation of power which enables high efficient power delivery and openness between utility and consumer. It has improved control, effectiveness, unwavering quality and protection in the distribution. The smart grid can be described as the transparent, seamless, and instantaneous bidirectional information exchange, enables better management in Power industry.

Secure data communication helps in smart grid communication to prevent the network attacks like message analysis, replay, injection and modification attacks [1]. The security characteristics like confidentiality, data protection and authentication in bidirectional communication must be considered [2–5]. As various attacks like DoS, replay attack and study of traffic may happen in smart grid data communication. Designing an efficient security protocol between nodes is much needed. Implementation of authentication process for both ends is essential in guaranteeing every correspondence association in the bidirectional data exchange [6–10]. In the wireless communication, the process of channel sharing is considered to be a challenging task. Various algorithms are introduced for information sharing in communication channel. Existing methods has shown the methods and process involved in employing a security algorithm in Smart grid, Neighbourhood gateways.

© Springer Nature Switzerland AG 2020
P. Karrupusamy et al. (Eds.): ICSCN 2019, LNDECT 39, pp. 469–477, 2020.
https://doi.org/10.1007/978-3-030-34515-0_49

1.1   Related Works

Al-Fares et al. [11] proposed an authenticated scheme on Diffie-Hellman key estab-lishment protocol, message integrity is achieved utilizing RSA and AES. This method is more advantage than Hash-MAC. This algorithm is lightweight and suited to the circumstances like smart meters in the SG environment. Liu et al. [12] designed a lightweight authenticated communication scheme using bitwise exclusive OR and Lagrange interpolation formula for performing encryption and authentication service This method has reduced the storage and communication cost achieving better secured bi-directional communication between SM and NG.

Abdrabou et al. [13] proposed an architecture with cellular frequency-reuse structure which benefits the multichannel ability of IEEE 802.11a. QoS-GPSR protocol was used for multi hop communications in each transformer location connected to the DCU through the Sensor/Control devices. Estimation of coverage space, density of nodes and data rates by analysing the network framework. Boustani et al. [14] focus on novel SGN architecture to enable better privacy for SMs. Privacy is preserved well in the utility because of two load predictions of energy level in Grid. At the SMs group level higher correlation is attained between predicted and consumed energy based on load pattern. Usage data of the consumer and channel usage of the SM in the grid are considered in the above methods.

Uludag et al. [15] developed an efficient protocol for Smart Grid data collection for a hierarchical architecture. The relay nodes collect and convey the data securely from SMs to Utility. Optimization problems for collection of data in the aspect of time using intermediary nodes are considered [16]. This method makes more expensive and over-lapping may occur in the implementation of security policies in a single device or multiple devices. The above communication model in SG is homogenous too facing security issues in wireless data communication between HAN-NAN, NAN-WAN and viceversa.

2   Proposed Method

Figure 1 depicts the proposed Smart Meter heterogeneous network. The nodes (SMs) have different types of specification, Bluetooth and Zigbee is used. The network model consists of Access point (Gateway) and DCU. The SMs communicates to the gateway, which in turn communicates to the DCU.

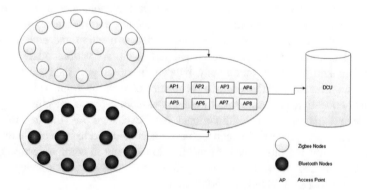

Fig. 1. Heterogenous SM to DCU communication environment

The proposed method more precisely address the HAN-NAN communication as Heterogenous at the SM end based on its type. Since the deployment of Smart meters is based on the type and capacity of the DCU deployed by the utility in the Grid. As vertical growth of SMs are estimated, heterogenous SM deployments will educe cost of the Utility. Our method uses Access point (AP) methods drive the HAN-NAN communication as a heterogenous enviornment. Local centralisation of data achieved through the AP. As such method ensure more accuracy in the identification of SMs based on IEEE 802.15.1 or IEEE802.15.4 standard. Making the HAN-NAN communication a heterogenous nature, incorporating security algorithm by accurate authentication process is developed in the proposed work. The above diagram shows that two SM groups selecting a single Access point.

In the proposed method the drawbacks of lightweight communication scheme in homogeneous SM environment is overcome by adopting a new methodology. The key exchange is performed by using Bilinear Diffie-Hellman method. The authentication is done automatically by enabling the key exchange, guarding the network from the attackers. The proposed method provides better network security and provides improvement in network efficiency. Intrusion Detection System (IDS) is proposed to avoid the attackers in network which detects different types of attackers inside the network. The IDS will identify and segregate the malicious nodes injected by the routing protocol.

IDS can detect the following attacks:

- Denial of service
- Man in middle attack.

This proposed methodology uses PSO(Particle Swarm Optimisation) in heterogeneous environment is compared with existing scheme which used AODV in homogeneous environment are analysed using Ns-2.

2.1 Proposed Algorithm

The Particle swarm optimisation is used to detect the intruders. PSO algorithm will help in finding the best agent (AP) to communicate. The PSO will help the best position of AP in the network by initializing a SM randomly. Based on the finding of the APs best position, the PSO algorithm modified for SM to DCU communication through gateway will check for data transfer by SMs in the network. As the Best agent moves to another position within the network, PSO will update the movement to the SM. The SM will identify that this agent (AP) is under his best point of communication based upon the boundary values it possess for communicating with AP.

Let S be number of Smart meters each having individual timeslot to communicate with DCU. Let p_i be the best known time of SM data i and let g be the best known time of data transmission. By fixing the minimum and maximum boundary for data transmission T_{lo} and T_{up}.

For each SM i = 1 ..., S do
Intialize SMs data transfer time S_i
  Intialize APs best position in the network p_i
   if $f(p_i) < f(g)$ then
      update the SMs best known timefor data transfer : $g \leftarrow p_i$
Intialize the APs time : $T_i\, U((-|T_{up}\text{-}T_{lo}|, |T_{up}\text{-}Tl_o|))$
Do till data transfer is completed
for each SM, i = 1, ..., S do
    for each time t = 1, ..., n do
      Pick random meters: $SM_p, SM_g \sim U(0,1)$
      Update the APs time: $t_i,T \leftarrow \omega\, t_i,T + \varphi p\; rp\; (p_i,d\text{-}x_i,t) + \varphi g\; rg\; (gd\text{-}x_i,t)$
    Update the APs position: $x_i \leftarrow x_i + t_i$
    if $f(x_i) < f(p_i)$ then
      Update the SMs best known time: $p_i \leftarrow x_i$
      if $f(p_i) < f(g)$ then
        Update the APsbest known position: $g \leftarrow p_i$.

3   Network Security

Bilinear Diffie-Hellman Key exchange ensures the security of the data best identification of which SM in the Network. The AP will generate the public and private key in the network and transfer it to other SMs. Grid. The SMs are authenticated based on the digital signature which is verified at DCU. The signature is validated based on the signature and the IDS is enabled to detect attacks.

The asymmetric key management is one of the cryptographic technique which consists of private and public key. The asymmetric key management techniques such as DES, AES, RSA and ECC are used for secure communication. In this method, for encrypting and decrypting purpose Bilinear Diffie-Hellman is used along with hash and exclusive OR operations. PSO is taken so that attackers can be detected and removed from the network.

4   Simulation Scenario

Using NS2, the network is simulated with the parameters listed in Table 1 in order to create the proposed heterogeneous network scenario. A heterogeneous network model has been created in the simulation using the Network simulator tool with the above specified simulation parameters. The traffic assumed here is the CBR traffic as the SMs send data at constant time interval (15 min)as the SM to DCU communication time is 15 min. The scenario has 1 DCU (sink), 8 gateways (AP) and 27 smart meters are taken at the network. In that 8 gateways all the gateways will not be active at once. Only one gateway will send the usage reports to the DCU (sink). The deployments of the nodes are shown in the Fig. 2. The DCU (sink) node is placed down, then comes gateways and grid.

Table 1. Simulation parameters

Total number of nodes	37 nodes	37 nodes
Network type	Wireless (Heterogenous)	Wireless (Homogenous)
MAC type	IEEE802.11	IEEE 802.11
Routing protocol	PSO	AODV
Max packets	100	100
IEEE 802.15.4	13 nodes	–
IEEE 802.15.1	14 nodes	27
Area considered	500*500	500 * 500
Time of simulation	10 s	10 s

The Fig. 2 also shows the naming of each nodes.. A and B (Blue colour node) act as a sensor nodes which is nothing but it will collect all the smart meter usage reports and send it to grid. In the Fig. 2 the node 16 is attacked by DOS and it is added to the malicious list. Now the node 19 is ready to launch the attack to node 2. Likewise node 22 which is red color is attacked by MIM attacked and added to the malicious Thus, when the node is attacked the it will indicate by changing its colour. Thus through the IDS techniques the DOS attack and MIM attack are detected and displayed as the output. When the malicious nodes are detected the communication of the node with the malicious nodes are stopped removal of such malicious nodes are done.

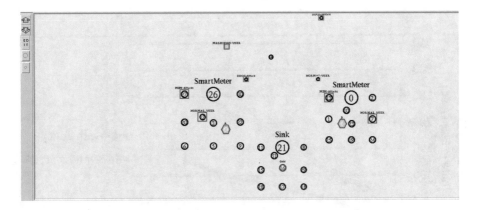

Fig. 2. Detection of attacks

4.1   Results and Discussion

Figure 3 shows the packet delivery ratio for the simulated existing and proposed scheme. Packet delivery ratio is proportional increasing towards time when compared to existing method. In case of AODV protocol the packet gets increased gradually from 0 to 12 s. In case of PSO, the packets sent gets increased from 0 to 4 s. In 6th s there is a drop due to the traffic and congestion in the system, the packets get dropped in the network. The packet delivery ratio's performance is increased in terms of percentage as 15%.

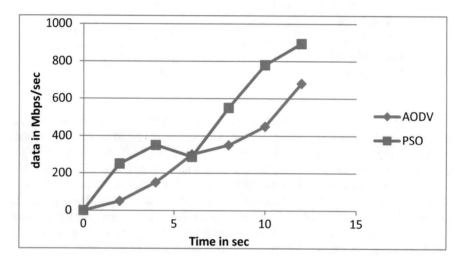

Fig. 3. Packet delivery ratio

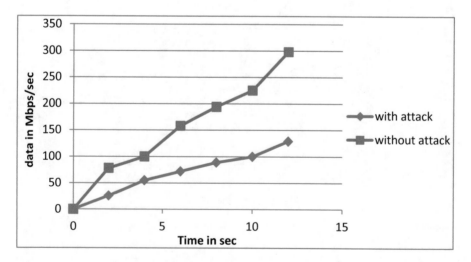

Fig. 4. Packet delivery ratio in terms of attacks

Figure 4 represents the packet delivery ratio with and without attacks. In the absence of attacks, the performance of the network is more better than the presence of attacks. IDS identifies and segregates the attacked nodes its reduces the channel usage by the utility.

Figure 5 depicts about the throughput in the network. As the Packet delivery ratio increase in a network by avoiding the error packets throughput of the network increases automatically. The existing methods in homogenous SM-DCU communication removes the attacked/altered data at DCU, our method removes the attacked/altered data before the DCU. In PSO after the time 8 s the throughput decreases slightly, due to accumulation of the packets generated the jitter in the communication link occurs and later it get increased.

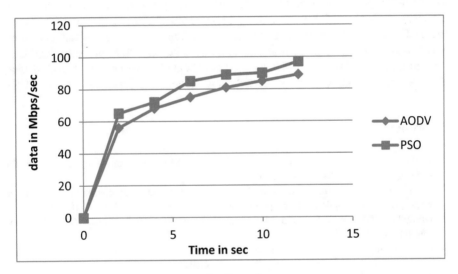

Fig. 5. Throughput

Comparison of packet delivery ratio, end to end delay and throughput energy consumption for proposed and existing scheme is tabulated in Table 2 which shows the better performance than the existing method.

Table 2. Comparison of performance analysis

Parameters	Performance in AODV	Performance in PSO
Packet delivery ratio	8%	15%
End to end delay	12%	7%
Throughput	14%	20%

Based on the displayed simulation results it is evident, the proposed method shows nearly 20% improvement in terms of throughput. Finally, the comparison shows that, the proposed method performs better in terms of throughput which in turn shows a considerable improvement in network performance.

5 Conclusion

Real-time monitoring and decision-making is possible in heterogeneous environment in the case of usage reports transmission in very short time. The proposed communication scheme, for the secure two-way communications of the SMs and the Gateways, which can adapt to fewer time duration of consumption data transmission. The SMs will undergo login phase followed by authentication phase, key management phase, data management phase and lastly the nodes get logged off. The key exchange method used here is Bilinear Diffie-Hellman algorithm which improves the nodes authentication process and PSO is used as a routing algorithm. As key exchange process takes place between the SM and the AP, SM is allowed to pass the information to the grid making the intruders in the network are identified using the Intrusion Detection System (IDS).

References

1. Li, H., Lu, R., Zhou, L., et al.: An efficient Merkle-Tree-based authentication scheme for smart grid. J. IEEE Syst. **8**, 655–663 (2014)
2. He, D., Wang, H., Khan, M.K., et al.: Lightweight anonymous key distribution scheme for smart grid using elliptic curve cryptography. J. IET Commun. **10**, 1795–1802 (2016)
3. Nordell, D.: Terms of protection: the many faces of smart grid security. J. IEEE Power Energy. **10**, 18–23 (2012)
4. Li, X., Liang, X., Lu, R., et al.: Securing smart grid: cyber attacks, countermeasures, and challenges. J. IEEE Commun. Mag. **58**, 38–45 (2012)
5. McDaniel, P., McLaughlin, S.: Security and privacy challenges in the smart grid. J. IEEE Secur. Priv. **7**, 75–77 (2009)
6. Ericsson, G.N.: Cyber security and power system communication essential parts of a smart grid infrastructure. J. IEEE Trans. Power Deliv. **25**, 1501–1507 (2010)
7. Lu, R., Li, X., Liang, X., et al.: GRS: the green, reliability, and security of emerging machine to machine communications. J. IEEE Commun. Mag. **49**, 28–35 (2011)
8. Zhu, H., Lin, X., Lu, R., et al.: SLAB: secure localized authentication and billing scheme for wireless mesh networks. J. IEEE Trans. Wirel. Commun. **7**, 3858–3868 (2008)
9. IEEE p 2030 draft guide. http://grouper.ieee.org/groups/scc21/2030/2030_index.html
10. Lin, X., Lu, R., Ho, P.H., et al.: TUA: a novel compromise-resilient authentication architecture for wireless mesh networks. J. IEEE Trans. Wirel. Commun. **7**, 1389–1399 (2008)
11. Al-Fares, M., Loukissas, A., Vahdat, A.: A scalable, commodity data center network architecture. SIGCOMM Comput. Commun. Rev. **38**, 63–74 (2008)
12. Liu, Y., Cheng, C., Gu, T., et al.: A lightweight authenticated communication scheme for smart grid. J. IEEE Sens. **16**, 836–842 (2010)
13. Khurana, H., Hadley, M., Lu, N., Frincke, D.A.: Smart grid security issues. J. IEEE Secur. Priv. **8**, 81–85 (2010)

14. Aravinthan, V., Namboodiri, V., Sunku, S., Jewell, W.: Wireless AMI application and security for controlled home area networks. In: IEEE Power and Energy Society General Meeting, pp. 1–8 (2011)
15. Hopkinson, K., et al.: Quality-of-Service considerations in utility communication networks. J. IEEE Trans. Power Deliv. **24**, 1465–1474 (2009)
16. Sherry, J., et al.: Making middleboxes someone else's problem: network processing as a cloud service. In: Proceedings ACM SIGCOMM, pp. 13–24 (2012)

Clustering Mixed Datasets by Using Similarity Features

Amir Ahmad[1]([envelope]), Santosh Kumar Ray[2], and Ch. Aswani Kumar[3]

[1] College of Information Technology, United Arab Emirates University, Al Ain, UAE
`amirahmad@uaeu.ac.ae`
[2] Department of Information Technology, Khawarizmi International College,
Al Ain, UAE
`santosh.ray@khawarizmi.com`
[3] School of Information Technology and Engineering, VIT University, Vellore, India
`aswanis@gmail.com`

Abstract. Clustering datasets consisting of numeric and nominal features is a challenging task as there are different similarity measures for numeric and nominal features. In the present paper, we propose a method to transform a mixed dataset to a numeric dataset. This method uses a similarity measure for mixed datasets and a randomly selected set of the data objects form the given mixed dataset and generate numeric similarity features. A clustering algorithm for pure numeric datasets is then applied on the newly generated numeric dataset to produce clusters. A comparative study with the other clustering algorithms demonstrated the superior performance of the proposed clustering approach.

1 Introduction

Clustering is a process to generate groups such that groups have similar data objects [14]. Many clustering algorithms are available for datasets consisting of only numeric features. Mixed datasets consist of both numeric and nominal features. Various clustering algorithms have been presented for these datasets [4]. However, the theoretical foundation of these algorithms are not as strong as that of clustering algorithms for numeric datasets. Attempts have been made to develop transformation techniques to generate a numeric dataset from a mixed dataset and then a clustering algorithm for numeric datasets is applied on this transformed dataset [4]. This approach has not been very successful as it is difficult to transform the mixed datasets without information loss. We present a transformation method to transform a mixed dataset to a numeric dataset. The proposed transformation method uses the technique developed to generate similarity features [5,6] with the similarity function for mixed datasets [1].

Following is the organization of the paper. Section 2 has related work. The proposed transformation method is presented in Sect. 3. Experiments and results are discussed in Sect. 4. Section 5 has conclusion and future directions.

© Springer Nature Switzerland AG 2020
P. Karrupusamy et al. (Eds.): ICSCN 2019, LNDECT 39, pp. 478–485, 2020.
https://doi.org/10.1007/978-3-030-34515-0_50

2   Related Work

In this section, we will discuss existing transformation techniques to covert mixed datasets to numeric datasets for clustering tasks.

1-in-q transformation is proposed [21] to transform a nominal feature to a numeric feature. The method increases the dimensionality of the data. The method considers that all the feature values have the same significance, which may not be the case. Spherical coordinates-based approach has been proposed to transform nominal features to numeric features [7]. Context-based approach has been used to transform mixed datasets to numeric datasets [22]. A mutual information-based transformation method has been suggested to convert nominal attribute to numeric attribute [23]. A fuzzy adaptive resonance theory-based method [8] has been proposed to transform mixed datasets to numeric features [17]. No theoretical results have been presented to show that the mixed dataset has been transformed into a numeric dataset with minimum information loss.

3   The Proposed Method

In the proposed approach, a mixed dataset is transformed to a dataset that has only numeric features. A clustering algorithm for the numeric dataset is applied on this new numeric dataset. The proposed clustering approach can be used with any clustering algorithm for numeric datasets. The proposed approach has two main components; the first is the methodology to convert a dataset to the other dataset using a similarity measure and the second is the similarity measure for the mixed datasets. We will discuss both of them in detail.

3.1   Similarity Features

A mapping is proposed to generate similarity features [6]. The mapping uses a kernel function and a set of data objects from the dataset. These data objects are selected randomly [6]. This mapping maintains the separability of the dataset for a given kernel.

Balcan *et al.* [6] suggest that using a similarity measure $K(x_i, x_j)$ between data objects x_i and x_j, a set of kernel features can be generated. To generate d features values of a data object x, they select d data objects $x_1, x_2, ..., x_d$ randomly, and use $K(x, x_i)$ (i = 1 to d) as the ith feature.

It is further suggested that the above approach can work with any similarity function [5]. In other words, a dataset can be transformed into another dataset by using a similarity function and some randomly selected data objects from the datasets with very small information loss. The steps of this approach are presented in Fig. 1.

Input- Dataset T with n data objects, K is a kernel function.
Select d data objects $x_1, x_2, ..., x_d$ randomly.
for i=1...n **do**
    For a data object x_i, the d similarity feature values will be $K(x_i, x_1)$, $K(x_i, x_2)$,...,$K(x_i, x_d)$.
end for
/*We have a new dataset with d similarity features*/

Fig. 1. The steps for creating kernel features.

3.2   Similarity Function for Mixed Datasets

Various similarity measures have been proposed for mixed datasets [4,10]. In the proposed method, we will use a dissimilarity measure that has been developed to compute the dissimilarity between two data objects of a mixed dataset [1]. This dissimilarity measure has been successfully used in various applications related with clustering [2,3].

This dissimilarity measure has two parts (Eq. 1). The first part is for numeric features and the second part is for nominal features. The Euclidean distance measure is used for numeric features. The important point is that the significance of a numeric feature is employed. Therefore, the contribution of each numeric feature is different in the dissimilarity measure. The significance of each feature is computed from the dataset. For the nominal features, instead of using the Hamming distance as the dissimilarity between a pair of feature values (0 if two feature values are the same, 1 they are different), the distances are computed from the datasets using co-occurrence based method. The dissimilarity, $S(x_i, x_j)$, between two objects x_i and x_j is defined in Eq. 1,

$$S(x_i, x_j) = \underbrace{\sum_{t=1}^{m_r}(w_t^r(x_{it}^r - x_{jt}^r))^2}_{numeric} + \underbrace{\sum_{t=1}^{m_c}(\Omega(x_{it}^c, x_{jt}^c))^2}_{nominal} \tag{1}$$

where w_{tt}^r is the weight of the tth numeric feature and x_{it}^r and x_{jt}^r are the values of the tth numeric feature of the ith data object and j^{th} data object respectively. Ω is the dissimilarity measure for a categorical attribute. x_{it}^c and x_{jt}^c are the values of the tth nominal feature of the ith data object and j^{th} data object.

For more detail of this similarity measure, readers are advised to refer the paper [1].

3.3   A Method to Transform Mixed Datasets to Numeric Datasets

Similarity features are numeric features. We can use the similarity function for mixed datasets [1] with the technique developed to create similarity features to generate numeric features for a given mixed dataset [5,6]. To generate m numeric features, we will select m data objects from the given mixed datasets, and use the method discussed in Fig. 1 to produce m numeric features. All the steps are presented in Fig. 2.

Input- A mixed dataset T with n data objects.
1- Select m data objects $x_1, x_2, ..., x_m$ randomly.
2- Calculate the distance between each pair of feature values for every nominal feature by using the method presented in [1]
3- Compute the weights of each numeric feature using the method presented in [1].
for i=1...n **do**
    For a object x_i, the m similarity feature values will be $S(x_i,x_1)$, $S(x_i,x_2)$,..,$S(x_i,x_m)$.
    /*Where S is the dissimilarity function discussed in Section 3.2 */
end for
/*We have a new dataset with m numeric features*/

Fig. 2. The steps for generating m numeric features for a mixed dataset

4   Experiments and Results

We created numeric datasets from mixed datasets using the proposed transformation method. Any clustering algorithm for numeric datasets can be applied on these datasets. In our experiments, we used the weighted k-means clustering algorithm [1]. k-means clustering algorithm works iteratively. In each iteration, cluster centres and cluster memberships are computed [14]. It has been shown that with weighted k-means clustering algorithm performs better than unweighted k-means clustering algorithm [1].

Four datasets were used in the experiment. Two of these datasets, Vote and Breast cancer, are pure nominal datasets, whereas two datasets, Heart (S) and Heart (C), are mixed datasets. The information on these datasets is provided in the Table 1. The number of the new numeric features was set to the number of the features in the dataset,

Table 1. Information on datasets.

Dataset	No. of data objects	No. of nominal features	No. of numeric features	No. of classes
Vote	435	16	0	2
Breast cancer	699	9	0	2
Heart (S)	270	7	6	2
Heart (C)	303	6	7	2

All these datasets have classes. These classes were used as the ground truth. In k-means clustering algorithm, the user has to decide the number of the desired clusters (the value of k). The number of classes was selected as the value of k. The mapping of clusters to classes was used to compute the clustering accuracy. The mapping is done such that the following clustering accuracy measure (AC) [1] was maximized;

$$AC = \frac{\sum_{i=1}^{K} G_i}{n} \tag{2}$$

Where G_i data objects in a cluster are correctly mapped to a class.

4.1   Results

Confusion matrix created using clustering results for Vote and Breast-Cancer datasets are provided in Tables 2 and 3 respectively. The comparative study with other state-of-the-art cluster algorithms ([11, 13, 16, 19, 20, 24]) is provided in the Table 4. Results show that the proposed clustering approach outperformed other cluster algorithms.

Confusion matrix created using clustering results for Heart (S) and Heart (C) are provided in Tables 5 and 6 respectively. The comparative study ([9, 12, 15, 18]) for mixed datasets presented in Table 7 suggests that the proposed clustering approach has the best performance.

Results demonstrate that the proposed clustering approach produces accurate clustering results. The accurate results indicate that the proposed transformation method is able to create numeric datasets from the nominal and mixed datasets with minimum information loss.

Table 2. Confusion matrix for Vote dataset.

Cluster no.	Number of republican	Number of democrat
1	155	36
2	13	231

Table 3. Confusion matrix for Breast-Cancer dataset.

Cluster no.	Number of benign	Number of malignant
1	447	14
2	11	227

Table 4. The comparative study for nominal datasets. "-" suggests that the result for the given clustering algorithm is not available in the literature. The numbers in bold represents he best clustering results.

Dataset	The proposed algorithm	[13]	[24]	[11]	[16]	[20]	[19]
Vote	**0.887**	0.497	–	–	0.850	0.869	0.856
Breast-cancer	**0.964**	0.836	0.911	0.911	0.913	0.933	-

Table 5. Confusion matrix for Heart (S) dataset.

Cluster no.	Number of Normal	Number of Heart Patient
1	132	23
2	18	97

Table 6. Confusion matrix for Heart (C) dataset.

Cluster no.	Number of Normal	Number of Heart Patient
1	140	26
2	24	113

Table 7. The comparative study for nominal datasets. "-" suggests that the result for for the given clustering algorithm is not available in the literature. The numbers in bold represents he best clustering results.

Dataset	The proposed algorithm	[12]	[15]	[18]	[9]
Heart (S)	**0.848**	0.770	–	–	0.814
Heart (C)	**0.834**	0.772	0.808	0.752	0.831

5   Conclusion

In the paper, we presented a method to transform a mixed dataset to a numeric dataset for the clustering task. A clustering algorithm for pure numeric dataset is then used on the generated dataset to create clusters. A detailed comparative study was carried out to show that the proposed clustering approach outperformed other clustering algorithms for nominal datasets. The similar superior performance was observed for mixed datasets. The excellent performance of the proposed clustering approach demonstrates that the proposed transformation technique is successful in transforming the mixed dataset into pure numeric datasets with minimum information loss.

In future, we will study the performance of other clustering algorithms for numeric datasets with the proposed transformation method.

References

1. Ahmad, A., Dey, L.: A k-mean clustering algorithm for mixed numeric and categorical data. Data Knowl. Eng. **63**(2), 503–527 (2007)
2. Ahmad, A., Dey, L.: A k-means type clustering algorithm for subspace clustering of mixed numeric and categorical datasets. Pattern Recogn. Lett. **32**(7), 1062–1069 (2011)
3. Ahmad, A., Hashmi, S.: K-harmonic means type clustering algorithm for mixed datasets. Appl. Soft Comput. **48**(C), 39–49 (2016)
4. Ahmad, A., Khan, S.S.: Survey of state-of-the-art mixed data clustering algorithms. IEEE Access **7**, 31883–31902 (2019)
5. Balcan, M.F., Blum, A.: On a theory of learning with similarity functions. In: Proceedings of the 23rd International Conference on Machine Learning (2006)
6. Balcan, M.F., Blum, A., Vempala, S.: Kernels as features: on kernels, margins, and low-dimensional mappings. Mach. Learn. **65**, 79–94 (2006)
7. Barcelo-Rico, F., Jose-Luis, D.: Geometrical codification for clustering mixed categorical and numerical databases. J. Intell. Inf. Syst. **39**(1), 167–185 (2012)
8. Carpenter, G.A., Grossberg, S., Rosen, D.B.: Fuzzy art: fast stable learning and categorization of analog patterns by an adaptive resonance system. Neural Netw. **4**(6), 759–771 (1991)
9. Cheung, Y.M., Jia, H.: Categorical-and-numerical-attribute data clustering based on a unified similarity metric without knowing cluster number. Pattern Recogn. **46**(8), 2228–2238 (2013)
10. Foss, A.H., Markatou, M., Ray, B.: Distance metrics and clustering methods for mixed-type data. Int. Stat. Rev. **87**(1), 80–109 (2018)
11. He, Z.: Farthest-point heuristic based initialization methods for k-modes clustering. CoRR, abs/cs/0610043 (2006)
12. Huang, Z.: Clustering large data sets with mixed numeric and categorical values. In: Proceedings of the First Pacific Asia Knowledge Discovery and Data Mining Conference, pp. 21–34. World Scientific, Singapore (1997)
13. Huang, Z.: A fast clustering algorithm to cluster very large categorical data sets in data mining. In: In Research Issues on Data Mining and Knowledge Discovery, pp. 1–8 (1997)
14. Jain, A.K., Dubes, R.C.: Algorithms for Clustering Data. Prentice Hall, Upper Saddle River (1988)
15. Ji, J., Pang, W., Zheng, Y., Wang, Z., Ma, Z., Zhang, L.: A novel cluster center initialization method for the k-prototypes algorithms using centrality and distance. Appl. Math. Inf. Sci. **9**(6), 2933 (2015)
16. Khan, S.S., Ahmad, A.: Cluster center initialization algorithm for k-modes clustering. Expert Syst. Appl. **40**(18), 7444–7456 (2013)
17. Lam, D., Wei, M., Wunsch, D.: Clustering data of mixed categorical and numerical type with unsupervised feature learning. IEEE Access **3**, 1605–1613 (2015)
18. Li, C., Biswas, G.: Unsupervised learning with mixed numeric and nominal data. IEEE Trans. Knowl. Data Eng. **14**(4), 673–690 (2002)
19. Lin, S., Azarnoush, B., Runger, G.: CRAFTER: a tree-ensemble clustering algorithm for static datasets with mixed attributes and high dimensionality. IEEE Trans. Knowl. Data Eng. (in Press)
20. Jiang, F., Liu, G., Du, J., Sui, Y.: Initialization of k-modes clustering using outlier detection techniques. Inf. Sci. **332**(C), 167–183 (2016)

21. Modha, D.S., Spangler, W.S.: Feature weighting in k-means clustering. Mach. Learn. **52**(3), 217–237 (2003)
22. Wang, C., Chi, C., Zhou, W., Wong, R.: Coupled interdependent attribute analysis on mixed data. In: Proceedings of the Twenty-Ninth AAAI Conference on Artificial Intelligence, AAAI 2015, pp. 1861–1867 (2015)
23. Wei, M., Chow, T.W.S., Chan, R.H.M.: Clustering heterogeneous data with k-means by mutual information-based unsupervised feature transformation. Entropy **17**(3), 1535–1548 (2015)
24. Wu, S., Jiang, Q., Huang, J.Z.: A new initialization method for clustering categorical data. In: Zhou, Z.-H., Li, H., Yang, Q. (eds.) Advances in Knowledge Discovery and Data Mining, Berlin, Heidelberg, pp. 972–980. Springer, Heidelberg (2007)

Analysis of Consumer Reviews by Machine Learning Techniques

Sourav Sinha$^{(\boxtimes)}$ and M. Sandhya

Department of Computer Science and Engineering, B. S. Abdur Rahman
Crescent Institute of Science and Technology, Chennai, India
`sourav_cse_phd_17@crescent.eudcation`

Abstract. The dynamic change in the usage of digital economy had drifted the customers towards using online shopping websites for their day to day purchase. Under such circumstances, customer reviews and ratings had been used by these online stores to generate trustworthy and esteemed brand in the digital market. Most of these reviews were unstructured text which had been widely used for sentiment analysis. This paper discusses about various machine learning algorithms proposed by different researchers to analyze the consumer reviews. Based on the existing work, we categorized three broad aspects in consumer reviews namely opinion mining, spam or fake review detection, collaborative filtering. In future work, we plan to propose customer review-oriented decision support system (CRDSS) which can help in consumer decision making process and thus improve the customer review helpfulness and rating prediction.

Keywords: eCommerce · Consumer review · Sentiment analysis · Machine learning

1 Introduction

Consumer reviews are of growing importance in everyday business activities in electronic commerce and retail industry under present scenario of internet dominated world. Customers are being motivated to enter their feedbacks voluntarily with most of the online e-commerce and shopping websites namely amazon, flipkart, snapdeal etc. These reviews are being used by other buyers to make their purchase both online through internet and offline from the retail stores. Most of these reviews are appropriate as well as helpful by other buyers while a few of them are being neglected for biased orientation in delivering the opinion.

According to the survey being conducted by Power-Reviews a leading cloud software-based service provider of review and ratings to global brands and retailers had conducted a study in 2018, 97% of consumers refers to the product reviews while 85% of consumers checks for the negative reviews before buying a product. Majority of them goes through the first ten reviews before making a purchase. About 50% of customers write their reviews for products which they have purchased as per 2018 with an increase from 42% in 2014 [16].

India is growing rapidly in eCommerce sector and will surpass US e-Commerce industry becoming the second largest Market in eCommerce sector in the world by

© Springer Nature Switzerland AG 2020
P. Karrupusamy et al. (Eds.): ICSCN 2019, LNDECT 39, pp. 486–496, 2020.
https://doi.org/10.1007/978-3-030-34515-0_51

2034. The e-Commerce Market revenues is expected to reach US $200 billion by 2027 as against US $38.5 billion in 2017. Online shoppers in India is likely to reach 220 million by 2025 as against 120 million in 2018 [17]. The Government of India's policies and its regulatory frameworks such as 100% foreign direct investment (FDI) in B2B e-commerce and 100% FDI under automatic route under the market place model of B2C e-commerce will catalyze the growth in e-commerce sectors [17].

As per the market share survey conducted by forrester an American market research company in 2017, the market share of various ecommerce sites namely flipkart, amazon, paytmmall, snapdeal, shopclues, pepperfry, bigbasket has been reported as 39.5%, 31.1%, 5.6%, 2.5%, 2.2%, 1.8%, 1.5% respectively [18].

Nowadays easy access to internet on getting public opinion reviews with effective utilization of technological advent as well made it possible to generate terabytes of data primarily in the form of unstructured data/text [21]. Reviews available on those e-commerce platforms might have played the role of guiding tool for the consumers on making informal decisions. However, it is more challenging task for the consumer to read through all the reviews for concrete knowledge extraction on their respective product. On the contrary selective readings of the product reviews may lead towards biased outlook on the product without proper consideration for its merits and demerits. Hence the needs are felt on emergent manner to make this unstructured data/text into some developed structured format with topic labeling under such instances [21]. To extract valuable thoughts from a large set of reviews, classification of reviews in positive and negative sentiments is needed. Segregation of reviews based on their sentiments can help buyers to establish feedback as merits and demerits in a constructive manner and reach to a better decision.

Customer review and rating are important measures which influence product buying decisions [19, 20]. As per the existing study, the product review can be helpful to the other consumers, but rating has its own limitations when it comes to comparison of different product reviews [19]. The reliability of the customer reviews is also an emerging challenge in the e-Commerce industry. The study also explains that there is breach between product review and its rating which heavily affects the productivity of decision-making process of the buyer [19]. The next portion in this paper contains the literature survey followed by conclusion and future work.

2 Literature Survey

Customer reviews from various sources are generally unstructured text of data [21]. Feature extraction and knowledge discovery is most widely being performed on this review comments to infer the consumers buying behavior patterns and their current emerging interest on marketed product and services. These features are being widely used along with various machine learning algorithms for further knowledge extraction and mining of these customer reviews.

An extensive literature review was performed on the field of analyzing various techniques being applied to analyze the consumer reviews and we broadly categorized the existing work on the consumer reviews based on machine learning techniques and methodology adopted:

(A) Use of opinion mining/sentiment analysis for Review Helpfulness/Classification
(B) Bias, Spam and Fake review detection from consumer reviews
(C) Collaborative Filtering in Recommendation Systems based on consumer reviews and rating.

We have summarized the brief overview of the related work based on various techniques used for the above-mentioned categories as presented below in tabulated form (Tables 1, 2 and 3).

Table 1. Techniques used in Sentiment Analysis for review helpfulness/classification

Ref no	Objective	Dataset & technology used	Outcome	Salient features	Conclusion/future work
[1]	To provide with feature-oriented opinion based on customer review for online products	Dataset used is obtained from amazon.com for 3 products - Nokia 6610, Canon G3 and DVD player Author uses two sentiment analysis system for comparison namely Recursive Deep Model & Senti-wordnet	Result gave high accuracy score when compared with Extended Page Ranking System (EXPS)	Recursive Deep Model is being implemented in order to explore the orientation of sentiments for reviews A review matrix has been computed to identify importance of product feature	Author presents with various feature extraction methods for products based on customer review The future work will be to improve the feature extraction by considering the implicit product feature and using machine learning technique
[2]	Categorize positive & negative reviews for different products and develop a supervised model to polarize reviews	Data used is taken from amazon product reviews based on three categories namely Electronics, Cellphone & Accessories and Musical Instruments In this paper, pool based active learning process is being used to reduce time complexity, as manual labeling of dataset is an difficult task. Author uses pipeline approach to apply TF-IDF, Chi-Square and another classifier	Machine learning algorithms like Support Vector Machine, Naive Bayesian, Stochastic Gradient Descent, Random Forest, Logistic Regression and Decision Tree being used for classification Various Information Retrieval metrics namely Accuracy, Precision, Recall and F1 Measure is being calculated to check performance of classification and system was evaluated with more than 90% accuracy	Pool Based Active Learning which is a semi-supervised learning approach is being used to label the datasets using oracle data source K-fold cross validation (resampling) procedure is being used to improve the accuracy and in this paper k = 5 and 10 are being used in the result for comparison and it was observed that 10-fold provides high accuracy	Author concludes that SVM classifier providing better classification result and high accuracy using 10-fold Future work will be to apply principal component analysis (PCA) for active learning process to automate labeling with less assistance of oracle data source
[3]	To Analyze and predict the open opinion customer reviews using probability-based classifier model	Data used is from Thai customer review's hotel from website hotel reservation service	Result was being compared between decision tree classifier and naive Bayesian classifier	An Advantages using naïve bayesian classifier is the ease of processing of data using	Paper explains the advantages of decision tree & naive Bayesian (NB) classifier. In NB classifier, system can summarize the rating

(continued)

Table 1. (*continued*)

Ref no	Objective	Dataset & technology used	Outcome	Salient features	Conclusion/future work
		Feature selection is performed based on frequent occurring positive and negative words in set of 10, 20, 36 respectively	Average accuracy of naive tree classifier was higher than decision tree classifier	bayes theorem. Probability value ranges from 0 to 1 Advantages using decision tree is due to ordering of its attributes. Information Gain is being used to classify data and its range is from 0 to 1. Highest value of gain is being for classification	which is useful for consumers to take decision rapidly Future Work is to automate the word extraction from sentence using machine learning to address different sentence polarity problem
[4]	To identify positive/negative sentiments from the Amazon's Product Review dataset	Data used is extracted from amazon product reviews Lexicon approach with n-gram along with 3 other machine learning approach namely the random forest learner with word-vector, random forest learner with n-gram, decision tree learner with document vector is studied and compared in this paper	Result presents that lexicon dictionary-based approach outperforms the other machine learning algorithms due to use of dictionary in accordance with the dataset	1. Lexicon Dictionary based Approach uses dictionary also known as opinion lexicon obtained online using wordnet 2. Random Forest with word vector uses random forest learner along with word vector to form a dictionary. Principal 3. Component Analysis is used to discard the irrelevant component Random Forest with N - grams approach is extension in which n-grams of word is used as vocabulary builder instead of word vector	This paper presents with relative study between lexicon dictionary-based approach and other machine learning approaches As part of future work, authors suggest giving different weights to reviews based on helpful parameter. Paper also suggests predicting the popularity of the product for a given time
[5]	To build an analysis system of e-commerce reviews based on Hadoop software framework for big data processing	Dataset used is a commodity review text from the chocolate brands Naive Bayesian (NB) sentiment classifier has been used along with map reduce parallel computing theory on hadoop platform	The experimental result suggest that accuracy rate of the classifier is more than 85% which indicates that system is effective in predicting the emotional tendency of review texts	Emotion Tendency classifier is based on NB classification algorithm and the results from the classifier is being integrated with map-reduce computing module to achieve parallel processing for emotional classification	Analysis of large volume of data from unstructured review comments helps merchants to alter its sales and marketing strategies Following Limitations needs to be addressed 1. Only NB classifier has been used, it cannot retrieve the attribute characteristics when text is large 2. Hardware limitations is a concern when large scale data is to be used on hadoop Platform

(*continued*)

Table 1. (*continued*)

Ref no	Objective	Dataset & technology used	Outcome	Salient features	Conclusion/future work
[6]	Unsupervised learning algorithms namely k-means clustering and peak searching clustering to group similar reviews	Data used is taken from amazon.com for the product (8" black Samsung Galaxy Tab 4) Before clustering the input data was converted to vector form using TF-IDF. Two flat clustering algorithms namely K-means clustering, & peak-searching clustering was applied	Two evaluation methods were used for evaluating outcome evaluation using Human Assessors Evaluation using purity 1.Based on human assessors, k-means clustering performed slighter better in terms of clustering of topics 2.Based on evaluation of purity, neither k-means nor peak search clustering algorithm performed better in terms of purity score	K-Means Algorithm used a set of input data and k set of desired output cluster. The approach used selected the seeds randomly Peak-Searching (PS) Algorithm used two parameters namely pre-processed set of reviews & weighted adjacency matrix which was created by counting number of terms in two document r_i A & r_j	Paper concludes that k-means clustering performs slightly better than peak-searching and clustering was done based on the major topic for each review Future work is to include other clustering algorithms like DBSCAN, k - Medoids & Fuzzy C-Means clustering
[7]	Use of deep neural network to process large scale of data & propose framework for sentiment analysis using deep learning	Dataset used is obtained from 3 sources of raw review data: (a) electronic products review from amazon.com (b) movie reviews from amazon.com and IMDB (c) Hotels review from TripAdvisor Hierarchical Deep Neural Network (DNN) framework and selection of hidden layer and nodes in DNN are proposed in this paper	Experiment was done with DNN and two other classical classifiers namely Naïve Bayesian and SVM with 80% of training data Result as per paper shows that accuracy of DNN become better as the dataset and number of the dimensions increases in size	A. In the proposed framework, feature vectors were constructed using 3 kinds of linguistic knowledge: 1. Feature based on word frequency 2. Feature based on contextual window 3. POS Tagging Feature B. In proposed HDNN, first neural network is being constructed based on feature vector to reduce dimensions and second target number of dimensions for independent feature vector is being set and subnets are constructed. Training separate sub-network for each feature vector in parallel improves the performance and reduces training time	Deep Neural Network has been used to address the high dimensional data problem and provides good results on dataset from different domains Future work is to apply the framework of sentiment analysis from document to aspect level to improve the system

(*continued*)

Table 1. (*continued*)

Ref no	Objective	Dataset & technology used	Outcome	Salient features	Conclusion/future work
[8]	To propose a lazy supervised learning model which can classify the given review sentences into two different classes namely – Subjective/Objective	Dataset consists of 5 different electronic product (i.e.) 2 digital camera, 1 cellular phone, 1 mp3 player & 1 dvd player extracted from Amazon review website Author proposes with a lazy supervised learning using k-nearest neighbor (k-NN) algorithm to classify the review sentences	Average accuracy of the proposed model using K-NN classifier for the Training datasets is 81.37% (±0.77%) Proposed method shows better results in terms of the average precision, recall and f-score	Paper suggests that most common used opinion words are generally adjectives (JJ) or adjectives (JJ) along with adverbs (AB). Only few of opinion bearing words are verbs and nouns Review are transformed into numeric vector to detect if the given word is a subjective or an objective Senti-wordnet is being used to detect opinion polarity of sentences using polarity score obtained from these opinion words	Paper presents with a lazy supervised learning approach using syntactic rules for linguistic patterns identification and subjectivity/objectivity analysis Future work is to include the implicit features for improving the existing approach and also determine the strength of the opinions
[9]	To propose two level hierarchical hidden Markov model (HHMM) for product feature extraction	Dataset used has been taken from Amazon reviews for three different kinds of digital products	In this paper, author proposes an approach based on HHMM to extract product features & also implements steps to improve the viterbi algorithm Experimental results for given proposed approach in this paper shows that precision and recall average value can reach 90% and also proves that HHMM can obtain high accuracy in product feature extraction	Paper presents with approach in which segment tags to divide comment text into Feature and Non-Feature Contained segment in HHMM1 and product feature in Non-Feature Contained segment in HHMM2 Method proposed in HHMM1 only extract features from Feature Contained segment and ignore the Non-Feature Contained segment to avoid time wastage	Proposed two level HHMM approach is more innovative and have high accuracy as compared to traditional extraction method based on probability Future work will be to improve the proposed model by marking training set artificially, use of nouns, adjectives and adverbs and ignore other part of speech (POS)

Table 2. Techniques used for Bias, Spam & fake review detection of consumer reviews

Ref no	Objective	Dataset & technology used	Outcome	Salient features	Conclusion/future work
[10]	To detect review spam using ensemble learning by combining active and supervised learning	Dataset used is gold standard dataset for labeled data and for unlabeled data, it uses the yelp academic dataset challenge Author proposes an ensemble approach named as Hybrid approach to Detect Review Spam (HDRS)	Best Result was achieved using naïve Bayesian classifier by using the 10-fold cross validation According to the paper, naïve Bayesian (NB) classifier using Bi-gram features had precision of 95%, accuracy of 88%, recall of 95% and f-measure score of 94.99%	HDRS model used in this paper uses three steps to detect spam review: 1. Detecting duplicate reviews 2. Creating Hybrid dataset for active learning 3. Ensemble method applied to detect fake reviews	Linear SVM was best in active learning model NB classifier model was over best among all classifier using the supervised learning model Future work is to experiment on larger dataset from different domains from different languages
[11]	Identify the spam reviews based on opinion spam detection's identification indicators by using spammer behavior features	Dataset used mobile phone reviews obtained from amazon.com Author proposes below algorithms to identify spam reviews: 1. Conversion matching algorithm (CMA) obtains conversion times from two reviews 2. Non-recursive LCS algorithm to detect longest common subsequence a. Calculate Similarity algorithm measures review similarity b. Word segmentation algorithm to retrieve relevancy between topic and review of product	Experiment contain 3 parts: 1. Similarity Experiment As per paper, proposed similarity algorithm performs more pruning than traditional methods Recursive LCS algorithm performance was very not good to applied on large data processing 2. Relevancy Experiment Proposed algorithm detects 46% of the irrelevant mobile phone buyer reviews Approximately 3. Human Evaluation As per paper, relevancy algorithm detected 28% more fake reviews on comparing with traditional algorithm	Paper describes eight spammer behavior features namely 1. Star User 2. Deviation Rate 3. Bias Rate 4. Review Similarity Rate 5. Content Length 6. Illustration 7. Burst review Relevancy between the topic and review of products is being graded by ranking in range of 0 to 3. Rank 0 being irrelevant and Rank 3 being completely relevant	Paper presents with eight spammer identification indicators to detect product spam review and present with two algorithms to check for similar and relevant reviews The Future work includes improving relevancy algorithm to find more irrelevant reviews
[12]	To propose a partial supervised model to identify the spammer groups	Dataset used is collected from Amazon China Author proposes a partially supervised model (PSGD) where PU labeling has been used for reliable negative set extraction and train semi supervised classifier on naïve Bayesian model & Expectation - Minimization (EM) algorithm	The result in this paper suggests that proposed partially supervised learning model performs effectively on spammer group detection	PSGD model uses the frequent item mining (FIM) to discover spammer group candidate from the review data PSGD model uses the NB model and EM algorithm on positive, negative and neutral instances to build classifier as spammer group detector	Future Work will be to use classification models like neural network, semi-supervised SVM & ensemble techniques to improve the performance of PSGD model

Table 3. Technique used for Collaborative Filtering/Recommendation based on user reviews

Ref no	Objective	Dataset & technology used	Outcome	Salient features	Conclusion/future work
[13]	To propose a hybrid collaborative filtering recommender system that enhances the traditional collaborative filtering by user free text reviews	Dataset used is obtained by extraction of user ratings and reviews from Internet Movie Database (IMDB) Author proposes two hybrid similarity collaborative filtering approach namely 1. User Hybrid Similarity Collaborative Filtering (UHSCF) 2. 2. Item Hybrid Similarity Collaborative Filtering (IHSCF)	Author analyzed the review through topic modeling and then result are being combined with UCF and ICF to improve the recommender system Root Mean Square Error (RMSE) was used for measuring the performance and as per result from the paper, UHSCF outperforms IHSCF based on the given dataset	1. User-based collaborative filtering (UCF) is being used to predict the rating based on previous ratings by other user & Item based collaborative filtering (ICF) is being done on items 2. User Topic Feature based collaborative featuring (UTFCF) is measure of topic review similarity from different users 3. Item Topic based collaborative Featuring (ITFCF) is like UTFCF done on items 4. UHSCF, IHSCF are the hybrid of UCF and UTFCF, ICF and ITFCF respectively	Experiment done by the author shows that UHSCF and IHSCF bring significant improvement in accuracy of the recommendation Future work is to test the algorithms on dataset which is good for item-based algorithms
[14]	To present with novel approach to enhance standard collaborative filtering (CF) model by utilizing LDA (Latent Dirichlet Allocation) to learn latent properties from textual description	Dataset used is taken from Movielens 1 M and Netflix datasets Author proposes Hybrid user neighborhood-based recommender (HUNR) to improve the quality of the recommendations	Precision Analysis shows that both standard Item CF performed, and standard user CF performed the worst. Proposed method HUNR performs the best Recall Analysis indicates that HUNR outperforms other approaches used	HUNR approach utilizes LDA to obtain the latent properties of items and then compute the hybrid user similarity score and using their rating overlap-based similarity to refine neighborhood formation	Author proposed a hybrid recommender approach using LDA which enhances the neighborhood formation and query recommendations Future work is to implement taxonomy induction approach to improve the quality in recommendation
[15]	To propose collaborative filtering approach to infer the numeric rating from text reviews	Dataset used from the Amazon instant video data with total of 37126 reviews for 1686 videos User Review Incorporated Collaborative filtering (URICF) algorithm was proposed which takes user item matrix and user reviews as input and the output is recommendation for video which is likely to watched by active user	Experiment was performed to measure the performance 1. Mean Absolute Error (MAE) was computed when neighbor size is changing. Proposed URICF performs slightly better than existing UCF (user based collaborative filtering) 2. URICF performs better than UCF in terms of similarity measure like cosine similarity and Pearson correlation	Rating inference process used in this paper contains 5 important steps: 1. 1. Data Acquisition 2. 2. Gauging Reviews 3. 3. Review Analysis 4. 4. Constructing Opinion Dictionary 5. 5. Numeric Rating Inference	Paper concludes that numeric rating computed from the user reviews are better than the rating provided by user Future work is to include feature engineering to improve the accuracy of recommendation process

3 Conclusion

In current digital era, there is an emerging need for analysis of the consumer review not just by the buyer but also for the manufacturer. The consumer review helps to influence the decision-making process which in turn drives the sales and marketing of the product. The objective of this paper was to investigate in detail about the different existing machine learning techniques being used for the evaluation of consumer reviews for its helpfulness.

In this literature survey, we classified the existing machine techniques being applied on different consumer reviews into three broad domains namely opinion mining/sentiment analysis for review helpfulness& rating prediction, Bias, spam and fake review detection, collaborative filtering for recommendation system.

During the survey of existing work, it was studied that most of them used supervised learning approach to classify and summarize the reviews using labeled data and naive Bayesian (NB) classifier has been most eminently used classifier for training and testing the proposed model. Certain researchers had suggested use of ensemble approach to take advantage of best classifier using majority voting mechanism for mining of opinions and fake review detection. Few of the researchers have proposed framework using the active and semi-supervised learning for training and testing their system with minimal training data. But as the volume of data in terms of number of customer reviews increases, performance is a serious concern for the storage, processing and analysis of these consumer reviews. Deep learning is an evolving machine learning technique simulating the neural network present in human brain which is capable to process data with large dimensions and also support parallel processing. Hence there is a shifting paradigm towards use of deep learning algorithm to design and implement an efficient and effective process for mining of these consumer reviews and achieve high performance results on large-scale data.

4 Future Work

The current trends with the substantial growth in number of online post transaction reviews in the e-Commerce industry, leads to an increasing demand for designing an effective and efficient Decision Support System which can influence the consumer buying decision and increases the sales and productivity for the manufacturer. The system should be able to extract the desired features which are specified by the consumers and these desired product features are to be communicated to the manufacturer to incorporate it in future product release. Thus, the proposed system will help the manufacturers to promote their products with these customer recommended features in the market. As a result, it will help to build a strong bonding between consumer and manufacturer.

In our future work, we plan to propose a consumer review-based Decision Support System (CRDSS) to address the consumer's and manufacturer's needs. We will also use the deep learning approaches such as Convolution Neural Network (CNN) combined with Long Short-Term Memory (LSTM) neural network to intelligently learn and analyze the consumer reviews.

References

1. Devasia, N., Sheik, R.: Feature extracted sentiment analysis of customer product reviews. In: International Conference on Emerging Technological Trends (ICETT), pp. 1–6. IEEE (2016)
2. Haque, T.U., Saber, M.N., Shah, F.M.: Sentimental analysis on large scale Amazon product reviews. In: International Conference on Innovative Research and Development, pp. 1–6. IEEE (2018)
3. Songpan, W.: The analysis of prediction of customer review rating using opinion mining. In: IEEE 15th International Conference on Software Engineering Research, Management and Applications (SERA), pp. 70–77 (2017)
4. Ejaz, A., Turabec, Z., Rashmi, M., Khoja, K.: Opinion mining approaches on Amazon product reviews a comparative study, pp. 173–179. IEEE (2017)
5. Zu, Q., Wu, J.: Big data analysis of reviews on e-commerce based on hadoop. In: HCC 2017. LNCS, vol. 10745, pp. 492–502. Springer (2018)
6. Fry, C., Manna, S.: Can we group similar Amazon reviews-a case study with different clustering algorithm. In: IEEE Tenth International Conference on Semantic Computer, pp. 324–329. IEEE (2016)
7. Hu, Z., Hu, J., Ding, W., Zheng, X.: Review sentimental analysis based on deep learning. In: IEEE 12th International Conference on e-business Engineering, pp. 87–94 (2015)
8. Khan, J., Jeong, B.S.: Summarizing customer review based on product feature and opinion. In: Proceedings of 2016 International Conference on Machine Learning and Cybernetics, pp. 158–165. IEEE (2016)
9. Wang, X., Lu, Z.: Extracting product features from online reviews based on two-level HHMM, pp. 1–4. IEEE (2014)
10. Ahsan, M.N.I., Nanian, T., Kofi, A.A., Hossain, M.I., Shah, F.M.: An ensemble approach to detect review spam using hybrid machine learning techniques. In: 19th International Conference on Computer and Information Technology, pp. 388–394. IEEE (2016)
11. Liu, P., Xu, Z., Ai, J., Wang, F.: Identifying indicators of fake reviews based on spammer's behavior features. In: IEEE International Conference on Quality Reliability and Security (Companion Volume), pp. 396–403 (2017)
12. Zhang, L., Wu, Z., Cao, J.: Detecting spammer group from product reviews: a partially supervised learning model. IEEE Open Access J. **6**, 2559–2568 (2012)
13. Wang, H., Lue, L.: Collaborative filtering enhanced by user free text reviews topic modelling. In: International Conference on Information and Communications Technologies, pp. 1–5. IEEE (2014)
14. Wilson, J., Chaudhury, S., Lall, B.: Improving collaborative filtering based recommenders using topic modelling. In: IEEE/WIC/ACM International Joint Conference on Web Intelligence (WI) and Intelligent Agent Technologies (IAT), pp. 340–346 (2014)
15. Mahadevan, A., Arock, M.: Credible user review incorporated collaborative filtering for video recommendation system. In: Proceedings of ICISS 2017 IEEE Xplore Compliant, pp. 375–379. IEEE (2017)
16. https://www.powerreviews.com/events/consumers-depend-on-reviews/. Accessed 02 Apr 2019
17. https://www.ibef.org/industry/ecommerce-presentation. Accessed 02 Apr 2019
18. https://economictimes.indiatimes.com/small-biz/startups/newsbuzz/amazon-india-reportedly-closing-in-on-flipkart/articleshow/63407725.cms. Accessed 02 Apr 2019
19. Lackermair, G., Kailar, D., Kanmaz, K.: Importance of online product reviews from a consumer prospective. Adv. Econ. Bus. **1**(1), 1–5 (2013)

20. Manuel, T., Benedikt, B.: Analyzing online customer reviews – an interdisciplinary literature review and research agenda. In: Proceeding of the 21st European Conference on Information Systems, pp. 1–12 (2013)
21. Ahmad, T., Doja, M.N.: Opinion mining using frequent pattern growth method from unstructured text. In: International Symposium on Computational and Business Intelligence, pp. 92–95. IEEE (2013)

Fuzzy Based Latent Dirichlet Allocation in Spatio-Temporal and Randomized Least Angle Regression

D. Nithya$^{(\boxtimes)}$ and S. Sivakumari

Department of Computer Science and Engineering, School of Engineering,
Avinashilingam Institute for Home Science and Higher Education for Women,
Coimbatore, India
nithya.apcse@gmail.com, prof.sivakumari@gmail.com

Abstract. Due to the emerging growth of Social Networks and web blogs, several news providers share their news articles on different web sites and web blogs. It is also used to get public opinion about the articles. Twitter is one of the popular microblogs which act as intermediate for publics to distribute their thoughts. Our intention is to find Twitter data related to the news in the web news articles are also used to enhance the performance of Evolving Fuzzy System-Penguins Search Optimization Algorithm (EFS-PeSOA) based web news mining. In this paper, Latent Dirichlet Allocation (LDA) is used to model the topics within the text of tweets. Twitter-specific tokenizer, part-of-speech tagger, and snowball stemmer are used to generate terms from the twitter data. Term-Frequency and Inverse-Document-Frequency (tf-idf) of each term in tweets are calculated along with the terms in web news articles for the creation of evolving fuzzy rules. Based on the evolving fuzzy rules, web news articles are categorized. In order to enhance the efficiency of categorization of web news articles, a Spatio-Temporal Generalized Additive Model (STGM) is developed where the spatial and temporal information of the tweets are considered for categorization. However, it generates different terms in tweets. So a Randomized Least Angle Regression (RLAR) is used to choose the most significant terms in tweets and only the selected term's tf-idf values are used in EFS for categorization of web news articles.

Keywords: Web news articles · Evolving Fuzzy System ·
Latent Dirichlet Allocation · Spatio-Temporal Generalized Additive Model ·
Randomized Least Angle Regression

1 Introduction

Text classification [1, 2] is the process of assigning a text document to one or more decided categories. This will help the users to find desired information easily by searching only the relevant categories. The importance of text classification is even more apparent when the information space is large. Because of increasing the rate of web pages in the World Wide Web (WWW), the classification of text documents is more difficult. So, machine learning methods have been introduced for automating the

© Springer Nature Switzerland AG 2020
P. Karrupusamy et al. (Eds.): ICSCN 2019, LNDECT 39, pp. 497–508, 2020.
https://doi.org/10.1007/978-3-030-34515-0_52

classification process. Online news articles represent a type of web information that is frequently referenced. Nowadays, online news is provided by many dedicated news-wires such as PR Newswires and Reuters. It will be useful to gather news from newswires sources and classify the news accordingly for ease reference.

A method based on Evolving fuzzy system [3] was proposed to categorize the web news articles according to the text content of the articles. The most relevant terms of each article were obtained by using tokenization, stopword elimination and stemming processes. Based on the relevance value of each term, a large number of terms were pruned by a user-specified pruning threshold. Then, evolving fuzzy rules were created based on the tf-idf value of terms. A Gaussian membership function was utilized in EFS to define the nearness of the article to the prototype. However, the problem of selection of the pruning threshold was solved by using a Penguins Search Optimization Algorithm (PeSOA) [4]. Instead of Gaussian membership function, a bell-shaped membership function was utilized to describe the nearness of the article to the proto-type. Because sometimes the Gaussian membership function is hard to justify. Clas-sification accuracy of web news mining is improved by utilizing PeSOA.

In this paper, the web news mining based on EFS is improved by analyzing the Twitter data which are related to the news articles. The topics of the Twitter data are modeled by LDA [5] model. Then the terms in tweets are generated by using twitter-specific tokenizer, part of speech tagger and stemming. The Term Frequency and Inverse Document Frequency (tf-idf) are calculated for each generated terms and prune the terms which have tf-idf value lesser than the optimized pruning threshold. Then, Evolving Fuzzy Rules are created based on the tf-idf of web news article terms and tf-idf of twitter terms which effectively categorize the web news articles. In addition to this, a Spatio-Temporal Generalized Additive Model (STGM) is developed to analyze the Twitter terms based on user's location and time when they tweet about the news. It generates different terms, so the most important terms are selected by using RLAR. Thus, the tf-idf value of the most important twitter terms is used in the creation of evolving fuzzy rules which improves the web news articles categorization accuracy.

2   Literature Survey

A news article classification [6] method was proposed to classify the news articles based on the vector representation of articles. The pre-processing process was enhanced by introducing words which reduced the word space without compromising the information value. A vector representation was introduced to improve the classification accuracy and reduce the computational complexity of the classification process. After the preprocessing process, the news articles were classified using Naïve Bayes, K-Nearest Neighbor (K-NN) and decision trees.

A Support Vector Machine (SVM) [7] was introduced for twitter news classifica-tion. Initially, small messages were taken out from the Twitter microblog by choosing different active newsgroups. Then the short messages were manually classified into 12 groups. SVM got words of each small message as features and it was trained by using the manually classified news data. Based on the trained data, the SVM classified the news into groups. The disadvantage of using the SVM algorithm is that it needs more memory for classification in many cases and involves high complexity.

Based on random forests and multimodal features, a news article classification framework [8] was introduced. In this framework, both visual features and textual features were used for category-based classification of news articles. The textual features and visual features were taken out from the textual part and image of the news articles respectively. Then, these two textual and visual features were combined using a late fusion strategy. Random forest classifier got fused features as input and classified the web news articles.

For web news classification, a deep learning technique [9] was proposed. Initially, a training database was created in which the class of each data was known. It was used to predict the class of online web news. Then, a neural network was employed for the classification of online web news. A goal was set for the neural network by the user. If the goal was not met, then change the weight of the neural network otherwise the training sides were reselected. Online web news was classified based on the weights. However, the user-specified goal greatly influences the online web news classification.

A method [10] was proposed for the classification of tweets about the newspaper reports. Initially, a dataset was formed by using the content of an article and the tweets having a link to the news. The unnecessary symbols and words in the collected data were removed in the pre-processing phase. Then, the stemming process was applied through morphological analyzers. A textual similarity approach was exploited to find the similarity between the tweets. Finally, a binary classifier called Naïve Bayes was applied to classify the tweets about the newspaper reports. The disadvantage of using Naive Bayes requires more data to get better results.

A Fuzzy Algorithm for Extraction, Monitoring, and Classification of Communicable Diseases (FAEMC-CD) [11] model was proposed based on evolving fuzzy model. This model extracts the information about communicable diseases from Twitter and news websites. Twitter and data in news websites were pre-processed using tokenizing, stemming, stop word filtering and term filtering. Then fuzzy rules were developed using fuzzy rule-based classification package.

3 Proposed Methodology

In this section, the proposed method for web news mining using web news articles and Twitter data is described in detail. The topics within the twitter data are modeled by the Latent Dirichlet Allocation (LDA). Then a Twitter-specific tokenizer and part of speech tagger are used to split the tweets into tokens and label the tokens. The terms of each tweet under different topics are pruned by a term reduction technique. If the terms have a relevance value more than the optimized pruning threshold, then those terms are removed from the twitter data. Based on the Term Frequency and Inverse Document Frequency (tf-idf) of terms in web news articles and twitter data, Evolving fuzzy rules are created where the web news articles are categorized. A Spatio-Temporal Generalized Additive Model (STGM) is developed where the location and time of the tweets are considered for efficient categorization of web news articles. For the efficient selection of dynamically changing terms of tweets, a Randomized Least Angle

Regression (LAR) is used. The tf-idf of selected terms of Twitter data is used in the Evolving fuzzy rules which enhance the accuracy of web news article categorization.

3.1 LDA Based Topic Modeling of Twitter Data

The news articles of different topic areas are extracted from the web and tweets containing a link to the news are collected from Twitter. The topics within the text of tweets are modeled by LDA model. A twitter dataset of unlabelled tweets is given as input to the LDA model. It found the hidden topics as distributions over the words in the vocabulary. The words are represented as observed random variables and the topics are represented as latent random variables. Initially, a topic weight vector is drawn that is modeled by a Dirichlet random variable. It finds out which topics are almost certainly to appear in the Twitter data. From the topic weight vector, a single topic was chosen for each word that is to appear in the Twitter data. By a different, randomly chosen topic each word in the Twitter data is generated. The available choices of topics for each tweet are, in order, drawn from a smooth distribution over the space of all possible topics. The LDA process is described as follows:

For each tweet indexed by $m \in \{1, 2, \ldots M\}$ in a twitter data:

1. Select a K-dimensional topic weight vector θ_m from the distribution $p(\theta|\alpha) = Dirichlet(\alpha)$. The Dirichlet distribution is given as follows:

$$p(\theta|\alpha) = \frac{\Gamma\left(\sum_i \alpha_i\right)}{\prod_i \Gamma(\alpha_i)} \prod_i \theta^{\alpha_i - 1} \tag{1}$$

Equation (1), Γ is the gamma function and α is a K-dimensional parameter.

2. For each word indexed by $n \in \{1, 2, \ldots N\}$ in a tweet:
   a. Select a topic $z_n \in \{1, 2, \ldots K\}$ from the multimodal distribution $p(z_n = k|\theta_m) = \theta_m^k$.
   b. Given the chosen topic z_n, draw a word w_n from the probability $p(w_n = i|z_n = j, \beta) = \beta_{ij}$, where β is a K-dimensional parameter.

3.2 Term Generation

The news articles undergo the term generation process where the tokenization, stop word elimination and stemming processes are carried out. A twitter-specific tokenizer is used to split the tweets into symbols, phrases and words are called tokens. In the Twitter-specific tokenizer, emoticons are treated as separate tokens. Since the emoticons in twitter data hold some semantic content that describes the user's emotional state. A part of speech tagger is used to label the tokens of twitter data as a common noun, proper noun, pronoun, proper noun + possessive, nominal + possessive, interjection, verb, adverb, adjective, hashtag, emoticon, proper noun + verbal, nominal + verbal and existential 'there' + verbal. The list of labelled tokens is typical to stop word elimination. Finally, the stemming process is carried out by Snowball Stemmer where the words are reduced into stems. It reduces the number of distinct terms in twitter.

3.3   Term Filtering and Web News Classification

After modelling the topics of twitter data, a term reduction technique is applied to reduce the number of terms associated with each tweet. The terms are pruned based on their frequencies of occurrence throughout the collection. A relevance value of each term is calculated by the Term Frequency and Inverse Document Frequency (tf-idf) metric. Remove the words from the dataset of news articles and Twitter when tf-idf value of those words is less than an optimized pruning threshold [4]. This process is called term filtering. The efficiency of EFS based web news mining approach is improved by including twitter feature with news article feature for classification of web news articles. Based on the tf-idf values of each term in news articles and twitter data, categorize the news articles based on the following fuzzy rules:

$$Rule_i = IF(A_1 \& T_1 \sim Prot_1) \, AND \, (A_2 \& T_2 \sim Prot_2) \ldots AND \, (A_n \& T_n \sim Prot_n) THEN \, Category$$
$$= Category_j$$

$$(2)$$

Equation (2), i denotes the number of rules, n denotes the number of terms in the collection of news articles and tweets, the A_i stores the tf-idf of the term i of the article A, the T_i stores the tf-idf of the term i of the tweet T, the $Prot_i$ stores the tf-idf value of the term i of one of the prototypes of the corresponding class and $Category_m \in \{set \, of \, different \, categories\}$. The category of new news articles A_z is identified by comparing the article A_z to all the prototypes using the cosine distance and it is given as follows:

$$Class(A_z) = Class(Prot^*)$$

$$(3)$$

$$Prot^* = Min_{i=1}^{NumP}(cosDist(Prot_i, A_z))$$

$$(4)$$

where, A_z denotes the non-categorized web news article to classify, $NumP$ denotes the number of existing prototypes, $Prot_i$ denotes the ith prototype and $cosDist$ denotes the cosine distance between two news articles.

3.4   Spatio-Temporal Generalized Additive Model for Web News Classification

The terms in tweets are changing over the time they tweets and their locations and a large number of terms are generated from the Twitter dataset. So a Spatio-Temporal Generalized Additive Model (STGM) is developed which considers both the location of user and time they tweets. The STGM model is built on spatial information of users with the temporal information encoded by dummy variables. This model is given as follows:

$$logit\left(p\left(Tweet_{s_i, t_j}\right)\right) = \sum_{n=1}^{N} f_n\left(Term_{n, s_i, t_j}\right) + \kappa_{s_i, t_j - t^0}$$

$$(5)$$

Equation (5), $p(Tweet_{s_i, t_j})$ is the probability of tweet in the spatial s_i at time t_j, $logit(p) = log\left(\frac{p}{1-p}\right)$ is a logit link function, N denotes the number of terms in a $Tweet$,

$Term_{n,s_i,t_j}$ is the nth term in tweet with location s_i and time t_j, f_n is the smooth function of the nth feature and $\mathcal{K}_{s_i,t_j-t^0}$ denotes the dummy variable representing the length of the continuous zeros that precede the current observation location s_i and time t_j. However, the collection of tweets based on time and location, generate different terms for tweets.

In order to choose the most important terms in the tweets, a Randomized Least Angle Regression (RLAR) is used. RLAR is the extension of Least Angle Regression (LAR) [12] method which is a feature selection model. The RLAR method introduced randomness into the dataset to enhance LAR when the relationship between twitter terms might be nonlinear. RLAR calculates the order of terms going into the regression model and selects the best m number of terms based on criteria like mean squared error. In the RLAR method, the T Twitter Dataset $T^s_{m \times q}$ from $T_{n \times p}$, where $m < n$ (n is the number of terms) and $q < p$ (p is the number of predictors). LAR is applied to each T^s to rank the terms. The term priority for T^s is $rank_s = \langle term_{s1}, term_{s2}, \ldots term_{sq} \rangle$. Finally, term priorities are voted on by $\{rank_s\}$. The overall process of RLAR is given as follows:

RLAR Algorithm:

Input: Twitter $T = \{term, y\}$

Output: Ranked terms $vote_i$

1. for $s = 1$ to S

2.   sample D^s from D

3.   Get ranked features $rank_s = \langle term_{s1}, term_{s2}, \ldots term_{sp} \rangle$ by applying LAR on D^s, $term_{sp}$ is added in the ith step of LAR algorithm

4. End for

5. for $i = 1$ to p

6.   for $s = 1$ to S

7.   if $term_i \in rank_s$

8. $vote_i = vote_i + r_{si}$, where $rank_s[r_{si}] == term_i$

9. $sample.time_i = sample.time_i + 1$

10.   End if

11.   End for

12. $vote_i = \dfrac{vote_i}{sample.time_i}$

13. End for

The smaller *vote*$_i$ is the more important term *term*$_i$ of tweets. The top most terms are considered for calculation of tf-idf. Based on the calculated tf-idf of twitter data and web news articles, web news articles are categorized by evolving fuzzy rules.

4 Result and Discussion

The efficiency of the News Article Categorization using Optimized Evolving Fuzzy System (NAC-OEFS), Twitter enriched News Article Categorization using Optimized Evolving Fuzzy System (TNAC-OEFS) and Spatio Temporal Twitter enriched News Article Categorization using Optimized Evolving Fuzzy System (STTNAC-OEFS) techniques is analyzed in terms of accuracy, precision, and recall. For the experimental purpose, 4000 tweets from the Twitter dataset are collected. The web news article is created from five different sets of data which is briefly explained in [4].

4.1 Accuracy

Accuracy is the percentage of web news articles that are predicted with the correct category.

$$Accuracy = \frac{True\,Positive + True\,Negative}{True\,Positive + True\,Negative + False\,Positive + False\,Negative} \quad (6)$$

Table 1 tabulates the accuracy of NAC-OEFS, TNAC-OEFS, and STTNAC-OEFS for five different sets of data.

Table 1. Accuracy model

Methods	Dataset				
	H-Sc	Sc-Te	H-Sc-Sp	B-He-Sc-Sp	A-B-H-Sc-Sp-Tr
NAC-OEFS	80	97.27	70.45	60.7	42.45
TNAC-OEFS	84	91.24	76.12	65.98	51.78
STTNAC-OEFS	90	94.56	80.87	71.32	60.85

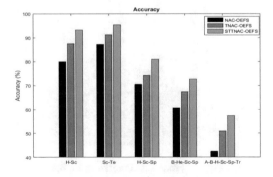

Fig. 1. Accuracy comparison model

Figure 1 shows the comparison of accuracy between existing News Article Cate-gorization using Optimized Evolving Fuzzy System (NAC-OEFS), proposed Twitter enriched News Article Categorization using Optimized Evolving Fuzzy System (TNAC-OEFS) and Spatio Temporal Twitter enriched News Article Categorization using Optimized Evolving Fuzzy System (STTNAC-OEFS) methods for five different sets of data. The different sets of data are taken in X-axis and the accuracy value in % is taken in Y-axis. For Health vs. Science set of data, the accuracy of STTNAC-OEFS is 12.5% greater than NAC-OEFS and 7.14% greater than TNAC-OEFS method. By using the spatial information of Twitter users and temporal information of tweets for web news categorization, the accuracy of STTNAC-OEFS method is greater than the other web news categorization methods. From this analysis, it is known that the pro-posed STTNAC-OEFS has high accuracy than the other methods.

4.2   Precision

$$Precision = \frac{True\ Positive}{(True\ Positive + False\ Positive)} \tag{7}$$

Table 2 tabulates the precision of NAC-OEFS, TNAC-OEFS, and STTNAC-OEFS for five different sets of data.

Table 2. Precision model

Methods	Dataset				
	H-Sc	Sc-Te	H-Sc-Sp	B-He-Sc-Sp	A-B-H-Sc-Sp-Tr
NAC-OEFS	81	86.97	69.96	61.13	43.29
TNAC-OEFS	87.5	91.24	74.15	67.29	50.87
STTNAC-OEFS	93.28	95.38	80.96	72.69	57.36

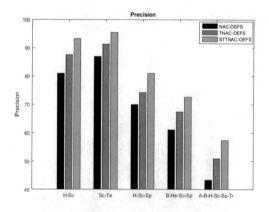

Fig. 2. Precision comparison model

Figure 2 shows the comparison of precision between existing NAC-OEFS, proposed TNAC-OEFS and STTNAC-OEFS methods for five different sets of data. The different sets of data are taken in X-axis and the precision value is taken in Y-axis. For Health vs. Science set of data, the precision of STTNAC-OEFS is 15.2% greater than NAC-OEFS and 6.6% greater than TNAC-OEFS method. By using twitter data, spatial and temporal information along with the web news articles, the true positive prediction of STTNAC-OEFS based web news categorization is high which also increase the precision value of STTNAC-OEFS method. From this analysis, it is proved that the proposed STTNAC-OEFS has high precision than the other methods.

4.3 Recall

$$Recall = \frac{True\ Positive}{(True\ Positive + False\ Negative)} \tag{8}$$

Table 3 tabulates the recall of NAC-OEFS, TNAC-OEFS, and STTNAC-OEFS for five different sets of data.

Table 3. Recall differentiation

Methods	Dataset				
	H-Sc	Sc-Te	H-Sc-Sp	B-He-Sc-Sp	A-B-H-Sc-Sp-Tr
NAC-OEFS	83.45	85.87	73.14	64.21	45.78
TNAC-OEFS	87	89.21	77.62	70.13	51.27
STTNAC-OEFS	91	94.36	82.41	76.34	57.36

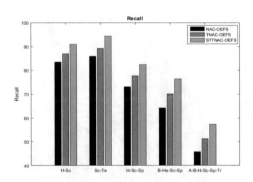

Fig. 3. Recall results for different methods

Figure 3 shows the comparison of recall between existing NAC-OEFS, proposed TNAC-OEFS and STTNAC-OEFS methods for five different sets of data. The different sets of data are taken in X-axis and the recall value is taken in Y-axis. For Health vs.

Science set of data, the recall of STTNAC-OEFS is 9.04% greater than NAC-OEFS and 4.6% greater than TNAC-OEFS method. From this analysis, it is known that the proposed STTNAC-OEFS has high recall than the other methods. The recall of STTNAC-OEFS is high because it categorizes the web news articles with the consideration of twitter data, spatial information of Twitter users and temporal information of tweets.

4.4   F-Measure

F-measure is the external measure for measuring goodness or accuracy of web news categorization methods. It depends on two factors are precision and recall. It is calculated as,

$$F - measure = \frac{2 \times Precision \times Recall}{Precision + Recall} \tag{9}$$

Table 4 tabulates the f-measure of NAC-OEFS, TNAC-OEFS, and STTNAC-OEFS for five different sets of data.

Table 4. F-measure analysis

Methods	Dataset				
	H-Sc	Sc-Te	H-Sc-Sp	B-He-Sc-Sp	A-B-H-Sc-Sp-Tr
NAC-OEFS	82.21	86.42	71.51	62.63	44.5
TNAC-OEFS	87.25	90.21	75.85	68.68	51.07
STTNAC-OEFS	92.13	94.87	81.68	74.47	57.36

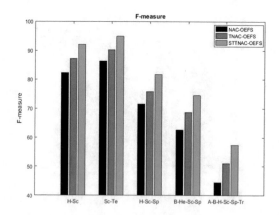

Fig. 4. F-measure analysis

Figure 4 shows the comparison of f-measure between existing NAC-OEFS, proposed TNAC-OEFS and STTNAC-OEFS methods for five different sets of data. The different sets of data are taken in X-axis and the f-measure value is taken in Y-axis. For Health vs. Science set of data, the f-measure of STTNAC-OEFS is 12.07% greater than NAC-OEFS and 5.59% greater than TNAC-OEFS method. From this analysis, it is known that the proposed STTNAC-OEFS has high f-measure than the other methods. By using efficient term filtering and efficient web news classification in STTNAC-OEFS, the f-measure of STTNAC-OEFS method is high.

5   Conclusion

In this paper, EFS based web news mining is improved by using Twitter data related to news in the web news articles. Initially, Twitter data containing a link to the news are collected along with news articles. The topics within the texts of tweets are modeled by using the LDA model. Then the terms in tweets under different topics are generated by twitter-specific tokenizer, part of speech tagger and Snowball Stemmer. A STGMmodel is developed which consider the location of user and time they tweets and the most important terms are selected by RLAR method. The tf-idf values of selected terms are calculated and it is used along with the tf-idf values of terms in the web news article to create the evolving fuzzy rules. Based on these rules, the web news articles are categorized. Finally, the efficiency of the proposed method has proved in terms of accuracy, precision and recall.

References

1. Mirończuk, M.M., Protasiewicz, J.: A recent overview of the state-of-the-art elements of text classification. Expert Syst. Appl. **106**, 36–54 (2018). https://doi.org/10.1016/j.eswa.2018.03.058
2. Thangaraj, M., Sivakami, M.: Text classification techniques: a literature review. Interdiscip. J. Inf., Knowl. Manag. **13**, 118–135 (2018) https://doi.org/10.28945/4066
3. Iglesias, J.A., Tiemblo, A., Ledezma, A., Sanchis, A.: Web news mining in an evolving framework. Inf. Fusion **28**, 90–98 (2016). https://doi.org/10.1016/j.inffus.2015.07.004
4. Nithya, D., Sivakumari, S.: Categorizing online news articles using Penguin search optimization algorithm. Int. J. Eng. Technol. **7**, 2265–2268 (2018). https://doi.org/10.14419/ijet.v7i4.15607
5. Bastani, K., Namavari, H., Shaffer, J.: Latent Dirichlet allocation (LDA) for topic modeling of the CFPB consumer complaints. arXiv preprint arXiv:1807.07468 (2018)
6. Kompan, M., Bieliková, M.: News article classification based on a vector representation including words' collocations. In: Third International Conference on Software, Services and Semantic Technologies, S3T 2011, pp. 1–8. Springer, Heidelberg (2011). https://doi.org/10.1007/978-3-642-23163-6_1
7. Dilrukshi, I., De Zoysa, K., Caldera, A.: Twitter news classification using SVM. In: IEEE 2013 8th International Conference on Computer Science & Education, pp. 287–291 (2013). https://doi.org/10.1109/iccse.2013.6553926

8. Liparas, D., HaCohen-Kerner, Y., Moumtzidou, A., Vrochidis, S., Kompatsiaris, I.: News articles classification using random forests and weighted multimodal features. In: Information Retrieval Facility Conference. Springer, Cham, pp. 63–75 (2014). https://doi.org/10.1007/978-3-319-12979-2_6

9. Kaur, S., Khiva, N.K.: Online news classification using Deep Learning Technique. Int. Res. J. Eng. Technol. (IRJET) 3, 558–563 (2016)

10. Demirsoz, O., Ozcan, R.: Classification of news-related tweets. J. Inf. Sci. 43, 509–524 (2017). https://doi.org/10.1177/0165551516653082

11. Jahanbin, K., Rahmanian, F., Rahmanian, V., Jahromi, A.S., Hojjat-Farsangi, M.: Application of Twitter and web news mining in monitoring and documentation of communicable diseases. J. Int. Transl. Med. 6, 167–175 (2018). https://doi.org/10.11910/2227-6394.2018.06.04.03

12. Khan, J.A., Van Aelst, S., Zamar, R.H.: Robust linear model selection based on least angle regression. J. Am. Stat. Assoc. 102, 1289–1299 (2007). https://doi.org/10.1198/016214507000000095

13. Malhotra, S., Dixit, A.: An effective approach for news article summarization. Int. J. Comput. Appl. 76(16), 5–10 (2013)

Designing Safe and Secure Land Registration- Ownership Using Blockchain Technology with a Focus on Mutual Authentication

B. R. Arun Kumar[✉] and B. Rohith

Department of MCA, BMS Institute of Technology and Management,
Bengaluru, India
arunkumarbr@bmsit.in, rohith.naidu.1009@gmail.com

Abstract. This paper presents a scheme for electronic land registration using blockchain technology. A blockchain is a comprehensive approach in industries and academia which need to be protected. The objective is to maintain an open ledger (blockchain) to record the land registration data. Idea is to offer the right person the right ownership. This paper highlights the need of blockchain technology in an application and addresses the need for adopting the mechanisms for confidentiality, authentication, integrity, and non-repudiation which are basics of any security systems referred to as CIA-R in short. This paper discusses the implementation of three blockchains namely People, Land and Transaction with more emphasis on mutual authentication. The application uses the following tools/technology for implementation.

Keywords: Blockchain technology · Fraudulent land registration · Transaction · Ownership · Co-ownership

1 Introduction

A blockchain is an open ledger where the list of records keeps growing called blocks, which are linked using cryptography. Each block contains a cryptographic hash of the previous block a timestamp, and transaction data (generally represented as a Merkle tree). A blockchain is designed in such a way that blocks are immutable, verifiable, efficient and permanent. By implementing the blockchain in land registration can create an impact on the present systems of land registration. It can strengthen the hands of government and right owners to enforce their rights, avoid corruption, bring integrity and enable practicing ethics. Several researchers have expressed that blockchain may redefine many areas including cybersecurity. At a rapid pace of development since the past two years, the market for blockchain technology worldwide is forecast to grow to 2.3 billion U.S. dollars by 2021. A shared ledger reduces the need for third parties thereby reducing complexity and time taken to processes the transaction. Further, it offers enhanced network resiliency, minimizes network attack increases accuracy and transparency [1–3]. Blockchain may be open to everyone or restricted to a set of users of an organization or coming under the single umbrella of a purpose (Public or private).

© Springer Nature Switzerland AG 2020
P. Karrupusamy et al. (Eds.): ICSCN 2019, LNDECT 39, pp. 509–516, 2020.
https://doi.org/10.1007/978-3-030-34515-0_53

Further, it can be thought of as permission or permission-less, meaning that either anyone can offer their services to add blocks to the chain or only a restricted group of users can do so [4]. Any application which implements blockchain technology needs to provide techniques to store the records securely and globally in a decentralized manner. Such applications can add values to the society and industry. As mentioned in [6], across the world land registration is improperly managed and in some cases corrupts. It is high for all countries to take this problem to vary seriously and offer technical solutions. Land reform is a major problem which seeks a solution with proper policies and technology so as to have land ownership as an individual affair or co-affair in case of co-ownership, it is very important to offer rights to right individual(s) so that family, organization and government property to be protected.

In this application, a new perspective of the complete land registration process is introduced and information is made available on-demand in a registration. In the literature survey, it is found that only a few research works have addressed this problem of national importance [6]. In the world of technical development, digital disruptive technology like blockchain can solve enormous problems making the system safe and secure. The blockchain acts as a ledger that stores all the transactions that have been performed during the land registration process. This means that the blockchain grows continuously, adding new blocks after every successful transaction. Across the world, there are lakhs of complaints of fraudulent land registration. In Tamil Nadu, India, as on July 2018 nearly 1,700 complaints reported were fraudulent land registrations. It is found that in some cases concerned authorities also involved illegally in the land registration process. Though the respective owners have lodged the complaint, it takes years together to solve the problem. Hence electronic innovative solutions with advanced technology are necessary. The scope of blockchain technology is not necessarily be associated with digital or virtual currency schemes, payments and financial services. The notion of block chain introduced by Satoshi Nakamoto's white paper in 2008 can be theoretical extended to a large number of sectors of applications such as trade and commerce, healthcare, governance. Its applications can have impact on the registration of shares, bonds and other assets such as Land/the transfer of property titles [7–16].

1.1 Organization of the Paper

The introduction section articulates the area of the work and literature survey briefly. The overall architecture of the proposed system is presented in Sect. 2. The technical solution proposed, data structure used and language used for implementation are explained in Sect. 3. The results obtained along with a summary and future work are presented in Sect. 4.

2 The Overall Architecture of the Proposed System

In a real scenario, the buyer and seller mutually come to an agreement and approaches the sub-register office for registration of their sale deed. In this proposed system during the preliminary investigation, mutual authentication among the buyer, seller, sub-register or government representative and Attorney is enforced. After, authentication actual transaction happens as per the processes defined by the concerned government. The successful transaction leads to node creation and gets added into the immutable register. The following high-level architecture depicts the mutual authentication among the stakeholders (Fig. 1).

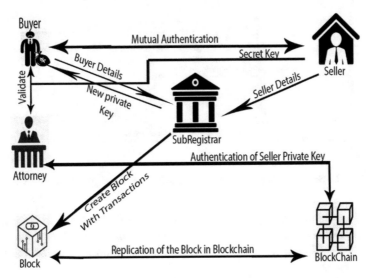

Fig. 1. High-level architecture of Land-ownership stakeholders with block chain

Fake registration is unethical imposing a huge loss to owners, especially a common man suffers a lot, in some cases fake registration may lead to litigation problem which consumes many days to be solved and the precious time of the legal authorities solving it. Fraudulent registration with the intention of grabbing government and private land is a global problem. Illegal transfer of property by forgery and impersonation may be another problem. As per the rule, it is mandatory to do registration sale and purchase of immovable property and ensures the conservation of evidence, prevention of fraud and assurance of title. Fake registration is unethical imposing a huge loss to owners, especially a common man suffers a lot, in some cases fake registration may lead to litigation problem which consumes many days to be solved and the precious time of the legal authorities solving it. Fraudulent registration with the intention of grabbing government and private land is a global problem. Illegal transfer of property by forgery and impersonation may be another problem.

2.1 Sequence Diagram of Registration Processes

The above figure explains the involvement of stakeholders in the sequence of mutual authentication before the actual transaction to be set. In this project work, the solution to avoid fake land registration is proposed, mutual authentication of stakeholders is

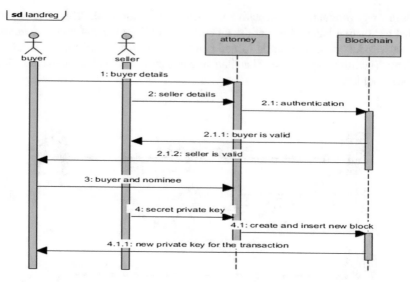

Fig. 2. Sequence diagram involving stakeholders for registration processes.

ensured, transaction integrity is maintained and the solution looks towards nonrepudiation of the stakeholders. In other words, using blockchain technology, land property gets a binder with the right owner and avoids unauthorized one encroaching the property (Fig. 2).

3 Technical Solution Proposed

The blockchain proposed here consist of 3 important data elements: People, Land, and Transaction. People element contain a list of blocks which contain the Aadhaar details of the people, Aadhaar number is considered as the unique identity of an individual. The land element contains a list of blocks which contain the details of every land and uniquely identified by Survey number and Katha number. Transaction element contains a list of a block containing the land registration which also has a private secret key which will only be available with the owner of the land Initially, people should register for the people block chain with their Aadhaar number. Land should be registered with its uniquely identified Survey and Katha number. The land and its owner should be bounded by making a transaction which will contain the land identity, owner identity, and nominee identity or co-owner identity, it also contains a secret private key which authenticates the transaction block and should only be available with the owner and kept confidentially to the outside world. Buyer of the land, and the seller of the land initially goes to the attorney, who is responsible for validating and authenticating the buyer and seller by their Aadhaar number. Further, the presence of the land is authenticated by considering the survey number, both buyer and seller also authenticates the attorney, after this procedure the secret key is given to the seller of the land while registering the land transaction is getting validated and new block is created with new owner and nominee with newly generated private key which would be given to the buyer and it is his responsibility to maintain it secretly.

3.1 Language and Data Structures Used

This application is developed using the Python 3.7 programming language with flask version 0.12.2 micro web framework. The interaction is done by HTTP request and parsing JSON object. The data from the front end web page goes through ajax call passing JSON object. The server responds for the request by sending the JSON object back to the web page. There are 3 blockchains in this single application which are implemented using 3 lists called People, Land, and Transaction. As soon as the instance of the class is created genesis block gets generated with no previous hash for all the 3 blockchains. All the data of an individual record are stored in a dictionary and that dictionary is appended to the respective list, each dictionary will have a field called previous hash which will hold the hash value of the previous dictionary present in that particular list. Dictionary contains the basic data of the blockchain like index, timestamp, proof, previous hash, and transaction. These elements are common for all the three lists, but the content in the transaction will vary between various list. Initially, genesis block is created with no previous hash, further created blocks have the previous hash which is the hash key generated by SHA256 algorithm taking the previous block (dictionary) as input (Fig. 3).

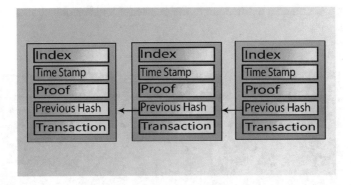

Fig. 3. The data structure used in blockchain

3.2 Algorithm for Creation of a Block in Land Blockchain

Start
Step 1: Web interface requests Node to create a block and send survey number of the land proposed for registration and survey number is defined as identification of the land.
Step2: Node sends survey number to the function, "add_transaction_land" invoked by it.

Step3: In the mechanism of add_transacion_land function following functionality is implemented which defines a dictionary with land no as a key and considering value as land chain length + 1, and survey number is passed as an argument by Node to add_transaction_land and append this dictionary to transaction_land list.

Step4: Node invokes "land_mine_block" function which does the job of solving the consensus problem, which results in defining proof followed by generation of the hash of the previous block.

Step5: Node invokes" land_create_block" function and sends proof and previous hash to it.

Step6: In land_create_block function has a mechanism to define a dictionary named includes index, timestamp, proof, previous hash and transaction as keys with their corresponding values and append this block to the land blockchain.

Stop.

4   Results and Conclusion

In this work, land registration system with a new perspective is partially implemented. The chains created enables participants to view, verify, validate, and commit permanently leading to avoid fake registration or multiple registrations. In the future work, it is planned to implement mechanisms for authentication of the sub-registrar entity and attorney, generate multiple nodes, validation of documents online including the Encumbrance Certificate, notification to the owner for any fraudulent attempt (Figs. 4, 5, 6 and 7).

Fig. 4. Initial registration of the land with the owner

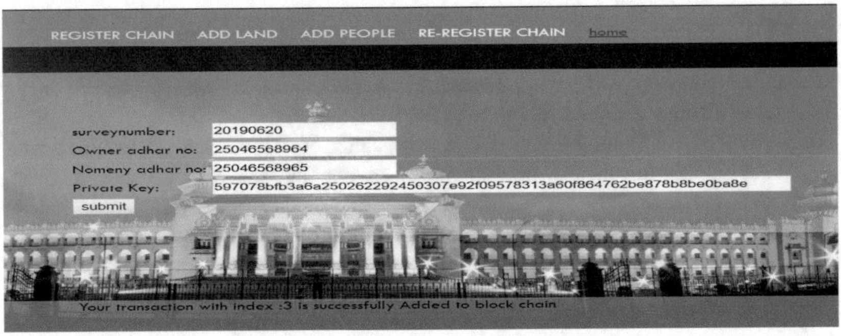

Fig. 5. When all the fields are correct while reregistration

Fig. 6. When tried to use the previously used key for resale

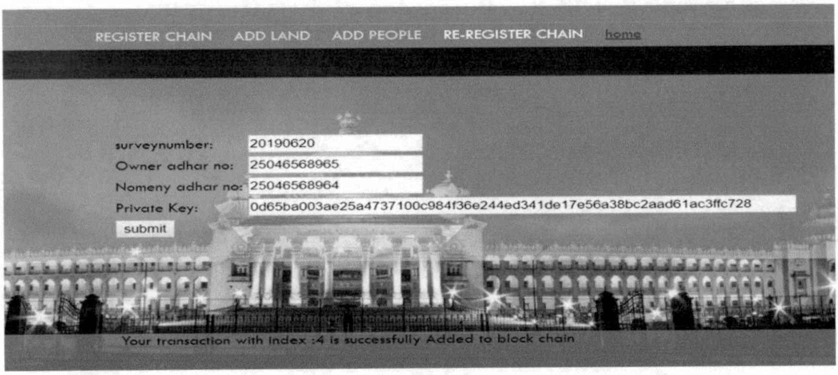

Fig. 7. Resale of land with the newly generated key

References

1. Fernández-Caramés, T.M., Fraga-Lamas, P.: A review on the use of blockchain for the Internet of Things. IEEE Access **6**, 32979–33001 (2018)
2. Vishwakarma, P., Khan, Z., Jain, T.: A brief study on the advantages of blockchain and distributed ledger in financial transaction processing. Int. J. Latest Technol. Eng. Manag. Appl. Sci. **VII**(I), 76–79 (2018). ISSN 2278-2540
3. https://housing.com/news/laws-related-registration-property-transactions-india
4. https://www.aciworldwide.com/-/media/files/collateral/trends/unlocking-benefits-of-blockchain-tl-us.pdf
5. http://www.cs.tau.ac.il/ ∼ msagiv/courses/blockchain/overview.pdf
6. https://www.elra.eu/wp-content/uploads/2017/02/10.-Jacques-Vos-Blockchain-based-Land-Registry.pdf
7. Houben, R., Snyers, A.: Policy department for economic, scientific and quality of life policies. http://www.europarl.europa.eu/supporting-analyses
8. http://www3.weforum.org/docs/WEF_Building-Blockchains.pdf
9. https://nvlpubs.nist.gov/nistpubs/ir/2018/NIST.IR.8202.pdf
10. Sagiv, M.: The block chain Technology. http://www.cs.tau.ac.il/ ∼ msagiv/courses/blockchain/overview.pdf
11. https://chromaway.com/papers/Blockchain_Landregistry_Report_2017.pdf
12. https://www.metropolis.org/sites/default/files/metobsip5_en_1.pdf
13. Saranya, A., Mythili, R.: A survey on blockchain based smart applications. Int. J. Sci. Res. (IJSR) **8**(1), 450–455 (2019)
14. Sravani, C., Murali, G.: Usage of blockchain-based provenance enabled technology to process and track land transactions. Int. J. Res. Advent Technol. Special Issue, 1–5, March 2019
15. Uzair, M.M., Karim, E., Sultan, P., Ahmed, S.S.: The Impact of Blockchain Technology on the Real Estate Sector Using Smart Contracts. https://mpra.ub.uni-muenchen.de/88934
16. Bocek, T., Rodrigues, B.B., Strasser, T., Stiller, B.: Blockchains everywhere - a use-case of blockchains in the pharma supply-chain. In: IEEE International Symposium on Integrated Network Management, IFIP, pp. 773–777 (2017). 978-3-901882-89-0

Intellectual Radio Architecture with Software Defined Radio and Its Functionalities

Reshmi Krishna Prasad$^{(\boxtimes)}$ and T. Jaya

Department of ECE, Vels Institute of Science,
Technology and Advanced Studies (VISTAS), Chennai, India
reshmiideepak@gmail.com, jaya.se@velsuniv.ac.in

Abstract. The tremendous growth of wireless digital communications and additive use of users has raised spectrum shortage and security issues in the last decade. The main obstacle for the growth of wireless digital communications is the fixed spectrum allocation and non-availability of new bands in the spectrum. For sharing the fixed spectrum allocation it makes spectrum inflexible. The inefficient utilization of spectrum bands was the result of pre-allocation of spectrum bands. The spectrum scarcity is not the real problem whereas inefficient spectrum allocation and its usage lead to the scarcity. For network implementation, Cognitive radio or Intellectual radio is believed to be the key enabling technology. Recent spectrum sensing algorithm-binary hypothesis detecting model was reviewed in this paper.

Keywords: Cognitive Radio · Intellectual radio · Software Defined Radio · Spectrum handoff

1 Introduction

Intellectual Radio (IR) is a form of wireless communication throughout which a device termed as as the transceiver senses channels associated with communication to determine which in turn channels are empty and even occupied. The cognitive radio's primary characteristics are:

- The user can be identified and authorized.
- Encryption and decryption can be executed.
- Capable of feeling wireless equipment nearby.
- It utilizes less interference with the radio frequency spectrum.
- A transceiver can determine its location [2].

Intellectual Radio is a paradigm of wireless communication in which a network or a wireless node changes its transmission or reception parameters to communicate effectively to avoid interference with licensed or unlicensed users. This particular parameter change is structured on the active security of several variables throughout the external and interior radio setting of radio frequencies. It absolutely considered an optimal aim for a Software Defined Radio (SDR) platform to produce a fully reconfigurable cellular black box that changes the communication variables automatically. SDR can execute beneficial tasks helping the consumer, helping the network, and

© Springer Nature Switzerland AG 2020
P. Karrupusamy et al. (Eds.): ICSCN 2019, LNDECT 39, pp. 517–522, 2020.
https://doi.org/10.1007/978-3-030-34515-0_54

minimizing spectral congestion. A few main applications that improve the capacity of the SDR and make this a cognitive radio:

1. Management of spectrum in addition to optimizations.
2. Interface having a broad range of systems and network resource optimization.
3. A human user interface and the provision associated with electromagnetic resources to support the particular human being.

2 Intellectual Radio Requirements

One of the primary aims of cognitive radio is to make the most effective use of current radio resources. Cognitive radio needs a number of circumstances to be fulfilled to guarantee optimum usage. The main demands for cognitive radio are:

(a) Significant interference with licensed systems
(b) The ability to adapt to different characteristics of the connection
(c) Capability of sensing and measuring critical environmental, channel
(d) Ability to take advantage of a range of spectral opportunities
(e) Flexible band width and pulse shape
(f) Adjustable data rates, adaptive strength of transmission, safety of information and restricted costs.

Cognitive Radio's objective is to use frequency bands owned by its registered consumers. Therefore, one of the cognitive radio's most important criteria is that the interference caused to licensed consumers by cognitive machines remains negligible. One of the cognitive radio concept primary characteristics is that the targeted frequency spectrum is scanned regularly to verify its accessibility for opportunistic use. Bands to be used for cognitive communication are determined according to the outcomes of this spectrum scan. Since the available groups may vary at unmistakable minutes and spots, it is foreseen that intellectual radio will be exceedingly adaptable in deciding the range it involves.

3 Software Defined Radio and Its Relation With Intellectual Radio

An SDR is a radio in which software defines the characteristics of the carrier frequency, signal bandwidth, modulation, and network access. SDR is a general purpose device where numerous wave forms are actualized at different frequencies utilizing a similar radio tuner and processors. This procedure has the advantage of making the gear increasingly adaptable and cost effective. The cognitive radio's primary features are the adaptability where the radio parameters (including frequency, energy, modulation, bandwidth) can be altered based on the radio setting, user status, and network condition. By avoiding the use of application specific fixed analog circuits and elements, SDR can provide very versatile radio features. Figure 1 shows the relation between SDR and CR.

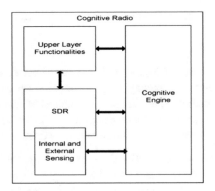

Fig. 1. Relation between SDR & IR

4 CR Architecture

In terms of both spectrum policy and communication techniques, the current wireless network environment uses heterogeneity. For the growth of communication protocols, a clear description of the cognitive radio network architecture is therefore essential [3].

The elements of the architecture of the cognitive radio network, as shown in Fig. 2.

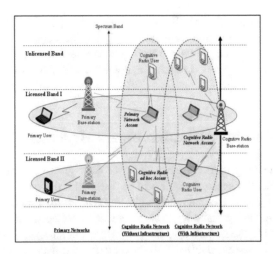

Fig. 2. CR network architecture [SSN ONLINE(2278-8875) PRINT (2320-3765)]

Primary_User

In a certain spectrum band, the primary user has a permit to function.

Essential_ Base-Station

Primary base station is a component of a fixed system foundation with a range license. In substance, there is no intellectual radio limit in the fundamental base station to impart range to clients of subjective radio. However, for cognitive radio users' main network access, it may be necessary to have both heritage and cognitive radio protocols.

Cognitive-Radio-User

There is no spectrum license for cognitive radio users. Therefore, access to spectrum is only permitted in an opportunistic way. Cognitive radio user capabilities include spectrum sensing, choice spectrum, spectrum handoff, and MAC/routing/transport cognitive radio protocols [1, 8]. It is presumed that the cognitive radio user has the ability to interact not only with the base station but also with other users of cognitive radio.

i. **Cognitive-Radio-Base-Station:** Cognitive base_station radio is a fixed element of infrastructure with cognitive radio capacities. Cognitive base_station radio offers a single hop link without a spectrum access permit to cognitive radio users. Intellctual_radio users can either interact with each other in a multihop way or access the base station [6]. Thus, there are three distinct kinds of access across heterogeneous networks in our cognitive radio network architecture that demonstrate distinct execution criteria as follows:

   a. **Cognitive-Radio-Network-Access:** Cognitive radio users in licensed and unlicensed spectrum bands can access their own cognitive radio base station. Since all interactions happen within the cognitive radio network, it is independent of the main network's medium access system [10].
   b. **Cognitive-Radio-Ad-Hoc-Access:** CR users can interact both licensed and unlicensed spectrum bands [7].
   c. **Primary-Network-Access:** If the main network is permitted, the cognitive radio user can access the main base station [4] via the licensed band. Unlike other kinds of access, main network medium access technology should be supported by cognitive radio users. In addition, cognitive radio skills should be supported by main base station [9].

5   Spectrum-Sensing Algorithm Binary Hypothesis Detecting

The spectrum sensing's is to correctly discover the "spectrum holes." Compared to demodulation, spectrum sensing requires to be a sensed if there are main users, and the initial information_bearing signal need not be extracted from a modulated carrier wave [11]. In practice, to ensure that normal communication between authorizing main users is not interfered with, the spectrum sensing algorithms in the Cognitive Radio Network 133 sensing sensitivity need 30–40 dB beyond the primary user receiver because cognitive users can not directly detect the channel scenario between primary users and transmitters and, at the same moment, the wireless spectrum. The binary hypothesis detection model is the easiest range sensing model on the AWGN (Additive White Gaussian Noise) channels.

$$H0 : X[n] = Y[N], \; H1 : X[n] = Z[n] + Y[N], n = 1, \ldots .N$$

Where H0 and H1 convey the distinct hypotheses of the main user signal exists or does not exist in the AWGN channels; where X[n] is the signals obtained by unlicensed users(cognitive users); where Z[n] is the signals transmitted by registered users (main users); where Y[N] is the White Gaussian Noise. The statistical T can be achieved by X[n], and the sensing outcomes can be achieved on the basis of the choice theorem.

According to the sensing algorithms.

$$H0 : T < \lambda, H1 : T > \lambda$$

Where λ is the decision threshold. The detailed process is shown as Fig. 3.

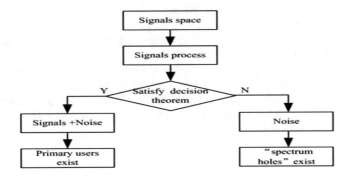

Fig. 3. Binary hypothesis detecting model

6   Conclusion

Cognitive radio is a solution for spectrum scarcity which results from inefficient spectrum allocation and its usage. Spectrum sensing is a significant feature of cognitive radio to avoid spectrum scarcity with licensed users and to define the spectrum available to improve the use of the spectrum. The binary hypothesis detection model is the earliest range sensing model in which the sensing outcomes can be achieved on the basis of the choice theorem on the AWGN channel.

References

1. Shakeel, A., Hussain, R., Iqbal, A., Latif, I., Malik, S.A., Omoniwa, B.: Adaptive spectrum handoff scheme in cognitive radio ad-hoc networks. In: 2018 International Conference on Advances in Computing and Communication Engineering (ICACCE), pp. 127–132. IEEE (2018)
2. Kumar, K.: A spectrum handoff scheme for optimal network selection in NEMO based cognitive radio vehicular networks: a cost function MADM method based on grey theory based approach. Procedia Comput. Sci. **125**, 275–281 (2018)
3. Koushik, A.M., Bentley, E., Hu, F., Kumar, S.: A hardware testbed for learning-based spectrum handoff in cognitive radio networks. J. Netw. Comput. Appl. **106**, 68–77 (2018)

4. Hoque, S., Arif, W.: Impact of secondary user mobility on spectrum handoff under generalized residual time distributions in cognitive radio networks. AEU-Int. J. Electron. Commun. **86**, 185–194 (2018)

5. Chen, A.-Z., Shi, Z.-P., He, Z.-Q.: A robust blind detection algorithm for cognitive radio networks with correlated multiple antennas. IEEE Commun. Lett. **22**(3), 570–573 (2018)

6. Singh, T., Ali, K.A., Chaudhary, H., Phalswal, D.R., Gahlaut, V.: Design and analysis of reconfigurable microstrip antenna for cognitive radio applications. Wirel. Pers. Commun. **98** (2), 2163–2185 (2018)

7. Carie, A., Li, M., Anamalamudi, S., Reddy, P., Marapelli, B., Dino, H., Khan, W., Jamal, W.: An internet of software defined cognitive radio ad-hoc networks based on directional antenna for smart environments. Sustain. Cities Soc. **39**, 527–536 (2018)

8. Hoque, S., Sen, D., Arif, W.: Impact of residual time distributions of spectrum holes on spectrum handoff performance with finite switching delay in cognitive radio networks. AEU-Int. J. Electron. Commun. **92**, 21–29 (2018)

9. Gupta, N., Dhurandher, S.K., Woungang, I., Obaidat, M.S.: Proactive decision based handoff scheme for cognitive radio networks. In: 2018 IEEE International Conference on Communications (ICC), pp. 1–6. IEEE (2018)

10. Liu, X., Li, B., Liu, G.: Simultaneous cooperative spectrum sensing and energy harvesting in multi-antenna cognitive radio. Mob. Netw. Appl. **23**(2), 263–271 (2018)

11. Koushik, A.M., Hu, F., Kumar, S.: Intelligent spectrum management based on transfer actor-critic learning for rateless transmissions in cognitive radio networks. IEEE Trans. Mob. Comput. **17**(5), 1204–1215 (2018)

Design and Development of Self-sovereign Identity Using Ethereum Blockchain

Amrita B. Chavan[(✉)] and K. Rajeswari

Pimpri Chinchwad College of Engineering, Nigdi, Pune, India
camritab@gmail.com, kannan.rajeswari@pccoepune.org

Abstract. Meaning of an identity is the fact of being who or what a person or thing is Now a days India moving towards digitization. Therefore, all the things becoming digital like bank transaction, education, government IDs, etc. We all are so much active on social media also. Hence we are having tremendous amount of login IDs and passwords. Through these kind of social media or Internet surfing we are unnecessarily reveling our personal information. In this way we aren't having control on our own identity. Consequently security of identity is becoming very serious issue, it becomes more hard to authorize a level of trust between multiple parties. On this issue Sovereign Identity will be the best solution. Sovereign means Unrestricted. There is no restriction on our identity and it will be completely under the control of the user. By using blockchain technology we can develop this kind of identity. This paper focusing on various aspects of blockchain technology like ethereum platform, smart contracts, ganache server, truffle framework, metamask, Zero knowledge proof (ZKP) and also the implementation part of development of sovereign identity using blockchain.

Keywords: Sovereign identity · Blockchain · Identity · Smart contracts · Ganache · Metamsk · Digital ids · ZKP

1 Introduction

Self-Sovereign Identity is an identity concept where people and businesses store and control their data on their own devices; providing this data when someone needs to validate them. This is all done without relying on a centralized database. These are concepts and projects like Civic (https://www.civic.com/solutions/kyc-services/) [1]. Employing blockchain for representing identities offers many applications, as listed in [2]. Creating self sustainable (self dependent) individuality utilizing blockchain offers both new avenues for start-ups as well as endorsement by government agencies. In case of the physical identities, governments issue identities. In India, probably an identity represented by an ID card or passport. Various providers attach claims to this identity, for instance, the driving license provider certifies that you are allowed to drive. This license, or claim verification, might be repeal in some situations. any a time, physical ID cards are issued. For instance in India, Government agencies require such physical IDs (e.g. passport), for claiming any right. For example, a driving license is required to certify that one is eligible and has the right to drive. Yet, several times, this right is repealed. As additional guarantee, blockchain can be employed to verify claims of any

P. Karrupusamy et al. (Eds.): ICSCN 2019, LNDECT 39, pp. 523–531, 2020.
https://doi.org/10.1007/978-3-030-34515-0_55

given person by checking whether his declarations have been digitally signed by other authoritative entities. Driving license issuing entities/authorities can digitally sign the concerned person's claim to eligibility to drive. Such examples are illlustrated in [2]. The main reasons for creating identity systems are trust and accountability. Precisely the role of identity online should make interactions smooth, energetic, and most particularly, secure and private. Today, with the details of our personal lives spread across the web, self sovereign identity seems like a isolated dream. When peer-to-peer (p2p) networks are employed for regulating the growth of data amongst institutions or establishments, creating in effect, decentralized databases, it is called blockchain. This technology is also referred to as Distributed Ledger Technology (DLT). In this case the verifier can see only the minimal information rest of information should hide this concept known as zero knowledge proof (ZKP). A ZKP is a method where one party can prove (prover) to another party (verifier) without revealing knowledge of secret itself. The three conditions that must be satisfied for a Zero Knowledge protocol are: • Completeness: Given a true statement AND that the same protocol is followed by both parties, then the verifier will be undoubtedly convinced. • Soundness: Given a false statement, a verifier will not be convinced. • Given a true statement, any verifier will learn only that the given statement is true, and will not learn anything else [22]. Zero knowledge proof explained by far has been proving itself to be capable of handling enterprise level businesses. Not everyone is a fan of the public ledger system where everyone can see your transactions. Yes, you do get the anonymity with the help of addresses, but still, people can track addresses too. Also, when it comes to storing additional sensitive information blockchain isn't the best idea. Enterprises deal with a lot of private info, and the existing privacy protocol isn't enough. Zero knowledge proofs explained only can improve blockchain, but it can also get rid of all the negative issues. Many enterprises aren't interested in blockchain even though; its a beautiful invention. But with the help of Zero Knowledge proof explained, now everyone can start to utilize it. [23] All details regarding the smart contract in Ethereum's blockchain model are publicly made available on their network. All transactions and codes are also visible [24] (Figs. 1 and 2).

Fig. 1. Current identity system [20]

Fig. 2. Self-sovereign identity system [20]

2 Review of Literature

Most architectures don't allow the user to store their own data. Presently, a single decentralized system is accessible, which has not yet gained wide spread recognition. A case study on a solution has been conducted in this paper. The solution allows trade of KYC attributes, which are outcomes of Customer Due Diligence(CDD). This is often seen during opening of bank accounts. Now, mortgage lenders and insurance companies can make use of these attributes to make the induction processes easier for their customers. This is so because the customers won't have to repeat the same documentation process again. Costs and errors too can be minimized if institutions outsourced their CDDs. Though this idea had potential, it didn't perform very well. The organization which performed the solution was small, and the process hugely complex, which was also proprietary. This dependence on vendor restricted reproducibility [1]. In the nascent phases of every technology, trials and errors abound. Interfaces and protocols modify over time. However, at some point, all this needs to be standardized. An inter-operable manner of accessing information is necessary.

Ethereum's ERC-725 vand blockchain startup uPort's ERC1056 are examples of such standards. uPort's standard is compliant with the proposals for Decentralized Identifiers put forward by W3C Verifiable Claims Groups. This standardization effort expands beyond the Ethereum ecosystem. uPort's alternative is also suited to IETF standard and JSON web tokens [2]. A budding application of blockchain is the decentralized application (dApp). It might even completely replace platforms that are operated by centralized third party entities. Managing USer Identity however still remains critical and is regulated by third party entities. This is aimed to be addressed by ERC725 standard. The creator of the ERC20 token standard, which is widely used in the blockchain technology - Fabian Vogelsteller - also created ERC725, which has been chosen by several institutions aiming to include in peer to peer market places decentralized identity standards [3]. Self-sovereign (Self dependent) identity gives the user fully autonomous digital identities. This means that a person can control their digital selves as much as their physical selves. They need not depend upon third parties such as Facebook, Instagram, Twitter etc. to issue an identity that they may use. They may create completely unique identifiers with the power to regulate what information is shared with whom, and under what king of circumstances. We don't need to be liable to a third party's terms and agreements. The new self-dependent identity system will help shift the balance [4]. Distributed ledgers are utilized so that decentralized identifiers can be accessed sans a centralized directory, in self-sovereign identity systems. Blockchain technology offers a possible solution to address the identity problem, in the form of decades of unused cryptography knowledge. This may allow people to prove their personage using an authentic online decentalized system. For instance, consider a place you wish to visit and you require to show your age proof. You may simply present your digital driver's license which is digitally signed by license issuing authorities. Anyone can search or decentralized identifiers and access publicly associated keys [5]. In light of what kind of blockchain show these DApps use, they are categorized into three classes. Several current applications are based over these categories.

- **Type1:** Such dApps have their own blockchain viz. bitcoin. Falling into the same category are other altcoins.
- **Type2:** Such dApps utilize blockchains of by type 1 dApps. They are protocols that have tokens that are necessary for their function. An example of a type 2 dApp is the omni protocol.
- **Type3**: Such dApps utilize type 2 dApps protocols. For example, in the SAFE network, distributed file storage are build using SafeCoins, which uses the Omni protocol [18].

3 System Architecture/System Overview

A certifiable audit trail is created of this transfer, which is facilitated by an thereum smart contract. A back-end interface (Javascript, Web3) and a front end interface (ReactJS, web browser) composes the structure of a dApp (Fig. 3).

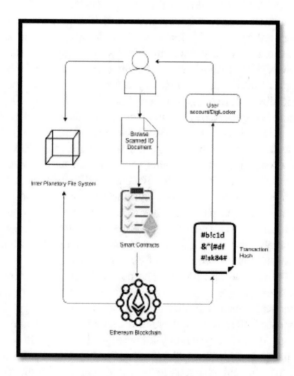

Fig. 3. System architecture

4   Proposed Methodology

In this system, smart contracts are applied to make documents tamper-proof and immutable. So it will overcome the theft of ID proofs or any confidential document. For this ethereum has been used as a blockchain platform because it is a public blockchain. One more advantage of this system is minimal or only required information will be disclose from our document, nobody can hack it. And for store those documents this system uses file system called Inter Planetory File System (IPFS).

5   Result and Discussion

DataSet: Real world documents like land contracts, Aadhar Card, Driving Licence, Voter ID etc. (Fig. 4).

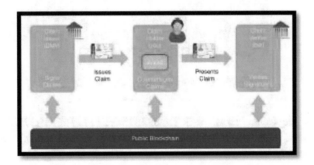

Fig. 4. Working of SSI dApp [5]

Self-sovereign Identity Dapp Performance and Transaction Status
Firstly we will compile our solidity smart contract which is the basis of transaction between two accounts. Tomorrow's web browser can be visited in a browser today using a bridge named MetaMask. Running a complete Ethereum node is unnecessary even though MetaMask will run Ethereum dApps in your browser. MetaMask also accomodates a wallet for secure identity, regulating your online identities and signing blockchain transactions via a user-interface. Metamask is used to choose rinke by test network and show the interaction between accounts below figure shows the contract deployment status (Fig. 5):

Fig. 5. Contract deployed on metamask

Contract is deployed from **0x7f1be5223136751111167d1b57c67eccd08976ba** account address to **0x83207c910b6af6a56ce0df9ef8b3278f8175d187** and the whole transaction summary i.e. its successful can be seen in below Fig. 6:

Fig. 6. Transaction successful

Our block status gas price and other statistics can be seen in below ether scan window Fig. 7:

Fig. 7. Statistics

Serverless dApps can be designed using Ethereum's blockchain technology. Users may store personal information. All this information is encrypted and stored in a decentralized form. This information can be shared with other user's at the owner's discretion.

6 Conclusion

A practical implementation of sharing objects using a dApp is shown in this work. This study is conducted with the ultimate aim to make a Decentralized dApp on which user must have their own control on their documents. This experiment is conducted using Smart contracts in solidity, Ganache Server, Metamask (Ethereum based DApp), ReactJS (Frontend), and IPFS (Inter Planetory File System). The Blockchain technology plays very important role to make our Identity decentralized. This dApp is useful to secure our documents and can store it into hash format so that no one hack it. In our smart contract demonstration, we have hash value of the document to the web app, also the smart contracts for zero knowledge proof (ZKP).

7 Future Work

This work can be done more effective on permissioned blockchain i.e. on 'Hyperledger Indy' we can make it more valuable. Also there is separate platform for sovereign identity namely 'Sovrin'. So by using sovereign it'll be more easy to implement unrestricted identity.

References

1. Baars, D.: Towards Self-Sovereign Identity using Blockchain Technology. University of TWENTE. https://essay.utwente.nl/71274/1/Baars-MA-BMS.pdf
2. BlockTelegraph, Dr Stefan Beyer Dr Stefan Beyer Self-Sovereign Identity: Blockchains Real Killer Application? Block Telegraph (blog), 28 September 2018. https://blocktelegraph. io/self-sovereign-identityblockchains-real-killer-application/

3. What is Ethereum ERC725? Self-Sovereign Identity Management on the Blockchain. CryptoSlate (blog), 8 August 2018. https://cryptoslate.com/what-is-erc725-self-sovereign-identitymanagement-on-the-blockchain/

4. Theres A Facebook Alternative. Its Called Self-Sovereign Identity - CoinDesk. https://www.coindesk.com/theresalternative-facebook-called-self-sovereign-identity. Accessed 28 June 2019

5. Windley, P.: How Blockchain Makes Self-Sovereign Identities Possible. Computerworld, 10 January 2018. https://www.computerworld.com/article/3244128/how-blockchain-makesself-sovereign-identities-possible.html

6. Jurowiec, P.: This Thing Called Blockchain... [Beginners Guide]. Piotr Jurowiec (blog), 20 August 2018. https://medium.com/@piotr-61543/this-thing-called-blockchain-beginners-guide-1849d79f1c99

7. How Blockchain Can Solve Identity Management Problems. https://www.forbes.com/sites/forbestechcouncil/2018/07/27/how-blockchain-can-solve-identity-managementproblems/2436b8e813f5. Accessed 1 Feb 2019

8. Azure Blockchain Content and Samples. Contribute to Azure-Samples/Blockchain Development by Creating an Account on GitHub.HTML. 2018. Reprint, Azure Samples (2019). https://github.com/Azure-Samples/blockchain

9. Bogner, A., Chanson, M., Meeuw, A.: A decentralised sharing app running a smart contract on the ethereum blockchain. In: Proceedings of the 6th International Conference on the Internet of Things - IoT16, 17778. Stuttgart, Germany. ACM Press (2016). https://doi.org/10.1145/2991561.2998465

10. Blockchain and Digital Identity A Good Fit? Internet Society (blog), 13 March 2018. https://www.internetsociety.org/blog/2018/03/blockchain-digital-identity-good-fit/

11. Ahammad, R.: Aivon Artificial Intelligence Image Identifier. Raju Ahammad (blog), 30 November 2018. https://medium.com/@rajuahammad1981/aivon-artificial-intelligenceimage-identifier-3371250080e5

12. Decentralized Digital Locker Blockchainerz. https://articles.abilogic.com/310377/decentralized-digitallocker-blockchainerz.html. Accessed 27 Feb 2019

13. Digital Identity Management for NGOs Using Blockchain (HackSociety 2018 Entry). Dynaquest (blog), 22 October 2018. http://www.dqtsi.com/2018/10/22/digital-identity-management-for-ngosusing-blockchain-hacksociety-2018-entry/

14. How Blockchain Can Solve Identity Management Problems - One Kosmos (BlockID). https://onekosmos.com/blog/how-blockchain-can-solve-identitymanagement-problems/. Accessed 1 Feb 2019

15. What Is DApps (Decentralised Apps)? BTC Wires (blog). https://www.btcwires.com/glossary/what-isdapps-decentralised-apps. Accessed 1 Feb 2019

16. South African Cryptocurrency Magazine. Cryptocurrency Yes or Now - Read More Here! http://technomagazine.net/image-post46.html. Accessed 1 Feb 2019

17. Ruff, T.: The Three Models of Digital Identity Relationships. Evernym (blog), 24 April 2018. https://medium.com/evernym/the-threemodels-of-digital-identity-relationships-ca0727cb5186

18. Step by Step Guide to Build a Dapp. Heptagon (blog),Paul, Moses Sam, 8 March 2018. https://medium.com/heptagon/step-by-step-guideto-build-a-dapp-a-homo-sapiens-2-day-love-affair-with-ethereum-dappde2b0dea12f1

19. What are DApps (Decentralized Applications)? The Beginners Guide. CoinSutra - Bitcoin Community, 24 July 2017. https://coinsutra.com/dapps-decentralized-applications

20. lmasi, P.P.: The Identity Revolution Self Sovereign Powered by Blockchain. Good Audience, 7 February 2019. https://blog.goodaudience.com/how-blockchain-could-become-theonramp-towards-self-sovereign-identity-dd234a0ea2a3

21. James: The Basics of Decentralized Identity. UPort (blog), 27 June 2018. https://medium.com/uport/the-basics-of-decentralized-identityd1ff01f15df1

22. Shish: Introduction to Zero Knowledge Proof: The Protocol of next Generation Blockchain. Medium (blog), 8 October 2018. https://medium.com/coinmonks/introduction-to-zero-knowledge-proofthe-protocol-of-next-generation-blockchain-305b2fc7f8e5
23. What Is ZKP? A Complete Guide to Zero Knowledge Proof. 101 Blockchains (blog), 29 November 2018. https://101blockchains.com/zero-knowledge-proof/
24. What Zero-Knowledge Proofs Will Do for Blockchain. VentureBeat (blog), 16 December 2017. https://venturebeat.com/2017/12/16/whatzero-knowledge-proofs-will-do-for-blockchain/

Analysis of Dispersion Compensation in Wavelength Division Multiplexing System for 8 * 20 Giga Bits Per Second Optical Fiber Link

Pragya Purohit$^{(\boxtimes)}$, M. L. Meena, and J. B. Sharma

Department of Electronics Engineering, Rajasthan Technical University,
Kota, India
pragyapurohit2172@gmail.com,
madan.meena.ece@gmail.com, jbsharma@rtu.ac.in

Abstract. Optical Fiber Communication is the rapidly evolving form of communication for long haul applications. Due to its compact size, low losses and versatile application, it is significantly used in thousands of kilometers of distance. But the major limiting factor of using fiber technology is Dispersion. It causes pulse broadening due to multiple signals propagation in Wavelength Division Multiplexing (WDM) network and leads to inter-symbol interference. To overcome the effects of dispersion, many dispersion compensation technologies has been adopted. Dispersion Compensation Fiber (DCF) is one of the compensation scheme used in fiber optics. In this paper, a WDM channel having 8 * 20 Giga bits per second link capacity has been analyzed using pre compensation, post compensation and symmetric compensation with different encoding schemes(NRZ and RZ). The results are investigated in terms of Quality factor (Q-factor) and Bit Error Rate (BER). The simulation is carried out in OptiSystem 7 simulator.

Keywords: Optical Fiber Communication · Dispersion Compensating Fiber · WDM · Q-factor · Ber · Optisystem 7 simulator

1 Introduction

Optical communication system has been in the development of high speed, large bandwidth and the long haul applications, but the loss and dispersion of the optical fiber limits the optical fiber transmission system [1]. With the development of erbium-doped fiber amplifier (EDFA) [2], optical signal transmission loss problem has been resolved. The optical signal can be transmitted farther. It can be used in many configurations like pre amplifier, inline amplifier and boost amplifier [3]. But the accumulation of dispersion is more considerable with the distance increasing. The resulting of dispersion pulse broadening has generated serious Inter-symbol Interference (ISI) [4], which limits the transfer rate and distance relay, so the dispersion problem is more prominent.

© Springer Nature Switzerland AG 2020
P. Karrupusamy et al. (Eds.): ICSCN 2019, LNDECT 39, pp. 532–541, 2020.
https://doi.org/10.1007/978-3-030-34515-0_56

OptiSystem Simulator Software is an advanced, comprehensive software design suite that enables users to plan, test, and simulate optical links in the transmission layer of modern optical networks. OptiSystem offers optical transmission system design and planning from component to system level and present the analysis and scenarios visually [5].

With the invention of Wavelength division multiplexing (WDM), signals with different wavelengths can be transmitted onto a single fiber, which increases the capacity of the system [6, 7]. AWDM model consists of a multiplexer where optical carriers of different wavelengths are combine and the resultant wave is fed to a de-multiplexer which split the optical carriers at their central frequencies. By employing this system, multiple signals can be transmitted with minimum bandwidth required and high capacity can be obtained [8, 9] (Fig. 1).

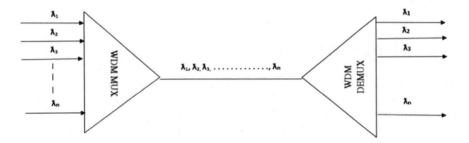

Fig. 1. Basic WDM model

In WDM network optical transmission, the factor of pulse broadening is arises because of multiple signals propagating inside the fiber which affects the quality of the signal and due to this at receiver, the probability of getting error is high [10]. To resolve this problem of overlapping of signals, many compensation schemes have been developed. The one of the most efficient technique is Dispersion Compensating Fiber (DCF) which cancels out the positive dispersion value and improve the quality of received signal [11].

2   Dispersion Compensating Fiber (DCF)

Chromatic Dispersion (CD) in a fiber comprising of Material and Waveguide Dispersion resulting from variation in group velocities associated with different spectral components of the signal is the major type of dispersion occurring in single mode fiber (SMF) [12].

Dispersion compensation fibers (DCF) are designed fibers possess high negative dispersion value in the range of −70 to −90 ps/(km-nm) [13]. They cancel out the positive dispersion value produced by single mode fiber along the fiber length. For compensating the effects of pulse broadening in SMF, the length and dispersion constant value of both fibers (SMF and DCF) must satisfy the equations [14]:

$$D_S L_D + D_D L_S = 0 \tag{1}$$

$$L_D = -(D_S/D_D)L_S \tag{2}$$

Here L_S and L_D are the lengths of SMF and DCF, D_S and D_D are the dispersion constants of SMF and DCF.

To compensate the effects of dispersion on receiving signal, DCF can be allocated at three different locations with respect to SMF [15]. The compensation techniques are defined below:

- **Pre Technique:** In this technique, the compensating fiber is placed before the SMF.
- **Post Technique:** In this technique, the compensating fiber is placed after the SMF.
- **Symmetric Technique:** It is the combination of pre and post technique. In this, compensating fiber is placed before and after the SMF.

3 Proposed Simulation Model

The basic blocks of fiber communication consists of a transmitter section, optical fiber channel, and a receiver section. The transmitter section includes Pseudo-random bit sequence generator (PBRS), NRZ encoder, C/W laser source and Mach-Zehnder modulator. PBRS generator produces a stream of random sequences at 20 Gbps which is encoded by NRZ encoder. The optical source used for generation of optical wave is continuous wave laser at input power of 5dBm and it is modulated by MZ modulator having extinction ratio of 30 dB.

The modulated wave propagates through channel comprises of optical fibers and amplifiers. At receiver, p-i-n photodiode is used to convert the optical signal into electrical pulses, then it is passed through low pass Bessel filter to filter unwanted frequencies and the received signal is analyzed through BER analyzer (Fig. 2).

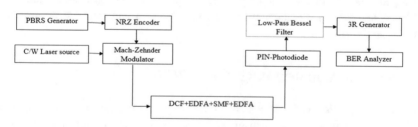

Fig. 2. An overview of fiber communication

In this study, analysis of dispersion compensation in WDM network is investigated using dispersion compensating fiber. In Fig. 3, the proposed model for 8 channel WDM network fiber link is given. At transmitter, 8 laser sources producing optical signals of power 5 dBm and the spacing between channels used is 0.1 THz. The WDM mux combine these signals of different wavelengths and transmit it over fiber link consisting of SMF, DCF (operating at 1550 nm optical window) and an inline Erbium Doped Fiber Amplifier (EDFA). The noise center frequency for EDFA in pre-compensation is chose to be 193.5 THz and in post-compensation, the noise center frequency is kept at 193.6 THz.

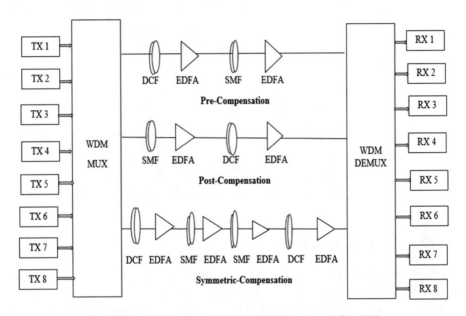

Fig. 3. Block Diagram of 8 channel WDM network fiber link

The variation in position of SMF and DCF is chosen according to the compensation scheme. The length of SMF and DCF in each compensation technique is kept same. At receiver, WDM de-mux is used to separate the received signal into 8 individual channel frequency. The center frequency of p-i-n diodes is set according to the individual frequencies at transmitter. The simulation parameters are tabulated in Table 1 and the optimized parameters of SMF and DCF are tabulated in Table 2.

Table 1. Simulation components parameters

Data source	Data rate: 20 Gbps
Encoding schemes (RZ and NRZ)	duty cycle (RZ): 0.5
C/W laser	Output Power: 5dBm FWHM line width: 10 MHz 1st–8th channel frequencies 193.1 THz–193.8 THz Channel spacing: 0.1 THz
Mach-Zehnder modulator	Extinction ratio: 30 dB
EDFA length	Variable gain(10–30 dB)
Receiver	Photodetector: p-i-n photodiode Sensitivity: −100dBm Probability of error: 10^{-9}
Low pass Bessel filter	−3 dB Bandwidth: 10 GHz

Table 2. SMF and DCF components values

Components	SMF	DCF
Length (km)	180	36
Dispersion value (ps/km.nm)	16	−80
Dispersion slope (ps/nm^2/km)	0.075	−0.475
Attenuation constant value (dB/km)	0.15	0.4

4   Results and Discussion

The parameters measured to analyze received signal quality in fiber communication is Q-factor and BER. Higher value of Q-factor and lower value of BER shows less noisy and better communication system. For enhanced optical communication, the value of Q-factor must be greater than 6 and for BER, the acceptable range is 10^{-9} to 10^{-12} or lower than this. The simulation design of the optical link is done by Optisystem 7 simulator. To compensate the effects of dispersion, pre, post and symmetric compensation DCF technique is used with NRZ and RZ modulation format having EDFA amplifier for SMF of length 180 km and DCF of length 36 km. For simplicity, the eye diagrams are given for 193.1 THz, 193.5 THz and 193.8 THz central frequencies. The output readings of compensation schemes are tabulated in Tables 3, 4 and 5. The graphs are also plotted for Q-factor and BER given in Figs. 5, 6, 7 and 8. From the eye diagrams given in Fig. 4, the received signal quality is best achieved in symmetric compensation with RZ modulation in terms of maximum Q-factor, min. BER and eye opening.

Table 3. Output readings in terms of Q-factor and BER for Pre-Compensation DCF

Channel frequency (THz)	PRE compensation			
	NRZ scheme		RZ scheme	
	Q-Factor	BER	Q-Factor	BER
193.1	37.1442	$1.89776e^{-302}$	36.6825	$4.83787e^{-295}$
193.2	28.951	$8.84996e^{-185}$	32.3947	$1.12726e^{-230}$
193.3	24.9698	$4.24563e^{-138}$	30.527	$4.02939e^{-205}$
193.4	31.2893	$2.13619e^{-215}$	33.714	$1.25043e^{-249}$
193.5	25.8723	$4.37323e^{-148}$	31.1494	$1.80508e^{-213}$
193.6	31.6092	$9.41259e^{-220}$	32.5838	$2.13947e^{-233}$
193.7	27.6757	$4.3387e^{-169}$	35.6917	$1.86216e^{-279}$
193.8	32.162	$2.04539e^{-227}$	35.3866	$9.50904e^{-275}$

Table 4. Output readings in terms of Q-factor and BER for Post-Compensation DCF

Channel frequency (THz)	POST-Compensation			
	NRZ scheme		RZ scheme	
	Q-Factor	BER	Q-Factor	BER
193.1	33.66	$7.6550e^{-249}$	37.809	$1.65038e^{-313}$
193.2	27.1446	$1.03616e^{-162}$	23.2946	$1.85177e^{-120}$
193.3	31.4205	$3.70004e^{-217}$	26.7691	$2.65944e^{-158}$
193.4	31.7924	$2.9987e^{-222}$	26.5105	$2.68492e^{-155}$
193.5	28.6078	$2.08693e^{-180}$	24.2812	$1.1573 \, le^{-130}$
193.6	30.0646	$5.08586e^{-199}$	24.3638	$1.41715e^{-131}$
193.7	30.4107	$1.46162e^{-203}$	21.1853	$4.51908e^{-100}$
193.8	34.075	$6.67567e^{-255}$	31.6745	$1.34148e^{-220}$

Table 5. Output readings in terms of Q-factor and BER for Symmetric Compensation

Channel frequency (THz)	SYMMETRIC-Compensation			
	NRZ scheme		RZ scheme	
	Q-Factor	BER	Q-Factor	BER
193.1	21.0408	$9.38535e^{-099}$	52.2396	0
193.2	20.6042	$8.72558e^{-095}$	59.34	0
193.3	19.8842	$1 \, 98101e^{-088}$	44.4776	0
193.4	21.0933	$3.17551e^{-099}$	51.1984	0
193.5	19.8559	$3.34729e^{-088}$	51.6194	0
193.6	19.0869	$1.04954e^{-081}$	54.2363	0
193.7	$19.3 \, 139$	$1.01284e^{-088}$	50.4826	0
193.8	22.4373	$5.90201e^{-112}$	47.624	0

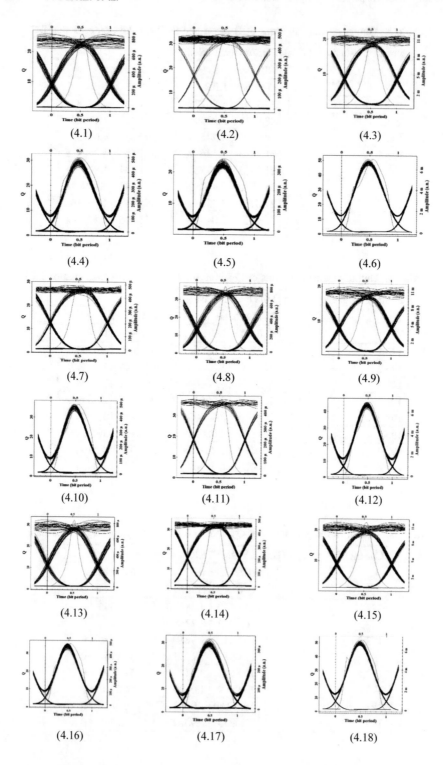

(4.1)   (4.2)   (4.3)

(4.4)   (4.5)   (4.6)

(4.7)   (4.8)   (4.9)

(4.10)   (4.11)   (4.12)

(4.13)   (4.14)   (4.15)

(4.16)   (4.17)   (4.18)

◄ **Fig. 4.** (4.1–4.3) Eye diagrams at 193.1 THz with NRZ scheme for pre, post and symmetric compensation. (4.4–4.6) Eye- diagrams at 193.1 THz with RZ scheme for pre, post and symmetric compensation. (4.7–4.9) Eye-diagrams at 193.5 THz with NRZ scheme for pre, post and symmetric compensation. (4.10–4.12) Eye diagrams at 193.5 with RZ scheme for pre, post and symmetric compensation. (4.13–4.15) Eye-diagrams at 193.8 with NRZ scheme for pre, post and symmetric compensation. (4.16–4.18) Eye-diagrams at 193.8 THz with RZ scheme for pre, post and symmetric compensation

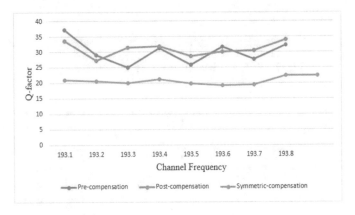

Fig. 5. Q-factor vs. channel frequency with NRZ format

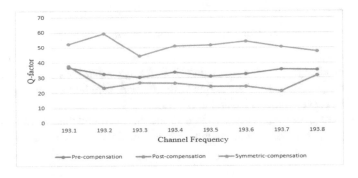

Fig. 6. Q-factor vs. channel frequency with RZ format

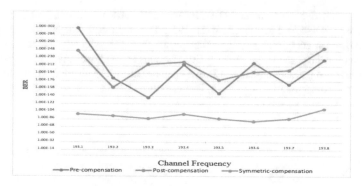

Fig. 7. BER vs. channel frequency with NRZ format

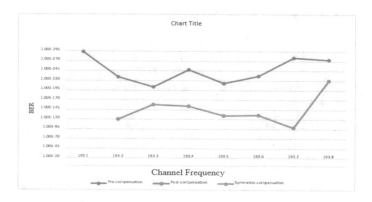

Fig. 8. BER vs. channel frequency with RZ format

The received signal is analyzed using BER Analyzer and the eye diagrams for the pre, post and symmetric DCF technique with different encoding schemes are given in Fig. 4 at 193.1 THz, 193.5 THz and 193.8 THz.

By comparing the graphs given in Figs. 7 and 8, it is found that the minimum BER is obtained in symmetric compensation with RZ modulation format at each channel frequency.

5 Conclusion

In this study, an optical link of length 216 km (SMF of 180 km and DCF of 36 km) has been analyzed for 8 channel Wavelength Division Multiplexing (WDM) network using Dispersion Compensation Fiber (DCF) as compensation scheme. The performance of designed optical link is done in Optisystem 7 simulator and the results are investigated in terms of parameters like Q-factor, BER and eye-diagram with RZ and NRZ modulation formats. Further the comparative analysis of pre, post and symmetric-compensation DCF technique has been evaluated and by comparing the received signals quality, it is found that max. Q-factor and min. BER is best achieved in symmetric-compensation scheme with RZ modulation format whereas with respect to NRZ modulation, pre and post compensation gives better results.

References

1. Mohammadi, S.O., Mozaffari, S., Shahidi, M.M.: Simulation of a transmission system to compensate dispersion in an optical fiber by chirp gratings. Int. J. Phys. Sci. **6**(32), 7354–7360 (2011)
2. Keiser, G.: Optical Fiber Communications, 4th edn, pp. 255–412. Tata McGraw Hill Education Private Limited (2008)
3. Zaman, M.U., Iqbal, S., Isha: Performance analysis of EDFA Amplifier for DWDM system. IEEE (2014)
4. Senior, J.M., Jamr, M.Y.: Optical Fiber Communications: Principles and Practices. Pearson Education (2009)
5. Dar, A.B., Jha, R.K.: Design and comparative performance analysis of different chirping profiles of tanh apodized fiber Bragg grating and comparison with the dispersion compensation fiber for long-haul transmission system. J. Mod. Opt. **64**(6), 555–566 (2017)
6. Zhang, P.: Study on dispersion compensation and modulation in DWDM system. Taiyuan University of Technology, Shan Xi (2013). (in Chinese)
7. Gopika, P., Thomas, S.A.: Performance analysis of dispersion compensation using FBG and DCF in WDM systems. Int. J. Adv. Res. Comput. Commun. Eng. **4**(10), 1–4 (2015)
8. Yadav, M., et al.: Design performance of high speed optical fiber WDM system with optimally placed DCF for dispersion compensation. Int. J. Comput. Appl. **122**(20), 1 (2015)
9. Patel, G.H., Patel, R.B., Patel, S.J.: Dispersion compensation in 40 Gbps WDM network using dispersion compensating fiber. J. Inf. Res. Electron. Commun. Eng. **2**, 662–665 (2013)
10. Meena, M.L., Meena, D.: Performance analysis of DWDM optical network with dispersion compensation techniques for 4 × 8 Gbps transmission system. ICTACT J. Microelectron. **4**(2), 613–617 (2018)
11. Shukla, R.P., Kumar, M.: Performance analysis of dispersion in optical communication link using different dispersion compensation fiber (DCF) models. IJRTE **1**(2), 161–163 (2012)
12. Moses, B.D., Lakshmy, G.B.: Analysis of intensity modulation techniques in DWDM transmission systems. IJSRD **3**, 114–117 (2015)
13. Tosson, M., El-Deeb, W.S., Abdelnaiem, A.E.: Dispersion compensation techniques for DWDM optical networks. IJARCCE **4**, 1–6 (2015)
14. Neheeda, P., Pradeep, M., Shaija, P.J.: Analysis of WDM system with dispersion compensation schemes. Procedia Comput. Sci. **93**, 647–654 (2017)
15. Sharma, M., Raghav, P.K., Chaudhary, R., Sharma, A.: Analysis on dispersion compensation in WDM optical network using pre, post, symmetrical DCF based on optisystem. MIT Int. J. Electron. Commun. Eng. **4**(1), 58–63 (2014)

A Smart Assistive Stick Guide for the Visually Impaired with a Crisis Alert and a Virtual Eye

Adwayt Pradeep Nadkarni[✉], Jaisagar Ahuja, Rahul Upadhyay,
and Budhaditya Bhattacharyya

School of Electronics Engineering, VIT, Vellore, Tamil Nadu, India
{adwaytpradeep.nadkarni2015, jai.sagar2015,
rahul.upadhyay2015, budhaditya}@vit.ac.in

Abstract. The conventional stick guide used by a visually impaired person helps detecting objects in their immediate vicinity, but it proves highly inefficient in detecting objects or obstructions that are not in the physical range of the stick. There is no provision made within the stick to help the visually impaired people from impending danger, nor there is no provision made to even contact a guardian when lost or injured. In this paper, a cane model is designed to exploit the sense of hearing and touch for the visually impaired person. The design also endows a visually impaired person with auxiliary "senses" through means of sensors and camera modules to percieve different kinds of obstacles, effectively increasing the perception range. The functionalities hailed in this paper include sensor-based close range detection, camera based traffic sign mid-long range detection and a GPS-based location deduction is developed to send alert messages.

Keywords: Arduino · Raspberry pi · Ultrasonic sensor · IR(infrared) sensor · Water level sensor · Camera module · GSM (global system for mobile) module · GPS (global positioning system) module · microSD card reader module · CNN (convolutional neural networks) · Google maps · SMTP(simple mail transfer protocol) query

1 Introduction

As per recent tabulations, World Health Organization (WHO) declared that 285 million people are visually impaired worldwide, of which, 39 million are blind and 246 have low vision [1]. Another estimate states that 90% of the visually impaired people around the world live in developing countries [2]. Locally, we see that within India, there are 12 million legally blind people, making India home to close to a third of the world's blind population [2].

The conventional assistive cane used by a visually impaired person has several drawbacks, and may not guarantee a collision less journey. Much work on Electronic Travel Aids (ETAs) devices has already been done, majority of which focuses on alerting the visually impaired person through means of sound and vibration [3, 4].

One such design is the voice operated outdoor navigation system for visually impaired persons [5], which uses a stick equipped with ultra-sonic sensors, GPS,

© Springer Nature Switzerland AG 2020
P. Karrupusamy et al. (Eds.): ICSCN 2019, LNDECT 39, pp. 542–554, 2020.
https://doi.org/10.1007/978-3-030-34515-0_57

buzzer, vibration motor and miroSD card module. Pre-recorded longitude and latitude data are stored in a SD card, which guides the visually impaired person to his/her destination. The ultra-sonic sensors detect any obstacle directly the buzzer will activate the vibration motor. This system can be classified as a low cost system affordable by the user. The issue with this design, however, is it does not take into consideration that the path pre-recorded may not be feasible which could lead to fatal accidents.

Another such design is the arduino based automated stick guide was made that incorporates ultrasonic sensors, IR sensors, GSM module and GPS module [3]. This design works efficiently for short range detection only and it also provides a crisis alert feature to contact a guardian which makes up for the short comings in [5]. However, the message alert provided are raw longitude and latitude data which, by themselves do not provide the location and leads to unnecessary delay. Also, this design heavily relies on the proper functioning of the guardian's cellular phone and does not make any provision if the guardian's phone stopped working.

Our assistive stick guide design incorporates this model and enhances it by adding components like a camera module which performs long-range to mid-range detection. Using a CNN based real time image detection algorithm discussed in later sections, an alert message is deployed based on the emergency signs. The image dataset includes images like slippery road sign, no entry sign and construction at work sign amongst 43 other signs [6] which if ignored, could pose a danger to a visually impaired person. Complementing the detection algorithms, the design also provides a feature to send a SMS and an e-mail to the guardian(s) in cases of emergency or if the visually impaired person feels lost [3]. This module was upgraded by providing a visually pleasing map in contrast to just the latitude and longitude coordinates.

2   Proposed Methodology

2.1   Architecture

The proposed architecture of the design is depicted in the block diagram given Fig. 1. For the ease of understanding the roles and requirements of each component, the design is split into three units.

- **The Sensor Unit** - consists of IR sensor, water level sensor, ultrasonic sensor and an arduino. Role: short distance detection [8].
- **The Virtual Eye Unit** – consists of camera module, microSD card reader module, speaker/headphones, Wif-Fi enabled raspberry pi and arduino. Role: long and mid range detection.
- **The Location Deduction Unit** – gsm module, gps module, push button, Wi-Fi enabled raspberry pi and arduino. Role: alert messages [6].

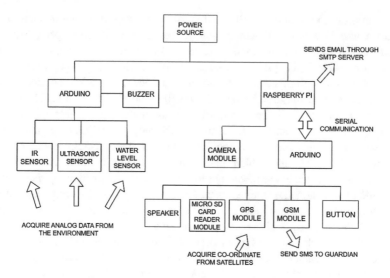

Fig. 1. Block diagram for the proposed design

2.2   Components Required

2.2.1   Software Requirements

- **Raspbian:** this software is Debian based and it is the official operating system for Raspberry Pi.
- **Python:** An open source programming language that is flexible and highly robust. In addition to this, it is compatible with the Raspberry Pi.
- **C ++:**A middle level language compatible with Arduino, which is normally used in hardware communication.

2.2.2   Hardware Requirements

- **Raspberry Pi with Wi-Fi module:** a single board mini-computer developed in the United Kingdom. It is the only component connected to the internet, hence empowering it to send emails.
- **Arduino:** a single board microcontroller which allows one to read analog data provided by the sensors and control other hardware modules.
- **Water level Sensor:** a single point level sensor which is connected at the end of the stick to sense a spillage or puddle of water on the street.
- **Ultrasonic Sensor:** This sensor is connected in the mid region of the to detect obstacles in front of the visually impaired person.
- **IR Sensor:** this sensor is placed at the end of the stick to measure depth of the elevated or depressed feature [8].
- **Camera Module:** a camera module with 3.85 mm focal length is placed on the top of the cane for the ease of image acquisition.

- **Buzzer:** an acoustic que, that deploys sounds of different frequencies depending on the sensor triggering it. It is placed on the top of the cane.
- **Speaker:** an over the ear 0.5 W, 8 Ω headphone speaker that deploys an audio file depending on the sign detected by the raspberry pi.
- **MicroSD card reader module:** stores the audio file.
- **GPS Module:** a delicate module which acquires the longitude and latitude information necessary to track the visually challenged person's location.
- **GSM Module:** this module uses a SIM card and its services to send an alert SMS to a guardian.

2.3   Implementation

2.3.1   Sensor Unit

- The sensor unit connections are as illustrated in Fig. 2. These sensors that are attached to the cane acquire signals from the real world and generate an analog values.
- These acquired readings forms the basis for decision making that takes place in the microcontroller.

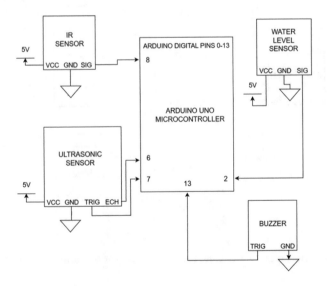

Fig. 2. Sensor unit hardware connections

- The acquired inputs are then compared with a set threshold value for each sensor such that, the tone of the buzzer, which is unique for every sensor, gets progressively sharper, the closer the obstruction.

2.3.2   Location Deduction Unit

- Connections with the GSM module, GPS module, a push button and a speaker to the Arduino and establish a serial connection with the Raspberry pi establishes the location deduction unit. The connections are illustrated in Fig. 3.
- In regards to the flowchart given in Fig. 4, when the press button is pushed, a flag is raised that triggers the GPS to collect location information from satellites. The data recorded from the GPS contains 19 types of free flowing NMEA sentences, of which GPGGA sentences are of importance to the design as they hold latitude and longitude information.
- This latitude and longitude information is forwarded to GSM module which sends SMS and the data is sent to Raspberry Pi via serial communication to send an email as illustrated in Figs. 6(a), (b) and Fig. 7.

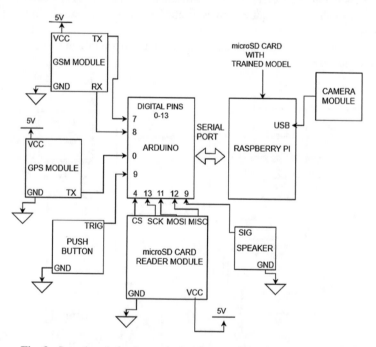

Fig. 3. Location deduction and virtual eye unit hardware connections

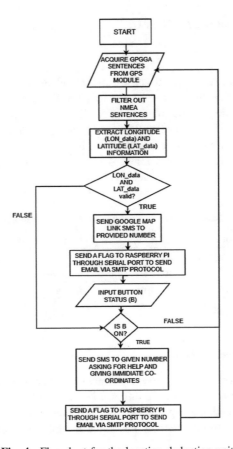

Fig. 4. Flowchart for the location deduction unit

2.3.3 Virtual Eye Unit

- To the given connections to the location deduction unit in, a camera module is connected via a USB connection to the raspberry pi, a microSD card reader module is connected to the arduino and 0.5 W, 8Ω speaker was connected to an arduino. This as shown in Fig. 3.
- Pre-made audio files corresponding to each class or sign are recorded, sampled at 16000 Hz and converted into an 8-bit WAV file.
- A dataset of 50,000 images with 43 classes illustrated in Fig. 5 with varying sizes of 15×15 to 222×192 were acquired [7]. 34,799 of these images were randomly selected as training images and were fed to a convolutional neural network to train the model.
- The image frames captured by the camera are resized to $32 \times 32 \times 3$ and pre-processed to fit the model's criteria and is fed to the model.
- The predicted result sends a flag to the arduino via serial communication which deploys the appropriate audio stored in microSD reader module to the speaker.

3 Proposed Design and Testing

3.1 Sensor Unit

Different sensors that are used to sense different elements of the environment are attached to the assistive stick and are used to sense the surroundings throughout it's functioning. This collected information forms the basis to produces analog signals which are sent to the microcontroller. This analog information is then used to compare against a set threshold value on basis of which the visually impaired person is made aware of his surroundings. Tables 1 and 4 provides details regarding tone and frequencies corresponding to the obtained readings.

Table 1. Sensor readings on test observations

Case type	Type of obstacle	Ultrasonic sensor	Infrared sensor	Water level sensor
TYPE I	Incoming Car	Active	Inactive	Inactive
TYPE II	Tree	Active	Inactive	Inactive
TYPEIII	Staircase	Inactive	Active	Inactive
TYPE IV	Water Puddle	Inactive	Inactive	Active

3.2 Location Deduction Unit

The location deduction unit was tried in different weather conditions and locations to realize the true potential of the unit. Table 2 shows that the location deduction unit works best outdoors and gets a good fix. The GPS however takes a long time to get a fix indoors, especially in rainy conditions where it hardly ever gets a good fix.

Table 2. Time of GPS fix in an urban environment

Case type	Type of obstacle	Location	Types of region	Time to get a GPS fix (minutes)
TYPE I	Sunny	Indoor	Urban	8–15
TYPE II	Sunny	Outdoor	Urban	2–3
TYPE III	Rainy	Indoor	Urban	>90
TYPE IV	Rainy	Outdoor	Urban	30–40

3.3 Virtual Eye Unit

We used a system with a Core i7-7700HQ (2.8 GHz), 16 GB DDR and a GTX 1050Ti graphics cards to train the model. The convolutional neural network model is a feed-forward neural network architecture that is trained using gradient descent. Weights of the neural network model were initialized by uniform random distribution.

Table 3. Architecture of the convolutional neural network

Layer (type)	Output shape	Parameters
Convolution 2D (1)	(None, 28, 28, 60)	1560
Convolution 2D (2)	(None, 24, 24, 60)	90060
Max pooling 2D (1)	(None, 12, 12, 60)	0
Convolution 2D (3)	(None, 10, 10, 30)	16230
Convolution 2D (4)	(None, 8, 8, 30)	8130
Max pooling 2D (2)	(None, 4, 4, 30)	0
Flatten	(None, 480)	0
Dense (1)	(None, 500)	240500
Dropout (1)	(None, 500)	0
Total parameters		378023
Trainable parameters		378023

Fig. 5. Image dataset with 43 classes (Source: Stallkamp J, Schlipsing M, Salmen J, Igel C. The German Traffic Sign Recognition Benchmark: A multi-class classification competition. In IJCNN 2011 Jul 31, p. 3)

The model discussed in Sect. 2.3.3 accepts images with a sized of $32 \times 32 \times 3$. A set of 12,630 images, which were not fed to network were used as validation set and the remaining 4,410 images were used as test dataset. The architecture of the convolutional neural network is depicted in Table 3 and the model's response to test images is graphed in Figs. 8 and 9. The 43 unique images focused in this paper is shown in Fig. 5.

4 Results and Discussion

4.1 Sensor Unit

The sensors provide real time data to a microcontroller which is compared against the parameters shown in Table 4 to deploy a unique buzzer tone which alerts the visually impaired person. The design was run on sample test obstacles shown in Tables 1 and 4.

Table 4. Response of test cases to sensor parameters

Obstacle	Type of obstacle	Parameters	Status	Frequency of tone (Hz)
Approaching Car	Dynamic	20 < D < 50	Reasonably far	8000
		10 < D < 20	Approaching	4000
		D < 10	Dangerously close	1000
Tree	Static	20 < D < 50	Reasonably far	8000
		10 < D < 20	Approaching	4000
		D < 10	Dangerously close	1000
Staricase	Elevated	5 < D < 9	Elevated feature	2000
	Depressed	D > 15	Depressed feature	2700
Water Puddle	Wet	V > 3.3	Possible chance of slipping	6000

where D is the experimental distance in centimetres acquired from ultrasonic and ir sensors, and A is the voltage reading recorded from water level sensor when it comes in presence with water.

4.2 Location Deduction Unit

When the proposed design was tested in an urban environment and indoor location with clear skies, the results are as illustrated in Figs. 6(a), (b) and 7. As seen in Fig. 7, this link is a query containing the cane's longitude and latitude coordinates acquired from a GPS module. When the link is pressed open, it takes the guardian to the Google maps app where the location of visually impaired person and the cane be seen graphically. Finally, a flag and the coordinates are sent from Arduino to Raspberry Pi via serial communication which through a simple SMTP query to send an email to the guardian.

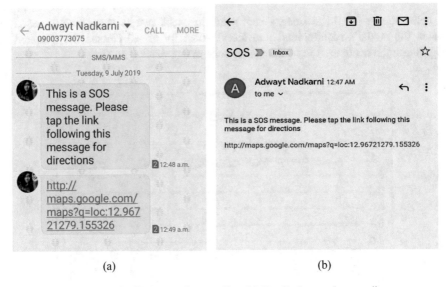

(a)   (b)

Fig. 6. (a) SMS alert to the guardian (b) Email alert to the guardian

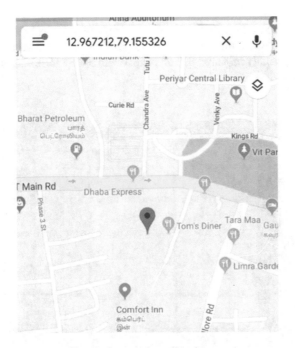

Fig. 7. Location on Google map

4.3   Virtual Eye Unit

A collection of 4410 test images kept aside to check results were pre-processed to match the model's requirements and the f1 score for all 43 classes was graphed in Fig. 8 and the accuracy model against epochs in Fig. 9.

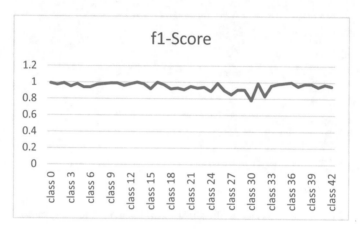

Fig. 8. Location on Google map

From Fig. 8, the plot shows that the f1 score for every class ranges between 0.8 and 1. As the f1 score is high, the chances of false positives and false negatives is low. This ensures that the visually impaired person is not disturbed by false alarms and can feel at ease while navigating.

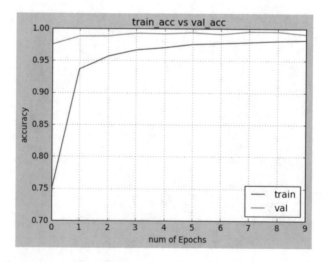

Fig. 9. Accuracy of training set and validation set against number of epochs

From Fig. 9, the plot shows that the model could not trained any more as the trend for accuracy on both datasets have become parallel to each other in the last few epochs. We can also see that the model has not over-learned the training dataset, and showing comparable skill on both datasets.

5 Conclusion

It is worth mentioning at this point that developing such an empowering design for the visually impaired people was on top of our priorities. Hence, our primary goal of collision avoidance, danger prediction and rescue in the guide stick commits to the same. The results, having successfully implemented, tabulated and tested show promise in practical scenarios. This design, though requires hard wiring several sensors and their associated modules, is not very bulky and can be carried around with ease. This paper is but a stepping stone to more efficient and reliant technologies that will improve independent navigation experience amongst the visually impaired people. One such way is incorporating artificial intelligence concepts to help the visually impaired person with route navigation and ground texture recognition. Also, the model can be made eco-friendly by adding solar cells so that the battery does not have to be changed repeatedly. The model's camera has been restricted to only emergency signs in this paper, but it can truly become a virtual eye if it can be successfully used to detect humans, features and provide more information on fast moving objects and vehicles. Nonetheless, this design has shown merits in short and medium range collision avoidance, long range precautionary caution and an improved visually pleasing alert feature in a hope that they prove beneficial in making the visually impaired person self-reliant in navigation.

References

1. World Health Organization: Universal eye health: a global action plan 2014–2019. World Health Organization (2014)
2. http://www.who.int/mediacentre/factsheets/fs282/en/
3. Gayathri, G., Vishnupriya, M., Nandhini, R., Banupriya, M.M.: Smart walking stick for visually impaired. IJECS 3(3), 4057–4061 (2014)
4. Mahmud, M.H., Saha, R., Islam, S.: Smart walking stick-an electronic approach to assist visually disabled persons. Int. J. Sci. Eng. Res. 4(10), 111–114 (2013)
5. Koley, S., Mishra, R.: Voice operated outdoor navigation system for visually impaired persons. Int. J. Eng. Trends Technol. 3(2), 153–157 (2012)
6. Swain, K.B., Patnaik, R.K., Pal, S., Rajeswari, R., Mishra, A., Dash, C.: Arduino based automated stick guide for a visually impaired person. In: 2017 IEEE International Conference on Smart Technologies and Management for Computing, Communication, Controls, Energy and Materials (ICSTM), 2 August 2017, pp. 407–410. IEEE (2017)

7. Stallkamp, J., Schlipsing, M., Salmen, J., Igel, C.: The German traffic sign recognition benchmark: a multi-class classification competition. In: IJCNN, 31 July 2011
8. Nada, A.A., Fakhr, M.A., Seddik, A.F.: Assistive infrared sensor based smart stick for blind people. In: 2015 Science and Information Conference (SAI), 28 July 2015. IEEE (2015)

Discovering Frequent High Average Utility Itemset Without Transaction Insertion

Pramod Singh Negi[(⊠)], Samar Wazir, and Md. Tabrez Nafis

Department of Computer Science and Engineering, Jamia Hamdard,
New Delhi 110062, Delhi, India
pramodsinghnegi75@gmail.com,
samar.wazir786@gmail.com, tabrez.nafis@gmail.com

Abstract. Data mining is a technique through which we can find interesting data and sequence from the available wide range of data source. Incremental high-average utility pattern mining (IHAUPM) algorithm is represented to manage the incremental database with transaction insertion. IHAUPM algorithm basically follows the comparison to the original database and newly inserted database itemset if itemset has High Average Utility Upper Bound Itemset (HAUUBI) in the initial database as well as new transaction database then the item always frequent. Second situation itemset has non-High Average Utility Upper Bound Itemset (non-HAUUBI) in the initial database as well as new transaction database then the item always not frequent. Otherwise, the itemset is recurring or not is identified by the given information. This new algorithm is represented in this paper to generate expected high utility frequent item set to form a new transaction database; this algorithm is much faster than the existing algorithm.

1 Introduction

Data mining is a technique through which we can find an interesting pattern of data from the huge Database. i.e. Data mining is the process of discovering patterns in large data sets involving methods at the intersection of machine learning, statistics, and database systems [3, 18]. There are some different requirement come inside the various application. Here we can show the knowledge differently as par requirements. Apriori [1] is an algorithm which is come first to mine the correlation between the purchase items collection (Sets). i.e. the Apriori algorithm [1] is represented to frequent itemset mining (items occurs in how many time within a transaction) and association rule that applied on the transactional database. It is processed through the identifying the frequent independent items inside the existing database. This algorithm also has association rule mining (ARM) [3]. Association mining is generally used to identify the purchase of one item is effect to purchase of another item. Association rule mining has followed two steps that is (1) Frequent Itemset Generation: that Generate the itemsets which follow some following condition support $\geq$ minsup and another step (2) Generation rule:- every frequent item set follow generate high confidence rule. Frequent itemset has binary partitioning of each rule. The Apriori algorithm represents the set of frequent item (FI) [8] set. Infrequent pattern (FP)-growth [20, 21] algorithm is used for

© Springer Nature Switzerland AG 2020
P. Karrupusamy et al. (Eds.): ICSCN 2019, LNDECT 39, pp. 555–569, 2020.
https://doi.org/10.1007/978-3-030-34515-0_58

searching the frequent itemset inside the transaction database without candidate generation. FP (frequent pattern) [21] growth is denoted the recurring items set in the form of hierarchical structure that is known as Frequent pattern tree (FP-tree) [21]. The database size is dynamically changed in the real world when a new transaction is merged to an original database then the representation of information is necessary to update the original database then maybe some information is lost at that time. FIM [8, 9] or ARM [3, 4], an algorithm can only, however. We have handled the database that has no change in the future i.e. static database. But we update the database; batch manner must be happening in this algorithm, that indicates the already represented information is not useable and it is necessary to update database it scanned and study to mine the necessary meaningful group of data (i.e. information) handle this situation of transaction insertion of the database. Here introduced a Fast Update Algorithm that contains the pruning mechanism that used inside the DHP (Direct Hashing and Pruning) algorithm is used for updating the already existing information. It is divided the initial database and new transaction inserted into the four variations and each variation act by the own algorithm to update the essential information. Some items that are not satisfied with the condition of FUP [22] that are eliminated inside the FUP. Significantly improve the updating performance when it can do. In data mining, Frequent pattern mining (FPM) have some problem, that contains to finding the recursive pattern inside a transactional database. main purpose of frequent mining itemset to search the frequent itemsets. These patterns have taken as input a transaction database and an argument that is 'minsup' i.e. the 'minsup' is a minimum support threshold. In this algorithm, produced a group of items that have must contain minimum support in transactions. Frequent pattern mining (FPM) have popular issue. But frequent pattern mining has some limitation when we are observing the customer transaction. It has some significant restriction is that have no mentioned purchase quantities into the account. i.e. in the whole transaction the purchase items can become visible once or zero time. Suppose that a buyer purchase six pieces of bread, twenty pieces of bread or thirty pieces of bread, these wholes of thing are considered as same. Another important constraint is that every item is taken as same importance the utility of weight suppose a buyer buys a very costly bottle of wine or a cheap piece of bread, then both of are considered as equal importance (Table 2).

Table 1. Transaction Items

	Transaction items
T_1	{2, 3, 1, 4, 5}
T_2	{6, 2, 1}
T_3	{5, 2, 1}
T_4	{1, 8, 5, 6, 3}
T_5	{4, 3, 1}

Table 2. Minimum support: 3

If minsup = 3 then	
Itemset	Support
{1}	5
{2}	4
{3}	4
{5}	3
{1, 2}	3
{1, 3}	3
{1, 5}	3

The limitation of frequent itemset mining FIM [8] has been redefining in High utility itemset mining (HUIM). In this issue, every transaction is stored into the transactional database along with selling quantities, and each item has per unit profit. e.g. we have taken the following transaction database (Table 3).

Table 3. Transactional database with quantity

Transaction	Items
T_1	1{2}, 2{1}, 3{7}, 4{1}
T_2	2{2}, 4{5}, 6{2}
T_3	1{4}, 5{2}
T_4	3{2}, 4{6}, 6{2}

where: A{B}: transaction has contained an item A and {B} are denoted the quantity of item A in that transactional database.

Table 4. Frequent item in transactional database

Item	Unit profit
1	5 $
2	2 $
3	1 $
4	3 $
5	2 $
6	4 $

Now we are taken transaction T3; that denotes the corresponding buyer taken 2 units of an item '3', 6 unit of item '4' and 2 units of item '6'. We easily look at above the transactional table that has represented the T3 transaction and the respective per unit item profit is denoted in Table 4. The item profit in the unit "3", "4" and "6" are respectively 1$, 3$ and 4$, i.e. for example, that every unit of item "6" generates the 4$ profit for per unit selling. HUIM [22, 23] is used to search an itemset (collection of items) that produced a high profit in a database if both of items are sold together. In high utility, itemset has contained a number of items and per unit value. This value is the weight of item in a transactional table. The user provides a minimum support value. This minimum support value is called the minimum utility threshold that denoted by 'minutil'. A HUIM algorithm is used to generates all the utility itemset. That means the utility value of the item in itemset is more than equal to the minimum defining threshold 'minutil'.

For Example, suppose we have defined a 'minutil' 25$ for the user. Then high utility itemset has considered the only itemset that has minutil equal and less than the utility of itemset taken that suppose minimum utility is equal to 25$ for the item in itemset. That the conclusion of the high utility itemset Mining algorithm is shown in below side Table 5.

Table 5. High utility itemset

High utility itemset	
{1}	30 $
{4}	36$
{3, 4}	30$

Here, I have to find a high frequency set inside a new transactional database and the original transactional database. That approach is used to calculate the which items in the new database are frequent with respect to original database. So without merging the new database into the original database, we have easily got it the Frequent item that exists in the new database.

2   Related Work

Here, we have to do some related work with the merging model of both databases (original database and new database). Both high utility itemset mining (HUMI), frequent item set, and respectively discussed the HAUIM. This concept is used in this paper. That is the key part of this paper.

2.1   Data Mining

Data mining is techniques that are used to filter a deserving data and engaging pattern or rules from a Hugh amount of Transactional database. The significant basic approach is known as association rule mining (ARM) [3, 4].that is used to find the combination of the item that has an impact on each other. That ARM is approach in the form of different level it first the frequent itemset (FIs) [16, 17] mines based on minimum threshold support and applies the grouping concept from the minded FIs [17] to get the ARs based on minimum threshold confidence. That means if an item is purchased with another item, then that represents the confidence. And the Apriori algorithm [1] is generally applied in the form of level wise approach. Thus a large amount of memory and runtime usage is required. To solve this problem through Frequent Pattern Tree structure (FP tree) [22] and FP growth algorithm for efficiently mining FPs without generation of candidate itemsets. In further, some improvement is done inside the FP growth and proposed the FP-array technique that has decreased. The FP-tree traversing time of node to node. Some other thing is NFP- tree structure that has also reduce the number of node in the tree and also reduced Header_table size that has to take counts two for every node of the tree. Likewise, the small node in the tree is traversed in the structure of the tree to nine the FIs [17].

These algorithms have to process fixed transactional database. In the real environment, the transaction database usually grows over time, and after we mine data, then we again re-evaluated. Like that association rules mining (ARM) [3, 4], some new ARs may be generated, and some old one may become invalid when new transaction inserted into the original Transactional database. But the traditional algorithm is processed in the form of batch manner to again processed the Updated Whole database. Now transaction inserts, delete or update forms the original Transactional database. That process has many computation operations by which discovered knowledge increase a cost and waste. Fast Updated (FUP) [25, 26] concept is very effectively managing the incremental transactional database. FUP [25, 26] concept of the old database and newly inserted transaction are categories into 4 part and every case is performed by its own process to manage and modify the information in the original transaction database.

2.2 Apriori Algorithm [1]

Apriori algorithm [1] is introduced by R. Agrawal and R. Srikant in 1994 for searching the itemset that is frequent in a dataset through the Boolean association rule. Name of Apriori is coming from prior knowledge of properties of a frequent itemset. That algorithm depends on the level-wise finding technique were to suppose K-frequent itemsets are used to search K + 1 itemset. To enhance the efficiency of level wise generated frequent itemset. So we are used here some important characteristic of Apriori that help to search space reducing This algorithm has followed some property. Subset of frequent itemset of all non-empty itemset is always. That property is called closer property. It assumes that the subset of a frequent itemset is must be frequent. If itemset may be infrequent then the superset of this itemset is always infrequent. We have already discussed all important definition and which are explained in above the side.

Suppose we have to Consider a transactional database, and we have to search the frequent itemset and generate association rules on this (Table 6).

Table 6. Minimum support count is 2, minimum confidence is 60%

Transaction_ID	Items
T1	{1, 2, 5}
T2	{2, 4}
T3	{2, 3}
T4	{1, 2, 4}

size of item k = 1

Compare itemset that considered minimum support (min_support = 2) for that transactional database to find the frequent itemset in it. if minimum support of itemset is less than 2 then that will be removed from the frequent itemset like that 3, 1 (Table 7).

Table 7. Frequent itemset count for size 1

Itemset	Sup_count
{1}	2
{2}	4
{4}	2

Itemset size K = 2

Generate Itemset I_2. In which item set have contained two items in each itemset. Now we will check all subset, frequent or not. If the itemset is frequent then that have stayed in frequent itemset. Otherwise, that have removed (Table 8).

Table 8. Frequent itemset count for size 1

Itemset	Sup_count
{1,2}	2

In size 3(K = 3) no them itemset satisfied the minimum support. So we no need to Processed the other itemset. So in above the table size, k = 1 and k = 2 have followed the minimum support, so we are only k = 1, and k = 2 are considered here.

$$\text{Confidence}(P \rightarrow Q) = \text{Support\_count}(P \cup Q)/\text{Support\_count}(P)$$

2.3   Frequent Itemset Mining

Frequent itemset mining [13, 17] is a technique through which we can search the item frequency in itemset of each transactional database. Now an item is purchased with another item also, so the Association Mining [6] searches for frequent items in the dataset. Infrequent mining usually, the correlations and interesting associations between item sets in transactional and relational databases are found in short, Frequent Mining shows which items appear together in a transaction or relation. We have required the association mining for predicate the purchase of one item is an impact on the purchase to another item in the transactional database. Frequent mining is the generation of association rules from a Transactional Dataset. If there are two items, X and Y frequently purchased then its good to put them together in stores or provide some discount offer on one item on purchase of another item. This can really increase sales. For example, it is likely to find that suppose a buyer buy a tea and biscuit, he also buys milk.

Important Definitions:

Support: It is one of the measures of interestingness. This tells about the certainty and usefulness of rules. 5% Support means a total of 5% part of that transaction inside the database are follows the rule.

$$support(P \to Q) = support\_count(Q \cup P)$$

Confidence: If let we consider a 60% confidence, that means 60% customer the have to buy tea and biscuit also bought milk.

$$Confidence(P \to Q) = Support\_count(P \cup Q)/Support\_count(P)$$

If both minimum support and minimum confidence are satisfies this rule. Then it is a strong rule.

Support_count(X): Total number of transaction on which the X is discovered. Suppose that X is P U Q(P union Q) then that has to provide the number of transaction on which both P and Q are lies.

Maximal Itemset: A set of items is called maximal frequent if all the superset is not frequent.

Closed Itemset: An itemset is closed if none of its immediate supersets has same support count same as Itemset.

K- Itemset: Itemset contain K items is called K-itemset. So we can say this itemset is frequent if the corresponding support count is greater than the minimum support count.

3   Problem Statement

In traditional, we have to find the high utility item in the new database and the old transactional database. First, we have to merge both database and after that, we have to find which item is either frequent or not. in this paper, we have introduced the way by which we have easily find the high average utility item set (HAUI) [22] before merging both databases under some specific condition. Suppose we have two databases first is new database d. that have some items set $d_I = \{i_1, i_2, i_3, i_4........\}$ And second is old transactional database D. that have also some items set $D_I = \{i_1, i_3, i_5, i_6...........\}$.So we need not to merge both transaction of the database (d + D) for finding which item is HAUI [22] with respect to old transactional database D.

4   Proposed Discover High Average Utility Itemsets with Frequent Itemset

In this post, we have to find the high average utility item set without merging both databases (old transactional database and new transactional database) inside to it. So by the help of this algorithm, we can easily identify which itemset in HAUI(High average utility itemset) [22]. That whole is categories into four-part. That parts are mentioned into the below side (Table 9).

Table 9. Cases for the transactional database(original and new)

Case	Itemset	Old (original) transactional database	New transactional database	Conclusion
Case 1	Utility itemset	High	High	High
Case 2	Utility itemset	High	Low	High
Case 3	Utility itemset	Low	High	The conclusion depends on that given information
Case 4	Utility itemset	Low	Low	The conclusion depends on the given information

In above Table 10, there are four cases is demonstrated here. Initial two cases of the table are indicating the high utility itemset in a transaction. A remaining case has no final conclusion because the whole calculations are depending on the given information of a new database.

Case 1:

If an itemset $I_D = \{i_1, i_2, i_3, i_5, \ldots\ldots.in\}$ in old transactional database 'D' with high average utility itemset and itemset $I_d = \{i_1, i_2, i_3, i_5, \ldots\ldots.in\}$ in new transactional database 'd' then we can easily say these itemset have high average utility after merging the both database.

That means

$$I_D \geq threshold(Itemset\ in\ old\ Transactional\ Database)$$
$$I_d \geq threshold(Itemset\ in\ new\ Transactional\ Database)$$

Before merging both of those transactional databases. We get conclusion the itemset has high average utility in Old transactional database.

So that

$$I_D + I_d \geq U_D\ Or\ I_D + I_d \geq (U_D + U_d)$$

Where U_D is denoted the minimum threshold in the old transactional database. U_d is denoted the minimum threshold in the new transactional database.

Case 2:

Both transactional database (T_D, T_d), here T_D represents the Original Transactional database and T_d is a new transactional database. In this case, the average utility of itemset in the Original Database is high, and the average utility of itemset in the new database is low. Then we can identify the average utility of that itemset before the database merging. So we need not be necessary to combine the transaction of both databases and calculate after that. We can easily say that itemset is high average utility itemset after margining

That means

$$I_D \geq \text{threshold} (\text{Itemset in old Transactional Database})$$
$$I_d \leq \text{threshold} (\text{Itemset in new Transactional Database})$$

Before merging both of those transactional databases. We get conclusion the itemset has high average utility in Old transactional database.

So that

$$I_D + I_d \geq U_D$$

Where U_D is denoted the minimum threshold in the old transactional database

U_d is denoted the minimum threshold in the new transactional database.

Case 3:

If in Original Transactional database itemset have low average utility and in new Transactional database itemset have high average utility then we don't say the itemset of new Transactional database also has high utility. But after merging the new Transactional database in the original database, we can say it. In which have decided based on the given information. That means the low average utility itemset of original database is make high utility after merging the high utility itemset of new Transactional database.

That means

$$I_D \leq \text{thresholds} (\text{Itemset in old Transactional Database})$$
$$I_d \geq \text{thresholds} (\text{Itemset in new Transactional Database})$$

Before merging both of that transactional database. We get conclusion the itemset has high average utility in Old transactional database.

So that

$$\text{If } I_D \leq \text{threshold and } I_d \geq \text{threshold}$$

then $I_D + I_d \geq U_D$ or $I_D + I_d \leq U_D$ {Depend on given information}

That conclusion depends on the given information of new transactional database and the original transactional database. That itemset may be either high average utility or low average utility. But that have decided on the base of the given information.

Where:

U_D is denoted the minimum threshold in the old transactional database

U_d is denoted the minimum threshold in the new transactional database.

Case 4:

If the average utility of an itemset is low in the new transactional database and average utility of the item set is also low in the original transactional database then. We don't say that itemset has high average utility in original database. i.e. it depends on the given information if that have satisfied the given information then we can say that itemset has high average utility itemset in the original database.

That means

$$I_D \leq threshold(\text{Itemset in old Transactional Database})$$
$$I_d \leq threshold(\text{Itemset in new Transactional Database})$$

Before merging both of that transactional database. We get a conclusion the itemset has high average utility in Old transactional database or not.

So that

If $I_D \leq threshold$ and $I_d \leq threshold$

then $I_D + I_d \geq U_D$ or $I_D + I_d \leq U_D$ {Depend on given information}

That conclusion depends on the given information of new transactional database and the original transactional database. That itemset may be either high average utility or low average utility. But that has decided on the based on the given information.

Algorithm 1:

Input: O_D is an original transactional database, and N_d is a new transactional database a quantitative database. Here profitable, a profit table have some utility TU_D in original database and T_{MU} Is indicate minimum average utility threshold,

1 for every $Td \in d$ do

2 populate $tmu(Td)$;

3 for every item kj in d do

4 populate $auub(kj)$;

5 get perform total utility in new transactional database N_d as TUNd ;

6 update and calculate the total utility of both Transactional database $TU_{ONU} = TUO_D + TU$Nd ;

7 for every item that has $kj \in$ indexTable in htree do

8 for every index item that has kj in d do

9 if $auub(kj)$Nd + $auub(kj)$OD $\geq T_{MU} \times TU_{ONU}$ then

10 $auub(kj)_{ONU} = auub(kj)Nd + auub(kj)OD$;

11 if $TU_{OD} > T_{MU}$ AND $TU_{Nd} > T_{MU}$

12 then given itemset is frequent as per this condition.

13 Refine (k) $\in$ indexTable as $auub(kj)_{ONU}$;

14 _item_set $\leftarrow$ insert_item_set $\cup$ kj ;

15 else if $TU_{OD} < T_{MU}$ AND $TU_{Nd} < T_{MU}$

16 itemset is not frequent so ignore it.

17 else if $TU_{OD} > T_{MU}$ AND $TU_{Nd} > T_{MU}$

18 _item_set $\leftarrow$ insert_item_set $\cup$ kj ;

19 else if $TU_{OD} < T_{MU}$ AND $TU_{Nd} > T_{MU}$

20 parentNode(kj).child$\leftarrow$ childNode(kj).parent ;

21 else if $(kj)OD + auub(kj)Nd < T_{MU} \times TU_{ONU}$ then

22 parentNode(kj).child$\leftarrow$ childNode(kj).parent ;

23 remove kj from htree;

24 remove kj from indexTable;

25 else

26 if ( )Nd $\geq TUNd \times T_{MU} \wedge kj \notin index\_table$ then

27 rescan O_D to obtain (kj)OD;

28 if $(kj)OD + auub(kj)Nd \geq TU_{ONU} \times T_{MU}$ then

29 rescan_item_set $\leftarrow$ rescan_item_set $\cup$ k ;

30 itemset rescan_item_set and sort in current auub-descending order;

31 insert items in rescan_item_set to the end of indexTable;

32 UPdateNodeBranch(rescan_item_set);

33 UPdateNodeBranch(insert_item_set);

Algorithm 2:

Input: provide the itemset as an argument.

Output: update the branch to the tree.

1 for every item set k_j in the *set* do

2 for every itemset $k_j \subseteq T_d \wedge T_d \in O_D \| T_d \in N_d$ do

3 if $k_j \in$ tree. branch $\wedge k_j \in$ insertItemSet then

4 every itemset node itemsetNode(k_j).*auub*+ = *tmu*(T_d);

5 itemsetNode (k).qun_ary_Ary += T_d .*quantity*[1..*j*];

6 else

7 if k*j* $\in$ insertItemSet $\wedge ij \in$ rescanItemSet then

8 find the k .*position* in *T*d $\in O_D$;

9 for every itemset itemsetNode(*ij*).*auub* = *tmu*(*Tq*);

10 itemsetNode(*ij*).quan_Ary = *T*d .*quantity*[1..*j*];

5   An Illustrated Example

In this section, we have to demonstrate the whole flow to that algorithm and also mentioned how these algorithms are worked here. We have to use two Transactional databases, first is the original database and the second is a new transactional database. In original Transactional table we have considered two transaction T_1 and T_2 both have respectively contained {1, 2, 3, 4, 6} and {2, 3, 5}, and new transactional database also have contain two Transaction that is T_1 {1, 2, 3, 4, 6} and T_2 {2, 3, 5} both Transactions have different minimum average utility suppose that we have declared minimum average utility for original transactional database is 10 and average utility for new transaction database is 6 then we get the conclusion over different itemset is comes under the high average utility or not.

Let we consider minimum utility is 15 for the original transactional database when were minimum utility is 6 for a new Transactional database. an itemset {1}, {2}, {3}, {4}, {5}, {6}, {1, 2}, {2, 3}, {4, 2} have contain average utility respectively 5, 8, 12, 9, 8, 5.5, 10, 15 in original Transactional database. But only {4, 2} itemset has satisfied the given minimum utility condition. But when we consider the new transactional database then some conclusion will be changed like that {4} have low average utility in original Transactional database when we consider the new transactional database then {4} average utility is an increase that has 18 average utility.

So that in above the case we have got the conclusion on the bases of Transactional database. So we can easily find out which itemset in high utility itemset and which is not.

Table 10. Original transactional database

Transaction	Items	Transaction utility	Item utilities for this transaction
T1	{1,3 2 4 6}	32	{9 6 6 5 6}
T2	{3 2 5}	16	{2 8 6}

Table 1 it denotes only the original transactional database and that is an old database which has some itemset. These itemsets have different average utility. In Table 11 we have mentioned only two transactions T_1, and T_2, both of these, have item set. Table 11 have contained only six items.

Table 11. New transactional database

Transaction	Items	Transaction utility	Item utilities for this transaction
T1	{1 4 2 3 6}	32	{5 6 9 6 6}
T2	{2 5 3}	10	{2 2 6}

6 Conclusion

In this paper, we can find the high utility of item set without merging both databases. i.e. we can find easily which item is more frequently occurred in every transaction without merging that process are basically save the memory so if have less memory then this method is more useable. That have considered basically 4 cases. That is (1) that have consider if items in both side (Old Transactional Database and New Transactional Database) are high frequency, then this item is must be frequent after merging both databases. (2) in the second case the items set are highly frequent in old Transaction database but low frequent in new Transactional database that itemset is frequent after merging both databases. (3) in the third case the itemset is a low frequency in the old transactional database but that have high in an old transactional database the item must be in high frequency after merging both databases. (4) in the last case both databases have an itemset in low frequency then itemset must be low.

References

1. Agrawal, R., Srikant, R.: Fast algorithms for mining association rules. In: Proceedings of the 20th VLDB Conference, Santiago, Chile, pp. 487–499 (1994)
2. Agrawal, R., Shafer, J.C.: Parallel mining of association rules. IEEE Trans. Knowl. Data Eng. 8(6), 962–969 (1996)
3. Han, E., Karypis, G., Kumar, V.: Scalable parallel data mining for association rules. In: Proceedings of 1997 ACM-SIGMOD International Conference on Management of Data, Tucson, Arizona (1997)
4. Han, E., Karypis, G., Kumar, V.: Scalable parallel data mining for association rules. IEEE Trans. Knowl. Data Eng. 12(3), 337–352 (2000)
5. Joshi, M.V., Han, E., Karypis, G., Kumar, V.: Efficient parallel algorithms for mining associations. In: Zaki, M., Ho, C.-T. (eds.) Large-Scale Parallel and Distributed Data Mining. Lecture Notes in Computer Science/Lecture Notes in Artificial intelligence (LNCS/LNAI), vol. 1759. Springer (2000)
6. Cheung, D.W., Han, J., Ng, V.T., Fu, A.W., Fu, Y.: A fast distributed algorithm for mining association rules. In: Proceedings of 4th International Conference on Parallel and Distributed Information System, Los Alamitos, California. IEEE Computer Society Press, pp. 31–42 (1996)
7. Cheung, D.W., Ng, V.T., Fu, A.W., Fu, Y.: Efficient mining of association rules in distributed databases. IEEE Trans. Knowl. Data Eng. 8(6), 911–922 (1996)
8. Bernecker, T., Cheng, R., Cheung, D.W., Kriegel, H.P., Lee, S.D., Wang, L.: Model-based probabilistic frequent itemset mining. Knowl. Inf. Syst. 37, 181–217 (2013)
9. Bernecker, T., Kriegel, H.P., Renz, M., Verhein, F., Zuefle, A.: Probabilistic frequent itemset mining in uncertain databases. In: Proceedings of the 15th ACM SIGKDD Conference on Knowledge Discovery and Data Mining (KDD 2009), Paris, France (2009)
10. Cheng, R., Kalashnikov, D., Prabhakar, S.: Evaluating probabilistic queries over imprecise data. In: SIGMOD (2003)
11. Chui, C.K., Kao, B., Hung, E.: Mining frequent itemsets from uncertain data. In: 11th Pacific-Asia Conference on Advances in Knowledge Discovery and Data Mining, PAKDD 2007, Nanjing, China (2007)
12. Zhang, Q., Li, F., Yi, K.: Finding frequent items in probabilistic data. In: SIGMOD (2008)
13. Wang, L., Cheng, R., Lee, S.D., Cheung D.W.-L.: Accelerating probabilistic frequent itemset mining: a model-based approach. In: CIKM (2010), pp 429–438
14. Sun, L., Cheng, R., Cheung, D.W., Cheng, J.: Mining uncertain data with probabilistic guarantees. In: SIGKDD (2010)
15. Chui, C.K., Kao, B.: A decremental approach for mining frequent itemsets from uncertain data. In: PAKDD, pp. 64–75 (2008)
16. Tong, Y., Chen, L., Cheng, Y., Yu, P.S.: Mining frequent itemsets over uncertain databases. In: VLDB 2012 (2012)
17. Tong, Y., Chen, L., Yu, P.S.: UFIMT: an uncertain frequent itemset mining toolbox. In: KDD 2012, 12–16 August 2012, Beijing, China (2012)
18. Cheung, D.W., Xiao, Y.: Effect of data distribution in parallel mining of associations. Data Min. Knowl. Discov. 3, 219–314 (1999)
19. Frequent Itemset Mining Implementations Repository http://fimi.ua.ac.be/
20. SPMF An Open-Source Data Mining Library. http://www.philippe-fournier-viger.com/spmf/index.php?link=datasets.php
21. Han, J., Pei, J., Yin, Y., Mao, R.: Mining frequent patterns without candidate generation: a frequent-pattern tree approach. Data Min. Knowl. Discov. 8, 53–87 (2004)

22. Hong, T.P., Lin, C.W., Wu, Y.L.: Incrementally fast updated frequent pattern trees. Expert Syst. Appl. **34**, 2424–2435 (2008)
23. Liu, Y., Liao, W., Choudhary, A.: A fast high utility itemsets mining algorithm. In: The International Workshop on Utility-Based Data Mining, pp. 90–99 (2005a)
24. Liu, Y., Liao, W.K., Choudhary, A.: A two-phase algorithm for fast discovery of high utility itemsets. Lecture Notes in Comput. Sci. **3518**, 689–695 (2005)
25. Agrawal, R., Srikant, R.:. Fast algorithms for mining association rules in large databases. In: The International Conference on Very Large Databases, pp. 487–499 (1994)
26. Cheung, D., Han, J., Ng, V., Wong, C.: Maintenance of discovered association rules in large databases: an incremental updating approach. In: IEEE International Conference on Data Engineering, pp. 106–114 (1996)
27. Ahmed, C.F., Tanbeer, S.K., Jeong, B.S., Lee, Y.K.: Efficient tree structures for high utility pattern mining in incremental databases. IEEE Trans. Knowl. Data Eng. **21**(12), 1708–1721 (2009)
28. Lan, G.C., Hong, T.P., Tseng, V.S.: Efficiently mining high average-utility itemsets with an improved upper-bound strategy. Int. J. Inf. Technol. Decis. Mak. **11**(5), 1009–1103 (2012)

Augmented Reality Based Visual Dictionary for Elementary Students

Sudhir Bagade, Aparna Lidhu, Yamini Manral$^{(\boxtimes)}$,
and Jidnyasa Vartak

Usha Mittal Institute of Technology, SNDT Women's University, Mumbai, India
bsudhiran@ieee.org, aparnlidhu@gmail.com,
manralyamini@gmail.com, jidnyasavartak1997@gmail.com

Abstract. The contemporary world lives on information. Data must be related to the user and its mannerism of presentation should be as such it is easily understood by the user. Developing technologies, like Augmented reality (AR), will be widely accepted rapidly. AR is an outstanding technology that is already used in some original and artistic applications. For example, Snapchat, Pokemon Go, Skymap etc. This paper discusses the new opportunity for improving the quality of education for the elementary students with the assistance of AR technology. We are implementing Visual Dictionary, with two modules namely, alphabets and occupations. These AR applications could lead students through learning process in an improved way, as AR is an advancement of traditional books with a digital touch. With such an innovative learning technique, education process can be made more fun and interesting.

Keywords: Augmented reality · Smart learning · 3D objects ·
Unity3D · Vuforia · Digital education

1 Introduction

Augmented reality (AR) [1] is a collaborative experience of a real-life setting where the objects that reside in the real-life are "augmented" by computer-generated information. The meaning of augmented reality in simple terms is overlaying digital objects on participant's display (screen) which shows a clear view of real world.

Augmented Reality (AR) is blending of virtual and physical information in real-time using camera and appropriate software [2]. AR could provide users technology-mediated immersive experiences. By adding AR into lessons and lectures, it may help educators directly involve students in the studying process by using interactive 3D models. Such activities help in motivating students, capture their attention in the class and helps them understand the concept deeper using augmented reality. This kind of learning can also be used at home and for the revisions.

A perspective of using smartphone application using augmented reality for education process is remarkable [2], but in meantime only a few educators really use it. Such an application helps the students in various ways to provide tempting, interactive digital information about any topic. Reading and teaching in 2D can be difficult in some concepts, however, complex information is easier to understand or represent in

P. Karrupusamy et al. (Eds.): ICSCN 2019, LNDECT 39, pp. 570–581, 2020.
https://doi.org/10.1007/978-3-030-34515-0_59

3D format. This visualization can be better explained in Fig. 1. In Fig. 1, a mobile device is used to project a 3D model of the monument in the history book. Thus, using AR can change the way we perceive information [3].

Fig. 1. Example of augmented reality application [3]

Niantic's Pokémon Go [4, 5] uses GPS to spot your location and move your in-game avatar, while your smartphone camera is used to show Pokémon in the real world. Using Snapchat [6, 7], one can create their own miniature version and watch it come to life—in hundreds of stickers, and via augmented reality projections.

Various educationalists are still using outdated education methods to conduct the teaching and learning activities in the classroom via verbal communication [2]. However, this learning milieu has grown beyond classroom. The addition of technology in the teaching and learning procedure is the new trend in education, with encouraging results [2, 3].

In this paper, we are developing a mobile application for elementary students using Augmented Reality. The application consists of two modules, namely (1) Alphabet recognition (2) Occupation detection. To develop these modules, we have used Unity 3D as the Integrated Development Environment [8] and Vuforia as Software Development Kit [9]. The application detects images from the book and plays appropriate informative videos in augmented 3D format. Learning in this way is an enjoyable process.

Our paper is organized as follows. In Sect. 2, we discuss the literature survey. In Sect. 3, we present our two proposed modules in detail. Section 4 contains the methodology, and Sect. 5 discusses the implementation details and tools used for building the project. Lastly, we conclude our paper in Sect. 6 and state the future work.

2 Literature Survey

Traditional learning based on paper material is changing by enabling 3D features using AR applications. AR technologies urge the engagement of students in the virtual illustration of real objects or situations, even from historical environments.

Following are a few surveyed examples where companies are trying to collaborate immersive technology for educational purposes:

Google Earth [10] **Virtual Reality (VR)** enables mapped location to be experienced in virtual reality, transporting the user straight to the area they want to be, around the globe. This technology helps user to see the world how it looks like in reality, and not just through maps and pictures. It is currently compatible with the Oculus Rift [11] and HTC Vive [12] virtual reality headsets [13].

Google Expeditions takes students on field trip [14] via technologies of AR and VR, so they never have to leave the classroom and also get a better view and understanding of their place of interest. Technologies like these are becoming more and more affordable so that educators can bring them up for educational purposes [13].

Unity's technology [8] can help in converting original ideas into realities. It is a platform for creating games and simulations for educational purposes. Unity's AR/VR can be used to train medical students, create enjoyable and interactive history lessons and lab simulations. Unity is a cross platform game engine developed by Unity Technologies. Unity is developed in C++ [12].

Labster VR [15] offers an accurate yet virtual lab experience that anyone can access to perform experiments in a risk-free setting. The VR labs have prospects not available in real labs like, (1) facility to zoom in to view life science at a molecular level; (2) tasks that connect the science to real life situations; and (3) ability to adjust time to make experiments faster, or go back in time to fix a mistake. It has a direct VR mode to preview work in an HMD (Head Mount Display) [11] [12].

Byju's - The Learning App [16] offers inclusive learning programs in Math and Science for students of class 4th to 12th. It also has preparation courses for competitive exams like CAT, NEET & JEE, IAS, GRE & GMAT etc.

DragonBox [17] offers a powerful, unique, interactive digital learning experience. All the apps offered by DragonBox are paid and are being used by institutions to include digital elements in their curriculum.

Based on above surveyed techniques, we can list some of the limitations below:

1. Expensive hardware requirements.
2. Limited VR [14] applications are available due to slow acceptance by the users.
3. Training of teachers to handle the hardware and to execute session in a safe environment is required.
4. Latest equipment (processors, memory) is required.

The importance of AR/VR in education [18] is summarized below:

1. Maximizes student enthusiasm and involvement by fitting in game elements in learning atmospheres.
2. Interactive situations for generating and directing replicated experiments based on real-life phenomena.

3. Keeps their engagement levels high and inspires them to learn more and better.
4. AR facilitates teachers and trainers in performing tasks in a safe environment [2].

In the next section, we describe the proposed architecture.

3 Proposed Architecture

We are proposing a mobile application for kindergarten children, for interactive learning using the Augmented Reality technology. This application can be used by teachers, tutors, and parents for better learning of their child. It is an easy to handle application that a child can learn to operate after a few usages. The proposed system will be helpful for the students to gain concentration on what is being taught since the application includes videos and voices which amuses kids.

Following are the two modules that we propose:

3.1 Alphabet Recognition

The child will point the camera to a given Alphabet, for e.g. letter 'A'. Then, the application will play a video of objects starting with letter 'A'. The mobile application contains information of all the alphabets, from A to Z.

3.2 Occupation Detection

The child will point at an image of let's say a 'doctor'. The app will play video of a doctor doing his routine. Such representation will allow a child to learn about the occupation related to the image, more than just its pronunciation and spellings. The app contains audio-visual information for the occupations.

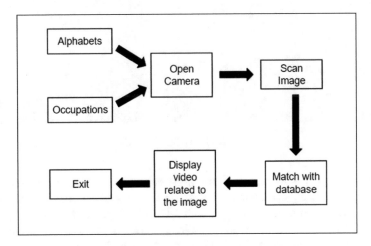

Fig. 2. Block diagram of visual dictionary using AR

Figure 2 shows the block diagram of visual dictionary using AR. In this Fig. 2, we have represented the workflow of the application. The Android application consists of an introductory page that shows the name of the application, with two buttons, leading to two different pages called (1) Alphabets (2) Occupations. After choosing one of the options, the application opens up the camera. The child will point at the image from the book, an augmented 3D video will pop up over the target image and gives appropriate information. This process can be repeated for all the images from the book. The images along with respective video and audio file are stored in databases. Application will fetch results from its database to scan the word and give appropriate videos as a 3D augmented object.

In this section we presented the architecture and working of the application in detail, in the next section we discuss the methodology used.

4   Methodology

Before we mention how the project is built, first we will describe the tools used in developing the application:

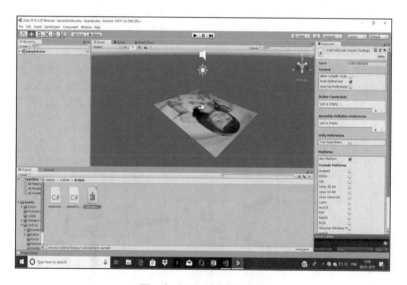

Fig. 3. Unity 3D IDE [8]

Unity3D [8, 19] is a cross-platform game engine used popularly for developing animations including games. Figure 3 displays the view of an opened Unity project. The Unity engine can be used to create both 3-D and 2-D games as well as replications for many platforms such as Windows, Linux, Android etc. We can achieve some special functions of the scenes by scripting in C#. When the scenes are created, assets and scripts are added to the scenes and image targets are created, we save the scenes, scripts, audios, attribute data, pictures, textures etc. in a "project" in Unity3D. After saving the project, the asset (3D data models) in a Unity3D Project is automatically imported.

Fig. 4. Online database on Vuforia [20]

Vuforia is an augmented reality software development kit (SDK) [9, 20] for mobile devices that allows the conception of augmented reality applications. It uses computer vision technology to identify and trail planar images (Image Targets) and simple 3D objects, for example *boxes*, in real time. It lets us create databases, as shown in Fig. 4, of images in Unity format. In the Developer's portal of Vuforia (Fig. 4) we created a database which consists of images, type, quality rating, status and date and time of image modification. The database can be downloaded if and when needed. We can upload the image database later and map the respective videos to the images in Unity3D.

The AR application based on Vuforia SDK can create better user experience, in terms of following four aspects:

(1) Faster targets (i.e. 2D and 3D objects) recognition;
(2) Can recognize millions of targets;
(3) Highly-robust target tracking that will not be affected easily by mobile devices;
(4) More effective than other AR SDKs and supportive of recognition of low-light and partly-covered target [9].

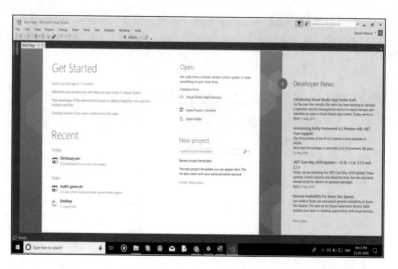

Fig. 5. Visual studio IDE [22]

Microsoft Visual Studio (Fig. 5) is an integrated development environment (IDE) from Microsoft. It is used to develop computer programs, as well as websites, web apps, web services and mobile apps. For our application development, we used Microsoft Visual Studio for scripting in C#.

Now, we will explain how the application is developed using the above discussed tools. The creation of the application starts by the Unity3D software installed on the desktop with enabled packages of Vuforia and Android SDK.

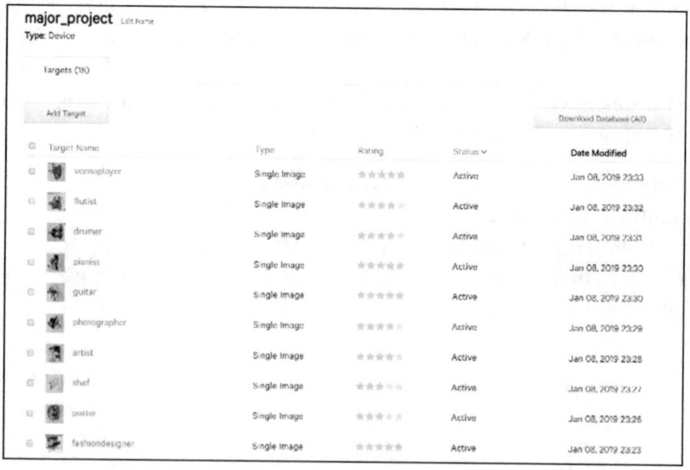

Fig. 6. Online database on Vuforia [20]

The database shown in Fig. 6, for the application, is made on Vuforia by creating an ID on Vuforia's developer portal. The database contains pictures that will be used for scanning.

After the completion of database, it is imported on the Unity 3D for further use. A license key is generated online on Vuforia, which is used to link the Vuforia ID and the Unity project. One License key per project is created.

Scenes are created for every image and videos are linked. A video clip is added on 3D object. The corresponding 3D object is linked with the image target.

Every time an image from the database is scanned, the app plays a video linked with the particular image. The database scans the data points on the image and depending on the quality of image as per rating, the recognition of the image is determined.

In this section, we stated the tools and methodology for the design of our application. In the following section, we present the implementation details and results.

5 Implementation and Result Analysis

The application is developed on Unity3D for Windows. The Unity project is built as an Android Package Kit (APK) and the application is executed on an Android device [21]. It is highly user-friendly as it is aimed at encouraging interactive learning with children of all ages.

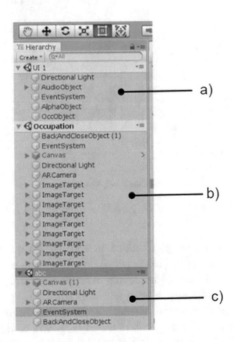

Fig. 7. (a) UI scene; (b) Occupation scene; (c) Alphabet scene [8]

Once License key for creation of scene is made, we copy paste it in the software and the new scene window appears. The main camera is replaced with AR camera.

This application consists of 3 scenes, namely User Interface (UI), Alphabets and Occupations, as elaborated in Fig. 7. Each Scene Contains information on what goes inside it and how will it be operated. The UI Scene contains the opening Activity of the Android Application. It has 2 buttons which point to the other 2 scenes: Alphabets and Occupations. When either one is pressed, the Activity and Scene is changed and AR Camera appears.

Figure 8a shows the application's first page. Figure 8b shows the AR Camera page. AR Camera can be opened by clicking one of the two buttons on the first page.

Both the Scenes that are, Alphabets and Occupations, have AR Camera component added to them. Each of these Scenes have Image Targets that have video and audio components attached to them. These Image Targets tells the application on how to react when certain image is scanned.

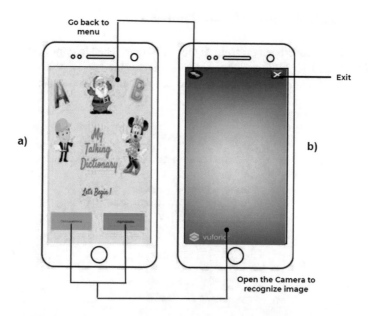

Fig. 8. The application view (a) User Interface; (b) AR Camera

After the event handler script is added to specific Image Target and UI (User Interface) Component, the application is tested on a Windows platform and is built for Android API Level 25 [21] and higher levels. Afterwards, the final APK was installed on an Android based mobile device such as Nokia 6.1 Plus.

In Fig. 9, we have shown the screenshots from the resulting application. For example, when an image of a singer is detected (Fig. 9a), a video corresponding to singer is played. Similarly, when an Alphabet 'C' is detected, a video corresponding to the alphabet plays (Fig. 9b).

(a)    (b)

Fig. 9. (a) Recognition of alphabet 'C' (b) Recognition of occupation

In this way, the application scans the images and retrieves the respective videos. Thus, our application leaves a great impact on students and teachers in terms of education, entertainment, grasping power and better imagination which leverages their understanding skills.

5.1   Comparison with Other Applications

After the discussion of implementation and its results, we compare our application with the existing applications for example, Byju's [16] and DragonBox [17]. Below, we will give the comparison of these application along with our application.

(A) **Byju's: The Learning App**
  1. Used for kids of age 6 and above. However, our application is for Elementary students of age 3–6 years.
  2. BYJU's provides study material in subjects like Mathematics, Science for classes 4th to 12th and for competitive exams. Our application provides 3D augmented videos for alphabets and occupations.
  3. Their study material is in form of videos, presentation and notes. Our application provides information in form of augmented videos in a fun-loving way.

(B) **DragonBox: Empower Kids!**
  1. DragonBox is an application for mathematical functions. Our application focuses on more basic concepts on understanding of alphabets and occupations.
  2. DragonBox's applications are proprietary, and target students of ages between 4 to 9. Our application is for Elementary kids and could be available for free of cost.

In this section, we discussed the implementation of our application and comparison with other applications in the market. In the next section, we conclude our paper and discuss the scope for the future work.

6   Conclusion and Future Work

In this paper, we implemented AR based visual dictionary for the students to learn via augmented 3D videos and audios. This smart application has two modules, Alphabets and Occupations which scans the images and augments the related 3D videos. For the successful implementation of application, we have used tools like Unity3D and Vuforia. The programming was mainly done in C# and some libraries were used exclusively.

However, our application mostly retrieves its visual data from locally stored devices and once the application is downloaded, it does not require internet connectivity. Our application has advantages over other e-learning applications like Byju's [16] in terms of augmented experience, entertainment and interactivity.

In future, we aim to shift the whole backend visual data to cloud for faster and wider access. Our application can be used not just for kids but also can be extended to other more complex subjects for tutorials.

References

1. Azuma, R.T.: A survey of augmented reality. Presence: Teleoper. Virtual Environ. **6**(4), 355–385 (1997)
2. Al-Azawi, R., Shakkah, M.D.S.: Embedding augmented and virtual reality in educational learning method: present and future. In: 9th International Conference on Information and Communication System, Gulf College, Oman (2018)
3. Paredes Velastegui, D., Lluma Noboa, A., Olmedo-Vizueta, D., Avila-Pesantez, D., Hernandez Ambato, J.: Augmented reality implementation as reinforcement tool for education in ecuador. In: IEEE Global Engineering Education Conference (EDUCON), Riobamba, Ecuador, April 2018
4. Muhammad, S.A.: Playing Pokémon GO in a public park in malaysia: a survey and analysis (2019). https://doi.org/10.1007/978-3-030-15616-9_11
5. Karkera, A., Dhadse, S., Gawde, V., Jain, K.: Pokemon fight augmented reality game. In: Proceedings of the 2nd International Conference on Inventive Communication and Computational Technologies (ICICCT 2018) (2018)
6. Lao, C., Mao, C., Sy, A.: Security analysis on snapchat (2019)
7. Vaterlaus, J., Barnett, K., Roche, C., Young, J.: "Snapchat is more personal": an exploratory study on snapchat behaviors and young adult interpersonal relationships. Comput. Hum. Behav. **62**, 594–601 (2010). https://doi.org/10.1016/j.chb.2016.04.029
8. Wang, S., Mao, Z., Zeng, C., Gong, H., Li, S., Chen, B.: A new method of virtual reality based on Unity3D. IEEE, Beijing
9. Amin, D., Govilkar, S.: Comparative study of augmented reality SDK's. researchgate.net (2015)
10. Patterson, Todd C.: Google earth as a (not just) geography education tool. J. Geogr. **106**(4), 145–152 (2007). https://doi.org/10.1080/00221340701678032

11. Desai, P.R., Desai, P.N., Ajmera, K.D., Mehta, K.: A review paper on oculus Rift-A virtual reality headset. Int. J. Eng. Trends Technol. (IJETT) **13**(4) (2014)
12. Kelly, J.W., Cherep, L.A., Siegel, Z.D.: Perceived space in the HTC vive. ACM Trans. Appl. Percept. **15**(1), Article 2 (2017)
13. Sheppard, S.R.J., Cizek, P.: The ethics of Google earth: crossing thresholds from spatial data to landscape visualization (2008)
14. Wickens, C.D.: Virtual reality and education. Aviation Research Laboratory, University of Illinois at Urbana-Champaign (1992)
15. Kaufmann, H., Schmalstieg, D.: Collaborative augmented reality in education. Comput. Graph. **27**(3), 339–345 (2003)
16. https://byjus.com/. Accessed 23 May 2019
17. https://dragonbox.com/. Accessed 23 May 2019
18. India Today: Role of AR/VR in the education system: how technology is shaping India's learning space (2018). [https://www.indiatoday.in/education-today/featurephilia/story/role-of-augmented-virtual-reality-in-education-1417739-2018-12-26
19. Kuang, Y., Bai, X.M.: The research of virtual reality scene modeling based on unity 3D. In: The 13th International Conference on Computer Science and Education (ICCSE 2018), 8–11 August 2018, Colombo, Sri Lanka (2018)
20. Peng, F., Zhai, J.: A mobile augmented reality system for exhibition hall based on Vuforia. In: 2nd International Conference on Image, Vision and Computing (2017)
21. https://developers.google.com/ar/develop/java/quickstart. Accessed 23 May 2019
22. https://visualstudio.microsoft.com. Accessed 23 May 2019

Security Algorithms for Cloud Based Applications

Bushra Shaheen[✉] and Farheen Siddiqui

Department of CSE, SEST, Jamia Hamdard, New Delhi, India
bushrashaheenll@gmail.com,
fsiddiqui@jamiahamdard.ac.in

Abstract. The concept of cloud computing provides various scalable resources and dynamic capacity of storage which performs operations on the internet. The main benefit from using the cloud computing is the reduction in the economic expenditure and ease in accessibility of data. Data protection is also one of the major concern of cloud computing and in order to provide the security to it's users, cloud computing uses different cryptographic encryption algorithm. The main concern of this paper is to analyze which encryption algorithm will be most efficient and convenient to be used in cloud computing to secure data as well as less time consuming algorithm. Analysis have been done on different criteria like block size, key length, security and speed, encryption flow chart and decryption flow chart.

Keywords: Cloud computing · Data security · Encryption algorithm · AES · DES · RSA · Public key · Private key

1   Introduction

Cloud computing is a technology which permit the user the approach to the multiple resource, and provides services which can be approached by the service providers with adequate amount of effort. As per NIST definition, the main mechanism of cloud computing by which it completes its execution are software as a service, platform as a service and infrastructure as a service [8].

The main aim of this paper is to completely understand the cloud computing mechanism and examine that which encryption algorithm is the most productive to be used in evolution of cloud computing [4, 16] (Fig. 1).

From the above survey report of both the years, we can clearly notice the increase in the adoption of cloud services by the several companies [18]. These companies have increased the usage in cloud services because cloud services provides several ease in the protection of data protection as well as platform, infrastructure and software as their service.

This Fig. 2 explains problems faced during the adoption of cloud computing services. The particular problems are: integrating with the existing system; changes in the IT policy degradation in the performance; modification in the workflow etc.

© Springer Nature Switzerland AG 2020
P. Karrupusamy et al. (Eds.): ICSCN 2019, LNDECT 39, pp. 582–588, 2020.
https://doi.org/10.1007/978-3-030-34515-0_60

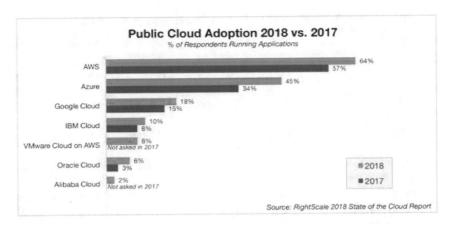

Fig. 1. Survey report by International Data Corporation (IDC) of the year 2018 and 2017 [18].

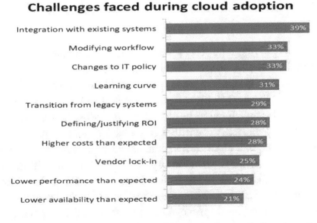

Fig. 2. Shows the challenges which the companies had to face during the adoption of cloud environment.

The details about rest of the paper are as follows:

Section 2 talks about the different reviews from the humongous papers which focus on the issues of the cloud security.

Section 3 throws light on the comparison between the different algorithms which have been described on the basis of architecture, on the basis of scalability, on the basis of flexibility, on the basis of security level.

Section 4 concludes the paper.

Table 1: This table shows the comparison between the encryption algorithms on the basis different perspective used in cloud based applications.

Table 1 speaks about the various algorithms and their performance on the basis of block size, key length it requires, the algorithm's processing speed and its performance on how much security can it provide. DES uses 64 bits of block size and 56 bits key

Table 1. Comparison between various encryption algorithms.

S.no	Algorithm	Developed	Block size (bits)	Key length (bits)	Security	Speed
1.	DES	1997	64	56	Proven inadequate	Very slow
2.	AES	2000	128,192,256	128,192,256	Highly secure	Very Fast
3.	TDES	1998	112,168	112,168	Considered secure	Slow
4.	BLOWFISH ALGORITHM	1993	64	32,448	Considered secure	Fast
5.	IDEA	1991	128	128	Inadequate	Slow
6.	HOMOMORPHIC ENCRYPTION (Modified RSA)	1978	128	1024,4096	Secure	Fast
7.	RSA	1977	128	1024	Considered secure	Very slow

length but unable to provide required security and it's speed is also slow. AES supports block size ranging from 128 bits to 256 bits and it also uses key length if 128 to 256 bits, it provides perfect amount of security required and it also provides good speed. Triple DES requires 112 and 168 bits of block size and key length as same bits as block size. This algorithm is considered to be secure but it is slow hence it is not efficient for encrypting the cloud data. RSA algorithm requires 128 bits of block size and 1024–4096 bits of key length, this algorithm is considered to be secure but its speed is slow compared to other algorithms. Homomorphic encryption algorithm is one of the modification of RSA algorithm which is also known as MREA (Modification in RSA Encryption Algorithm). This algorithm uses 128 bits of block size and 4096 bits of key length additionally it uses large prime numbers which enhances the security of this algorithm and also provides the processing speed as required. In order to maintain the security as well as the speed, we can clearly observe that AES algorithm and Modification in RSA algorithm which is Homomorphic Encryption are the most appropriate and efficient one to use.

2   Literature Review

In this section, short description of cloud computing and its security issues is given and the main focus is to analyse which encryption algorithm will be best to use in order the secure the cloud data more appropriately.

According to the NIST, the Cloud Computing definition is "Cloud Computing is a model for enabling omnipresent, efficient, on-demand access to the network to a large pool of extensible computer resources (e.g. networks, servers, storage, applications and services) that could be rapidly offered and issued with negligible governance attempt" [11]. According to A. Venkatesh, the stellar threat to the cloud computing is its data security issue. Hu and A. Klein has provided a standard for cloud based data in transit security approach. In order to guard data during migration, a benchmark for encryption was discussed. For robust security, additional encryption is required as well as additional computation must be executed. The baseline explored in their research shows a balance for security and overhead encryption [4]. M. A. Mushtaque, H. Dhiman, they both has centered their focus to discuss the limitations and opportunities of deploying data management issues on these emerging cloud computing platforms [3].

3 Descriptive Analysis on the Above Security Algorithms

3.1 Based on Architecture

Algorithms have been compared on structure, key size, block size and number of processing rounds in the architecture section.

i. DES (Data Encryption Standard)
   DES is an algorithm for symmetric key block encryption. DES uses 64-bit block text and 56-bit key length. The overall number of 16 handling rounds in DES is operated to authenticate the data.
ii. 3DES (Triple DES)
   This algorithm is the Data Encryption Standard's advanced version. It has 64-bit block size and 192-bit key size [19].
iii. AES (Advanced Encryption Standard)
   For the key length of 128 bit AES performs 10 processing rounds, 12 rounds for the key length of 192 bit and 14 rounds for the key length of 256 bit.
iv. RSA
   It is an algorithm for the encryption of asymmetric key streams. RSA has 128 bits block size with 1024 to 4096 bits key length [8] (Table 2).

Table 2. Comparison of algorithms on architecture level

Algorithm	Symmetric/Asymmetric	Block size (in bits)	Key length (in bits)
DES	Symmetric	64	56
3DES	Symmetric	64	192
AES	Symmetric	128	128,192,256
RSA	Asymmetric	128	1024,4096

3.2 Based on Scalability

Based on the performance such as key scheduling, data encryption and the encryption algorithm's requirement for space, if any encryption algorithm requires fewer amount of memory then it will definitely be more efficient (Fig. 3).

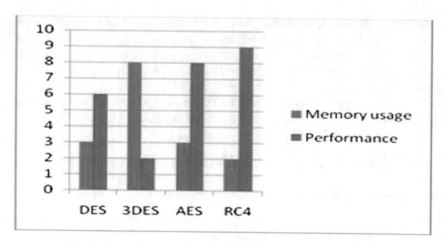

Fig. 3. Comparison of different algorithms based on scalability [8]

3.3 Based on Versatility

Table 3: This table shows the challenges the versatility of the following encryption algorithms used in cloud computing.

Table 3. Comparison between encryption algorithms used in cloud computing on the basis of versatility

Algorithm	Versatility	Alteration	Comment
DES	No	No	Does not support any Modification
3DES	Yes	128 bits	3 DES Performs DES Operation
AES	Yes	128,192,256 bits	It's structure is flexible to the multiple of 64
RSA	Yes	RC4, RC4C and many More	Supports key size Ranging from 8 to 256 bits

In this Table 3, we can observe that DES algorithm is restricted for the modifications, as well as 3DES algorithm also doesn't support modification. AES algorithm is flexible for the modification up to the multiple of 64 bits and the algorithm RSA supports the modification ranging from 8 bits to 256 bits. So, we can state that among above all the algorithms, the RSA is the most appropriate one in order to meet up with any modification.

3.4 Based on Security Level

i. DES

It is the only drawback of DES algorithm. It fails to provide better security because of its key length of 56 bits.

ii. 3DES

It provides high level of security.

iii. AES

It is the advanced version of the DES

iv. RSA

RSA also fails to provide high level of security and it has also very slow speed of execution. The reason behind the weak level of security is the poor key scheduling [8] (Table 4).

Table 4. Comparison of algorithms on security level

Algorithm	Security
DES	Not secure
3DES	Highly secure
AES	Highly secure
RSA	Not secure

4 Conclusion and Future Scope

- DES can be widely used in financial applications in the form of encryption schemes.
- AES can be ideal in encrypting the messages via chat or can be used to ensure the monetary transactions.
- RSA has the good speed and it can be used in different commercial software packages.

After analysing the performance mechanism of every encryption algorithm mentioned in this paper, conclusion can be stated that AES algorithm is well secured and provides high speed for the processing of algorithm which enhances the efficiency of the algorithm. As well as Homomorphic Encryption algorithm which is one of the modification in RSA algorithm is also very efficient one as it is also highly secured and have high speed in processing and moreover it uses large prime numbers which makes the data very difficult to decrypt.

So both these algorithms can be used in cloud computing.

Cloud computing provides several services which proves to be of benefit to various organisations and business but these services are useless unless it provides the reliable security to the data. There are various encryption algorithms which can be used in cloud computing in order to provide better level of security. As per analysis done in this paper, Advance Encryption Algorithm (AES) is the most efficient one algorithm to use. But the security can be damaged by anyone. So it is very mandatory to make the algorithm more strong and secure [19].

References

1. Ganapathy, S.: A secured storage and privacy-preserving model using CRT for providing security on cloud and IoT-based applications. Comput. Netw. **151**, 181–190 (2019)
2. Rath, M.: Resource provision and QoS support with added security for client side applications in cloud computing. Int. J. Inf. Technol. **11**(2), 357–364 (2019)
3. Agarwal, R., Pranay, S.S., Rachana, K., Parveen Sultana, H.: Identity-based security scheme in internet of vehicles. In: Smart Intelligent Computing and Applications, pp. 515–523. Springer, Singapore (2019)
4. Manzoor, A., Liyanage, M., Braeke, A., Kanhere, S.S., Ylianttila, M.: Blockchain based proxy re-encryption scheme for secure IoT data sharing. In: 2019 IEEE International Conference on Blockchain and Cryptocurrency (ICBC), pp. 99–103. IEEE (2019)
5. Arulmozhi, P., Rayappan, J.B.B., Raj, P.: A lightweight memory-based protocol authentication using radio frequency identification (RFID). In: Advances in Big Data and Cloud Computing, pp. 163–172. Springer, Singapore (2019)
6. Sahu, M.T.S.N.K., Pal, A.K.: A Review on Analysis of Data Search Scheme for Secure Information Retrieval in Cloud Computing (2019)
7. Zhu, L., Wu, Y., Gai, K., Choo, K.K.R.: Controllable and trustworthy blockchain-based cloud data management. Future Generation Computer Systems **91**, 527–535 (2019)
8. Roy, M., Mali, K., Chatterjee, S., Chakraborty, S., Debnath, R., Sen, S.: A study on the applications of the biomedical image encryption methods for secured computer aided diagnostics. In: 2019 Amity International Conference on Artificial Intelligence (AICAI), pp. 881–886. IEEE (2019)
9. Stergiou, C., Plageras, A.P., Psannis, K.E., Gupta, B.B.: Secure Machine Learning Scenario from Big Data in Cloud Computing via Internet of Things Network (2019)
10. Gupta, V., Johari, R., Gupta, K., Bhatia, R., Seth, S.: LBCLCT: location based cross language cipher technique. In: Smart Cities Performability, Cognition, & Security, pp. 221–234. Springer, Cham (2020)
11. Naseri, A., Navimipour, N.J.: A new agent-based method for QoS-aware cloud service composition using particle swarm optimization algorithm. J. Ambient Intell. Hum. Comput. **10**(5), 1851–1864 (2019)

An Empirical Model for Thyroid Disease Diagnosis Using Data Mining Techniques

Umar Sidiq[1(✉)] and Syed Mutahar Aaqib[2]

[1] Department of Computer Science and System Studies,
Mewar University, Chittorgarh, Rajasthan, India
umarsidiq67@gmail.com
[2] Department of Computer Science, Amar Singh College, Srinagar,
Jammu and Kashmir, India
syed.auqib@gmail.com

Abstract. Utilization of data mining in healthcare sectors showing great role in effectiveness of treatment, healthcare administration, finding of fraud and abuse and customer relationship management but besides that it is also used for diagnosis of diseases. In this work, our focus is on diagnosis of thyroid diseases by using three classification models like K-Nearest Neighbor (K-NN), Decision Tree and Naïve bayes based on certain clinical thyroid attributes like Age, Gender, TSH, T3 and T4. The entire research work is to be conducted with RapidMiner version 8.1, an open source tool under Windows 10 environment. The Experimental study shows that decision tree outperformed over other models.

Keywords: Thyroid disease · K-nearest neighbor · Decision tree · Naïve Bayes

1 Introduction

Data mining acts as crucial role in detection and analyzing survivability of diseases by reducing the payment bills to the patients. In this modern world, it became an important to diagnose the diseases like diabetes, kidney disease, heart diseases etc. because of its highly occurrence in mankind and it is also found that thyroid disease also become a common endocrine health disease and in India, 42 million people is expected that are suffered from thyroid disease [1–3]. Hyperthyroidism and Hypothyroidism are main two diseases of thyroid occurs due to excess secretion and deficient secretion of thyroid hormones producing from the thyroid gland [4, 5]. The T3 and T4 are the two types of thyroid hormones that help in maintain the balance of proteins, temperature and energy production of the body [6]. Hypothyroidism and hyperthyroidism causes various symptoms including weight loss, weight gain, nervousness, tiredness, weakness etc. The underestimated thyroid disease causes thyroid storm, graves' disease and myxedema which causes death [7]. In this research work, three separate experiments were carried out by applying K-NN, Decision tree and Naïve bayes through k-fold cross test method on thyroid dataset.

The rest of the paper is followed as: Sect. 2 shows related work in diagnosis of thyroid diseases. Section 3 contains dataset and methods. Section 4 shows construction

© Springer Nature Switzerland AG 2020
P. Karrupusamy et al. (Eds.): ICSCN 2019, LNDECT 39, pp. 589–597, 2020.
https://doi.org/10.1007/978-3-030-34515-0_61

of model. Section 5 represents the results and discussion. Section 6 shows the conclusion of the work and references are mentioned at last.

2 Related Work

By studying literature, it seems that Data mining is extensively used in medical care for the diagnosis of thyroid diseases. Ozyilmaz et al. in 2002 proposed three commonly used methods of neural network (MLP, RBF and CSFNN) for thyroid diseases diagnosis and achieved accuracy of 89.80%, 79.08% and 91.14% respectively [8]. Polat et al. in year 2007 proposed artificial immune- recognition system (AIRS) classification method for thyroid disease diagnosis and achieved 81% accuracy on experiment. AIRS was combined with Fuzzy weighted pre-processing showed 85% on another separate experiment [9]. Temurtas in 2009 have implemented MLNN, PNN, LVQ algorithms of neural network to estimate the accuracy performance of each algorithm on 3 fold cross validation and 10 fold cross validation test methods. On experiment, it was calculated that 94.81% was highest accuracy found on experiment [7]. Saravana Kumar et al. in year 2014 proposed two classification techniques (K-NN and SVM) for diagnosis of thyroid disease. Experimentally, K-NN showed better accuracy than SVM [13]. Khushboo Chandel et al. in year 2017 used KNN and Naïve Bayes for analysis of diagnosis of thyroid diseases implemented in RapidMiner tool. On comparison of these classifiers, the KNN shows 93.44% of accuracy and Naïve bayes shows accuracy of 22.56% [22].

3 Dataset and Methods

The clinical data of 807 patients was taken from the diagnostic laboratory in Kashmir. There are five input attributes as: age, gender, TSH, T3, T4 and one output attribute called as class label having three class values as normal, hypothyroidism and hyperthyroidism. The description of our dataset is shown in Table 1.

Table 1. Attributes used in our dataset.

Serial No.	Attribute name	Description	Value
1	Age	Age in years	Numeric
2	Gender	M-Male F- Female	Nominal
3	TSH	Continuous	Numeric
4	T3	Continuous	Numeric
5	T4	Continuous	Numeric
6	Results	Normal Hyperthyroidism Hypothyroidism	Nominal

3.1 Classifiers Used in this Research Work

3.1.1 K Nearest Neighbor

K-nearest neighbor or KNN is a supervised and non-parametric method used to classify the unknown instances on the basis of K closest instances in the feature space [18, 19]. The prediction attribute of the most similar instances is summarized and returned as the prediction for the undetected instance [20].

3.1.2 Decision Tree

Decision Tree is tree like graph that splits the dataset into parts by the process of making series of decisions [10]. In decision tree, each internal node represents a test, each branch denotes the outcome of that test and each leaf node has a class label and the path where a particular test data is to classify from root to leaf represent classification rules based on maximum information gain [11].

3.1.3 Naïve Bayes

Naïve Bayes is a simple classification algorithm for predictive modeling with clear semantics, representing and the probabilistic learning method based on Bayesian theorem [12]. Naive Bayes classifier assumes that the values of attributes are independent upon other and presence or absence of particular attribute doesn't affect the overall prediction process. Suppose there are m classes say C1, C2....Cn having a unidentified data sample X, Naive Bayes classifier will predict the X to the class Ci because of having highest probability [21].

$$P(Ci|X > P(Cj|X) \text{ for } 1 \leq j \leq m, j \neq i$$

4 Model Construction

To build a predictive model for thyroid disease diagnosis, the required dataset was loaded into the RapidMiner version 8.1 tool. After loading it successfully, the dataset is need to be preprocessed, so that dataset become error free by removing all unwanted data and by which performance is enhanced. After preprocessing, it is necessary to divide the dataset into training and testing. In this work, the whole thyroid dataset was divided by K fold cross validation method (10 fold) into K equal subsets and among one subset or 10% act as testing data and remaining K − 1 or 90% acts as training data [7]. After cross validation, the K-NN, Decision tree and Naïve bayes are applied in three separate experiments to construct the models. Evaluation of developed model is to calculate confusion matrix for measuring certain performance vectors like accuracy, precision, recall and classification error.

The accuracy is ratio of true predictions by total number of predictions or ACC = tp + tn/tp + tn + fp + fn.

The precision is ratio of true positive prediction by total positive predictions or P = tp/(tp + fp).

Recall is ratio of true positive prediction by sum of true positive and false negative predictions or R = tp/(tp + fn).

The Classification error or misclassification rate is measured by 100% − the achieved accuracy or ERR = 100% − achieved accuracy.

Figure 1 shows the proposed flow chart for diagnosis of thyroid diseases. Figures 2, 3 and 4 shows the training and testing phase of K-NN, Decision tree and Naïve bayes respectively.

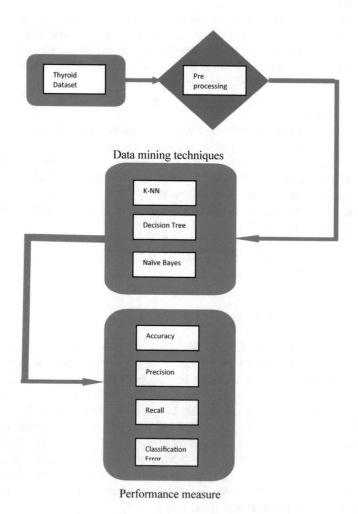

Fig. 1. Shows proposed flow chart for diagnosis of thyroid diseases

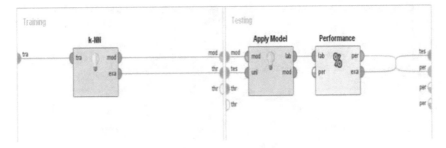

Fig. 2. Shows training and testing process in KNN

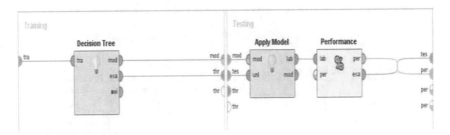

Fig. 3. Shows training and testing process in Decision Tree

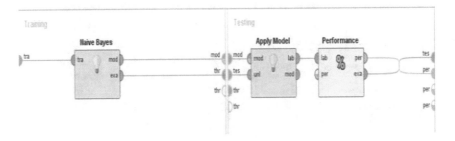

Fig. 4. Shows training and testing process in Naïve Bayes

5 Results and Discussion

In this work, decision tree achieved highest accuracy of 99.38%, followed by K-nearest neighbor having accuracy of 93.80% and naïve bayes achieved least accuracy of 91.70%. The decision tree also achieved highest values of precision, recall and least value of classification error. The decision tree that was build after experiment is shown in Fig. 5. The achieved accuracy of K- nearest neighbor, Decision tree and Naïve bayes are shown in Figs. 6, 7 and 8 respectively. Comparative analysis among the classifiers is shown in Table 2, the error report that was generated of each classifier is shown in Tables 3 and 4 shows accuracy of our model higher the existing models available in literature.

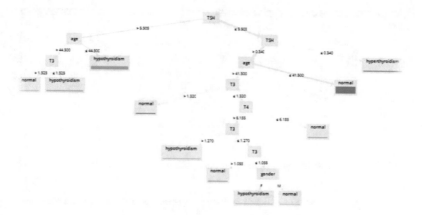

Fig. 5. Shows decision tree that was build after experiment

accuracy: 93.80% +/- 2.09% (mikro: 93.80%)

	true normal	true hypothyroidism	true hyperthyroidism	class precision
pred. normal	523	14	6	96.32%
pred. hypothyroidism	21	204	0	90.67%
pred. hyperthyroidism	9	0	30	76.92%
class recall	94.58%	93.58%	83.33%	

Fig. 6. Shows accuracy with confusion matrix for K-nearest neighbor

accuracy: 99.38% +/- 0.62% (mikro: 99.38%)

	true normal	true hypothyroidism	true hyperthyroidism	class precision
pred. normal	551	3	0	99.46%
pred. hypothyroidism	2	215	0	99.08%
pred. hyperthyroidism	0	0	36	100.00%
class recall	99.64%	98.62%	100.00%	

Fig. 7. Shows accuracy with confusion matrix for Decision Tree

accuracy: 91.70% +/- 2.34% (mikro: 91.70%)

	true normal	true hypothyroidism	true hyperthyroidism	class precision
pred. normal	549	59	4	89.71%
pred. hypothyroidism	4	159	0	97.55%
pred. hyperthyroidism	0	0	32	100.00%
class recall	99.28%	72.94%	88.89%	

Fig. 8. Shows accuracy with confusion matrix for Naïve Bayes

Table 2. Shows Comparative analysis among the classifiers

Technique used	K-Nearest Neighbor	Decision Tree	Naïve Bayes
Accuracy	93.80%	99.38%	91.70%
Recall	90.76%	99.42%	87.14%
Precision	89.95%	99.53%	95.88%
Classification error	6.20%	0.62%	8.30%

Table 3. Shows error report of each classifier.

Classifier used	Classification error	kappa	Absolute error	Relative error	Root mean squared error	Root relative squared error
K-NN	6.20%	0.866	0.062	6.20%	0.245	0.383
Decision tree	0.62%	0.986	0.008	0.85%	0.057	0.089
Naïve bayes	8.30%	0.803	0.117	11.65%	0.248	0.388

Table 4. Shows our model achieved higher accuracy than existing models available in literature.

Study	Method	Highest accuracy
Polat et al. [9]	AIRS on 10-fold cross validation	85.00%
	AIRS with Fuzzy on 10-fold cross validation	
Keles et al. [14]	ESTDD	95.33%
Temurtas [7]	MLNN with LM (3-fold cross validation)	94.81%
	PNN (3-fold cross validation)	
	LVQ (3-fold cross validation)	
	MLNN with LM (10 fold cross validation)	
	PNN (10-folds cross validation)	
	LVQ (10-fold cross validation)	
Dogantekin et al. [15]	GDA-WSVM	91.86%
Gopinath [16]	SVM	96.30%
	FKNN	
	Decision Tree	
Sidiq et al. [17]	KNN	98.89%
	SVM	
	DT	
	NB	
Our model	KNN	99.38%
	Decision Tree	
	Naïve bayes	

6 Conclusion

The main aim of this experimental study is to evaluate KNN, Decision tree and naïve bayes based on RapidMiner tool for thyroid diseases diagnosis. Experimentally, decision tree was proved as best algorithm because it obtained highest performance metrics in terms of accuracy, precision and recall and least classification error.

Acknowledgment. There is no conflict of interest. We used our own data.

References

1. Sehgal, M.S.B., Gondal, I.: K-ranked covariance based missing values estimation for microarray data classification. IEEE (2004)
2. Bonner, A.: Comparison of discrimination methods for peptide classification in tandem mass spectrometry. IEEE (2004)
3. http://www.ncbi.nlm.nih.gov/pmc/articles/PMC3169866/. Accessed Dec 2015
4. Shen, X., Lin, Y.: Gene expression data classification using SVM–KNN classifier. IEEE (2004)
5. Xia, C., Hsu, W.: BORDER: efficient computation of boundary points. IEEE (2006)
6. Kodaz, H., et al.: Medical application of information gain based artificial immune recognition system (AIRS): diagnosis of thyroid disease. Expert Syst. Appl. **36**(2), 3086–3092 (2009)
7. Temurtas, F.: A comparative study on thyroid disease diagnosis using neural networks. Expert Syst. Appl. **36**, 944–949 (2009)
8. Ozyılmaz, L., Yıldırım, T.: Diagnosis of thyroid disease using artificial neural network methods. In: Proceedings of ICONIP 2002 9th International Conference on Neural Information Processing, pp. 2033–2036. Orchid Country Club, Singapore (2002)
9. Polat, K., Sahan, S., Gunes, S.: A novel hybrid method based on artificial immune recognition system (AIRS) with fuzzy weighted preprocessing for thyroid disease diagnosis. Expert Syst. Appl. **32**, 1141–1147 (2007)
10. http://en.wikipedia.org. Accessed on 24 Dec
11. Apte, C., Weiss, S.M.: Data Mining with Decision Trees and Decision Rules. T. J. Watson Center (1997). http://www.research.ibm.com/dar/papers/pdf/fgcsapteweissue_with_cover.pdf
12. John, G.H., Langley, P.: Estimating continuous distributions in Bayesian classifiers. In: Proceedings of the 11th Conference on Uncertainty in Artificial Intelligence, San Francisco, pp. 338–345 (1995)
13. Saravana Kumar, K., Manicka Chezian, R.: Support vector machine and k-nearest neighbor based analysis for the prediction of hypothyroid. Int. J. Pharma Bio Sci. **2**(5), 447–453 (2014)
14. Keles, A., Keles, A.: ESTDD: Expert system for thyroid diseases diagnosis. Expert Syst. Appl. **34**(1), 242–246 (2008)
15. Dogantekin, E., Dogantekin, A., Avci, D.: An expert system based on generalized discriminant analysis and wavelet support vector machine for diagnosis of thyroid diseases. Expert Syst. Appl. **38**(1), 146–150 (2011)
16. Gopinath, M.P.: Comparative study on classification algorithm for thyroid data set. Int. J. Pure Appl. Math. **117**(7), 53–63 (2017)

17. Sidiq, U., et al.: Diagnosis of various thyroid ailments using data mining classification techniques. Int. J. Sci. Res. Comput. Sci. Eng. Inf. Technol. (IJSRCSEIT) **5**(1), 131–136 (2019). ISSN 2456-3307
18. Roychowdhury, S.: DREAM: diabetic retinopathy analysis using machine learning. IEEE (2014)
19. Chetty, N., Vaisla, K.S., Patil, N.: An improved method for disease prediction using fuzzy approach. IEEE (2015)
20. Jacob, J., et al.: Diagnosis of liver disease using machine learning techniques. IRJET **05**(04), 4011–4014 (2018)
21. Srinivasan, B., Pavya, K.: Diagnosis of thyroid disease using data mining techniques: a study. Int. Res. J. Eng. Technol. **3**(11), 1191–1194 (2016)
22. Chandel, K., Kunwar, V., Sabitha, S., Choudhury, T., Mukherjee, S.: A comparative study on thyroid disease detection using K-nearest neighbor and Naive Bayes classification techniques. CSI Trans. ICT **4**(2–4), 313–319 (2016)

Design and Implementation of Smart Vehicles for Smart Cities

P. S. Sheeba$^{(\boxtimes)}$, Kanhaiya Kumar Sah, Snehal Chavan,
Rushabh Pawar, and Megha Chand

Department of Electronics Engineering, Lokmanya Tilak College of Engineering,
Navi Mumbai, India
sheebaps@gmail.com

Abstract. Over-speeding and accidents caused due to sudden braking are two of the leading causes of road accidents, which is increasing every year. To monitor every stretch of road 24/7, manpower or resources are required for traffic enforcement officials which is not sufficient as of now. This leads to people getting away with speeding in speed zones. During manufacturing, the proposed system may be installed into a vehicle, to automatically apply brakes, when the vehicle ahead applies brakes, to avert a collision and to detect if a vehicle is maintaining the speed limit and if not, the RTO will instantly send a notification of the fine via SMS to the driver. Another major problem to be tackled is noise pollution caused by constant honking of vehicles. By using inter-vehicular communication unnecessary honking noise can be reduced to a certain extent. Inter vehicle communication can also be used to avert a collision by applying brakes when the vehicle in the front applies brakes suddenly.

Keywords: Inter-vehicular communication · Automatic braking · Noise pollution · Road safety · Vehicle safety

1 Introduction

The current world's daily dilemma is to bear with the unnecessary havoc created by the blowing of car horns on the roads. According to a journal article, which takes a look at India's noise pollution policy, road traffic sounds are often higher than industrial and even construction noise [1]. The main reason for the noise pollution in cities is vehicle density and traffic congestion. Drivers use horns frequently in their journey due to improper traffic systems and indiscipline, increased number of vehicles on road and traffic congestion. As per the noise pollution (regulations and control) rules, noise level for ambient air quality has been set at 55 dB for residential areas and 65 dB in commercial areas [2]. These levels of noise are appropriate for the human condition, but according to data from India's central pollution control board (CPBC) and independent reports, traffic sounds averages around 100 dB in some streets. Noise pollution is disturbing and harming the human health and if not controlled can lead to serious consequences. Noise pollution can be responsible for severe health risks like hearing loss, high stress levels, sleep disturbance, hypertension and other harmful effects.

© Springer Nature Switzerland AG 2020
P. Karrupusamy et al. (Eds.): ICSCN 2019, LNDECT 39, pp. 598–603, 2020.
https://doi.org/10.1007/978-3-030-34515-0_62

Along with increasing noise levels in residential areas, rules are not being followed in silent zones like hospitals and schools, which cause a lot of inconvenience to patients and students.

One of the leading problems caused by heavy traffic and high vehicle density is accidents caused due to sudden braking. When a driver suddenly applies brakes, due to a number of reasons, there isn't a sufficient amount of reaction time for the driver who is behind the suddenly stopped vehicle, this leads to a collision which can cause injuries and might even lead to fatal accidents. In India, over-speeding is one of the major causes of road accidents which keeps on increasing day by day [4]. These accidents are caused due to people speeding way over stipulated limits [5]. In [11], authors discuss about how a group of vehicles can communicate by adhoc networks by means of on-board sensors. How the information about traffic density and weather conditions are communicated to the drivers by means of intelligent system is discussed in [12]. A new concept of social internet of vehicles is explored in [13]. Drivers are aware of their rule breaking but since they get away with it most of the time, people tend to ignore the speed limits and over speed. This paper addresses all these issues and presents an inter-vehicular communication system design using IrDA and speed detection using active RFID. The proposed system may be incorporated during manufacturing into the vehicle, to implement inter vehicular communication for honking inside the vehicle itself instead of the environment along with a automatic braking to avoid collision and, speed detection and SMS intimation or "Speeding ticket".

2   Background

2.1   Inter-vehicle Communication

Currently some research has been done in the area of inter-vehicle communication. Some studies show, inter-vehicle communication using Vehicular Adhoc Networks or VANET's, a car would be able to communicate and apply emergency braking to avoid collision [6]. Although VANETs can be used, it relies on vehicle itself to provide network functionality as these networks does not have fixed infrastructure. It is difficult to manage the rapid changes in the VANETs topology. The network topology changes rapidly due to high relative speed between cars.

Ultrasonic sensors for detection of proximity between two vehicles has been implemented earlier [7]. IR sensors are necessarily compatible with digital circuits and are more reliable, reproducible and precise than ultrasonic sensors. Ultrasonic require electrical and mechanical circuitry, IR sensors require electronics circuitry as shown in Fig. 1.

2.2   Speed Detection and Ticketing

Current methods used by law enforcement for speed detection is Light Detection and Ranging (LIDAR) and traffic cameras [8]. Speeding vehicle is detected using line of sight and/or constant manual operator. Speed detection using RFID tag is also possible is what is used in our system [9]. Majority of the over-speeding accidents happens

Fig. 1. Range of infrared sensor

when there is less police manpower and the roads are relatively empty which generally happens between 12 am and 5 am. To combat India's propensity to overspeed and lack of manpower to enforce regulations can be overcome by leveraging this technology which is stated on multiple occasions by the Minister for Road Transport and Highways. This system will be highly beneficial in overcoming these problems faced by the Indian citizens and law enforcement.

Reduced Inertial Sensors have also been used in prior research and study [10], the problem with these is, the reliability of the inertial sensor data for the proposed use based on the degree of variations in measurements for intended application is questionable. Small errors in the measurement of acceleration and angular velocity are integrated into progressively larger errors in velocity, which are compounded into still greater errors in position.

3 Detailed Design

3.1 Overview

The proposed system is based on three major features such as Hornless Honking, and Automatic Braking, which will be implemented using inter-vehicle communication, and Real time speed detection and RTO intimation which will use Active RF transmitter and receiver and GSM. IR transmitter, transistors, IR receiver, RF Active Transponder, LCD, buzzer, LED, GSM module, Relay, Arduino, are the components used in the proposed system. The proposed system's block diagram representation is shown in Fig. 2.

3.2   Hardware Systems

In this paper microcontroller unit consist of AVR ATMEGA328 from AVR family or AT89S52 from MCU 8051 family as a main controlling core. Basically, system is based on the embedded platform in which multiple sensors are actuators are connected to the system.

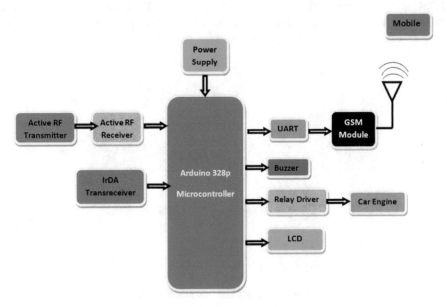

Fig. 2. Block diagram

IR transmitter and receiver pair act as primary sensors in the system to transmit and receive the modulated data. Modulation of the signal will take place at 38 kHz Frequency. Also, RF transmitter and receiver pair act as secondary sensors used to transmit the fixed signal at a frequency of 433 MHz in the certain area of coverage and receive the signal by the receiver which is integrated in the vehicle system. Relay driver and motor resembles the car engine. Also, LCD is used for the current system status indication. Horn switch and brake switch is used for actuation of the horn and car brake respectively.

When there is horn initialization from car unit 1, a signal will be transmitted from one IR transmitter and it strikes to the car unit 1 receiver section. This signal is then decoded and horn indicator will be raised high.

Similarly, when there is an emergency condition at car unit 2 with emergency braking system, a high priority signal is transmitted on the reverse channel which makes the car unit 1 brakes activated and the tires or engine will stop. Buzzer will give the audio indications for the different conditions of the system.

Similarly, RF transmitter is installed to the speed limit sign board which transmits the fixed or constant signal i.e., Speed limit of that region in the form of binary digits which will be decoded by the RF receiver which is integrated in the system of vehicle.

3.3   Software Systems

When the device is switched ON, the hardware devices connected to the microcontroller is initialized with default values. When the IRDA transmits IR rays from vehicle 1, it is sensed by the IR receiver in vehicle 2 switching on the buzzer. If the "Horn Switch" is set at digital LOW, the vehicle 1 transmits the IR signal say 'A' and "Horn pressed by Car 1" is shown on the LCD, If the vehicle 2 receives this signal 'A' the Buzzer is set to HIGH and "Horn received by Car 1" is shown on the LCD. Since the Buzzer is placed inside the vehicle, close to the driver, the horn is heard only by the people inside and thus unnecessary noise can be reduced from the environment (as shown in Fig. 3).

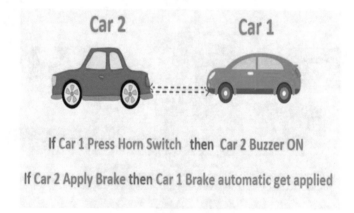

Fig. 3. Inter-vehicle communication

The inter-vehicle communication of automatic braking happens in a similar way, when the vehicle 2 applies the brakes suddenly the "Brake Switch" is set to digital LOW and an IR signal 'B' is transmitted to the vehicle 1, which is behind vehicle.

This is also indicated on the LCD as "Brakes applied by Car 2". If vehicle 1 receives this signal the relay is set to HIGH and brakes are automatically applied.

Also "Brakes applied by Car 2" is shown on the LCD (as shown in Fig. 2).

An RF transmitter which is installed on the speed limit sign board will constantly transmit the desired speed limit, in the form of binary digits, within its range. When a vehicle passes through this range, RF receiver integrated in the system of the vehicle will receive the binary digit signal, then the system will decode and compare it with the instantaneous speed of the vehicle. If the speed is more than the decoded value, a notification will be displayed on the LCD asking to reduce the speed. After a delay of few seconds, if the driver fails to follow the notifications that are transmitted to him then there is a provision to automatically generate an e-ticket. This will be implemented using a GSM module, that will send an SMS intimation to the RTO database. A database connection is created to insert necessary details such as the vehicle registration number, registered mobile number, the "fine" generated, etc. After creating the database connection, required SQL queries will be performed, the data will be retrieved and displayed on the page in table format.

4 Conclusion

Over-speeding and accidents caused due to sudden braking are two of the leading causes of road accidents, which is increasing every year. Manpower is not sufficient to monitor the traffic. To overcome these problems, the paper presented an inter-vehicular communication system design using IRDA and speed detection using active RFID. By successful implementation, the proposed system may be used to counter noise pollution at a certain level and help the environment as well as human health and ensure better travelling experience along with road safety, as it will prevent fatal road accidents resulting in severe loss. This system will act as a major deterrent to rash and undisciplined driving as the violators are immediately penalized by this speed detection and ticketing system.

References

1. Garg, N., Maji, S.: A retrospective view of noise pollution control policy in India: status, proposed revisions and control measure. Curr. Sci. **111**(1), 10 (2016)
2. The noise pollution (regulation and control) rules. Published in the Gazette of India, Extraordinary, Pt. II, Sec. 3 (ii) dated 14th February 2000 (2000)
3. Sharma, V., Aware, S., Goel, P.: Soundless horn using ad-hoc network (VANET). Int. Res. J. Eng. Technol. (IRJET) **04**(06), 2837–2840 (2017)
4. Kambadkone, P.R., Hancke, G.P., Ramotsoela, T.D.: Real time speed detection and ticketing system. In: 2017 IEEE AFRICON, Cape Town, pp. 1593–1598 (2017)
5. Sridharamurthy, K., Govinda, A.P., Gopal, J.D., Varaprasad, G.: Violation detection method for vehicular adhoc networking. Secur. Commun. Netw. **9**(3), 201–207 (2016)
6. Abdi, L., Takrouni, W., Meddeb, A.: Inter-vehicle communications over vehicular ad hoc networks. In: 13th International Wireless Communications and Mobile Computing Conference (IWCMC), Valencia, Spain, 26–30 June 2017
7. Kuchimanchi, C., Garimella, S.: Collision warning with automatic braking system for electric cars. Int. J. Mech. Eng. Res. **5**(2), 153–165 (2015). ISSN 2249-0019
8. Adnan, M.A., Sulaiman, N., Zainuddin, N.I., Besar, T.B.H.T.: Vehicle speed measurement technique using various speed detection instrumentation. Business Engineering and Industrial Applications Colloquium (BEIAC), pp. 668–672. IEEE (2013)
9. Al-Shabibi, L., Jayaraman, N., Vrindavanam, J.: Automobile speed violation detection system using RFID and GSM technologies. Int. J. Appl. Inf. Syst. **7**(6), 24–29 (2014)
10. O'Kane, T., Ringwood, J.V.: Vehicle speed estimation using GPS/RISS (reduced inertial sensor system). In: 24th IET Irish Signals and Systems Conference (ISSC 2013) (2013)
11. Sawant, H., Tan, J., Yang, Q.: A sensor networked approach for intelligent transportation systems. In: IEEE International Conference on Intelligent Robots and Systems (IROS), 28 September–2 October (2004)
12. Barba, C.T., et al.: Smart city for VANETs using warning messages, traffic statistics and intelligent traffic lights. In: IEEE Intelligent Vehicle Symposium, 3–7 June 2012
13. Maglaras, L.A., et al.: Social internet of vehicles for smart cities. J. Sens. Actuator Netw. **5**(1), 3 (2016)

RETRACTED CHAPTER: Standard Weight and Distribution Function Using Glowworm Swarm Optimization for Gene Expression Data

K. Sathishkumar[1]($\boxtimes$), E. Balamurugan[1], Jackson Akpojoro[1], and M. Ramalingam[2]

[1] University of Africa, Toru-Orua, Arua, Bayelsa, Nigeria
sathishmscgasc@gmail.com
[2] Gobi Arts and Science College, Gobi, TN, India

Abstract. This work shows an examination of swarm insight based grouping calculations to manage the quality articulation information successfully. In this work, a quality bunching strategies have been proposed to improve the looking and the grouping execution in genomic information. Also, through execution probes genuine informational collections, the proposed strategy Fuzzy Possibilistic C-Means Algorithm utilizing Expectation Maximization Algorithm is appeared to accomplish higher productivity, bunching quality and mechanization than other grouping technique.

To keep up bond between the areas in territory, Locality Sensitive Discriminant Analysis is utilized and a production meta experimental advancement calculation named Modified Artificial Bee Colony utilizing Fuzzy C Means grouping known as MoABC for bunching the quality articulation dependent on the example. At that point effective Standard Weight and Distribution Function - Glowworm Swarm Optimization (SWDF-GSO) grouping is utilized for bunching the quality articulation dependent scheduled on example. The trial results demonstrates that proposed calculation accomplish a higher grouping exactness and proceeds slighter fewer bunching period once contrasted and standing calculations.

Keywords: Clustering · LSDA · MABC · Fuzzy C Means · Swarm intelligence

1 Introduction

Grouping is the most famous methodology of dissecting quality articulation information and has demonstrated fruitful in numerous applications, for example, finding quality pathway, quality grouping, and capacity expectation. There is an extremely enormous territory of writing on bunching when all is said in done and on applying grouping strategies to quality articulation information specifically. A few agent algorithmic systems have been created and tested in grouping quality articulation

---

The original version of this chapter was retracted: The retraction note to this chapter is available at https://doi.org/10.1007/978-3-030-34515-0_85

information, which incorporate yet are not constrained to various leveled bunching, Self-Organizing Maps (SOMs), and realistic theoretic bunching.

In the ongoing work, dimensionality decrease is performed by utilizing Locality Sensitive Discriminant Analysis (LSDA). An ABC utilizing Fuzzy C Means (FCM) grouping remains utilized aimed at bunching the quality articulation dependent scheduled the example. So as to improve the exactness of the bunching, the dimensionality decrease of microarray quality articulation information is completed utilizing Neuro-Fuzzy Discriminant Analysis. To keep up link among the areas trendy territory, Glowworm Swarm Optimization (GSO) grouping with Standard Weight and Distribution Function (SWDF-GSO) is utilized for bunching the quality articulation dependent on the example.

2   Literature Review

Various grouping calculations have been created to improve the previous bunching calculations, unwinding the issues and fit for explicit fields [1]. Deng et al. proposed an improved fluffy grouping content bunching technique dependent on the FCM calculation and the alter separate calculation [2]. Vignes and Forbes [3] proposed a probabilistic model that has the bit of leeway to represent singular information and pairwise information at the same time. Shanthi and Uganya [4–7] proposed compelling bunching procedures named EMFPCM (Enhancement Maximization of Fuzzy Possibilistic C-Means) Algorithm utilizing Enhancement Maximization. The calculation was executed and the test result demonstrates that this technique is exceptionally powerful in anticipating client conduct. Grouping precision, Mean Squared Error (MSE), Implementation Time and Union contact were the parameters utilized for assessment [8].

3   Proposed Methodology

The proposed methodology comprises of three phases to be specific verdict of co-regulated bi-clusters utilizing Bimax calculation, dimensionality decrease utilizing Neuro-fuzzy Discriminant Analysis (NFDA) and bunching utilizing Standard Weight and Distribution Function (SWDF-GSO).

3.1   Proposed Enhanced Bimax Algorithm

This work proposed EB (Enhanced Bimax) Algorithm is associated data withdrawal framework organized gathering. In gathering similar precedents and tantamount quality tests are dealt with in a way with the objective would story close to each other. It contains three techniques. Three techniques are Enhanced Bimax, Breath first search (BFS) and BSP. Starting advance is institutionalization framework which is used to oust the dreary data and for social affair characteristics in the specific conditions.

Second step Breath First Search (BFS) is a count and it is proposed for exploring or glancing through tree or outline data structures. Finally Binary space partitioning (BSP) is proposed for again splitting a universe into angled.

3.2 Dimensionality Reduction – Neuro Fuzzy Discriminant Analysis (NFDA)

The essential stage concentrated on overhauling the handiness of MLFNN (Multilayer Feed-Forward Neural Networks) playing out the examination [11]. As referenced previously, another one example dedicated on Fisher's Discriminant analysis the situation and mixing this system to play out the nonlinear mapping.

3.3 Differential Evolution Established Loads Optimization

Differential Evolution (DE) is a straightforward enhancement method consuming similar, straight hunt, simple to utilize, great union, and quick usage properties [12, 13]. The essential supposed overdue DE is another plan for producing preliminary limit vectors by including the subjective contrast route amid two populace individuals $x_{r1.g}$ and $x_{r2.g}$ to a last group $x_{r0.g}$ at generation 'g'. The succeeding equation protests how to syndicate three dissimilar, arbitrarily selected routes to make a distorted vector, $v_{i,g}$ from the present group g:

$$v_{i,g} = x_{r0.g} + F \times \left(x_{r1} - x_{r2.g} \right) \tag{1.1}$$

Where $F \in (0, 1)$ is a measure aspect that panels the degree at which the populace changes. The index 'g' designate the group.

3.4 Glowworm Swarm Optimization (GSO) Clustering

This work presents a GSO calculation with upgrading multi-modular capacities, streamlining multi-modular capacities, physical specialists $i(i = 1, \ldots, n)$ are at first haphazardly sent in the target capacity space. Every specialist in the swarm chooses its bearing of development by the quality of the sign got from its neighbors. In this way, the glowworm allegories use to speak to the hidden standards of the GSO calculation [10]. GSO calculation to advance the multi-modular capacity incorporates five noteworthy advances: The Fig. 1 demonstrates the GSO calculation portrayal.

Input : Dimensionality reduced gene data

Output : Clustering results

Step 1 : Start the clustering process

Step 2: Fuzzy clustering for gene expression data

Step 3: Describe fuzzy membership value

Step 4: Random selection of **centroid** value

Step 5: Optimal selection of centroid value by GSO

Step 6: Each glowworm i converts the impartial purpose value $J(x_i(t))$ at its present location $x_i(t)$ into a luciferin value $l_i(t)$ by formula (1.2),

$$l_i(t) = (1 - \rho)l_i(t - 1) + \gamma J(X_i(t))$$

(1.2)

Step 7: Constructing neighborhood set $N_i(t)$;

Step 8: each glowworm i calculate move toward j probability $p_{ij}(t)$ by formula (1.3),

$$p_{ij}(t) = \frac{l_j(t) - l_i(t)}{\sum_{k \in N_i(t)} l_k(t) - l_i(t)}$$

(1.3)

Step 9: Select the touching substances j^*, using the formula (1.4) compute the novel local $x_i(t - 1)s$, is the moving step;

$$X_i(t + 1) = X_i(t) + s^* \left(\frac{X_j(t) - X_i(t)}{\|X_j(t) - X_i(t)\|} \right)$$

(1.4)

Step 10: Update the range of the active choice area by formulation (1.5)

$$r_d^i(t + 1) = \min \left\{ r_s, max\{0, r_d^i(t) + \beta(n_t - |N_i(t)|)\} \right\}$$

(1.5)

Step 11: Updating of optimal centroid value.

Step 12: Minimization of clustering objective function

Step 13 : Clustered results

Fig. 1. GSO Algorithm for gene expression data

In relations of the entire amount of summits apprehended, they propose the limit assortment is exposed in Table 1. Thus, only n and s r essential to be designated. These limitations cost takes more suitability to persons to smear the GSO algorithm.

Table 1. The GSO algorithm parameter selection

ρ	γ	β	n_t	S	l_0
0.4	0.6	0.08	5	0.03	5

The earlier segment transitory presented the GSO procedure, the GSO gathering procedure has been projected in this segment, contraction GSOCA.

3.5   Standard Weight and Distribution Function (SWDF-GSO)

GSO algorithm has some of the disadvantages:

- Faintness in universal pursuit which gives short clustering correctness and higher computation.
- It is easily to reduction into indigenous finest.
- Early conjunction and hunt correctness isn't enough all, also has the little competence in the advanced repetitions.

In instruction to overawed overhead difficulties of GSO, this work obtainable a Standard Weight and Distribution Function (SWDF-GSO) algorithm, which weights values are introduced for each feature based on the standard deviation and new position update based on the distribution function. The Fig. 2 shows the proposed algorithm representation.

The desirability of data object X is designed by the formula (1.2)

$$f(X) = -\ln\left(\frac{1}{num_g}\right) + \ln(d(X)) \tag{1.2}$$

where $d(X)$ the local space relative density of is X, ln() is the natural logarithm. The more X similarity with the data containing in its local space has more attraction.

Input : Dimensionality reduced gene data

Output : Clustering results

Step 1 : Start the clustering process

Step 2: Fuzzy clustering for gene expression data

Step 3: Describe fuzzy membership value

Step 4: Random selection of *centroid* value

Step 5: Optimal selection of centroid value by **SWDF-GSO**

Step 6: Each glowworm i codes the impartial purpose value $J(x_i(t))$ at its present position $x_i(t)$ into a luciferin value $l_i(t)$ by formula (1.6),

$$l_i(t) = (1 - \rho)l_i(t - 1)sw + \gamma J(X_i(t)) \qquad (1.6)$$

Step 7: Constructing neighborhood set $N_i(t)$;

Step 8: Generate weight value via the use of the standard deviation from equation (1.7)

$$x_i(t + 1) = DF(0, sd(t) * x_i(t)) \qquad (1.7)$$

In projected algorithm, glowworm is required to Distribution Function rendering to the formulation (1.8) if its site is altered in any group. $DF(0, sd(t))$Produce a normal distribution random number which means 0 standard deviation is sd(t). t is the stage scope. The standard deviation sd(t) is given by the equation (1.8)

$$sd(t) = \sqrt{\frac{\sum(x - x')}{n}} \qquad (1.8)$$

At the same time weight values are generated for each samples based on the attributes.

Fig. 2. SWDF-GSO Algorithm for gene expression data

Step 9: Each glowworm $i$calculate moves toward $j$probability $p_{ij}(t)$by formula (1.9),

$$p_{ij}(t) = \frac{l_j(t)-l_i(t)}{\sum_{k \in N_i(t)} l_k(t)-l_i(t)} \qquad (1.9)$$

Step 9: Select the affecting substances j^*, using the formula (1.10) compute the new location $x_i(t-1)s$,is the moving step;

$$x_i(t+1) = DF\big(0, sd(t) * x_i(t)\big) \qquad (1.10)$$

Step 10: Update the range of the active conclusion area by formula (1.11)

$$r_d^i(t+1) = \min\left\{r_s, max\left\{0, r_d^i(t) + \beta * w(n_c - |N_i(t)|)\right\}\right\}$$

(1.11)

Where sw is denoted as the weight value of the attributes generated from gene samples.

Step 11: Updating of optimal centroid value.

Step 12: Minimization of clustering objective function

Step 13 : Clustered results

Fig. 2. (*continued*)

4 Outcomes and Discussion

The estimated method for microarray quality grouping consumes remained actualized in the employed stage of tool (Matlab). For assessing the proposed strategy, has used the microarray quality examples of humanoid intense leukemia and colon disease information (Microarray quality) is talked about in Table 2. The high dimensional quality articulation information has been exposed to dimensionality decrease thus the dimensionality diminished quality information with measurements has been acquired. Consequently LPP strategy is connected to distinguish educational qualities and diminish quality dimensionality for bunching tests to recognize their phenotypes [9].

Table 2. Microarray gene data measurement applied for the assessment procedure

Categories of genetic factor data	Number of examples	Number of genes	Dimensionality reduction data using LPP
ALL	41	7139	41×41
AML	36	7128	36×40
COLON	68	3000	62×42

An example of microarray quality dataset of three classes that has been utilized for testing is given in the Table 3. Bunching for microarray quality articulation information whose sum is huge can be completely determined by deciding the limit of the groups.

Table 3. A sample of the microarray gene data

Class	ALL		AML		COLON	
Gene	Sample					
	ALL 16125TA-Norel	ALL 23368TA-Norel	AML SH5	AML SH1	AFFX-murIL2	AFFX-murIL10
AFFX-Crex-5 at (endogenous control)	−172 A	−93 A	−2 A	−11 A	20.6	−16
AFFX-Crex-3 at (endogenous control)	52 A	10 A	−12 A	112 A	−8.7	41.2
AFFX-BioB-3_st (endogenous control)	−134 A	59 A	−104 A	−176 A	4880	26.2

From the Table 4, it tends to be seen that the proposed SWDF-GSO system has given more exactness, connection and less separation and blunder rate instead of other grouping strategies. More precision and less slip-up rate prompts fruitful gathering of the given microarray quality data to the veritable class of the quality.

While testing, when a quality dataset is given, the proposed system needs to distinguish its having a place bunch. Bunching calculations, for example, Fuzzy C-implies and SWDF-GSO Algorithm methodologies are connected both to gather qualities, to parcel tests in the beginning period and have demonstrated to be helpful. The exhibition of each grouping calculation may shift significantly with various informational indexes.

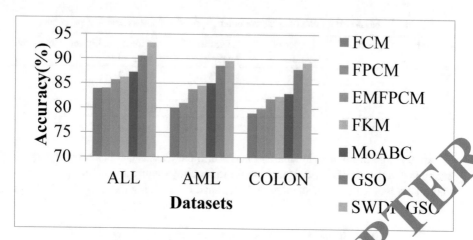

Fig. 3. Clustering accuracy vs. Clustering methods (Proposed SWDF-GSO technique)

Figure 3 shows the results it concludes that the proposed SWDF-GSO technique provides higher accuracy results of 93.25% respectively for ALL dataset samples (Table 4). From the Table 4, it can be understood that the projected SWDF-GSO technique has provided added accuracy, rather than existing clustering techniques.

Table 4. Accuracy Performance assessment between clustering algorithms (Proposed SWDF-GSO technique)

Types of data	Accuracy (%)						
	FCM	FPCM	EMFPCM	FKM	MoABC	GSO	SWDF-GSO
ALL	83.9	83.98	81.69	86.23	87.25	90.54	93.25
AML	80.06	02	83.84	84.56	81.12	88.63	89.62
COLON	79	79.9	81.96	82.45	83.04	87.98	89.25

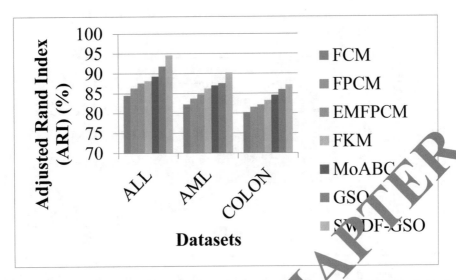

Fig. 4. Adjusted Rand Index (ARI) vs. Clustering methods (proposed SWDF-GSO technique)

Figure 4 shows the results it concludes that the proposed SWDF-GSO technique provides higher Adjusted Rand Index (ARI) results of 94.56% respectively for ALL dataset samples (Table 4). From the Table 5, it can be understood that the proposed SWDF-GSO method has providing more ARI, rather than the other clustering techniques.

Table 5. Adjusted Rand Index (ARI) comparison between clustering algorithms (Proposed SWDF-GSO technique)

Type of gene data	Adjusted Rand Index (ARI) (%)						
	FCM	FPCM	EMFPCM	FKM	MoABC	GSO	SWDF-GSO
ALL	84.5	86.32	87.56	88.12	89.25	91.79	94.56
AML	82.25	83.69	84.79	86.21	86.98	87.52	90.21
COLON	80.18	81.57	82.14	83.25	84.56	81.96	87.15

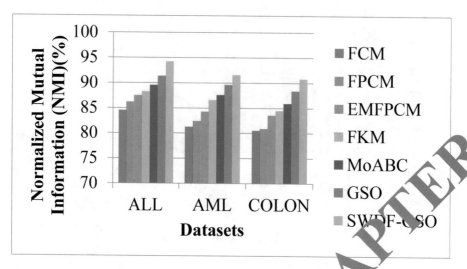

Fig. 5. Normalized Mutual Information (NMI) vs. Clustering methods (Proposed SWDF-GSO technique)

Figure 5 shows the results it concludes that the proposed SWDF-GSO technique provides higher Normalized Mutual Information (NMI) results of 94.21% respectively for ALL dataset samples (Table 5). From the Table 6, it can be understood that the proposed SWDF-GSO technique has providing more Normalized Mutual Information (NMI), rather than the other clustering techniques.

Table 6. Normalized Mutual Information (NMI) comparison between clustering algorithms (Proposed SWDF-GSO technique)

Type of gene data	Normalized Mutual Information (NMI) (%)						
	FCM	FPCM	EMFPCM	FKM	MoABC	GSO	SWDF-GSO
ALL	84.56	86.21	87.51	88.26	89.56	91.36	94.21
AML	81.25	82.41	84.28	86.58	87.63	89.63	91.58
COLON	80.52	80.89	83.58	84.49	81.93	88.41	90.81

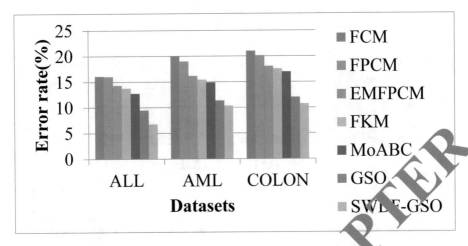

Fig. 6. Error Rate vs. Clustering methods (Proposed SWDF-GSO technique)

Figure 6 shows the error rate comparison result of three different datasets under seven clustering methods such FCM, FPCM, EMFPCM, FKM, MoABC and GSO algorithm. From the results it concludes that the proposed SWDF-GSO technique provides lesser error rate results of 6.75 respectively for ALL dataset samples (Table 6). From the Table 7, it can be understood that the proposed SWDF-GSO method has providing lesser error rate, rather than the rest of the clustering techniques.

Table 7. Error rate comparison between clustering algorithms (Proposed SWDF-GSO technique algorithm)

Type of gene data	Error rate (%)						
	FCM	FPCM	EMFPCM	FKM	MoABC	GSO	SWDF-GSO
ALL		16.02	14.31	13.77	12.75	9.46	6.75
AML	19.94	18.98	16.16	11.44	14.88	11.37	10.38
COLON	21	20.1	18.04	17.55	16.96	12.02	10.75

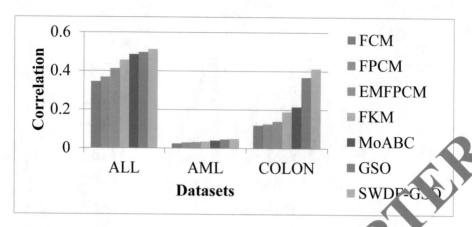

Fig. 7. Correlation measure vs. Clustering methods (Proposed SWDF-GSO technique)

Figure 7 shows the results it concludes that the proposed SWDF-GSO technique provides higher correlation results of 0.5125 respectively for ALL dataset samples (Table 6). From the Table 8 it can be understood that the proposed SWDF-GSO technique has providing supplementary correlation value than the other clustering techniques. It concludes that the proposed SWDF-GSO technique has higher correlation results.

Table 8. Correlation comparison between clustering algorithms (Proposed SWDF-GSO technique)

Type of gene data	Correlation						
	FCM	FPCM	EMFPCM	FKM	MoABC	GSO	SWDF-GSO
ALL	0.345	0.368	0.412	0.456	0.4852	0.4962	0.5125
AML	0.024	0.029	0.0315	0.0345	0.0396	0.0462	0.0489
COLON	0.119	0.125	0.139	0.189	0.215	0.368	0.4125

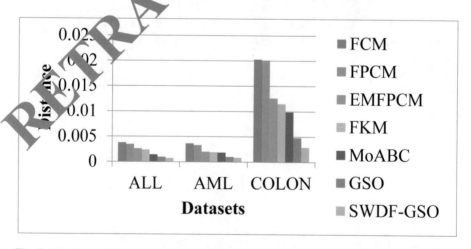

Fig. 8. Euclidean distance measure vs. Clustering methods (Proposed SWDF-GSO technique)

Figure 8 shows the results it concludes that the proposed SWDF-GSO technique provides lesser euclidean distance value of 0.00072 respectively for ALL dataset samples (Table 7). From the Table 9, it can be perceived that the proposed SWDF-GSO technique has providing lesser distance between the two samples than the other clustering techniques. It concludes that the proposed SWDF-GSO technique has higher clustering efficiency.

Table 9. Distance comparison between clustering algorithms (Proposed SWDF-GSO technique)

Types of data	Distance						
	FCM	FPCM	EMFPCM	FKM	MoABC	GSO	SWDF-GSO
ALL	0.00379	0.00346	0.00263	0.0024	0.00142	0.00098	0.00072
AML	0.00364	0.00331	0.00201	0.00192	0.00185	0.00101	0.00085
COLON	0.02029	0.02011	0.0126	0.0115	0.0099	0.00098	0.00289

5 Conclusion

In the proposed work, microarray quality information grouping strategy is carryout with three noteworthy advances: Bimax calculation, NFDA and SWDF-GSO. It tends to be seen that the proposed SWDF-GSO approach yields similarly great outcomes for the whole useful classification. The near outcomes have been demonstrated that the proposed procedure has better exactness, relationship and lesser separation, mistake rate than FCM, FPCM quality grouping systems. Trial results demonstrate that SWDF-GSO has higher grouping exactness when contrasted with GSO calculation.

References

1. Huber, W., von Heydebreck, A., Vingron, M.: Analysis of microarray gene expression data. Max-Planck Institute for Molecular Genetics, Berlin, 2 April 2003
2. Deng, J., He, LL., Chi, H., Wu, J.: An improved fuzzy clustering method for text mining. In: Second International Conference on Networks Security Wireless Communications and Trusted Computing (NSWCTC), vol. 1, pp. 65–69 (2010)
3. Vignes, M., Forbes, F.: Gene clustering via integrated Markov models combining individual and pairwise features. IEEE/ACM Trans. Comput. Biol. Bioinform. (TCBB) **6**(2), 260–270 (2009)
4. Shanthi, R., Suganya, R.: Enhancement of fuzzy possibilistic C-means algorithm using EM algorithm (EMFPCM). Int. J. Comput. Appl. **61**(12), 10–15 (2013). 0975–8887
5. Wang, G., Cui, W., Shao, Y.: Discriminant locality preserving projection. Res. J. Appl. Sci. Eng. Technol. **4**(24), 5572–5577 (2012)
6. Holland, S.M.: Principal components analysis (PCA). Department of Geology, University of Georgia, Athens, GA (2008)
7. Kumar, C.A., Srinivas, S.: Concept lattice reduction using fuzzy K-means clustering. Expert Syst. Appl. **37**(3), 2696–2704 (2010)

8. Khushaba, R.N., Al-Jumaily, A., Al-Ani, A.: Dimensionality reduction with neuro-fuzzy discriminant analysis. Int. J. Comput. Intell. **5**, 225–232 (2009)
9. Sathishkumar, K., Narendran, P.: An efficient artificial bee colony and fuzzy C means based co-regulated biclustering from gene expression data. In: Mining Intelligence and Knowledge Exploration, LNCS, vol. 8284, pp. 120–129. Springer (2013)
10. Sathishkumar, K., Ramalingam, M., Thiagarasu, V.: Biclustering of gene expression using glowworm swarm optimization and neuro-fuzzy discriminant analysis. Int. J. Adv. Res. Comput. Sci. Softw. Eng (IJARCSE) **4**(1), 188–196 (2014). ISSN 2277 128X
11. Casasent, D., Chen, X.: Radial basis function neural networks for nonlinear fisher discrimination and Neyman-Pearson classification. Neural Netw. **16**(5–6), 529–535 (2003)
12. Price, K.V., Storn, R.M., Lampinen, J.A.: Differential Evolution: A Practical Approach to Global Optimization. Springer, Berlin (2006)
13. Wua, H.X., Zhoua, J.J.: Fuzzy discriminant analysis with kernel methods. Pattern Recogn. **39**(11), 2236–2239 (2006)

Performance Analysis of Deep Learning Based Video Face Recognition Algorithm

Shahzadi Asra[✉] and Holambe Sushilkumar Nathrao

TPCT's College of Engineering, Osmanabad, Maharastra, India
shahzadiasra123@gmail.com

Abstract. The identities verification in videos has many applications in area of surveillance, social media and law enforcement. The existing algorithms have obtained higher verification accuracies at equal error rate but it is very difficult to achieve higher accuracy at low false accept rate and this has become major research challenge. An efficient video face recognition system has to develop and the performance is carried out to obtain accurate face recognition from video. We propose a novel algorithm for face verification from video signal and MATLAB is used to implement and simulate proposed algorithm. The performance analysis of proposed algorithm is carried out using databases such as YouTube faces and point and shoots challenge.

Keywords: Face · Features · Deep learning · Verification · Image · Recognition · Extraction

1 Introduction

With the advances in information and communication technology and huge demand for usage of smart mobile phones and surveillance camera for capturing the video have made face recognition from video highly significant [1]. The face is most extensively accepted means of person authentication and verification. However, the task for biometric authentication is based on face images and these images obtained using mobile devices such as smart phone, PDA and laptops in uncontrolled surroundings is extremely challenging [2, 3]. The performance of Face verification can be improved using multiple samples. The face recognition from video is process involving face verification and face identification and it can be classified into single face to video matching [3] which performs matching single face with various frames of image contain in a video recording and image set (video) to video matching which matching image set with video footage of an individual and this type of matching is also used for learning the model of person's [4, 5]. The main objective of the image set (represents variation in a face appearance) method is to classify an unidentified set of vectors to single training classes, each also characterized by more than a few image sets. The existing algorithms [6–10] have achieved high accuracies of verification at equal error rate, but to obtain low false accept rate with high performance is a still challenging issue [11–15]. We propose a algorithm based on deep learning. Deep learning intends

© Springer Nature Switzerland AG 2020
P. Karrupusamy et al. (Eds.): ICSCN 2019, LNDECT 39, pp. 619–625, 2020.
https://doi.org/10.1007/978-3-030-34515-0_64

to extract/remove hierarchical depictions from large scale images and video by employing deep learning architecture model called convolutional neural networks and it uses non linear transformations (multiple layers). Better performance can be achieved by utilizing deep learning technique. The principle concept of deep learning is to separate different levels of abstractions(concepts) entrenched in experiential data by highly designing the depth of layer and width of layer, and correctly selecting image features which are useful for learning tasks. We propose an algorithm to identify and verify face in video signal to obtain the best results at low false accept rate. The principle concept of this algorithm is to use frame selection (video frame) and feature representation based on deep learning. This algorithm performs feature rich frame selection from input videos by employing decomposition of wavelet and also entropy. The selected frames features are extracted using deep learning framework. The feed forward neural network is used for matching of extracted features representation from two videos. The database such as Point and Shoot Challenge and YouTube Faces are considered for performance analysis of proposed algorithm. Video face recognition system will employ verification and identification process to recognize face from video. Video face verification process is performed to obtain 1:1 matching of face images in video with a template face images whose identity being maintained. By employing the face verification process we can find whether the given two faces from video belong to the same person. Video face identification process is carried out to obtain 1: N difficulties matching of query video face image with all face image templates in a video face database. This step is carried out to get the identities of the given video faces from the known face set. The video face verification and identification process will be carried out based on deep learning technique and the expected results will be very high accuracy and efficient. The objective of this research is to assess the performance of proposed face recognition algorithm and to validate the effectiveness of image related quality measures for face video matching. The rest of the paper is organized as below. We present the proposed algorithm for face recognition in Sect. 2. In Sect. 3, we present the simulation results and discussions. Finally, the last section concludes the paper.

2   Proposed Face Recognition Algorithm

The proposed algorithm performs three steps for face recognition from video signal: the first step is frame selection, second step is feature extraction based on deep learning and third step is face verification by employing learnt representations. Figure 1 illustrates the steps carried out in video face recognition.

2.1   Frame Selection

The proposed algorithm utilizes feature richness of image as criteria for frame selection from input video signal. The feature richness aids in frames extraction with higher possibility of having biased features. To compute feature richness of image the following steps are carried out Preprocessing: The preprocessing of Input image (detected

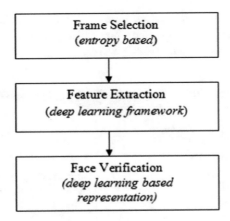

Fig. 1. Steps for video face recognition

face) is to have standard size and to convert the detected image to grayscale image. From the input image only facial region is considered by performing face detection without interfere of non-face content of frame using proposed algorithm/technique image normalization: This step is carried out using image mean and standard deviation. Also the Discrete wavelets transform (DWT) is applied to processed image.

2.2 Feature Extraction

After completion of process of obtaining feature rich frames i.e. step 1, the next step 2 called feature extraction and matching process is carried out using stacked de-noising auto encoders (SDAE) and DBM (deep Boltzmann machine) based algorithm. This algorithm utilizes limited training data and gives good results. In this step a deep learning framework is employed to perform the feature extraction and matching.

2.3 Face Verification

This step is carried out after obtaining the feature and matching from all the frames present in two videos. Face verification is performed using deep learning based representation. This step is carried out to verify the face from video.

3 Simulation Results and Discussion

The simulation analysis results for proposed algorithm are shown in Figs. 2, 3, 4, 5 and 6.

The Fig. 2 shows the plot for genuine accept rate v/s false accept rate. This illustrates the verification performance comparison of proposed algorithm using Receiver Operating Characteristic (ROC) curves.

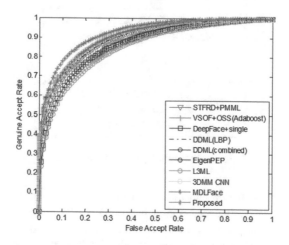

Fig. 2. ROC curves comparing, the verification performance of proposed algorithm

The Fig. 3 shows the plot for genuine accept rate v/s false accept rate of proposed algorithm simulation using YouTube faces database in the form of ROC curves.

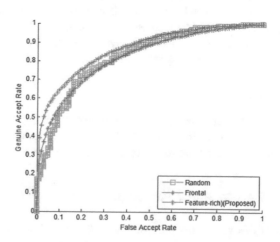

Fig. 3. ROC curves for YouTube faces database

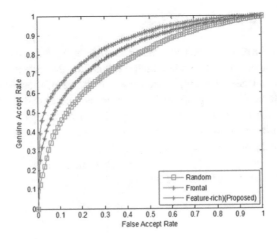

Fig. 4. ROC curves for handheld subset of PaSC database

The Fig. 4 illustrates the plot for genuine accept rate(GAR) v/s false accept rate (FAR) of proposed algorithm simulation using hand-held sub-set of PaSC, database in terms of ROC curves.

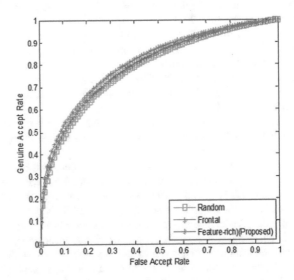

Fig. 5. ROC curves for control subset of PaSC database.

The Fig. 5 shows the plot for genuine accept rate v/s false accept rate of proposed algorithm simulation using control subset of PaSC database in terms of ROC curves.

4   Conclusion and Future Work

İn this paper, we have discussed the face recognition algorithm using the principle concept of deep learning. The video face verification and identification process is carried out based on deep learning technique and the results obtained are very high accuracy and efficient. The proposed algorithm is successfully implemented and simulated using MATLAB software and its performance analysis is carried out. The performance result of proposed algorithm shows that it has achieved the higher verification accuracy at low false accept rate.

As a future research work, we extend the implementation and simulation of proposed algorithm using VHDL and Xilinx tool.

References

1. Barr, J.R., Bowyer, K.W., Flynn, P.J., Biswas, S.: Face recognition from video: a review. Int. J. Pattern Recogn. Artif. Intell. **26**(5), 1266002 (2012)
2. Beveridge, J., et al.: The challenge of face recognition from digital point and-shoot cameras. In: Proceedings of IEEE Conference on Biometrics Theory, Applications and Systems, pp. 1–8, October 2013
3. Wolf, L., Hassner, T., Maoz, I.: Face recognition in unconstrained videos with matched background similarity. In: Proceedings of IEEE Conference Computer Vision and Pattern Recognition, pp. 529–534, June 2011
4. Li, H., Hua, G., Lin, Z., Brandt, J., Yang, J.: Probabilistic elastic matching for pose variant face verification. In: Proceedings of IEEE Conference on Computer Vision and Pattern Recognition, pp. 3499–3506, June 2013
5. Méndez-Vázquez, H., Martínez-Díaz, Y., Chai, Z.: Volume structured ordinal features with background similarity measure for video face recognition. In: Proceedings of International Conference on Biometrics (ICB), pp. 1–6, June 2013
6. Wolf, L., Levy, N.: The SVM-minus similarity score for video face recognition. In: Proceedings of IEEE Conference on Computer Vision and Pattern Recognition, pp. 3523–3530, June 2013
7. Khan, N.M., Nan, X., Quddus, A., Rosales, E., Guan, L.: On video based face recognition through adaptive sparse dictionary. In: Proceedings of IEEE International Conference and Workshops on Automatic Face and Gesture Recognition, pp. 1–6, May 2015
8. Li, H., Hua, G., Shen, X., Lin, Z., Brandt, J.: Eigen-PEP for video face recognition. In: Proceedings of Asian Conference on Computer Vision, pp. 17–33 (2014)
9. Li, H., Hua, G.: Hierarchical-PEP model for real-world face recognition. In: Proceedings of IEEE Conference on Computer Vision and Pattern Recognition, pp. 4055–4064, June 2015
10. Goswami, G., Bhardwaj, R., Singh, R., Vatsa, M.: MDLFace: memorability augmented deep learning for video face recognition. In: Proceedings of IEEE International Joint Conference on Biometrics (2014)
11. Hu, J., Lu, J., Tan, Y.: Discriminative deep metric learning for face verification in the wild. In: Proceedings of IEEE Conference on Computer Vision and Pattern Recognition, pp. 1875–1882, June 2014
12. Sun, Y., Wang, X., Tang, X.: Deeply learned face representations are sparse, selective, and robust. In: Proceedings of IEEE Conference on Computer Vision and Pattern Recognition, pp. 2892–2900, June 2015

13. Ding, C., Tao, D.: Trunk-branch ensemble convolutional neural networks for video-based face recognition, July 2016. https://arxiv.org/abs/1607.05427
14. Yang, J., Ren, P., Chen, D., Wen, F., Li, H., Hua, G.: Neural aggregation network for video face recognition, March 2016. https://arxiv.org/abs/1603.05474
15. Tran, A.T., Hassner, T., Masi, I., Medioni, G.: Regressing robust and discriminative 3D morphable models with a very deep neural network, December 2016. https://arxiv.org/abs/1612.04904

Computational Measure of Cancer Using Data Mining and Optimization

Ashutosh Kumar Dubey$^{(\boxtimes)}$, Umesh Gupta, and Sonal Jain

Institute of Engineering and Technology,
JK Lakshmipat University, Jaipur, India
ashutoshdubey123@gmail.com,
{umeshgupta, sonaljain}@jklu.edu.in

Abstract. In this paper approaches for cancer prediction through computational measures has been discussed and analyzed. This paper provides the basis of worldwide cancer impact, methodological study with discussion, attributes and parametric impact, gaps analyzed, and the suggested computational solutions. This paper also explores the impact and the association measures of the influencing factors. The methods covered in this study are from data mining and optimization. The latest trends in the methods used and applicability have been discussed with the gaps.

Keywords: Computational measure · Data mining · Optimization · Cancer prediction

1   Introduction

According to the latest report of GLOBOCAN, cancer is the leading cause of death worldwide [1–3]. The impact can be understood easily from the cancer incidence and mortality rates of 2018 as shown in Fig. 1. It is clear from Fig. 1 that the highest incidence has been observed in case of breast cancer and the lowest in the case of ovarian cancer. The highest mortality has been observed in case of lung cancer and lowest in cases of thyroid. Incidence shows the measure of new occurrences of cancer cases in a specific population and time period [4, 5]. Mortality shows the measure of cancer death in a specific population and time period [4, 5]. The current worldwide statistics show the alarming state.

This paper's aim is to analyze the methods in the direction of cancer detection and prediction through computational variability and suggest some approaches which can be helpful in the future detection and prediction framework development. There are several works have been published in case of breast cancer [6–9], lung cancer [10–12] and oral cancer [13, 14]. We have considered some of the latest papers as the number of research papers published in the cancerous area are too wide and the work is already being in progress due to the need of betterment in the detection and prediction system. The main problems in the development of the efficient cancer detection or prediction framework are the attribute association, data unavailability, demographic regions and the parameter variations. The above facts make the scenario typical when the factors

© Springer Nature Switzerland AG 2020
P. Karrupusamy et al. (Eds.): ICSCN 2019, LNDECT 39, pp. 626–632, 2020.
https://doi.org/10.1007/978-3-030-34515-0_65

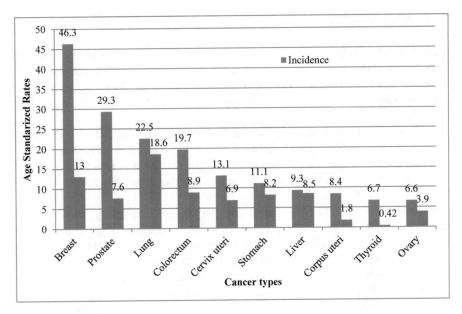

Fig. 1. Worldwide incidence and mortality rates 2018 (age standardized) [1–3]

and their associations are unavailable from the sources. Then there are other methodical challenges which make it complex. The main challenges are as follows:

1. How to determine the combinations of method to fit in the problem area or with the given dataset.
2. If the framework will extend with the attributes and the parameters, how it can be handled?
3. Then the impact and the association's measures are also a challenging issue.
4. To handle the new issues and prospect generates a requirement for the automatic enhancement in the methods and their experimentation ranges.

The above challenges show the issues which are almost same in every cancer research irrespective of the method and its applicability. The main methods used in the cancer analysis in general are data mining and optimization [4, 5]. The use of clustering and classification techniques are the most in cancer detection and prediction based on the literatures.

So, the main objectives of this paper are:

1. To explore the statics in case of cancer worldwide
2. To explore the opportunity in case of different cancer.
3. To review and analyses the latest trends in terms of the method uses and applicability.
4. To highlights the gaps and provides the suggested solutions.
5. To suggest the computational analysis approaches for cancer prediction through data mining and optimization.

2   Related Works

In 2017, Kawashima et al. [15] discussed about the biomedical literatures. The main focus of this paper is the automatic extraction of the related genes. For the relation extraction they have used text mining and pattern clustering. They have extracted the candidate genes first and then clustering algorithms have been used for the candidate association.

In 2017, Lemsara et al. [16] discussed about the determination of the patient subtypes. They have considered multi-omics datasets for the experimentation. They have proposed a refined multi-view clustering algorithm for the multi-view datasets in order to ensure patient subtypes with better quality. The approach performs the initial clustering on the data.

In 2017, Li et al. [17] discussed about the subspace clustering. Firstly, affinity matrix has been observed, and then the spectral clustering has been applied. They have proposed a joint optimization framework that is structured sparse subspace clustering (S3C). It is used for the learning. Their results support the approach.

In 2017, Behera et al. [18] proposed an evolutionary clustering algorithm. It has been proposed for the clustering of genes having similar expression profiles. They have used the combination of clustering algorithm and evolutionary computation. In case of evolutionary computation, they have used mutation, crossover, and natural selection. Based on the mutual information they have found the genetic distances. Then, based on the similarity clusters have been formed. They have applied this approach to gastric cancer.

In 2017, Govinda et al. [19] discussed the expert systems for medical applications. They have applied machine learning approach to the cancer diagnosis. The detection is applied using the artificial neural network. Backpropagation algorithm has been used for the network training. Fuzzy-c-means algorithm has been used for the membership value calculation. They have developed a neuro-fuzzy system based on the mamdani model. For the cancer identification rule pruning has been used.

In 2018, Hossain and Rahaman [20] discussed about the bone cancer. They have suggested the dangerous impact due to this cancer because of the early death. So, they have suggested the need for early detection. They have applied FCM clustering. They have considered 120 verified MRI images for the experimentation. They have used adaptive Neuro fuzzy inference system (ANFIS) for the classification purpose. They have extracted gray level co-occurrence matrix (GLCM) features of the training and testing. They have calculated accurately, sensitivity and specificity. They have achieved 93.75% classification accuracy.

In 2018, Fathurahman et al. [21] discussed the Identification of cancer subtypes identification. They have discussed the problems in the case of the analysis of gene expression data. The problem they identified is higher dimensional attributes, missing values and sparse data problem. They have proposed an iterative scaling fuzzy additive spectral clustering (is-FADDIS) for the above discussed problems. Their results support the approach.

In 2018, Gupta and Malhi [22] discussed about the head and neck cancer detection. For experimentation they have considered 26019 CT scan images from cancer imaging archive (TCIA). They have used deep learning framework. They have used FCM for the segmentation. They have used gray level co-occurrence matrix (GLCM) for the feature extraction. They have achieved the accuracy of 98.8%.

3 Problem Statements

The following problem statements have been identified based on the literature explored.

1. Traditional methods may fail in finding association from multiple relations that are independent and identically distributed.
2. If the data preprocessing is needed for biological sequencing, clustering and classification algorithms alone may not provide better results, considering that the functional properties and structure are different.
3. The membership value affects the data point's value and it may affect the next cycle, and, consequently, the results. There are also chances to confine it to local optima, as it may be influenced by initialization.
4. The influence of an individual parameter inside each attribute is overlooked; thus, it is important to examine each parameter separately.
5. The influence of the combined parameters of the attributes may be considered to improve the prediction accuracy.
6. To handle the non-clustered data efficiently is a difficult problem.
7. One method may not be sufficient and successful for all purposes, necessitating the need for a hybrid framework.

4 Computational Analysis

This section provides an overview of the way to apply clustering to map and scale the clustering approach to lung cancer, oral cancer and bone cancer based on their attributes. The approach for such clustering based on each attribute is mentioned below:

Cancer Type: Lung Cancer
1. Age: This attribute is diagnosed mostly in elderly age [4]. Therefore, the scaling should be considered based on the age. The following scaling can be used for the calculation of computational predictive accuracy and clustering. For the ages (30–40)→3, (40–50)→5, (50–60)→7, (60–70)→9, (>70)→10. As the chances of lung cancer are low under the age of 30, so the scaling can be considered from age 30.
2. Gender: This attribute is common in both males and females.
3. Alcohol Usage: It is categorized generally in two parts alcoholic and non-alcoholic. Alcohol can be measured by ml or g. It can be calculated based on the unit consumption in terms of weeks or month [23]. So, the scaling can be considered from 1 for non-alcoholic to 10 for high consumption per week.
4. Obesity: Obesity may affect the insulin resistance, so it can be scaled in terms of resistance.

5. Smoking: People (age > 50) in the habit of smoking and using tobacco come under high risk of lung cancer. In India deaths between 25–69 age group people are more due to cigarette or beedi [24, 25]. So, it can be categorized age wise habit of smoking for the better risk prediction. The scaling can be provided based on two factors first is age and the second is frequency of smoking in a day. For age, the categories may be (10–20), (20–30), (30–40), (40–50) and (>50). In case of frequency it can be measured with the average number of cigarettes smoked per day. Then it is also important to consider that the habit is how much older. It also affects badly. Then another condition is if he/she left smoking, then from how many days or months or years. So, the combination of these attributes can form the proper association between the attributes and provide the scaling weight for the clustering.

6. Chest pain: It can be scaled on the basis of the pain conditions like regular and irregular along with the duration of continuing pain. Then the associated factors in pain are acute pain, chronic pain, breakthrough pain, soft tissue pain, nerve pain, referred pain, phantom pain and bone pain. So, it can be scaled on the basis of 1–10 with the associated and affected attributes in the combination.

7. Coughing of blood: The first attribute associated for computational measurement in case of coughing of blood is smoking behavior, past history of smoking and age. Then the second factor associated with this is persistent coughing, wheezing and shortness of breath. Then the frequency of regular coughing is also considered for checking the complete impact.

8. Weight loss: It can be scaled for the computational calculation based on the weight loss percentage and the duration of the weight loss. Muscle mass loss can also be considered. The weight loss consideration also accounts the pre-chemotherapy and post- chemotherapy status.

9. Shortness of breath: The attributes associated for computational measurement in case of shortness of breath are smoking behavior, past history of smoking, age, persistent coughing, wheezing and the frequency of regular coughing is also considered for checking the complete impact.

Cancer Type: Oral Cancer

1. Difficulty in chewing: It can be scaled for the computational calculation based on the smoking behavior, tobacco and gutkha usage. Then the associated factors in pain are acute pain, chronic pain, breakthrough pain, soft tissue pain, nerve pain, referred pain, phantom pain and bone pain. So, it can scale on the basis of 1–10 with the associated and affected attributes in combination.

2. Painless ulceration: It can be scaled based on the number of days and the affected area with the chewing habits of tobacco and gutkha. More than three weeks should scale on higher side.

3. Bleeding: The scaling for this attribute can be considered based on blood or bruises in the mouth, rash on tongue, under the tongue, on roof of mouth, and/or on inside of cheeks and blood oozing from the mouth. Then the number of days may be considered as the associated and influencing parameter.

4. Hoarseness of voice: The voice changes like hoarseness may be considered here if that does not improve within 15–20 days. So, the scaling should be done on the basis of a number of days.

5. Swelling: It is considered based on the growth of the tumor inside and outside of the mouth. So, in this case the scaling is on the basis of tumor size.

Cancer Type: Bone Cancer
The attributes of bone cancer are weight loss, bone pain and swelling. Therefore, similar clustering approach is applicable to them as above.

In this section we have discussed several attributes with the suggested computational analysis. These factors are also common in other cancers so this computational approach may be helpful in other cases also.

5 Conclusion

This paper discusses and analyzes the computational approaches used for the cancer detection with the prospective of different attributes and parametric impacts. This study explores the areas where the traditional methods fail and the gap analysis has been done with the discussion. The main focus of this paper is to highlight the methods, discuss the impact, gap analysis and the computational discussion based on the attributes. Finally, based on the discussion suggested computing solutions have been suggested.

References

1. Bray, F., Ferlay, J., Soerjomataram, I., Siegel, R.L., Torre, L.A., Jemal, A.: Global cancer statistics 2018: GLOBOCAN estimates of incidence and mortality worldwide for 36 cancers in 185 countries. CA Cancer J. Clin. **68**(6), 394–424 (2018)
2. Ferlay, J., Colombet, M., Soerjomataram, I., Mathers, C., Parkin, D.M., Piñeros, M., Znaor, A., Bray, F.: Estimating the global cancer incidence and mortality in 2018: GLOBOCAN sources and methods. Int. J. Cancer **144**(8), 1941–1953 (2019)
3. Ferlay, J., Ervik, M., Lam, F., Colombet, M., Mery, L., Piñeros, M., Znaor, A., Soerjomataram, I., Bray, F.: Global Cancer Observatory: Cancer Today. International Agency for Research on Cancer, Lyon (2018). https://gco.iarc.fr/today. Accessed 20 Mar 2019
4. Dubey, A.K., Gupta, U., Jain, S.: Epidemiology of lung cancer and approaches for its prediction: a systematic review and analysis. Chin. J. Cancer **35**(1), 71 (2016)
5. Dubey, A.K., Gupta, U., Jain, S.: Breast cancer statistics and prediction methodology: a systematic review and analysis. Asian Pac. J. Cancer Prev. **16**(10), 4237–4245 (2015)
6. Dubey, A.K., Gupta, U., Jain, S.: Analysis of k-means clustering approach on the breast cancer Wisconsin dataset. Int. J. Comput. Assist. Radiol. Surg. **11**(11), 2033–2047 (2016)
7. Dubey, A.K., Gupta, U., Jain, S.: A survey on breast cancer scenario and prediction strategy. In: Proceedings of the International Conference on Frontiers of Intelligent Computing: Theory and Applications, pp. 367–375. Springer, Cham (2015)
8. Dubey, A.K., Gupta, U., Jain, S.: Comparative study of K-means and fuzzy C-means algorithms on the breast cancer data. Int. J. Adv. Sci. Eng. Inf. Technol. **8**(1), 18–29 (2018)
9. Wang, Z., Li, M., Wang, H., Jiang, H., Yao, Y., Zhang, H., Xin, J.: Breast cancer detection using extreme learning machine based on feature fusion with CNN deep features. IEEE Access **7**, 105146–105158 (2019)

10. Hussain, L., Aziz, W., Alshdadi, A.A., Nadeem, M.S., Khan, I.R.: Analyzing the dynamics of lung cancer imaging data using refined fuzzy entropy methods by extracting different features. IEEE Access **7**, 64704–64721 (2019)

11. Wu, J., Guan, P., Tan, Y.: Diagnosis and data probability decision based on non-small cell lung cancer in medical system. IEEE Access **7**, 44851–44861 (2019)

12. Delen, D.: Analysis of cancer data: a data mining approach. Expert Syst. **26**(1), 100–112 (2009)

13. Chan, C.H., Huang, T.T., Chen, C.Y., Lee, C.C., Chan, M.Y., Chung, P.C.: Texture-map-based branch-collaborative network for oral cancer detection. IEEE Trans. Biomed. Circ. Syst. **13**(4), 766–780 (2019)

14. Gawade, P., Chauhan, R.P.: Detection of lung cancer using image processing techniques. Int. J. Adv. Technol. Eng. Explor. **3**(25), 217 (2016)

15. Kawashima, K., Bai, W., Quan, C.: Text mining and pattern clustering for relation extraction of breast cancer and related genes. In: International Conference on Software Engineering, Artificial Intelligence, Networking and Parallel/Distributed Computing, pp. 59–63. IEEE (2017)

16. Lemsara, A., Ouadfel, S., Batouche, M.: Multi-view clustering with local refinement for cancer patient stratification. In: Intelligent Systems and Computer Vision, pp. 1–5. IEEE (2017)

17. Li, C.G., You, C., Vidal, R.: Structured sparse subspace clustering: a joint affinity learning and subspace clustering framework. IEEE Trans. Image Process. **26**(6), 2988–3001 (2017)

18. Behera, N., Sinha, S., Gupta, R., Geoncy, A., Dimitrova, N., Mazher, J.: Analysis of gene expression data by evolutionary clustering algorithm. In: International Conference on Information Technology, pp. 165–169. IEEE (2017)

19. Govinda, K., Singla, K., Jain, K.: Fuzzy based uncertainty modeling of cancer diagnosis system. In: International Conference on Intelligent Sustainable Systems, pp. 740–743. IEEE (2017)

20. Hossain, E., Rahaman, M.A.: Bone cancer detection classification using fuzzy clustering neuro fuzzy classifier. In: International Conference on Electrical Engineering and Information & Communication Technology, pp. 541–546. IEEE (2018)

21. Fathurahman, M., Veritawati, I., Wasito, I.: Experimental analysis of iterative-scaling fuzzy additive spectral clustering (is-FADDIS) for cancer subtypes identification. In: International Conference on Advanced Computer Science and Information Systems, pp. 435–440. IEEE (2018)

22. Gupta, P., Malhi, A.K.: Using deep learning to enhance head and neck cancer diagnosis and classification. In: International Conference on System, Computation, Automation and Networking, pp. 1–6. IEEE (2018)

23. Mons, U., Gredner, T., Behrens, G., Stock, C., Brenner, H.: Cancers due to smoking and high alcohol consumption: estimation of the attributable cancer burden in Germany. Deutsches Ärzteblatt Int. **115**(35–36), 571 (2018)

24. Thankappan, K.R., Thresia, C.U.: Tobacco use & social status in Kerala. Indian J. Med. Res. **126**(4), 300 (2007)

25. Gajalakshmi, V., Peto, R., Kanaka, T.S., Jha, P.: Smoking and mortality from tuberculosis and other diseases in India: retrospective study of 43 000 adult male deaths and 35 000 controls. The Lancet **362**(9383), 507–515 (2003)

RETRACTED CHAPTER: Gene Expression Analysis Using Clustering Methods: Comparison Analysis

K. Sathishkumar[1(✉)], E. Balamurugan[1], Jackson Akpojoro[1], and M. Ramalingam[2]

[1] University of Africa, Toru-Orua, Bayelsa, Nigeria
sathishmscgasc@gmail.com
[2] Gobi Arts and Science College, Gobichettipalayam, Tamil Nadu, India

Abstract. A microarray measures the articulation dimensions of thousands of qualities. Meanwhile, Grouping examines microarray quality articulation information. In this paper, have executed a biclustering calculation to distinguish subgroups of information which shows corresponded conduct under explicit test conditions. During the time spent for discovering bi-clusters, Fuzzy K-implies grouping is utilized to bunch the qualities and tests with most extreme enrolment work. Both dimensionality and lessening the quality sieving are finished utilizing LFDA and quality sifting with the capacity separately. From the outcomes it presumes that the proposed work performs better when contrasted with other existing bunching calculations, for example, FCM, FPCM and EMFPCM.

Keywords: Clustering · Fuzzy K means · Classification · Cancer data · LFDA

1 Outline

A rise of microarray innovation has made it conceivable to screen the articulation dimensions of thousands of qualities at the same time. The Challenge is to adequately investigate and decipher is enormous volume of information. Two measurable tasks normally connected to microarray information are arrangement and grouping however the most critical region is bunching microarray information and examination. Microarray so that remains supportive for study varieties of the numerous qualities by methods for all the while. With advancement of the microarray approaches numerous works are done on examination of the quality articulation information. Late work could present the Fuzzy Possibilistic C-Means Algorithm with the assistance of Expectation Maximization Algorithm (EMFPCM) [1–3].

2 Literature Review

Liu et al. [1] work has focusing on finding the subset of explicit qualities in which those could display a comparative articulation designs and alongside subset of the conditions in quality articulation network. In explicitly, this work is searching for an Order

---

The original version of this chapter was retracted: The retraction note to this chapter is available at https://doi.org/10.1007/978-3-030-34515-0_85

© Springer Nature Switzerland AG 2020, corrected publication 2024
P. Karrupusamy et al. (Eds.): ICSCN 2019, LNDECT 39, pp. 633–644, 2020.
https://doi.org/10.1007/978-3-030-34515-0_66

Preserving groups (OP Cluster), by methods for every subset of the qualities that instigating the comparable straight requesting with subset of a conditions.

Gerstein et al. [2] presented variants of an approach, which is based on whether the specific normalization gets over the genes and conditions is been done by independently or in coupled fashion. Yang et al. [3] generalize a model of the bi-cluster that incorporate the null values and also propose probabilistic algorithm which could discovering set of the k possibly available overlapping bi-clusters in simultaneously.

In this review, the bi-clustering algorithm has proposing in the intent of identify the subgroups of a data that shows the correlated behaviour by means of specific experimental conditions that under. In process of find the bi-clusters, the Fuzzy K-Means (FKM) clustering is found as usable one for clustering out the genes and particular samples along maximum membership function. The results are been measured based on clustering metrics namely specific accuracy, the error rate, the correlation and the distance measure.

3   Local Fisher's Discriminant Analysis (LFDA)

Fisher Discriminant Analysis (FDA) [4] is one of prominent methodology particularly for the straight dimensionality decrease by which it expands in the middle of the both class dissipate and furthermore limits inside the class disperse. Conventional FDA is additionally known for functioning admirably and furthermore for all intents and purposes valuable for even at this point. In any case, it likewise tends in give an undesired yield and if the examples are in some of class that structure a surely isolated bunches to be specific multimodal [5]. The multimodality has found as regularly see in a significant number of down to earth applications.

Fisher Discriminant Analysis (FDA) is one of prominent decision for lessen a dimensionality of a unique informational index accessible. It additionally amplifies in the middle of class dissipate and could limit in inside class disperse. It may fill in as truly well by methods for training, be that as it may, it needs on certain contemplations of multimodality. Multimodality is along inside the huge numbers of uses to be specific the ailment determination, though they might be a different explanation behind specific sickness. In this specific situation, the FDA couldn't catch multimodal qualities of bunches. In managing the multimodality, the Local-Preserving Projection (LPP) may assume indispensable job explicitly in protecting neighborhood structure of information [5].

LPP would keep close-by accessible information combines in unique space close in installing space, in that multimodal information can implanted by without lose its very own neighborhood structure. In other hand, the LPP is one of an unsupervised dimensionality decrease approach that doesn't take name data in to a record [5]. Proposing new dimensionality decrease system and calling as Local Fisher Discriminant Analysis (LFDA), is confined variation of the Fisher discriminant examination. LFDA could take the neighborhood structure of information in record thus that multimodal information could implant in suitably. Furthermore, LFDA has given more independent implanting than the FDA. To clarify a reason naturally, would initially take note of a FDA that can viewing as expanding of between-class dissipate which is under requirement by keeping out inside class disperse for certain dimension [5]. Because of this specific solid imperative, the less level of an opportunity has left for the expanding detachability, thus FDA yields in extremely less discrete installing. On other hand, LFDA isn't requiring the multimodal tests in the feeling of falling into an individual group.

4 Bi Clustering and K Means Clustering

The Bi-bunching, the square grouping the co-bunching, or the two-mode bunching is information mining approach in which it permits the concurrent grouping of lines and the segments of lattice. Giving arrangement of the m tests are been speaking to by methods for n-dimensional component based vectors; a whole dataset may peak to m pushes in the n number of segments [7–11].

In explicit quality articulation based information examination, this point is constantly went with an extra prerequisite of a huge line change by sensibly. The method of reasoning is for low mean squared build-up that can just demonstrates the quality articulation levels vacillate by around in the harmony, which incorporates steady bi-groups though there has no or by little change among all. Contrastingly bi-bunching calculations have shifting meanings of the bi-group. The are:

1. Bi-bunch with a steady qualities (a),
2. Bi-group with the consistent qualities on explicit lines (b) or on sections (c),
3. Bi-group alongside intelligent qualities (d, e).

4.1 Bicluster with Consistent Qualiti

While bi-bunching calculation would attempts in discovering consistent bi-group, typical path for reorder lines and the sections of grid thus that could gather by together comparably profiting lines/segments and furthermore discover the bi-bunches along the comparative qualities. This technique has found as OK when information is as clean.

4.2 Bi-Clusters with Steady Qualities on Lines or Sections

This kind of the bi-bunches couldn't assess by difference of its own qualities. To complete recognizable proof, segments and lines must get standardized by first. There are just different calculations, are without the standardization step, discover the bi-groups having lines and the sections along changing methodologies.

4.3 Bi-Clusters with Reasonable Qualities

For bi-bunches with the reasonable qualities on segments and lines, whole improvement over calculations particularly for the bi-groups with a steady an incentive on the lines or on the segments must got considered. This implies modern calculation is a required one.

4.4   K-Means Clustering

K-Means Clustering is one of an unconfirmed knowledge calculation which attempts in grouping the information that relies upon their own comparability. Unsupervised adapting likewise implies, there are no more results for anticipated, and a calculation additionally simply giving a shot in discovering designs in information. In explicit 'k' signifies bunching, having indicates measure of the groups, need the information to get assembled. This calculation could arbitrarily allot perception to group, and furthermore discovers centroid to every one of bunch. At that point, calculation could emphasize by means of two stages:

1. Reassigning the information focuses to group which are centroid is as near st.
2. Calculate the new centroid of an each group.

These are two stages rehashed till inside the group variety which ca diminish any of further. At that point inside group variety has been determined as aggregate of explicit Euclidean separation in the middle of information focuses and furthermore their own particular bunch based centroids. In the feeling of improving an exactness of k means grouping calculation fluffy has consolidated and furthermore increment the precision of bunching.

5   Proposed Methodology

The proposed approach contains three phases specifically finding of co-regulated bi-clusters expending Bimax algorithm, dimensionality decrease consuming LFDA and clustering using Fuzzy K-Means Clustering Algorithm (FKM).

5.1   Enhanced Bimax Algorithm

The graph for discovery the co-regulated groups utilizing calculation is appeared in Fig. 1. The info is quality grouping f the smaller scale cluster information. Improved Bimax calculation is utilized to show a best bi-clusters esteem and shows a co-regulated bi-cluster. The Enhanced Bimax calculation is utilized to quantify a specific quality is available or not. Standardization system cast-off to designate makings are displayed in the specific gathering it or not. The income is demonstration the conversion features [17].

5.2   Bimax Algorithm

The Bimax calculation needs to ensure the main ideal, incorporation maximal bi-clusters which is to be produced. It is utilized to indicate the examination of DNA chips and quality systems. The calculation understands the gap and-vanquish procedure. Figure 1 portrays a unique Bimax calculation. It contains of three strategies. They are Enhanced Bimax, Conquer and Divide. Overcome capacity is sound and checked the complaint is in the event that the qualities and conditions are equivalent, at that point the dividing is start, else it halt the procedure. Another step is part the information and

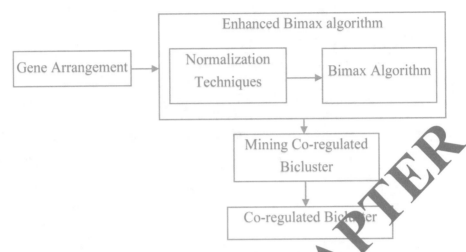

Fig. 1. Excavating Co-regulated Bicluster

standardization procedure is utilized to amass the splited information. It is utilized to discover entirely include the most extreme gatherings when all is said in done quality articulation information. Each co-regulated qualities are gathering the specific articulation esteem and the specific circumstance.

5.3   Suggested Enhanced Bimax Algorithm

In this work proposed Enhanced Bimax calculation is connected information mining strategy on grouping. In the grouping comparative examples and comparable quality tests are composed in a manner so they would untruth near one another. Initial step is standardization procedure which is utilized to evacuate the repetitive information and for gathering qualities in the particular conditions. Standardization is the way toward segregating factual blunder in rehashed estimated information. The objectives in doing dispense with all the excess information and guarantee information conditions. Breath First Search (BFS), Binary Space Partitioning (BSP) capacities are called and check the condition is in the event that the qualities and conditions are equivalent, at that point the dividing is start. Breath First Search (BFS) is a calculation for crossing or looking through tree or chart information structures [12].

Else it stops the procedure. It determines that a specific quality is available in the given gathering then it is speaks to a one. With these most extreme gatherings when all is said in done quality articulation information can be found. Every co managed qualities are gathering the specific articulation esteem and the specific circumstance. Generally the quality is absent in the given gathering it is speaking to as zero. Figure 2 depicts a proposed Enhanced Bimax calculation [13].

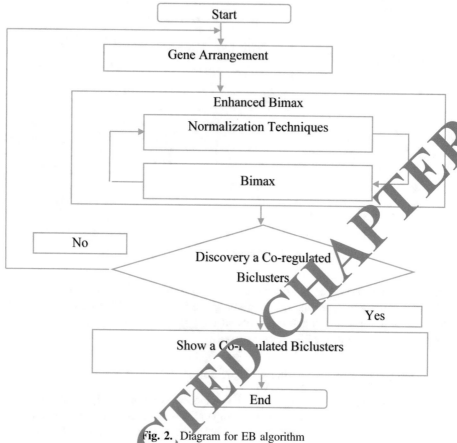

Fig. 2. Diagram for EB algorithm

5.4 LFDA (Local Fisher's Discriminant Analysis)

LFDA is an ongoing expansion to LDA that can adequately deal with the multi-modular/non-Gaussian issue. It is an administered highlight projection system that adequately joins the belongings of LDA and an unverified multifaceted knowledge method – LPP (Locality Preserving Projection) [14, 15]. The overall thought of LFDA is to acquire a decent division of tests from various modules though defensive the neighborhood building of opinion exhausts of each lesson [13, 17]. The native inside method throw matrix $S^{(lw)}$ and the local between class scatter matrix $S^{(lb)}$ used in LFDA are distinct as monitors:

$$S^{(lb)} = \frac{1}{2}\sum_{i,j=1}^{n} W_{i,j}^{(lb)}(X_i - X_j)(X_i - X_j)^T \tag{1}$$

$$S^{(lw)} = \frac{1}{2} \sum_{i,j=1}^{n} W_{i,j}^{(lw)} (X_i - X_j)(X_i - X_j)^T \tag{2}$$

where $W^{(lb)}$ and $W^{(lw)}$ are $n \times n$ conditions definite as

$$W_{i,j}^{(lb)} = \begin{cases} A_{i,j}\left(\frac{1}{n} - \frac{1}{n_l}\right), & \text{if } y_i = y_j = l, \\ \frac{1}{n} & \text{if } y_i \neq y_j, \end{cases} \tag{3}$$

$$W_{i,j}^{(lw)} = \begin{cases} \frac{A_{i,j}}{n_l} & \text{if } y_i = y_j = l, \\ 0, & \text{if } y_i \neq y_j, \end{cases} \tag{4}$$

The attraction matrix $A_{i,j}$ hand-me-down in this effort is distinct as

$$A_{i,j} = \exp\left(-\frac{\|X_i - X_j\|^2}{\Upsilon_i \Upsilon_j}\right) \tag{5}$$

where $\Upsilon_i = \left\|X_i - X_j^{(k_{nn})}\right\|$ signifies the native climbing of information examples in the locality of X_i, and X_i^k is the adjacent national of X_i.

5.5   FKM Algorithm

The fuzzy k-means clustering (FKM) algorithm does iteratively the divider stage and novel bunch illustrative group stage until junction. The fuzzy k-means clustering algorithm partitions data points looked on k clusters $S_l(l = 1, 2, \ldots k)$ and clusters S_l are related with senates C_l. The association amid a information fact and bunch typical is fuzzy [17]. The association $u_{i,j} \in [0, 1]$ is cast-off to signify the grade of data point X_i and cluster midpoint C_j. Denote the set of data points as $S = \{X_i\}$. The algorithm is founded on falling:

$$J = \sum_{j=1}^{k} \sum_{i=1}^{N} u_{i,j}^m d_{ij} \tag{6}$$

Through esteem to the cluster congresses C_j and associations $u_{i,j}$, where N is the number of data points; m is the fuzzier limitation; k is the number of clusters; and d_{ij} is the squared Euclidean reserve amid data point X_i and cluster typical C_j. It is noted that $u_{i,j}$ should content the next limitation:

$$\sum_{j=1}^{k} u_{i,j} = 1, \text{ for } i = 1 \text{ to } N \tag{7}$$

The significant procedure of FKM is charting a assumed arrangement of agent paths into an better one through parcelling material efforts. It jumps with a ration of commencement collection focuses and rehashes this charting process till a stopping instruction is fulfilled. For the situation that two bunch focuses correspond, a group focus ought to be irritated to dodge happenstance in the iterative procedure. If $d_{ij} < \eta$, then $u_{i,j} = 1$ and $u_{j,l} = 1$ for $l \neq j$, where η is a actual minor optimistic amount. The FKM clustering algorithm is now offered as:

1. Input a set of early cluster centers $SC_o = \{C_j(0)\}$ and the value of ε. Set p = 1.
2. The set of cluster centers SC_p, compute d_{ij} for i = 1 to N and j = 1 to k. Inform associations $u_{i,j}$ using the next equality:

$$u_{i,j} = \left((d_{ij})^{\frac{1}{m-1}} \sum_{l=1}^{k} \left(\frac{1}{d_{il}} \right)^{\frac{1}{m-1}} \right)^{-1} \tag{8}$$

If $d_{ij} < \eta$, set $u_{i,j} = 1$, where η is a very small optimistic amoun.
3. Compute the center for each cluster using Eq. (9) to obtain a new set of cluster councils SC_{p+1}.

$$C_j(p) = \frac{\sum_{i=1}^{N} u_{i,j}^m X_i}{\sum_{i=1}^{N} u_{i,j}^m} \tag{9}$$

4. If $\left\| C_j(p) - C_j(p-1) \right\| < \varepsilon$ for j = 1 to k then stop, where $\varepsilon > 0$ is a small optimistic amount. Else set $p + 1 \rightarrow p$ and drive to stage 2.

The main computational multi-layered landscape of FKM is after stages two and three. In any case, the computational multi-layered nature of stage three is substantially less than that of stage two. In this manner the computational multifaceted nature, as far as the quantity of separation figurings [16].

6 Outcomes and Discussion

The projected method for microarray quality bunching takes persisted executed in the employed stage of MATLAB. For assessing the projected scheme, have used the microarray quality examples of human intense leukemia and colon disease information. Table 1 it very well may be seen that the proposed FKM method has given more precision, and lesser mistake rate instead of the other bunching procedures.

Table 1. Performance comparison of accuracy and error rate metrics

Type of gene data	Accuracy (%)				Error rate (%)			
	FCM	FPCM	EMFPCM	FKM	FCM	FPCM	EMFPCM	FKM
ALL	83.9	83.98	85.69	86.23	16.1	16.02	14.31	13.77
AML	80.06	81.02	83.84	84.56	19.94	18.98	16.16	15.44
COLON	79.0	79.9	81.96	82.45	21	20.1	18.04	17.55

Figure 3 appearances the accuracy comparison results of three different datasets under four clustering methods such as FCM, FPCM, EMFPCM and FKM. From the outcomes concludes the suggested FKM technique provides higher accuracy results of 86.23% whereas other existing FCM, FPCM, and EMFPCM clustering algorithms provides only 83.9%, 83.98% and 85.69% respectively for ALL dataset samples (see Table 1).

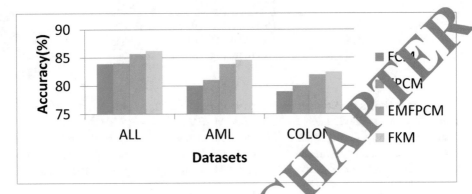

Fig. 3. Clustering accuracy vs. Clustering methods (three different datasets)

Figure 4 appearances the error rate comparison results of three different datasets under four clustering methods such as FCM, FPCM, EMFPCM and FKM. From the outcomes concludes the projected FKM technique provides lesser error rate results of 13.77% whereas other existing FCM, FPCM, and EMFPCM clustering algorithms provides higher error rate results of 16.1%, 16.02% and 14.31% respectively for ALL dataset samples (See Table 1).

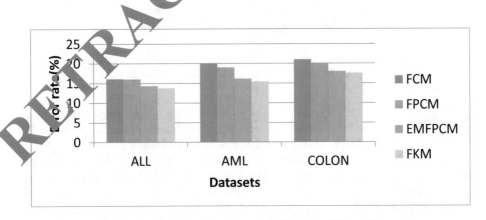

Fig. 4. Error Rate vs. Clustering methods (three different datasets)

From the Table 2, it very well may be seen that the proposed FKM method has given more relationship esteem than the other bunching strategies. It presumes that the proposed work has higher connection results.

Table 2. Correlation comparison between clustering algorithms

Type of gene data	Correlation			
	FCM	FPCM	EMFPCM	FKM
ALL	0.345	0.368	0.412	0.456
AML	0.024	0.029	0.0315	0.0345
COLON	0.119	0.125	0.139	0.189

Figure 5 shows the correlation comparison results of three different datasets under four clustering methods such as FCM, FPCM, EMFPCM and FKM. From the outcomes concludes the projected FKM technique provides higher correlation results of 0.456 whereas other existing FCM, FPCM, and EMFPCM clustering algorithms provides lesser correlation of 0.345, 0.368 and 0.412 respectively for ALL dataset samples (See Table 2).

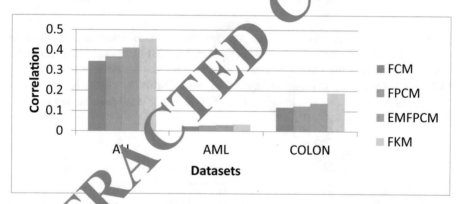

Fig. 5. Correlation measure vs. clustering methods

From the Table 3, it very well may be seen that the proposed FKM strategy has given lesser separation between the two examples than the other bunching systems. It presumes that the proposed work has higher bunching proficiency.

Figure 6 shows the euclidean distance comparison results of three different datasets under four clustering methods such as FCM, FPCM, EMFPCM and FKM. From the outcomes concludes the projected FKM technique provides lesser euclidean distance value of 0.0024 whereas other existing FCM, FPCM, and EMFPCM clustering algorithms provides higher euclidean distance value of 0.00379, 0.00346 and 0.00263 respectively for ALL dataset samples (See Table 3).

Table 3. Distance comparison between clustering algorithms

Type of gene data	Distance			
	FCM	FPCM	EMFPCM	FKM
ALL	0.00379	0.00346	0.00263	0.0024
AML	0.00364	0.00331	0.00201	0.00192
COLON	0.02029	0.02011	0.0126	0.0115

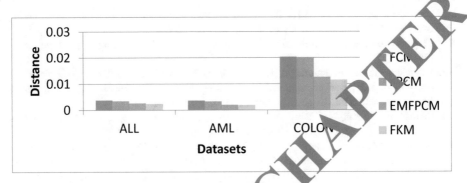

Fig. 6. Euclidean distance measure vs. Clustering

7 Conclusion

So as to pick up a profound knowledge into the malignancy arrangement issue, it is important to investigate the issue, the proposed arrangements and the related issues all together. The proposed methodology has three phases specifically finding of co-managed bi-clusters utilizing Bimax calculation, dimensionality decrease utilizing LFDA and bunching utilizing Fuzzy K-Means Clustering Algorithm (FKM). From the outcomes it reasons that the proposed work performs better when contrasted with other existing bunching calculations, for example, FCM, FPCM and EMFPCM.

References

1. Liu, J., Yang, J., Wang, W.: Bi clustering in gene expression data by tendency. In: Null, pp. 182–193 (2004)
2. Gerstein, M., Chang, J., Basri, R., Kluger, Y.: Spectral bi clustering of microarray data: co clustering genes and conditions. J. Genome Res. **13**(4), 703–716 (2003)
3. Yang, J., Wang, W., Haixun, W., Yu, P.: Enhanced bi clustering on expression data. In: 3rd IEEE Conference on Bioinformatics and Bioengineering, pp. 321–327 (2003)
4. Belkin, M., Niyogi, P.: Laplacian eigenmaps for dimensionality reduction and data representation. Neural Comput. **15**(2003), 1373–1396 (2003)
5. Fukunaga, K.: Introduction to Statistical Pattern Recognition, 2nd edn. Academic Press, Inc., Boston (1990)
6. He, X., Niyogi, P.: Locality preserving projections. In: Advances in Neural Information Processing Systems 16. MIT Press, Cambridge (2004)

7. Govaert, G., Nadif, M.: Block clustering with bernoulli mixture models: comparison of different approaches. Comput. Stat. Data Anal. **52**(6), 3233–3245 (2008)

8. Balamurugan, R., Natarajan, A.M., Premalatha, K.: Stellar-mass black hole optimization for biclustering microarray gene expression data. Appl. Artif. Intell. **29**(4), 353–381 (2015)

9. Govaert, G., Nadif, M.: Co-Clustering: Models, Algorithms and Applications. Wiley, Hoboken (2013)

10. Balamurugan, R., Natarajan, A.M., Premalatha, K.: A modified harmony search method for biclustering microarray gene expression data. Int. J. Data Mining Bioinform. **16**(4), 269–289 (2016)

11. Van Mechelen, I., Bock, H.H., De Boeck, P.: Two-mode clustering methods: a structural overview. Stat. Methods Med. Res. **13**(5), 363–394 (2004)

12. Graph500 benchmark specification (supercomputer performance evaluation). Graph500.org (2010)

13. Sathishkumar, K., Balamurugan, E., Narendran, P.: An efficient artificial bee colony and fuzzy C means based co-regulated bi clustering from gene expression data. In: Mining Intelligence and Knowledge Exploration, pp. 120–129 (2013)

14. Shamir, R., Sharan, R., Tanay, A.: Discovering statistically significant biclusters in gene expression data. Bioinformatics **19**(1), 136–144 (2002)

15. Wang, H., Wang, W., Haixun, W., Yu, P.: Clustering by pattern similarity in large data sets. In: ACM SIGMOD International Conference on Management of Data, pp. 394–405 (2002)

16. Madeira, S.C., Oliveira, A.L.: Bi clustering algorithms for biological data analysis: a survey. IEEE/ACM Trans. Comput. Biol. Bioinform. (TCBB) **1**(1), 24–45 (2004)

17. Sathishkumar, K., Thiagarasu, V., Ramalingam, M.: Gene expression analysis using fuzzy k-means clustering. Int. J. Adv. Innov. Res. (IJAIR) **5**(3), 173–177 (2016). ISSN 2278-7844

RETRACTED CHAPTER: Automated Investigation of Power Structures Annoyance Data with Smart Grid Big Data Perception

R. Lavanya[1(✉)] and V. Thanigaivelan[2]

[1] Department of Computer Science and Engineering, SRM Institute of Science and Technology, Chennai, India
lavanya27382@gmail.com
[2] Department of Mechanical Engineering, SRM Institute of Science and Technology, Chennai, India
thanigailav@gmail.com

Abstract. Scrutiny of liability and instability play vital part in protected and trustworthy electrical power supply. Digital error recorders (DER) facilitate digital tracking of the power schemes transitory action with elevated excellence and huge extent. Though, conversion of statistics to information, expectedly in a computerized way, is a big confront for the power utilities universal. This is a primary focus for apprehending the 'Smart Grid'. In this paper, the structural design and stipulation for the principal and the derivative information for the computerized schemes are portrayed. This affords qualitative and quantitative strategy about the information to obtain a way of the annoyance data. An enumerated approximation of big data for the substations has been anticipated in the paper. Probable customs of dropping the big data by employing intellectual segmentation procedures are depicted, corroborated by factual instance. Deployment of centralized security and distant annoyance scrutiny for dropping big annoyance data are also conversed. Thus, in demand to form a precise real-time observing and predicting scheme, dual original ideas have been engaged into interpretation in the scheme proposal. First, all accessible statistics from diverse bases, has been combined, whereas a communiqué socket has been intended where numerous simulated specialists interrelate and mark conclusions on data.

Keywords: Smart grid · Digital tracking · Power system · Big data · Artificial experts

1 Introduction

Conventionally, the phrase framework is worn for an energy scheme that possibly will hold up each and every one or a few of the subsequent four procedures: power production, power conduction, power allocation, and power manage. A well turned-out framework as well described neat power framework, intellectual framework, intelligrid, future framework is an enrichment of the new generation power framework [1]. The conventional power frameworks are commonly worn to bear energy since little central

The original version of this chapter was retracted: The retraction note to this chapter is available at https://doi.org/10.1007/978-3-030-34515-0_85

producer to a huge amount of consumer or clients. In disparity, the SG employ mutual stream of power and information to generate a programmed and scattered superior power deliverance system. By make use of recent information knowledge, the SG is accomplished of distribute power in added proficient customs and reacting to extensive series circumstances and actions. Generally affirmed, the SG might react to actions which happen anyplace in the framework, like energy production, communication, sharing, and utilization, and accept the consequent approaches. Introspection of energy utilization information to accomplish experiences into customer use is the thing that various energy firms attempt to achieve for objective applications like units consumed per hour, units consumed in various domains etc.

The meter which is set up shall gather information either hourly or on a monthly basis. The nature of data processing knowledge fluctuates significantly with totally unique applications. Antiquated RDBMS of utility firms might be a hurdle in beating this approach. HDFS is an elixir to this problem and makes sure that the vast amount of data is made available that helps network administrators make precise determinations [2]. Apache Hadoop is an open-source software framework that accesses complicated data sets with aplomb. The technique that the file systems in Hadoop shall make use is MapReduce which processes and generates vast amount of data in a parallel and distributed manner. HDFS in our project shall help client to accomplish tasks like accurate data, breaking down customer's use of electricity, response by concerned authorities to take proactive measures etc. With the development of sensible meters for sensible circulation and utilization of energy, the power created should be used appropriately and make sure that everyone gets their share. As we progress and show more inclination towards new technologies, we must keep in mind that the environmental loss is minimal and should aim for sustainable development. The system's utilization can give helpful valuable yields that include: soliciting precision, time-from utilization levying charges etc.

2 Literature Work

2.1 Impacts of Raw Data Temporal Resolution on Residential Electricity Load Profiles

Here it's creating excitement for watching practices of force customers in both the private and business divisions. With the approach of high-assurance time-course of action control ask for informations over advance concept this data could be excessive from the processing assessment [3]. One of the notable frameworks is grouping, however liable upon computation course of action of the statistics with a basic impudence of consequent gatherings. This article exhibits that common assurance of drive mandate article influences way for gathering method, the consistency of pack support (profiles indicating near lead), and the efficiency of the gathering methodology [4]. The incentive of this effort will upgrade gathering of force freight subjects to aid perceive customer sorts aimed at collect diagram and trading, liability and blackmail recognizable proof, ask for side organization, and essentialness efficiency measures. The key control for mining extensive enlightening lists is the methods by which slight evidence to acquire a strong outcome, though keeping up assurance and safety.

2.2   Inconsistency and Inclination Support Global Decree Training Replica to NTL Discovery in Power Corporation

This paper proposes a far reaching system to identify non-specialized misfortunes (NTLs) and recoup electrical vitality (lost by variations from the norm or misrepresentation) by method for an information mining investigation, in the Spanish Power Electric Industry. It is partitioned into four areas: information determination, information preprocessing, enlightening, and prescient information mining. The creators demand the significance of the learning of the specific qualities of the Power Company client: the primary components accessible in databases are depicted [5].

The paper presents two inventive measurable estimators to append significance to changeability and pattern investigation of electric utilization and offers a prescient model, in view of the Generalized Rule Induction (GRI) show. This prescient investigation finds affiliation manages in the information and it is supplemented by a twofold Quest tree classification technique. The nature of this structure is delineated by a contextual investigation considering a genuine database, provided by Endues a Company.

2.3   Investigation and Grouping of Inhabited Clients Power Behavioral Stipulate with Elegant Meter Statistics

Bunching strategies are progressively being connected to private savvy meter information, which gives various vital open doors for circulation organize administrators (DNOs) to oversee and arrange low voltage systems. Grouping has various potential points of interest for DNOs, including the identification of appropriate contender for request reaction and the change of vitality profile displaying. In any case, because of the high stochastic city and inconsistency of family level request, definite investigation is required to define proper credits to bunch. In this paper, we show inside and out examination of client keen meter information to better comprehend the pinnacle request and real wellsprings of changeability in their conduct [6].

We find four key ways in which the information ought to be broke down, and utilize this to shape important characteristics for our grouping. We display a finite blend demonstrate based bunching, where we find ten particular conduct bunches portraying clients in view of their request and their inconstancy. At long last, utilizing a current bootstrap system, we demonstrate that the bunching is dependable. To the creators' learning, this is the first time in the power frameworks writing that the example strength of the bunching has been tried [7].

2.4   Enhancing WFA K-Means Procedure for Claim Retort Plans Requests

There are a few example based bunching strategies which are utilized for various applications, for example, design acknowledgment, information mining, and so on. As of late, some of these strategies are executed in power framework contemplates, particularly to cluster stack bends for planning appropriate duties, request reaction programs determination, and so forth. Decision of the best bunching technique for certain application is a standout amongst the most imperative issues which is case subordinate and ought to be considered in utilizing of grouping burden bends [8]. Request reaction projects are broadly utilized as a part of force framework for various applications, for example, crests cutting, request diminishment, and so on since request reaction projects are highlighted with various attributes. Along these lines, choice of appropriate projects for various client classes is of extraordinary significance. In this paper, an enhanced weighted fluffy normal (WFA) K-implies with the end goal of interest reaction programs applications is produced. This technique is actualized on 316 load bends of Tehran dispersion arrange and the outcomes are examined.

2.5 Small Tenure Capacity Predicting Centered on Big Data Knowledge

With the development of savvy network, loads of recharge capable vitality ssets, for example, wind and sunlight based are conveyed in power framework. It may make the power framework stack differed complex than before which will acquire difficulties here and now stack gauging territory. To beat this issue, this paper proposes another transient load gauging system in view of enormous information advancements. To start with, group an assumed names for each shaped to characterize every day stack designs for individual burdens utilizing keen meter information [1]. This is trailed by the utilization of a choice tree to set up classification rules. At that point, fitting estimating models are decided for various load designs. At long last, the determined aggregate framework load is acquired through a conglomeration of an individual load's anticipating comes about. Contextual analyses utilizing genuine load information demonstrate that the proposed new structure can enhance the precision of here and now stack anticipating inside required points of confinement.

3   System Methodology

Attempted idea concerns with altering database by utilizing hadoop tool with no barrier on information size and simply involving additional systems to the group provide outcomes with lower time, more throughput and less upkeep cost. Concepts like partitions and backup are too used which are useful to organize data depending on your needs. The projected strategy achieves fine in the all-encompassing communal and in sub-populated areas. The project aims to observe the number of elements disbursed in the recent times and how abundant sum the customers rewarded in the last 4 years. Professed "elegant" gauge and application include the probable to put aside power, to heighten max out power tradition, and to diminish menace of collapse [9]. Emblematic net indicator plan embrace intermittent communication of power, segment, and regularity statistics from the client to the power division corporation. Accessibilities will employ the statistics in invoice estimate beneath time-of-day charge, for stack managing investigate, to offer client criticism, and/or to regulate client domestic device.

Though, there has been small conversation of the packed methodical, financial, and momentous prospective of these statistics whose utility may expand further than the unique rationale for which the statistics will be composed and accumulate as given in Fig. 1.

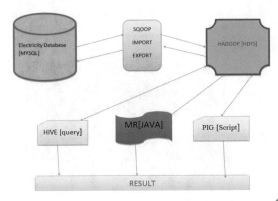

Fig. 1. Architecture for smart grid system analysis.

Power statistics are merely solitary instance wherever the administration has haggle energy to make easy admission to statistics by capable researchers, except captivating exploit is appropriate, since this statistics rebellion is in its extremely near the beginning phases. This editorial focuses on the comprehension of energy streak replica. Energy streak replica is the major element of intricate HPC statement scheme [10].

4 Results and Discussion

4.1 Data Preprocessing Phase

The smart substructure scheme is the drive, evidence, and communiqué substructure essential of the SG which can (1) progressive power cohort, distribution, and ingesting; (2) progressive evidence metering, checking, and organization; and (3) progressive communiqué tools. In this phase, It is to generate Data-set for Electricity Consumption [11].

4.2 Data Relation Segment by Sqoop

Here with dataset currently our goal aims to relocate the dataset to hadoop (HDFS). Sqoop is application for transporting data from database software to Hadoop. Sqoop can achieve the job, need to procure the particular column [12].

4.3 Data Systematic Segment by Hive

Hive is an information product line framework to Hadoop. It turns SQL based questions named HQL (Hive inquiry dialect) that get exclusive altered above to delineate jobs. Hive reinforces Data description Language analyzed in Fig. 2. With examination database HIVE employing HQL Language. Employing hive we achieve Tables appearances, connections, Divider idea. The MapReduce scheming covers double dynamic undertakings, in precise Map and Reduce. In this element similarly employed for dividing the informational catalog utilizing MAP REDUCE [13].

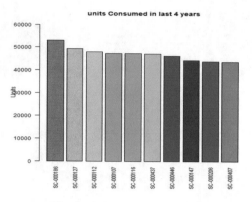

Fig. 2. Data analytic using R.

4.4   Data Analytic Module with R

R is a dialect and environment for factual processing and illustrations. It is a GNU venture which is like the S dialect and environment which was produced at Bell Laboratories (some time ago AT&T, now Lucent Technologies) by John Chambers and colleagues [14]. Pig grips together structured and formless language. The MapReduce scheming covers double dynamic undertakings, in precise Map and Reduce. In this element similarly employed for dividing the informational catalog utilizing MAP REDUCE. Outline Run by Java Program and is given in Fig. 3.

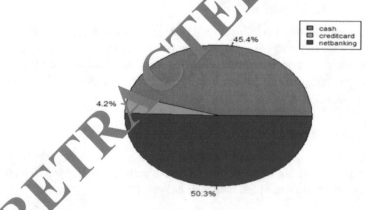

Fig. 3. Transaction type count

5   Conclusion

A technique for envisaging forthcoming consumption of electricity being established by structures mined as of client's previous ingesting. The objective of this training is to examine amount of elements are spent in the recent years and the volume has been compensated in the earlier four years which gives us a glimpse about future consumption Rational use of energy should be practiced everywhere ranging from a bigger cluster of firms, municipalities to public organizations and this shall create awareness among the masses about the importance of energy. A complete image of their energy use, potential for savings, in conjunction with prices is given to them by good meter knowledge analytics enabling effective energy management. Good meter sends energy consumption knowledge at tiny intervals leading to generating huge knowledge. Time and storage are vital factors that have an effect on lots on building any application. The answer for handling such huge knowledge is Hadoop.

References

1. Wang, Q.R.: Performance evaluation of smart grid based on optimal fuzzy algorithm method. Appl. Mech. Mater. **482**, 346–349 (2013)
2. Clerk Maxwell, J.: A Treatise on Electricity and Magnetism, vol. 2, 3rd edn. Clarendon, Oxford (1892)
3. Jacobs, I.S., Bean, C.P.: Fine particles, thin films and exchange anisotropy. In: Magnetism, vol. III, pp. 271–350 (1963)
4. Li, L., Liu, L., Yang, C., Li, Z.: The comprehensive evaluation of smart distribution grid based on cloud model. Energy Procedia **17**, 96–102 (201)
5. Young, M.: The Technical Writer's Handbook. University Science, Mill Valley (1989)
6. Ipakch, A., Albuyeh, F.: Grid of the future. IEEE Power Energy Mag. **7**, 52–62 (2009)
7. Lee, J., Song, H., Mok, A.K.: Design of reliable communication system for grid-style traffic control networks. In: Proceedings of the 16th IEEE Real-Time and Embedded Technology and Applications Symposium, pp. 133–142 (2010)
8. Luan, S., Teng, J., Chan, S., Hwang, L.: Development of a smart power meter for AMI based on ZigBee communication. In: Power Electronics and Drive Systems, pp. 661–665 (2009)
9. Bonneville, E., Rialhe, A.: Demand side management for residential and commercial end-users (2006). http://www.leonardo-energy.org/Files/DSM-commerce.pdf
10. Abras, S., Pesty, S., Ploix, S., Jacomino, M.: An anticipation mechanism for power management in a smart home using multi-agent systems. In: Proceedings of the 3rd International Conference on From Theory to Applications, pp. 1–6 (2008)
11. Derin, O., Ferrante, A.: Scheduling energy consumption with local renewable micro-generation and dynamic electricity prices. In: Proceedings of the 1st Workshop on Green and Smart Embedded System Technology: Infrastructures, Methods, and Tools (2010)
12. Katsigiannis, Y.A., Georgilakis, P.S., Karapidakis, E.S.: Multiobjective genetic algorithm solution to the optimum economic and environmental performance problem of small autonomous hybrid power systems with renewables. IET Renew. Power Gener. **4**(5), 404–419 (2010)
13. Facchinetti, T., Bibi, E., Bertogna, M.: Reducing the peak power through realtime scheduling techniques in cyber-physical energy systems. In: Proceedings of the 1st International Workshop on Energy Aware Design and Analysis of Cyber Physical Systems (2010)
14. Lee, J., Park, G., Kang, M., Kwak, H., Lee, S.: Design of a power scheduler based on the heuristic for preemptive appliances. In: Proceedings of Asian Conference on Intelligent Information and Database Systems, pp. 396–405 (2011)

Secure Data Integrity in Cloud Storage with Multi-level Hash Indexing Technique

T. P. Kavyashree[✉] and A. S. Poornima

Department of CSE, Siddaganga Institute of Technology, Tumkur, India
kavyashreetp@gmail.com, poornimaarvind9@gmail.com

Abstract. A growing technology which offers storage, compute and network resources over the internet on a pay-for-use basis can be named as cloud computing. Cloud Storage is one of computing and service model in which the system allows you to store, manage, maintained and backed up remotely over a network and made available to users to use. Data integrity implies that the accuracy, validity of data over its Life cycle has to be maintained and any violations are to be detected like if data is loss, altered or compromised. User money and time can be saved by using cloud services. So user loses control over the data as there are maintained, stored by the cloud and it can be hacked, tampered by the attacker. Now integrity of data has to be maintained in a cloud where data is present in cloud. Here the aim is to provide data integrity for which the data will be stored in the cloud by user. Here paper says, we need to have an auditing technique which must be efficient and check the data integrity using Third party auditor (TPA), the auditing schemes uses Advanced Encryption Standard (AES) and Secure hash algorithm for encryption and to generate verification meta data respectively and using multi-level hash indexing technique to eliminate duplicate files, reduce storage consumption of file which is stored in the cloud.

Keywords: Cloud storage · Data Integrity · AES algorithm · Secure Hash algorithm · Multi-level Hash Index Technique

1 Introduction

Cloud computing is referred as a model for on-demand availability of hardware and software resources, convenient network access and delivery model which deliver the hosted services and resources over the internet. The cloud computing technology provides pay as you go basis where you need to use the services or the technology of what you want and can pay for that only and services like applications, storage, processing power. It can manage the large amount of information. Cloud computing is consistent and reliable, due to which the organizations and many companies use this infrastructure and no need to maintain, build their own. It is a platform which provides service as computing resources. The services which are based on cloud consist of (SAAS) Software as a service, (PAAS) Platform as a service and (IAAS) Infrastructure as a service. Many Internet service Providers and cloud computing infrastructure are Microsoft Windows Azure storage services, Amazon's EC2, Google app engine and

© Springer Nature Switzerland AG 2020
P. Karrupusamy et al. (Eds.): ICSCN 2019, LNDECT 39, pp. 652–660, 2020.
https://doi.org/10.1007/978-3-030-34515-0_68

S3, etc. A Large amount of storage space, resources are computed and customizable are provided by them. The local machines responsibility can be eliminated by computing platforms so that the maintenance and local data storage burden can be relieved by the user.

Although cloud computing has many uses, it also has some disadvantages too, and the most important is security issue. It has many security related issues such as identity management, resource management, integrity, access control, privacy, confidentiality, personal security, availability and so on. In cloud, if any user or organization wants to store their data, they should provide it to the utility provider or to cloud server. Therefore the data must be secured and maintained in the cloud. Even though there are many advantages of cloud and we also find some disadvantages too where user feel, hesitate to use this technology because of data loss, altered, compromised and hacked by the hacker. So there are many problems facing the cloud in which we cannot know that the problems from an outsider or from insider which can use cloud vulnerabilities to do harm. These problems may cause many treats and results in integrity, privacy, confidentiality of data. Hence, there is most probability of misusage of confidential data or the sensitive user data may be altered intentionally by hackers or accidentally by other users. So this lead to data integrity, confidentiality and consistency breach.

By considering every one of the issues with respect to security of the data, it is considered that data integrity is one of security problem in cloud computing. And many approaches are proposed and various techniques are used to check data integrity. Cryptography is widely approved method for ensuring the information security. This mechanism will secure the information by changing it to the non-readable form.

2   Related Works

Govinda Ramaiah and Vijaya Kumari [1] has proposed a strategy where they used the homomorphic encryption technique for combining both confidentiality and integrity of data to give a solution which is impending in factor. Computations are performed with the help of encrypted data and it provides block less verification where the blocks of data is not downloaded. It considered as stateless verification in which there is no state verified by verifier.

Awad, Kadry, Lee, Maddodi and O'Meara [2]. Has used both hard and soft attestations for a new integrity assurance system has been proposed. The changes are allowed to data artifacts if any detection is found. They communicate results with cloud users in which it is tested on the open stack and trust protocol. The target measurements or changes are checked to verify in system uses a TPM, property based attestations to given a certification system.

Rukavitsyn, Borisenko, Holod, Shorov [3] has used a method which describes the separate services outside cloud. In order to avoid the unauthorized usage, it uses metadata for controlling of integrity. And as data stored in encrypted form, it cannot communicate with the database.

Suganya, Durai Raj Vincent [4] has proposed a cloud auditing technique in which the data auditing is verified with help of protocol where it checks the integrity of data to provide certificates to the user uninterrupted and also service for the secure cloud, efficient process.

Bansal, Agrawal [5] proposed a new method which uses the cloud storage and also ECC algorithm for integrity and security purpose where they have used very less CPU power, processing time is said to be efficient in terms. In order to enhance security further they used the meta data in which it is encrypted

Chen, Li, Chen [6] used a RSA encryption technique where it also checks with the integrity of data and to know the comprehensive analysis of static data performed repeatedly for the verification purposes. It also verifies the replay attack and MitM.

Mahajan and Shetty [8] used a method to minimize the storage space so that performance also increases. This system is developed in web based technology and tested with three servers. Where one is the gateway server and other two are community servers. The system tested with large number of files and the results are shown.

3   Proposed System

In this System architecture it consists of many actors like Admin, User, TPA and Cloud Storage to secure data integrity in the cloud. And also how the process can be done is viewed in the figure and can know the actor responsibilities also.

Fig. 1. System architecture

The above Fig. 1 shows you the diagram of system Architecture where it includes all the actors in order to secure data integrity. The admin has User details, TPA details, Transaction details, cloud details and hash code details. User can view his profile, edit his profile, can upload and download the file and can also make a request to TPA to check his data. Auditor will look for message and Integrity check request in portal if the user requested him. From his portal he can upload and download. Data will be uploaded to cloud storage. Once user authenticates with his username and password, then he logged in to his account and it is provided by the admin. The user can view, edit his profile so that he can change his information according to him and when the users wants to upload a file he must selects the file to be uploaded to cloud from his system. The Fig. 2 shows you the diagram of user uploading a file where he selects the file and the file splits in to block and then the file is encrypted using AES algorithm and uploads this file to cloud storage and a meta data is also generated and it can be sent to the TPA for verification purpose using SHA2 algorithm and the file undergoes to multi-level-hash indexing technique. It will check if the block is present or not and it can be explained as follows.

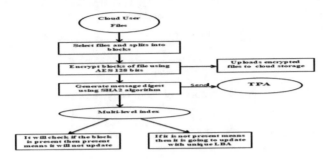

Fig. 2. User uploading the file

Figure 3 shows multi-level hash indexing technique in which the file is uploading, the whole file is divided into multiple blocks according to the size and for each block hash code is generated. With the help of hash code the system will check for the presence of the block in the cloud. If the block is present then it will not update. If the block is not present means then it is going to update.

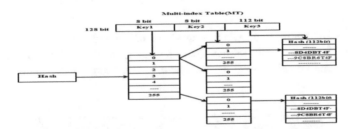

Fig. 3. Multi-level hash indexing

The Fig. 4 shows you the diagram of the working of TPA where users send the request to auditor in order to check the integrity of his data that is uploaded to cloud. So the auditor will get the message to verify the user data. Now he will check integrity of data by comparing the old hash code of the file to the newly created hash code of the file and if it matched then the integrity of data is maintained as if it is not matched then integrity of data is not maintained. The results are send to the user email id who requested for the integrity check result.

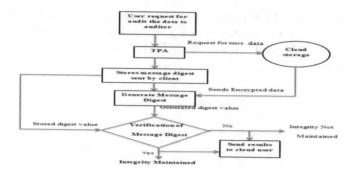

Fig. 4. Working of Third party auditor

3.1 Implementation and Results

Fig. 5. Admin login

The Fig. 5 show admin login where it needs admin id and password.

The Fig. 6 shows admin profile where he can view his profile, transactions, user details, hash code details, cloud details where he can add or delete the user.

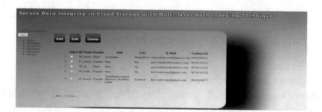

Fig. 6. Admin profile

The Fig. 7 show user login page where he needs to provide user id and password.

Fig. 7. Login page of user

The Fig. 8 show user profile in which user is going to upload a file. Here he can view, edit his profile.

Fig. 8. User profile

The Fig. 9 show user 011 where he will uploaded a file name hello1.

Fig. 9. user 011 profile

The Fig. 10 show nishi user also uploads with different file name with the same content and it is also saved successfully but not stored in cloud (Fig. 11).

Fig. 10. User Nishi profile

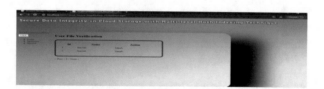

Fig. 11. Auditor login page.

The Fig. 12 shows auditor page where checking message for verification.

Fig. 12. Auditor profile

The Fig. 13 show Auditor will check for the details and it shows that the file is verified.

Fig. 13. Verification page

The Fig. 14 shows the auditor has sent the message to the user mail id if it is verified or not.

Fig. 14. Verification message

4 Conclusion

Users store their information in cloud storage where information can be hacked by any others so there is need for security and also data must be maintained. The main issue to check for integrity of data stored. So data is uploaded in encrypted form as confidentiality is achieved and with the help of TPA we can check the data integrity is maintained or not. The result is send to the user mail id. By using the multi-level Hash index technique it removes redundant data, reduces the storage space in the cloud.

References

1. Govinda Ramaiah, Y., Vijaya Kumari, G.: Complete privacy preserving auditing for data integrity in cloud computing. In: 12th International Conference on Trust, Security and Privacy in Computing and Communications, pp. 1559–1566. IEEE (2013)
2. Awad, A., Kadry, S., Lee, B., Maddodi, G., O'Meara, E.: Integrity assurance in the cloud by combined PBA and provenance. In: 10th International Conference on Next Generation Mobile Applications, Security and technologies, pp. 127–132. IEEE (2016)
3. Rukavitsyn, A.N., Borisenko, K.A., Holod, I.I., Shorov, A.V.: The method of ensuring confidentiality and integrity data in cloud computing, pp. 272–274. IEEE (2017)
4. Suganya, S., Durai Raj Vincent, P.M.: Improving cloud security by enhancing remote data integrity checking algorithm. In: International Conference on Innovations in Power and Advanced Computing Technologies, pp. 1–6. IEEE (2017)
5. Bansal, A., Agrawal, A.: Providing security, integrity and authentication using ECC algorithm in cloud storage. In: International Conference on Computer Communication and Informatics (ICCCI), pp. 1–5. IEEE (2017)
6. Chen, Y., Li, L., Chen, Z.: An approach to verifying data integrity for cloud storage. In: 13th International Conference on Computational Intelligence and Security, pp. 582–585. IEEE (2017)
7. Jianhong, Z., Hua, C.: Security storage in the cloud computing: a RSA-based assumption data integrity check without original data. In: International Conference on Educational and Information Technology, pp. 143–147. IEEE (2010)
8. Mahajan, M.J., Shetty, C.: In-line de-duplication technique using multi-level hash indexing. Imp. J. Interdiscip. Res. **3**(7) (2017)
9. Morea, S., Chaudhari, S.: Third party public auditing scheme for cloud storage. Int. J. Procedia Comput. Sci. **79**, 69–76 (2016)
10. Amazon.com. Amazon Web Services (AWS) (2008). http://aws.amazon.com
11. Allen, W., Aldossary, S.: Data security, privacy, availability and integrity in cloud computing. Int. J. Adv. Comput. Sci. Appl. **7**(4), 485–498 (2016)

12. Yang, N.K., Jia, X.: Data storage auditing service in cloud computing: challenges, methods and opportunities. World Wide Web **15**(4), 409–428 (2012)
13. Wang, C., Chow, S.S.M., Wang, Q., Ren, K., Lou, W.: Privacy preserving public auditing for secure cloud storage. IEEE Trans. Comput. **62**(2) (2013)
14. Wang, C., Wang, Q., Ren, K., Lou, W.: Privacy-preserving public auditing for data storage security in cloud computing. In: 2010 Proceedings IEEE INFOCOM, pp. 1–9 (2010)
15. Tejaswini, K.S., Prashanth, S.K.: Privacy preserving public auditing service for data storage in cloud computing. Indian J. Res. PARIPEX **2**(2), 131–133 (2013)

Energy Efficient Wireless Communication System

Pamarthi Supriya, Y. Mohana Roopa$^{(\boxtimes)}$, and Ankam Sudheera

Computer Science and Engineering, Institute of Aeronautical Engineering,
Hyderabad, India
pamarthisupriya1998@gmail.com,
ymohanaroopa@gmail.com, ankamsudheera@gmail.com

Abstract. Now-a-days energy efficient wireless communications had been paid developing attention under the background of constrained energy aid and environmental-friendly transmission behaviour. Energy efficiency is important for developing wireless networks [1]. Battery technology in wireless networks is low but the consumption of energy is excessive. In order to improve the battery era, energy saving is needed and this will be accomplished via extracting energy from greenhouse gases. The energy extracted from greenhouse gases can be stored in the battery and converts into dc current. This dc current can be further used for the development of battery technology.

Keywords: Energy consumption · Energy efficiency · Wireless communication · Greenhouse gases · Battery

1 Introduction

Wireless communication includes the transmission of data over a distance without the assist of wires, cables or some other types of electrical conductors. It is a extensive time period that incorporates all tactics and styles of connecting and speaking among two or more devices using wireless signal through wireless communication technology and devices. Wireless communication is, through any measure, the quickest growing section of the communications industry. As such, all the media attention is towards this wireless communication and the creativeness among society. Wireless communication has observed sudden change in growth during the last decade and at present there are billion customers across the world. Indeed, wireless communication have turn out to be a vital commercial enterprise device and part of everyday lifestyles in maximum advanced nations, and are rapidly supplanting antiquated wireline systems in many growing nations. Along with this, wireless communication system is mostly used to update stressed out networks in lots of areas, campuses, and corporations [2]. Cellular system, satellite, wireless networking, wi-fi, wimax, wireless router, Bluetooth technology, zigbee, wireless adaptor, repeater, microwave, infrared, radio are some of the wireless communication devices that allow users to communicate even from remote areas [3] (Fig. 1).

© Springer Nature Switzerland AG 2020
P. Karrupusamy et al. (Eds.): ICSCN 2019, LNDECT 39, pp. 661–668, 2020.
https://doi.org/10.1007/978-3-030-34515-0_69

Fig. 1. Wireless networks

- Wireless networking technologies connect multiple systems together and provide the required communication among the various networks.
- Mobile communication systems are used to communicate over a single frequency band.
- Satellites are used to communicate directly with the orbiting satellites via radio waves.
- Wi-fi is a low power wireless communication system allows user to connect only within small proximity.
- Bluetooth allows to connect different electronic devices wirelessly to transfer and share the data.
- Wireless repeater is a device used to extend the range of a wireless router [4].
- Wireless router is used to provide security access among the network, in order to increase speed of communication between two networks.

2 Wireless Communication System Architecture

- **Application Layer:**
  This layer is accountable for traffic management system and propose software with plentiful applications. This layer is mainly used to transmit the data into crystal form to find absolute information. The sensors used in this network are arranged in the form of large applications in different fields such as agricultural, military, environment, medical, etc. [5] (Fig. 2).

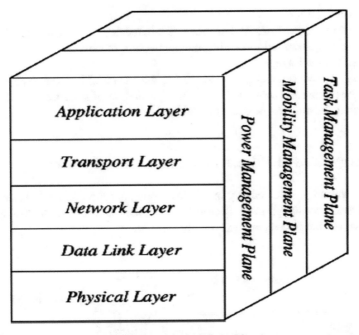

Wireless Sensor Network Architecture

Fig. 2. Wireless network architecture

- **Transport Layer:**
  The main task of this layer is to provide surplus escape and reliability where many protocols are proposed to deliver this function on the upstream. These protocols requires opposite mechanisms for less recognition and less recovery. This layer is exactly needed when a system is planned to contact with other systems. Providing a dependable low recovery is more efficient and is one of the main reason why TCP is not fit for WSN. In general, this layer can be split into different forms such as Packet driven and Event driven. Some of the popular protocols in this layer are namely STCP (Sensor Transmission Control Protocol), PORT (Price-Oriented Reliable Transport Protocol and PSFQ (pump slow fetch quick).

- **Network Layer:**
  The essential feature of this layer is routing, it posses few obligations which are primarily depended on utility. But clearly, the main obligations are inside the strength protecting, partial memory, buffers, and sensor. These obligations need to no longer own any ordinary ID and ought to be Independent. The simple concept of the routing protocol is to give an explanation for a safe and inordinate lanes, according to a satisfied scale known as metric. This generally varies from one protocol to another protocol. There are numerous present protocols for this network layer, among these which may be separated into flat routing and hierarchal routing or may be separated into time driven, question-pushed & occasion pushed.

- **Data Link Layer:**
  This layer is accountable for numerous data frame detection, data streams, MAC, & error control, and used to confirm the point–point reliability.
- **Physical Link Layer:**
  This layer produce an edge for converting the stream of bits above physical medium. It is important for opting frequency, generation of a carrier frequency, signal detection, Modulation & data encryption [6].

3 Energy Used in Wireless Communication Systems

The total consumption of energy in wireless communication network exceeds 3% of the worldwide consumption of electric energy. There is also a scope to increase the percentage of energy consumption rapidly (Fig. 3).

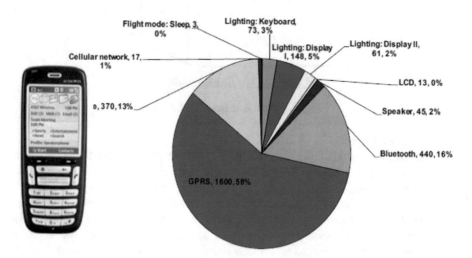

Fig. 3. Energy consumed in mobile networks

The main source of energy intake in cellular mobile network is the base station (BS) sites, that are chargeable for kind of two-thirds of the total carbon dioxide emissions radio access networks. Mainly, In the overall network the blue-tooth consumes 16% of energy among the complete energy in the cellular network. In surrounding networks the speaker technology requires 2% of energy consumption. Among the cellular networks, the display content takes 5% of energy that differs from different mobile networks. Keyboard lighting requires for about 3% of energy and it depends upon the various mobile networks. Including all these features, the GPRS acquires excess amount of energy i.e., 58% of energy consumption among the cellular mobile networks [6].

3.1   Techniques for Wireless Communication

Current evaluation through tactics and network operators has shown that modern day wireless networks are not very power green, specifically the lowest stations via which terminals access offerings from the network.

- Green radio i.e., radio strategies to permit electricity-green wireless networks.

    1. Radio transceivers – The system for generating transmit indicators to and decoding signals from cell terminals.
    2. Power amplifiers-these devices enlarge the transmit signals from the transceiver to a excessive sufficient electricity stage for transmission commonly around 5 – 10 watts.
    3. Transmit antennas – A major possibility to obtain the power discount goals of the program lies in developing techniques to enhance the performance of base station hardware [7].

- MIMO (multi input multi output) - A cross-layer technique to electricity performance for adaptive exploiting spare capability. More than one enter gives a couple of output and multiple antennas are used in MIMO [8].
- SISO (single input single output) – In SISO one antenna is used at transmitter and the other antenna is used at receiver [9].

4   Battery Technology

Batteries are getting more extensively used. As the use of transportable and mobile devices increases, so does the use of battery technology. The growing needs being positioned on batteries has supposed that the era has evolved notably within the beyond few years, and more development can be predicted inside the future. With the large demand of batteries, there is a extensive variety of different battery and mobile technology available. With a massive want for batteries, there is a large quantity of battery era improvement underway and new sorts of cells and batteries will no doubt end up to be had presenting even better ranges of overall performance.

- **Primary batteries:** These are essential batteries that cannot be recharged. They irreversibly remodel chemical power to electrical energy. When the chemical substances within the battery have all reacted to produce electric strength and they are exhausted, the battery or cell cannot be effectively restored by way of electric manner.
- **Secondary batteries:** These cells are extraordinary to primary ones in that they may be recharged. The chemical reactions within the mobile or battery may be reversed with the aid of imparting electricity to the cell, restoring their authentic configuration [10].

    In batteries the power of chemical compounds behave like a storage medium. In this process a type of chemical system occurs which generates power that may be drawn from the battery within the form of an electric powered cutting-edge at a certain voltage. The cell reaction is a chemical reaction that differentiates the battery. When the battery is discharged, chemical compounds of high energy are converted into low energy by using cell reaction. Generally, the released energy is treated as heat. But in a battery, the cell reaction is divided into two electrode reactions among which one releases the electrons and the other absorbs the electrons. This flow of electrons forms current which can be drawn from the battery. By this process, the consumption of energy that is connected to the cell reaction is directly converted into electric current.

- **Lithium ion battery:** They are widely used in transportable client electronic gadgets, and in electric vehicles starting from full sized cars to radio managed toys. The time period of "lithium battery" point to a circle of relatives for various lithium-metallic chemistries, composing many kinds of cathodes and electrolytes but all with metal lithium as anode but the battery required is from 0.15 to 0.3 kg of lithium per kWh. As per the idea of this model these primary systems use a charged cathode, which is an electro-energetic fabric with diffraction vacancies that are filled moderately throughout discharge. Lithium batteries are light weight and powerful and are at risk of leaking and catching fire [11].
- **Electric battery:** It is a tool consisting of one or more electrochemical cells with external connections provided to energy electric gadgets together with flash lights, smart phones, and electric motors. When a battery is presenting electric strength, its tremendous terminal is cathode and its poor terminal is anode. When a battery is connected to external electric powered load, a redox reaction converts excessive-power reactants to low-strength products, and the unfastened-electricity distinction is introduced to the outside circuit as electrical power [12].
- **Rechargeable battery:** It is a power garage gadget that can be charged again even after discharging from DC current to its terminals. They allow for various usages of a cell, lowering waste and provide a better long term investment. It is commonly a more realistic and can also be replaced by reuse of batteries, they produce electricity by using a chemical reaction in which a reactive anode is consumed. The anode in a rechargeable battery gets consumed as well but a low rate, allowing for many charges and discharges. However, after discharge the batteries are positioned in charger or, insert in batteries, an analog current and digital current adapter is hooked up [13].

4.1   Problem and Solution

In the present scenario, battery technology is low whereas usage of battery energy is very high. Energy consumption is excessive when compared to the existing energy in the battery. In order to improve the battery technology, excess of energy is required. This can be done by collecting and storing the energy emitted by greenhouse gases like carbon dioxide, water vapour, and nitrous oxide. The energy emitted by this greenhouse gases are stored in the battery and are further used to improve the battery technology.

Greenhouse gases emit radiant energy that produce heat, Despite the fact that you usually can't see it, all objects provide off radiant energy and you may from time to time feel this energy. For example, if there's a pot of hot water on your range, you may experience the radiant energy it offers off without touching it. You usually name what you sense "heat," but it is more accurate to consider it as a form of invisible light called "infrared radiation" that warms your skin, much like the daylight. The amount of infrared radiation energy a warmed item offers off depends on its temperature—the higher the temperature, the more energy is given off [14].

Here, the main problem is that the energy present in the battery is less but the consumption of energy from the battery is excess. The solution for this problem is to extract the energy from the greenhouse gases and store this energy in the battery. The energy stored in the battery is converted to dc current and this current is used to improve the battery technology. Finally the battery performance can be improved by using the energy emitted by greenhouse gases. As the battery technology improves, the energy in wireless communication can be effectively used.

5 Conclusion

The main problem in wireless communication is, battery technology is very much low but the consumption of energy from the battery is excessive. Energy present in the battery is less whereas the energy consumed from the battery is more. In order to overcome this problem, performance of battery should to improved. Battery technology can be improved by collecting and using the radiant energy emitted by some of the greenhouse gases. As the battery technology is improved, excessive consumption of energy in wireless communication can be effectively done.

References

1. Energy efficient wireless communications: tutorial, survey and open issues
2. Linebaugh, K.: Medical Devices in Hospitals go wireless. Wall Street J. (2010), Web 27 October 2013. Online.wsj
3. Stallings, W.: Wireless Communications and Networks, 2nd edn. Prentice Hall, Upper Saddle River (2005)
4. Poole, I.: What exactly is.. ZigBee? IEEE Commun. Eng. 2, 44–45 (2004)
5. Porcino, D., Hirt, W.: Ultra-wideband radio technology: potential and challenges ahead. IEEE Commun. Mag. 41, 66–74 (2003)
6. Akyildiz, I.F., McNair, J., Martorell, L.C., Puigjaner, R., Yesha, Y.: Medium access control protocols for multimedia traffic in wireless networks. IEEE Netw. 13, 39–47 (1999)
7. Mohana Roopa, Y., et al.: Component-based self-adaptive middleware architecture for networked embedded systems. Int. J. Appl. Eng. Res. 12, 3029–3034 (2017)
8. Holma, H., Toskala, A.: LTE for UMTS. Wiley, Hoboken (2009)
9. Kim, H., et al.: A cross-layer approach to energy efficiency for adaptive MIMO systems exploiting spare capacity. IEEE Trans. Wirel. Commun. 8(8), 4264–4275 (2009)
10. Labiod, S., Guerra, T.M.: Adaptive fuzzy control of a class of SISO nonaffine nonlinear systems. Fuzzy Sets Syst. 158, 1126–1137 (2007)
11. Hu, Y., Li, X., Wang, J., Li, R., Sun, X.: Free-standing graphene–carbon nanotube hybrid papers used as current collector and binder free anodes for lithium ion batteries. J. Power Sources 237, 41–46 (2013)

12. Nykvist, B., Nilsson, M.: Rapidly falling costs of battery packs for electric vehicles. Nat. Clim. Change **5**, 329 (2015). nature.com
13. David, L., Bhandavat, R., Singh, G.: MoS2/graphene composite paper for sodium-ion battery electrodes. ACS nano **8**, 1759–1770 (2014)
14. Hu, R.Q., Qian, Y.: An energy efficient and spectrum efficient wireless heterogeneous network framework for 5G systems. IEEE Commun. Mag. **52**(5), 94–101 (2014)

ESP Module Based Student Attendance Automation System

V. Linga Bharath[(⊠)], K. Surya, P. Varatharajan,
G. K. D. Prasanna Venkatesan, and Kamalraj Subramaniyam

Department of ECE, KAHE, Coimbatore, India
lingabharath387@gmail.com,
suryakathirvell998@gmail.com,
varatharajan2910@gmail.com,
prasphd@gmail.com, kamalrajece@gmail.com

Abstract. The student attendance automation system plays a key role in eliminating the disadvantages of manual attendance management system. To present added features like manual override. Which are essential for seem-less automated system in recent years these systems can be comprised into proxy-defining, passive tags and biometric recognition systems, where the calculated error rate was high with some wait-in-line delay. The proposed work eradicates the erroneous attendance marking and simplifies the delay process by using ESP modules (when staff requires to get the class count) which are connected to network and to count the list of established networks from the active tag and the innovative manual override keeps the malfunctions or user problems such as on-duty. As a result our project is very less time consuming with high transparency to the users.

1 Introduction

The concept of collecting attendance to mark a presence of particular individual has taken its own value in the logistical process of educational institutes industries and corporate. Several methods in the past decades have been involved to make an automated attendance system which has turned out to be presumptuous, as the requirements eventually found some errors and malfunctions. The faults such as wait-in-line delay commences only when the data traffic is high or function which displays number count would be decreased to minimum individuals. The debugging of minimal error would require tedious efforts of programming the overall process. This featured system flawlessly by assigning the IP addresses for individual students or employees through the ESP tags which are connected to the router controlled Raspberry pi and enables us to simultaneously monitor the presence without any delay. This monitoring system with integral support to manually override the entries provides us with vast implementing possibilities.

2 Literature Review

Chaudhari et al., a paper described an idea about RFID based Attendance System [1]. Islam et al., a paper described an idea about Smartphone based application Attendance system [2]. Rahman, et al., a described about Attendance automation system using

© Springer Nature Switzerland AG 2020
P. Karrupusamy et al. (Eds.): ICSCN 2019, LNDECT 39, pp. 669–678, 2020.
https://doi.org/10.1007/978-3-030-34515-0_70

Fingerprint sensor [3]. Yadav and Nainan a paper described an idea about Attendance system notification to parents [4]. Gagare et al., a paper described an idea about Enrolling Attendance system with the help of barcode Reader [5].

Nandya, et al., a paper described an idea about Two way attendance systems with the help of RFID and Fingerprint Sensor [6]. Kumar et al., a paper described an idea about Attendance Automation System using Raspberry Pi [7]. Prince et al., the paper described an idea about Automatic Attendance System with the detection of Web Server [8]. Sapkal et al., a described an idea about Attendance system with the help of Internet Of Things [9]. Praveen Kumar and Mani Kumar, a paper described an idea about RFID based attendance monitoring system using IOT [10].

Lodha, Gupta, Jain, Narula, a paper describe that, Bluetooth Smart Based Attendance Management System [11]. Agarwal and Bansal a paper describe that, Online Attendance Management System Using RFID With Object [12]. Chaudhari et al., in which described about attendance system where the students data are used in full campus area [13]. Yadav et al., the paper described about the attendance status will share with the parents using gsm module and the taken was taken by rfid [14]. Nandyal et al., the paper described about the attendance system using the combination of both rfid and internet of things [15].

Prof. Kanawade et al., the author described about the new type of attendance system [16]. Sutar et al., the paper described about the attendance system using the combination of rfid and internet of things [17]. Jadhav et al., in which the author described about attendance system using internet of things [18]. Chintalapati et al., in which the author described the about face recognition process which is used in attendance system [19]. Akinduyite et al., in which the author has described the idea abundance management system by fingerprint [20].

Ashwin, Kamalraj, Azath, made a study on the trust based clustering algorithm on MANETS [21]. Gowtham, Subramaniam, proposed an algorithm using TCP and max-min fairness algorithm to reduce congestion and CCAODV protocol to control routing [22]. Ashwin, Kamalraj, Azath, proposed a particle swarm optimization algorithm for efficient and secure communications [23]. Gowtham, Subramaniam, proposed an algorithm to transfer the data in the channel without congestion and also to recover the lost packet [24]. Ramachandran, Subramaniam, presented a secure and efficient data forwarding algorithm for forwarding and recovering information from the untrusted cloud environment [25].

Swaminathan, Prasanna Venkatesan, designed an control system using mobile adhoc sensor [26]. Annamalai, Kumaresan, Prasanna Venkatesan, a low profile tri band antenna is designed to exhibit good return loss VSWR and moderate gain [27]. Jegadeesan, Shanmugapriya, et al., presented a system to increase the target secrecy rate by reducing the security performance of full duplex jamming communication relay model [28].

3   Proposed System

For the Proxy, manual and recognition attendance taking problem, the best solution is provided by in this proposed system. The system concept is made with the ESP8266 Module along router and Internet Protocol address to the database for making the attendance for the students (Fig. 1).

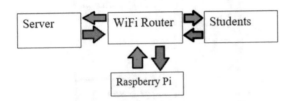

Fig. 1. Block diagram

First the students IP address and their details are stored in the data server. And all the details are stored in the database with the help of Wi-Fi modem or router for each classroom. This proposed system has limited range of the Wi-Fi signals like in the classroom. If the student enters the class room then only it receives the signal from the router or Wi-Fi modem after the student in the signal range and connect the Wi-Fi with the help of the specific router. The process is initiated by admin by means of web analysis Raspberry Pi search specific IP address that are stored in database server by transmitting an acknowledgement through router to ESP Module. The ESP module receive the signal from the raspberry pi and then it will generate acknowledgement for the same signal. When the Raspberry Pi Gets an acknowledgement from ESP Module with Specific IP Address, it is considered to be present for the specific student with the particulars details that are previously stored in the database server. The student must be connected in the Wi-Fi modem for making the attendance this saves the time for the staffs and students for taking the attendance by giving the timesheet or signature register or else saying the names of the students one by one. If the student leaves the class room then the signal is lost and the data stored as present or absent when the admin refers the attendance through web server with the help of IP All the details are stored in the database server and this data can be viewed by the staffs.

4   Methodology Steps

The methodological feature adds up the connections between router and the various functions which include the server and the ESP tag modules. The overall system is designed to automate and monitor the system at any point of time.

5   Flow Chart

See Fig. 2.

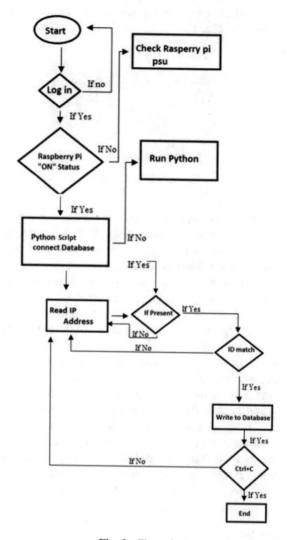

Fig. 2. Flow chart

6   Flowchart Explanation

The starting process and Raspberry Pi turning ON phase happens after the Login Page was successfully shifted to the attendance Homepage. The python script runs only when a user logs in and it matches the IP address of the individual ESP tags and if the access was successfully established, every IP might have been added to the report page and by fetching the details with the help of IP address the attendance is displayed. If the access establishment or any other process is collapsed the process loops and starts again till the resulting access turns null. Whenever a resulting loop finds the result zero it will automatically end up in the adjusting quantified place and restarts the process again.

7   Server

This can be calculated as the various usable function's, the server maintains and unfolds the data's required to the concerned partition from where the request has been received. Every single client requests makes use of the server to fetch data and unload it to the needed place.

8   Raspberry-Pi

This can be considered as a Mini CPU as it provides all the necessary functions of a well-built processing unit, the detailed specification differs by versions. Raspberry-Pi is a defined controller, for accessing all the activated elements in the automated system. Software can be thought of as a program or a collection of programs that instruct a computer on what to do and how to do it. It uses 8 GB SD for installing the Raspbian OS and for storage. The USB port is used for connecting keyboard, mouse, dongle and pen drive. The power supply is given through USB connector.

9   Esp Module (Node MCU, Student)

These modules have extremely small form factor, and are the most commonly used Wi-Fi module. They can be easily interfaced with any microcontroller and operate strictly on 3.3 V. This acts as the tag for individual students which are set to have different IP addresses; this will complete the edging closure of the receiving end.

10   Wi-Fi Router

This provides us various possibilities of fetching the data, in this case governing over the raspberry Pi and following the acknowledgement from the ESP tags and it flexes the part where the need to monitor at desired intervals of time, even frequently.

11   Results and Discussion

The further process efficiently discloses the steps involving the online server Entry level page and the varied accessing difference provided for staffs and students (Fig. 3).

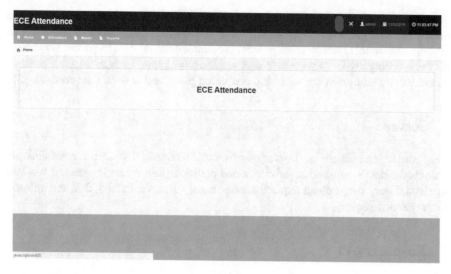

Fig. 3. Online server entry level page

Home button page through which the entire process has its start, the green light says the active status of the server. It is the display page after login (Fig. 4).

Fig. 4. Basic outlook page-Student Id logged on

This page is the logged-on display for the ADMIN user, it has specifications for filtering the attendance by period, ESP tags (students) etc. It can maintain the previously saved data as it principle report. The staff has the option to mark 'present' or absent on the template. Based on the record, the current database will be created so the total number of students present or absent for a particular date and at particular time will be evaluated (Fig. 5).

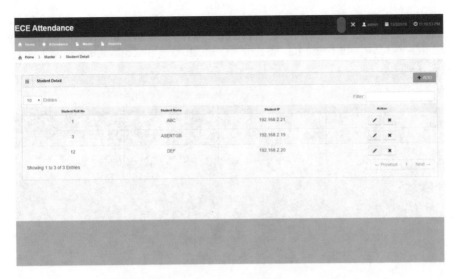

Fig. 5. Customization page-staff user access

This is the detailed information of each entered IP of every individual, this can be a major time saving effort when the immediate requirement of information takes place. The number of Ip addresses can be added and removed whenever necessary. The staff can edit the number of entries in the class separately in this page.

12 Manual Override

The overlapping of any previous process by which the individual ESP tags are not present in this work area or anywhere is considered to be at ease due to the manual override function which confirms the alteration of the attendance through Admin access (Fig. 6).

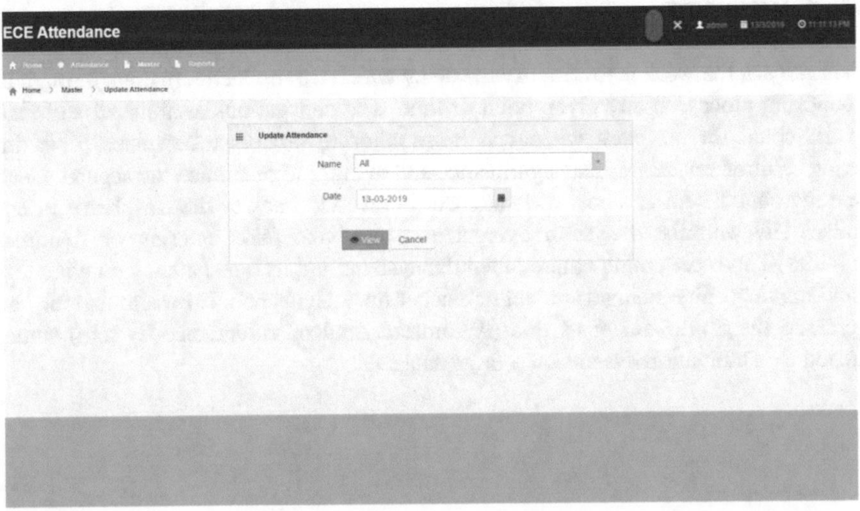

Fig. 6. Manual override function

We can view the attendance at any given point of time, this can result in the observant end through which the time and complexity are differed through the display, and this window will fixate on the date for which then override function is required (Fig. 7).

Fig. 7. System declared function for manual override

Once the changes are made through the override function, it can be saved as the last updated entry and the update can only be made for the particular date. It differs from the server entry which will be having one or more added manual entry.

13   Conclusion

The Entity of the work is to simplify the delay which was collectively done through the monitoring process at any given point of time. The connections established will make up the count for the class whenever the monitoring process takes place. This time saving solution renders a great significance and in order to maximize the accuracy, each time the data is sent and acknowledgement is reverted back to the Raspberry Pi controller. This automated system overcomes issues like false entering or fraudulent activities, Time consuming nature of similar systems and its transparency on whomever could monitor either being a student or teacher marks its bench. Error simulations have increased the improvement of this (Automated Student Attendance System) student attendance automation system on a large scale.

References

1. Chaudhari, P., Rane, S., Ahire, N., Shinde, H., Patil, A.: Implementation of RFID based campus management system. Int. J. Innov. Res. Comput. Commun. Eng.
2. Islam, Md.M., Hasan, Md.K., Billah, Md.M., Uddin, Md.M.: Development of smartphone-based student attendance system. In: IEEE Region 10 Humanitarian Technology Conference (R10-HTC), December 2017
3. Rahman, S., Rahman, M., Rahman, Md.M., et al.: Automated student attendance system using fingerprint recognition. Edelweiss Appl. Sci. Technol. (2018)
4. Yadav, R., Nainan, S.: Design of RFID based student attendance system with notification to parents using GSM. Int. J. Eng. Res. Technol. (IJERT) **3**, 1406–1410 (2014)
5. Gagare, P.S., Sathe, P.A., Pawaskar, V.T., Bhave, S.S.: Smart attendance system. Int. J. Recent Innov. Trends Comput. Commun.
6. Nandya, S.K., Gadgey, B., Pujari, V.: RFID and fingerprint based smart attendance system. Int. J. Res. Appl. Sci. Eng. Technol. (IJRASET) **5**(VI) (2017)
7. Kumar, P., Suresh, K., Indumathi, T., Kumar, K.: Smart attendance system using raspberry Pi. Int. J. Trend Sci. Res. Dev. (IJTSRD)
8. Prince, N., Sengupta, A., Unni, K.: Implementation of IoT based attendance system on a dedicated web-server. Int. J. Sci. Eng. Res. **7**(6), 351–355 (2016)
9. Sapkal, P., Jaiswal, K., Gohil, H., Thakkar, K., Lopes, A.: IoT based smart attendance system. Int. Res. J. Eng. Technol. (IRJET)
10. Praveen Kumar, M., Mani Kumar, B.: RFID based attendance monitoring system using IOT with TI CC3200 launchpad. Int. J. Mag. Eng. Technol. Manag. Res.
11. Lodha, R., Gupta, S., Jain, H., Narula, H.: Bluetooth smart based attendance management system. In: International Conference on Advanced Computing Technologies and Applications (ICACTA-2015) (2015)
12. Agarwal, A., Bansal, A.: Online attendance management system using RFID with object counter. Int. J. Inf. Comput. Technol.
13. Chaudhari, P., Rane, S., Shinde, N., Shinde, H., Patil, A.: Implementation of RFID based campus management system. Int. J. Innov. Res. Comput. Commun. Eng.
14. Yadav, R., Nainan, S.: Design of RFID based student attendance system with notification to parents using GSM. Int. J. Eng. Res. Technol. (IJERT)
15. Nandyal, S.K., Gadgey, B., Pujari, V.: RFID and fingerprint based smart attendance system. Int. J. Res. Appl. Sci. Eng. Technol. (IJRASET)
16. Kanawade, S.Y., Shinde, S., Matale, K., Shelar, M.: Modern attendance system. Int. J. Adv. Res. Comput. Commun. Eng. (IJARCCE)
17. Sutar, M., Patil, M., Waghmare, S.: Smart attendance system using RFID in IoT. Int. J. Adv. Res. Comput. Eng. Technol. (IJARCET)
18. Jadhav, V., Lakshman, K.: Office automation & attendance system using IoT. Int. Res. J. Eng. Technol. (IRJET)
19. Chintalapati, S., Raghunadh, M.V.: Automated attendance management system based on face recognition algorithms. In: 2013 IEEE International Conference on Computational Intelligence and Computing Research (2013)
20. Akinduyite, C.O., Adetunmbi, A.O., Olabode, O.O., Ididunmoye, E.O.: Fingerprint-based attendance management system. J. Comput. Sci. Appl. **1**, 100–105 (2013)
21. Ashwin, M., Kamalraj, S., Azath, M.: Weighted clustering trust model for mobile ad hoc networks. Wirel. Pers. Commun. **94**, 2203–2212 (2017)
22. Gowtham, M.S., Subramaniam, K.: Power aware cross layer approach in MANETs using advanced congestion control approach. **9**, 1600–1615 (2017). Special Issue 18-Special Issue

23. Ashwin, M., Kamalraj, S., Azath, M.: Multi objective trust optimization for efficient communication in wireless M learning applications. Cluster Comput., 1–9 (2017)
24. Gowtham, M.S., Subramaniam, K.: Congestion control and packet recovery for cross layer approach in MANET. Cluster Comput., 1–8 (2018)
25. Ramachandran, B., Subramaniam, K.: Secure and efficient data forwarding in untrusted cloud environment. Cluster Comput., 1–9 (2018)
26. Swaminathan, S., Prasanna Venkatesan, R.: Embedded traffic control system using wireless ad hoc sensors. Middle-East J. Sci. Res. **20**(2), 225–227 (2014)
27. Annamalai, S., Kumaresan, M., Prasanna Venkatesan, G.K.D.: A low profile higher band IOT antenna for security applications. In: AIP Publishing - AIP Conference Proceedings, vol. 2039, no. 1, p. 020048 (2018)
28. Jegadeesan, S., Shanmugapriya, S.S., Azees, M., Dhamodaran, M., Prasanna Venkatesan, G.K.D., Kumar, R.P.: An efficient technique to improve the reliability and security of relay networks in physical layer. J. Web Eng. **17**(6), 2341–2359 (2018)

Cassava Disease Prediction Using Data Mining

Amal Anand[✉], Merin Joseph, S. K. Sreelakshmi, and G. Sreenu

Department of Computer Science and Engineering,
Muthoot Institute of Technology and Science, Varikoli, Kochi, India
2014amalanand@gmail.com, ponnythankuz@gmail.com, sksreelakshmi97@gmail.com,
sreenug@mgits.ac.in

Abstract. Pesticides are used abusively mainly used to control plant diseases and pests, which lead to reduced quality of vegetables and endangering the life of the living beings. So we propose a model that can predict the presence of diseases with the fulfillment of speed and accuracy. And here for the successful predictions it mainly depends on the selected parameters which we use for the prediction. The parameters we use are predictable. Predictable computer applications that predicts diseases under favourable conditions will be of great help to all farmers. Such applications would reduce problems related to plant protection. In order to create this prediction model we need to consider different prediction variables. Prediction is done using the weather variables such as humidity, temperature and soil conditions such as soil type and data that represents specific disease characteristics. Through this model we check the presence of diseases present in Cassava plants. The proposed solution will take current weather details and soil conditions as input and would predict the diseases, if present any, along with some suggestions to overcome or suppress these diseases.

Keywords: RFT · Disease prediction · WEKA · Machine learning

1 Introduction

The fruit/vegetables protection is a complex and difficult task. We need ensure the production of healthy food and make sure that it does not leave any harmful residue behind. Farmers faces huge losses due to this. These conditions leads to quality deterioration of fruits and vegetables. It also effects the exporting and financial rate of the country. These are the reasons that inspired us to propose a plant disease prediction system using data mining. In this proposed project we are trying to develop a system that can predict the presence of possible diseases of Cassava in certain given conditions and also show the control measures that have to be taken for its prevention. The project is based on data mining techniques. For this the data sets of Cassava will be obtained from ICAR-Central Tuber Crops Research Institute, Trivandrum. The existing system of detecting diseases

© Springer Nature Switzerland AG 2020
P. Karrupusamy et al. (Eds.): ICSCN 2019, LNDECT 39, pp. 679–686, 2020.
https://doi.org/10.1007/978-3-030-34515-0_71

in plant is either using image processing or naked eye observation. But these methods are time consuming and expensive in large farms. They will try to treat these diseases using pesticides and other harmful chemicals in a huge amount. Also this would lead to quality deterioration of plants. So, our proposed project would be very helpful for the farmers and consumers in future. Our project would also tell the farmers about the possible measures to be taken in order to avoid or control the disease. So the usage of harmful chemicals and pesticides can be reduced to some extent.

Cassava is the third-biggest wellspring of carbohydrates in the tropics, after rice and maize [11]. Cassava is a noteworthy staple sustenance in the creating scene, giving an essential eating routine to over a large portion of a billion people. It is a standout amongst the most dry spell tolerant yields, fit for developing on minor soils. Nigeria is the world's biggest maker of cassava, while Thailand is the biggest exporter of dried cassava.

Around the world, 800 million individuals rely upon cassava as their essential nourishment staple.

2 Literature Review

Early cherry fruit pathogen disease detection based on data mining [1]. This is the base paper which explains about the pathogen disease detection of Cherry plant based on data mining. This paper proposes a system where we predict the disease of cherry plant using data mining. The proposed system predicts two main diseases of cherry fruit. The prediction model proposed in this system make use of six weather variables and one variable represents the month in the year. Data sets are organized in matrix forms, each row representing the instance and columns representing the attributes.

Data mining based tool for predicting the possible fruit pathogen infections [2]. In this paper data mining tool are used for predicting the fruit pathogen infections. An open source application is named WEKA is used for this. The GUI created in C-sharp along with WEKA . Prediction accuracy is 89%. The most important condition needed for predicting the infectious pathogen fruits are the appropriate weather data and presence of an active pathogen spores. Prediction is performed on 'Hayward' Kiwi fruit crops in New Zealand. Main goals of this paper was to do a comparison of achieved results and then is model is verified against the new data. Weather variables are min temperature, max temperature, avg temperature, wind speed, rainfall, avg humidity. Information about spores are gathered - whether they are passive or active. Each row represents instances and column represents attributes. The class attribute values are - Monilia, Cocomyces, nothing, both. After that a file is created with all the attributes, using the software tools prediction model is trained. It is based on concept of decision tree. It is designed so that it can be used by the user, whose has access to the weather data and information about pathogen spore activities. An user whose is an educated farmer or a certified service can also lead as the software users. This tool is not just designed for only one type of data file, neither for a predefined set of diseases.

A review of neural networks for disease detection of plants with the help of hyper spectral data [3]. In This paper Neural-Network (NN) technique are used for computing available hyper-spectral informations, giving main importance to detect the plant diseases. Hyper-spectral imaging is an special technique which is used for remote sensing. Hyper-spectral sensors collects the information from the visible through the Near-InfraRed (NIR) in ranges of the spectrum, and obtain the spectral informations from hundred of narrow spectral band. These benefits of hyper-spectral imaging has created exactness plat protection even a lot realizable. Some techniques such as Principal-Component Analysis (PCA), Support-Vector Machine (SVM), Fuzzy logic, Cluster Analysis (CA), Neural Networks (NNs) and Partial Least- Square (PLS) have been used for hyper-spectral spectroscopy. The most promising tool used in hyper-spectral analysis is NN. The mechanisms of NN are based on the human nervous system. Basically, NN are very helpful for the pattern recognition, despite any explicit recognition rule. NN are mathematical models which is used for data mining. This paper has extensively reviewed on SDI.

A novel Neurofuzzy classification technique for data mining [4]. This paper proposes a novel Neurofuzzy classifications techniques which are used for data mining. The planned technique uses a fuzzification matrix's in-which the input pattern are associated with a degree of membership to completely different class. Artificial-neural network (ANN) or simple neural-network (NN) are the some of the best data modeling tools that will be used for intellectual task which has similarity to the human brain. NNs are well known-ed for its prime preciseness and high learning abilities even if a small amount of data is offered. This paper proposes a Novel Neuro fuzzy classification techniques for data mining which use a combinations of MLPBPN and fuzzy-set theory approaches. This technique extract a feature wise data about couple of input pattern and fuzzifies its corresponding patterns using its membership-function (MF), and it also provide a degree of the memberships for several categories to many class. The main three classification techniques used are ANFIS, NFS and RBFNN. These are well trained and tested.

Prediction of Plant Disease from Weather Forecasting using Data Mining [5]. This paper develops model that predicts disease of orange plants using segmentation technique like k-means clustering, SGLDM and deep neural network to detect the disease of the orange plant based on available data. The model is developed by using MATLAB software. The data-set includes the disease image of orange plant and also the weather feature such as temperature, rainfall and humidity. The data-set used the seven categories of disease namely blackspot, greening, Melanosa, Greasy spot, Scab disease, Alternaria brown spot and Cenker disease. The segmentation technique such as K-means clustering divide the image into specific part and show the disease part of orange plant. Then apply the deep neural network learning. This is used to extract the automatic feature without human conflict and then spatial grey level dependence matrix. This system can only be used for predict the seven categories of diseases found in Orange. And also the users have to give image of the disease affected part as input. These are some disadvantages of this paper.

Prediction of Potato Late Blight Disease is Based upon Environmental Factors in Faisalabad, Pakistan [6]. This paper deals with Potato blight disease, caused by Phytophthora-infestans which is a water mold lower plant life that infect the potato crop in both cool and wet climate. Because of its lack of resistances. Within the potato plant the disease is managed through fungicides by the growers. It effect the environment and thus they developed a model on 2 years information about potato blight diseases severity and epidemiologic factor with step-wise regression analysis. Five varieties of potatoes like Desiree, Diamant, SH-5, SH-339 and FD 35-36 were planted for collecting data-set and it is recorded on weekly basis used a 1–9 Henfling scales until the end of the season. Minimum and maximum temperatures, rainfall, relative-humidity and wind speed are the considered environmental factors. Environmental factors exhibits a important relationship with disease severity is plotted graphically and critical ranges of environmental variables contributive for potato late blight disease were determined. Those factors are examined for five varieties in different seasons and determined the favourable range of factors, that were responsible for disease. It depend on the location where data collection is done. So it may not work in other location. Accuracy enables the farmer to implement a timely disease management plans with reduced use of pesticides.

Development of a web-based disease prediction system for strawberries [7]. This paper is for strawberries in Florida. In Florida about sixteen millions flats of strawberries produced each year. This model utilised wetness of the leaf and temperature throughout the wet amount to predict outbreak. This forecasting system predicts these diseases, Strawberry-Advisory-System (SAS), was implemented on the Agro climate web site using weather data from the Florida agricultural weather networks. A database is required to store information derived from completely different source about weather. It is enforced using the R statistical analysis software, which is a language and environment for graphics generation and statistical computing. The C++language and fortran was used for implementing the specific code functions for certain task which requires a high computing power.These functions were integrated using the R and were called throughout the execution of these model. Results from the models layers are static information, primarily formed by, HTML, texts, XML and graphics. This system recommends the spray for Botrytis and Anthracnose based on questionnaire. Model outputs can be in table formats or graphic. This is used to indicate the disease risk levels based on weather information gathered at the station and forecast for the next three day. And it also send Emails and SMS messages to registered users whenever their occurs a high risk. This can be used for only two diseases. And the weather data are collected from another website. So any problem in data updation will affect the prediction model. This is an disadvantage of this model.

Detecting Bakanae disease in rice seedlings by machine vision [8]. In this paper combination of both machine learning algorithms and image processing techniques. Using these me hods we can easily distinguish infected from the healthy seedlings. Bakanae disease, also known as The 'foolish seedling', is a rice seed borne disease. The images of infected as well the controls seedling were acquired exploitation and the flat-bed scanner are used to quantify its morphological and colour trait. Using Support vector machine (SVM) classifier we can distinguish the infected seeds from the healthy seedling. A genetic algorithmic rule were used for choosing essential trait and optimum model parameter for the SVM classifiers. Flatbed color scanners used in order to acquire the images of seedlings. For identifying the anatomical points in the images, Image processing algorithm were used. Morphological and colour trait of the seedling can taken as the attribute for distinguishing the seedling with different inoculation treatment. Support vector machine (SVM) classifier were developed for distinguishing the health status of the seedling. The first classifiers SVM1, were used for distinguishing the seedlings which are controlled ed and inoculated. The second classifiers SVM2, for distinguishing the seedlings at different inoculation level.

Wheat Leaf Disease Detection Using Machine Learning Method [9]. A formal image recognition method includes segmentation, image pre-processing, feature extractions and pattern recognitions. Image segmentation is the key step, and therefore used for exactness of the segments directly influence the dependability of feature extraction and therefore the recognition accuracy. Clustering and segmentation method are based on statistical pattern recognition. This paper present, detection of wheat leaf using SVM image processing techniques. The fundamental step to implement machine learning strategies is for the detection of wheat disease. The input are pre-processing of image, use of machine learning algorithms. Support vector machine (SVM) is used for classification in disease detection in the leaves. Neural network is one amongst the known strategies for regression which used for disease detection in leaves. The main important characteristics of disease detection of wheat leaf are speed and accuracy. It permits the turnout and detection in wheat plant diseases. Wheat disease is detected by the SVM classifiers.

A Hybrid of Plant Leaf Disease and Soil Moisture Prediction in Agriculture Using Data Mining Techniques [10] This paper focuses on monitoring the plant health, evaluating the water content level for different plants by analyzing the plant and soil images Initially, plants and soil images are captured through digital camera with a required resolution. In addition, texture and color features are extracted from the soil images. By mistreatment the contour options of the plant pictures, the plant type is identified through the botanical plant species dictionary. The leaf diseases are predicted by Transductive Support Vector Machine classification. The causes for the specific plant disease are identified supported the Latent Dirichlet Allocation and Artificial Neural Network classification technique through the options of soil pictures and also the morbid plant pictures. The soil moisture prediction technique was proposed by using improved BP algorithm.

3   Design and Methodology

The proposed solution is designed using Random Forest Tree algorithm which is a collection of many decision trees. The main advantage of this algorithm is that it can be used for both classification and regression. Rather than utilizing information gain or gini index for computing the root node, the way toward finding the root node and splitting the nodes will happen haphazardly in Random Forest. There is a root node and a split function (chosen randomly) at each node and leaf nodes that denotes the target/goal value which is the possible diseases for Tapioca in this case.

The user has to give the weather details and the soil conditions (mainly pH level of soil) as input and the system would provide the disease which is most likely to occur and provides control measures in order to control the disease. The system will start from the root node and goes to the subsequent child node according to the split value obtained and ultimately reaching one among the many leaf nodes which will denote the disease that will be present. The best feature in a random subset of features is selected for splitting a node. The system will be trained first and then tested before making it available to all. The tree is traversed from the root node for each input and gradually reaching the target. The tree construction happens randomly adding more accuracy and correctness to the results obtained. The Random Forest preparing calculation begins with developing various trees. In the literature review, a few techniques were utilized, for example, arbitrary trees, C4.5, J48, CART and so on. In this paper we are utilizing the random trees in building the Random Forest classifier with no pruning, which makes it light, from a computational point of view. The following stage is setting up the preparation set for each tree, which is shaped by haphazardly sampling the training dataset (Fig. 1).

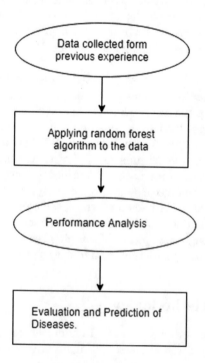

Fig. 1. Data Flow diagram

4   Implementation

There are some basic steps in RFT algorithm From dataset, take n random records. Then create decision tree on this random records. In case of regression problem, each tree in forest predict a value for output and the final result is calculated by taking the average of all the value is predicted by all trees. In case of classification drawbacks, each tree in the forest predict the category to which the new record belong.

For our system initially we collected data containing attributes like climate, weather, diseases from sources. The data should be given to system with all values. Then this data is divided as training and test data. First training data is used to train the system and after that system can predict whether the plant has chance to occur disease or not. Algorithm over fitting can be reduced by pruning the dataset. By using random forest algorithm we can predict all tapioca plant disease with highest accuracy. Implementation of the Random Forest Tree is done in Python. There are many Python libraries available which helps in implementing machine learning algorithms in a more easier way. One such library of use is ScikitLearn. It has many inbuilt modules that helps in building the tree, testing and training the data and so on. It can be used for both data mining and data analysis. The sklearn ensemble module contains the Random Forest tree. Each tree is built using Bootstrapping which is data sampling with replacement.

5   Results and Conclusion

This paper proposes a system that can accurately predict the diseases found in Cassava plants and guide the farmers in taking appropriate counter measures to control the diseases. The proposed system uses Random Forest Tree algorithm for accurate prediction and takes the actions accordingly. Prevention is better than cure. So through the proposed solution the farmers can prevent diseases that are most likely to happen in future. The agricultural field will be the most benefited and this system will also encourage freshers into the field of agriculture.

6   Future Scope

The paper proposes a system that will increase the yield in Cassava in the years to come. The Cassava farmers will be so relieved that they can finally treat the plants right. The best counter measure against anything is to treat the problem right. Right treatment itself will solve majority problems concerning the yield. In future the system can be modified to predict not just the diseases in Cassava but also among a variety of other fruits and vegetables too. The User Interface can be modified for much easier use of the users.

References

1. Ilic, M., Ilic, S., Jovic, S., Panic, S.: Early cherry fruit pathogen disease detection based on data mining prediction. Comput. Electron. Agric. **150**, 418–425 (2018)
2. Predic, B., Ilic, M., Spalevic, P., Trajkovic, S., Jovic, S., Stanic, A.: Data mining based tool for early prediction of possible fruit pathogen infection. Comput. Electron. Agric. **154**, 314–319 (2018)
3. Golhani, K., Balasundram, S.K., Vadamalai, G., Pradhan, B.: A review of neural networks in plant disease detection using hyperspectral data. Inf. Process. Agric. **5**, 354–371 (2018)
4. Ghosh, S., Biswas, S., Sarkar, D., Sarkar, P.P.: A novel Neuro-fuzzy classification technique for data mining. Egypt. Inform. J. **15**, 129–147 (2014)
5. Kaur, K., Kaur, M.: Prediction of plant disease from weather forecasting using data mining. Int. J. Future Revolut. Comput. Sci. Commun. Eng. (2018)
6. Ahmed, N., Khan, M.A., Khan, N.A., Ali, M.A.: Prediction of potato late blight disease based upon environmental factors in Faisalabad, Pakistan. J. Plant Pathol. Microbiol. (2015)
7. Pavan, W., Fraisse, C.W., Peres, N.A.: Development of a web-based disease forecasting system for strawberries Comput. Electron. Agric. **75**, 169–175 (2011)
8. Chung, C.-L., Huang, K.-J., Chen, S.-Y., Lai, M.-H., Chen, Y.-C., Kuo, Y.-F.: Detecting Bakanae disease in rice seedlings by machine vision. Comput. Electron. Agric. **121**, 404–411 (2016)
9. Dixit, A., Nema, S.: Wheat leaf disease detection using machine learning method. Int. J. Comput. Sci. Mob. Comput. **7**, 124–129 (2018)
10. Sabareeswaran, D., Guna Sundari, R.: A hybrid of plant leaf disease and soil moisture prediction in agriculture using data mining techniques. Int. J. Appl. Eng. Res. **12**, 7169–7175 (2017)
11. Predic, B., Ilic, M., Spalevic, P., Trajkovic, S., Jovic, S., Stanic, A.: Data mining based tool for early prediction of possible fruit pathogen infection. Comput. Electron. Agric. **154**, 314–319 (2018)
12. Dhomse Kanchan, B., Mahale Kishor, M.: Study of machine learning algorithms for special disease prediction using principal of component analysis. In: International Conference on Global Trends in Signal Processing, Information Computing and Communication (2016)
13. Kiania, E., Mamedovba, T.: Identification of plant disease infection using soft-computing: application to modern botany. Faculty of Engineering, Near East University, Nicosia, North Cyprus, National Academy of Science, Azerbayjan, Mardakan Dendrary
14. Sabareeswaran, D.: A hybrid of plant leaf disease and soil moisture prediction in agriculture using data mining techniques. Research Scholar, Department of Computer Science, Karpagam Academy of Higher Education (KAHE), Karpagam University, Coimbatore, Tamilnadu, India
15. Sonare, Ms.B., Zarkar, S., Talele, P., Deshmukh, R., Shelake, U.: Review on crop pests forewarning with weather factors using machine learning. Department of Information Technology, Pimpri Chinchwad College of Engineering

Online Attendre Tracking System

M. Suguna[1]([⊠]), D. Prakash[2], and G. Shobana[1]

[1] Department of CSE, Kumaraguru College of Technology,
Coimbatore, Tamilnadu, India
{suguna.m.cse, shobana.g.cse}@kct.ac.in
[2] Anna University, Chennai, Tamilnadu, India
pp_d@rediffmail.com

Abstract. The Educational institutions much more focus on their academic and business motives were, the students are not comfortable with the stand-alone applications provided to them for their privilege to have a connectedness within themselves and the management. This is because those applications are comprised with much less features which failed to have such an interactivity among the students. This derivate the objective of the proposed system which, is to provide students, the utmost interactive experience to be socially connected and to make the users to be socially responsible. Here, the student can use Blood-Bank portal to be the donor or beneficiary by providing their non-confidential health details. And, The Institution can make better decisions by conducting an online election/poll with the students as the voters. Also the students can share their ideas and opinions with other members through the Blog. Portal, where the students can even use this as a platform to show their support for the polling. The students can also keep a track of their presence in the institution and they will get notified on insufficient attendance. Admin is also provided with the facility to post some prominent issues like circulars, new regulations, events etc. Which shall be shared with all the students for their kind references. Overall, the application is developed for a group of people to make them to be socially connected and to be socially responsible. The entire control of the application lies in the hand of the administrator and the admin possess the power to remove the user in case of any misconduct or breach of any of the terms and conditions laid down by the application.

Keywords: ERP · Poll Portal · Database · Data-mart · Blood bank portal

1 Introduction

The use of mobile devices was tremendous for the last couple of years, from which the users can perform almost all tasks in the mobile context within the very small span of time [3]. Hence, the users always need the solutions to be handy, reliable and fast as its performance. Mainly, the apps under the category of "Utilities and productivities", had shown a huge impact among the users, which majorly concentrates on their own field of work and always need to reduce their load in terms of automation. One among them to be noted is ERP. As of now, almost all ERPs are much more focused on their business and the users who use those ERPs are less benefited apart from their

© Springer Nature Switzerland AG 2020
P. Karrupusamy et al. (Eds.): ICSCN 2019, LNDECT 39, pp. 687–693, 2020.
https://doi.org/10.1007/978-3-030-34515-0_72

commercial use [1]. This is because those are developed in such a way that is mainly dedicated to the business and there exists no functionality that can provide interactivity amongst the users.

Interaction in the sense, it majorly lags to have something that can unite them for a cause and to be socially responsive. Major institutions failed to have such a scenario and let them use only to fulfil for the management's cause. And especially in the case of educational institutions, most of their students need such a thing to express themselves and to share their thoughts with all their fellow students and to the management. The future completely relies on the students for own institution, which can't be given by anyone else. Hence, their own confidential data and their suggestions for the betterment of an institution is something priceless. So, it should be received and well organized in the best way to have a good relationship between the students and the institution. Since, it could make an open system among the institution, then it will increase the potential of every student to perform well in their overall academics as well as their carrier. To proposed a progressive Android-web application which includes many modules to engage students [4]. Among which the Blood bank, the Polling and the blog portal available in the system allows the students to interact and share their ideas and opinions with others along with their admin to overview it. The students can also use the Attendance portal to keep track of their attendance in the academics and will get notified later in case of insufficient attendance. In a nutshell, the application is developed for a group of users to be socially connected and to be socially responsible.

2 Literature Survey

Survey monkey is an online survey development cloud-based software that provides free and platform allows the users to tailor your surveys according to tailor the surveys according to a defined target audience. It can provide an impetus for making smart and wise decisions related to the different areas of the enterprise. The survey Monkey app is used to collect various opinions from the respondents and allows and encourages the poll organizer to make an important decision [8]. If anyone needs emergency bloods for a specific group, just make a search on Digital Blood Bank (DBB) database by putting Blood Group Name and community name that is needed. DBB is the most valuable project in Health and Treatment. The Digital Blood Bank App allows the users to receive the volunteers of blood from a nearby location. The limitation with this application is that it does not allow the user to be the blood donor [9]. Online Educational Enterprise resource planning (ERP) implementation approving as the mandatory software for all technical institutions [10].

3 Proposed System

Most of the utilities-built applications are built in order to reduce the work of their customers in terms of their own workload, and this system mainly concentrates with the student's activities and their tasks. And so, it is built with the architecture of modules providing the students with the features of attendance, blog, poll, data-mart, blood bank. Attendance is being controlled by the students for their own sake, of calculation to trace back their presence and the poll is being controlled by the admin to ask the students regarding their works and for the betterment to be made in the institution in the future. Blood-bank is one of the feature made in-order to collect the student's blood related details to safeguard their own students in future with those details which, furthermore helps us to co-relate them. Blog is something always important for the students to share their posts with all the fellow students in the network along with the overview of admin provided to delete some awkward posts. Datamart helps the institution to share some important notice to their students, which are like the events, circulars, results, etc. And this feature can also be used in best way by sharing some good books and the other portable documents to increase the betterment of the education. And much more manipulation processes, including the student creation, manage the student list, poll results, data-mart upload and attendance report can be viewed and updated via admin panel seems the biggest advantage to control the students and their activities (Fig. 1).

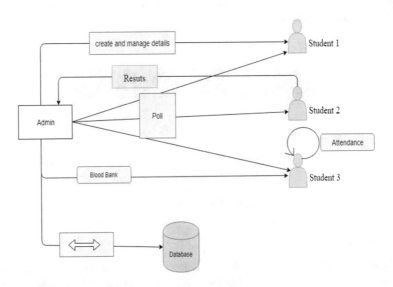

Fig. 1. Architecture diagram

The entire system control lies with the control of the Institution, the admin of application. The Institution can conduct one or more polls at a time and each and every student can vote for the posted questions in the Poll Portal. Once the time allotted for the Poll gets over, the results will be published to the participants if it is needed. The main purpose of the Poll portal is to enable the Institution to make management or academic decision by taking into account the opinion of the students. Also, if any student is in a medical emergency and requires blood donation, the students with the same blood group can be found. The Blood Bank Portal gives the basic non-confidential students of the students. Also the students can use this app to keep track of his attendance during a particular term.

The students can use the Blog-Portal as the discussion forum to share their opinions with other students. The Institution can use the Data-Mart Portal to post the details about the Institution decision and other confidential details.

4 List of Modules

4.1 Blood Bank Portal

The Institution will collect the necessary personal and health details of the students at the time of admission. This Blood-Bank Portal provides the feature that when one needs blood, this portal shows the person who have same blood group. If A needs a blood group of AB+, then A can search on the portal for the persons with the blood-group of AB+. The Portal will provide the basic details of the persons such as name, roll number, phone number (Fig. 2).

Fig. 2. Attendre entry details

4.2 Poll Portal

The Institution when need to make an important decision either in academics or management side may consider that the collective opinions of the students almost matters the most. At such situation, they can conduct a poll and the students can vote in the poll. The students can vote only for the allotted time period after which the poll will get deleted. The Institution can then make a decision depending upon the majority opinions known through the election. The Institution can conduct the poll for a particular year or department, particular course and can set the time-period of the poll (Fig. 3).

Home
Create Student
Manage Student
Reports
Polling
View Data-mart & Blog
Logout
Create Polling
View Result
Question *
1)*
2)*
3)
4)
Select Dept
CSE IT ECE EEE MECH CIVIL
Select Course
BE ME
Select Year
1 2 3 4
April 2018
Su Mo Tu We Th Fr Sa
Select End Date
Send
Cancel

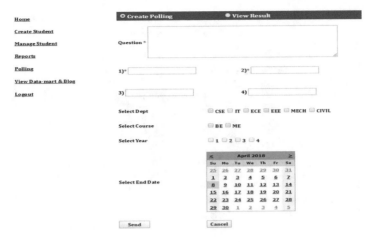

Fig. 3. Create polling form

4.3 Blog and Data-Mart

Data-Mart feature allows the students to share the syllabus books, journals, magazines and other technical-news with other students. The right-to-delete option is available to the admin and the post in Data-Mart can be removed in-case of any complaint in terms of misconduct or mis-behaviour or breach of any terms and conditions laid down by the institution. And the Blog Portal can be used as the canvas campaign for the election being conducted by the Institute. The students can also share their thoughts with others and the students can feel the response received from the blogs (Figs. 4 and 5).

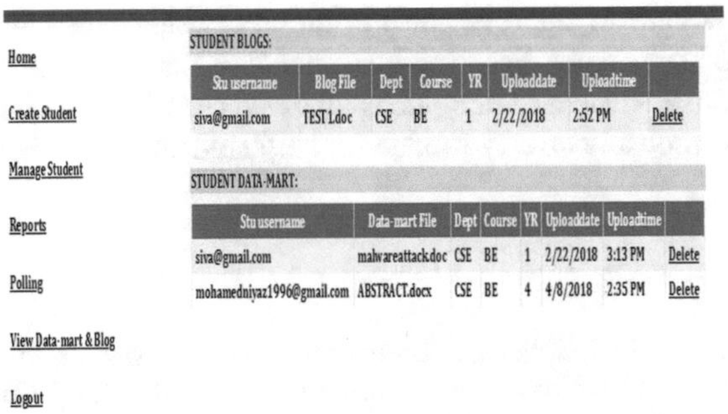

Home
Create Student
Manage Student
Reports
Polling
View Data-mart & Blog
Logout
STUDENT BLOGS:
Stu username Blog File Dept Course YR Uploaddate Uploadtime
siva@gmail.com TEST1.doc CSE BE 1 2/22/2018 2:52 PM Delete
STUDENT DATA-MART:
Stu username Data-mart File Dept Course YR Uploaddate Uploadtime
siva@gmail.com malwareattack.doc CSE BE 1 2/22/2018 3:13 PM Delete
mohamedniyaz1996@gmail.com ABSTRACT.docx CSE BE 4 4/8/2018 2:35 PM Delete

Fig. 4. Students Blog and Data-Mart

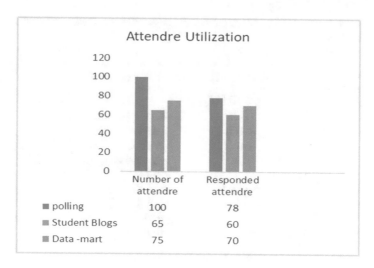

Fig. 5. Number of Attendre Utilization chart

5   Conclusion

Utility based initiatives can always make sure about their user's usage and their work load which dignifies that Attendre will definitely play an important role in the student's community to co-relate, organize, update and express themselves. It can be available anywhere and anytime to access the updates and emergencies to help others or to get sorted. On the other hand, it is well designed to enhance the communication between the student and their college management to improvise the betterment of the student community. Hence, it is easily used to extract and track the failures in the student network by the admin to improve the institution's vision along with the student's mission. Some other features like chat-bot, attendance notifications and alerts, voice recognition and other advanced encryptions involved security can be made in the phase 2 to get the most probable quality product to help students in the beneficial way.

References

1. Othman, M., Ismail, S.N., Noradzan, H.: An adaptation of the web-based system architecture in the development of the online attendance system. In: 2012 IEEE Conference on Open Systems, pp. 1–6. IEEE, October 2012
2. Zhang, Z., Gong, P., Hao, C.: Educational administration attendance system based on XML and Web service (2008)
3. Oliver, E.: A survey of platforms for mobile networks research. ACM SIGMOBILE Mob. Comput. Commun. Rev. **12**(4), 56–63 (2009)
4. Sidi, J., Junaini, S.N., Lau, S.L.: ISAMS: tracking student attendance using interactive student attendance management system (2007)
5. Hornback, G., Babu, A., Martin, B., Zoghi, B., Papu, M., Singhal, R.: AAS: automatic attendance system (2009). Accessed 4 October 2009

6. Misra, H.: How relevant is E-Governance measurement? Experiences in Indian scenario. In: Proceedings of the 2014 Conference on Electronic Governance and Open Society: Challenges in Eurasia, pp. 1–6. ACM, November 2014

7. Kumar, A., Tadayoni, R., Sorensen, L.T.: Metric based efficiency analysis of educational ERP system usability-using fuzzy model. In: 2015 Third International Conference on Image Information Processing (ICIIP), pp. 382–386. IEEE, December 2015

8. https://play.google.com/store/apps/details?id=com.surveymonkey

9. https://play.google.com/store/apps/details?id=com.maktra.digitalbloodbank

10. Kassim, M., Mazlan, H., Zaini, N., Salleh, M.K.: Web-based student attendance system using RFID technology. In: 2012 IEEE Control and System Graduate Research Colloquium, pp. 213–218. IEEE, July 2012

11. Patel, U.A., Priya, S.: Development of a student attendance management system using RFID and face recognition: a review. Int. J. Adv. Res. Comput. Sci. Manag. Stud. 2(8), 109–119 (2014)

12. Deugo, D.: Using QR-codes for attendance tracking. In: Proceedings of the International Conference on Frontiers in Education: Computer Science and Computer Engineering (FECS), p. 267. The Steering Committee of the World Congress in Computer Science, Computer Engineering and Applied Computing (World Comp) (2015)

13. Srinidhi, M.B., Roy, R.: A web enabled secured system for attendance monitoring and real time location tracking using Biometric and Radio Frequency Identification (RFID) technology. In: 2015 International Conference on Computer Communication and Informatics ICCCI), pp. 1–5. IEEE, January 2015

Congestion Control for Improved Cluster-Head Gateway Switch Routing Protocol Using Ad-hoc On-demand Distance Vector Algorithm

U. B. Mahadevaswamy[(⊠)] and M. P. Jananitha

Department of ECE, JSSS&TU, Mysuru, India
ubms_sjce@yahoo.co.in

Abstract. MANETs are emerging as the torrid research topic in the wireless communication domain. These MANETs are used in emergency cases like military, flood, earthquake, etc. to set-up a fast, easy and feasible network. Proactive is a table-driven protocol, consists of routing table information before setting up a connection. Reactive protocols setup path only when it is demanded by the network. In this paper, we have proposed a protocol that combines the advantages of both cluster-head gateway switch routing protocol (CGSR) and adhoc on-demand distance vector (AODV) to minimize the traffic congestion in ad-hoc wireless network. The results of performance metrics like routing overhead, end-end delay, and packet delivery ratio are enhanced and compared with other clustering protocols like CGSR, LEACH, and DUCS.

Keywords: MANET · AODV · CGSR · Cluster-head · Congestion control

1 Introduction

Mobile ad-hoc network (MANET) is a group of mobile nodes or moving nodes which establishes a communication network voluntarily through a shared wireless channel. MANETS have no pre-existing infrastructure and minimal central administration. In MANETs, every node acts as a router to initiate the connection between the source and the destination. The Mobile ad-hoc network is more feasible and easy maintenance when compared to the traditional communication method. For a MANET to be established, we need a node ready to send data to a node that is ready to accept the data. These networks often use in emergency and rescue operations like military, floods and other natural calamities to establish the network. Since MANETs don't have any infrastructure, hence the network security is poor, congestion is more, and the probability of link failure is high.

There are three types of routing protocols in MANETs. Proactive Routing protocols: These are well known as table-driven protocols as each mobile node maintains routing table which is useful for data traffic, path information, power consumption, etc. These protocols are not appropriate for large networks. Examples of proactive routing

P. Karrupusamy et al. (Eds.): ICSCN 2019, LNDECT 39, pp. 694–703, 2020.
https://doi.org/10.1007/978-3-030-34515-0_73

protocols are DSDV, WRP, CGSR, etc. Reactive Routing protocols: These are well known as demand routing protocol as nodes create routing table only when they are needed. These are best suited for larger networks. Example of reactive routing protocol is AODV, DSR, etc. Hybrid routing protocols: This protocol combines the advantage of proactive and reactive protocols. An example of Hybrid routing protocol is ZRP [12].

In this paper, the method of controlling congestion in the CGSR protocol using AODV is implemented. AODV is a reactive protocol which establishes connection only on demand. CGSR is a proactive protocol that contains routing table information before the connection is established. The proposed method combines the advantages of the pro-active and reactive protocols to enhance the performance of the network [9]. AODV is used as an algorithm to ICGSR which leads to the improvisation of performance of the CGSR protocol. By using AODV algorithm, CGSR establishes path only when demanded by the network according to AODV method, which in-turn reduces congestion in terms of routing overhead, end-end delay, and packet delivery ratio.

2   Recent Works

Routing process can be controlled by allowing hops with maximum possible distances on the received signal strength in a route based protocol at each node by using topology-aware and power-aware AOMDV based on maximum transmission range [12]. A node waits for acknowledgment for the threshold period. If the acknowledgment is not received within the threshold period then the node broadcasts again to select an alternate path. By using this method, congestion can be controlled in AODV [4]. To achieve better route cost and smaller delay, piggyback mechanism and weighted neighbor stability algorithm respectively can be used [8]. Better route maintenance can be achieved by providing alternate routes by an extra broadcast in case of route failure. These alternate routes reduce the number of control packets in AODV for different time pause and source [1].

The clustering and queuing techniques strengthen the mobility effect and ad-hoc nature by controlling traffic congestion [10]. Improving wireless cluster algorithm and other similar algorithms with some analytical models and clustering schemes to get a better quality of clustering [7]. A dynamic list of nodes is maintained which helps in forming in clusters and electing the cluster-head faster [5]. If the frequency of cluster-head changing by applying a different range of speed and pause time for all mobile nodes is reduced, then the performance of cluster-based routing will improve and makes the cluster more stable [11].

3 Methodology

The objective of this paper is to control the congestion of the CGSR protocol by using an enhanced AODV algorithm. In CGSR, the rerouting process occurs often and often because of the mobility of the nodes [1]. This leads to change in cluster-head which in-turn maximizes the traffic congestion in the network. This proposed method minimizes the congestion which is proved by comparing some performance metrics like end-end delay, routing overhead, and packet delivery ratio with the existing methods [9].

3.1   Improved Cluster-Head Gateway Switch Routing Protocol (ICGSR)

In Improved cluster-head gateway switch routing protocol (ICGSR) the path discovery is implemented using the Ad-hoc On-demand Distance vector (AODV) algorithm. CGSR being a proactive routing protocol it has a routing table set up before the path establishment. CGSR gives higher performance when compared to other protocols because of gateway-switch type traffic redirections [5]. Clusters provide a powerful membership of nodes for connectivity. Each node transmits data based on its sequence number. This is to assure the updates and new path discovery. Each node maintains tree topology of cluster members that narrate the next hop to find the cluster head of the destination. In clustering the nodes are assigned with groups based on distance, energy etc. there are three types of nodes in the CGSR protocol [11].

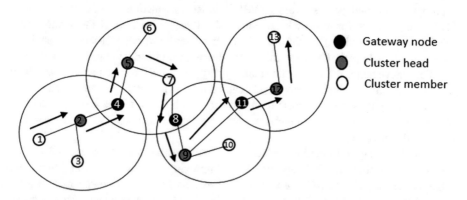

Fig. 1. CGSR routing from node 1 to 13

Cluster head: These are like the brain of the network and it works as a leader to nodes present in their cluster. Every node communicates with other nodes through the cluster-head. Its functions are inter-clustering, data packet forwarding, intra-clustering, and maintenance of the whole network. This node contains routing table information of its cluster members. The algorithm for the cluster-head selection method is shown in Algorithm 1.

Gateway Node: It acts as a communication medium between one clusters to another. Data packets transmission between one nodes to another is done via this node.

```
// convey the network 'n' number of nodes
// 'k' is a variable
//Compare the mobile nodes based on capacity of node
BEGIN {
Compare the mobile nodes and assign rank R_k
}
//compare the distances of all mobile nodes from
sink {
Allocate the rank D_k for each mobile node
}
//assign rank for each mobile nodes based on energy
level {
Assign rank based on energy level E_k
}
//evaluate the weight of mobile nodes {
W_k = R_k + D_k +E_k
}
//select cluster head based on the weights {
Assign cluster head with highest weight to each
clusters
}
END
```

Algorithm. 1. Cluster-head selection method

Cluster members: These are the common nodes present in the cluster. It doesn't have any inter-cluster links.

ICGSR uses DSDV as an underlying protocol and LCC algorithm for the clustering process. Most of the clustering process in many protocols is periodical. Re-clustering happens often and often to know some specific features of cluster-heads. LCC algorithm has two steps in the clustering process that is cluster formation and cluster maintenance.

Cluster member finds the nearest cluster-head along the path to the destination according to the routing tree and the cluster member tree on receiving a packet as shown in Fig. 1. Source Cluster-head searches for destination cluster-head and also finds the next hop in the routing table for the destination path. Assigned cluster member transmits a packet to the destination cluster head [7]. If the source nodes move out of its cluster head range, then it will rediscover the route for the destination node and then transmit the data packet [6].

3.2   Enhanced AODV Mechanism

Ad-hoc On-demand Distance Vector (AODV) protocol is a reactive protocol which establishes connection only when demanded by the network. It uses a route discovery and route maintenance process as in DSR and uses periodic beacons and sequence number for the transmission of data packets as in DSDV [13]. In AODV, when a node needs route information to a destination it broadcasts a Route Request message to its neighbors with a last known sequence number [4]. This message is then flooded through the network until it reaches the node which has information about the route to the destination. This node generates a Route Reply message which contains the sequence number and number of nodes to the destination as shown in Fig. 2. Each node that crosses the Route Request then participates in forwarding this Route Reply message towards the originator of the Route.

Request message and makes a 'forward route' entry in its routing tables pointing to the destination node. To keep the routing table updated, each node periodically sends a 'Hello" message once in a second. When a link goes down, the upstream node is notified through an "Unsolicited Route Reply" containing an infinity metrics for the destination. Whenever the route fails the RREP discards the old path and shortens the new path to the destination [2]. The AODV algorithm is shown in Algorithm 2. By following this method, the cluster-head establishes path whenever it is demanded by the network. This in turn reduces traffic congestion and gives better performance in the network.

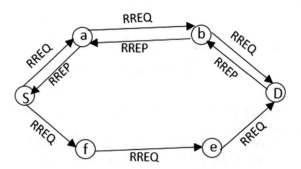

Fig. 2. Route discovery method in AODV

```
// Broadcast RREQ messages
1. BEGIN
2. Send RREQ (nodeP)
3. {SET sqn#_rq=1, Hop_Count_rq = 0
4. BROADCAST RREQ to Neighbors}
5. END
// Handling RREQ Messages
6. BEGIN
7. Delay time = ((current tie - timestamp)/
(Hop_Count))
8. IF (delay time > threshold) THEN
9. DISCARD THE RREQ message
10. END IF
11. END
// Broadcast RREP messages
12. BEGIN
13. SendRREP (nodeP, RREQ)
14. {SET sqn#_rp = sqn#_rq, Hop_Count_rp = 0
15. BROADCAST RREP to Neighbors}
16. END
// Handling RREP Messages
17. BEGIN
18. Receive RREP (RREP, nodeP)
19. {{IF (nodeP == source) UPDATE Route, DATA
20. IF (node x 1 = Destination) {
21. IF (sqn#_rp> sqn#_tb) OR (sqn#_rp == sqn#_tb) AND
(Hop_Count_rp < Hop_Count_tb) UPDATE, FORWARD RREP
22. ELSE FORWARD RREP, UPDATE RREP: sqn#_tb = sqn#_rp,
Hop_Count_tb = Hop_Count_rq+1}
23. UPDATE Route, Sqn#_tb = sqn#_rp, Hop_Count_tb =
Hop_Count_rp}
24. END IF
25. END IF
26. END
```

Algorithm 2. Modified AODV algorithm

4   Simulation Results and Discussions

The proposed work is simulated using the tool Network Simulator 2 (NS2). The obtained results of ICGSR are compared with CGSR, LEACH, ECGSR, and DUCS protocols. The simulation parameters are shown in Table 1. The performance of the protocols in terms of end-end delay, packet delivery ratio, and routing overhead are compared and plotted in graphs. NS-2 is an open-source discrete event simulator used for simulation of both wired and wireless network designs. The back-end programs in NS2 is written in C++ and the front-end program is written in Tool Command Language (TCL). When we run a TCL program, it generates two types of files namely, Network Animator (NAM) file and Trace file. These files define the behavior of the nodes and keep the record of connection type, number of packets send and received, number of hops between any two nodes, etc. at any instance of time.

Table 1. Simulation Parameters

Parameters	Values
Simulation area	1000 m^2
Number of nodes	100,150,200,250
Size of packets	5000
Simulation time	175 ms
Packet rate	250 kb
Initial energy	100 J

4.1   Routing Overhead

In MANETs, link breakage occurs frequently due to the mobility of nodes, which leads to frequent route discovery. In MANETs, nodes move in the selected path for packet transmission and that might get interrupted [6]. This, in turn, leads to the rerouting of the route to the destination whenever a node moves from one cluster to another. This increases the traffic on the channel which leads to routing overhead.

The routing overhead is defined as the total number of routing packets transmitted through a channel at one instance of time. It is measured using Eq. 1.

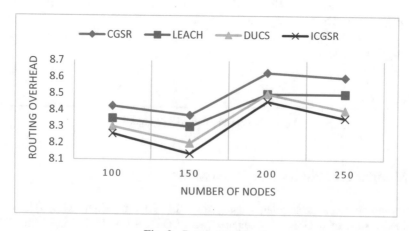

Fig. 3. Routing overhead

$$\text{Routing overhead} = PL + SN + TR \tag{1}$$

PL indicates Packets lost before reaching the destination, SN indicates Messages contains serial numbers, and TR indicates Triggered messages. The simulation result of the routing overhead is shown in the Fig. 3. The routing overhead of the proposed protocol is less by when compared to the other protocols.

4.2 End-End Delay

The end-end delay is defined as the time taken by the packet to travel from source to destination across a network. Since the nodes are mobile in MANETS, it leads to frequent route discovery & queue in the transmission of a data packet and re-routing of the data packets [3]. As a result, a delay is developed in transmitting the packet from source to destination. This delay in transmission is called end-end delay. This end-end delay in a network is calculated using Eq. 2.

$$\text{End - End Delay} = \frac{\sum(\text{arrival time} - \text{transmitted time})}{\sum \text{Number of connections}} \tag{2}$$

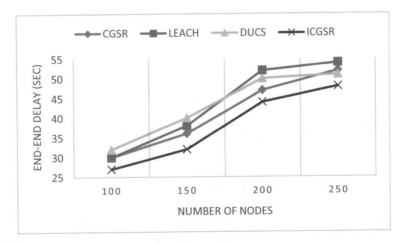

Fig. 4. End-End Delay

As the end-end delay decreases the throughput of the network increases, this leads to better performance of the protocol. The end-end delay of the ICGSR protocol is shown in the Fig. 4. The end-end delay of the ICGSR is less than other protocols. As a result, network performance gets better.

4.3   Packet Delivery Ratio

It is the ratio of the number of the data packets sent to the number of data packets received by the destination as shown in Eq. 3. PDR also represents the minimization of the packet drop in the network.

$$\text{Packet Delivery Ratio} = \frac{\sum(\textbf{\textit{Number of data packets received}})}{\sum \textbf{\textit{number of data packets sent}}} \tag{3}$$

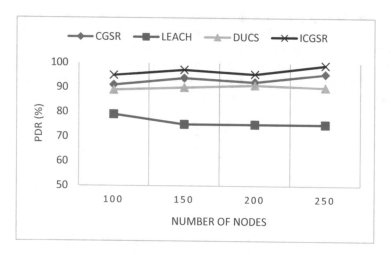

Fig. 5. Packet delivery ratio

The packet delivery ratio of ICGSR compared with other protocols is shown in Fig. 5. The result shows ICGSR PDR is better than other protocols.

5   Conclusion and Future Work

In this paper, an Improved CGSR protocol is implemented using an enhanced AODV algorithm to minimize the congestion in the network. AODV algorithm is modified in such a manner that, it helps in path establishment in CGSR protocol and minimizes the link failure, which leads to minimization of traffic congestion. Hence, the routing overhead and end-end delay decreases with increasing the packet delivery ration compared to the other clustering protocols like CGSR, LEACH, and DUCS.

Since two protocols are used in the network the energy consumption is pretty much high. Hence, the future scope of this work is minimizing the energy consumption and calculate the throughput for the same by modifying the AODV algorithm.

References

1. Singh, A.P., Bhukya, S.N.: HP-AODV: an enhanced AODV routing protocol for high mobility MANETs. In: Advanced Computing and Communication Techniques for High Performance Applications, pp. 0975–8887 (2014)
2. Anju, S.S.: Modified AODV for congestion control in MANET. Int. J. Comput. Sci. Mob. Comput. **4**(6), 984–1001 (2015)
3. Abu-Ein, A., Nader, J.: An enhanced AODV routing protocol for MANETs. Int. J. Comput. Sci. Issues **11**(1), 54 (2014)
4. Heena, Goyal, D.: Congestion control using enhanced AODV routing mechanism in MANET. Int. J. Comput. Sci. Mob. Comput. **3**(8), 555–565 (2014)
5. Sathiamoorthy, J., Ramakrishnan, B.: Energy and delay efficient dynamic cluster formation using hybrid AGA with FACO in EAACK MANETs. Wirel. Netw. **23**, 371–385 (2017)
6. Devarajan, K., Padmathilagam, V.: An enhanced cluster-head gateway switch routing protocol (ECGSR) for congestion control using AODV algorithm in MANET. Int. J. Comput. Appl. **123**(3), 37–42 (2015)
7. Aissaa, M., Belghith, A.: Quality of Clustering in mobile Ad Hoc networks. In: 5th International Conference on Ambient Systems, Network and Technologies, pp. 245–252 (2014)
8. Mungare, S.A., Patil, R.S.: Review of enhanced AODV routing protocol for mobile adhoc wireless networks. Int. J. Adv. Res. Electron. Commun. Eng. **1**(5), 1–6 (2012)
9. Phate, N., Saxena, M., Rizvi, M.A.: Minimizing congestion and improved QoS of AODV using clustering in MANET. In: IEEE International Conference on Recent Advances and Innovations in Engineering (2014)
10. Sharma, S., Jindal, D., Agarwal, R.: An approach for congestion control in mobile ad hoc networks. Int. J. Emerg. Trends Eng. Dev. **7**(3), 2249–6149 (2017)
11. Pathak, S., Dutta, N., Jain, S.: An improved cluster maintenance scheme for MANET. In: International Conference on Advances in Computing, Communications and Informatics (2014)
12. Lalitha, V., Rajesh, R.S.: Power aware and topology aware ad-hoc on demand multipath distance vector routing for MANET. In: IEEE 8th International Conference on Intelligent Systems and Control (2015)
13. Rishiwal, V., Yadav, M., Singh, O., Yadav, M.: Analysis of ADHOC routing protocols: a retroscopic view. In: International Conference on Advances in Computing, Communication, & Automation (2016)

Improvement in Massive MIMO System Performance Through Caching Technique

Ankita Singh[✉], Saurabh M. Patel, and Kirit Kumar R. Bhatt

E&C Department, SVIT, Vasad, Gujarat, India
24mayankita@gmail.com

Abstract. In this work we study the system performance of massive multiple input multiple output (MIMO) system by applying the wireless technique that is caching. As we know the current mobile cellular network are getting better towards 5G wireless network to withstand the huge rise of connected devices and data peckish applications of mobile users and we know this very well that spectrums are limited, so we have to improve the system performance for meeting the higher demand in wireless system. Among the possible solutions, caching at base station is a encouraging plan that can provide improved quality of service. This work is target for Cache-aided system which enhances the system performance by decreasing the congestion of network. Here we show the massive MIMO system with cache outperforms by reducing the outage probability.

Keywords: Multiple input multiple output · Cache-aided system · Outage probability

1 Introduction

As per current scenario, the new wireless devices is become the fact of appearing everywhere or of being common. There is rapid increase of demand of smartphones which is easily accessible due to which social networking applications or bandwidth intensive applications increase rapidly. Therefore due to proliferation the network load increases exponentially as wireless data traffic is increases tremendously. Through rich media application, video streaming contribute maximum in data traffic as it account for almost 50% of mobile data traffic which is expected to increase tremendously in coming future. Also by the survey, the social networking also contribute second largest traffic volume. Therefore it is high demand to evolve the traditional cellular networks towards the next generation network which is referred as 5G networks for satisfying the high demand of data or fulfill the data crunch. So this data tsunami need to develop mobile operators in more advanced way to increase the coverage area of network and to boost the capacity of network more efficiently without much altering the cost and efficiency of system. As per current demand it is required to achieve better data transmission rate or provide maximum content to user in less time. So among all the demanding efforts caching the content of users' at the end of transmitter of network emerge as a possible solution for satisfying the mobile data tsunami [1, 2].

© Springer Nature Switzerland AG 2020
P. Karrupusamy et al. (Eds.): ICSCN 2019, LNDECT 39, pp. 704–710, 2020.
https://doi.org/10.1007/978-3-030-34515-0_74

By this promising content caching strategy, high demanding files are stored (cache) in the midway of servers or routers or base station (BSs) of antennas due to this the demands from users for same contents can be fulfilled easily without any delay from the servers and hence the superfluous traffic can be eliminated significantly. Hence in 5G mobile networks caching emerge as a potential technique for achieving the better and fast network condition [3, 4].

As we know that the high data traffic is due to the facsimile downloads of a few popular files (e.g. viral videos or popular news) with hefty size. Therefore researchers and engineers have been studying or searching operative way to reduce the facsimile content transmission by putting into use storage related techniques referred as caching at the edge of wireless network namely at the level.

For reducing the multimedia traffic, the knowledge of caching at the transmitter end of network, namely at the level of base stations and user terminals has been drawing special attention in many works which includes edge caching, femto-caching [5] and proactive caching [6] which has capabilities to alleviate the backhaul loads and satisfy the users' demand.

The key restriction in the backhaul is the capacity and delays. Many literature works exist which focus on to mitigate the backhaul crunch by reducing the backhaul payload which is quibbling issue in 5G networks. Caching at the edge of BSs before transmission can significantly reduce the demand for backhaul resources by network. However for caching the memory of cache size is limited at the BSs. Therefore it is not realistic to store all the files at the BSs, as it is resource consuming, due to this cache placement approach must be carefully design to set of scales among the resource consumption and data transmission improvement. Making use of information of popular content for storing the data at the edge of network is utilized in the design of caching scheme would therefore develop a better system and conclude at improved system performance by analyzing the content popularity and demand of the user equipment which is requested most probably in peak duration, as the demand of the user depends on content which is popular or viral and the peak time of a day [7].

2 Edge Caching

2.1 System Model

In this system, we study a massive MIMO system having L number of transmitter and receiver pairs where base stations (BS) employ a uniform linear array. A ULA pattern having N_t antenna elements, which are evenly placed in a straight line with a gap of d. For simplicity, here we adopt a homogeneous network where transmitter (having uniform linear array) have N_t antennas and all receivers (users) consist of N_r antennas. From transmitter k to its paired receiver k, the data streams are transmitted, these number of independent data streams is symbolized by d_k, with $d_k \leq \min(N_t; Nr)$.

There is a channel between a single-antenna user and BS which can be characterized from the antenna array theory as [8].

$$\mathbf{h_k} = \int_{\theta \in A_k} \alpha_k(\theta) a(\theta) d\theta$$

Where θ represent the direction of arrival (DOA) of each ray which is inside the incident signal and A_k is the angular spread (AS) of the incident signal from user-k. Here $\alpha_k(\theta)$ is the complex gain of the incident ray at DOA θ and $a(\theta)$ is the array manifold vector (AMV), whose expression is dependent on the array structure. When a uniform linear array is put into use,

$$\mathbf{a}(\theta) = \left[1, e^{j\frac{2\pi d}{\lambda} \sin \theta}, \ldots \ldots, e^{j\frac{2\pi d}{\lambda}(M-1)\sin\theta}\right]^{\mathbf{T}}$$

In ULA, the inter antenna spacing i.e. d has a linear relationship with λ (i.e. $d = \lambda\mu$) and λ is referred as carrier wavelength. We assume that the users are unsystematically spread inside the cell, and hence azimuth angles track a uniform distribution in the [0, 2π] interval. Then the probability density function (p.d.f.) of the users' azimuth angles ϕ_k ($\forall k = 1 \ldots, K$) can be describe as

$$f(\phi_k) = \frac{1}{2\pi}.$$

Here ϕk represent the azimuth angle of the AoD of the k-th user.

2.2   Performance Analysis

The system is said to be in outage, when the selected transmission rate R is larger than the maximal rate for reliable communication R_{max} for the given block, the system is said to be in outage. While the outage probability is defined for a targeted transmission rate, the capacity with outage is defined for a targeted outage probability p_o. It is the maximum transmission rate for which the outage probability is smaller than p_o.

As we observed the downloading of same content multiple times or streaming of some popular content on high demand create the backhaul crunch. This backhaul crunch can be effectively reduced by increasing the transmission time and reducing the outage event by applying the caching at the edge of network. We observed that the effectively cached memory provide better cache placement. Effective cost of storage device open up new opportunities of accessing the storage memory more easily in wireless communication. The optimal cache placement is effecting in alleviating the heavy traffic from network. For reducing the backhaul crunch and latency, we utilizes wireless local caching pre-fetch the files on the basis of its popularity profile in storage.

As we describe in above section, we take two type of condition for analyzing the outage performance. For this we divide the cached files in two sets on the basis of its popularity profile. First distribution is based on the high popularity profile of files (i.e. high demand by user equipment) which is represented by F and second distribution is based on less popular files which are represented by (K − F). K represents the all files requested by user equipment.

As we know there is a limitation of cache resources as the cache storage memory is finite. Here the base station can cache (store) up to C files which must be less than or equal to all set of files i.e. K, which is requested by the user equipment. The popular files has given priorities in transmission, therefore the popular files is to be cached by all local base station as it is most likely to be requested by many user equipment by multiple times. Due to the limitation in cache storage resource, we do not cache less popular files to all the base station. Instead of this, less popular files are cached at only part of the base station. Therefore the part of station is divided into L clusters which are denoted as $L = \frac{(K-F)}{(C-F)}$. Less popular files should not be ignored as it is demanded by users but not multiple times. Files cached in the l-th cluster $(1 \leq l \leq L)$ form the set $\{K - F\}_l$.

On request, all base station serves all user equipment therefore the transmission outage probability for the popular file [9] is denoted by P_1^{out} i.e.

$$P_1^{out} = P_e^{FC}(e_o) \sum_{c=1}^{F} f_c \tag{1}$$

Where fc denotes popularity profile of c-th file and $P_e^{FC}(e_o)$ denotes the outage probability regarding the bit error rate (BER).

By Zipf Probabilities,

$$\sum_{c=1}^{F} f_c \approx \frac{F^{(1-\alpha)} - 1}{K^{(1-\alpha)} - 1}. \tag{2}$$

Here α is the popularity parameter defined by Zipf distribution [10].
$P_e^{FC}(e_o)$ can be described as,

$$P_e^{FC}(e_o) \approx Pr\left\{\frac{1}{12}exp\left(-\frac{c_2 c_1^2}{2}\frac{W}{R}SINRi\right) + \frac{1}{4}exp\left(-\frac{2c_2 c_1^2}{3}\frac{W}{R}SINRi\right) > e_o\right\} \tag{3}$$

Here W is bandwidth and R is code rate c_1 and c_2 [11] are constants which is dependent on understanding of modulation and coding scheme. For this system for 64 QAM we take $c_1 = 1$ and $c_2 = \frac{3}{4}$ [12]. R is code rate and W is bandwidth. For ideal condition we take $\frac{W}{R} = 1$.

Now next we consider the performance of less popular file i.e. (K − F) and consider its outage performance. Here for less popular files transmission $\overline{P}_2^{out}$ is an average outage probability which is represented as,

$$\overline{P}_2^{out} = \overline{P}_e^{PC}(e_o) \sum_{c=F+1}^{C} f_c. \tag{4}$$

Where C is the files cached by the base station in storage unit (cache). Here $\overline{P}_e^{PC}(e_o)$ can be expressed as,

$$\overline{P}_e^{PC}(e_o) = \frac{1}{L}\sum_{l=1}^{L} P_{e,l}^{PC}(e_o). \tag{5}$$

Where $P_{e,l}^{PC}(e_o)$ is,

$$P_{e,l}^{PC}(e_o) \approx Pr\left\{\frac{1}{12}exp\left(-\frac{c_2c_1^2}{2}\frac{W}{R}SINR_{i,l}\right) + \frac{1}{4}exp\left(-\frac{2c_2c_1^2}{3}\frac{W}{R}SINR_{i,l}\right) > e_o\right\} \tag{6}$$

Here $P_{e,l}^{PC}(e_o)$ denotes the outage probability regarding the BER (bit error rate).

Outage performance of both popular and less popular file is explained by the average of both the probability equation of popular and less popular set which is better for the system than other performances. For our system, we take K = 500 and F = 400.

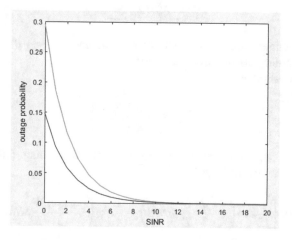

Fig. 1. A Outage performance versus SINR

2.3   Conclusion

In MPC strategy based on to cache the most popular content with all storage capacity according to the popularity profile of the content. In Fig. 1, green line which has higher outage probility shows the outage performance of MPC as it shows that outage probability reduces with signal to noise ratio. The outage probability performance is equal to the P_1^{out}. The blue line which has lower outage probability than another one shows the outage performance of average of both popular and less popular files, which is reduces with signal to noise ratio.

Here it shows average caching scheme which include both popular and less popular file is much better than caching of only most popular file as it has lower outage probability. Cache placement minimizes the outage probability with increase of the SINR (dB).

Caching schemes for improving the traffic in massive MIMO with uniform linear array is utilised. As we observed the downloading of same content multiple times or streaming of some popular content on high demand create the backhaul crunch. This backhaul crunch can be effectively reduced by reducing the outage event by applying caching at the edge of network. We observed that the effectively cached memory provide better cache placement. We can divide cached files in two file sets according to their popularity profile i.e. popular file(cached at all base station) and another is less popular file (cached at only part of the base station). By this content placement we utilise the storage more efficiently. With cache-enabled system, outage probability decreases with increase of signal to noise ratio.

In future work, we can extend the work by utilising the multiple sub array system [13] and work on the different parameters [14] of caching including hybrid cache placement [15].

References

1. Ylitalo, J., Juntti, M., Matsumoto, T.: Overview and recent challenges of MIMO systems. IEEE Commun. Mag. **50**(2), 4–9 (2003)
2. Larsson, E.G., Edfors, O., Tufvesson, F., Marzetta, T.L.: Massive MIMO for next generation wireless systems. IEEE Commun. Mag. **52**(2), 186–195 (2014)
3. Andrews, J.G., Buzi, S., Choi, W., Hanly, S., Lozano, A., Soong, A.C., Zhang, J.C.: What will 5G be?. arXiv preprint: arXiv:1405.2957 (2014)
4. Wang, X., et al.: Cache in the air: exploiting content caching and delivery techniques for 5G systems. IEEE Commun. Mag. **52**(2), 131–139 (2014)
5. Shanmugam, K., Golrezaei, N., Dimakis, A.G., Molisch, A.F., Caire, G.: FemtoCaching: wireless content delivery through distributed caching helpers. IEEE Trans. Inf. Theory **59** (12), 8402–8413 (2013)
6. Bastug, E., Bennis, M., Debbah, M.: Living on the edge: the role of proactive caching in 5G wireless networks. IEEE Commun. Mag. **52**(8), 82–89 (2014)
7. Ji, M., Caire, G., Molisch, A.: Wireless device-to-device caching networks: basic principles and system performance. IEEE JSAC **34**(1), 176–189 (2016)
8. Tatar, A., et al.: A survey on predicting the popularity of web content. J. Internet Serv. Appl. **5**(1), 1–20 (2014)
9. Zhou, F., Fan, L., Wang, N., Luo, G., Tang, J., Chen, W.: A cache aided communication scheme for downlink coordinated multipoint transmission. IEEE Access **6**, 1416–1427 (2018)
10. Breslau, L., Cao, P., Fan, L., Phillips, G., Shenker, S.: Web caching and Zipf-like distributions: evidence and implications. In: Proceedings of IEEE INFOCOM, New York, NY, USA, pp. 126–134 (1999)
11. Mahmod, A., Jäntti, R.: Packet error rate analysis of uncoded schemes in block-fading channels using extreme value theory. IEEE Commun. Lett. **21**(1), 208–211 (2017)
12. Gallager, R.G.: Error probability for orthogonal signal sets in Principles of Digital Communication, 1st edn., pp. 294–298. Cambridge University Press, New York, NY, USA (2008)

13. Tan, W., Smith, P.J., Suraweera, H.A., Matthaou, M., Jin, S.: Spectral efficiency of multiuser mm Wave systems with uniform linear arrays and MRT. In: Proceedings of IEEE VTC-Spring, pp. 150–156 (2016)
14. Liu, D., Chen, B., Yang, C., Molisch, A.F.: Caching at the wireless edge - design aspects, challenges and future directions. IEEE Commun. Mag. **54**(9), 22–28 (2016)
15. Zheng, G., Suraweera, H.A., Krikidis, I.: Optimization of hybrid cache placement for collaborative relaying. IEEE Commun. Lett. **21**(2), 442–445 (2017)

Distributed Energy Efficiency for Fault Tolerance Using Cuckoo Search in Wireless Sensor Networks

N. Priya[✉] and P. B. Pankajavalli

Department of Computer Science, Bharathiar University, Coimbatore, India
npriyaresearch@gmail.com, pankajavalli@buc.edu

Abstract. A Wireless Sensor Network (WSN) is a self-organized network, which consists of thousands of in-expensive and low powered devices and these devices are highly energy constrained. Therefore, energy plays a vital role in communication between the sensors nodes as the livelihood of the nodes may get affected with lack of energy. Hence energy efficiency is mandatory for maintaining the network longevity. The proposed Distributed Energy Efficient (DEE) fault tolerance clustering using cuckoo search (CS) is having the ability of tolerating the failure of sensor nodes that will ensure the uninterrupted communication. The performance evaluation of the DEE with the existing algorithms such as Particle swarm optimization (PSO), Bee colony optimization (BCO), and Fire fly optimization (FFO) in terms of alive node identification, dead node identification and network life time. The comparison result shows that DEE outperforms the existing algorithms.

Keywords: Wireless sensor network · Cuckoo search · Energy efficiency · Fault tolerance

1 Introduction

Wireless Sensor Network (WSN) is a collection of sensor nodes which convenes information from physical and ecological conditions such as temperature, motion, vibration finally the convene information's are forwarded to the sink node [1]. Particularly sensor nodes are armed with batteries with limited amount of energy. Renewing or swapping those batteries with other batteries is impossible and fixed energy can also be support the network only with particular amount of time. Therefore, minimizing the energy for information transmission becomes ultimate consideration when sensor nodes are positioned in an inaccessible circumstance which impacts the lifetime of a network.

Clustering is one of the approaches that are used to evenly distribute energy to each and every sensor node for extending the lifetime of a network. In clustering, sensor nodes can be partitioned into several groups of nodes called clusters; in each cluster it can have a single Cluster Head (CH). CH gathers information from member of clusters, process the information and forward it towards the sink node [2].

© Springer Nature Switzerland AG 2020
P. Karrupusamy et al. (Eds.): ICSCN 2019, LNDECT 39, pp. 711–719, 2020.
https://doi.org/10.1007/978-3-030-34515-0_75

Further, using huge amount of sensor nodes also might leads to failure [3]. Sensor nodes may get failure due to depletion of energy, failure of hardware, communication link errors and malicious attacks [4]. If sensor nodes are positioned in the harsh environment, sensors may get failure due to energy depletion. Hence, any good clustering algorithm should have the capability to tolerate the fault [5].

Discovering the optimal routes and computational difficulty of cluster formation for large-scale WSNs is extremely high [6]. Therefore, a nature inspired meta-heuristics algorithms are used to resolve clustering problems [7]. Cuckoo Search (CS) is one of the effective meta-heuristic algorithms used to solve the clustering problem. It can satisfy the global convergence necessities and thus has guaranteed global convergence belongings [8]. The other meta-heuristic algorithms which solve the optimization problems are Particle swarm optimization (PSO), Bee colony optimization (BCO), and Fire Fly Optimization (FFO). However, CS algorithm does global search and local search which are controlled by switching or discovery probability and global search uses Levy flights path identification strategy method instead of standard random walk approaches. Because Levy flights method can have boundless mean and variance, CS can discover the search space more efficiently than algorithms using standard Gaussian procedures [9].

The proposed DEE algorithm addresses the key issue like energy efficiency and energy balanced clustering for full heterogeneous WSN is proposed and it applies to fault tolerance for sensor nodes. The rest of the paper is organized as follows:

A comprehensive survey of energy efficiency techniques in WSNs is presented in Sect. 2. The proposed DEE is presented in Sect. 3. The simulation results are presented in the Sect. 4 and finally Sect. 5 concludes the work.

2 Related Works

Now a days, energy efficiency in WSN is an operational sustainability concern for all the domains of computer research. Generally, energy efficiency and energy balancing are two important parameters for the data communication to be effective. This section briefly review the existing energy efficiency approaches in WSN as well as the optimization algorithms that have been extensively studied to enhance the process of energy efficiency in WSN.

There are plenty of clustering algorithms developed for WSNs. Kaur et al., has proposed an energy efficient protocol in hybrid meta-heuristic optimization, which is a tree-based routing protocols in contributing techniques such as hybrid Ant colony Optimization (ACO) and PSO. This hybrid ACO, PSO protocol does energy efficient clustering for data aggregation [10]. Azharuddin et al. proposed a distributed fault-tolerant clustering algorithm for WSN. For the cluster formation the algorithm uses a cost function of the CHs. However the algorithm cannot deal with the sensor nodes residual energy, as well as energy balancing of CHs for balancing the energy they have used the PSO-based approach. For both clustering and routing algorithms authors arise a multi objective fitness function called Particle-encoding scheme and the algorithm is only capable for tolerating the failure of CHs. [11]. Ali Ghiasiana et al. proposed a cuckoo based clustering algorithm for WSN. Three variants of cuckoo algorithm in which the energy of path length is considered as one important factor in CH selection.

To prevent quick energy resolve of the CHs, the role of CH should be circulated among different nodes. However, the algorithms do not consider the fault tolerance issue for the sensor nodes which is a serious drawback [12].

Yang et al. and Anuja et al. addresses the problem of hatching possibility, i.e., cuckoos who put their eggs in a communal roost they may detach other birds eggs to extend the hatching possibility of their own eggs. Rather an amount of species involves in oblige brood parasitism by laying their eggs in the roosts of other host birds [13, 14]. Das et al. proposed and clustering and routing for WSN using cuckoo search. In their approach the random walk was computed in which the step-lengths were according to a heavy-tailed probability distribution. The detachment from the beginning of the random walk inclines to a firm distribution after huge number of steps [15].

3   Methodology

3.1   Cluster Setup

The sensor nodes form together in order to complete the task is called as cluster. Based on the size, the cluster is classified into two types: (i) even (ii) uneven. Inter cluster and intra cluster are the two types of techniques involved in cluster formation. Intra cluster communicates between cluster member and CH and inter cluster communicates between CH and Base Station (BS). Cluster approach is used to increase the energy consumption and network lifetime. The other issue to be discoursed is, if a single CH is proceeding at all times as the head node, it will die because of losing its energy level after certain amount of time. In this situation, a new sensor node is mandatory to be selected as the CH. The choice of choosing a new CH will be done by the DEE algorithm.

The CH selection depends on residual energy and energy consumption rate. CHs act as a gateway among cluster members and BS. Here CH selection is based on maximum energy of a sensor node and it senses the data efficiently and transmits to the BS. Normal node energy consumption depends on the distance among the node from the BS. Mostly, entire clustering procedures involve four main steps and two parts. The four steps are: selection of CH, formation of cluster, data aggregation, and data communication. The two parts are the setup and steady state. In the preliminary stage of each and every setup phase, all sensor nodes send the data directly to the sink node about their present energy level and positions. Based on the obtained data, the sink node then computes the average energy level of entire nodes. In every single round, only the higher energy sensor nodes are selected as CH of that certain cluster. This procedure guarantees that only nodes with an acceptable energy level are nominated as CH. Here the sensor node energy is considered as $E(N)$ and N represents the Nth sensor node and N lies between 300 to 700 ($Min(N) = 300$ and $Max(N) = 700$).

If $E(N)$ = initial energy (α), then the sensor node will be considered as a normal sensor node.

$$(i.e) \qquad N_{norm} = E(N_{norm}) = \alpha \qquad (1)$$

The sensor node will be considered as an advanced sensor node if the energy of the sensor node having the initial energy plus the residual energy (β). The advanced sensor node is represented as follows.

$N_{adv} = E(N_{adv}) = $ Initial energy + Residual Energy

$$(i.e) \qquad E(N_{adv}) = \alpha + \beta \tag{2}$$

CS uses levy flights strategy for path identification and a new solution are obtained according to the Eq. (3).

$$x_i(t+1) = E(N_i) \otimes (m \oplus l) \tag{3}$$

Where m is the packet size and l provide randomness in gaining new solutions. l is a period of delta - gamma steady functions, which takes up unlike other disseminations based on the value of delta and gamma.

CHs distribute a packet based on the request to attract other normal sensor nodes in the network. Distance of that particular CH with BS and fitness_number value can be comprised with the packets which are going to broadcast. In this case sensor nodes must have the knowledge about CHs and which are non-CH nodes in the network. Subsequently the normal sensor nodes can treat CH sensor nodes as the nests and they try to pick a foremost CH node for laying their eggs, then the sensor node will send JOIN_REQUEST for an elected CH. JOIN_REQUEST is preserved as the cuckoo egg. The finest CH is elected based on the distance (d) and energy level of the neighbor sensor nodes and fitness function.

$$d(CH, N_i) = |p(CH) - p(N_i)|$$

$$\text{Shortest Path} = Min\{d(CH, N_1), d(CH, N_2), \ldots, d(CH, N_i)\} = d(CH, N_s)$$

$$(i.e) \qquad \text{finest CH} = N_s \tag{4}$$

$$\text{Fitness} = \text{fitness\_number}/[(\delta * N_{BS}) + (\gamma * d\ (CH, N_{BS}))] \tag{5}$$

Where *Fitness* is the fitness of a CH (nest), calculated on the basis on the fitness_number, based on its remaining energy in that exact round an integer value is allotted to a CH. δ and γ are the unchanging factors which portraying the desirability of cuckoo bird on the way to a specific roost, distance among the CH and the sink node is signified as N_{BS}, and distance among the normal node and the CH node is signified as d (CH, N_{BS}). $d(CH, N_{BS})$ is computed using Eq. (4). With the appropriate values predetermined for δ and γ, entirely the normal sensor nodes choose finest high-quality CH, contingent on any CH obtains uppermost value for the objective function *Fitness* using Eq. (5) the normal node transmits JOIN_REQUEST for its elected CH node. CHs on the response of the invocation, and produces a random possibility of whether to admit or discard the invocation based on the threshold value which varies for each and every round and whether the refusal possibility cross the threshold (t) or not. This case can be plotted to the refusal or affirmation of cuckoo eggs by the host bird with some possibility say p. In this DEE work, a JOIN_REQUEST is examine only if p > t (where the threshold value is considered as t), otherwise the node is disallowed by CH. The whole nodes in the conspiracy, separate itself into cluster formation, where the communication of detected data occurs among the nodes and the sink node. Pseudo code for cluster formation using CS algorithm has shown in Table 2.

When the energy level of the sensor node does not receive heartbeat message within a particular time limit the energy level of the node is considered to be low which leads to failure of sensor node. This consequence is resumed by swapping another node with the low energy node by using dynamic nodes swapping which replaces the lower energy with high energy node to avoid sensor node failure. The replaced sensor node IDs are communicated with the neighboring nodes. The information retained in the sensor nodes which is swapped will not be lost in the system which results in system efficiency improvement (Figs. 1 and 2).

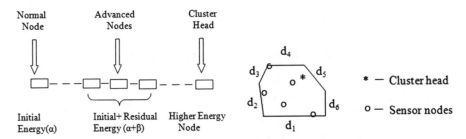

Fig. 1. Sensor Nodes in cluster   **Fig. 2.** Path identification for neighboring

4   Simulation Results

4.1   Simulation Setup

The execution of the DEE algorithm is simulated in matlab. The simulations parameters and parameter values taken for the CS algorithm are shown in Table 1. To assess the performance of the DEE algorithm, the experiment is carried out. In this scenario the position of the sink node lies with the coordinate (250, 250) and the number of nests is kept 50. Probability value pa = 0.25 gives the best result among selected values, since the DEE algorithm consist of clustering algorithm. Figure 3 shows the cluster formation of 500 sensor nodes using CS.

Table 1. Simulation parameters

Parameter	Value
Area	500*500
Number of sensor nodes (n)	300–700
Number of CHs	60–90
Sensor nodes initial energy	0.5 J
Amount of simulation iterations	200
Communication range of sensor nodes	60 m
Communication range of CHs	120 m
E_{ele}	50 nJ/bit
E_{fs}	10 pJ/bit/m2
E_{mp}	0.0013 pJ/bit/m4
D_0	30.0 m
E_{DA}	5 nJ/bit
Size of the packet	4000 bits
Size of the message	200 bits
Number of Nests	500
Pa	0.25
Lower bound	−100
Upper bound	100

Table 2. Pseudo Code for clustering using CS

```
Input    :  Number of nodes
Output   :  Cluster formation
Begin
for a=1: Normal node
for b=1: Advanced node with highest residual energy
Select CH by CS Algorithm
end
end
CH selection and cluster creation ();
for m=1: Entire cluster
for k=1: Assigning cluster member
Formation of clusters
end
end
while (Entire Nodes)
Send data packets to the sink node
Calculate the energy consumption value;
end
End
function CH selection and cluster creation ()
begin
for j=1: Entire Nodes
if Node is not a CH
Selecting the shortest distance between CH and sensor node;
Sends the join request;
end
end
end
end
```

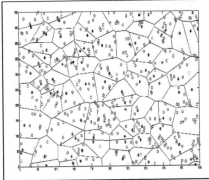
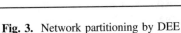

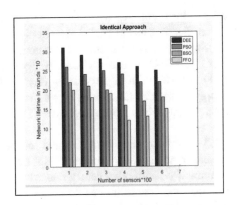

Fig. 3. Network partitioning by DEE   **Fig. 4.** Comparison of network lifetime

4.1.1 Performance with Respect to Network Lifetime

The result of the DEE algorithm is compared in terms of the network lifetime with respect to expiry of first CH. The existing algorithms like PSO, BCO and FFO were compared with DEE. Figure 4 is an experimental snapshot which depict that DEE outperforms the PSO, BCO and FFO in terms of network lifetime. This is due to the usage of energy balancing techniques in the clustering phase. Additionally, it considers the CHs whose energy depletes frequently by assigning less number of sensor nodes in the cluster. This reduces the frequent death of CH and increases network lifetime. However, the existing algorithms do not consider the energy balance concept of the CHs due to data forwarding and data routing, which are the main reason for the rapid decaying of CHs which are in close proximity to the sink node.

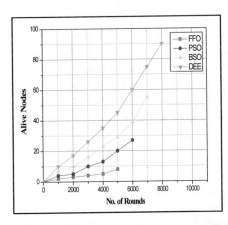

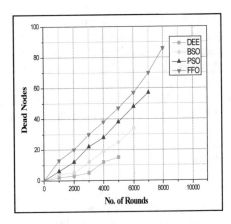

Fig. 5. Comparison of algorithms with respect to alive nodes   **Fig. 6.** Comparison of algorithms in terms of inactive nodes

4.1.2 Performance with Respect to Alive Nodes

Figure 5 represents that DEE algorithm has significant amount of alive sensor nodes to improve the network lifetime when compared to the existing algorithms such as PSO, BCO, and FFO. Because The DEE algorithm will take the advantage of selecting the node as a CH with higher energy level by taking into contemplation the outstanding energy of the CH candidates as well as sensor nodes and also the minimum distance between the sensor nodes and their CHs.

4.1.3 Performance with Respect to Number of Inactive Nodes

A sensor node which can have highest residual energy but it has no alive CH to communicate with other sensor nodes are called inactive sensor nodes. Figure 6 shows that the DEE algorithm has less number of inactive sensor nodes per round than the existing algorithms. The reason behind this is that the CHs energy consumption is minimized at clustering phase and CHs lifetime is maximized that helps sensor nodes to be active for the longer period.

The fault tolerance of the sensor node can also be the reason for CH node failure so that the failed CH node is combined with another CH node based on the communication range in order to minimize the energy consumption. The failed CH members are allocated to its neighboring nodes whereas the frequent failure of CH and its member sensor node are not considered as an issue of fault tolerance.

5 Conclusion

Nowadays wireless sensor networks (WSN) emerge as an active research area due to the challenging topics involving energy consumption, fault tolerance, routing algorithms. The proposed distributed energy efficiency (DEE) algorithm is used to create the energy efficient and energy balanced clusters, which is carried out at various levels to enhance the network lifetime and improve the fault tolerance. In this article energy depletion was identified at both node level and cluster head level. In sensor fields the nodes can be deployed in a random manner, CH can be selected based on the CS algorithm via levy flights strategy using remaining unconsumed energy of sensor nodes. The experimental results showed that DEE is efficient in comparison with the other existing algorithms and Fault tolerance of the sensor nodes are done using DEE algorithm, which reduces the expiry nodes in every single round.

References

1. Kuila, P., Jana, P.K.: Energy efficient load-balanced clustering algorithm for wireless senso networks. Procedia Technol. **6**, 771–777 (2012)
2. Azharuddin, M., Jana, P.K.: PSO-based approach for energy-efficient and energy-balanced routing and clustering in wireless sensor networks. Soft Comput. **21**(22), 6825–6839 (2017)
3. Ghiasiana, A., Hosivandib, M.: Cuckoo based clustering algorithm for wireless sensor network. Int. J. Comput. (IJC) **27**(1), 146–158 (2017)

4. Mannan, M., Rana, S.B.: Fault tolerance in wireless sensor network. Int. J. Curr. Eng. Tech. **5**(3) (2015). E-ISSN 2277-4106, P-ISSN 2347-5161

5. Singh, S.S., Jinila, Y.B.: Sensor node failure detection using checkpoint recovery algorithm. In: Fifth International Conference on Recent Trends in Information Technology (2016)

6. Kuila, P., Jana, P.K.: A novel differential evolution-based clustering algorithm for wireless sensor networks. Appl. Soft Comput. **25**, 414–425 (2014)

7. Sharawi, M., Emary, E., Saroit, I.A., El-Mahdy, H.: Bat swarm algorithm for wireless sensor networks lifetime optimization. Int. J. Sci. Res. (IJSR) **3**(5), 654–664 (2014)

8. Yang, X.-S.: Nature-Inspired Optimization Algorithm. Elsevier Inc., Amsterdam (2014)

9. Yang, X.-S., Deb, S.: Engineering optimisation by cuckoo Search. Int. J. Math. Model. Numer. Optim. **1**(4), 330–343 (2010)

10. Kaur, S., Mahajan, R.: Hybrid meta-heuristic optimization-based energy efficient protocol for wireless sensor networks. Egypt. Inform. J. **19**(3), 145–150 (2018)

11. Azharuddin, M., Kuila, P., Jana, P.K.: A distributed fault-tolerant clustering algorithm for wireless sensor networks. In: International Conference on Advances in Computing, Communications and Informatics (ICACCI), pp. 997–1002. IEEE (2013)

12. Bhatti, G.K., Raina, J.P.S.: Cuckoo based energy effective routing in wireless sensor network. Int. J. Comput. Sci. Commun. Eng. **3**(1), 92–95 (2014)

13. Yang, X.-S., Deb, S.: Cuckoo search via levy flights. In: Proceeding of World Congress on Nature & Biologically Inspired Computing (NaBIC 2009), pp. 210–214. IEEE Publications, USA (2009)

14. Joshi, A.S., Kulkarni, O., Kakandikar, G.M., Nandedkar, V.M.: Cuckoo search optimization-a review. In: ICAAMM, vol. 4, no. 8, pp. 7262–7269 (2016)

15. Das, S., Barani, S., Wagh, S., Sonavane, S.S.: Optimal clustering and routing for wireless sensor network based on cuckoo search. Int. J. Adv. Smart Sens. Netw. Syst. (IJASSN) **7**(2/3), 1–13 (2017)

A Literature Survey on Various Classifications of Data Mining for Predicting Heart Disease

Divya Singh Rathore[✉] and Anamika Choudhary

Computer Science and Engineering, JIET, Jodhpur, India
singhrathorediya@gmail.com,
anamika.choudhary@jietjodhpur.ac.in

Abstract. The heart is a very important hard working organ in the human body, which pumps blood to supply nutrients and oxygen throughout the whole body. The prediction of the occurrence of heart disease in the medical area is an important task. Algorithm of data mining are very helpful in the detection of Cardiovascular disease. In this paper, a survey has been provided for data mining classification techniques, in which health professionals have been offered to help in diagnosing cardiovascular diseases. We start by over-viewing the data mining techniques and describing various classification models used in for the earlier detection of heart diseases. Then, we review the proposed research works on using data mining classification techniques in this area.

Keywords: SVM · Back propagation · Prediction

1 Introduction

Heart disease causes millions of death worldwide every year. According to a report of World Health Organization, 17.9 million people died from Cardiovascular disease in 2016. If heart disease could be predicted earlier, then many patients would be prevented from dying. Identifying disease is a very challenging task in the field iatrology. Heart disease is one of the most critical issues faced by people. The demonstration of the proposed work is tested in the dataset of heart disease collected from University of California, Irvine's repository. The dataset consists of 76 attributes. However, all published experiments require a subset of 14 amongst them. Specifically, machine learning researchers have used Cleveland database, particularly at all times.

1.1 Data Mining

The process of exploration of large dataset in order to extract hiddeb interesting facts and patterns to analyze data is known as data mining. Different data mining tools are available to analyze various types of data in data mining. Education systems, customer retention, scientific discovery, production control, market basket analysis and decision making are some of the data mining application used to analyze the gathered information [1]. The multimedia, relational, object-relational and data warehouses are some of the databases for which data mining has been studied. As shown in Fig. 1, the complete procedure of data mining has been explained.

© Springer Nature Switzerland AG 2020
P. Karrupusamy et al. (Eds.): ICSCN 2019, LNDECT 39, pp. 720–726, 2020.
https://doi.org/10.1007/978-3-030-34515-0_76

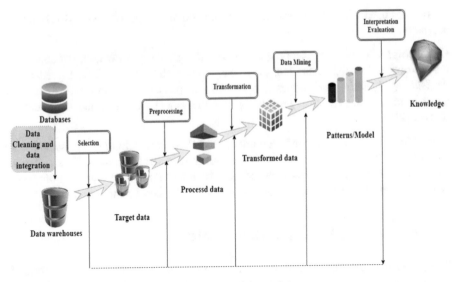

Fig. 1. Process of data mining

Data Cleaning. To ensure the accuracy and correctness of data by altering it within the given storage resource is known as a data cleaning process.

Data Transformation. The process through which data can be modified from one form to another is known as data transformation. Mainly, a new destination system is generated from the source system's format.

Pattern Evaluation. Truly interesting patterns that represent the knowledge on the basis of interesting measures are identified through pattern evaluation.

Data Integrity. In case, if the data stored within the database or data warehouse is accurate and consistent then the data is known as integral. The state, process or function of data can be known through data integrity and it can also be known as data quality.

Data Selection. For the determination of appropriate source of data and its type, the data selection process is applied. Further, the appropriate measures through which data can be gathered can be known here. The actual practice of data collection is preceded by the data selection process.

1.2   Techniques of Data Mining

Association: It is used to predict heart disease which is used to provide information about the relationship of various characteristics and to analyze.

Classification: A classic approach based on machine learning is known as the classification of data mining techniques. Within the predefined set of groups or classes, each item in the data set is classified by a classification approach. Different mathematical approaches used within the classification system are linear programming, decision trees and so on.

Clustering: Clustering is a technique of data mining, in which clustering of objects are done using automatic technique, as it has similar characteristics. The clusters are defined by clustering techniques and objects are placed in the cluster, which is in contrast to the classification process, where the objects are assigned in predefined classes [2].

Prediction: In the data mining technique, prediction is a technique that searches for relationships between dependent variables and independent variables. This technique can be utilized in various fields such as in sale in order to predict profit for future hence, profit is referred as a dependent variable and sale as independent variable.

2 Classification Models of Data Mining

The group membership for data instances can be predicted with the help of classification technique within data mining. Classification models that can be used in data mining for predicting heart disease are given below:

- Decision trees
- Neural networks
- Naive Bayes Classifier
- SVM (Support Vector Machine) classifier.

Decision Tree
In the classification of data mining, the decision tree approach is considered as the most powerful technique. In this method, all the models build in the form of a tree structure. Datasets are breaks into small sets and help in the formulation of an associated decision tree. Both the numerical data and categorical data are handled by the decision trees. In the different attributes, the order is determined by the decision trees for the medical purpose and on the basis of attributes a decision is taken [3].

Advantages

- Interpretation of this method is easy to understand.
- Generation of rules are easy
- Feature selection is performed by the implicit fuction.
- A large number of new data can be added using this algorithm

Disadvantaged

- Over-fitting is the drawback
- It is difficult to handle non-numeric data.
- It is difficult to understand the trees with many branches.
- This process consumes a lot of time [4].

Neural Networks

In the Neural network, a large number of elements is organized in a different number of layers that are interconnected to each other. Through this approach, the adaptive non-linear data processing algorithms have applied that help in integrating all the multi-processing units. Three layers are present in the artificial neural network these are input layer, hidden layer and output layer. A hidden layer also termed as an extra layer is present between the input layer and output layer [5].

Advantages

- Any kinds of noise present within the data can be handled easily through the application of neural networks.
- There is no need for re-programming once the data is trained.
- Within huge datasets, it is easy to perform this approach.

Disadvantaged

- Proper training is required for the neural network in order to operate in an effective manner. It is difficult to handle non-numeric data.
- Large networks consume High processing time.
- The retraining of the neural network is impossible as no modification of data can be done in the existing network.

Naive Bayes Classifier

The simple probabilistic classifier that depends on Bayes' theorem is known as Naive Bayes classifier which is a strong independence Naive assumption. This algorithm is also known as the independent feature model [6]. Naive Bayes classifier based on the assumption that features of a particular class present is unrelated to the present feature of any other class. Naive Bayes classifiers are trained to work in supervised learning.

Advantages

- Handling a large amount of data easily.
- The higher the speed of training and classification can be done.
- This approach provides no sensitivity for irrelevant features.
- Handling of real and discrete data.
- It is possible to manage data streaming [7].

Disadvantages

- Accuracy loss
- Dependencies among variables are present in this method however the classifier does not manage these dependencies.
- The independence of features is assumed to be provided in this approach.

SVM Classifier

A binary classifier through which the margin is increased is known as SVM classifier. This algorithm helps in performing classification in which all the data points present in individual class are separated by the best hyperplane. The best hyperplane of SVM can be presented on the basis of the highest margin present in the two classes [8]. Margin is defined as the largest width that exists amongst slabs that are parallel to the hyperplane that has no interior data points. The SVM algorithm is used to isolate the maximum margin in a hyperplane. The margin planes determined using the point from each class are called support vectors (SVs). It has many applications such as text and image recognition, bioinformatics etc. It becomes popular due to its success in handwritten digit recognition. Several operations are being performed within this algorithm to perform several tasks efficiently.

Advantages

- Easily extended to perform numerical calculations
- Used for pattern classification [9].

Disadvantages

- It is computational inefficiency.

3   Literature Review

A study related to the growing medical field was presented in paper [10]. In this area, large quantities of data have been generated every single day. Therefore this data need to be handled properly, which is not an easy task. For handling this much data different technologies need to be used. After that, data need to be mined in order to get useful patterns. The optimal results of the medical line prediction-based system are produced by the techniques of medical data mining. The human effects and cost effectiveness has been reduced using proposed prediction system based on data mining.

The paper [11] has proposed a new convolutional neural network based multi model disease risk prediction (CNN-MDRP) algorithm. Both structured and unstructured types of data was gathered from a hospital. A latent factor model was used in this method so that the incomplete types of data contained within the collected data can be reconfigured. To assess the performance of the proposed method, various experiments have been done on a regional chronic disease of cerebral infraction. It was seen through various comparisons between existing and proposed techniques that none of the pre-existing works focused on both types of data, which were gathered from the medical areas. In comparison to other similar algorithms, 94.8% accuracy was predicted with high convergence speed.

In paper [12], the author has perform various experiments for the investigation of heart disease. J48, KStar, Bayes Ne, Multilayer Perceptronand SMO were used for this purpose that can be possible through software called Weka. The performance of data mining technique is compared with the standard data setin terms of roc curve auc value and predictive accuracy. The Bayes Net and SMO technique shows the optimal performance as compared to the other techniques.

The paper [13] discussed the use of data mining techniques in the medical system. These techniques provides the idea to the doctor weather the patient is suffering from any heart disease or not, using Hidden Naive Bayes. The conditional independence assumption of traditional method, in the data mining is relaxed by using this model. On the basis of the performed experiments, this paper concluded that Hidden Naive Bayes (HNB) performs better in comparison to Naive Bayes in terms of optimal accuracy.

The paper [14] has discussed various data mining techniques which have been utilized to detect the rate of heart disease. On the basis of the number of attributes taken different accuracy was provided by the different technologies. Using ID3 and KNN algorithm, the risk rate of heart disease was estimated and it also provides the accuracy level for different attributes. It is concluded from the observation that using new algorithms the numbers of attributes could be reduced that increase the accuracy for the detection of the heart disease.

The paper [15] has proposed a technique I order to enhance the performance of k-Nearest Neighbor (kNN) algorithm with the integration of Ant Colony Optimization technique. With the help of this method prediction of the heart disease becomes easy. In this technique there are two different phases. kNN algorithm was utilized in the initial phase for the classification of the test data. For the optimized solution, The ACO technique was utilized as it initializes the population and search to get desired result.

The paper [16] has proposed a framework for predicting cardiovascular disease at the early stage using dataset of heart. All datasets are based on Association Classification techniques and this technique has been applied to the Cleveland Heart Disease dataset to check data technologies. According to the experiments conducted it has been concluded that a hybrid technology has been used to take care of classification associative rules which provides optimum accuracy.

4 Conclusion

The study surveyed some data mining classification techniques to predict heart disease at early stages. These techniques are decision tree, Naive Bayes classification, Neural Network and SVM classifier. Among all these algorithms, decision tree's algorithm having less error rate and it is the easier algorithm as compared to Naive Bayes. Each and every algorithm having their own merits and demerits. Decision tree is good at dealing with irrelevant features, Naive Bayes is good for handling multiple classes and so on. Their performance is also affected by the size, type and quality of data set, time duration etc.

References

1. Dey, M., Rautaray, S.S.: Study and analysis of data mining algorithms for healthcare decision support system. Int. J. Comput. Sci. Inform. Technol. 6(3), 234–239 (2014)
2. Beant, K., Singh, W.: Review on heart disease prediction system using data mining techniques. Int. J. Recent Innov. Trends Comput. Commun. 2(10), 3003–3008 (2014)
3. Bellaachia, A., Guven, E.: Predicting breast cancer survivability using data mining techniques, Washington DC 20052, vol. 6, no. 3, pp. 234–239 (2010)
4. Osamor, V.C., Oyelade, J.O., Adebiyi, E.F., Doumbia, S.: Reducing the time requirement of K-means algorithm. PLoS One 7(4), 56–62 (2012)

5. Gupta, P., Kaur, B.: Accuracy enhancement of heart disease diagnosis system using neural network and genetic algorithm. Int. J. Adv. Res. Comput. Sci. Softw. Eng. **103**(13), 11–15 (2014)

6. Agrawal, K.C., Nagori, M.: Clusters of ayurvedic medicines using improved k-means algorithm. In: International Conference on Advances in Computer Science and Electronics Engineering, vol. 23, no. 4, pp. 546–552 (2013)

7. Oliver, D., Martin, F.C., Daly, F., McMurdo, M.E.: Risk factors and risk assessment tools for falls in hospital in-patients: a systematic review. Age Ageing **33**(2), 122–130 (2004)

8. Ravi Kumar, G., Nagamani, K., Ramchandra, G.A.: An efficient feature selection system to integrating SVM with genetic algorithm for large medical dataset. Int. J. Adv. Res. Comput. Sci. Softw. Eng. **4**(2), 272–277 (2014)

9. Rosalina, A.H., Noraziah, A.: Prediction of hepatitis prognosis using support vector machine and wrapper method. In: 7th IEEE International Conference on Fuzzy System and Knowledge Discovery, pp. 2201–2211 (2010)

10. Al-Radaideh, Q.A., Assaf, A.A., Alnagi, E.: Predicting stock prices using data mining techniques. In: The International Arab Conference on Information Technology (ACIT 2013), vol. 23, no. 17, pp. 32–38 (2013)

11. Chen, M., Hao, Y., Hwang, K., Wang, L., Wang, L.: Disease prediction by machine learning over big data from healthcare communities, vol. 15, no. 4, pp. 215–227. IEEE (2017)

12. Sultana, M., Haider, A., Uddin, M.S.: Analysis of data mining techniques for heart disease prediction. In: 3rd IEEE International conference on Electrical Engineering and Information Communication Technology (ICEEICT), vol. 14, no. 1, pp. 123–138 (2016)

13. Jabbar, M.A., Samreen, S.: Heart disease prediction system based on hidden naive bayes classifier. In: International Conference on Circuits, Controls, Communications and Computing, vol. 4, no. 11, pp. 23–48 (2016)

14. Princy, R.T., Thomas, J.: Human heart disease prediction system using data mining techniques. In: International Conference on Circuit, Power and Computing Technologies [ICCPCT], vol. 4, no. 1, pp. 23–48 (2016)

15. Rajathi, S., Radhamani, G.: Prediction and analysis of rheumatic heart disease using kNN classification with ACO, vol. 4, no. 7, pp. 223–248. IEEE (2016)

16. Singh, J., Kamra, A., Singh, H.: Prediction of heart diseases using associative classification, vol. 7, no. 9, pp. 23–48. IEEE (2016)

RETRACTED CHAPTER: A Performance Analysis of Novel Classification Algorithm (C18) for Credit Card Fraud Detection

S. Subbulakshmi[✉] and D. J. Evanjaline[✉]

PG and Research Department of Computer Science,
Rajah Serfoji Government Arts College (Autonomous),
Thanjavur, Tamilnadu, India
07subbu.lakshmi@gmail.com

Abstract. Payment by credit card is becoming very common today. Credit card is the simplest way to use your bank account. Despite the hype, it also has some security challenges. Fraudsters are the primary intruder in the event of loan card. Some unauthorized transactions can be accessed by these intruders. In this paper an efficient classification algorithm is developed for different datasets, to handle noisy data and produce results with more accuracy. The result of these proposed algorithm compared with Naïve Bayes and AdaBoost. We observe that the proposed C18 algorithm gives better result accuracy for the classification of credit card fraud detection.

Keywords: Naïve Bayes · AdaBoost · C18 algorithm · Credit card

1 Introduction

The merchants are affected by the loss from credit card fraud, where they have to bear all costs, including card issuer fees, charges, and administrative charges. Therefore, it is very important to decrease the loss. To overcome these challenges, a helpful fraud detection system to reduce fraud cases should be proposed. From the beginning of banking, the bank's major job was to protect the money of their client, to safely move money from one city to a new and to certify/assure its customer payment. In this way, bank business has always been related to fraud prevention. New technologies such as automatic teller machines (ATMs), credit cards and internet banking have enlarged the number of connections banks have to deal with [2].

In Today's reality high reliance on web innovation has appreciated expanded Visa exchanges however credit card misrepresentation had additionally quickened as on the web and disconnected exchange. A credit card fraud is a developing worry with extensive outcomes in the administration, corporate associations, fund industry, as credit card exchanges turn into a boundless method of installment, center has been given to later computational philosophies to deal with the charge card misrepresentation issue. There are numerous extortion location arrangements and programming which anticipate fakes in organizations, for example, credit card, retail, online business,

The original version of this chapter was retracted: The retraction note to this chapter is available at https://doi.org/10.1007/978-3-030-34515-0_85

protection, and ventures. It is difficult to be sheer sure about the genuine goal and legitimacy behind an application or exchange. Information mining strategy is one outstanding and prevalent techniques utilized in taking care of credit misrepresentation identification issue [3]. Truly, to search out conceivable confirmations of extortion from the accessible information utilizing scientific calculations is the best successful alternative. Misrepresentation identification in credit card is the genuinely the way toward distinguishing those exchanges that are deceitful into two classes of genuine class and extortion class exchanges, a few procedures are planned and executed to comprehend to credit card misrepresentation location, for example, hereditary calculation, fake neural system visit thing set mining, machine learning calculations, relocating winged animal improvement calculation, near examination of strategic lapse SVM, choice tree and arbitrary woodland is completed.

A credit card extortion discovery is an exceptionally famous yet in addition a troublesome issue to comprehend. Right off the bat, because of issues of having just a restricted measure of information, credit card makes it trying to coordinate an example for dataset. Besides, there can be numerous passages in dataset with truncations of fraudsters which likewise will fit an example of genuine conduct. Additionally, the issue has numerous limitations. Right off the bat, informational collections are not actually available for open and the aftereffects of inquiries about are regularly covered up and blue-penciled, making the outcomes difficult to reach and because of this it is trying to benchmarking for the models fabricated. Datasets in past looks into with genuine information in the writing is no place referenced [4]. Furthermore, the enhancement of techniques is progressively troublesome by the way that the security concern forces a confinement to trade of thoughts and strategies in extortion location, and particularly in credit card misrepresentation discovery. In conclusion, the informational indexes are consistently advancing and changing making the profiles of ordinary and deceitful practices constantly extraordinary that is the genuine exchange in the past might be an extortion in present or the other way around. This research evaluates C18 classification algorithm compared with Naïve Bayes and Ada boost algorithm.

2   Related Work

Misrepresentation recognition depends on the rule that each client displays a specific example of conduct in his online exchanges. The examples are made according to his past online buys and it might fluctuate from client to the client, for example the examples are one of a kind for each online client or client. These examples are produced from the current online exchange information store, in view of the highlights of the online exchange of the clients. The chose highlights are decreased utilizing and just the hugest highlights were utilized for the preparation of individual clients utilizing oneself arranging map calculation. At whatever point any new exchange is on the pipe to get the endorsement from the bank, it ought to be checked with the prepared framework. On the off chance that there is a match in the examples, the exchange will be endorsed by the bank and on the off chance that it is found not coordinating with the example, a trigger will be raised. In the event that the trigger is raised, the exchange won't be endorsed and the comparing card will be blocked. The general work is partitioned into four sections, in particular the information gathering, preprocessing, preparing and mapping. The online exchange information store contains a huge number of exchanges of clients over some stretch of time. This information will be put away in various bunches and there ought to be a system accessible to gather these terabytes of information for preparing. This uses the current Hadoop dispersed File System or the SPARK group design which underpins capacity and recovery of the tremendous volume of information. The information put away in these bunches are dispersed and recreated in nature to stay away from the information misfortune.

The Naïve Bayes machine culture classifier attempts to expect a class which is known as result class reliant on probabilities, and in addition preventive probabilities of its occurrence from the training information [5]. This kind of knowledge is remarkably productive, quick and high in accuracy for genuine situations, and in addition this learning kind is known as administered learning. KNN go under an extremely unusual sort of classification of machine learning calculations that are known as 'Languid Learners' since this estimation adapts steadily than contrasted with different calculations. KNN pursues a procedure to learn in which it continues intent on putting away the information until the end when it is actually having the information whose mark or class is planned to be anticipated. KNN classifier predicts that how shut the anonymous tuple is to the K preparing set, and KNN does this by utilizing some division measure. In sensible terms, KNN is uncomplicated and still ready for tackling difficult issues as it can work moderately with little data as well. The Naïve Bayes machine learning classifier attempts to predict a class which is known as product class dependent on probabilities, and in addition dependent probabilities of how often it happened from the training information. This sort of knowledge is awfully dynamic, quick and high in exactness for authentic situations, and is known as keeping pace learning. Additionally, this is intensely proficient on the grounds that it gauges the parameters by utilizing little preparing information which is utilized for classification and depends on word freedom. Despite the fact that Naïve Bayes is very easy to execute and understand and utilizes solid uncertainties. It gives actually exact outcomes and in addition it has been established again and again the time that Naïve Bayes works sufficiently in dissimilar zones recognized with machine learning [6].

For framing cross breed models an AdaBoost and casting a ballot strategy are connected. To furthermore review the power and steadfast eminence of the models, clamor is added to this present realism informational collected works [1]. The key assurance of this research is the assessment of a variety of machine learning models with an authentic Via informational collected works for extortion acknowledgment. While dissimilar specialists have utilized dissimilar techniques on openly available informational indexes, the informational index utilized in this research are extricated from authentic charge card switch over data more than three months. The Feed-Forward Neural Network (NN) utilizes the back-propagation calculation for preparing too. The relations connecting the units don't frame a synchronized cycle, and data just pushes ahead from the info nodes to the yield nodes, through the hidden nodes. Deep knowledge depends on an MLP systematize equipped utilizing a stochastic slope plunge with back transmission. Each node catches a duplicate of the worldwide model parameters on nearby information, and contributes intermittently toward the worldwide model utilizing model averaging [2].

3   Existing Methodology

3.1   Naïve Bayes

The Naive Bayes categorization is a natural technique that utilizes the dependent probabilities of each credit having a place with each category to make an anticipatic It utilizes Bayes' Theorem, a technique that computes a probability by examination the reappearance of character and blends of character in the genuine in order. Parmete estimation for Naive Bayes models utilizes the strategy for maximum probability. In bitterness over-disentangled doubts, it regularly performs better in frequent random genuine situation. One of the important preferences of Naïve Bayes suggestion is that it requires a little measure of preparing information to review the parameters. The Bayesian Classification speaks to an administered learning policy ast as a practical method for classification. Expect a basic probabilistic replic and it enables us to catch weakness about the model righteous by deciding probabilities of the results. It can take care of logical and discerning issues [7] (Fig. 1).

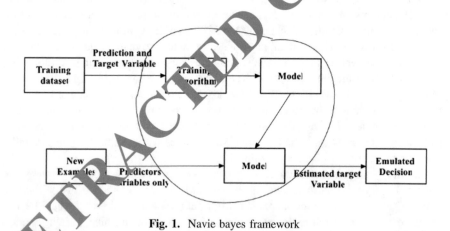

Fig. 1. Navie bayes framework

3.2   Adaboost Algorithm

AdaBoost can be utilized to support the execution of any machine learning calculation. It is best utilized with powerless students. These are models that accomplish precision simply above irregular possibility on a characterization issue. The most suited and accordingly most normal calculation utilized with AdaBoost are choice trees with one dimension. AdaBoost is best used to support the execution of choice trees on parallel grouping issues. All the more as of late it might be alluded to as discrete AdaBoost in light of the fact that it is utilized for arrangement as opposed to relapse.

AdaBoost can be utilized to support the execution of any machine learning calculation. It is best utilized with powerless students. These are models that accomplish exactness simply above irregular shot on a grouping issue. The most suited and along these lines most regular calculation utilized with AdaBoost are choice trees with one dimension. Since these trees are so short and just contain one choice for grouping, they are regularly called decision stumps (Fig. 2).

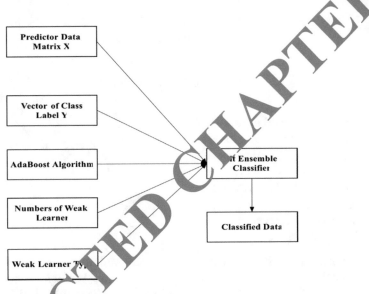

Fig. 2. Adaboost frame work

4 Proposed Methodology

C18 is the firmly related issue of multi-class characterization are variations of the arrangement issue where different marks might be allocated to each case. C18 classification is a speculation of multiclass order, which is the single-mark issue of arranging examples into correctly one of multiple classes; in the multi-name issue there is no requirement on what number of the classes the occasion can be allocated to. Formally, multimark characterization is the issue of finding a model that maps inputs x to paired vectors y (out an estimation of 0 or 1 for every component (name) in y). Each return neuron is assigned the task of recognizing a given class. The result code for that class ought to be 1 at this data, and 0 for the others. Along these lines, it will require N = K input in the result layer, where K is the quantity of classes. When testing an obscure model, the data giving the most extreme yield is viewed as the class name for that precedent (Fig. 3).

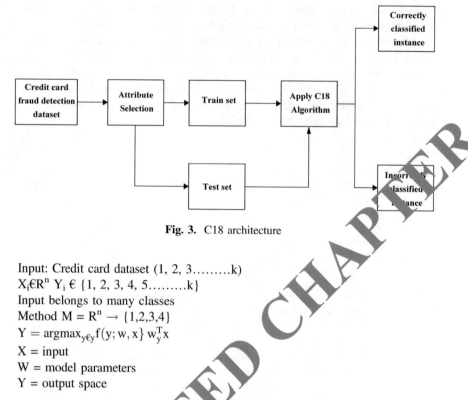

Fig. 3. C18 architecture

Input: Credit card dataset $(1, 2, 3\ldots\ldots\ldots k)$
$X_i \in R^n$ $Y_i \in \{1, 2, 3, 4, 5\ldots\ldots\ldots k\}$
Input belongs to many classes
Method $M = R^n \to \{1,2,3,4\}$
$Y = \text{argmax}_{y \in y} f(y; w, x\} w_y^T x$
X = input
W = model parameters
Y = output space

For class k, construct a binary classification work as:

Initially, the size of the certainty steems may vary between the double classifiers. Second, regardless of whether the class dissemination is adjusted in the preparation set, the parallel order students the unequal circulations on the grounds that regularly the arrangement of negatives they see is a lot bigger than the arrangement of positives.

5   Experimental Result

In this research work, taking a vulnerability of 1000 clients holding credit card and making a comparison with three classification algorithms: Naïve Bayes, AdaBoost and C18. The dataset assembles individuals showed by a lot of qualities as great or horrible credit dangers. It Comes in two organizations (one all numeric) with a cost framework. Calculations are connected to the dataset the disarray grid is created. Analyses are performed with 10 overlap cross approval. Ten times cross approval has been ended up being factually adequate in assessing the execution of the classifier. The initial step is to locate the quantity of occurrences of Credit card dataset utilizing Naïve Bayes, Ada-Boost and Novel characterization calculation. In the subsequent stage of the investigation will total the characterization exactness and cost examination. Perplexity framework depicts the data about real and anticipated grouping, figured in the last.

Confusion matrix are.

(1) True positive – if the output of estimate is p and the real rate is also p than it is called true positive (TP).
(2) False positive-if real rate is n than it is false positive (FP)
(3) Precision – precision is quantity of accuracy and quality-Precision = tp/(tp + fp)
(4) Recall- measure of comprehensiveness and quantity Recall = tp/(tp + fn) (Fig. 4 and Table 1).

Table 1. Comparison table of precision, recall, F-measure

Algorithm	Precision	Recall	F-measure
C18 classification	0.747	0.755	0.75
Naïve Bayes	0.743	0.754	0.746
Ada boost	0.661	0.695	0.665

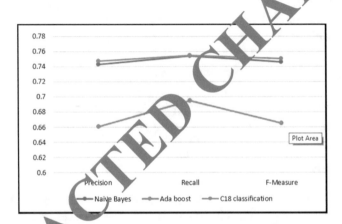

Fig. 5. Comparison chart for precision, recall, F-measure

Our experiment results show that, Naïve Bayes algorithm provide a 75.4% of correctly classified instance and 24.6% incorrectly classified instance. Ada boost algorithm provide a 69.5% of correctly classified instance and 30.5% incorrectly classified instance. The proposed C18 classification algorithm provide a 75.5% of correctly classified instance and 24.5% incorrectly classified instance. It gives better accuracy (Fig. 5 and Table 2).

Table 2. Comparison result of correctly classified instance and incorrectly classified instance

Algorithm	Correctly classified instances	Incorrectly classified instances
C18 classification	75.50%	24.50%
Naïve Bayes	75.40%	24.60%
Ada boost	69.50%	30.50%

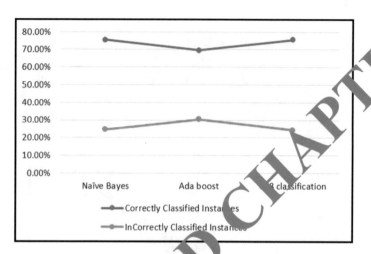

Fig. 5. Comparison output of properly classified instance and wrongly classified instance.

6   Conclusion

The proposed work is detecting the frauds in credit card effectively for different data sets. The result of proposed work compared with Naïve Bayes and AdaBoost, and found that it gave the accuracy for correctly classified instance as 75.50% and incorrectly classified instance as 24.50%. It gives better results compare to Naïve Bayes, AdaBoost.

References

1. Roy, A., Sun, J., Mahoney, R., Alonzi, L., Adams, S., Beling, P.: Deep learning detecting fraud in credit card transactions. IEEE (2018). 978-1-5386-6343
2. Randhawa, K., Loo, C.K., Seera, M., Lim, C.P., Nandi, A.K.: Credit card fraud detection using adaboost and majority voting. IEEE Access **6**, 14277–14284 (2018)
3. Carneiro, N., Figueira, G., Costa, M.: A data mining-based system for credit-card fraud detection in e-tail. Decis. Support Syst. **95**, 91–101 (2017)
4. Scholar, M.A., Ali, M., Fellow, P.: Investigating the performance of smote for class imbalanced learning: a case study of credit scoring datasets, vol. 13, no. 33, pp. 340–353 (2017)

5. Yassin, W., Udzir, N.I., Muda, Z., Sulaiman, M.N.: Anomaly-based intrusion detection through kmeans clustering and naives bayes classification. In: Proceedings of the 4 th International Conference on Computing and Informatics, ICOCI 2013, pp. 298–303 (2013)
6. Int. J. Eng. Res. Technol. www.ijert.org. Int. J. Adv. Res. Ideas Innov. Technol. (ISSN 2454-132x)
7. Research India Publications. www.ripublication.com

Novel Approach for Stress Detection Using Smartphone and E4 Device

Tejaswini Panure[(✉)] and Shilpa Sonawani

School of Computer Science and Engineering,
Dr. Vishwanath Karad MIT World Peace University, Pune, India
tejaswinipanure@gmail.com,
shilpa.sonawani@mitpune.edu.in

Abstract. Stress reduction is important for maintaining overall human health. There are different methodologies for detecting stress including clinical tests, traditional methods and various sensors and systems developed using either a smartphone, wearable devices or sensors connected to the human body. In this paper, a novel methodology is proposed by creating a personalized model from the generalized model because stress differs from person to person for the same work profile. A generalized model for stress detection is developed from smartphone and E4 device data of all the available individuals. A generalized model is used to build a personalized model that is a person-specific model and will be build up over a time of time when enough amount of person-specific data gets collected. This proposed methodology intends to give more accuracy as two devices are used with the novel approach of model building. Various machine learning algorithms such as ANN, xgboost, and SVM are implemented with the E4 device dataset while the LASSO regression model is used for smartphone data. ANN worked best than xgboost and SVM with 93.71% accuracy. In LASSO, 0.6556 RMSE is achieved.

Keywords: Stress · E4 device · Artificial Neural Network (ANN) · LASSO regression · Xgboost · Support Vector Machine (SVM)

1 Introduction

As per information by the World Health Organization (WHO), the count of individuals experiencing depression is almost 350 million. Stress can prompt sadness and further leads to suicide. Every year the count of suicide is about 1 million [31], so it is important to distinguish stress at an essential level. Positive stress resembles an inspiration while negative stress seriously influences human wellbeing as it is enduring [34]. Being a long haul it antagonistically influences the human body, so it is important to recognize negative stress. It is generally found in the all-inclusive community at working zones where the rest of the job needing to be done is more, close by due dates and weight of perfection of the work. In this investigation, utilizing information gathered from different inbuilt sensors of convenient gadgets, for example, smartphone and E4, stress examination will be done, for example, at a particular time, paying little

© Springer Nature Switzerland AG 2020
P. Karrupusamy et al. (Eds.): ICSCN 2019, LNDECT 39, pp. 736–745, 2020.
https://doi.org/10.1007/978-3-030-34515-0_78

mind to whether the individual is stressed or not, his pressure examples like area, time and forecast of stress later on. By combining data from smartphone and E4 device, a model will be developed which will take data from all available individuals. While using this model, newly arriving data will be stored in parallel on cloud and later after enough amount of data is gathered, personalized model will be trained using this data. This model is person specific. For this research, we used E4, a wearable device of an individual, along with his smartphone. Both of these devices will be capturing data to gain accurate results. We used WESAD [2] as E4 device data and StudentLife dataset [1] for smartphone data with features as mentioned in below sections. The objective of this study is to develop a personalized model. The paper is organized into following sections. Section 1 presents the introduction to the novel approach of stress detection system. Section 2 presents the work done till now, Sect. 3 includes proposed methodology for E4 device using ANN, xgboost and LASSO regression model for smartphone data. Section 4 presents the in depth implementation details of system and Sect. 5 conclude the paper.

2 Related Work

Stress detection systems can be extensively ordered into three sorts as indicated by the kind of information accumulation. First type includes medical tests. There are various medical tests available to know an individual's stress. Pathologist and a doctor examine samples and reports, and for this individual need to visit the clinic. The second type includes carrying different sensors attached to the human body to capture different types of physiological data. In any case, this isn't helpful for everyone in day by day life. Third is to capture data from sensors which are already available in handy devices of the individual. Third technique proves the automatic and efficient way of data collection. Almost everybody utilizes smartphone much of the time. It has different inbuilt sensors for various purposes as clarified in Table 1. More or less of these sensors are accessible in wearable gadgets like Fitbit, E4 and so forth. Uses of these sensors are explained in Table 1. Additionally, no uncommon treatment of these sensors is required. Example of utilizing smartphone and detecting information impacts a great deal on an individual's stress [15]. Close by this, dozing example and social association is likewise considered for better outcomes [4]. Information received from these sensors is utilized for stress detection by different analysts, as clarified beneath. Accelerometer sensor in cell phone gathers physical movement of the client. When a cell phone is kept still (someplace amid rest, meeting, etc.) makes obstacles in gathering physical movement information. This doesn't happen when wearable gadgets like smart watches are utilized in light of the fact that equivalent sensors are accessible in the smart watch. Here cell phone and smart watch are utilized for information accumulation [14]. Introduction and position of these sensors in a smartphone or other wearable gadgets significantly impact the achievement of this framework. But in real time application, the variability in the placement and orientation of these devices should be handled properly. The fusion of sensors is a method in which information from sensors is fused together to get increasingly exact outcomes [14]. As per Thapliyal H., utilizing IoT, various gadgets can be created for automatic stress detection. Different sensors that gather distinctive kind of information required to distinguish pressure are clarified. A versatile application is important to work these sensors and break down information. Here five wearable gadgets are taken for this examination; these are HeartMath Inner Balance, Spire, WellBe, Zensorium Being, Zensorium Tinke. All of these contraptions is annexed with no short of what one, either Android or iOS application moving toward the location and calendar with the customers' assent. As indicated by this examination, HRV strategy is essential to recognize pressure. Gjoreski M. demonstrated that utilization of cell phone only makes numerous challenges. In spite of the fact that the work profile is the same for the general population, stress observed is distinctive for every individual. Here individual wise different model worked better using the random forest. Alongside the utilization of sensors that distinguishes physiological signs, voice examination should likewise be possible [12]. Personspecific model works better [13], but for this, more labeled data is required and in case of the general model, no labeling is required. Garcia- Ceja used the only accelerometer is used for sensing data. Even for mood prediction, blood pressure, heartbeats, blood pressure etc. can be considered. SMS, Email contents also gives a basic idea of the mood of a person, but it requires the permission of the user to his/her highly private and sensitive contents. Data from sensors of a smartphone such as a microphone, accelerometer, electronic compass, light sensor and app usage history is taken from the user for this study. Mood-Explorer is an android app which collects sensing information. This app sends notifications to the user and reports their emotional state. Around 76% accuracy is taken from the ML model [15]. Jin Lu's combination of smartphone and Fitbit along with questionnaire data is used. Toward the start of this study, the clinical diagnosis of each participant is done. From a smartphone, the location and physical activity of the user is taken. For clustering, DBSCAN is used. Activity data, heart rate, sleeping patterns are extracted from Fitbit. Here multi-task learning is a proposed method [16]. All the studies above include either a smartphone or wearable devices or sensors attached to the body. It is not convenient to handle sensors attached to the human body every time. Considering this in this study, both smartphone and wearable devices (E4) are used for stress detection. During stress, the human body changes its external as well as internal behavior. These changes in internal behavior can be captured through physiological signals. E4 device gives information about the physiological flag, for example, pulse, the electro-dermal activity of the skin, blood volume pulse, temperature. All these signals contribute to determining the stress of a person effectively. Every-time it is not possible to handle these sensors. E4 device gives all of this data automatically. Just like this, the smartphone has inbuilt sensors that give data as discussed below. In this paper, data is collected from a smartphone as well as E4 device to gain more insights and achieve high accuracy of the system.

Table 1. Different Parameters of Smartphone and E4 devices

Smartphone data parameters	E4 device parameters
Phone usage	Activity data
Call log	Heart rate
Phone lock	Blood volume pulse
Audio data	Electro-dermal activity
Sleep data	Respiration
	Body temperature

3 Proposed Methodology

In this paper, a novel methodology is proposed for stress identification in the wake of learning numerous commitments in this area. Among every one of the commitments referenced above, just Lu J. has utilized a blend of smartphones and wearable devices, for example, Fitbit, yet the smartphone is utilized just for location information and heart rate is taken from Fitbit as a physiological parameter [16]. Beneath, we clarified how the blend of smartphone and wearable devices can be utilized to augment the exactness of stress prediction and addition more experiences from the information. Despite the fact that the work profile is the equivalent for all, stress differs from individual to individual [12]. To choose the stress of a specific individual, there ought to be a personalized model that will get prepared from the information taken from a person's devices. A personalized model is a person-specific model. In this methodology, two distinctive datasets of smartphones and E4 devices are consolidated. This dataset is utilized to build a generalized model. The generalized model development process is shown in Fig. 2. This information contains different stress patterns. While utilizing this model, after a specific period, an adequate measure of information will be produced which is individual explicit (Fig. 1).

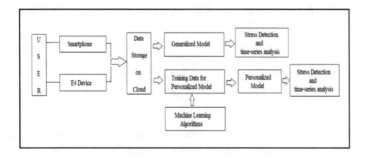

Fig. 1. Development of personalized model while using generalized model

This produced information will be utilized to prepare another model called the personalized model. Changing from a generalized model to the personalized model is the key idea of this novel system. The main distinction between these two is that a generalized model is constructed utilizing a dataset that is a mix of accessible smartphone and E4 device data of the significant number of individuals, while the personalized model will be constructed utilizing information taken from smartphones and E4 device of a specific person.

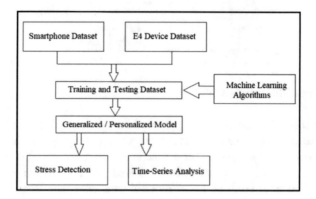

Fig. 2. General model development process

Here, we developed two models, one for each device and tested on new unseen data. Details of these models and their accuracies are discussed in sections below (Fig. 3).

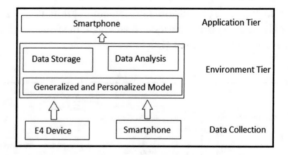

Fig. 3. Architecture of the system

The architecture of the system is as shown above partitioned into three tiers. Data is collected from E4 and smartphone devices of an individual and stored in the cloud. Using ANN and regression models for E4 and smartphone devices respectively, predictions are made on this new data. Predictions, whether an individual is stressed or not along with various analysis results taken out of this data are sent as notifications to the user's smartphone. For a new individual, first he/she will be using the generalized model to know their stress, after a particular period of time, when enough amounts of his own data will be generated, a new personalized model will be developed and further predictions will be done by this new personalized model. So basically there will be a model for every individual who will be trained and tested on his own data.

3.1 Data Collection

As we talked about above, for this study we utilized two datasets one is E4 device and the other is a smartphone that is publically accessible on the web. Student Life dataset [1] contains information gathered from smartphones of 49 students. For E4 gadget information, we utilized the WESAD dataset [2]. This contains information gathered from the E4 device of 15 people.

3.2 Data Pre-processing

In WESAD, data collected from E4 and raspiban devices are stored in separate folders, one per individual. This raw data is converted into SI units, along with feature engineering, filling missing values. Information caught is every second and it is changed over into five minutes of information. Also, normalization of the data is done using the min-max method that re-scaled data to fit within the 0 to 1 range. In the StudentLife dataset, each sensors data is stored in different folders, so gathering all this data is an important step. Missing qualities are supplanted with a mean value of a respective attribute. Here also feature engineering is done.

3.3 Artificial Neural Network

The artificial neural network worked best for this study as compared to the xgboost and SVM algorithm which resulted in lesser accuracy. We created our own neural network and trained it using a batch gradient descent algorithm for WESAD. Gradient descent algorithm works with very little memory. There are two layers in this network except for the input layer with activation function, Rectified Linear Unit in the hidden layer and sigmoid in the output layer. The input layer has 8 units because we have selected 8 features from WESAD for this study and the next layers have 15, 1 and 1 neurons. This network performs back-propagation and gradient descent. 65% of data is used for training while remaining is used for testing purposes. We run a number of iterations, performing forward and in reverse goes to refresh weights. Now, this neural network is applied to WESAD pre-processed data. Predictions are made on new data to check the accuracy of the model. Also, deep exploration is done to know how back-propagation and gradient descent works.

3.4 LASSO Regression

This analysis is accomplished for the StudentLife dataset. We have actualized a Lasso Linear Regression model on account of the idea of our dataset. LASSO (least total shrinkage and determination administrator) is a relapse investigation strategy that performs both variable choice and regularization so as to upgrade the forecast exactness and interpretability of the measurable model it produces. As the dataset is exceedingly conflicting, we have created every one of the highlights by condensing existing qualities.

4 Performance Analysis

Performance analysis of various algorithms used for classification such as ANN, Xgboost, SVM for E4 device data are shown in the Fig. 4. Highest accuracy on validation dataset is achieved by ANN, followed by xgboost and SVM.

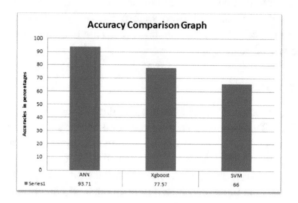

Fig. 4. Comparative analysis of validation accuracy for ANN, xgboost and SVM

In ANN, first we instantiate our network and after that run various emphases along with backward and forward passes and updating weights. After every iteration, loss value is calculated which and stabilizes at a low level. This chart appears in Fig. 5 and how precision varies as per the number of cycles is appeared in Fig. 6.

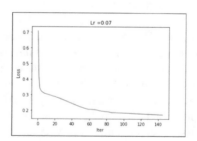

Fig. 5. Graph showing stabilization of loss values

In Fig. 6 stress level of students during various activities is shown such as stress level is highest during academic and lowest while watching movies and TV. Here, the correlation between the activity of a student and his stress level is shown after processing. Such different correlation heat maps for activity data, SMS data, and sleep data are shown in Figs. 7, 8 and 9 respectively.

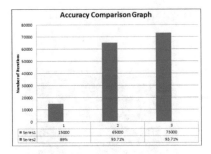

Fig. 6. Accuracy of ANN with number of iterations to stabilize loss values

In the StudentLife dataset, various types of data are extracted from a smartphone. Connection between each of these data as a feature and level of stress experienced by that student is shown below.

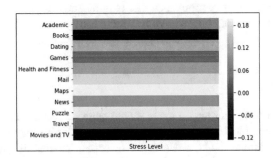

Fig. 7. Correlation heat map between stress level and activity data

In the Fig. 7, stress level during various activities of a student is shown. It is in the highest range during academics, books etc., while during watching movies and TV, it is very low.

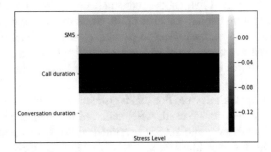

Fig. 8. Correlation heat map between stress level and SMS data

Same thing is shown in Fig. 8, people engaged in long-duration phone calls are less stressed.

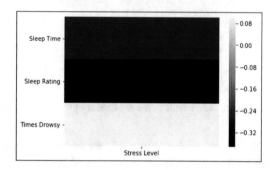

Fig. 9. Correlation heat map between stress level and sleep data

Using these heat maps, various analyses is done such as people engaged in long-duration phone calls, read books and/or watch Movies and TV are less stressed. Stressed individuals don't rest for a longer length of period and so forth (Fig. 9).

5   Conclusion

As most of the work for automatic stress identification is done till now, utilizes either smartphones or wearable devices like Fitbit. Fitbit doesn't give numerous physiological parameters of the human body as E4 does. In certain studies, it happened to utilize direct sensors appended to the human body. It isn't advantageous to deal with these numerous sensors appended to the body in everyday life. In this work, the advantages of the devices, smartphone and wearable device E4 are used to structure a proficient, advantageous and increasingly accurate stress detection framework. No additional treatment of sensors would be required. In this investigation, the generalized model is created utilizing publically available information of a considerable number of people.

This model will keep running until enough information about a specific individual is accumulated with the goal that another personalized model would be prepared. This new model would require enough patterns of the stress of a specific person recorded to get proficiency. As stress varies person to person, a general model will not be able to achieve the required accuracy. To shift from generalized to personalized model, the model is trained using a particular individual's data. During the collection of such person-specific data, the general model will be used. Here we developed two generalized models, one for each device. For this a different algorithms like ANN, xgboost, SVM and LASSO regression are used. ANN achieved the highest accuracy of 93.71% using the WESAD dataset while LASSO resulted in 0.6556 as RMSE (Root Mean Squared Error) with the best alpha value as 0.001. Both of these models are tested against new unseen data i.e. prediction is done.

References

1. Wang, R., Chen, F., Chen, Z., Li, T., Harari, G., Tignor, S., Zhou, X., Ben-Zeev, D., Campbell, A.T.: StudentLife: assessing mental health, academic performance and behavioral trends of college students using smartphones. In: Proceedings of the ACM Conference on Ubiquitous Computing (2014)
2. Schmidt, P., Reiss, A., Duerichen, R., Marberger, C., Van Laerhoven, K.: Introducing WESAD, a multimodal dataset for Wearable Stress and Affect Detection. In: 2018 International Conference on Multimodal Interaction (ICMI 2018), 16–20 October 2018, Boulder, CO, USA, 9 p. ACM, New York (2018)
3. Mitrpanont, J., Phandhu-fung, J., Klubdee, N., Ratanalaor, S., Pratiphakorn, P., Damrong-vanakul, K., Chuanvaree, P., Mitrpanont, T.: iCare-stress: caring system for stress. IEEE Conference (2019)
4. Akmandor, A., Jha, N.: Keep the stress away with SoDA: stress detection and alleviation system. IEEE Trans. Multi- Scale Comput. Syst. 3(4), 269–282 (2017)
5. de Santos Sierra, A., Avila, C.S., Casanova, J.G., BailadordelPozo, G.: A stress detection system based on physiological signals and fuzzy logic. IEEE Trans. Ind. Electron. 58(10), 4857–4865 (2011). 2018 IEEE International Conference on Applied System
6. Ramteke, R., Thool, V.: Stress detection of students at academic level from heart rate variability. In: 2017 International Conference on Energy, Communication, Data Analytics and Soft Computing (ICECDS) (2017)
7. Gjoreski, M., Luštrek, M., Gams, M., Gjoreski, H.: Monitoring stress with a wrist device using context. J. Biomed. Inform. 73, 159–170 (2017)
8. Thapliyal, H., Khalus, V., Labrado, C.: Stress detection and management: a survey of wearable smart health devices. IEEE Consum. Electron. Mag. 6(4), 64–69 (2017)
9. Ollander, S., Godin, C., Campagne, A., Charbonnier, S.: A comparison of wearable and stationary sensors for stress detection. In: 2016 IEEE International Conference on Systems, Man, and Cybernetics (SMC) (2016)
10. Gjoreski, M., Gjoreski, H., Lutrek, M., Gams, M.:. Automatic detection of perceived stress in campus students using smartphones. In: 2015 International Conference on IntelligentEnvironments (2015)
11. Garcia-Ceja, E., Osmani, V., Mayora, O.: Automatic stress detection in working environments from smartphones' accelerometer data: a first step. IEEE J. Biomed. and Health Inform. 20(4), 1053–1060 (2016)
12. Vaizman, Y., Ellis, K., Lanckriet, G.: Recognizing detailed human context in the wild from smartphones and smartwatches. IEEE Pervasive Comput. 16(4), 62–74 (2017). Zhang, X., Li, W., Chen, X., Lu, S. (2018)
13. MoodExplorer: Proceedings of the ACM on Interactive, Mobile, Wearable and Ubiquitous Technologies, vol. 1, no. 4, pp. 1– 30
14. Lu, J., Bi, J., Shang, C., Yue, C., Morillo, R., Ware, S., Kamath, J., Bamis, A., Russell, A., Wang, B.: Joint modeling of heterogeneous sensing data for depression assessment via multi-task learning. Proc. ACM Interact. Mob. Wearable Ubiquit. Technol. 2(1), 1–21 (2018)
15. Xia, L., Malik, A., Subhani, A.: A physiological signal- based method for early mental-stress detection. Biomed. Signal Process. Control 46, 18–32 (2018)
16. Sioni, R., Chittaro, L.: Stress detection using physiological sensors. Computer 48(10), 26–33 (2015)
17. Sriramprakash, S., Prasanna, V., Murthy, O.: Stress detection in working people. Proc. Comput. Sci. 115, 359–366 (2017). Kostopoulos, P., Kyritsis, A., Deriaz

Blockchain Based Electronic Voting System Using Biometric

Resham Nimje$^{(\boxtimes)}$ and D. M. Bhalerao

Sinhgad College of Engineering, Pune, India
nimjeresham94@gmail.com, dmbhalerao.scoe@sinhgad.edu

Abstract. Electronic voting system aims to eliminate the possibility of human error, manipulation and tamplering which occurs in traditional voting system. Electronic voting systems require heavy encryption to ensure transparency, security, auditability and secrecy for a vote. The proposed system modeled utilizing the information block of currently available framework to join them to another increasingly secure framework. This uses decentralized distributed database called blockchain. This is highly cost efficient and helps to increase the trust in the voting process. For encrypting data, we have used Elliptic Curve Digital Signature Algorithm (ECDSA) and hash code. Fingerprint module is interfaced with Raspberry Pi to access finger impression of user.

Keywords: Blockchain · E-voting · Biometric · Fingerprint module · Encryption · Digital signature · Raspberry pi

1 Introduction

A blockchain is a distributed database, where the total information is shared between all members in the system. Information, which should be stored in this database, is bundle into packages with a characterized most extreme size and confirmed with a particular hash. Blockchain innovation was first used inside Bitcoin which is a public record all things considered. A blockchain stores all the transactions in a block; the block gets completed as more transactions are executed. After the task is completed it is added in a linear, sequential order to the blockchain as seen in Fig. 1.

The blockchain is an add-only framework, where information is put away in a dispersed record that can't be altered or erased making the record unchanging. The squares are binded with the goal that each square has a hash that is a segment of the past square, and by including the aggregate past chain, it gives the ensurity of perpetual nature. This new innovation has three essential highlights: Immutability, Confirmable, Distributed oneness.

© Springer Nature Switzerland AG 2020
P. Karrupusamy et al. (Eds.): ICSCN 2019, LNDECT 39, pp. 746–754, 2020.
https://doi.org/10.1007/978-3-030-34515-0_79

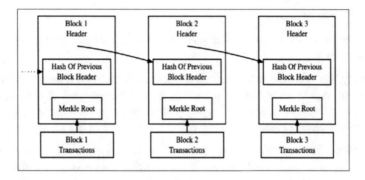

Fig. 1. Simplified bitcoin blockchain [1]

1.1 Basic Principles of Blockchain

Decentralization: There is no focal expert, with no single purpose of defenselessness or disappointment.

Trustlessness: A blockchain does not require trust in any authority or any participant.

Consensus network: A process allows participants to come to an agreement over what is true or false. For a cryptocurrency, it would typically concern the validity of a transaction.

Transaction transparency: The validity of all transactions is available to everyone on the network.

Transaction immutability: Once added to the blockchain, an exchange.

Pseudonymous: Transactions are unknown (in that they don't require individual data) however can be followed back to a public key.

1.2 List of Requirements in E-Voting System

- A decision framework ought not empower pressured casting a ballot.
- A decision framework ought to permit a technique for secure confirmation by means of a character check administration.
- A decision framework ought not enable detectability from votes to particular voters.
- A decision framework ought to give straightforwardness, as a verifiable confirmation to every voter that their vote was tallied, effectively, and without taking a chance with the voter's security.
- An ballot system should not allow anyone to mislead the votes.
- A race framework ought not bear the cost of any single element command over counting cast a ballot and deciding the afteref-fect of a race.
- A decision framework should just enable qualified people to cast a ballot in a race.

1.3   Preliminaries

A few necessities must be satisfied to make a casting a ballot framework pertinent for this present reality.

Availability. An e-casting a ballot framework should remain out there all through the full race and should serve voters interfacing from their gadgets. Particularly, the e-casting a ballot framework ought to be prepared for prime work, because of can |there'll be periods any place numerous natives will put their vote in the meantime.

Qualification. Only qualified voters are allowed to make a choice, while only one decision in favor of each voter counts. It is allowed to cast a vote on different events (furthermore called re-vote), the most recent count will be tallied and all others must be discarded.

Trustworthiness. The reasonableness of the vote ought to be verified. decision frameworks must ensure that the tallies aren't modified all through any progression of the race. Else we won't confide in this procedure.

Anonymity and Election Secrecy The affiliation between the vote of a user and therefore the user herself should not be reconstruct in a position while not her facilitate.

Fairness. The framework should probably endure (a few) broken votes. Aggressors would perhaps attempt and strong vindictive tallies, anyway these tickets ought to be identified. A constituent framework must recognize these tickets to thwart vote-control or assaults on the servers.

Definiteness. The election results should be summed up correctly and printed fairly.

Power. The framework ought to endure (a few) broken votes. Aggressors may attempt to cast malevolent votes; however, these tallies must be distinguished. A casting a ballot framework needs to perceive these polls to avoid vote-control or assaults on the servers.

Universal Verifiability. After the counting procedure, the outcomes are distributed and should be irrefutable by everyone. The casting a ballot framework must give systems to confirm the race's result.

Voter Verifiability. The balloter most likely confirm that her tally touched base in the polling booth. This guarantees the voter is certain her vote was tallied and was not changed.

2   Related Work

The utilization of the grouping planned within the blockchain creation method during this framework takes in account that in an exceedingly constituent framework negligible for mining as within the crypto-currency framework in lightweight of the actual fact that the elector info and digits are clear and don't seem to be permissible to decide on over once, the planned succession guarantees that everyone hubs that is lawfully associated and may maintain a strategic distance from impact in transportation [1].

As so much as expense will likewise be more practical on the grounds that it does not need hardware that's faithfully modified in each race did. In lightweight of the structure and therefore the consequences of analysis led, its o.k. is also probable that the framework is effective usefulness of chronicle the e-casting a ballot framework relies on Blockchain innovation [2].

A blockchain may be a common, taken over, and permanent record. The blockchain has made serious enthusiasm for use in an assortment of enterprises and areas, going from saving cash, back, and protection to medicative services, government, retailing, and collection. Associations are utilizing blockchains to grow new applications that are additional solid and practiced [3].

We tend to initial direct a organized investigation of uses and problems known with blockchain innovation, and subsequently distinguish a number of problems that need in addition consider with the top goal to be licitly cared-for. we tend to in addition examine the potential use of blockchain in instruction, and the way coaching frameworks will profit by the approaching of blockchain innovation [4].

The bitcoin [5], or, in alternative words and most acknowledge digital currency, has been acceptive a good deal of thought. one among its specialized highlights is that it empowers solid exchanges while not an incorporated administration element notwithstanding whether or not their area unit untrustworthy members within the system, and this component is gotten by the innovation of blockchain innovation.

With the increasing bitcoin exchanges attending to be larger and greater, talk about blockchain applications apart from those for money has been pushed [6] into the spotlight. this is often on account of the unwavering quality of the innovation has been preserved withal once it's place to use on a considerable scale.

With the development of advancement, the utilization of advancement in beating the issues that happen terrains up urgent, and furthermore the complexities of the development methodology [7].

Security is steadfastly the best stress for an e-throwing a vote structure. There should be no e-throwing a ticket system to grapple information and need the capacity to face up to potential ambushes. Blockchain advancement is one course of action that might be utilized to diminish the issues that occur in throwing a poll. Blockchain has been utilized in Bitcoin trade data structures [8].

Blockchain is spread, unchangeable and simple record who can't deny reality [9].

3   Proposed Methodology

This paper proposes an information maintaining framework on e-voting implemented in block chain. This technique aims to take care of the information integrity, that is shielded from manipulation that happen within traditional election (Fig. 2).

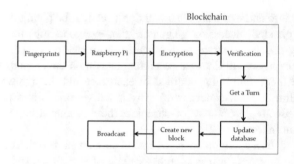

Fig. 2. Proposed system

Fingerprint model is used for making database for individual who wants to cast a vote. Further procedure starts once the choice procedure at each hub has been finished. Before the ballot casting procedure starts, every single hub creates a personal and non-secret key. Non-secret key of each hub is distributed to any or all nodes recorded within the ballot procedure, so each node has AN non-secret key of all nodes. Purpose once the choice happens, each node assembles the choice results from each citizen. At the purpose once the election procedure is finished, the nodes can wait their communicate create the square. Blockchain procedure is pursued. once the block of every node is arrived, the verification of every block is completed to examine whether the square is valid or not. If the square is valid, the info is enclosed with the knowledge within the block.

3.1   Block Creation

Confirmation and Update: The check strategy begins from the getting of a data square containing the ballot result, the past hash of the hash worth beginning from the previously mentioned significant square, and subsequently the modernized imprint. These two hash values block measure at that point looked at, in the event that the value is that the equivalent, at that point the advanced mark is substantial and in this way the technique proceeds, notwithstanding if the value isn't rise to it is viewed as invalid and hence the framework can deny the block to proceed with the strategy.

Get a Flip: The determination time can initialize and complete in the meantime. when the choice time has been finished, each hub can hold up it intercommunicate produce a square. The framework can constantly communicate the data pursued by the ID of a given hub. The hub ID is a token, in the event that a hub distinguishes that the distributed ID has a place with its ID, at that point it's the hub's go to deliver another square. nonetheless, to make another square it's important to explain that the sender of the square might be a sustantial sender and a piece of the decision, at that point the check technique is done.

Make New Square and Broadcast: Nodes accumulate cast a ticket from each selector, by then decided and united with the past hash as an electronic record in the system. The electronic chronicle is set up with a hash ability to make a message digest. It encodes the hash worth using the private key ECC. The proposed square suggests the examination implied involving an id hub, a timestamp, and three approval sections also in this examination notwithstanding an id hub of the hub that earned a next turn.

4   Implementation

The hardware implementation of the proposed system contains the Raspberry pi which is connected to the fingerprint module through the USB to TTL. The voting process can be seen on the computer or mobile screen which in turn is connected to Raspberry pi (Fig. 3).

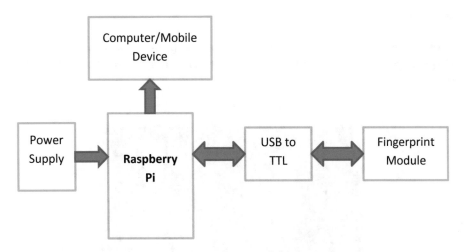

Fig. 3. Block diagram of proposed system

4.1   Hardware Components

Raspberry Pi b
The Raspberry Pi is a cost-efficient, small-sized computer that can be plugged into display devices, and can be interfaced with a keyboard and mouse. It is a more powerful processor (BCM2387 chipset), 10x faster (1.2 GHz Quad-Core ARM Cortex-A53) than the first generation Raspberry Pi. Introduction.

Fingerprint Module
This is an optical biometric unique mark peruser/sensor (R305) module with TTL UART interface for an immediate association with microcontroller UART.

4.2   Steps for Voting Process

Registration. The initial phase in the casting a ballot procedure is to enlist the fingerprints of the qualified voters. The voter needs to put his/her unique mark on the unique mark module so as to select his/her character in the database.

Initialization. The subsequent advance is to put the finger on the unique mark module which confirms the contribution with accessible database and produces three cases i.e., (i) confirmed (ii) Not checked/Thumbprint not discovered (iii) Your can proceed.

Verification. After the voter is checked, he/she is permitted to decide in favour of their preferred gatherings by tapping on the separate catch which updates the tally.

Acknowledgement. The voter will be recognized that their vote has been casted.

5   Result

The hardware implementation of the e-voting system using biometric contains the Raspberry pi b module with the fingerprint module attached with the USB to TTL (Fig. 4).

Fig. 4. Hardware model

When a ballotor wants to give vote, he/she puts the finger on the fingerprint module, then the fingerprint is initialised, finger is detected, the hash values are generated with the help od ECDSA algorithm, verify's the identity and the block is generated. Finally, if the fingerprint match is found the voter is able to choose the party he/she wants to vote and the vote can be casted (Fig. 5).

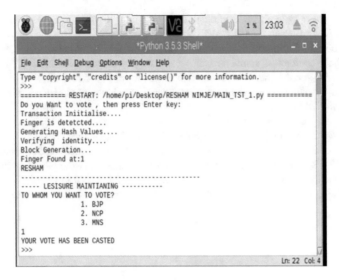

Fig. 5. Voting process

6 Conclusion

It is essential for a majority rules system to have a straightforward casting a ballot framework that must have minimal number of impediments for a voter to cast a ballot. The proposed framework handles voter protection and auditability as well as gives a straightforward framework to confirmation of the race. The arrangement of blockchain casting a ballot will affect society in an exceptionally positive manner. The framework will build accommodation for voters. It will make it simple for individuals with incapacities or who experience difficulty moving around to cast a ballot. It is brisk and private approach to cast a ballot. This will expand the quantity of voters since the procedure does not take up a lot of their time. It will help increment the trust of the general population in the administration since it is more straightforward than the present tally framework.

References

1. Hanifatunnisa, R., Rahardjo, B.: Blockchain based e-voting recording system design (2017). 978-1-5386-3546-9/17/$31.00 ©2017 IEEE
2. Kamboj, D., Yang, T.A.: An exploratory analysis of blockchain: applications, security, and related issues. In: International Conference on Scientific Computing, CSC (2018)
3. Shah, S., Kanchwala, Q., Mi, H.: Block chain voting system (2016)
4. Christian: Desain dan implementasi visual cryptography padasistem e-voting untuk meningkatkan anonymity. Institut Teknologi Bandung (2017)
5. Nakamoto, S.: Bitocoin: a peer-topeer electronic cash system (2008)
6. Bonneau, J., et al.: SoK: research perspectives and challenges for Bitcoin and cryptocurrencies. In: 36th IEEE Symposium on Security and Privacy (2015)

7. Dougherty, C.: Vote chain: secure democratic voting (2016)
8. Wijaya, D.A.: Bitcoin tingkat lanjut (2016)
9. Watanabe, H., Fujimura, S., Nakadaira, A., Miyazaki, Y., Akutsu, A., Kishigami, J.J.: Blockchain contract: a complete consensus using blockchain. In: 2015 IEEE 4th Global Conference on consumer electronics, GCCE 2015, pp. 577–578 (2016)
10. Kulkarni, A.: How to choose between public and permissioned-blockchain for your project. In: Chronicled (2018)
11. Weaver, N.: Secure the vote today
12. Agora: Agora: bringing our voting systems into the 21st century (2017)
13. Barnes, A., Brake, C., Perry, T.: Digital voting with the use of blockchain technology (2016)
14. Frecè, J.T., Selzam, T.: Tokenized ecosystem of personal data exemplified on the context of the smart city. JeDEM – eJ. e-Democr. Open Govern. **9**, 110–133 (2017)
15. Stanchfield, D.: Blockchain technology can make voting systems more secure (2016)

Proof of Game (PoG): A Game Theory Based Consensus Model

Adarsh Kumar and Saurabh Jain[(✉)]

School of Computer Science, University of Petroleum and Energy Studies,
Bidholi, Dehradun, India
{adarsh.kumar, saurabh.jain}@ddn.upes.ac.in

Abstract. Now-a-days, Blockchain networks are widely accepted in various
applications for its enhanced security levels. Blockchain characteristics like:
peer-to-peer, decentralized and immutable distributed ledger makes this technology acceptable to academia, research and industry communities. This work
proposes 'proof-of-game (PoG)' consensus algorithm suitable for resourceful
and resource-constrained devices. Heavy computational challenge in block
structure protects the blockchain network from selfish miners and majority
attacks. PoG consensus algorithm is suitable for both single and multi-player
challenges. It is observed that single and multi-bit challenges increases the
resource consumption and makes it difficult for resource constrained device to
confirm block in stipulated time period. However, a multi-round multi-bits
challenge makes it feasible for resource constrained devices to provide high
security within specified time period. In implementation, it is observed that
mined blocks indicates the chances of attacks. Large number of blocks are mined
if block miner is honest, computational challenge is high and number of participants associated with block is large. Similar scenario is possible with transactions. In results, it is observed that presence of large selfish miners decreases
the blocks mined exponentially with increase in computational challenge.

Keywords: Blockchain · Game theory · Proof of concepts · Selfish miner ·
Miner algorithm · Cryptocurrency

1 Introduction

In 2008, Satoshi Nakamoto introduces the Blockchain concepts an immutable timestamp ledger of blocks [1]. These blocks contains transactions records in a distributed
manner. These transaction recors includes payment history, personal data, challenges,
information about consensus mechanisms etc. In recent years, blockchain has attracted
tremendous importance in various industry domains. This importance can be realized
from the number of cryptocurrencies adopted day-by-day. Presently, there are more
than 2200 cryptocurrencies [1, 2]. In Blockchain, a distributed digital ledger of
transactions is shared with all participants which in-turn eliminates the chances of
centralized authority. New blocks can be added to existing Blockchain if user solves a
computationally hard and easily verifiable challenge. This challenge is backbone of
consensus algorithm used in Blockchain network, which restricts the users from
malicious activities. This restriction is resource dependent and existing Blockchain

© Springer Nature Switzerland AG 2020
P. Karrupusamy et al. (Eds.): ICSCN 2019, LNDECT 39, pp. 755–764, 2020.
https://doi.org/10.1007/978-3-030-34515-0_80

implementation uses various consensus algorithms: Proof of Stake Model (PSM), Proof of Elapsed Time (PET), Byzantine Fault Tolerance Algorithm (BFTA), Practical Byzantine Fault Tolerance Algorithm (PBFTA), SIEVE Consensus Protocol (SCP), Cross Fault Tolerance (XFT), Federated Byzantine Agreement (FBA), Ripple Consensus Protocol (RCP), Stellar Consensus Protocol (SCP) etc. Resource-intensive consensus algorithm makes it impossible for a resource constrained block generator in selecting nodes randomly and/or limiting a number of blocks it can generate. Presence of multiple block generators can create duplicate blocks with a pool of transactions. To protect the Blockchain network from a large number of malicious block generator, block and fake transactions, trustworthy consensus algorithm is required such that once block can be generated for a Blockchain network within specified-period.

In this work, "Proof-of-Game (PoG)" for resource dependent and computational independent consensus algorithm is introduced. PoG restricts the block generator in generating the number of blocks based on the level of game it can play and score it can earn. PoG concept can be applied for single or a group of block generators. Single block generator uses one-player and group block generator uses multi-player PoG concept. Single player PoG put challenge individually for users interested to add new blocks in existing Blockchain. Whereas, multi-player PoG allows multiple block generators to combine and play (put challenge) with user. This scheme is useful for both resourceful and resource constrained environments because game difficulty levels, number of game levels and scoring levels are proportionate to resource availability. High difficulty level and/or earning high game or scoring levels indicates trustworthiness which in-turns increases the importance of Blockchain network.

The rest of this work is organized as follows: State-of-the-art consensus algorithms are explained in Sect. 2. Section 3 proposes the single and multi-player PoG concept for Blockchain network. Experimental analysis of proposed consensus algorithm in Blockchain network is performed in Sect. 4. Finally, Sect. 5 concludes the work and outlines the future work.

2   Literature Survey

As discussed earlier, extensive work has been done [3, 4] to design a consensus algorithm for Blockchain network. Bach et al. [5] performed a comparative analysis of blockchain consensus algorithms. This comparative analysis is performed on parameters like: scalability, method of rewarding validators, security risks etc. In observations, it is found that Proof of Work (PoW) is well accepted consensus algorithm now a days. In another observation, it is found that there are blockchain networks, which uses other blockchain networks like: Proof of Stake (PoS), Proof of Luck (PoL), Proof of eXercise (PoX) etc. According to [3], a blockchain network should include incentive-compatible sub-stages with tolerance to Byzantine and unfaithful faults, attackers with cumulated computational power should be identifiable and isolated a balance between processing throughput and network scalability etc. Game theory is also widely acceptable in consensus mechanisms. Various game theory based consensus algorithms are explored as follows:

Jiang and Wu [6] uses game-theoretic model to study impact of block size in miner's payoff. Here, miner's strategy in varying block sizes for earning profits is analyzed. Miner's varies block sizes to optimize payoff. An experimental analysis to identify equilibrium where miners have no incentive to misbehave provide the guidelines on the default block size. Results show that a block size of 4 MB should be the default block size.

Swanson [7] argues that a permissionless set of nodes in blockchain network must require some game-theoretic approach for security. Blockchain network protocols versus trust tradeoff is a major challenge and various attempts are made to advance different sections of both sides. Cryptoeconomic game theory such as block reward halving i.e. rewards of miners are split into half after every four years is not fruitful for participants because they want to depend on a network for more than a decade. This would either be possible with significantly higher transaction fees or lower security thresholds from miner (or both). Another possible solution is to design an efficient and advanced game-theory model with less fee, high security and long participant dependency.

Dey [8] proposes a methodology with software agents to monitor blockchain network's stakeholders. This monitoring is allowed to detect anomalies using supervised machine learning algorithm and algorithm game theory. Supervised machine learning for anomaly detection is helpful in identifying various attacks like: majority population size attack, DoS, DDoS etc.

Liu et al. [9] performed a comprehensive survey over applications of game theory in blockchain. According to [9], various options available for game theory models are: non-cooperative game, extensive-form game, stackelberg game and stochastic game. Game theory can be applied in detecting various attacks like: selfish mining attack, majority attack, Denial of Service attack, false data sharing, distrustful goods trading and cyber-insurance. According to Nakamoto protocol [4], anyone in blockchain network can use token supply and transaction tipping while playing the role of blockchain mining. This process inculcates the profit process and maximize the payoff with effective application of game theory. Additionally, game theory can be integrated with two ways of mining management: individual mining and pool mining. In individual mining process, dominant strategy of each miner is either computational power or fork a longest chain. In addition to this, block size setting, pool selection and reward allocation are other action in pool mining based game theory model for payoff.

Dimitri [10] proposes a mining game theory based on block size. A large block size indirectly increases the power computation cost for a miner in generating a hash value. Further, author identifies that bitcoin mining process is profitable if fewer independent entities are actively mining, and the rationality in miners is profitable for all as large payoffs provide support to earn additional computational power.

Stone [11] proposes a game theoretical mode with payoff strategy related to block size. Large block size means large payoff price. Here, a threshold block size is analyzed for real scenario. Larger block size adds additional delay, which in-turns put the block into isolation. Isolation is a major cause for Denial of Service (DoS) or Distributed Denial of Service (DDoS) attack. It is also bserved that physical structure of blockchain network does not allow a large gas price for large block size.

Various other game theory based models are adopted for blockchain mining [12–14]. Among those evolutionary game models, hierarchical game and auctions are widely explored. However, single game model for both resourceful and resource constrained device's network is not explored much. Major challenge in designing this model is resource dependent and challenge variated rewarding point based game theory approach.

3 Proof of Game

In this work, PoG incorporating game theoretical model and mining process is proposed for blockchain network. Algorithm 1 explains the regular blockchain construction process and specific block entities. Various functions are written to generate genesis block and blockchain's subsequent blocks. Each of these blocks are linked with predecessor and successor blocks through a calculated hash value. As desired, this model of blockchain network construction has the provision of adding blocks of any size at any location.

In order to protect a blockchain network from attacks, computational challenges are put forward to node. If a node is able to solve the challenge within stipulated time period then node is allowed to add new block in existing blockchain network. A hash based computational challenge is implemented using algorithm 2. Computational challenges in Proof of Work (PoW) can be implemented using various ways [6, 7]. Today, more than seven hundred cryptocurrencies are using PoW concept in building consensus. Algorithm 2 is an extention of Algorithm 1 integrating SHA256 based

---

Algorithm 1 Regular Blockchain

Goal: To create a blockchain without proof of game

```
 1  Index <- BlockCounter()
 2  BBlock <- null
 3  UData <- UserData()
 4  PHash <- PreviousHash(BBlock)
 5  BTime <- CurrentTimestamp()
 6
 7  BlockCounter <- function() {
 8    index <- index+1
 9  }
10
11  UserData <- function() {
12    # User entered data
13  }
14
15  GBlock <- function(block) {
16    BBlock <- block
17    BTime <- TTime()
18    pHash <- null
19    Hash <- SHA256(BBlock)
20  }
21
22  PreviousHash <- function(block) {
23  if (block=="Genesis" || Index=="0"){
24    return "0"
25  else if (block == "null") {
26    return "null"
27  else{
28    return Hash(PreviousBlock)
29    }
30    }
31  }
32
33  Hash <- function(block) {
34    return SHA256(block)
35  }
36
37  Blockchain <- function(block) {
38  Index <- BlockCounter()
39  UData <- UserData()
40  PHash <- PreviousHash(BBlock)
41  BTime <- CurrentTimestamp()
42  }
43
44  AddBlock <- function(block) {
45    Index <- block.index
46    UData <- block.data
47    PHash <- block.PreviousHash(BBlock)
48    BTime <- CurrentTimestamp()
49    return Blockchain((Index, UData, PHash, BTime))
50  }
51
52  GBlock(0,CurrentTimestamp(),"Genesis","0",SHA256(0,CurrentTimestamp(),"Genesis","0")) #
        =>Creates Genesis Block
53  FirstBlock = new Blockchain(Index,UData,PHash,BTime,SHA256(Index,UData,PHash,BTim))
54  Blockchain(FirstBlock) # =>Creates First Block
55  SecondBlock = new Blockchain(Index,UData,PHash,BTime,SHA256(Index,UData,PHash,BTim))
56  Blockchain(SecondBlock) # =>Creates Second Block
57  ThirdBlock = AddBlock(Index,UData,PHash,BTime),SHA256(Index,UData,PHash,BTime))
```

computational challenges for blockchain network. In this implementation, hash computational challenger is given to node and it has to compute the hash within stipulated time period. If it is able to generate the hash then node is allowed to insert new block in blockchain network else it can be put under scrutiny. Line 35–39 shows the process of add new block if challenge is verified.

Algorithm 2 Extending a regular blockchain with 'Proof of Work'

Goal: To create a blockchain with 'Proof of Work' and without 'Proof of Game'

```
 1  Index <- BlockCounter()
 2  BBlock <- null
 3  UData <- UserData()
 4  Challenge <- GenerateChallenge()
 5  PHash <- PreviousHash(BBlock)
 6  BTime <- CurrentTimestamp()
 7
 8  GenerateChallenge <- function(Value, Operation) {
 9  Value <- GenerateValue()
10  Operation <- GenerateOperation()
11  ChallengeTime <- CurrentTimestamp()
12  }
13
14  ValidateChallenge <- function(TimeThreshold) {
15  CurrentTime <- CurrentTimestamp()
16
17  if {parent != SHA256(previousBlock) ||  !ValidTransaction(transaction) || (CurrentTime -
         ChallengeTime >= TimeThreshold)}{
18      return "False"
19  else{
20      return "True"
21      }
22  }
23  }
24
25  GBlock(0,CurrentTimestamp(),"Genesis","0", SHA256(0,CurrentTimestamp(),"Genesis","0",)) #
         =>Creates Genesis Block
26  NewBlock = new Blockchain(Index,UData,PHash,BTime,SHA256(Index,UData,PHash,BTim))
27  GenerateChallenge(Value, Operation) # =>Creates a Challenge
28  ValidateChallenge(TimeThreshold) # =>Validate a Block
29  Blockchain(NewBlock) # =>Creates New Block
```

Algorithm 3 extendeds Algorithms 1 and 2 with Proof-of-Authority (PoA), Proof-of-Ownership (PoO) and Proof-of-Stake (PoS). In PoA, one or group of selected nodes in the blockchain network is given the authority to put challenge rather than anyone put a challenge. Thus, trusted nodes are given priority using this concept. In order to implement PoA for single node, highly trusted node is selected. Whereas, a common signature of group of nodes is taken as consent of authority for group based PoA. PoA is additional overhead over PoW. However, it is useful in avoiding various attacks and providing higher security. PoO is an extension of PoA. In PoA, one or set of active nodes can be selected for challenge generation whereas, PoO is node specific. One or group of Owners need to be active for challenge generation and addition of new blocks in blockchain network. Apart from challenge generation, node with ownership right can verify any node's credentials before allowing it to add new block. Now, if a group of nodes are interested in creating a blockchain network based on node's investment then PoS is the best option. In PoS, those nodes are allowed to participate in the network which are having stake greater than a specific amount, within a specific range or

---

Algorithm 3 Proof of Game (PoG)

Goal: To identify and validate blockchain authority using 'Proof of Game'

```
PoW <- function(player) {
  if (player == 'single') {
    Lightweight_Hash_Computation(Time_
    Period, Hash_Algorithm, Data)
  }
  else if (player == 'multi') {
    While(Participant)) {
      group_sign=SIGNATURE(Identity,
      stakeholderList)
      Lightweight_Hash_Computation(Time_
      Period, Hash_Algorithm, Data, Group_
      Sign)
    }
  }
}

PoA <- function(player) {
  if (player == 'single') {
    Compute_Authority_Decision(Player_
    Signature, Player_Credentials)
  }
  else if (player == 'multi') {
    While(Participant)) {
      group_sign=SIGNATURE(Identity,
      stakeholderList)
      Compute_Authority_Decision(group_
      sign, Player_Credentials)
    }
  }
}

PoO <- function(player) {
  if (player == 'single') {
    Compute_Authority_Decision(Player_
    Signature, Owner_Creteria)
  }
  else if (player == 'multi') {
    While(Participant)) {
      group_sign=SIGNATURE(Identity,
      stakeholderList)
      Compute_Owner_Decision(group_sign,
      Owner_Creteria)
    }
  }
}
```

```
PoS <- function(player) {
  if (player == 'single') {
    Verify_Stake(Player_Stake)
  }
  else if (player == 'multi') {
    While(Participant)) {
      group_sign=SIGNATURE(Identity,
      stakeholderList)
      Compute_Owner_Decision(group_sign,
      Player_Stake)
    }
  }
}

PoG <- function(player) {
  if (player == 'single') {
    Execute PoW() || PoA() || PoO() || PoS
    ()
    Lightweight_Game_Single(bit_value,
    position_value)
  }
  else if (player == 'multi') {
    While(Participant)) {
      group_sign=SIGNATURE(Identity,
      stakeholderList)
      Lightweight_Game_multi(bit_value,
      position_value,participant_list)
    }
  }
}

Lightweight_Game_Single <- function(player
    ) {
bit <- Evaluate_Challenge(position)
if(verify(bit,position)=='valid') {
  return(bit,position)
  }
else {
  return 'invalid'
  }
}

Lightweight_Game_multi(bit_value, position
    _value,participant_list){
bit <- Evaluate_Challenge(position)
While(Participant <- participant_list)) {
bit_p <- Participant.bit
if(verify(bit,bit_p, position)=='valid') {
  return(bit,position)
  }
else {
  return 'invalid'
  }
}
```

interested to invest a minimum amount in business. Algorithm 3 explains the integration of PoW, PoA, PoO and PoS in blockchain network. This implementation is game theory based rewarding system. Initially, all participants are considered to be honest and share their resources. Computational resources are identified and verified by challenger before allowing any node to add new block. If a node is having sufficiently high computational resources then heavyweight computational challenge is put which in-turn ensures high security. Whereas, if node is resource constrained then lightweight computational challenges are used. Although lightweight computational challenges provide comparatively lighter security as compare to heavyweight computational challenges but security level can be enhanced using multi-rounds verification process explained in Algorithm 3. In multi-round verification process, bit value and bit position are verified at both challenge generation and challenge solver ends. If a threshold number of bit value and positions are verified then node is allowed to add new blocks. Challenger's trust over node increases with increase in number of new blocks. This trust is computed using rewarding points.

4 Analysis

This section explores the possibilities of various attacks in PoG algorithm for blockchain network.

- **Selfish Mining and Majority-Attack Detection:** Selfish mining and majority attacks in blockchain network is explored by various authors [12]. Improper incentive-compatibility and selfish miners may compromise the game theory process and gain higher profits than their due shares. In nutshell, selfish miners act as attacks and produce a tampered chain. This can be detected as follows:

Let there are two sides of miners: honest miner (H_M) and selfish miner (S_M). Each of these sides have their own population sizes. Let M represents the population size of H_M playing PoG, and m represents the population size of S_M interested in breaking PoG and earn higher profit. A miner competition is played between both sides to earn maximum profit, and perform or avoid majority attacks.

$$Chance_M^{Success} =$$
$$Maximize\left\{Height\ of\ honest\ miner - \frac{Sum\ of\ transactions\ secuted\ successfully\ at\ respected\ height}{total\ transactions\ of\ honest\ miner}\right\}$$

$$(i)$$

$$Chance_m^{Success} =$$
$$Maximize\left\{Height\ of\ selfish\ miner - \frac{Sum\ of\ transactions\ secuted\ successfully\ at\ respected\ height}{total\ transactions\ of\ selfish\ miner}\right\}$$

$$(ii)$$

Now, success probability of PoG is calculated as:

$$P_{Success}^h = \sum_{i=1}^h Chance_M^{Success} - Chance_m^{Success} \tag{iii}$$

From Eqs. (i), (ii) and (iii), it can be easily predicted that if $Chance_M^{Success} < Chance_m^{Success}$ then new block is considered to be created by selfish miner and can be rejected. After this check, transaction and block sizes are also compared before accepting or rejecting the new block. Eqs. (iv), (v) and (vi) explains the block size comparison.

$$Block_{size}^{selfish\_miner} = \frac{Sum\,of\,the\,block\,sizes\,created\,by\,selfish\,miner}{total\,number\,of\,blocks\,created\,by\,selfish\,miner} \tag{iv}$$

$$Block_{size}^{selfish\_miner} = \frac{Sum\,of\,the\,block\,sizes\,created\,by\,honest\,miner}{total\,number\,of\,blocks\,created\,by\,honest\,miner} \tag{v}$$

If $Block_{size}^{honest\_miner} \leq Block_{size}^{selfish\_miner} \leq Block_{size}^{honest\_miner}$ then
Now, Eq. (vi) evaluates the block future state in ideal condition

$$Block^{future\_state} = Height\,of\,new\,block\,in\,Merkle\,Tree$$
$$- \frac{Sum\,of\,Expected\,height\,of\,blocks\,in\,Merkle\,Root\,Tree}{total\,number\,of\,transaction\,expected\,to\,be\,executed\,by\,honest\,block\,miners} \tag{vi}$$

If $Block_{size}^{honest\_miner} \mp \delta < Block_{size}^{selfish\_miner} < Block_{size}^{honest\_miner} \mp \delta$ then block miner with new block is considered to be honest else miner is considered as selfish and new block is put in majority block attack category. Here, δ is acceptable minor variation in block size. δ varies with challenge. A larger challenge results in larger block size and a smaller challenge results in smaller block size. In this work, 1-bit challenge to Hash plus exponentiation challenge is analyzed for proposed work. In analysis, it is observed that the success probability of selfish miner and majority attack decreases with increase in honest blocks and heavy computational challenge.

An experimental analysis of blockchain network creation is performed over Intel Core i5-7200U CPU@2.5 GHz, 4 GB RAM and 64-bit operating system. In this experimental analysis, variation in blocks mined per second is observed with increase in challenge. A challenge variation from 1-bit to Hash with exponentiation is analyzed and results are shown in Fig. 1. Results shows that there is exponential decrease in number of blocks mined with increase in challenge. 1-bit challenge to 35-bit challenge verification process is 'match and miss' process. In this process single or multiple players has to predict, compute or guess bits at specific positions. Increase in number of bits verified increased the trust.

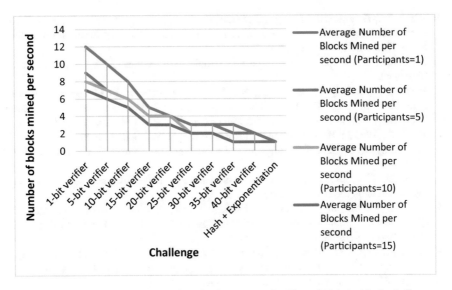

Fig. 1. Impact on number of blocks mined per second with variation in block challenge.

5 Conclusion

This work proposes a PoG consensus algorithm for blockchain network. It is observed that blockchain network can be protected from selfish miner and majority attack with increase in computational challenge. A heavyweight computational challenge consumes resources but provides higher security against selfish miner and majority attack. Whereas, lightweight computational challenge are suitable for resource constrained devices but provides comparatively lighter security against said attacks. However, a multi-round multi bit PoG concept for single or multiplayer game enhances the security for resource constrained devices as well. Thus, challenge variation from 1-bit to 35-bits is considered for analysis. Results shows that there is exponentiation decrease in blocks minded with increase in computational challenge but computational heavy challenge provides higher security. In single or multi-players PoG concept selfish and honest miners are identified through proper verification process. This verification process identifies the position of block in Merkle root tree and compares the block and transaction sizes. A large variation in block or transaction sizes indicates the chances of selfish miners. In future, single and multiplayer PoG will be extended for various lightweight cryptographic primitives and protocol based challenges in order to enhance the security levels.

References

1. All Cryptocurrencies. https://coinmarketcap.com/all/views/all/
2. Zheng, Z., Xie, S., Dai, H., Chen, X., Wang, H.: An overview of blockchain technology: Architecture, consensus, and future trends. In: 2017 IEEE International Congress on Big Data (BigData Congress), pp. 557–564. IEEE, June 2017

3. Wang, W., Hoang, D.T., Xiong, Z., Niyato, D., Wang, P., Hu, P., Wen, Y.: A survey on consensus mechanisms and mining management in blockchain networks. arXiv preprint arXiv:1805.02707, pp. 1–33 (2018)
4. Nakamoto, S.: Bitcoin: a peer-to-peer electronic cash system (2008). http://bitcoin.org/bitcoin.pdf
5. Bach, L.M., Mihaljevic, B., Zagar, M.: Comparative analysis of blockchain consensus algorithms. In: 2018 41st International Convention on Information and Communication Technology, Electronics and Microelectronics (MIPRO), pp. 1545–1550. IEEE, May 2018
6. Jiang, S., Wu, J.: Bitcoin Mining with transaction fees: a game on the block size. In: Proceedings of the 2nd IEEE International Conference on Blockchain (Blockchain 2019), May 2019
7. Swanson, T.: Consensus-as-a-service: a brief report on the emergence of permissioned, distributed ledger systems. Report, April 2015
8. Dey, S.: Securing majority-attack in blockchain using machine learning and algorithmic game theory: a proof of work. In: 2018 10th Computer Science and Electronic Engineering (CEEC), pp. 7–10. IEEE, September 2018
9. Liu, Z., Luong, N.C., Wang, W., Niyato, D., Wang, P., Liang, Y.C., Kim, D.I.: A survey on applications of game theory in blockchain. arXiv preprint arXiv:1902.10865 (2019)
10. Dimitri, N.: Bitcoin mining as a contest. Ledger **2**, 31–37 (2017)
11. Stone, A.: An examination of single transaction blocks and their effect on network throughput and block size. Self-published Paper, June 2015. http://ensocoin.org/resources/1txn.pdf
12. Liu, X., Wang, W., Niyato, D., Zhao, N., Wang, P.: Evolutionary game for mining pool selection in blockchain networks. IEEE Wirel. Commun. Lett. 1 (2018)
13. Xiong, Z., Feng, S., Niyato, D., Wang, P., Han, Z.: Optimal pricingbased edge computing resource management in mobile blockchain. In: 2018 IEEE International Conference on Communications (ICC), Kansas City, Kansas, May 2018
14. Jiao, Y., Wang, P., Niyato, D., Xiong, Z.: Social welfare maximization auction in edge computing resource allocation for mobile blockchain. In: 2018 IEEE International Conference on Communications (ICC), Kansas City, Kansas, May 2018

Line of Sight (LOS) Stabilization Using Disturbance Observer (DOB)

Akansha Rautela$^{(\boxtimes)}$ and R. P. Chauhan

Department of Physics, National Institute of Technology Kurukshetra,
Kurukshetra, India
akansharautela.16@gmail.com, chauhanrpc@gmail.com

Abstract. The Line Of Sight (LOS) electro-optical sighting system based on gyroscopic stabilized gimbal staging is difficult to control because of the non-linearities, friction, disturbances and many other unknown parameters associated with the outside environment. Disturbance Observer (DOB) provides an estimate of the disturbance, which is used to perform disturbance compensation by using negative feedback. To meet advanced LOS disturbance rejection requirement, in this work, compensator and PI controller with DOB are designed for an electromechanical gimbal. The system modelling and control design are carried out in MATLAB/Simulink. Simulation results show the performance enhancement of control structure with DOB in the presence of disturbance and measurement noise. The work focuses on command following, disturbance-rejection specifications as well as robustness of the system to be met.

Keywords: Line of Sight · Stabilization · Disturbance observer · PI · Gimbal · Robustness

1 Introduction

In fire control systems (FCS), performance improvement of LOS stabilization sub-system enhances its mission capabilities [1]. The main function of LOS stabilization systems can be defined briefly as maintaining or controlling the LOS of one object relative to another one. The stabilization sub-system minimizes the jitter on the LOS, thereby enhancing the image quality for an electro optical (EO) sighting system. EO imaging sensors mounted on a mobile platform usually require some form of control to stabilize the sensor pointing vector along the target LOS [2, 3]. The present work deals with the stabilization of the platform for mobile land vehicle. In land vehicles, typical disturbance spectrum of the platform is about 20 deg/sec at low frequencies (1 to 3 Hz) in yaw and pitch axis. To achieve the desired specifications, PI controller and compensators are designed along with disturbance observer to reject disturbances to increase the robustness of the design. DOB based control design provides an alternate way to improve disturbance rejection and robustness against uncertainties without sacrificing nominal control performance [4].

The remnant of the paper is structured as follows. Section 2 introduces the system description of the gimbal system. The problem statements have been formulated in

© Springer Nature Switzerland AG 2020
P. Karrupusamy et al. (Eds.): ICSCN 2019, LNDECT 39, pp. 765–772, 2020.
https://doi.org/10.1007/978-3-030-34515-0_81

Sect. 3. Sections 4 deals with the control strategy, while the simulation results have been presented and compared in Sect. 6. Finally, Sect. 8 concludes the paper.

2 System Description

Figure 1 shows the block diagram of gimbal system. The system under consideration is dual axis stabilized platform that consists of inner and outer gimbals [6]. In this work, the stabilization in only one axis is demonstrated. The plant consists of permanent magnet DC (PMDC) motor-driven gimballed payload followed by power amplifier and controller. Microelectromechanical system (MEMS) gyro is used to measure angular rate whereas encoder is used to measure position of the axis.

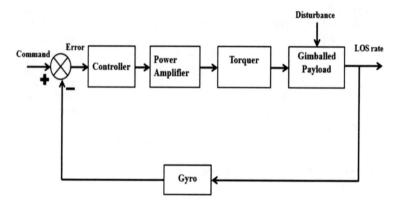

Fig. 1. Block diagram of gimbal system

3 Problem Formulation

The disturbances in a system can be due to bearing and motor friction, unbalanced aerodynamics, vibration force [5]. Due to these factors resolution of the imaging system degrades. By using high performance, low power consumption MEMS gyro, these disturbances can be sensed to generate appropriate control actions, so that the effect of these disturbances on the system performance can be significantly minimized. The disturbance rejection ratio and command following performance are of primary concern. Design specifications are given in Table 1:

Table 1. Problem statement

Specification	Required
Gain margin	≥ 6 dB
Phase margin	$\geq 30°$
Steady state error	0.00
Disturbance freq.	1 Hz and 3 Hz
LOS jitter	$\leq 100(1\sigma)$

The parameters of the system are given in Table 2:

Table 2. Plant parameters

Parameter	Description	Unit
$J = 0.325$	Moment of inertia of the inner gimbal	[Nm]
$B = 0.56$	Viscous friction coefficient of motor	[Nm/rad/sec]
$K_t = 0.8$	Torque constant	[Nm/A]
$K_b = 0.8$	Back EMF constant	[V/rad/sec]
$K_g = 0.1$	Rate gyro scale factor	[V/(rad/sec)]
$W_g = 628$	Natural frequency of the rate gyro	[rad/sec]
$L = 0.005$	Armature inductance	[Henry]
$R = 12$	Armature resistance	[Ohms]
$K_a = 10$	Power amplifier gain	–
$W_{r2} = 1884$	Gimbal resonance frequency	[rad/sec]
$W_{r1} = 1550$	Gimbal anti-resonance frequency	[rad/sec]
$\xi_r = 0.07$	Damping coefficient at resonance	–
$\xi_g = 0.7$	Damping coefficient of the rate gyro	–

4 Controller Design

4.1 Compensator

Here, bode plot approach has been utilized for designing compensators. The design of compensators has been done using SISOTOOL in MATLAB. The transfer function in Eq. (1) represents a phase lead-lag compensator.

$$G_c(s) = \frac{(1 + aTs)}{(1 + Ts)} \frac{(1 + bTs)}{a(1 + Ts)} \tag{1}$$

Here, a > 1 for lead compensation and b < 1 for lag compensation.

4.2 PI Controller

The basic architecture of PI controller is shown in Fig. 2.

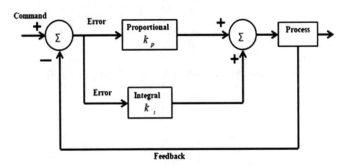

Fig. 2. Block diagram of PI controller

The proportional term results in a fast response, but generates offset error. The integral term removes offset error, but makes the response slow. Thus, a proper tuning is required for the efficient working of the PI controller. This is achieved by adjusting the controller parameter variables. The LOS jitter at the position output for 1 Hz and 3 Hz is shown in Figs. 3 and 4.

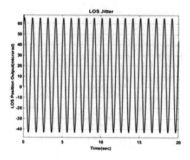

Fig. 3. LOS jitter of compensator at 1 Hz   **Fig. 4.** LOS jitter of compensator at 3 Hz

5   Proposed Robust Design

In the presence of uncertainties, the design of a controller to provide desired performance of the system is quite a demanding task [9]. To solve this problem, robust control methods have been developed [7]. In DOB-based control system design, robust stability of the DOB loop and over-all control system are analyzed [8, 10]. DOB uses nominal plant model and a low pass Q-filter to estimate the disturbance acting on the system. Control based on Disturbance observer is an effective method to achieve robustness against disturbances and model uncertainties. DOB utilizes simple control configuration. Basic architecture of DOB is shown in Fig. 5 and simplified model of DOB is shown in Fig. 6. The basic structure of Q-filter is given in Eq. (2):

$$Q(s) = \frac{1 + \sum\limits_{k=1}^{N-r} a_{Nk}(\tau s)^k}{1 + \sum\limits_{k=1}^{N} a_{Nk}(\tau s)^k} \tag{2}$$

Where,

$a_{Nk} = \frac{N!}{k!(N-k)!}$, N = Order of the filter, r = Relative degree of the filter.

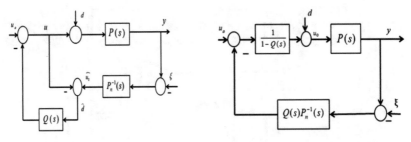

Fig. 5. Basic architecture of DOB   **Fig. 6.** Plant with DOB

6 Simulation Results

The gimbal system performance using different controllers with DOB and without DOB has been analyzed in this section. For platform disturbance of 0.3 rad/sec at 1 Hz and 0.1 rad/sec at 3 Hz, the closed loop performance of stabilized platform is discussed for four different cases, namely, compensator, PI, compensator with DOB and finally, PI with DOB. The proposed control scheme for system stabilization in closed-loop for compensator and PI is shown in Figs. 7 and 8 respectively.

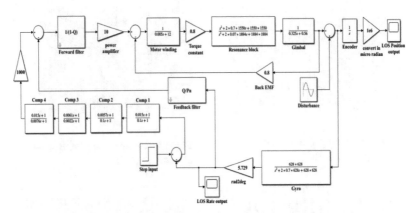

Fig. 7. Block diagram of Compensator with DOB

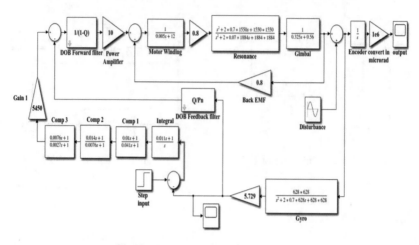

Fig. 8. Block diagram of PI with DOB

Figures 9, 10, 11 and 12 shows the residual jitter in LOS position output of controllers with DOB for 0.3 rad/sec at 1 Hz and 0.1 rad/sec at 3 Hz.

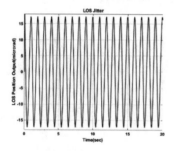

Fig. 9. Compensator-DOB residual jitter at 1 Hz

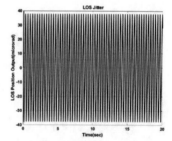

Fig. 10. Compensator-DOB residual jitter at 3 Hz

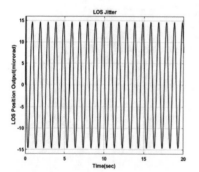

Fig. 11. PI - DOB disturbance at 1 Hz

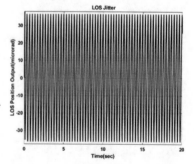

Fig. 12. PI - DOB disturbance at 3 Hz

7   Comparative Analysis for Performance Evaluation

The characteristics of the responses of the proposed DOB design with two controllers are compared with the existing controllers for different frequencies in Table 3.

Table 3. Comparison of different control techniques

Specifications	Disturbance	Units	Lead-lag compensator		PI controller	
			Without DOB	With DOB	Without DOB	With DOB
Settling time	–	sec	0.09	0.09	1.2	1.2
Gain Margin	–	dB	7.5	7.5	8	8
Phase Margin	–	deg	50	50	42	42
Steady state error	–		0.00	0.00	0.00	0.00
Residual jitter in LOS	0.3 rad/sec at 1 Hz	μ rad	38.5	12.36 (Q-filter freq. 3 Hz)	32.9	10.54 (Q-filter freq. 3 Hz)
Residual jitter in LOS	0.1 rad/sec at 3 Hz	μ rad	38.1	26.79 (Q-filter freq. 3 Hz)	36.7	25.85 (Q-filter freq. 3 Hz)

8   Conclusions

The study has been done such that the system must be command following and attenuates disturbance. This work mainly focuses on the design algorithm of DOB for the gimbal system. For this purpose various control methods have been designed. DOB has been included with compensator and PI controller. DOB is used to reject disturbance at particular frequency by increasing the gain at that frequency. Comparison of different control techniques has been obtained for gimbal system for desired specifications. The PI controller achieves the specifications readily and with inclusion of DOB in feedback, it significantly attenuates the disturbance robustly as compared with other techniques. Whereas, compensator and PI alone cannot achieve disturbance rejection specification, as for disturbance rejection high gain is required at given frequency. Results proclaim that the performance of PI with DOB is better as compared to other three cases namely, compensator, PI controller and compensator with DOB.

References

1. Singh, S., Marathe, R.: SMC based LOS stabilization of electro-optical sighting system. In: 1st IEEE International Conference on Power Electronics, Intelligent Control and Energy Systems (ICPEICES) (2016)
2. Singh, A., Thakur, R.: Design and optimal control of line of sight stabilization of moving target. IOSR, J. Electr. Electron. Eng. **9**, 27–32 (2014)

3. Ansari, Z.A., Nigam, M., Kumar, A.: Quick reaction target acquisition and tracking system. In: Proceedings of International Conference on Computer Vision and Image Processing, CVIP 2016, vol. 2. Springer, Roorkee (2016)
4. Choi, Y., Yang, K., Chung, W.K., Kim, H.R., Suh, I.H.: On the robustness and performance of disturbance observers for second-order systems. IEEE Trans. Autom. Control **48**, 315–320 (2003)
5. Jia, R., Nandikolla, V.K., Haggart, G., Volk, C., Tazartes, D.: System performance of an inertially stabilized gimbal platform with friction, resonance, and vibration effects. Hindawai J. Nonlinear Dyn. **2017**, 1–20 (2017)
6. Habashi, A.G., Ashry, M.M., Mabrouk, M.H., Elnashar, G.A.: Controller deign for line of sight stabilization system. Int. J. Eng. Res. Technol. (IJERT) **4**(11), 650–658 (2015)
7. Kim, B.K., Chung, W.K.: Advanced disturbance observer design for mechanical positioning systems. IEEE Trans. Ind. Electron. **50**, 1207–1216 (2003)
8. Ansari, Z.A., Nigam, M.J., Kumar, A.: Improved scheme for quick positioning of dual gimbal sight. In: Third International Conference on Advances in Control & Optimization of Dynamical Systems (ACODS 2018), IFAC-PapersOnLine, Hyderabad, vol. 51, no. 1, pp. 686–690 (2018)
9. Zhou, P., Dai, W., Chai, T.Y.: Multivariable disturbance observer based advanced feedback control design and its application to a grinding circuit. IEEE Trans. Control Syst. Technology **22**, 1474–1485 (2013)
10. Briat, C.: Linear Parameter-Varying and Time-Delay Systems. Analysis, Observation, Filtering & Control. Springer, Heidelberg (2015). ISBN 978-3-662-44049-0

Design Challenges for 3 Dimensional Network-on-Chip (NoC)

N. AshokKumar[1]([✉]), P. Nagarajan[1], SathishKumar Selvaperumal[2], and P. Venkatramana[1]

[1] Centre for VLSI and Embedded Systems,
Department of Electronics and Communication Engineering,
Sree Vidyanikethan Engineering College, Tirupati, India
ashoknoc@gmail.com
[2] School of Engineering, Asia Pacific University of Technology and Innovation,
Kuala Lumpur, Malaysia

Abstract. NoC (Network-on-Chip) is a technology is expected for resolve the problem of short imminence of buses. This technology approach is to design the information exchange between the subsystem of IP cores. The usage of common buses, which have the problem that they cannot scale in concern fixed and also the number of resources grows. NoC is proposed for resolving this short coming and especially implementing for switches/micro routers and resources in network strategy. The 3 Dimensional - NoC design paradigm is anticipated for ASIC design. The significant main impetus at the back move to 3D-NoC based arrangements is the deficiency of VLSI between chip correspondence plan strategy for the profound sub-micron chip producing innovation. Planning a proficient Network-on-Chip (NoC) engineering, while fulfilling the application execution imperatives is an intricate procedure. The outline issues traverse a few deliberation levels, going from abnormal state application displaying to physical design level execution. Probably the most critical stages in outlining the 3D-NoC include: Analyzing and portraying application activity. Combining the NoC topology for the application. Mapping and official of the centers with the NoC parts. Discovering ways for the activity streams and holding assets over the NoC. Picking 3D-NoC design parameters, for example, the information width of the connections, cradle sizes and recurrence of operation. Checking the outlined NoC for accuracy and execution.

Keywords: 3D NoC (3 Dimensional Network-on-Chip) · TSV (Through Silicon Via) · Topology · Routing

1 Introduction

A network-on-chip is framed by the basic building blocks of Physical links, router and interface topologies. The first is the links that physically make the contact between the nodes. The second is router, that is for information exchange protocol (the decentralized logic behind the communication protocol) strategy. Building reproduction [5], combination and copying models for the NoC Guarantee solid framework operation from untrustworthy parts keeping in mind the end goal to handle the configuration

© Springer Nature Switzerland AG 2020
P. Karrupusamy et al. (Eds.): ICSCN 2019, LNDECT 39, pp. 773–782, 2020.
https://doi.org/10.1007/978-3-030-34515-0_82

intricacy and meet the tight time-to-business sector limitations, it is vital to robotize the vast majority of these NoC outline stages. To accomplish outline conclusion, the diverse stages ought to be incorporated likewise in a consistent way (Fig. 1).

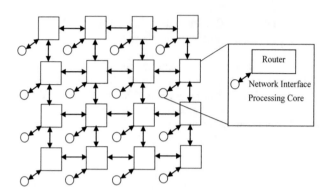

Fig. 1. A mesh based NoCs architecture

SoCs that utilize a NoC [7] base are financially attainable in the event that they can be utilized as a part of a few item variations, and if the outline can be reused in various application ranges. Then again, effective items must give great execution attributes, in this way requiring committed arrangements that are custom-made to particular needs. As an outcome, the NoC outline challenge lies in the capacity to plan equipment advanced, adjustable stages for every application area. PC helped amalgamation of NoCs is especially vital on account of utilization particular frameworks on chip, which more often than not contain registering and capacity varieties of different measurements and in addition joins with different limit prerequisites. Additionally, originators may utilize NoC combination as a method for developing arrangements with different qualities that can be thought about viably just when a point by point model is accessible. Therefore combination of NoCs can be utilized for looking at models. Obviously, amalgamation may likewise be exceptionally proficient for outlining NoCs [7] with general topologies as, for instance, multi-preparing frameworks with homogeneous centers.

Configuration and advancement can be accomplished by encouraging the coordination of domain specific calculation assets in a fitting and-play outline style. Standard interface attachments, for example, the Open Core Protocol (OCP) have been produced for this reason and backing the utilization of a typical NoC as the premise for framework coordination. A pertinent assignment of these interfaces is to make the NoC versatile to the distinctive elements of the coordinated centers (e.g., information and location transport width). NoC structures are pushing the advancement of customary circuit outline procedure to bargain viably with utilitarian assorted qualities and unpredictability. At the application level, the key outline test is to uncover undertaking level parallelism and to formally catch simultaneous correspondence in models of calculation. At that point, abnormal state simultaneous assignments must be mapped to the basic correspondence and calculation assets. At this level, a theoretical model [8] of the equipment engineering is normally presented to the mapping device, so that region and force assessments can be given in the early outline stage, and diverse target capacities (e.g. minimization of correspondence vitality) can be considered to assess the plausibility of option mappings. In this setting, a basic stride in correspondence mapping is the NoC engineering union for its noteworthy effect on general framework execution, which is progressively correspondence, overwhelmed. At last, it is imperative to accomplish a dependable NoC operation by giving strength from perpetual and transient postponement and rationale mistakes in the framework. With a specific end goal to shield the framework from blunders that happen in the correspondence sub-framework, we can utilize mistake recuperation components that are utilized as a part of conventional large scale systems. As the blunder location/redress capacity, zone control overhead and execution of the different mistake recognition/remedy plans vary, the decision of the mistake recuperation plan for an application includes various force execution dependability exchange offs that must be investigated.

2   Motivation

Inspirations of the proposition are the idea of VLSI design and execution of on-chip switches with worthwhile elements and attributes to create networks on-chip for multiprocessor frameworks. Since the primary center of the exploration is the NoC switches plan idea, then this proposal will talk about some issues and parts of the NoC switch design and its supporting particular segments. Hence a few subjects, for example, exchanging strategy, directing calculation, system stream control, and the inside NoC switch pipeline micro architecture including its pipeline control are the primary extents of this theory. The examination investigates the NoC-based multi-processor frameworks outfitted with a programming model, and application programming interface (API) of the multiprocessor framework with disseminated memory design are a piece of the exploration premiums led in our foundation. Nonetheless, the plans of NoC-based multiprocessor frameworks, on-Chip Network Interface (OCNI), and parallel programming models have passed the extent of this postulation. The general goal of this doctoral proposition is to show a configuration idea and nonex-clusive design of a NoC model with particular elements supporting particular admin-istrations. The particular destinations of this theory are proposed to enhance the current procedure, outline ideas and qualities of NoC switches that have been created so far in the NoC research range. The particular targets are:

1. To exhibit another wormhole changing strategy and to demonstrate hypothetically the invaluable qualities contrasted with conventional wormhole exchanging, in which the head-of-line-blocking issue is settled without utilizing virtual channels.
2. To introduce another hypothesis for gridlock free multicast directing calculation and to demonstrate the profitable attributes and VLSI usage contrasted with existing halt free multicast steering techniques, in which the multicast reliance (conflict) issue is understood without the utilization of virtual channels.
3. To display another way to deal with outline runtime versatile directing determi-nation systems in view of dispute and data transfer capacity data or mix of both

data, and to demonstrate their worthwhile execution attributes contrasted with other versatile steering choice methodologies exhibited in the written works.

4. To exhibit another and more adaptable Switched-Virtual Circuit (SVC) arrangement strategy to outline a NoC switch with association situated ensured data transfer capacity benefit and to demonstrate the favorable VLSI design and procedure to join the ensured throughput administration with the connectionless best-exertion administration contrasted with existing approaches displayed so far in the NoC research territory.

5. To present an adaptable VLSI miniaturized scale design of a NoC correspondence framework that can adaptable bolster the previously stated novel hypothesis and techniques.

3 Design Problems 3D Router

One of the configuration issues in the switch outline by Kumar [1], is that it utilizes 37-bit wide information lines for all the info/yield ports. This implies that each port comprises of 37 lines for the information and 37 lines for the yield. On each clock cycle of the switch, one piece is sent through an information line. Since this information line has a little defer contrasted with the clock cycle on which the switch is working, a bit can be sent at a much quicker interim, contrasted with the clock cycle of the switch. Along these lines the limit of an information line is all the more effectively used. In the event that the interim for sending somewhat through an information line would be diminished, more bits could be transported, which implies less information lines would be required so as to transport an equivalent measure of bits. Another issue in the switch outline by S. Kumar is that it just works in one clock space (Fig. 2).

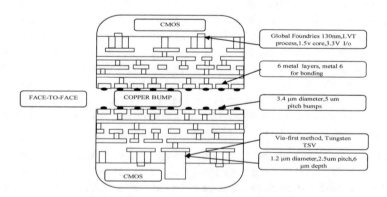

Fig. 2. 3 Dimensional IC technology with consequent parameters

This implies that when utilized as a part of an unmodified NoC setup, every one of the switches ought to work on the same clock recurrence. This is difficult to accomplish, particularly between two switches sitting on various levels. An answer must be found which empowers the operation in various clock areas. The physical configuration issues that emerge when utilizing the switch outline of S. Kumar, are the requirement for Trough Silicon Vias (TSVs) for each information line in the vertical bearing. As clarified before, there are 37 info and yield information lines in every bearing. Since a TSV has an expansive impression contrasted with a transistor, a great deal of chip territory will be required so as to suit all the TSVs which will bring out cost increment [6]. On the off chance that the quantity of required TSVs could be diminished, less region would be taken by them, bringing about lower costs. This theory is attempts to tackle the aforementioned issues by utilizing another configuration system. This theory attempts to solve the configuration issues of the switch by Kumar [1] through the following steps:

1. Low use of the limit in the information lines
2. Communication among various clock spaces
3. Large impression of required TSVs.

4 Network on-Chip for FPGAs

Originators of NoCs for FPGAs [2] need to choose whether to execute a hard (silicon) system of restricted configurability or a delicate (configurable) system utilizing the valuable reconfigurable assets. Numerous hard and delicate systems have been proposed and they exchange off between the two methods of execution in various ways.

1. Hard Routers
2. Soft Routers

4.1 Deadlock Avoidance

Keeping in mind the end aspiration to sustain a strategic distance from system halt all cyclic conditions must be disposed of. The West-First, North-Last calculation is regularly used to force an aggregate request on the system channels of cross section systems. This is actualized by compelling bundles to travel west before voyaging north or south and east before north. This is a straightforward arrangement, yet expels some way differing qualities from the system.

4.2 Topology

The topology of a system is the course of action and availability of hubs and channels inside a system. System hubs may have neighborhood associations with permit information on and off the system, which are known as terminal hubs and structure a subset of the arrangement of hubs. Non-terminal hubs are frequently alluded to as switch hubs. The level of the hub is the quantity of channels associating with it. Out degree and in degree can vary, however these are thought to be equivalent unless generally expressed. Channels interface with correctly two hubs and are directional [7]. They ought not to be mistaken for steering channels on a FPGA. The reasonableness of a topology for a system relies upon various exchange offs. When all is said and done the execution of a system is characterized by the throughput, inertness and way differing qualities. Dormancy: The inertness of a system relies on the quantity of bounces an information must go between its source and destination and the inactivity of every jump. A jump is typically taken to be a switch or switch and an area of interconnection wire (Figs. 3 and 4).

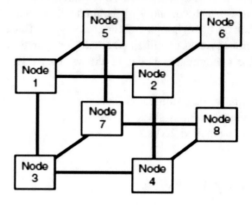

Fig. 3. 3-D Hypercube

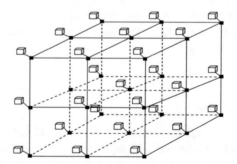

Fig. 4. 3D-Noc 3 × 3 × 3 mesh topology

The quantity of jumps can be improved for the normal case or the most pessimistic scenario. The idleness of every bounce will rely on the multifaceted nature of the switch. The quantity of bounces between two terminals in a system may shift contingent upon the way picked. The base jump includes two hubs the system is the base number and of hubs information must pass however to get from the info hub to the yield hub over every single conceivable way between the two hubs. The distance across of a system is the most astounding least jump number over all the conceivable information yield hub sets in the system. Transmission capacity: The transmission capacity of a channel is the most extreme rate of information move in that channel. It can be figured by increasing the channel bit width with the most extreme working recurrence. The transfer speed of channels need not be uniform over the system. Data transmission regularly given is a measure of system execution, however ought not to be mistaken for throughput. The throughput of a system is the rate of information exchange at run-time.

The throughput will rely on upon the hub engineering and the movement designs, and the topology. Way Diversity: The way differences is the quantity of ways between any two terminal hubs in a system. Way differing qualities in NoCs are utilized for adaptation to non-critical failure and blockage administration by element steering calculations. A circuit can be set up through a system by allotting an arrangement of hubs and directs in a way between an information hub and an yield hub. These assets have to be adequate to permit information to be exchanged between the information hub and yield hub and ought not to be utilized for some other correspondence whilst the circuit is set up.

4.3 Datapackets

System information is assembled into parcels. Bundles can be a small amount of the channel width or be particularly bigger. A system may have an altered or variable bundle length. Ordinarily a bundle will be prefixed with data about its destination. In a system with variable length parcels the length must be incorporated or a footer marker annexed to the end of the bundle. Parcels are typically partitioned into bounces [9]. A bounce is generally the same size as the channel width and is the biggest measure of information that can be moved in parallel. It is in this way the least measure of information to which assets can be allotted.

4.4 Routing

The strategy by which bundles are guided from source to destination can be deterministic, unaware or versatile. Systems with no way differences are compelled to utilize deterministic directing techniques; different topologies may utilize unaware or versatile calculations. Unaware calculations course bundles with no learning of the condition of the system. Deterministic [2] calculations are a subset of unaware calculations [2]. Unmindful calculations are anything but difficult to execute and make gridlock free. Versatile calculations may utilize data about deficiencies or clog to course parcels or circuits (Fig. 5).

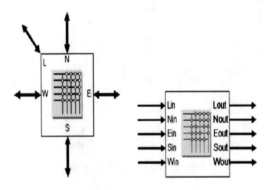

Fig. 5. Router view

4.4.1   Source Routing

Source steering [7] requires the way through changing hubs to be pre-figured at source. This requires a worldwide information of the system state and the condition of the system, however can decrease idleness however switch hubs [3].

4.4.2   Table-Based Routing

Table based steering can be actualized at the source or at switch hubs. Just the segment of the table relating to a specific hub should be put away at that hub. The table may store various ways for a specific destination from a given hub.

4.4.3   Algorithmic Routing

Algorithmic steering [7] can be utilized to register the yield port from the destination ID. For instance, in a cross section to organize the hub needs just decide the relative bearing of the destination from the destination ID and the present hub ID if outright tending to is utilized. Generally the bearing can be processed from a relative location and the header bundle upgraded for hub.

5   Comparisons of NoC and 3D-NoC

The stacking of two or more bites the dust utilizing vertical interconnects is known as 3-D [11] Network-on-chip. In order to meet the disadvantage of 2-D NoC approach, another system methodology is produced which is called 3-D NoC. By stacking of passes on with system approach, the measure of the chip is decreased furthermore. It lessens the postponement and it just uses a few quantities of bounces. The length of the correspondence channel is likewise diminished by stacking the 2-D layers. The two 2-D layers [9] are associated utilizing Through-Silicon-vias (TSV) or essentially we can call it vertical interconnects.

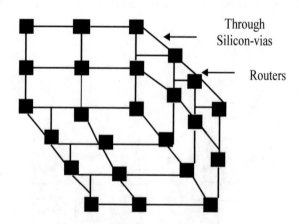

Fig. 6. 3D Network-on-Chip

In 3-D NoC, the engineering comprises of vertical and level interconnects which is utilized to navigate the bundles between various handling components in various layers. In this methodology additionally, the switch comprises of switches, registers and cradles and it likewise has six info ports and six yield ports. These ports are north, south, east, west, here and there. All the ports are utilized to navigate the parcels through vertical interconnects to its neighboring here and there layers. The basic 3-D NoC engineering appears in the Fig. 6.

Three-dimensional NoCs are normal augmentations of 2D [9] outlines. 3D system structures [4] give a superior execution contrasted with customary, 2D NoC designs. Both work and tree-concerned NoCs are equipped for accomplishing better execution in a 3D IC atmosphere than the customary 2D usage. The cross section based structures show noteworthy execution picks up as far as throughput, inertness, and vitality dissemination with a little region overhead. Then again, the 3D tree-based NoCs accomplish huge addition in vitality dissemination and territory overhead with no adjustment in throughput and idleness. Be that as it may, if the NoC changes are intended to be as quick as the interconnect, even the 3D tree-based NoCs will display execution advantages as far as inactivity and transfer speed are concerned. The NoC worldview keeps on drawing in critical examination consideration in both the educated community and industry. With the approach of 3D [10] ICs, the achievable execution profits by NoC system will be much. Thus, this will encourage reception of the NoC model as a standard configuration answer for bigger multicore framework chips than before.

6 Advantages of 3D-NoC

The advanced incorporated circuit patterns of today are toward very prominent levels of reconciliation. With apparently ceaseless gadget scaling and the presentation of 3D [4] advances, framework architects have extra gadgets available to them with which they can consolidate more usefulness into a solitary computerized chip. Such frameworks, with a developing number of modules, have come to require between part I/O data transmission which must be met by system on-chip (NoC) interconnection fabrics: this extraordinarily entangles the compositional arranging of such frameworks. Superior on-chip systems are pictured in three measurements, and ordinarily they have complex topologies in which the system hubs have high degrees of network. An essential disadvantage to VLSI NoC usage is that extremely customary, effective high radix system topologies lose quite a bit of their advantage when smoothed, and their physical outline is limited to a 2D plane.

Research studies in the field have recommended that there is a critical advantage to 3D NoC [8] executions where a hub may have six closest neighbors, over ones in two measurements with only four. Preparatory examinations demonstrate that relying on system topology, we may see NoC joins length diminishments of 30%–46% in a three level execution, versus a solitary level. This work explores plan tradeoffs in application particular system on-chip outline, and measures the advantages of a 3D NoC as far as execution, region, and force utilization when contrasted and a practically proportional 2D NoC.

For each of these network configurations we can quantify the differences between a 2D and 3D NoC, also varying the number of allowed 3 Dimensional IC tiers. The different network configurations will be evaluated using the NoC simulator in stipulations of power consumption and area, so that we may determine which networks benefit the most from 3DIC technologies and find the magnitude of that benefit for a realistic application.

7 Conclusion

Finally, 3D-NoC test system is produced utilizing "C" programming language with the blending parameter. A novel 3D-NoC that is intended to enhance the parameters of throughput, delay, latency of the on-chip network. 3D NoC is reserved small by stacking the dynamic layer vertically. Elevated asset limitation is accomplished, since solitary chip is utilized to execute the equipments of all used in NoC. High versatility, decreased force utilization and low inertness are a portion of the empowering zone of NoC. Dependable exchanging model, directing plan with decreased force alongside particular topologies are the convention followed in Network on Chip.

References

1. Roy, S., Kumar, R., Prvulovic, M.: Improving system performance with compressed memory. In: Proceedings of 15th International Parallel Distributed Processing Symposium, pp. 630–636 (2001)
2. Nallathambi, G., Rajaram, S.: A particle swarm optimization approach for low power very large scale integration routing. J. Math. Stat. **10**, 58 (2014)
3. Murugappa, P., Al-Khayat, R., Baghdadi, A., Jezequel, M.: A flexible high throughput multi-ASIP architecture for LDPC and turbo decoding. In: Proceedings of the Design, Automation and Test in Europe Conference and Exhibition (TECE), pp. 1–6 (2011)
4. Jacob, P.: Predicting the performance of a 3D processor-memory stack. IEEE Des. Test Comput. **22**(6), 540–547 (2005)
5. Kao, Y.H., Chao, C.J.: 5th Conference of IEEE/ACM International Symposium on Networks on Chip, vol. 81 (2011)
6. Viswanathan, N., Paramasivam, K., Somasundaram, K.: Exploring optimal topology and routing algorithm for 3D network on chip. Am. J. Appl. Sci. **9**, 300–308 (2012)
7. Zhang, W., Hou, L., Wang, J., Geng, S., Wu, W.: Comparison research between XY and odd-even routing algorithm of a 2-dimension 3X3 mesh topology network-on-chip. In: WRI Global Congress on Intelligent Systems, GCIS 2009, vol. 3, pp. 329–333 (2009). https://doi.org/10.1109/GCIS.2009.110
8. Topol, A.W.: Three-dimensional integrated circuits. IBM J. Res. Dev. **50**(4/5), 491–506 (2006)
9. Seo, D.: Near-optimal worst-case throughput routing for two-dimensional mesh networks. In: International Symposium on Computer Architecture, pp. 278–287 (2004)
10. Ashok Kumar, N., Nagarajan, P., Vithyalakshmi, N., Venkataramana, P.: Quad-rail sense-amplifier based NoC router design. In: Hemanth, J., et al. (eds.) ICICI 2018, LNDECT 26, pp. 1449–1454. Springer (2019)
11. Ashokkumar, N., Kavitha, A.: A novel 3D NoC scheme for high throughput unicast and multicast routing protocols. Tech. Gazettee **23**(1), 215–219 (2016)

R-Assistant Robots for Unstable Emotions

Ayngaran Krishnamurthy$^{(\boxtimes)}$ and N. Pughazendi

Department of Computer Science and Engineering,
Panimalar Engineering College, Chennai, India
ayngaranrk@gmail.com, pughazendi@gmail.com

Abstract. Robots are becoming more sociable nowadays. Recently, socio robots are used to mimic the human-like intelligence by taking human emotions, posture and so on into consideration. The designed R-assistant is the real-time robotic assistant which accompanies humans to eliminate their stress outburst and loneliness. In this scenario, R-Assistant with its underlying Machine Learning architecture with data analytics model perceives human emotions with a highly precise technique called Image processing, which can appropriately sense the facial expression of humans and delivers potential insights to predict human emotions.

Keywords: Assistant · Real time · Machine Learning · Emotions · Loneliness

1 Introduction

From the space stations to a friendly human assistants, robots are playing a predominant in almost all of our day-to-day activities. The research potential on robotic applications are significantly increasing to make them feel, understand and react to the human emotions and emerge as a perfect humanlike partner to the evolving smart human population.

The project R-assistant as you might have guessed it from the name is a robot it is not just any mere robot you can see elsewhere, the robot has a touch of human emotions, that is being built with the idea of how to react and respond to any human reactions, and help their owners when they are in need.

2 Literature Survey

Before making a solution for the problem the problem was studied much deeper in order to be sure about the needs that are to be met to make people either to feel less lonely or not lonely at all [1]. Here I have shortlisted some articles as a part of the literature survey for your convenience [2]:

The article [3–5] says that the loneliness epidemic is heightening the risks of premature deaths as linked by a new global study. In a conference organized by American Psychological Association Dr. Julianne Holt-Lunstad professor of Psychology at Brigham Young University presented two great studies, the study stated that people with deep social ties has 50% reduced risk of premature/early death? In the

© Springer Nature Switzerland AG 2020
P. Karrupusamy et al. (Eds.): ICSCN 2019, LNDECT 39, pp. 783–790, 2020.
https://doi.org/10.1007/978-3-030-34515-0_83

conference D. Julianne Holt-Lunstad stated that [6–12]," The project is being worked on since the last month. The robot is being designed with upcoming concepts and technologies to make sure that this can be the best companion for any human to have.

3   The R-Assistant

The R-assistant is everything you can imagine it to be, it sobs with you when you're sad or plays music for you when you are happy. A lot of similar papers on robotics, datasets and about the robots created were gone through, yet there is no idea/record that it can have its own emotions in the form of 0's and 1's. It is being built to make sure that it can meet the needs of every human out there. It is found to be very much helpful to those who feel lonely or insecure. The project is not yet finished but it is in its progressing stage. The robot is an android to make it easy to connect it to any required personal devices, due to the reason it is connected to your devices it acts as an interface between you and your personal devices thereby helping you access your devices without needing you to even touch your device. The robot traces the human facial deteriorations to determine their mood without the human actually telling the robot of his/her mood. The facial deteriorations are identified and the counteract fed for the emotion/mood will be executed. The robot is being fed with datasets of every human activity and their counteracts the project is of great interest involving Machine Learning in order to process the datasets, the branch of machine learning being used is the Deep Learning concept under the Image Processing technique, the algorithm used to process the real-time facial detections is the K-Nearest Algorithm. It also helps us to look over our loved ones whenever needed via its cameras in real time from anywhere at any time. Since it is still under the construction stage many efforts are being put to make sure that it is flamboyant among its family of robots. There is a safety dataset to help in times when there is an emergency situation, the dataset can be accessed by the user to add a few of their family or friend's numbers to the safety panel. This helps the robot contact them in case of emergency along with other hotlines

A. Methodologies Used
As said earlier to get the maximum potential the uprising and more advanced techniques are used to make it a complete companion, out of those the two of the major techniques are: the Machine Learning technique and Data Analytics. The Machine Learning techniques are implied to understand the user's needs, necessities, moods, favorites and also analyses the activities of humans to be a better companion. Now you might have two questions: 1. Why do you keep saying better companion, 2. What is so big about it the answers to your questions lie's in the next paragraph. It is called a better companion because, since the day the project started it was made that it must never feel less than human to make sure the robot is good at its work. Coming to the second question, the deal is that the robot is at first programmed with only the basic information of the client, the other actions and behavior of the user will be analyzed by the robot itself by the process of Deep Learning (a branch of machine learning that

capacitates the robot to learn from experiences without being explicitly programmed) under the Image Processing concept. The Data Analytics is used to guess the person's normal facial expressions from time to time and compare them to the real-time facial expressions to know of the person's mood and help accordingly. The robot is using machine learning as a basic step because only the reactions are fed as data sets into the robot but the actions and the according reactions are decided by the Machine Learning Technique. The algorithm used is the K-nearest algorithm to compare and contrast for better learning.

B. Equations and Process

(i) Image Processing: In this the picture is split up into a number of pixels each forming a form of matrix, these matrices are processed for the information. The common equations used for processing images are Laplace Transforms and Partial Differential Equations. Given below are some of the Partial Differential Equation used in image processing [1] (Pictures [7]). In Fig. 1 determines the feeling of the human based on their retinal changes in their eye (Fig. 2).

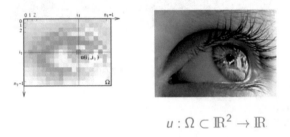

$$u : \Omega \subset \mathbb{R}^2 \to \mathbb{R}$$

Fig. 1. An image is seen as a function defined in continuous space

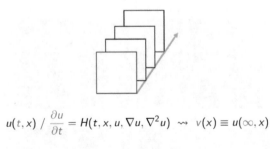

$$u(t, x) \ / \ \frac{\partial u}{\partial t} = H(t, x, u, \nabla u, \nabla^2 u) \ \rightsquigarrow \ v(x) \equiv u(\infty, x)$$

Fig. 2. A PDEs an evolution

Gaussian filtering
Let u_0 an image we defined (Fig. 3):

$$u_\sigma(x) = (G_\sigma * u_0)(x) \quad \text{avec} \quad G_\sigma(x) = \frac{1}{2\pi\sigma^2} \exp\left(-\frac{|x|^2}{2\sigma^2}\right)$$

$\sigma = 0$ $\sigma = 5$

$\sigma = 11$ $\sigma = 17$

Fig. 3. Gauss filtering

Heat equation
A linear PDE (Fig. 4)

$$\begin{cases} \frac{\partial u}{\partial t}(t, x) = \Delta u(t, x). & t \geq 0. \\ u(0, x) = u_0(x). \end{cases}$$

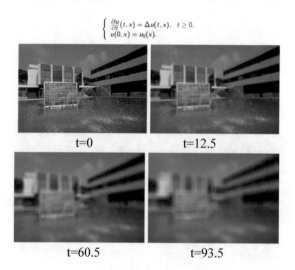

t=0   t=12.5

t=60.5   t=93.5

Fig. 4. Heat equation

A notion of class

Solution of the heat equation is a convolution (Fig. 5)

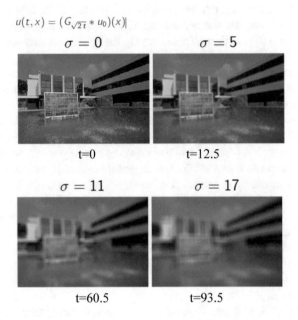

$$u(t,x) = (G_{\sqrt{2t}} * u_0)(x)|$$

$\sigma = 0$ $\sigma = 5$

t=0   t=12.5

$\sigma = 11$ $\sigma = 17$

t=60.5   t=93.5

Fig. 5. Gaussian filtering (One operation in a large neighbourhood) and Heat equation (A succession of local operations)

(ii) Neural Networks: The neural networks are the basis of Machine Learning the datasets given are processed by the networks. The neural network decided to be used is the multilayer neural network the process function of the neural networks is given below [2, 3].

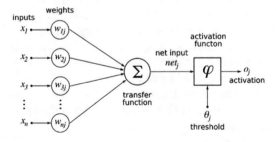

C. Datasets Used

Datasets are created for the robot to learn from them by the Machine Learning technique, every dataset is processed and it is made easily accessible. The datasets are made for the following categories they are: Medical, military, and various Speech and interaction models, A special dataset (of puns, jokes, metaphors, and cliché to make the interaction with the robot more realistic), Teaching, Safety Dataset.

(i) *Medical Dataset:* This dataset can be categorized according to the needs and age of the users, the dataset will come in three subcategories they are: child, adolescent, senior citizen. This is to help the robot act faster and efficiently to a close call medical emergency.

(ii) *Fight Dataset:* The fight dataset is common for all the fight dataset doesn't mean the robot can fight but the dataset is implemented so that it can counteract and defend their user from the people attacking them.

(iii) *Emotional Dataset:* The dataset can simply be defined as the heart of the robot, as we all know the truth is that robots don't have emotions or even understand our language, but what if I said that nowadays even you can make a robot weep, sob, cry and show anger?. Yes, it is possible what if we started to create the emotions in binary or hexadecimal codes. These codes are in association with the real-time facial recognition and corresponding binary/hexadecimal codes run to make sure that the robot counteracts or consoles the feelings of the user.

(iv) *Mood Dataset:* We all have a common problem that we face every day, the change of moods. There are various factors causing a change in mood the robot tries to talk out the issues if the user is ready to share the information, or else the robot plays music from set of genre to sooth the mood of the user to make them feel better

(v) *Speech and Interaction Dataset:* As seen everywhere these are like the usual speech to text and text to speech conversion techniques as used in Google Assistant, Siri, etc.

(vi) *Special Dataset:* The special data set is the magical dataset that makes the interaction feel more human-like, the dataset consists of normal puns (the funny little responses) with a set of examples where they can be used in order to help robot of their better usage similarly, jokes, metaphors, and cliché are used to make the user feel more comforting to that of conversing to a human being.

(vii) *Teaching Dataset:* When told that a separate dataset for teaching will be added quite a few were shook that why should such a dataset be added. But, consider a family where both of the parents are working, their child will be alone even though it can play songs and keep kids entertained there is also a part called learning. Henceforth the robot starts downloading the reference books based on the Board of Education and Standard at which they study.

(viii) *Safety Dataset:* The safety dataset tops all the other seven datasets, the safety dataset is like the adrenaline hormone of our body the dataset contains various safety measures and helplines in case of emergency situations, added that it even contacts the numbers registered to the safety panel of the robot.

The learning method used in machine learning technique is Reinforcement (In this type of learning the output turns out to be a reward or a punishment for the given input) learning.

D. Scope of R-Assistant

The project has a great scope as it involves human emotions and is an easier form of human-computer interaction. Moreover, in the upcoming generation, there are going to be many people who are likely to suffer a lot of issues due to less interaction. In the later times there are more possibilities for a reduction in human-human interaction due to their irregular schedules of work and after a tiring day, anyone would like to have someone to talk with someone who is ready to hear that's where R-Assistant is at its utmost potential. The project is expected to make people feel that they're not alone and the interaction is more human-like, which might help the people to have better communication with others. The safety panel provides a more secure environment for the user. Moreover, since the robot is an android it becomes more accessible and makes life simpler.

E. Design of R-Assistant

The design of the R-assistant is simple and compact, now the idea is to make an entire working robot alone, the looks of the robot like a human is to be decided later. The idea of design is expected to be more complex as it also must be safe for kids to use. The robot consists of digital eyes to make the expressions more realistic and look like real eyes, the eyes are the normal LED robotic eyes. The batteries to be used are High polymer Li-ion batteries for the greater life of the robot, the charging type is planned to be induction type so that the charging can be easy and efficient. Above all, due to its online service, the robot can even help us watch over our loved ones via its own eyes.

4   Conclusion

This research work aims to make people to be relaxed and stress-free. R-Assistant can be used to perceive emotional state just by sensing their facial reactions. This gives humans a satisfaction that someone, at last, can understand them. The innovativeness of the project is to help the users to make them feel secure. In the near future, R assistant can also be used for children with autism or other mental ailments, where we can understand them only through their facial expressions.

References

1. Mannella, C., Lederer, W.J., Jafri, M.S.: Understanding the role of mitochondrial cristae structure on energy metabolism through simulation
2. Jeong, S.Y., Lee, J.W., Choi, S.H., Kwon, S.W.: Single-incision laparoscopic cholecystectomy using instrumental alignment in robotic single-site cholecystectomy. Ann. Surg. Treat. Res. **94**(6), 291–297 (2018)
3. Assistant, Why Robotic: Control and obstacle avoidance of a mobile platform used as robotic assistant for elderly and disabled
4. Kumar Senthil, M., Vijayanandh, R., Ramganesh, T.:
5. Mohan, C.B., Baskaran, R.: Improving network performance using ACO based redundant link avoidance algorithm. Int. J. Comput. Sci. Issues (IJCSI) **7**(3), 27 (2010)

6. Raj, A.H.: Development of an intelligent sensor based inspection robot for closed environment. In: 2015 5th Nirma University International Conference on Engineering (NUiCONE), pp. 1–6. IEEE (2015)
7. IBM: Real-Time Analytics and Machine Learning. https://www.ibm.com/it-infrastructure/z/capabilities/real-time-analytics
8. Anderson, J.A.: An Introduction to Neural Networks. ISBN 978-02-623-1588-3
9. Fausett, L.: Fundamentals of Neural Networks. ISBN 978-81-317-0053-2, 813-17-0053-4
10. Duck, M., Read, R.: Data Communications and Computer Networks. ISBN 978-81-317-2698-2
11. The Hindu Website: https://www.thehindu.com/todays-paper/tp-opinion/loneliness-could-literally-kill-you/article19437568.ece
12. The WebPsychology Webpage. https://www.webpsychology.com/news/2015/12/14/severe-loneliness-leads-depression-248116
13. Image Processing using Partial Differential Equations. https://www.lri.fr/~gcharpia/VisionSeminar/slides/2014-04-02-kornprobst-vist.pdf

Synthesis of Low Sidelobe Radiation Patterns from Embedded Dipole Arrays Using Genetic Algorithm

G. Anjaneyulu[✉] and J. Siddartha Varma

MVGR College of Engineering, Vizianagaram, India
anjaneyulu.mvgr@gmail.com

Abstract. The main aim of this paper is to generate low sidelobe radiation patterns from array of horizontal dipole arrays embedded in different dielectric of different slab thickness using evolutionary algorithm Genetic Algorithm (GA) to make them suitable for high resolution radars, point to point communication and low EMI applications. Array element amplitude excitations are optimized using GA for horizontal dipole arrays embedded in alumina dielectric material with and without ground plane is considered.

Keywords: Embedded dipole arrays · Genetic Algorithm · High resolution radars

1 Introduction

Radiation pattern is used to identify the relative strength of the intensity field from the antenna in different direction with constant distance. In horizontal and vertical planes, radiation patterns are measured on two dimensional patterns. Mostly, format of pattern measurements in the order of rectangular or polar format. Figure 1 shows the different parts of radiation patterns. Main lobe is used to identify the direction of radiation field. Side lobe is a radiation lobe and it is adjacent to main lobe. It occupies the hemisphere in the direction of main lobe.

Radiation of minor lobes measured in undesired directions and side lobes will be greater than the minor lobes. Side lobes are calculated based on the ratio of lobes in the power density to the main lobe and it is defined as side lobe ratio (SLR) or side lobe level (SLL). Due to the transmission, there is a possibility of excessive side lobes and that leads to energy loss and high interference. Receiving side, side lobes can absorb the interfaced sign and it improves the noise level that leads to false detection of targets. Capacity of the communication system can be increased by the reduction of interference occurs due to transmitting and receiving antennas [1, 2].

Sidelobe levels of −20 dB or smaller are usually not desirable in most applications. Attainment of a sidelobe level greater than −30 dB usually requires very careful design and construction [3]. A method of sidelobe suppression of the sum pattern multiplication has been reported by Justice [4]. In this method, it has been shown that patterns of broadside and end-fire radiators can be used to minimize the sidelobe levels of the radiation patterns of uniformly excited arrays. King [5] carried out some useful studies

© Springer Nature Switzerland AG 2020
P. Karrupusamy et al. (Eds.): ICSCN 2019, LNDECT 39, pp. 791–797, 2020.
https://doi.org/10.1007/978-3-030-34515-0_84

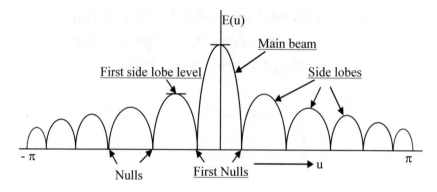

Fig. 1. Variation of E with u

on two element arrays of full wave elements and established relations for the currents, impedances and fields under different driving conditions. They include the broadside, bilateral end-fire, unilateral end-fire and for the case of parasitic elements. The analysis of linear arrays using mathematical approach is carried out by Cheng et al. [6]. In this paper, the current distribution of the elements of the linear array is considered. From the sampled values of a continuous function, closed form expressions are derived. In this approach, discrete linear array is treated as sampled data system. Known relations in z transforms are developed for sampled data systems and they are used to express the array polynomial in closed form. The same thing is valid for non uniform arrays. The techniques for determining the nulls, the principal maximum, the location of nulls and half power beamwidth are described.

The method of derivation of array factor in terms of composite array function is also present. In the paper reported by Ludwig [7] the source aperture distributions are investigated for low sidelobe radiation patterns. Here both peak and wide angle side-lobe levels are considered by modifying Taylor's technique. A method of reducing the sidelobes of narrow beams has been reported by Lo [8].

The method consists of randomly spaced elements in the array. It has been possible to reduce the number of elements while obtaining the desired patterns. It is found that number of elements depends upon required sidelobe level. The resolution of beamwidth is found to depend upon aperture dimension and to a lesser degree on the probability density function according to which the elements are placed. A method of generation of radiation pattern with desired beamwidth and sidelobe levels has been reported by Ishimaru et al. [9, 10]. The method consists of design of unequal spacing functions. The patterns are expressed in a series of angular functions and its sidelobe level is found to decrease approximately as $n^{-0.5}$. Where n being the total number of elements.

2   Formulation

Figures 2, 3, 4, 5 and 6 shows that the horizontal dipole array radiation pattern which is embedded with a dielectric slab materials in a both ground plane and non-ground plane field [11].

Therefore

$$E(u) = E_\theta(\theta)(AF)_n \tag{1}$$

(a) Far field with ground plane condition and it is a horizontal dipole which is embedded in a dielectric slab material [12–15].

$$|E_\theta(\theta, \phi)| = K \frac{N(1 + \rho_h) \sin[\beta N(z + d)] \cos(\phi_0 - \phi)}{\varepsilon_r \left[(1 + \rho_h)^2 \cos^2 \beta N d + (1 - \rho_h)^2 \sin^2 \beta N d \right]^{1/2}} \tag{2}$$

Where,

$$K = \frac{2\omega\mu_0 l_a i_a}{2\pi r}$$

(b) Far field without ground plane condition and it is a horizontal dipole which is embedded in a dielectric slab material,

$$|E_\theta(\theta, \phi)| = \frac{K}{2} \frac{N(1 + \rho_p) \cos(\phi_0 - \phi)}{\varepsilon_r} \frac{\left[(1 - \rho_p)^2 \cos^2 \beta N(z + d) + (1 + \rho_p)^2 \sin^2 \beta N(z + d) \right]^{1/2}}{\left[(1 - \rho_p)^2 \cos^2 \beta N d + (1 + \rho_p)^2 \sin^2 \beta N d \right]^{1/2}} \tag{3}$$

$$(AF)_n = \sum_{n-1}^{N} A(y_n) e^{j\left[\frac{2\pi d}{\lambda} u y_n + \phi(y_n) \right]} \tag{4}$$

3   Results

The amplitude excitation coefficients are optimized using GA, to produce low sidelobe radiation patterns. The amplitude excitation coefficients obtained by GA, for N = 20, and patterns are numerically computed and presented in Figs. (2, 3, 4, 5 and 6). The sidelobe levels of radiation patterns of array of embedded vertical and horizontal dipoles with and without ground plane from the optimization technique is presented in Tables (1 and 2).

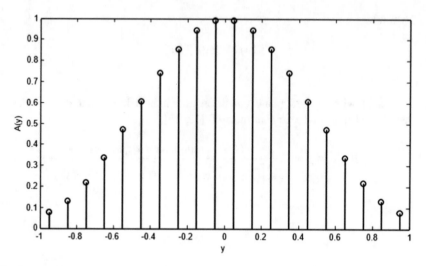

Fig. 2. Amplitude excitation coefficients obtained by GA method for N = 20 element array

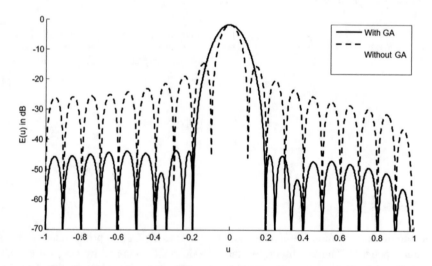

Fig. 3. Comparison of radiation patterns (N = 20) with ground plane.

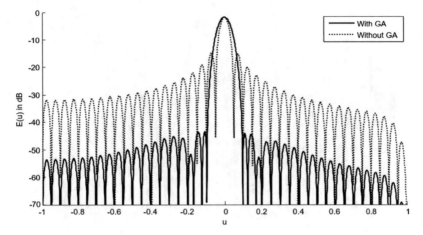

Fig. 4. Comparison of radiation patterns (N = 40) with ground plane.

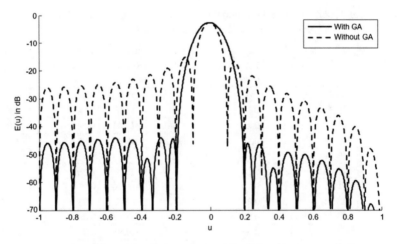

Fig. 5. Comparison of radiation patterns (N = 20) without ground plane.

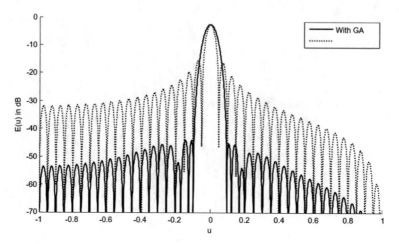

Fig. 6. Comparison of radiation patterns (N = 40) without ground plane.

Table 1. Comparison of embedded horizontal in a polystyrene ($\varepsilon_r = 2.54$) dielectric with ground plane and slab thickness d = $\lambda/2$, with and without GA.

No. of elements	With Genetic Algorithm			Without Genetic Algorithm			Central peak
	SLL(R) dB	SLL(L) dB	Beam width (rad)	SLL(R) dB	SLL(L) dB	Beam width (rad)	
10	−46.46	−41.26	0.68	−28.29	−13.22	0.40	−1.61
20	−50.37	−43.76	0.38	−19.54	−14.30	0.18	−1.62
30	−52.61	−44.51	0.26	−18.02	−14.04	0.13	−1.76
40	−49.89	−44.66	0.18	−17.10	−15.36	0.09	−1.77
50	−53.01	−43.17	0.14	−16.92	−14.79	0.08	−1.77
60	−47.92	−44.67	0.12	−16.75	−15.23	0.06	−2.35
70	−51.01	−48.20	0.10	−16.53	−16.16	0.05	−1.90
80	−46.52	−44.96	0.09	−17.39	−17.12	0.05	−1.72
90	−46.93	−44.81	0.08	−17.91	−17.37	0.04	−1.77
100	−49.13	−44.68	0.09	−20.82	−19.81	0.04	−1.70

Table 2. Comparison of radiation patterns of embedded horizontal dipole array in an alumina ($\varepsilon_r = 9.5$) dielectric without ground plane and slab thickness d = $\lambda/2$, with and without GA.

No. of elements	With Genetic Algorithm			Without Genetic Algorithm			Central peak
	SLL(R) dB	SLL(L) dB	Beam width (rad)	SLL(R) dB	SLL(L) dB	Beam width (rad)	
10	−49.96	−41.04	0.78	−17.51	−14.37	0.40	−2.66
20	−47.49	−44.08	0.40	−16.70	−15.29	0.18	−2.72
30	−49.19	−45.34	0.26	−16.87	−15.56	0.13	−2.74
40	−47.31	−45.74	0.18	−17.63	−16.10	0.09	−2.85
50	−49.88	−46.40	0.14	−17.37	−16.86	0.07	−2.88
60	−47.18	−46.02	0.12	−16.61	−17.13	0.06	−3.48
70	−50.53	−49.63	0.10	−18.22	−17.72	0.05	−3.16
80	−47.33	−48.62	0.12	−17.30	−18.41	0.05	−3.03

4 Conclusion

It is apparent from results that the patterns obtained by applying Genetic Algorithm are found better than that of radiation patterns without Genetic Algorithm in terms of sidelobe levels. It is clear that for small arrays, null-to-null beamwidth is more than that of large arrays. Moreover, for $N = 10$ elements, the beamwidth is about 0.40 rad for both vertical and horizontal dipole arrays embedded in a dielectric slab and for $N = 20$, the beamwidth is about 0.18 rad. As the number of elements increases in the array, beamwidth decreases that is presented in Tables (1 and 2). The results with very low sidelobe patterns synthesized by GA are useful in point to point communication and for high resolution radars and these results are unique.

Acknowledgement. This work (Major project) was funded by Science and Engineering Research Broad, the Department of Science and Technology (DST), Government of India, Sanction No. EEQ/2016/000396 and Order No. SERB/F/8020/2017-18.

References

1. Raju, G.S.N.: Antenna and Wave Propagation, 4th edn. Pearson Education, Singapore (2005)
2. Ma, M.T.: Theory and Applications of Antenna Arrays. Wiley, New York (1974)
3. Lavery, G.: Effect of antenna peak sidelobe levels on L-bandspace based GTMI radar performance. In: IEEE AerospaceConference 2005, pp. 2162–2169, pp. 5–12, March 2005
4. Justice, R.: Sidelobe suppression by pattern multiplication. IRE Trans. Antennas Propag. **4**(2), 119–124 (1956)
5. King, R.: Linear arrays: currents, impedances and fields. IRE Trans. Antennas Propag. **7**(5), 440–457 (1959)
6. Cheng, D.K., Ma, M.T.: A new mathematical approach forlinear array analysis. IRE Trans. Antennas Propag. **8**(3), 255–259 (1960)
7. Ludwig, A.C.: Low sidelobe aperture distributions forblocked and unblocked circular apertures. IEEE Trans. Antennas Propag. **30**(5), 933–946 (1982)
8. Lo, Y.T.: A mathematical theory of antenna arrays with randomly spaced elements. IEEE Trans. Antennas Propag. **AP-12**(3), 257–268 (1994)
9. Ishimaru, A., Chen, Y.S.: Thinning and broad banding antennaarrays by unequal spacing. IEEE Trans. Antennas Propag. **13**(1), 34–41 (1965)
10. Rajya Lakshmi, V., Raju, G.S.N.: Pattern synthesis usinggenetic algorithm for low sidelobe levels. IJCA **31**(4), 53–57 (2011)
11. Anjaneyulu, G., Varma, T.A.N.S.N., Siddartha Varma, J.: Radiation patterns generation of horizontal dipole array embedded in different dielectric slabs. In: IEEE Conference on Electronics, Communication and Aerospace Technology (ICECA), Coimbatore, 29–31 March 2018
12. Anjaneyulu, G., Raju, G.S.N.: The far-field patternanalysis of vertical and horizontal dipole embedded in a dielectric slab. IJEST **6**(8), 489 (2014)
13. Anjaneyulu, G., Raju, G.S.N.: Analysis of the far fieldpattern of ungrounded dipole in different dielectric slabs. IOSR_JECE **9**(6), Ver.I, 49–58 (2014)
14. Anjaneyulu, G., Raju, G.S.N.: Generation of radiationpatterns from array of vertical dipoles embedded in differentdielectric slabs. IJECT **6**(1), 410–412 (2015)
15. Varma, T.A.N.S.N., Anjaneyulu, G.: Design of uniform linear practical antenna arrays for ultralow side lobe sum patterns. In: Springer International Conference on Optical and Wireless Technologies (OWT 2018), Jaipur, 10–11 February 2018

Retraction Note to: Sustainable Communication Networks and Application

P. Karrupusamy, Joy Chen, and Yong Shi

Retraction Note to:
P. Karrupusamy et al. (Eds.): *Sustainable Communication*
Networks and Application, **LNDECT 39,**
https://doi.org/10.1007/978-3-030-34515-0

The Series Editor and the Publisher have retracted this chapter. An investigation by the Publisher found a number of chapters, including this one, with various concerns, including but not limited to compromised editorial handling, incoherent text or tortured phrases, inappropriate or non-relevant references, or a mismatch with the scope of the series and/or book volume. Based on the findings of the investigation, the Series Editor therefore no longer has confidence in the results and conclusions of this chapter. The authors have not responded to correspondence regarding this retraction.

The retracted version of these chapters can be found at
https://doi.org/10.1007/978-3-030-34515-0_2
https://doi.org/10.1007/978-3-030-34515-0_39
https://doi.org/10.1007/978-3-030-34515-0_46
https://doi.org/10.1007/978-3-030-34515-0_63
https://doi.org/10.1007/978-3-030-34515-0_66
https://doi.org/10.1007/978-3-030-34515-0_67
https://doi.org/10.1007/978-3-030-34515-0_77

Author Index

© Springer Nature Switzerland AG 2020
P. Karrupusamy et al. (Eds.): ICSCN 2019, LNDECT 39, pp. 799–801, 2020.
https://doi.org/10.1007/978-3-030-34515-0

Printed in the United States
by Baker & Taylor Publisher Services